FEATURES AND BENEFITS

Advanced Mathematics, A Precalculus Course, New Edition

All important precalculus topics are presented to lay the groundwork for further study of mathematics at the college level, pp. iii–ix (contents), 64, 132.

The **complete treatment of trigonometry** and its applications assures student understanding of both the concepts and the uses of this important topic, Ch. 6–10.

Independent chapters on **statistics, probability,** and **introductory calculus** provide a variety of options in structuring the course, Ch. 14, 15, 16.

Clear presentation of theory combined with **numerous worked-out examples** enable students to progress confidently and independently, pp. 149, 249.

Numerous graded exercises that range from straightforward practice to challenging problems and applications accommodate a wide range of student abilities and interests, pp. 55, 130, 283, 318.

Interesting and motivating **applications** are presented throughout to increase students' appreciation of the power and uses of mathematics, pp. 120, 173, 196, 198, 296.

Separate **Calculator and Computer Exercises, Trigonometric Table in decimal degrees,** and **computer use in some optional sections** enable students to apply new technology in a meaningful way, pp. 29, 80, 81, 156, 303–305, 487, 633.

Chapter Summaries, Chapter Tests, Cumulative Reviews, and **Cumulative Tests** provide continuing reinforcement of previously learned material, pp. 233, 271, 355–357, 602–604.

A **Geometry Review** and a list of **Trigonometric Formulas** provides students with a handy reference guide when needed, pp. 605–609.

A section on **College Entrance Examinations** gives students an excellent opportunity to prepare for these important examinations, pp. 610–623.

A comprehensive **Teacher's Manual with Solutions** and supplementary **Tests** (on duplicating masters) offer maximum teaching support.

Supplementary **Precalculus Computer Activities** are keyed to the text and allow students to work independently with the computer as they proceed through the course.

Advanced Mathematics

A Precalculus Course
REVISED EDITION

Richard G. Brown

David P. Robbins

EDITORIAL ADVISER

Andrew M. Gleason

HOUGHTON MIFFLIN COMPANY/BOSTON

Atlanta   Dallas   Geneva, Ill.   Princeton, N.J.   Palo Alto   Toronto

THE AUTHORS

Richard G. Brown, Mathematics Teacher, Phillips Exeter Academy, Exeter, New Hampshire. A teacher and author, Mr. Brown has taught a wide range of mathematics courses for both students and teachers at several schools and universities. His affiliations have included the Newton (Massachusetts) High School, the University of New Hampshire, Arizona State University, and the North Carolina School for Science and Mathematics during the school year beginning in 1983. He is an active participant in professional mathematics organizations and the author of mathematics texts and journal articles.

David P. Robbins, Institute for Defense Analyses, Princeton, New Jersey. Dr. Robbins has taught mathematics at the Phillips Exeter Academy, the Massachusetts Institute of Technology, the Fieldston School in New York City, and Washington and Lee University in Virginia. He is the author of several articles in professional mathematics journals.

EDITORIAL ADVISER

Andrew M. Gleason, Hollis Professor of Mathematics and Natural Philosophy, Harvard University. Professor Gleason is a well-known research mathematician and a member of the National Academy of Sciences. He has served as President of the American Mathematical Society.

Printed in U.S.A.

ISBN: 0-395-46137-5

DEFGHIJ-RM-9876543210

Contents

6 Trigonometric functions 189

7 Triangle trigonometry 237

16 Introduction to calculus 567

List of symbols

Symbol	Meaning	Page
$P(x, y)$	point P with coordinates x and y	2
$\sqrt{}$	square root	4
$\overline{AB}$	line segment with endpoints A and B	4
AB	the distance from point A to B	4
$\overline{x}$	x bar	5
$\angle$	angle	7
$\triangle$	triangle	7
m	slope of a line	8
$\Delta x, \Delta y$	The difference, $x_2 - x_1$, in the x-coordinates of two points, and the corresponding difference between their y-coordinates	8, 9
$\therefore$	therefore	16
i	$\sqrt{-1}$	21
$\overleftrightarrow{PB}$	the line through P and B	28
$\perp$	is perpendicular to	32
$\overrightarrow{OP}$	ray OP	36
x^n	the nth power of x	47
$f(x)$	f of x, or function of x	48
$\approx$	equals approximately	53
$|r|$	absolute value of r	56
V	volume	81
$\{t \mid 1 \leq t\}$	the set of all t such that t is greater than or equal to one	118
F	Fahrenheit temperature	122
$f(g(x))$	f of g of x	127
$f^{-1}(x)$	f inverse of x	140
$\log_b x$	logarithm of x to base b	160
e	base of natural logarithms	180
$\ln$	$\ln x = \log_e x$	180
RPM	revolutions per minute	189
$8°15'20''$	8 degrees, 15 minutes, 20 seconds	190
$\sim$	is similar to	225
$[x]$	greatest integer $\leq x$	292
$(r; \theta)$	polar coordinates	334
e	eccentricity	363
$\mathbf{v}$	vector quantity	400
$\overrightarrow{AB}$	velocity vector	400
$\mathbf{v}_1 \cdot \mathbf{v}_2$	dot product	422

Symbol	Meaning	Page
$\begin{vmatrix} a_1 & b_1 \\ a_2 & b_2 \end{vmatrix}$	determinant	438
lim	limit	466
∞	infinity	466
$\sum\limits_{k=5}^{10} 3k$	sum of $3k$ for values of k from 5 to 10	479
$n!$	n factorial	483
s^2	variance	503
s	standard deviation	503
$P(n, r)$	number of permutations of r things chosen from n things	540
$C(n, r)$	number of combinations of r things chosen from n things	540
$P(A)$	probability of event A	546
$P(B\,|\,A)$	the probability of B, given A	548
$f'(x)$	derivative of $f(x)$	580

Metric units

Length: meter (m)   *Mass:* kilogram (kg)*   *Capacity:* liter (L)
Time: second (s), minute (min)   *Temperature:* degree Celsius (°C)
Force: newton (N)   *Pressure:* pascal (P)   *Frequency:* hertz (Hz)
Electrical charge: coulomb (C)   *Luminous intensity:* candela (cd)

A prefix multiplies a unit by the factor given in the table. For example,

$$1 \text{ MHz} = 10^6 \text{ Hz} = 1,000,000 \text{ Hz}$$

Compound units may also be formed by multiplication or division. Examples are kilometers per hour (km/h) and cubic meters (m³).

Factor	Prefix	Symbol
10^6	mega	M
10^3	kilo	k
10^{-2}	centi	c
10^{-3}	milli	m

*Although the kilogram is defined as the base unit, the gram (g) is used with the prefixes to name other units of mass.

Greek letters

Letters		Names	Letters		Names	Letters		Names	Letters		Names
A	α	Alpha	E	ϵ	Epsilon	Π	π	Pi	Σ	σs	Sigma
B	β	Beta	M	μ	Mu	P	ρ	Rho	Θ	θ	Theta
Δ	δ	Delta							Ω	ω	Omega

CHAPTER ONE

Coordinate geometry

OBJECTIVES

1. To find the point of intersection of two lines by graphing or by solving a system of linear equations.
2. To find the length and midpoint of a line segment.
3. To find the slope of a nonvertical line.
4. To determine whether two lines are parallel, perpendicular, or neither.
5. To find an equation of a line given certain geometric properties of the line.
6. To add, subtract, multiply, and divide complex numbers.
7. To solve a quadratic equation for real and imaginary roots by factoring, completing the square, or the quadratic formula.
8. To find an equation of a circle given its center and radius.
9. To find the center and radius of a circle given its equation.
10. To find the intersection points of circles with lines or other circles.
11. To prove geometric theorems by coordinate methods.

Equations of lines

1-1/POINTS AND LINES

Each point in a plane can be associated with an ordered pair of numbers, called the **coordinates** of the point. Also, each ordered pair of numbers can be associated with a point. The association of points with ordered pairs is the basis for coordinate geometry. Coordinate geometry enables us to express geometric ideas like point, line, circle, and perpendicularity in terms of numbers, variables, and algebraic relationships.

The glass surface of this New York skyscraper suggests a coordinate plane.

1

Points

To set up a coordinate system, we choose two perpendicular lines as the **x-axis** and **y-axis** and designate their point of intersection as the **origin.** Using a convenient unit of measure, we mark off the axes as number lines with zero located at the origin. We usually show the x-axis as horizontal and the y-axis as vertical. The positive part of the x-axis is to the right of the y-axis, and the positive part of the y-axis is above the x-axis. The x-axis and y-axis divide the coordinate plane into the four quadrants shown in the diagram.

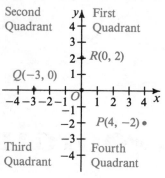

You can see in the diagram that point P has an **x-coordinate** of 4 and a **y-coordinate** of -2. We shall use the ordered pair $(4, -2)$ as another name for P and write $P = (4, -2)$. Frequently, we shall omit the equals sign and write $P(4, -2)$. Points with an x-coordinate of zero, such as R in the diagram, lie on the y-axis. Points with a y-coordinate of zero, such as Q in the diagram, lie on the x-axis. The coordinates of the origin are $(0, 0)$, usually shown as O.

Linear equations

A **solution** of the equation $2x - 3y = 12$ is an ordered pair of numbers that makes the equation true. For example, $(0, -4)$ is a solution because $2(0) - 3(-4) = 12$. Several ordered-pair solutions of this equation are plotted as points in the diagram. The set of all points in the plane corresponding to ordered-pair solutions of an equation is called the **graph** of the equation. The graph of $2x - 3y = 12$ is the line shown in the diagram. We

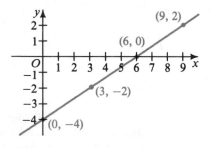

call -4 the **y-intercept** of the line because the line intersects the y-axis at $(0, -4)$. We call 6 the **x-intercept** of the line because the line intersects the x-axis at $(6, 0)$.

Any equation of the form $ax + by = c$ where a and b are not both zero is called a **linear equation** because its graph is a line. The graph is often referred to as "the line $ax + by = c$." Any line in the plane is the graph of a linear equation. We call the form "$ax + by = c$" the **general form** of the linear equation.

One way to draw the line that is the graph of a linear equation is to find the x- and y-intercepts of the line. Then draw the line passing through the points determined by the intercepts, as shown in the next example.

EXAMPLE 1. Draw the graph of $3x + 2y = 18$.

SOLUTION: To find the y-intercept, let $x = 0$ and solve for y. The line passes through $(0, 9)$, so the y-intercept is 9.

$$3 \cdot 0 + 2y = 18$$
$$y = 9$$

To find the x-intercept, let $y = 0$ and solve for x. The line passes through $(6, 0)$, so the x-intercept is 6.

The graph of $3x + 2y = 18$ is the line determined by the points $(0, 9)$ and $(6, 0)$, as shown in the diagram. To check that we have drawn the line correctly, we plot a third point whose coordinates satisfy the equation, in this case $(4, 3)$.

$$3x + 2 \cdot 0 = 18$$
$$x = 6$$

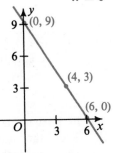

When one of the constants a, b, or c in the linear equation $ax + by = c$ is 0, you can draw certain conclusions about the graph of the equation. If $c = 0$, then the line passes through the origin [figure (a)]. If $a = 0$, then the line is horizontal [figure (b)]. If $b = 0$, then the line is vertical [figure (c)].

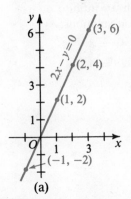

(a)

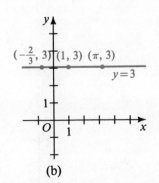

(b)

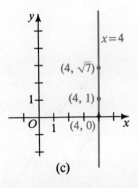

(c)

Intersection of lines

You can determine where two lines intersect by drawing their graphs or by solving their equations simultaneously. Consider the following pair of linear equations:

$$2x + 5y = 10 \quad (1)$$
$$3x + 4y = 12 \quad (2)$$

To solve the equations simultaneously, we multiply both sides (sometimes called members) of equation (1) by 3. Then we multiply both sides of equation (2) by 2 and subtract it from equation (1).

$$6x + 15y = 30$$
$$6x + 8y = 24$$
$$7y = 6$$
$$y = \tfrac{6}{7}$$

Next we substitute $\frac{6}{7}$ for y in equation (1) and solve for x.

$$2x + 5(\tfrac{6}{7}) = 10$$
$$x = \tfrac{20}{7}$$

Thus $(\frac{20}{7}, \frac{6}{7})$, or $(2\frac{6}{7}, \frac{6}{7})$, is the common solution of the two equations. It is also the point common to their graphs. Be aware that solutions found by graphing are not always exact. For example, you can tell by the location of the point of intersection in the diagram that the value of x is a bit less than 3 and the value of y is a bit less than 1, but you need the algebraic solution to determine the exact values.

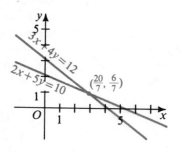

When two linear equations have no common solutions, their graphs are parallel lines [figure (a)]. Two linear equations may also have an infinite number of common solutions; two such equations have the same graph [figure (b)].

(a) A system with no common solutions:

$$6x + 4y = 8$$
$$3x + 2y = 1$$

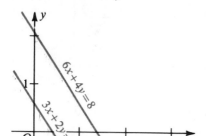

(b) A system with an infinite number of common solutions:

$$6x + 4y = 8$$
$$3x + 2y = 4$$

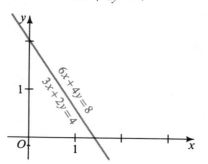

Distance and midpoint formulas

We denote the line segment with endpoints A and B as $\overline{AB}$. We denote the distance from A to B as AB. If you know the coordinates of A and B, then you can use the formulas below to find the distance AB and the coordinates of the midpoint of $\overline{AB}$. Derivations of these formulas are called for in Exercises 36 and 37, respectively.

Let $A = (x_1, y_1)$ and $B = (x_2, y_2)$.

DISTANCE FORMULA: $AB = \sqrt{(x_2 - x_1)^2 + (y_2 - y_1)^2}$

Let M be the midpoint of $\overline{AB}$.

MIDPOINT FORMULA: $M = \left(\dfrac{x_1 + x_2}{2}, \dfrac{y_1 + y_2}{2} \right)$

The midpoint formula can also be written as $M = (\bar{x}, \bar{y})$ where $\bar{x}$ (read "x bar") is the mean of the x-coordinates, $\dfrac{x_1 + x_2}{2}$, and $\bar{y}$ is the mean of the y-coordinates, $\dfrac{y_1 + y_2}{2}$.

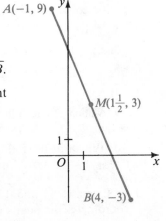

EXAMPLE 2. If $A = (-1, 9)$ and $B = (4, -3)$, find:
    a. the length of $\overline{AB}$.
    b. the coordinates of the midpoint of $\overline{AB}$.

SOLUTION: **a.** Finding the length of $\overline{AB}$ is equivalent to finding the distance from A to B.

$$AB = \sqrt{(4 - (-1))^2 + (-3 - 9)^2}$$
$$= \sqrt{25 + 144}$$
$$= 13$$

b. $M = \left(\dfrac{-1 + 4}{2}, \dfrac{9 - 3}{2}\right) = (1\tfrac{1}{2}, 3)$

ORAL EXERCISES 1-1

Use the coordinates that are given for C and D to find:
a. the length of $\overline{CD}$.   **b.** the coordinates of the midpoint of $\overline{CD}$.

1. $C(0, 0)$, $D(8, 6)$ **2.** $C(4, 2)$, $D(6, 6)$

3. $C(-3, 4)$, $D(3, -2)$ **4.** $C(7, -9)$, $D(7, -1)$

5. Which of the following points are on the line $2x + 3y = 15$?
    a. $(3, 3)$ **b.** $(9, -1)$
    c. $(2.5, 3.5)$ **d.** $(-10.5, 12)$

6. Give the coordinates of the points where the graph of $2x + 3y = 15$ intersects the x-axis and the y-axis.

7. The diagram at the right shows the graphs of the equations $x + y = 5$ and $2x - y = 1$.
    a. Use the graph to estimate the common solution of these equations.
    b. Solve the equations simultaneously and compare the solution with your estimate in part (a).

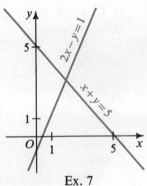

Ex. 7

8. The point $(8, 4)$ is on the horizontal line H.
    a. Name three other points of H.
    b. State an equation of H.

9. The point $(8, 4)$ is on the vertical line V.
    a. Name three other points of V.
    b. State an equation of V.

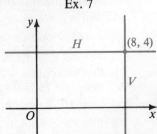

Ex. 8, 9

Coordinate geometry   **5**

Use the coordinates that are given for C and D to find:
a. the length of $\overline{CD}$.
b. the coordinates of the midpoint of $\overline{CD}$.

A
 1. $C(1, 0)$, $D(7, 8)$ **2.** $C(3, 3)$, $D(15, 12)$
 3. $C(-8, -3)$, $D(7, 5)$ **4.** $C(-2, -1)$, $D(4, 9)$
 5. $C(\frac{1}{2}, \frac{9}{2})$, $D(-2, -\frac{3}{2})$ **6.** $C(\frac{7}{2}, -1)$, $D(-\frac{5}{2}, \frac{7}{2})$
 7. $C(4.8, 2.2)$, $D(4.8, -2.8)$ **8.** $C(1.7, 5.7)$, $D(-2.3, 5.7)$

 9. Which of the following points are on the graph of $3x - 2y = 15$?
  a. $(9, 6)$ **b.** $(8, 4)$ **c.** $(-\frac{4}{3}, -\frac{19}{2})$ **d.** $(3.4, -3.2)$ **e.** $(-9, -22)$

 10. Which of the following points are on the graph of $-5x + 4y = 18$?
  a. $(-1.2, 3.0)$ **b.** $(3, -\frac{3}{4})$ **c.** $(-18, 24)$ **d.** $(-6, -3)$ **e.** $(3.6, 9)$

In Exercises 11 and 12 sketch the graph of the given equation. Label with coordinates the points P and Q where the graph intersects the x-axis and the y-axis, respectively. If $O = (0, 0)$, find the area of $\triangle OPQ$.

 11. $3x - 2y = 6$ **12.** $4x + 3y = 24$

 13. In one drawing, sketch the horizontal line through $(4, 3)$ and the vertical line through $(5, -2)$. What is the intersection point of these lines? What are the equations of these lines?

 14. Repeat Exercise 13 for the horizontal line through $(-2, -1)$ and the vertical line through $(-2, 3)$.

In Exercises 15–18 solve the given pair of equations simultaneously. Then sketch the graphs of the two equations and label the intersection of the graphs with the common solution of the equations.

 15. $3x - 5y = 9$ **16.** $2x + 3y = 15$
   $x + y = 3$ $4x - 9y = 3$
 17. $x - 3y = 4$ **18.** $-2x - 6y = 18$
   $5x + y = -8$ $x - 3y = 6$

 19. Plot the points $A(1, 7)$, $B(3, 5)$, $C(4, -1)$, and $D(2, 1)$. Use the distance formula to show that the opposite sides of quadrilateral $ABCD$ are equal. What kind of figure is $ABCD$?

 20. Plot the points $A(-6, 3)$, $B(-1, 6)$, $C(2, 1)$, and $D(-3, -2)$, and use the distance formula to show that quadrilateral $ABCD$ is a square. (*Hint:* Show that the four sides are equal and that the two diagonals are equal.)

 21. Plot the points $A(5, 1)$, $B(7, -1)$, $C(1, -3)$, and $D(-1, -1)$, and use the midpoint formula to show that the diagonals of quadrilateral $ABCD$, $\overline{AC}$ and $\overline{BD}$, have the same midpoint. What kind of quadrilateral is $ABCD$?

 22. Plot the points $A(2, 0)$, $B(4, -6)$, $C(9, 1)$, and $D(7, 7)$, and show that $\overline{AC}$ and $\overline{BC}$ bisect each other. What kind of quadrilateral is $ABCD$?

23. a. Given the three points $A(-3, 3)$, $B(1, 11)$, and $C(3, 15)$, show that B is on $\overline{AC}$ by showing that $AB + BC = AC$.

  b. What is the ratio of AB to BC?

24. Repeat Exercise 23 for the points $A(-3, 7)$, $B(-1, 4)$, and $C(3, -2)$.

B

25. a. Show that the point $P(4, 2)$ is equidistant from the points $A(9, 2)$ and $B(1, 6)$.

  b. If $(2, k)$ is equidistant from A and B, find the value of k.

26. a. Show that $P(1, 4)$ is equidistant from $A(-5, -3)$ and $B(-1, -5)$.

  b. If $(3, k)$ is equidistant from A and B, find the value of k.

27. P is a point on the x-axis that is 10 units from $(3, 8)$. Find all the possible coordinates of P.

28. P is a point on the x-axis that is 13 units from the point $(-3, 5)$. Find all the possible coordinates for P.

29. Q is a point on the y-axis that is 5 units from $(3, 6)$. Find all the possible coordinates of Q.

30. Q is a point on the y-axis that is $2\sqrt{10}$ units from $(6, 1)$. Find all the possible coordinates of Q.

31. a. Plot the points $A(-6, 7)$, $B(6, 3)$, and $C(-2, -1)$, and show that $(BC)^2 + (AC)^2 = (AB)^2$. What can you conclude about $\angle C$?

  b. Give the coordinates of the midpoint, M, of $\overline{AB}$. Verify these coordinates by showing that $CM = \frac{1}{2}AB$.

32. Show that the three lines $x + 3y = 19$, $2x - 5y = 5$, and $x - 2y = 4$ are concurrent. (*Hint:* Find the intersection of two lines and show that this point is on the third line.)

33. Determine whether or not the three lines $3x + 2y = 4$, $5x - 2y = 0$, and $4x + 3y = 3$ are concurrent. (*Hint:* See Exercise 32.)

C

34. Three vertices of a parallelogram have coordinates $(-3, 1)$, $(1, 4)$, and $(4, 3)$. Find the coordinates of the fourth vertex. How many possible answers are there?

35. In $\triangle ABC$, $D(7, 3)$ is the midpoint of side $\overline{AB}$, $E(10, 9)$ is the midpoint of side $\overline{BC}$, and $F(5, 5)$ is the midpoint of side $\overline{AC}$. Find the coordinates of A, B, and C.

36. In this proof, you may assume the following: The distance between two points on the same vertical line is the absolute value of the difference in y-coordinates; the distance between two points on the same horizontal line is the absolute value of the difference in x-coordinates.

Given: $A = (x_1, y_1)$, $B = (x_2, y_2)$

Prove: $AB = \sqrt{(x_2 - x_1)^2 + (y_2 - y_1)^2}$

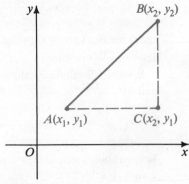

37. Given: $A = (x_1, y_1)$, $B = (x_2, y_2)$;
$M(x_3, y_3)$ is the midpoint of $\overline{AB}$.
P is the midpoint of $\overline{BC}$.
Q is the midpoint of $\overline{AC}$.

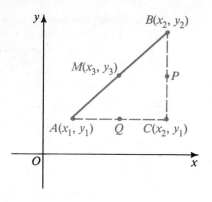

Prove: **a.** $P = \left(x_2, \dfrac{y_1 + y_2}{2}\right)$

b. $Q = \left(\dfrac{x_1 + x_2}{2}, y_1\right)$

c. M and P have the same y-coordinate.

d. M and Q have the same x-coordinate.

e. $M(x_3, y_3) = \left(\dfrac{x_1 + x_2}{2}, \dfrac{y_1 + y_2}{2}\right)$

CALCULATOR EXERCISES

The area of a triangle with sides a, b, and c units long can be found using *Hero's formula:*

$$\text{Area} = \sqrt{s(s - a)(s - b)(s - c)} \text{ where } s = \frac{a + b + c}{2}$$

Find to the nearest hundredth the area of the triangle with vertices $A(-13, 2)$, $B(5, 17)$, and $C(22, -4)$.

1-2/PARALLEL AND PERPENDICULAR LINES

The *slope* of a nonvertical line is a number measuring the steepness of the line relative to the x-axis. If (x_1, y_1) and (x_2, y_2) are any two points of a line, then the **slope** m of the line is defined by the equation:

$$m = \frac{y_2 - y_1}{x_2 - x_1}$$

Later we shall prove that the slope of a line is the same regardless of the points used to calculate it.

Remark: Sometimes the slope m of a line through (x_1, y_1) and (x_2, y_2) is written

$$m = \frac{\Delta y}{\Delta x} \quad \text{instead of} \quad m = \frac{y_2 - y_1}{x_2 - x_1}.$$

The symbol Δ is a Greek letter, pronounced "delta," which is used in mathematics to refer to a change, or *difference,* in the values of a variable. Thus, Δx refers to the difference, $x_2 - x_1$, in the x-coordinates of two points, and Δy

refs to the corresponding difference in the y-coordinates of the two points. Since Δy does not refer to the product of two numbers, Δ and y, it is incorrect to cancel the Δ's in the symbol $\dfrac{\Delta y}{\Delta x}$. That is, $\dfrac{\Delta y}{\Delta x} \neq \dfrac{y}{x}$.

Some important facts about the slope of a line are presented below.

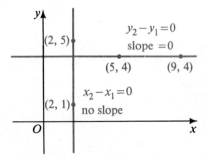

1. Horizontal lines have a slope of 0 because $\Delta y = 0$ for any two points of the line.
2. Vertical lines are considered to have no slope because $\Delta x = 0$ for any two points of the line. Consequently, you would have to divide by 0 to calculate the slope, and division by 0 has no meaning in our work. It should be apparent that having *no slope* is quite different from having a slope of 0.
3. Lines with positive slope rise to the right as you look at points with greater and greater x-coordinates. The greater the slope, the more steeply the line rises.

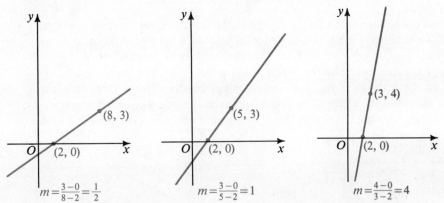

4. Lines with negative slope go down as you look at points with greater and greater x-coordinates. The greater the absolute value of the slope, the more steeply the line declines.

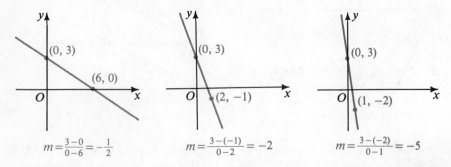

The following theorem provides an easy way to find the slope of a line from its equation.

Coordinate geometry   **9**

THEOREM 1. The slope of a line with equation $y = mx + k$ is m.

Proof: Let $P(x_1, y_1)$ and $Q(x_2, y_2)$ be any two different points $(x_1 \neq x_2)$ of the line with equation $y = mx + k$. Then:

$$y_1 = mx_1 + k$$
$$y_2 = mx_2 + k$$

Therefore, the slope is the following expression.

$$\frac{y_2 - y_1}{x_2 - x_1} = \frac{(mx_2 + k) - (mx_1 + k)}{x_2 - x_1}$$

$$= \frac{m(x_2 - x_1)}{x_2 - x_1} = m$$

COROLLARY 1. The slope of a line is constant and does not depend on the points used to calculate it.

When the equation of a line is written in the form $y = mx + k$, the numbers m and k should provide you with a mental picture of the line. According to the theorem, m is the slope of the line. The number k is the y-intercept of the line because the line intersects the y-axis at the point $(0, k)$. For these reasons, we refer to the form "$y = mx + k$" as the **slope-intercept** form. The diagrams below illustrate the effect of m and k on the graph of an equation written in slope-intercept form. Notice in the diagram that different lines with the same slope are parallel.

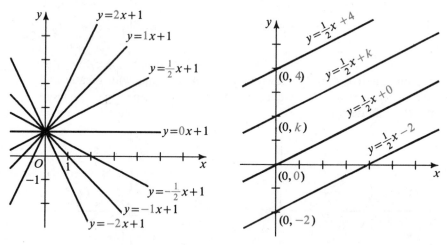

EXAMPLE 1. What are the slope and y-intercept of the line with equation

$$7x + 13y = 26?$$

SOLUTION: First rewrite the equation as shown on the following page.

$$7x + 13y = 26$$
$$13y = -7x + 26$$
$$y = -\tfrac{7}{13}x + 2$$

The slope is then $-\tfrac{7}{13}$, the coefficient of x. The y-intercept is 2.

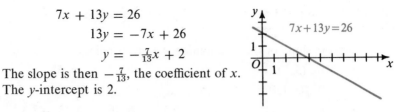

Theorems 2 and 3, which follow, show how the slope is related to the geometric ideas of parallel and perpendicular lines.

THEOREM 2.   **a.** If two nonvertical lines are parallel, then they have the same slope.

          b. If two (different) nonvertical lines have the same slope, then they are parallel.

    Proof: See Exercise 35 on page 13 for an algebraic proof and Exercises 37 and 38 on page 14 for a geometric proof.

In coordinate geometry it is sometimes convenient to consider lines that coincide to be parallel. If this is done, then the word "different" can be omitted from part (b) of Theorem 2.

Since vertical lines have no slope, they are not included in Theorem 2 above nor in Theorem 3. Note, however, that any two (different) vertical lines are parallel and any vertical line is perpendicular to any horizontal line.

THEOREM 3.   Given two lines with slopes m_1 and m_2:

          a. If the lines are perpendicular, then $m_1 = -\dfrac{1}{m_2}$.

          b. If $m_1 = -\dfrac{1}{m_2}$, then the lines are perpendicular.

    Proof: See Exercise 36 on page 14 for an algebraic proof and Exercises 39 and 40 on page 14 for a geometric proof.

EXAMPLE 2.   The equations of three lines are given. Which lines are parallel, and which are perpendicular?

    $l_1: y = \tfrac{3}{4}x - 7$ $l_2: 4x + 3y = 10$ $l_3: 3x - 4y = 11$

    SOLUTION: Find the slopes of the lines by rewriting their equations in slope-intercept form.

$$l_1: y = \tfrac{3}{4}x - 7 \qquad m_1 = \tfrac{3}{4}$$
$$l_2: y = -\tfrac{4}{3}x + \tfrac{10}{3} \qquad m_2 = -\tfrac{4}{3}$$
$$l_3: y = \tfrac{3}{4}x - \tfrac{11}{4} \qquad m_3 = \tfrac{3}{4}$$

Since both m_1 and m_3 are equal to $\tfrac{3}{4}$, lines l_1 and l_3 are parallel.

Since $m_2 = -\tfrac{4}{3} = -\dfrac{1}{\tfrac{3}{4}} = -\dfrac{1}{m_1} = -\dfrac{1}{m_3}$, line l_2 is perpendicular to l_1 and l_3.

Find the slope of the line joining the given points.

1. $(0, 0)$, $(5, 7)$ **2.** $(2, 0)$, $(3, 6)$

3. $(-2, 3)$, $(1, 0)$ **4.** $(2, -6)$, $(8, -3)$

5. $(-8, 2)$, $(8, 2)$ **6.** $(5, -3)$, $(5, -7)$

Find the slope and *y*-intercept of the line whose equation is given.

7. $y = 3x + 4$ **8.** $y = \frac{2}{5}x - 3$

9. $4x + 3y = 9$ **10.** $y = -2$

11. A line *l* has slope $\frac{4}{5}$. What is the slope of a line **(a)** parallel to *l*? **(b)** perpendicular to *l*?

12. A line *l* has equation $3x + 2y = 10$. What is the slope of a line **(a)** parallel to *l*? **(b)** perpendicular to *l*?

13. a. Show that opposite sides of quadrilateral *ABCD* are parallel.
    b. Show that adjacent sides of quadrilateral *ABCD* are perpendicular.

14. Explain why a vertical line has no slope.

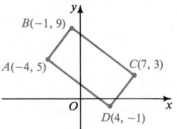

Find the slope of the line joining the points whose coordinates are given.

A   **1.** $(4, 2)$, $(9, 5)$ **2.** $(0, 4)$, $(12, 0)$

      3. $(-4, -2)$, $(2, -6)$ **4.** $(-2, 6)$, $(2, -2)$

      5. $(8, 5)$, $(-7, 5)$ **6.** $(-3, 8)$, $(-3, -2)$

      7. $(0.25, 1.5)$, $(0.5, 1)$ **8.** $(-\frac{1}{3}, 2)$, $(\frac{1}{4}, -\frac{1}{3})$

      9. (a, b), (b, a) **10.** $\left(a, \dfrac{a}{b}\right)$, $\left(b, \dfrac{b}{a}\right)$

11. Line *l* has slope $\frac{2}{7}$. What is the slope of a line **(a)** parallel to *l*? **(b)** perpendicular to *l*?

12. Line *l* has slope $-\frac{5}{4}$. What is the slope of a line **(a)** parallel to *l*? **(b)** perpendicular to *l*?

What are the slopes and the *y*-intercepts of the lines whose equations are given?

13. $y = 3x + 5$ **14.** $y = 4 - 5x$

15. $3x - 2y = 7$ **16.** $3x + 7y = 10$

17. $3x + 6y = 7$ **18.** $3y = 11x$

19. $y = -3$ **20.** $y = 7$

21. Line l has equation $3x - 7y = 9$. What is the slope of a line **(a)** parallel to l? **(b)** perpendicular to l?

22. A line l has equation $7x + 4y = 9$. What is the slope of a line **(a)** parallel to l? **(b)** perpendicular to l?

23. Show that the line through $(2, -3)$ and $(7, 2)$ is perpendicular to the line through $(-3, 7)$ and $(2, 2)$.

24. Show that the line through $(2, 3)$ and $(5, -2)$ is perpendicular to the line $3x - 5y = 15$.

25. Find the value of k if the line joining $(4, k)$ and $(6, 8)$ and the line joining $(-1, 4)$ and $(0, 8)$ are **(a)** parallel, **(b)** perpendicular.

26. Find the value of h if the line joining $(3, h)$ and $(5, 10)$ and the line $y = 3x + 4$ are **(a)** parallel, **(b)** perpendicular.

27. Tell which of the following equations have parallel line graphs and which have perpendicular line graphs:
    a. $y = \frac{5}{2}x - 8$ **b.** $-15x + 6y - 10 = 0$ **c.** $4x + 10y = 15$

28. Repeat Exercise 27 using the equations:
    a. $3y = 5x - 5$ **b.** $y = -\frac{3}{5}x + 4$ **c.** $10y = -6x - 7$

B 29. Given the points $A(-4, -6)$, $B(2, 4)$, $C(8, 6)$, and $D(2, -4)$:
   a. Show by using slopes that quadrilateral $ABCD$ is a parallelogram.
   b. Verify that both diagonals have the same midpoint.

30. Given the points $A(-4, 1)$, $B(2, 3)$, $C(4, 9)$, and $D(-2, 7)$:
   a. Show that quadrilateral $ABCD$ is a parallelogram with perpendicular diagonals.
   b. What special name is given to a quadrilateral like $ABCD$?

31. Show that the points $(-4, -5)$, $(-3, 0)$, $(0, 2)$, and $(5, 1)$ are the vertices of an isosceles trapezoid. (*Hint:* Find two pairs of points that determine parallel lines.)

32. Show that $(-1, -1)$, $(9, 4)$, $(20, 6)$, and $(10, 1)$ are the vertices of a rhombus. Find the area of this rhombus.

33. Given $A(-2, -4)$, $B(3, 1)$, and $C(4, -5)$, show that C is on the perpendicular bisector of $\overline{AB}$ by using two methods.
   a. Method I: Find the coordinates of the midpoint M of $\overline{AB}$ and show that $\overline{MC}$ is perpendicular to $\overline{AB}$.
   b. Method II: Show that $CA = CB$ and use a theorem from geometry.

34. Repeat Exercise 33 for $A(0, -2)$, $B(4, -4)$, and $C(7, 7)$.

35. **a.** Suppose lines l_1 and l_2 have equations $y = m_1x + k_1$ and $y = m_2x + k_2$, respectively. By solving these equations simultaneously, show that the lines intersect when $x = \dfrac{k_2 - k_1}{m_1 - m_2}$.
   b. What can you say about this value of x when $m_1 = m_2$? What can you say about the intersection of the lines when $m_1 = m_2$?

C　**36.** The lines $y = m_1x$ and $y = m_2x$ intersect the vertical line $x = 1$ at $A = (1, m_1)$ and $B = (1, m_2)$.

    a. Use the distance formula to find OA, OB, and AB.

    b. Use part (a) to show that if
$$(OA)^2 + (OB)^2 = (AB)^2,$$
then $m_1 = -\dfrac{1}{m_2}$. Can you prove the converse also?

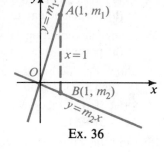

Ex. 36

37. Given: Slope of l_1 = slope of l_2
    Prove: **a.** $\triangle ABC \sim \triangle DEF$
          b. $l_1 \parallel l_2$

38. Given: $l_1 \parallel l_2$
    Prove: **a.** $\triangle ABC \sim \triangle DEF$
          b. Slope of l_1 = slope of l_2

39. Given: Line l_1 with slope $m_1 = -\dfrac{BC}{CA}$

    Line l_2 with slope $m_2 = \dfrac{DE}{AE}$

    $l_1 \perp l_2$

    Prove: **a.** $\triangle BAC \sim \triangle ADE$

          b. $\dfrac{BC}{CA} = \dfrac{AE}{ED}$

          c. $m_1 = -\dfrac{1}{m_2}$

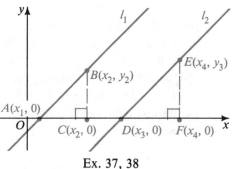

Ex. 37, 38

40. Given: Line l_1 with slope $m_1 = -\dfrac{BC}{CA}$

    Line l_2 with slope $m_2 = \dfrac{DE}{AE}$

    $m_1 = -\dfrac{1}{m_2}$

    Prove: **a.** $\triangle BAC \sim \triangle ADE$ **b.** $\angle BAD = 90°$

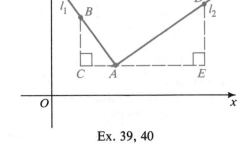

Ex. 39, 40

41. $ABCD$ is a square with vertices $A(0, 0)$ and $B(6, 8)$. Give the coordinates of C and D. (Two answers are possible.)

42. $PQRS$ is a rectangle with vertices $P(-4, -1)$ and $Q(-6, 5)$ and $PQ = 2(QR)$. Find the coordinates of R and S.

CALCULATOR EXERCISE

Given the points $A(-1.8, 2.3)$, $B(-0.4, 4.4)$, and $C(2.4, -0.5)$:
a. Verify that $\triangle ABC$ is a right triangle and find its area.
b. Show that the midpoint of the hypotenuse is equidistant from the three vertices.

1-3/FINDING EQUATIONS OF LINES

A linear equation can be written in several different forms, three of which are shown below.

The general form $\qquad ax + by = c$

The slope-intercept form $\quad y = mx + k$ $\qquad$ Line has slope m and y-intercept k.

The point-slope form* $\qquad \dfrac{y - y_1}{x - x_1} = m$ $\qquad$ Line has slope m and goes through (x_1, y_1).

Sometimes it is easier to use one form of a linear equation rather than another. The examples below illustrate this.

EXAMPLE 1. Find an equation of the line with x-intercept 8 and y-intercept 4.

SOLUTION: We use the points $(8, 0)$ and $(0, 4)$ to find the slope:

$$m = \frac{4 - 0}{0 - 8} = -\frac{1}{2}$$

Since the y-intercept is 4, the slope-intercept equation of the line is $y = -\dfrac{1}{2}x + 4$. An equation of the line in general form is $x + 2y = 8$.

EXAMPLE 2. Find an equation of the line through $(1, 4)$ and $(5, -2)$.

SOLUTION: First we find the slope of the line:

$$m = \frac{4 - (-2)}{1 - 5} = -\frac{3}{2}$$

Then we write an equation in point-slope form.

Using the point $(1, 4)$: $\quad$ or $\quad$ Using the point $(5, -2)$:

$$\frac{y - 4}{x - 1} = -\frac{3}{2} \qquad\qquad \frac{y + 2}{x - 5} = -\frac{3}{2}$$

Each equation can be written in general form as $3x + 2y = 11$.

EXAMPLE 3. Find an equation of the line through $(1, 2)$ and parallel to the line $y = 5x - 9$.

SOLUTION: The slope of $y = 5x - 9$ is 5. Thus, the slope of the parallel line is also 5. The point-slope form and the slope-intercept form of the equation are shown on the following page.

---

*The point-slope equation is sometimes written in the form $y - y_1 = m(x - x_1)$ to make it formally apparent that (x_1, y_1) satisfies the equation.

Point-slope form:

$$\frac{y-2}{x-1} = 5$$

Slope-intercept form:

$$y = 5x + k$$
$$2 = 5 \cdot 1 + k$$
$$-3 = k$$
$$\therefore y = 5x - 3$$

Each equation can be written in general form as $5x - y = 3$.

Example 4, which follows, shows two different ways of finding an equation of the perpendicular bisector of a line segment. Exercise 41 on page 20 shows a method for finding the equation of an angle bisector.

EXAMPLE 4. Write an equation of the perpendicular bisector of the segment joining $A(-2, 3)$ and $B(4, -5)$.

SOLUTION 1:   1. The slope of $\overline{AB}$ is

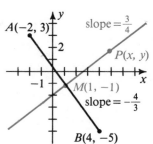

$$\frac{-5 - 3}{4 - (-2)} = \frac{-8}{6} = \frac{-4}{3}.$$

By Theorem 3 the slope of the perpendicular bisector is the negative reciprocal of $-\frac{4}{3}$, namely $\frac{3}{4}$.

2. The perpendicular bisector must pass through the midpoint of $\overline{AB}$. We use the midpoint formula to find the coordinates of this midpoint:

$$M = \left(\frac{-2 + 4}{2}, \frac{3 + (-5)}{2}\right) = (1, -1)$$

3. The equation of the line through $(1, -1)$ with slope $\frac{3}{4}$ is

$$\frac{y - (-1)}{x - 1} = \frac{3}{4},$$

which can be written in general form as $3x - 4y = 7$.

SOLUTION 2:   This method of solution uses the fact that $P(x, y)$ is on the perpendicular bisector of $\overline{AB}$ if and only if $PA = PB$. Using the distance formula, proceed as follows:

$$PA = PB$$

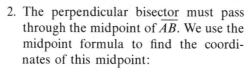

$$\sqrt{(x + 2)^2 + (y - 3)^2} = \sqrt{(x - 4)^2 + (y + 5)^2}$$
$$(x^2 + 4x + 4) + (y^2 - 6y + 9) = (x^2 - 8x + 16) + (y^2 + 10y + 25)$$
$$4x - 6y + 13 = -8x + 10y + 41$$
$$12x - 16y = 28$$
$$3x - 4y = 7$$

Notice that the sequence of equations just given is reversible. Thus, if $P(x, y)$ is on the line $3x - 4y = 7$, then it is equidistant from A and B.

Find an equation in slope-intercept form of the line described.

1. The line with a slope of 3 and a y-intercept of 7

2. The line with a slope of $\frac{5}{3}$ and a y-intercept of -2

Find an equation in point-slope form of the line described.

3. The line through $(-4, 6)$ with a slope of 3

4. a. The line through $(1, 2)$ and parallel to the line l
   with equation $y = 4x - 3$
   b. The line through $(1, 2)$ and perpendicular to l

5. Describe the steps in finding an equation of l,
the perpendicular bisector of $\overline{AB}$.

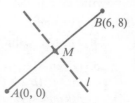

In Exercises 1-4, write an equation in slope-intercept form of the line described.

A **1.** The line with slope -2 and y-intercept 8

2. The line with slope $\frac{3}{5}$ and passing through the origin

3. The line with y-intercept 4 and x-intercept -2

4. The line with y-intercept -6 and x-intercept 7

In Exercises 5-10, write an equation in point-slope form of the line described.

5. The line through $(1, 7)$ with slope 4

6. The line through $(-2, 5)$ with slope 3

7. The line through $(2, 8)$ with slope $\frac{1}{2}$

8. The line through $(-3, 4)$ with slope $-\frac{2}{5}$

9. The line through $(-1, 4)$ and $(5, 8)$

10. The line through $(0, 5)$ and $(6, 1)$

In Exercises 11-16, write an equation in general form of the line described.

11. The horizontal line through $(5, -7)$ **12.** The vertical line through $(5, -7)$

13. The y-axis   **14.** The x-axis

15. The line through $(2, -7)$ and $(2, 3)$ **16.** The line through $(5, -3)$ and $(2, -3)$

In Exercises 17-24, write an equation in any form for the line described.

17. The line through $(4, 3)$ parallel to the line $5x + 4y = 1$

18. The line through $(-2, 4)$ parallel to the line through $(1, 1)$ and $(5, 7)$

19. The line through $(8, -2)$ perpendicular to the line $y = 7 - 2x$

20. The line through the origin perpendicular to the line $x - 3y = 9$

Write an equation in any form for the line described.

21. The perpendicular bisector of the segment joining $(0, 3)$ and $(-4, 5)$

22. The perpendicular bisector of the segment joining $(2, 4)$ and $(4, -4)$

In Exercises 23 and 24, refer to the diagram and write an equation for the line described.

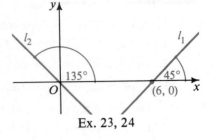

Ex. 23, 24

B **23.** The line l_1 through $(6, 0)$ with inclination $45°$

24. The line l_2 through $(0, 0)$ with inclination $135°$

25. Given $A(2, 0)$ and $B(8, 4)$, show that $P(3, 5)$ is on the perpendicular bisector of $\overline{AB}$ by these two methods:
Method 1. Show $\overline{AB} \perp \overline{PM}$ where M is the midpoint of $\overline{AB}$.
Method 2. Show $PA = PB$.

26. Repeat Exercise 25 using $A(0, 7)$, $B(2, -1)$, and $P(5, 4)$.

27. $\triangle PQR$ has vertices $P(4, -1)$, $Q(-2, 7)$, and $R(9, 9)$.
   a. Find an equation of the median from R.
   b. Find an equation of the altitude from R.
   c. Are your answers to (a) and (b) the same? Is $\triangle PQR$ isosceles?

28. $\triangle DEF$ has vertices $D(-2, 5)$, $E(6, -1)$, and $F(5, 6)$.
   a. Verify that $\triangle DEF$ is isosceles.
   b. Write an equation of the bisector of $\angle F$.

29. $\triangle ABC$ has vertices $A(-2, -1)$, $B(0, 7)$, and $C(8, 3)$.
   a. Write equations for the three medians of the triangle.
   b. Show that the medians are concurrent at a point G, called the *centroid* of the triangle.

30. $\triangle PQR$ has vertices $P(0, 4)$, $Q(8, -2)$, and $R(7, 5)$.
   a. Write equations for the perpendicular bisectors of the three sides of the triangle.
   b. Show that the perpendicular bisectors are concurrent at a point C, called the *circumcenter* of the triangle.
   c. Note that C is the midpoint of $\overline{PQ}$. What kind of triangle is $\triangle PQR$?

31. In $\triangle KLM$, $K = (0, 0)$, $L = (18, 0)$, and $M = (6, 12)$.
   a. Write equations for the altitudes to the three sides of the triangle.
   b. Show that the altitudes are concurrent at a point O, called the *orthocenter* of the triangle.

32. $\triangle RST$ has vertices $R(1, 2)$, $S(25, 2)$, and $T(10, 20)$.
   a. Find the centroid G of $\triangle RST$ using the method of Exercise 29.
   b. Find the circumcenter C of $\triangle RST$ using the method of parts (a) and (b) of Exercise 30.
   c. Find the orthocenter O of $\triangle RST$ using the method of Exercise 31.
   d. Show that G, C, and O lie on the same line. This line is called *Euler's line*.

33. Repeat Exercise 32 using $R(-3, 0)$, $S(17, 0)$, and $T(1, 8)$.

34. Each of the lines shown below passes through (2, 1) and forms with the axes $\triangle OPQ$. Which of these three triangles has the least area?

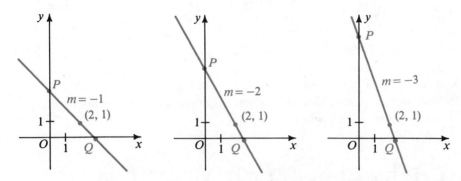

35. Consider the set of all lines through (2, 6) with negative slope. As in the drawings for Exercise 34, let P and Q be the points where the line intersects the y- and x-axes, respectively. Determine by experimentation the slope of the line that gives $\triangle OPQ$ the least area.

36. The vertices of $\triangle ABC$ are $A(8, 5)$, $B(0, 1)$, and $C(9, -2)$.
  a. Find the length and an equation of $\overline{BC}$.
  b. Find an equation of the altitude from A to $\overline{BC}$.
  c. Find the point where the altitude from A intersects $\overline{BC}$.
  d. Find the length of the altitude from A to $\overline{BC}$.
  e. Find the area of $\triangle ABC$.

37. Solve Exercise 36 using $A(2, -1)$, $B(-8, -11)$, and $C(-3, 4)$.

C **38.** Find the distance from the point (9, 5) to the line $4x - 3y = -4$.

39. Given the point $P(x_1, y_1)$ and the line l with equation $ax + by = c$:
  a. Write an equation of the line j that passes through P and is perpendicular to l.
  b. Show that l and j intersect at the point

  $$Q\left(\frac{b^2x_1 - aby_1 + ac}{a^2 + b^2}, \frac{a^2y_1 - abx_1 + bc}{a^2 + b^2}\right).$$

  c. Show that PQ, the distance between the point P and the line l, is given by this formula:

  $$PQ = \frac{|ax_1 + by_1 - c|}{\sqrt{a^2 + b^2}}$$

  d. Use the formula in part (c) to solve Exercise 38.

40. **a.** Use the formula derived in Exercise 39 to verify that the point (7, 9) is 3 units from the lines $12x - 5y = 0$ and $3x - 4y = 0$.
  b. Tell why $P(7, 9)$ is on the bisector of $\angle AOB$. Write an equation of this angle bisector and note that its slope is *not* the average of the slopes of lines OA and OB.

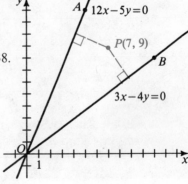

41. a. Suppose $P(x, y)$ is a point in the interior of $\angle AOB$, as shown in the diagram. Explain why P is on the bisector of $\angle AOB$ if

$$\frac{|a_1 x + b_1 y - c_1|}{\sqrt{a_1^2 + b_1^2}} = \frac{|a_2 x + b_2 y - c_2|}{\sqrt{a_2^2 + b_2^2}}$$

(*Hint*: See Exercise 40(a).)

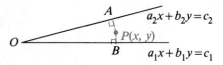

b. Lines l_1 and l_2 have equations $4x - 3y = -6$ and $3x - 4y = -4$. Use the equation of part (a) to write equations of the two lines that bisect the angles formed by l_1 and l_2.

Solutions to quadratic equations

1-4/THE COMPLEX NUMBERS

Throughout the history of mathematics, new kinds of numbers have been invented to fill deficiencies in the existing number system. In earliest times, there were only the counting numbers, that is, the positive integers 1, 2, 3 . . . . The ancient Greeks invented the *rational* numbers, so named because they are *ratios* of two integers. The rational numbers are used to represent fractional parts of a quantity. The Greeks also discovered that some numbers were not rational. For example, the ratio of a diagonal of a square to its side cannot be expressed as a quotient of two integers. We know that this ratio is $\sqrt{2}$, an irrational number.

The invention of zero came from India about A.D. 200. Zero was used to represent an empty column in a counting board similar to an abacus. The negative numbers were invented very much later in Renaissance Europe. One

The co-founders of coordinate geometry are the French mathematicians René Descartes (1596–1650), shown here, and Pierre de Fermat (1601–1655), shown on page 131. In honor of Descartes, the coordinate plane is often called the Cartesian plane, and rectangular coordinates are called Cartesian coordinates.

story tells that the plus and minus signs were first used in German warehouses to indicate whether crates of goods weighed more or less than some standard amount.

The number system consisting of zero and all positive and negative integers, rational numbers, and irrational numbers is called the *real number system*. One of the basic properties of real numbers is that their squares are never negative. In other words, the square root of a negative number cannot be a real number. Accordingly, the word "imaginary" gradually came to be used in the 16th and 17th centuries to describe such numbers as $\sqrt{-1}$ and $\sqrt{-15}$. The use of the word imaginary reflects some of the original uneasiness mathematicians had with these nonreal numbers. Today, the phrase "imaginary numbers" seems a little unfortunate since these numbers are firmly established in mathematics. They are routinely used in advanced mathematics, electrical AC circuits, map making, and quantum mechanics, to name just a few fields.

We introduce the **imaginary unit** i with the following properties:

$$i^2 = -1$$

We consider $\sqrt{-1}$ to be i and define the square root of any negative number as follows:

$$\text{If } a > 0, \ \sqrt{-a} = i\sqrt{a}.$$

EXAMPLE 1.   **a.** $\sqrt{-25} = i\sqrt{25} = 5i$
               b. $\sqrt{-7} = i\sqrt{7}$

EXAMPLE 2.   Simplify $\sqrt{-9} - 2\sqrt{-25}$.

     SOLUTION: $\sqrt{-9} - 2\sqrt{-25} = i\sqrt{9} - 2i\sqrt{25}$
$$= 3i - 2 \cdot 5i$$
$$= -7i$$

Any number of the form $a + bi$ where a and b are real numbers is called a **complex number.** For example, $5i$, $3 + 4i$, $5.2 - i\sqrt{7}$, 6, and 0 are all complex numbers. Notice that we usually write $5i$, 6, and 0 instead of, respectively, $0 + 5i$, $6 + 0i$, and $0 + 0i$. We call a complex number for which $b \neq 0$ an **imaginary number.** For example, $-2 - 4i$, $3i$, and πi are all imaginary numbers. The complex number system consists of the real numbers and the imaginary numbers.

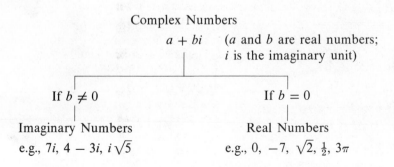

Complex Numbers

$a + bi$ (a and b are real numbers; i is the imaginary unit)

If $b \neq 0$ If $b = 0$

Imaginary Numbers   Real Numbers
e.g., $7i$, $4 - 3i$, $i\sqrt{5}$ e.g., 0, -7, $\sqrt{2}$, $\frac{1}{2}$, 3π

As the chart on the previous page suggests, you can think of the complex number system as consisting of two distinct sets, the real numbers and the imaginary numbers. Imaginary numbers for which $a = 0$, such as $3i$, $-i$, and $i\sqrt{7}$, are called **pure imaginary** numbers.

Two complex numbers $a + bi$ and $c + di$ are equal if and only if $a = c$ and $b = d$. The examples below show that when you add or multiply two complex numbers $a + bi$ and $c + di$, you proceed as if i were a variable. Just remember that $i^2 = -1$.

EXAMPLE 3. $(2 + 3i) + (4 + 5i) = 6 + 8i$

EXAMPLE 4. $(2 + 3i)(4 + 5i) = 8 + 10i + 12i + 15i^2$
$$= 8 + 22i + 15(-1) = -7 + 22i$$

The complex numbers $a + bi$ and $a - bi$ are called **complex conjugates.** Their sum is $2a$, a real number. Their product is $a^2 + b^2$, a nonnegative real number. The conjugate of the complex number $z = a + bi$ is often denoted by $\bar{z} = a - bi$. In the following example, we use the fact that the product of two complex conjugates is a real number. See Exercises 37 and 38 on page 23.

EXAMPLE 5. Divide $5 - 2i$ by $4 + 3i$ and express the answer in the form $a + bi$.

SOLUTION: We multiply the numerator and denominator of $\dfrac{5 - 2i}{4 + 3i}$ by the complex conjugate of the denominator.

$$\frac{5 - 2i}{4 + 3i} \cdot \frac{4 - 3i}{4 - 3i} = \frac{20 - 15i - 8i + 6i^2}{16 + 9} = \frac{20 - 23i - 6}{25}$$

$$= \frac{14 - 23i}{25} = \frac{14}{25} - \frac{23}{25}i$$

Warning: In algebra, you have used the fact that $\sqrt{a}\,\sqrt{b} = \sqrt{ab}$. But this statement is not true if a and b are *both* negative numbers. For example:

$$\sqrt{-4}\,\sqrt{-9} \neq \sqrt{(-4)(-9)}$$

$(2i)(3i)$	$\sqrt{36}$
$6i^2$	6
-6	6

ORAL EXERCISES 1–4

Simplify.

1. $(2 + 3i) + (5 + 4i)$
2. $(5 - 2i) - (6 + 3i)$
3. $(3 + i)(1 + i)$
4. $(9 + i)(3 - i)$
5. $(4 + i)(4 - i)$
6. $(3 + 5i)(3 - 5i)$
7. $(\sqrt{2} + i)(\sqrt{2} - i)$
8. $(a + bi)(a - bi)$

9. Find real numbers x and y such that $4 - 5i = 2x + yi$.

In Exercises 1-6, simplify the given expression.

A **1.** $\sqrt{-4} + \sqrt{-16} + \sqrt{-1}$ **2.** $\sqrt{-49} - \sqrt{-9} + \sqrt{-36}$

3. $\sqrt{-1}\sqrt{-9}$ **4.** $\sqrt{-2}\sqrt{-5}$

5. $\dfrac{\sqrt{-12}}{\sqrt{-3}}$ **6.** $\dfrac{\sqrt{-25}}{\sqrt{-50}}$

In Exercises 7-30, simplify the given expression; that is, write the given expression as a complex number.

7. $(4 - 3i) + (-6 + 8i)$ **8.** $(7 - 8i) - (6 + 2i)$

9. $4(3 + 5i) - 2(2 - 6i)$ **10.** $\frac{1}{6}(7 - 2i) + \frac{2}{3}(5 - 5i)$

11. $(6 - i)(6 + i)$ **12.** $(7 + 3i)(7 - 3i)$

13. $(5 + i\sqrt{5})(5 - i\sqrt{5})$ **14.** $(\sqrt{3} + 4i\sqrt{2})(\sqrt{3} - 4i\sqrt{2})$

15. $(8 + 3i)(2 - 5i)$ **16.** $(5 - 2i)(-1 + 3i)$

17. $(4 - 5i)^2$ **18.** $(4 + 7i)^2$

19. $\dfrac{1}{2 + 5i}$ **20.** $\dfrac{1}{4 - 3i}$

21. $\dfrac{5 + i}{5 - i}$ **22.** $\dfrac{3 - 2i}{3 + 2i}$

23. $\dfrac{3 + i\sqrt{2}}{7 - i\sqrt{2}}$ **24.** $\dfrac{2 + i\sqrt{5}}{3 - i\sqrt{5}}$

25. $\dfrac{5}{i}$ **26.** $\dfrac{i^2 + 2i^3}{i}$

27. $i + i^2 + i^3 + i^4 + i^5$ **28.** $i^{46} + i^{47}$

29. i^{-35} **30.** $(i^n)^4$, where n is any integer

31. Find real numbers x and y such that $(2x + y) + (3 - 5x)i = 1 - 7i$.

32. Find real numbers x and y such that $(3x - 4y) + (6x + 2y)i = 5i$.

B **33. a.** How could you show that 79 is a square root of 6241?
b. How could you show that $3 - i$ is a square root of $8 - 6i$?

34. Show that $4 - 3i$ is a square root of $7 - 24i$.

35. Show that $\dfrac{\sqrt{2}}{2}(1 + i)$ is a square root of i.

36. Show that $\dfrac{\sqrt{3} + i}{2}$ is a cube root of i.

37. Show that the sum of $a + bi$ and its conjugate is a real number.

38. Show that the product of $a + bi$ and its conjugate is a nonnegative real number.

C **39.** If $z = \overline{z}$, what special kind of number is z?

40. Show that $\overline{z_1 + z_2} = \overline{z_1} + \overline{z_2}$.

41. Show that $\overline{z_1 z_2} = \overline{z_1} \cdot \overline{z_2}$.

42. Show that $\overline{z_1 / z_2} = \overline{z_1} / \overline{z_2}$ if $z_2 \neq 0$.

43. Use Exercise 41 to show that $\overline{z^2} = (\overline{z})^2$.

44. a. Use Exercises 41 and 43 to show that $\overline{z^3} = (\overline{z})^3$.

    b. Make a generalization of Exercise 43 and based on part (a).

CALCULATOR EXERCISE

Find $(6 + 7i)^8$. Use the fact that $(x + yi)^2 = (x^2 - y^2) + (2xy)i$.

COMPUTER EXERCISE

Write a program that will compute $(a + bi)^n$ when you input the positive integer n and the real numbers a and b.

1-5/QUADRATIC EQUATIONS

Any equation that can be written in the form

$$ax^2 + bx + c = 0,$$

where $a \neq 0$, is called a **quadratic equation.** There are three common methods for solving quadratic equations:

(1) by factoring;

(2) by completing the square;

(3) by using the quadratic formula.

These methods are illustrated below.

Solution by factoring

Whenever the product of two factors is zero, at least one of the factors must be zero. For this reason you can just look at the equation $(3x - 2)(x + 4) = 0$ and immediately conclude that $3x - 2 = 0$ or $x + 4 = 0$. That is, $x = \frac{2}{3}$ or $x = -4$. A quadratic equation must be written in the **standard form** $ax^2 + bx + c = 0$ before it is solved by factoring.

EXAMPLE 1. Solve $(3x - 2)(x + 4) = -11$.

    SOLUTION:
$$3x^2 + 10x - 8 = -11$$
$$3x^2 + 10x + 3 = 0$$
$$(3x + 1)(x + 3) = 0$$

Therefore, $\quad\quad\quad 3x + 1 = 0 \quad$ or $\quad x + 3 = 0$

$$x = -\tfrac{1}{3} \mid \quad\quad x = -3$$

The solutions are -3 and $-\frac{1}{3}$.

Solution by completing the square

EXAMPLE 2. Solve $2x^2 - 12x - 7 = 0$.

SOLUTION:

1. Divide both sides by the coefficient of x^2 so that x^2 will have a coefficient of 1.

$$x^2 - 6x - \frac{7}{2} = 0$$

2. Subtract the constant term from both sides.

$$x^2 - 6x = \frac{7}{2}$$

3. Complete the square by adding the square of one half the coefficient of x to both sides.

$$x^2 - 6x + (-3)^2 = \frac{7}{2} + (-3)^2$$

$$(x - 3)^2 = \frac{25}{2}$$

4. Take the square root of both sides and solve.

$$x - 3 = \pm \frac{5\sqrt{2}}{2}$$

$$x = 3 \pm \frac{5\sqrt{2}}{2}$$

The solutions are

$$3 - \frac{5\sqrt{2}}{2} \text{ and } 3 + \frac{5\sqrt{2}}{2}.$$

Solution by the quadratic formula

The quadratic formula, shown below, is derived by completing the square. (See Exercise 56.)

QUADRATIC FORMULA: The solutions to the quadratic equation $ax^2 + bx + c = 0$ are given by:

$$x = \frac{-b \pm \sqrt{b^2 - 4ac}}{2a} \qquad (a \neq 0)$$

EXAMPLE 3. Solve $2x^2 + 7 = 4x$.

SOLUTION: Rewrite the equation in standard form:

$$2x^2 - 4x + 7 = 0$$

Substitute $a = 2$, $b = -4$, and $c = 7$.

$$x = \frac{-(-4) \pm \sqrt{(-4)^2 - 4 \cdot 2 \cdot 7}}{2 \cdot 2}$$

$$= \frac{4 \pm \sqrt{-40}}{4} = \frac{4 \pm 2i\sqrt{10}}{4} = 1 \pm \frac{i\sqrt{10}}{2}$$

The solutions are $1 - \dfrac{i\sqrt{10}}{2}$ and $1 + \dfrac{i\sqrt{10}}{2}$.

The discriminant $b^2 - 4ac$

The quantity that appears beneath the radical sign in the quadratic formula, $b^2 - 4ac$, can tell you whether the roots of a quadratic formula are real or imaginary. Because of this "discriminating ability," $b^2 - 4ac$ is called the **discriminant.** Also, if an equation with integral coefficients has any real roots, the discriminant can tell you whether or not the roots are rational.

THEOREM 4   Given the equation $ax^2 + bx + c = 0$, where a, b, and c are real numbers:

If $b^2 - 4ac < 0$, there are two conjugate imaginary roots.

If $b^2 - 4ac = 0$, there is one real root (called a *double root*).

If $b^2 - 4ac > 0$, there are two different real roots.

(See Oral Exercises 11 and 12.)

THEOREM 5   Given the equation $ax^2 + bx + c = 0$, where a, b, and c are integers: The discriminant $b^2 - 4ac$ is the square of an integer if and only if the equation has rational roots. (See Oral Exercise 13.)

The following is an equivalent statement to Theorem 5: the discriminant $b^2 - 4ac$ is the square of an integer if and only if the equation can be factored with integral coefficients.

Choosing a method of solution

Although you can use the quadratic formula to solve any quadratic equation, it is often much easier to factor or complete the square. In particular, if a, b, and c are integers and $b^2 - 4ac$ is a perfect square, then you can factor. If the equation has the form

$$x^2 + (\text{even number})x + \text{constant} = 0,$$

then completing the square is usually the easiest method. In the remaining situations, the formula is usually easiest, especially if you have to solve an equation like $px^2 + qx + r = 0$, where the coefficients are variables whose values are not known.

ORAL EXERCISES 1–5

What must be added to the following expressions to complete the square?

1. $x^2 + 8x +$ _?_   **2.** $x^2 - 10x +$ _?_   **3.** $x^2 + 7x +$ _?_   **4.** $x^2 + 2ax +$ _?_

Does factoring, completing the square, or using the quadratic formula seem to you the easiest method for solving the following equations?

5. $x^2 + 14x = 374$ **6.** $4x^2 - 5x = 0$ **7.** $3x^2 + 11x - 4 = 0$

8. $px^2 + qx + r = 0$ **9.** $4x^2 - x - 7 = 0$ **10.** $x^2 - 14x - 736 = 0$

Exercises 11–13 refer to the quadratic equation $ax^2 + bx + c = 0$, where a, b, and c are real numbers.

11. Explain why the equation has imaginary roots when $b^2 - 4ac < 0$.

12. Explain why the equation has only one root, and why this root is real, when $b^2 - 4ac = 0$. Does it seem reasonable to you to call this sort of root a *double root?*

13. If a, b, and c are integers and $b^2 - 4ac$ is the square of an integer, explain why the equation has rational roots.

14. The discriminant of the equation $x^2 - x\sqrt{5} + 1$ is the square of an integer, but the roots of the equation are not rational. Does the equation offer a contradiction of the result of Exercise 13?

15. Can a quadratic equation with irrational coefficients be solved by using the quadratic formula?

16. Can a quadratic equation with irrational coefficients be solved by completing the square?

WRITTEN EXERCISES 1-5

In Exercises 1–4, solve the equations by factoring.

A   **1.** $3x^2 - 4x - 7 = 0$ **2.** $4x^2 - 8x - 32 = 0$

     3. $(2x - 3)(x + 4) = 6$ **4.** $(3y - 2)(y + 4) = 24$

In Exercises 5–14 solve the equations by completing the square. Give both real and imaginary roots.

     5. $x^2 - 2x = 399$ **6.** $x^2 - 6x = 391$

     7. $x^2 - 10x = 1575$ **8.** $x^2 + 12x = 2464$

     9. $2z^2 - 16z - 1768 = 0$ **10.** $x^2 - 8x - 20 = 0$

    11. $x^2 + 6x + 10 = 0$ **12.** $y^2 + 10y + 35 = 0$

    13. $x^2 - 6x - 3 = 0$ **14.** $x^2 - 7x + 11 = 0$

In Exercises 15–20 solve the equations by using the quadratic formula. Give your answers in simplest radical form. Give both real and imaginary roots.

    15. $5x^2 + 2x - 1 = 0$ **16.** $4x^2 - 4x - 17 = 0$

    17. $3t^2 = 12t - 15$ **18.** $5u^2 + 2 = 5u$

    19. $\dfrac{4}{v} = \dfrac{v - 6}{v - 4}$ **20.** $\dfrac{4}{z} = \dfrac{3z}{z - 3}$

In Exercises 21–24 find the discriminant of the given equation. If the discriminant is the square of an integer, solve by factoring. Otherwise, solve by another method. Give both real and imaginary roots.

    21. $6x^2 - 5x + 3 = 0$ **22.** $8y^2 = 7 - 10y$

    23. $7 - 3y^2 = 5y$ **24.** $4t = 1 + 15t^2$

In Exercises 25–30 solve the equations by whichever method seems easiest. Be sure to rule out values of the variable that cause a denominator to be zero.

25. $(3x - 2)^2 = 121$

26. $\dfrac{v + 2}{v - 1} = \dfrac{3v}{10} + \dfrac{1}{v - 1}$

27. $\dfrac{1}{z - 2} - \dfrac{1}{z + 2} - \dfrac{2}{7} = 0$

28. $(4y + 4)^2 = -16$

29. $\dfrac{17}{n^2 - 25} + \dfrac{1}{n + 5} = \dfrac{n}{n - 5}$

30. $\dfrac{y^2 + 3}{y + 1} = \dfrac{y + 1}{3} + \dfrac{4}{y + 1}$

In Exercises 31 and 32 $\overline{DE}$ is parallel to $\overline{BC}$. Find the value of x.

31.

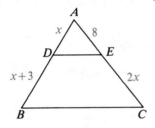

32.

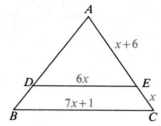

In Exercises 33 and 34, $\overset{\leftrightarrow}{PB}$ intersects the circle at points A and B, and $\overset{\leftrightarrow}{PD}$ intersects the circle at points C and D. By theorems from geometry, $PA \cdot PB = PC \cdot PD$ in each case. Find the value of x.

33.

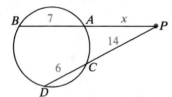

34.

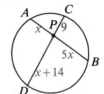

B   **35.** **a.** What is the discriminant of the equation $4x^2 + 8x + k = 0$?
　　b. For what value of k will the equation have a double root?
　　c. For what values of k will the equation have two real roots?
　　d. For what values of k will the equation have imaginary roots?
　　e. Name three values of k for which the given equation has rational roots.

36. Repeat Exercise 35 for the equation $5x^2 + 8x + k = 0$.

In Exercises 37 and 38, x_1 and x_2 are the roots of $ax^2 + bx + c = 0$. Thus

$$x_1 = \frac{-b + \sqrt{b^2 - 4ac}}{2a} \quad \text{and} \quad x_2 = \frac{-b - \sqrt{b^2 - 4ac}}{2a}.$$

37. Show that $x_1 + x_2 = -\dfrac{b}{a}$.

38. Show that $x_1 \cdot x_2 = \dfrac{c}{a}$.

Use the results of Exercises 37 and 38 to find the sum and product of the roots of the equations given in Exercises 39–42.

39. $3x^2 + 5x + 1 = 0$ **40.** $4x^2 + 9x + 2 = 0$

41. $5x^2 - 7x - 9 = 0$ **42.** $8x^2 - 6x = 0$

43. a. Note that the equation $ax^2 + bx + c = 0$ can be rewritten as the equivalent equation:

$$x^2 - \left(-\frac{b}{a}\right)x + \left(\frac{c}{a}\right) = 0$$

Using Exercises 37 and 38, explain why a quadratic equation can be written as:

$$x^2 - (\text{sum of its roots})x + (\text{product of its roots}) = 0.$$

b. Use part (a) to write a quadratic equation having roots 3 and 4.

In Exercises 44–49, use the result of Exercise 43(a) to write a quadratic equation in standard form that has the given roots.

44. 9, 6   **45.** 8, −2   **46.** $6 \pm \sqrt{3}$

47. $-4 \pm \sqrt{10}$ **48.** $3 \pm i$ **49.** $5 \pm 2i$

In Exercises 50–55, solve each equation.

50. $\sqrt{2}x^2 - 5x + \sqrt{8} = 0$ **51.** $4x^2 - 2\sqrt{5}x - 1 = 0$

52. $x^2 + \sqrt{6}x + \sqrt{2} = 0$ **53.** $x^2 + \sqrt{5}x = \sqrt{5}x^2 + x$

C   **54.** $x^2 - 6ix - 9 = 0$ **55.** $ix^2 - 3x - 2i = 0$

56. Derive the quadratic formula. (*Hint:* Solve the general quadratic equation $ax^2 + bx + c = 0$, $a \neq 0$, by completing the square.)

57. a. Suppose that a is a positive integer such that $ax^2 + x - 6$ can be factored. Find the five smallest values of a.
     b. Find two values of a that are greater than 100.

58. If a, b, and c are odd integers, can the roots of $ax^2 + bx + c = 0$ ever be rational? Prove that your answer is correct.

CALCULATOR EXERCISE

Solve for x to the nearest thousandth.

1. $(4 + \sqrt{5})x^2 - 2\sqrt{3}x - 3 = 0$ **2.** $(3 + 2\sqrt{2})x^2 + 5\sqrt{2}x - \sqrt{17} = 0$

COMPUTER EXERCISE

Write a computer program that will print the roots of $ax^2 + bx + c = 0$ when you input real numbers for a, b, and c. Have the program print REAL ROOTS, DOUBLE ROOT, or IMAGINARY ROOTS as appropriate.

Circles

1-6/EQUATIONS OF CIRCLES

The set of all points in the plane that are 5 units from the point $C(2, 4)$ is a circle. To determine an equation of this circle, we use the distance formula:

$$PC = 5$$
$$\sqrt{(x - 2)^2 + (y - 4)^2} = 5$$
$$(x - 2)^2 + (y - 4)^2 = 25$$

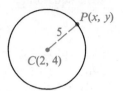

In general, if $P(x, y)$ is on a circle with center $C(h, k)$ and radius r, then:

$$PC = r$$
$$\sqrt{(x - h)^2 + (y - k)^2} = r$$
$$(x - h)^2 + (y - k)^2 = r^2$$

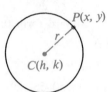

The steps leading to the last equation can be reversed to show that any point $P(x, y)$ satisfying the equation is on the circle with center (h, k) and radius r. Therefore, we have the following:

Equation of circle with
radius r and center (h, k): $(x - h)^2 + (y - k)^2 = r^2$

When the center of the circle is the origin, the form of the equation is:

Equation of circle with
radius r and center at origin: $x^2 + y^2 = r^2$

For brevity, we often refer to a circle with equation $x^2 + y^2 = r^2$, for example, as "the circle $x^2 + y^2 = r^2$."

EXAMPLE 1. Find the center and radius of each circle.
a. $(x - 3)^2 + (y + 7)^2 = 19$ **b.** $x^2 + y^2 - 6x + 4y - 12 = 0$

SOLUTION: **a.** The equation $(x - 3)^2 + (y + 7)^2 = 19$ is already in *center-radius form*. The center of the circle is $(3, -7)$, and the radius is $\sqrt{19}$.

b. We rewrite the equation in center-radius form by completing the squares in x and y.

Original equation: $x^2 + y^2 - 6x + 4y - 12 = 0$
$$(x^2 - 6x \quad\;) + (y^2 + 4y \quad\;) = 12$$
$$(x^2 - 6x + 9) + (y^2 + 4y + 4) = 12 + 9 + 4 = 25$$

Center-radius form: $(x - 3)^2 + (y + 2)^2 = 25$

Thus, the center of the circle is $(3, -2)$ and the radius is 5.

The graph of an equation in the form

$$x^2 + y^2 + Dx + Ey + F = 0,$$

where D, E, and F are constants, is usually a circle. The equation given in part (b) of Example 1 was in this form with $D = -6$, $E = 4$, and $F = -12$. The graph of an equation in the general form just given can also be a single point (the graph of $x^2 + y^2 = 0$) or no points at all (the graph of $x^2 + y^2 + 1 = 0$). Any circle, however, has an equation in this form. If you multiply out the squares and collect terms in the equation in part (a) of Example 1, you get the equation $x^2 + y^2 - 6x + 14y + 39 = 0$.

ORAL EXERCISES 1-6

Find the center and radius of each circle whose equation is given.

1. $x^2 + y^2 = 16$ **2.** $(x - 2)^2 + (y - 7)^2 = 36$

3. $(x - 4)^2 + (y + 7)^2 = 7$ **4.** $x^2 + (y^2 + 12y) = 0$

5. $(x^2 - 2x) + (y^2 - 6y) = 9$ **6.** $4x^2 + 4y^2 = 36$

Find equations of the following circles.

7. The circle with center $(7, 3)$ and radius 6

8. The circle with center $(-5, 4)$ and radius $\sqrt{2}$

9. The circle with center $(0, 0)$ that passes through $(-5, 12)$

WRITTEN EXERCISES 1-6

In Exercises 1–16 write an equation of the circle described. We use C to denote the center and r to denote the radius.

A   **1.** $C(4, 3)$, $r = 2$ **2.** $C(5, -6)$, $r = 7$

      3. $C(-4, -9)$, $r = 3$ **4.** $C(a, b)$, $r = f$

      5. $C(6, 0)$, $r = \sqrt{15}$ **6.** $C(-4, 2)$, $r = \sqrt{7}$

7. The center is $(2, 3)$; the circle passes through $(5, 6)$.

8. The center is $(4, -6)$; the circle passes through $(10, 6)$.

9. The points $(8, 0)$ and $(0, 6)$ are endpoints of a diameter.

10. The points $(-4, 4)$ and $(2, 8)$ are endpoints of a diameter.

11. The center is $(5, -4)$; the circle is tangent to the x-axis.

12. The center is in the third quadrant, the radius is 8, and the circle is tangent to both axes.

13. The center is $(6, 2)$, and the circle is tangent to the line $y = 3$.

14. The center is $(-3, 1)$, and the circle is tangent to the line $x = 4$.

15. The circle is tangent to the x-axis at $(4, 0)$ and has y-intercepts -2 and -8.

16. The circle passes through $(-2, 16)$ and has x-intercepts -2 and -32.

17. Show that a point (x_1, y_1) is inside the circle $x^2 + y^2 = r^2$ if $x_1^2 + y_1^2 < r^2$.

18. Given the circle $x^2 + y^2 = 40$, tell whether each of the following points is on the circle, inside the circle, or outside the circle.

    a. $(-6, -2)$ **b.** $(3, 5)$ **c.** $(-4, 5)$ **d.** $(3, -5.5)$

Rewrite each of the equations in Exercises 19–26 in center-radius form, and then give the center and radius of the graph of the equation.

19. $x^2 + y^2 - 2x - 8y + 16 = 0$ 20. $x^2 + y^2 - 4x + 6y + 4 = 0$

21. $x^2 + y^2 - 12y + 25 = 0$ 22. $x^2 + y^2 + 14x = 0$

23. $2x^2 + 2y^2 - 10x - 18y = 1$ 24. $2x^2 + 2y^2 - 5x + y = 0$

25. $3x^2 + 3y^2 - 6x + 3y = 4$ 26. $9x^2 + 9y^2 + 9x - 27y + 1 = 0$

In Exercises 27–32 find the distance between the centers of the two circles whose equations are given. By comparing this distance with the two radii, determine whether the circles intersect in two points, one point, or no points. Make a sketch of the two circles.

27. $x^2 + y^2 = 9$, $(x - 8)^2 + y^2 = 25$

28. $x^2 + y^2 = 4$, $x^2 + (y - 4)^2 = 9$

29. $x^2 + y^2 = 12$, $(x + 4)^2 + (y + 4)^2 = 3$

30. $x^2 + y^2 = 3$, $x^2 + y^2 - 4x - 4y = 4$

31. $x^2 + y^2 - 2x - 4y = -1$, $x^2 + y^2 + 2x + 2y = 14$

32. $x^2 + y^2 = 9$, $x^2 + y^2 + 2x + 4y = -1$

33. Show that the line $y = 2x + 8$ passes through the center of the circle $x^2 + y^2 + 6x - 4y + 8 = 0$.

34. Determine whether the line $3x + 2y = 6$ passes through the center of the circle $x^2 + y^2 + 4x - 12y + 24 = 0$.

B   35. A circle with center $C(2, 4)$ has radius 13.

    a. Verify that $A(14, 9)$ and $B(7, 16)$ are points on this circle.

    b. If M is the midpoint of $\overline{AB}$, show that $\overline{CM} \perp \overline{AB}$.

36. A circle with center $C(-4, 0)$ has radius 15.

    a. Verify that $A(8, 9)$ and $B(-13, 12)$ are points on this circle.

    b. Write the equation of the perpendicular bisector of $\overline{AB}$ and show that the coordinates of point C satisfy the equation.

37. A diameter of a circle has endpoints $A(13, 0)$ and $B(-13, 0)$.

    a. Show that $P(-5, 12)$ is a point on this circle.

    b. Show that $\overline{PA}$ and $\overline{PB}$ are perpendicular.

38. **a.** Find the coordinates of A and B if $\overline{AB}$ is a horizontal diameter of the circle $x^2 + y^2 - 34x = 0$.

    b. Show that $P(2, 8)$ is a point on this circle and that $\overline{PA} \perp \overline{PB}$.

39. Sketch the graph of the semicircle whose equation is given.

    a. $y = \sqrt{9 - x^2}$ **b.** $y = -\sqrt{9 - x^2}$

    c. $x = \sqrt{9 - y^2}$ **d.** $x = -\sqrt{9 - y^2}$

40. Sketch the graph of the semicircle whose equation is given.

    a. $y = \sqrt{16 - (x - 5)^2}$ **b.** $y = -\sqrt{16 - (x - 5)^2}$

    c. $x = \sqrt{16 - (y - 5)^2}$ **d.** $x = -\sqrt{16 - (y - 5)^2}$

41. Given $A(6, 8)$ and $B(-6, -8)$, write an equation in terms of x and y for all points $P(x, y)$ such that $\overline{PA} \perp \overline{PB}$. Simplify the equation and interpret your answer.

42. Given $O(0, 0)$ and $N(12, 0)$, find an equation for all points $P(x, y)$ in terms of x and y such that $\overline{PO} \perp \overline{PN}$. Simplify this equation and show that P is on a circle. What are the center and radius of this circle?

In Exercises 43–46, find an equation of the circle that passes through all the given points.

43. $A(0, 0)$, $B(2, 0)$, $C(2, 2)$, and $D(0, 2)$ **44.** $P(0, 0)$, $Q(6, 0)$, and $R(0, 8)$

45. $L(8, 2)$, $M(1, 9)$, and $N(1, 1)$ **46.** $D(7, 5)$, $E(1, -7)$, and $F(9, -1)$

Describe the set of points satisfying each equation.

47. $x^2 + y^2 + 2x + 2y + 2 = 0$ **48.** $x^2 + y^2 - 6x + 8y + 26 = 0$

C   **49.** $(x^2 + y^2 - 1)(x^2 + y^2 - 4) = 0$ **50.** $x^3y + xy^3 - xy = 0$

1-7 / INTERSECTIONS OF LINES AND CIRCLES

Intersection of a line and a circle

The figure shows a line and a circle with equations $y = 2x - 2$ and $x^2 + y^2 = 25$, respectively. To determine the coordinates of their intersection points, A and B, we solve these two equations simultaneously:

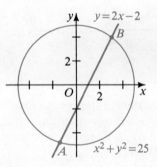

$$y = 2x - 2 \qquad (1)$$
$$x^2 + y^2 = 25 \qquad (2)$$

Substituting for y in equation (2) gives us:

$$x^2 + (2x - 2)^2 = 25$$
$$x^2 + 4x^2 - 8x + 4 = 25$$
$$5x^2 - 8x - 21 = 0$$
$$(5x + 7)(x - 3) = 0$$
$$x = -\tfrac{7}{5} \text{ or } x = 3$$

Substituting these values for x in equation (1) gives us

$$y = 2(-\tfrac{7}{5}) - 2 = -\tfrac{24}{5} \quad \text{and} \quad y = 2(3) - 2 = 4.$$

Thus, $A = (-\tfrac{7}{5}, -\tfrac{24}{5})$ and $B = (3, 4)$. We can check this solution by substituting the coordinates of A and B in the equations of the line and the circle.

TO FIND THE INTERSECTION OF A LINE AND A CIRCLE:

1. Solve the linear equation for y in terms of x (or x in terms of y).
2. Substitute this expression for y (or x) in the equation of the circle.
3. Solve the resulting quadratic equation.
4. Substitute each x-solution in the *linear* equation to get the corresponding value of y (or vice versa). Each point (x, y) is an intersection point.
5. You can check your result by substituting the coordinates of the intersection points in the two original equations.

It is, of course, possible that a line and a circle will not intersect. In this case, the quadratic equation solved in Step 3 of the procedure will have a negative discriminant and the equation will have only imaginary roots. If the discriminant is zero, then there is only one real root, indicating that the line intersects the circle in a single point. In such a case, the line is tangent to the circle.

Intersection of two circles

The figure shows circles with equations $x^2 + y^2 = 20$ and $(x - 5)^2 + (y - 5)^2 = 10$, respectively. To determine their points of intersection, we find the common solutions of their equations. We write both equations in expanded form and then subtract.

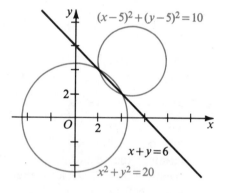

$$
\begin{aligned}
x^2 + y^2 &= 20 \quad &(1) \\
x^2 + y^2 - 10x - 10y &= -40 \quad &(2) \\
\hline
\text{By subtracting,} \quad 10x + 10y &= 60 \\
x + y &= 6 \quad &(3)
\end{aligned}
$$

Equation (3) represents the line through the intersection points of the circles. If we solve this equation simultaneously with equation (1), we can find the coordinates of these points. Solving (3) for y gives $y = 6 - x$. Then substituting in (1) gives:

$$
\begin{aligned}
x^2 + (6 - x)^2 &= 20 \\
x^2 + 36 - 12x + x^2 &= 20 \\
x^2 - 6x + 8 &= 0 \\
(x - 2)(x - 4) &= 0 \\
x = 2 \quad \text{or} \quad x = 4.
\end{aligned}
$$

When $x = 2$, | When $x = 4$,
$y = 6 - x = 4$ | $y = 6 - x = 2$

Thus the intersection points are $(2, 4)$ and $(4, 2)$. This result can be checked by substituting $(2, 4)$ and $(4, 2)$ in equations (1) and (2).

If the two original circles are tangent, then the linear equation derived by subtracting is the equation of their common internal tangent. [See figure (a) below.] If the two original circles do not intersect at all, then the derived linear equation represents the special line shown in figure (b).

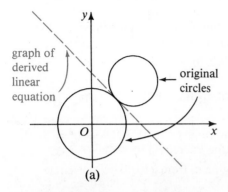

(a)

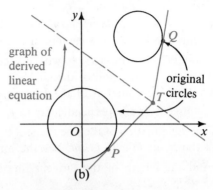

(b)

The original circles are tangent; the graph of the derived linear equation is their common internal tangent.

The original circles do not intersect. The graph of the derived equation consists of those points $T(x, y)$ for which tangents $\overline{TP}$ and $\overline{TQ}$ are equal.

WRITTEN EXERCISES 1-7

In Exercises 1–8 the equations of a line and a circle are given, and in Exercises 9–14 the equations of two circles are given. In each exercise solve the equations simultaneously to find the intersection points of their graphs. If the graphs are tangent or fail to intersect, say so. Illustrate graphically.

Line and Circle:

A   **1.** $x + y = 23,\ x^2 + y^2 = 289$ **2.** $9y - 8x = 10,\ x^2 + y^2 = 100$

     3. $2x - y = 7,\ x^2 + y^2 = 7$ **4.** $x + 2y = 10,\ x^2 + y^2 = 20$

     5. $5x + 2y = -1,\ x^2 + y^2 = 169$

     6. $y = 7,\ x^2 + y^2 - 4x - 6y = -9$

     7. $y = \sqrt{3}x,\ x^2 + (y - 4)^2 = 16$

     8. $x - y = 3,\ x^2 + y^2 - 10x + 4y = -13$

Two Circles:

     9. $x^2 + y^2 = 16,\ (x - 4)^2 + y^2 = 16$

   10. $x^2 + y^2 = 4,\ x^2 + (y - 6)^2 = 25$

   11. $x^2 + y^2 = 20,\ (x - 2)^2 + (y + 1)^2 = 13$

   12. $x^2 + y^2 - 4y = 0,\ x^2 + y^2 - 2x = 4$

   13. $x^2 + y^2 = 5,\ x^2 + y^2 - 12x + 6y = -25$

   14. $x^2 + y^2 = 4,\ x^2 + y^2 - 6x - 6y = -14$

B **15.** The line $x - 2y = 15$ intersects the circle $x^2 + y^2 = 50$ at the points A and B. Show that the line joining the center of the circle to the midpoint of $\overline{AB}$ is perpendicular to $\overline{AB}$.

16. a. Find an equation of the common chord of the circles $x^2 + y^2 = 16$ and $x^2 + y^2 - 6x - 8y = 0$.

b. Use the equation to show that the common chord is perpendicular to the line joining the centers of the two circles.

17. Point $P(2, 3)$ is on a circle with center $O(0, 0)$.

a. Write an equation of the circle.

b. Write an equation of the line tangent to the circle at P. (*Hint:* The tangent line is perpendicular to $\overrightarrow{OP}$.)

18. Find an equation of the line tangent to the circle $(x - 3)^2 + (y - 4)^2 = 5$ at the point $P(4, 2)$. (*Hint:* See the hint for Exercise 17.)

19. Find the length of the tangent line segment from $(10, 5)$ to the circle $x^2 + y^2 = 25$.

20. Find the length of a tangent line segment from $(2, 8)$ to $(x + 5)^2 + (y - 2)^2 = 25$.

21. Compare the graphs of the equations $x^2 + y^2 = 9$ and $(x + y)^2 = 9$.

22. Solve simultaneously and illustrate with a graph:

$$x^2 + y^2 = \frac{10}{9}$$

$$(x + y)^2 = \frac{16}{9}$$

C **23.** The lines $y = x$ and $y = 0$ form a $45°$ angle at the origin. The figure shows the ruler-and-compass construction marks used to obtain the bisector of this angle. If all the circular arcs have radius 3, find the coordinates of P and thus find an equation of the angle bisector $\overleftrightarrow{OP}$.

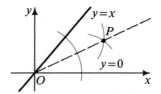

24. If the quadratic equation $x^2 + bx + c = 0$ has two real roots, they can be located graphically as follows: First, locate the points $P(0, 1)$ and $Q\left(\dfrac{-b}{2}, \dfrac{c + 1}{2}\right)$. With Q as center, draw a circle passing through P. This circle will intersect the x-axis in two points whose x-coordinates are the roots of the quadratic equation. Prove this. Also state what happens if the equation has **(a)** no real roots and **(b)** just one real root.

CALCULATOR EXERCISE

A circle circumscribes a rectangle with length 517 and width 238. Find the area of the shaded region to the nearest integer.

Coordinate proofs

1-8/PROOFS IN COORDINATE GEOMETRY

Now that we have seen how algebraic expressions can be used to describe geometric relationships, we can use these expressions to prove geometric theorems. If the theorem is about a right triangle, there are several ways that we can assign coordinates to the vertices of the triangle. For example, figures (1) and (2) below both illustrate a right triangle with legs a and b units long. Most people prefer to work with figure (2), because more of the coordinates are zero and this makes the work easier. Figure (3) shows that even if the right triangle is "tilted," we can choose the coordinate axes in such a way that the vertices of the triangle have several zero coordinates.

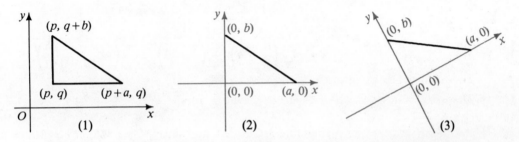

Similarly, if we are going to prove a theorem about a trapezoid or a parallelogram, we can always choose our axes so that one of the vertices of the figure is at the origin and one of its parallel sides lies on the x-axis. See figures (4) and (5) below.

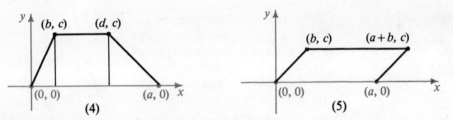

EXAMPLE 1. Prove that the midpoint of the hypotenuse of a right triangle is equidistant from the three vertices.

SOLUTION: **Step 1.** *First, we make a coordinate diagram of the triangle and note what we are given and what we must prove.* In this example we place the axes along the legs of the right triangle.

Given: $\angle C$ is a right angle.
M is the midpoint of $\overline{AB}$.
Prove: $MC = MA$. (We already know that $MB = MA$.)

Step 2. *Next, we use what is given, to add information to the diagram or to express algebraically any given fact not shown in the diagram.* In this example, we use the given fact that M is the midpoint of $\overline{AB}$ to find its coordinates.

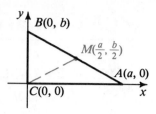

Step 3. *Finally, we use algebra to complete the proof.* To prove $MC = MA$:

$$MC = \sqrt{\left(\frac{a}{2} - 0\right)^2 + \left(\frac{b}{2} - 0\right)^2} = \sqrt{\left(\frac{a}{2}\right)^2 + \left(\frac{b}{2}\right)^2}$$

$$= \sqrt{\frac{a^2}{4} + \frac{b^2}{4}}$$

$$MA = \sqrt{\left(\frac{a}{2} - a\right)^2 + \left(\frac{b}{2} - 0\right)^2} = \sqrt{\left(-\frac{a}{2}\right)^2 + \left(\frac{b}{2}\right)^2}$$

$$= \sqrt{\frac{a^2}{4} + \frac{b^2}{4}}$$

Therefore $MC = MA$, and since $MA = MB$, $MA = MB = MC$.

EXAMPLE 2. (Converse of Example 1) If in $\triangle ABC$, the midpoint of side $\overline{AB}$ is equidistant from A, B, and C, then $\angle C$ is a right angle.

SOLUTION: **Step 1.** *We make a coordinate diagram and list what is given and what is to be proved.* We choose coordinate axes so that the x-axis contains one side of the triangle, and the y-axis passes through the opposite vertex.

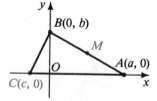

Given: (1) M is the midpoint of $\overline{AB}$.
       (2) $MC = MA$.
Prove: $\angle C$ is a right angle.

Step 2. *Next, we use what is given, to add information to the diagram or to express algebraically any given fact not shown in the diagram.* We use the first given item to label the diagram with the coordinates of M. The second given item, $MA = MC$, is expressed algebraically as shown at the top of the next page.

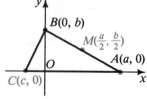

$$\sqrt{\left(\frac{a}{2}-c\right)^2+\left(\frac{b}{2}-0\right)^2}=\sqrt{\left(\frac{a}{2}-a\right)^2+\left(\frac{b}{2}-0\right)^2}$$

$$\left(\frac{a}{2}-c\right)^2+\left(\frac{b}{2}\right)^2=\left(\frac{a}{2}-a\right)^2+\left(\frac{b}{2}\right)^2$$

$$\left(\frac{a}{2}-c\right)^2=\left(-\frac{a}{2}\right)^2$$

$$\frac{a^2}{4}-ac+c^2=\frac{a^2}{4}$$

$$c^2-ac=0$$

$$c(c-a)=0$$

$$c=0 \quad \text{or} \quad c=a$$

Since $c \neq a$ (why not?), we have $c = 0$.

Step 3. *We reword what we are to prove in algebraic terms.* We can prove that $\angle C$ is a right angle by showing that $c = 0$ since this implies that $C = (0,0)$. Since we have already shown $c = 0$ in Step 2, we are done.

EXAMPLE 3. Prove that the median of a trapezoid is parallel to the bases and has length equal to the average of the lengths of the bases.

SOLUTION: Step 1. *We show a diagram, and the "Given" and "Prove."* We place the x-axis along one base of the trapezoid, with the origin at an endpoint of the base. Since the bases of a trapezoid are parallel, and base $\overline{OP}$ has been chosen to be horizontal, base $\overline{RQ}$ is also horizontal. Thus R and Q have the same y-coordinate.

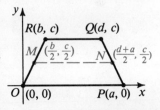

Given: $OPQR$ is a trapezoid. M and N are midpoints of $\overline{OR}$ and $\overline{PQ}$.

Prove: (1) $\overline{MN} \parallel \overline{OP}$ and (2) $MN = \dfrac{OP + RQ}{2}$.

Step 2. *We use the given information* that M and N are midpoints to label their coordinates in the diagram.

Step 3. *We reword what we are to prove in algebraic terms.*

(1) To prove $\overline{MN} \parallel \overline{OP}$, we must show that $\overline{MN}$ and $\overline{OP}$ have the same slope. A quick check shows that both slopes are zero, so this part of the proof is done.

(2) To prove $MN = \dfrac{1}{2}(OP + RQ)$ means to prove that

$$\left(\frac{d+a}{2}-\frac{b}{2}\right)=\frac{1}{2}(a+(d-b)).$$

But this last equation is an identity, so our proof is done.

EXAMPLE 4. Prove that the altitudes of a triangle are concurrent.

SOLUTION: **Step 1.** *We show a diagram and the "Given" and "Prove."*

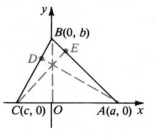

Given: $\triangle ABC$ with altitudes $\overline{AD}$, $\overline{BO}$, and $\overline{CE}$.

Prove: Lines AD, BO, and CE have a common point.

Notice that the axes are placed in such a way that one of the altitudes lies on the y-axis.

Step 2. *We use the given information to express algebraically the fact that $\overline{AD}$, $\overline{BO}$, and $\overline{CE}$ are altitudes.*

a. To find the slope of line AD, we note that the slope of line BC is $\dfrac{b-0}{0-c} = -\dfrac{b}{c}$, so that the slope of line AD is $\dfrac{c}{b}$.

Since line AD contains the point $(a, 0)$, its equation is

$$\frac{y-0}{x-a} = \frac{c}{b}, \quad \text{or} \quad cx - by = ca.$$

b. Likewise, an equation of the line CE is

$$\frac{y-0}{x-c} = \frac{a}{b}, \quad \text{or} \quad ax - by = ca.$$

c. An equation of the vertical line BO is $x = 0$.

Step 3. *We reword what we are to prove in algebraic terms.* To prove that lines AD, BO, and CE have a common point, we must show that their equations have a common solution. Using subtraction to solve

$$cx - by = ca$$
$$ax - by = ca,$$

gives us

$$cx - ax = 0,$$
$$x(c - a) = 0.$$

Since $c \neq a$, $x = 0$.

Thus, the first two altitudes intersect at a point on the y-axis, that is, on the third altitude, so we are done. The point of concurrency of the altitudes is called the *orthocenter* of the triangle.

SUMMARY OF METHODS COMMONLY USED IN COORDINATE PROOFS

1. To prove line segments equal, use the distance formula to show that they have the same length.

2. To prove lines parallel, show that they have the same slope.

3. To prove lines perpendicular, show that the product of their slopes is -1.

4. To prove two line segments bisect each other, use the midpoint formula to show that each segment has the same midpoint.

5. To show lines are concurrent, show that their equations have a common solution.

1. Study the coordinates of the vertices in the following diagrams and tell which figures represent isosceles triangles.

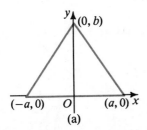

(a)

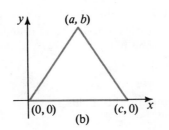

(b)

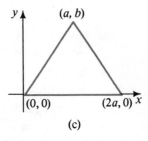

(c)

2. Which of the following diagrams represent parallelograms?

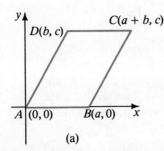

(a)

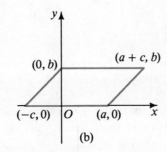

(b)

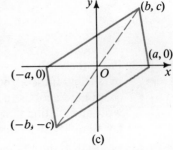

(c)

WRITTEN EXERCISES 1–8

In the proofs called for in the following exercises, do not use any geometric theorems other than the Pythagorean theorem. You may use the distance formula, the midpoint formula, Theorems 1 through 5, and results proved in earlier exercises of this section.

A

1. Prove that the medians to the legs of an isosceles triangle are equal. (*Suggestion:* Use figure (a) or (c) of Oral Exercise 1.)

2. Prove that if a triangle has two equal medians, then it is isosceles.

3. Prove that the line segment joining the midpoints of two sides of a triangle is parallel to the third side and has length half that of the third side.

4. Use either figure (a) or (b) of Oral Exercise 2 to prove that the diagonals of a parallelogram bisect each other. How does this result show that any parallelogram can be represented by figure (c) of Oral Exercise 2?

5. Prove that if the diagonals of a quadrilateral bisect each other, then the figure is a parallelogram.

6. Prove that the diagonals of a rectangle are equal.

Coordinate geometry   **41**

7. Study the figure for Example 3 on page 39 and write an equation that expresses the fact that the diagonals of a certain trapezoid are equal. Simplify this equation and prove that the trapezoid is isosceles.

8. Prove that the diagonals of an isosceles trapezoid are equal.

9. Prove that the line segments joining the midpoints of successive sides of any quadrilateral form a parallelogram.

10. Prove that the line segments joining the midpoints of successive sides of any rectangle form a rhombus.

11. Prove that an angle inscribed in a semicircle is a right angle.

   Given: $P(a, b)$ is a point on the circle
   $$x^2 + y^2 = r^2$$
   Prove: $\overline{PA} \perp \overline{PB}$

12. Prove that if $\triangle APB$ is a right triangle with the right angle at P, then P is on a circle with diameter $\overline{AB}$.

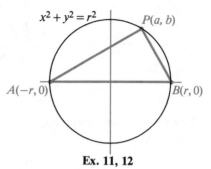

Ex. 11, 12

B 13. Prove that if the diagonals of a parallelogram are perpendicular, then the parallelogram is a rhombus. (*Suggestion:* As your first step, choose your coordinate axes as in figure (a) of Oral Exercise 2. As your second step, write an equation that expresses the fact that the diagonals are perpendicular. Simplify this equation to show that it implies $AD = AB$.)

14. Prove that the diagonals of a rhombus are perpendicular. (*Suggestion:* First, use the parallelogram of figure (a) of Oral Exercise 2. Next, express the fact that the parallelogram is a rhombus by writing an equation stating that $AB = AD$. Simplify this equation to show that it implies $\overline{AC} \perp \overline{BD}$.)

15. If P is any point in the plane of rectangle $ABCD$, prove that $(PA)^2 + (PC)^2 = (PB)^2 + (PD)^2$.

16. Use the diagram at the right to prove that P is on the perpendicular bisector of $\overline{AB}$ if $PA = PB$.

17. Prove that the medians of $\triangle PQR$ meet in a point G. This point is called the *centroid* of the triangle. [*Suggestion:* Use as vertices $P(0, 0)$, $Q(2a, 0)$, and $R(2b, 2c)$.]

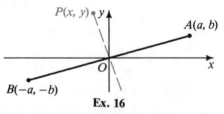

Ex. 16

18. Using Exercise 17, prove that the centroid G divides each median in a $2:1$ ratio.

19. **a.** Prove that the perpendicular bisectors of the sides of $\triangle PQR$ meet in a point C. This point is called the *circumcenter* of the triangle.
   b. Prove that C is equidistant from P, Q, and R.

C 20. Let G be the intersection point of the medians of a triangle, let C be the intersection of the perpendicular bisectors of the sides, and let H be the intersection of the altitudes. Prove that G, C, and H are collinear and that $GH = 2 \cdot GC$. (*Hint:* Use the results of Example 4 on page 40 and Exercises 17 and 19.)

21. Suppose that point $P(a, b)$ is any point on the circle with center $O(0, 0)$ and radius r and that line l is perpendicular to $\overline{OP}$ at P. Prove l is tangent to the circle as follows:

   a. Show that the equation of l can be written $ax + by = r^2$.

   b. Solve the equations of l and the circle simultaneously and show that there is only one solution.

Chapter summary

1. Given any two points $A(x_1, y_1)$ and $B(x_2, y_2)$:

   a. The distance $AB = \sqrt{(x_2 - x_1)^2 + (y_2 - y_1)^2}$.

   b. The *midpoint* of $\overline{AB}$ is $\left(\dfrac{x_1 + x_2}{2}, \dfrac{y_1 + y_2}{2}\right)$.

   c. The *slope* of $\overleftrightarrow{AB} = \dfrac{y_2 - y_1}{x_2 - x_1}$ if $x_2 \neq x_1$.

   d. If $x_1 = x_2$, then $\overleftrightarrow{AB}$ is a vertical line and consequently has no slope.

2. The graph of a linear equation $ax + by = c$, where a and b are not both zero, is a line. If the equation is written in the form $y = mx + k$, the slope is m and the y-intercept is k. The equation of a vertical line has the form $x = c$; vertical lines have no slope.

3. Two different nonvertical lines are parallel if and only if they have the same slope. Two lines with slopes m_1 and m_2 are perpendicular if and only if

$$m_1 = -\frac{1}{m_2}.$$

4. **a.** An equation of the line through (x_1, y_1) with slope m is

$$\frac{y - y_1}{x - x_1} = m.$$

   b. To find an equation of the line through (x_1, y_1) and (x_2, y_2), first find the slope m of the line and then use the equation in part (a).

5. **a.** The *imaginary unit* i is defined in such a manner that $i^2 = -1$. A *complex number* is a number of the form $a + bi$ where a and b are real numbers.

   b. The chart on page 21 shows the relationship of the complex numbers to the real and imaginary numbers.

   c. If $a > 0$, $\sqrt{-a} = i\sqrt{a}$.

6. A *quadratic equation*, $ax^2 + bx + c = 0$, can be solved by factoring, by completing the square, or by using the quadratic formula:

$$x = \frac{-b \pm \sqrt{b^2 - 4ac}}{2a}$$

If the *discriminant* $b^2 - 4ac > 0$, there are two real roots. If $b^2 - 4ac = 0$, there is one real root (called a double root). If $b^2 - 4ac < 0$, there are two imaginary roots that are complex conjugates. If a, b, and c are integers, then the equation has rational roots if and only if $b^2 - 4ac$ is the square of an integer.

7. The equation of a circle can be given in the general form

$$x^2 + y^2 + dx + ey + f = 0$$

or in the center-radius form

$$(x - h)^2 + (y - k)^2 = r^2.$$

The center-radius form shows that the center is (h, k), and the radius is $|r|$.

8. The procedure for finding the intersection of a line and circle is outlined on page 34. To find the intersection of two circles, first subtract their equations to get a linear equation. Then solve this linear equation together with an equation for one of the circles.

9. A summary of the methods used in coordinate proofs is given on page 40.

Chapter test

1. Let $A = (-2, -6)$ and $B = (-4, 2)$.   1–1
   a. Find the length of $\overline{AB}$.
   b. Find the coordinates of the midpoint of $\overline{AB}$.

2. Find the value of a if it is known that the point $(4, -2)$ lies on the line $2x + ay = 14$.

3. Solve the equations $2x + 3y = 2$ and $6x - y = -4$ simultaneously. Then sketch the graphs of the lines and label the intersection point with its coordinates.

4. Find the slope and the y-intercept of the line $4x - 2y = 7$.   1–2

5. Tell which of the following equations have parallel line graphs and which have perpendicular line graphs:
   a. $2x + 3y = 1$ b. $y = \frac{3}{2}x + 3$ c. $6x - 4y - 10 = 0$

6. Write an equation in slope-intercept form of the line through $(6, -2)$ and   1–3
   $(-3, 1)$.

7. Write an equation in general form of the line through $(5, 5)$ parallel to the line $4x + 3y = -2$.

8. Write an equation of the vertical line through $(4, -2)$.

9. Write an equation of the perpendicular bisector of the segment joining $(7, 0)$ and $(1, 8)$.

44   Chapter 1

Express each complex number in the form $a + bi$.

10. $\sqrt{-50} - \sqrt{-8}$ **11.** $(2 + 3i)^2$ **12.** $4(3 + 2i) - 5(1 - i)$ 1-4

13. $\dfrac{1}{2 + 3i}$ **14.** $\dfrac{\sqrt{3} + i}{\sqrt{3} - i}$ **15.** i^{17}

16. Solve for x.   1-5
     a. $7x^2 - 2 = 5x$ **b.** $x^2 - 4x = 9$

17. Find the discriminant of $3x^2 - 2x + 2 = 0$ and solve for x by the easiest method.

18. Find the center and the radius of the circle $x^2 + y^2 - 12x + 10y + 45 = 0$.   1-6

19. Find an equation of the circle with center $(-3, 4)$ that is tangent to the line $y = 8$.

20. Sketch the circles $x^2 + y^2 = 5$ and $x^2 + y^2 - 12x + 4y + 15 = 0$. Solve their equations simultaneously to find the intersection points of their graphs.   1-7

21. Prove by using coordinate geometry that the diagonals of a square are perpendicular and equal.   1-8

CHAPTER TWO

Polynomials

OBJECTIVES

1. To find the zeros of a polynomial function.
2. To solve second and higher degree equations by factoring.
3. To sketch the graphs of polynomial functions.
4. To apply the Remainder Theorem and the Factor Theorem.
5. To find the rational roots of a polynomial equation.
6. (Optional) To approximate the real roots of a polynomial equation using the bracket-and-halving method, linear-interpolation, or a computer.
7. To determine the maximum and minimum values of a polynomial function.
8. To find the sum and the product of the roots of a polynomial equation.

Factoring polynomial equations

2-1/POLYNOMIALS

A **polynomial** in x is an expression of the form

$$a_n x^n + a_{n-1} x^{n-1} + \cdots + a_2 x^2 + a_1 x + a_0$$

where n is a nonnegative integer. The numbers a_n, $a_{n-1}, \ldots, a_2$, a_1, a_0 are called **coefficients** of the polynomial. In this book, the coefficients represent real numbers, but the values of x may be real or imaginary numbers. (In more advanced courses, even the coefficients may be imaginary.) The highest power of x that has a nonzero coefficient is called the **degree** of the polynomial. The polynomial 0 has no degree. The coefficient of the term with the highest power of x is called the **leading coefficient** of the polynomial. Polynomials of the first few degrees have special names as shown in the examples that follow.

The path of the center of gravity of a leaping animal is shaped like part of a parabola.

Degree	Name	Example
0	constant	5
1	linear	$2 + 3x$
2	quadratic	$3 + x^2$
3	cubic	$x^3 + 2x + 1$
4	quartic	$-3x^4 + x$
5	quintic	$x^5 + \pi x^4 - 3.1x^3 + 11$

You can see that terms do not have to be written in any special order although they are usually written in order of increasing or decreasing powers of x.

Polynomials are examples of *functions,* an idea we shall be studying throughout this book. Basically, a function is nothing more than a correspondence or rule that assigns to each value of x a unique number. For example, consider the quadratic function $3x^2 - x + 7$.

When $x = 0$, the value of $3x^2 - x + 7$ is 7.
When $x = 1$, the value of $3x^2 - x + 7$ is 9.
When $x = -\frac{1}{2}$, the value of $3x^2 - x + 7$ is $8\frac{1}{4}$.

The work above can be written much more simply by using the symbol $f(x)$, called "f of x," to represent a "function of x." Thus, we write

$$f(x) = 3x^2 - x + 7.$$

Since the value of the function at 0 is 7, we write $f(0) = 7$. The previous results also show that $f(1) = 9$ and $f(-\frac{1}{2}) = 8\frac{1}{4}$.

A function of x can be represented by $f(x)$, $g(x)$, $h(x)$ or other similar expressions. For polynomial functions, the symbol $P(x)$ is often used.

One can substitute variables as well as constants for x. For example, if $g(x) = 4 + 2x^2$, then:

$$g(a + 1) = 4 + 2(a + 1)^2 \qquad g(3n) = 4 + 2(3n)^2$$
$$= 2a^2 + 4a + 6 \qquad\qquad = 18n^2 + 4$$

You can even replace x by another expression containing x:

$$g(x - 5) = 4 + 2(x - 5)^2 = 2x^2 - 20x + 54.$$

A function that can be defined by a linear polynomial is referred to as a **linear function.** For example, $g(x) = -2x + 3$ is a linear function. Constant, quadratic, cubic, quartic, and quintic functions correspond in a similar way to constant, quadratic, cubic, quartic, and quintic polynomials, respectively.

EXAMPLE 1. What is the change in value in the linear function $f(x) = \frac{3}{2}x + 2$ when the value of x increases by 6?

SOLUTION: When the value of x increases by 6, the value of $f(x)$ changes to $f(x + 6)$. We find the change in value of $f(x)$ as follows:

$$f(x + 6) - f(x) = [\tfrac{3}{2}(x + 6) + 2] - (\tfrac{3}{2}x + 2)$$
$$= (\tfrac{3}{2}x + 9 + 2) - (\tfrac{3}{2}x + 2)$$
$$= 9$$

A value of the variable that causes the value of the polynomial function to be zero is called a **zero** of the function. For example, 2 and -2 are zeros of $f(x) = x^2 - 4$. The zeros of $g(x) = x^2 + 9$ are the imaginary numbers $3i$ and $-3i$.

In later chapters we will present a general definition of a function and study many functions that are not polynomial functions. For example, we will study exponential functions, logarithmic functions, and trigonometric functions. Each of these functions, however, can be approximated using polynomial functions. Computers and calculators make frequent use of such approximations.

ORAL EXERCISES 2-1

Classify each polynomial function as linear, quadratic, cubic, quartic, or quintic. Give the leading coefficient of each.

1. $f(x) = 2 + 3x + 5x^2$ **2.** $g(x) = 5x^4 - x^2$

3. $f(x) = 7 - 4x^3$ **4.** $P(x) = x^4 + 9x^2 - 1 - 2x^5$

If $f(x) = 2x^2 + 5$, evaluate the following.

5. $f(3)$ **6.** $f(-1)$ **7.** $f(\sqrt{3})$ **8.** $f(i)$

9. $f(n + 1)$ **10.** $f(3x)$ **11.** $f\left(\dfrac{1}{x}\right)$ **12.** $f(a + b)$

Give the zeros of each function.

13. $f(x) = (x + 5)(3x - 9)$ **14.** $g(x) = (x^2 - 16)(x^2 + 25)$

15. $P(x) = x^2 - 6x + 9$ **16.** $h(x) = ax^2 + bx + c$

WRITTEN EXERCISES 2-1

In Exercises 1–12, state whether or not the function is a polynomial function. Give the zeros of each function, if they exist.

A   **1.** $f(x) = x^2 - 6x + 8$ **2.** $g(x) = 1 - 5x + 6x^2$

     3. $h(x) = 17 - 3x$ **4.** $r(x) = x - \dfrac{1}{x}$

     5. $f(x) = \dfrac{x^2 - 3x - 4}{x^2 + 1}$ **6.** $g(x) = px^2 + r + qx$

     7. $k(x) = 9$ **8.** $h(x) = (x - 7)^2(x^2 + 7)$

     9. $m(x) = x^3 + 2x^2 + x$ **10.** $p(x) = x^3 - 9x$

     11. $S(x) = \dfrac{1}{x^4 + 2}$ **12.** $f(x) = 2x^4 - x^3 - x^2$

In Exercises 13–18, find the requested values of the given functions. Simplify the answers.

13. $f(x) = 8x - 5$

     a. $f(7)$ **b.** $f(-4.5)$ **c.** $f(x + 2)$ **d.** $f\left(\dfrac{1}{2a}\right)$

Find the requested values of the given functions. Simplify the answers.

14. $g(x) = 12 - 4x$

    a. $g(\sqrt{2})$ **b.** $g(-3.25)$ **c.** $g(x - 1)$ **d.** $g(x + 3)$

15. $h(x) = 2x^2 - 5x + 6$

    a. $h(-1)$ **b.** $h(2i)$ **c.** $h(1 + i)$ **d.** $h(3a)$

16. $P(x) = 8x - 4x^2$

    a. $P(2\sqrt{3})$ **b.** $P(1 - \sqrt{2})$ **c.** $P(1 + 2i)$ **d.** $P\left(\dfrac{2}{x}\right)$

17. $f(x) = x^3 - 9x$

    a. $f\left(-\dfrac{\sqrt{2}}{3}\right)$ **b.** $f(i\sqrt{3})$ **c.** $f\left(\dfrac{x}{3}\right)$ **d.** $f(x - 3)$

18. $k(x) = x^2(x^2 + 16)$

    a. $k(4 - \sqrt{2})$ **b.** $k(1 + i)$ **c.** $k\left(\dfrac{p}{q}\right)$ **d.** $k(x^2)$

B

19. A quadratic polynomial $f(x)$ has a leading coefficient of -2, a constant term of 6, and no x term. Find the zeros of $f(x)$.

20. The leading coefficient of a cubic polynomial $P(x)$ is 2, and the coefficient of the linear term is -5. If $P(0) = 7$ and $P(2) = 21$, find $P(3)$.

21. If 4 is a zero of $f(x) = 3x^3 + kx - 2$, find the value of k.

22. If $2i$ is a zero of $f(x) = x^4 + x^2 + a$, find the value of a.

23. Sketch the graph of the linear function $f(x) = 4x - 8$. What are its x- and y-intercepts?

24. Sketch the graph of the linear function $g(x) = 3.6x + 18$. What are its x- and y-intercepts?

25. The graph of a linear function $f(x)$ is a line with slope 2 and y-intercept 5. What is the equation of $f(x)$?

26. The graph of a linear function $g(x)$ has y-intercept -3 and x-intercept 5. What is the equation of $g(x)$?

27. The graph of a linear function $h(x)$ passes through the points $(7, 1)$ and $(4, -3)$. What is the equation of $h(x)$?

28. The graph of a linear function $p(x)$ passes through the points $(-4, 2)$ and $(-1, -5)$. What is the equation of $p(x)$?

29. If $f(x) = 7x + 2$, evaluate: **a.** $f(9) - f(8)$ **b.** $f(x + 1) - f(x)$

30. If $g(x) = 3 - 8x$, evaluate: **a.** $g(6) - g(4)$ **b.** $g(x + 2) - g(x)$

31. If $f(x) = 4x + 6$, find the change in the value of $f(x)$ when x increases by 7.

32. If $f(x) = 9 - \frac{3}{2}x$, find the change in the value of $f(x)$ when x decreases by 10.

33. If $f(x) = mx + k$ and $h \neq 0$, show that the value of $\dfrac{f(x + h) - f(x)}{h}$ does not depend on x or h. Interpret this result graphically.

34. If $f(x) = x^2$ and $h \neq 0$, find the value of $\dfrac{f(x + h) - f(x)}{h}$. Is the value of this expression independent of the values of x and h, as it was for the function of Exercise 33?

35. For each function $f(x)$, determine whether or not $f(a + b) = f(a) + f(b)$.
  a. $f(x) = 3x$ **b.** $f(x) = 3x + 7$ **c.** $f(x) = x^2$ **d.** $f(x) = \sqrt{x}$

36. For each function in Exercise 35, determine whether or not $f(4x) = 4 \cdot f(x)$.

C   **37. a.** Consider the following table of values for the quadratic function $f(x) = x^2 + 2x + 3$.

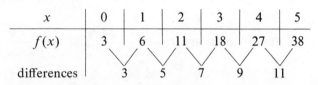

x	0	1	2	3	4	5
$f(x)$	3	6	11	18	27	38
differences		3	5	7	9	11

What pattern do you observe in the differences?
  b. Make a difference table similar to the one in part (a) for the function $g(x) = 2x^2 - 3x - 1$. What pattern do you observe in the differences?
  c. Do the differences in the values of *all* quadratic functions form a pattern like those you observed in parts (a) and (b)? Justify your answer with a difference table.

38. Experiment with the differences in the values of cubic functions. (See Exercise 37.) Can you detect any patterns?

CALCULATOR EXERCISE

If $f(x) = 21.3x - 173.4$, estimate the value of $f(49.6) - f(39.6)$ without using a calculator. Check your guess with a calculator.

2-2 / SOLVING HIGHER-DEGREE EQUATIONS

An equation that can be written in the form

$$P(x) = 0,$$

where $P(x)$ is a polynomial, is called a **polynomial equation.** Many algebraic problems in a single variable can be reduced to solving a polynomial equation. The difficulty of solving a polynomial equation depends primarily upon the degree of the polynomial. If the degree of the equation is 1 or 2, then the equation is easily solved using methods we already know. For degrees greater than 2, more advanced methods are usually needed, but there are certain higher-degree equations that can be solved using familiar methods such as factoring and the quadratic formula. We present some examples in this section. Other, more general, methods for solving polynomial equations will be presented in later sections of this chapter.

EXAMPLE 1. *Solving a cubic equation with no constant term*

Solve: $2x^3 + 6x^2 + 6x = 0$

SOLUTION: The first step in solving any polynomial equation is to look for a factor common to each term. If a polynomial in x has no constant term, then x will always be a factor. In this case, $2x$ is a common factor. Thus, we have:

$$2x(x^2 + 3x + 3) = 0$$

$2x = 0$	$x^2 + 3x + 3 = 0$
$x = 0$	$x = \dfrac{-3 \pm i\sqrt{3}}{2}$

The solutions are 0, $\dfrac{-3 - i\sqrt{3}}{2}$, and $\dfrac{-3 + i\sqrt{3}}{2}$.

EXAMPLE 2. *Solving a polynomial equation by grouping terms*

Solve: $x^3 + 5x^2 - 4x - 20 = 0$

SOLUTION: We group the first two terms and the last two terms.

$$(x^3 + 5x^2) - (4x + 20) = 0$$

Then we factor each group as follows:

$$x^2(x + 5) - 4(x + 5) = 0$$
$$(x^2 - 4)(x + 5) = 0$$
$$(x + 2)(x - 2)(x + 5) = 0$$

$x + 2 = 0$	$x - 2 = 0$	$x + 5 = 0$
$x = -2$	$x = 2$	$x = -5$

The solutions are -5, -2, and 2.

Sometimes a polynomial of higher degree has a **quadratic form.** For example,

$$7x^4 + 3x^2 - 2 \text{ can be rewritten as } 7(x^2)^2 + 3x^2 - 2.$$

If we let $y = x^2$, the quadratic form is apparent.

$$7x^4 + 3x^2 - 2 \text{ can be written as } 7y^2 + 3y - 2,$$

which is a quadratic polynomial in y. Another example is $8x^6 - 5x^3 + 1$. If we let $y = x^3$:

$$8x^6 - 5x^3 + 1 = 8(x^3)^2 - 5x^3 + 1 = 8y^2 - 5y + 1.$$

If $P(x)$ is a polynomial that has a quadratic form, then we say the equation $P(x) = 0$ has a quadratic form. The next example shows two methods for solving equations that have a quadratic form.

EXAMPLE 3. *Solving a polynomial equation that has a quadratic form*

Solve: **a.** $2x^4 - 8x^2 - 90 = 0$ **b.** $x^4 - x^2 - 1 = 0$

SOLUTION: **a.** Let $y = x^2$ and proceed as follows:

$$2x^4 - 8x^2 - 90 = 0$$
$$2y^2 - 8y - 90 = 0$$
$$2(y - 9)(y + 5) = 0$$

$y = 9$	$y = -5$
$x^2 = 9$	$x^2 = -5$
$x = \pm 3$	$x = \pm i\sqrt{5}$

The solutions are $-i\sqrt{5}$, $i\sqrt{5}$, -3, and 3.

b. In this solution we do not introduce another variable.

$$x^4 - x^2 - 1 = 0$$

We use the quadratic formula to solve for x^2.

$$x^2 = \frac{-b \pm \sqrt{b^2 - 4ac}}{2a}$$

$$= \frac{1 \pm \sqrt{5}}{2} \approx \frac{1 \pm 2.24}{2}$$

$x^2 \approx 1.62$	$x^2 \approx -0.62$
$x \approx \pm 1.27$	$x \approx \pm 0.79i$

To the nearest hundredth the solutions are $-0.79i$, $0.79i$, -1.27, and 1.27.

If polynomials occur on both sides of an equation or in a denominator in an equation, you must organize your solution carefully in order not to lose a root or gain a root.

1. **Losing a root.** Remember not to lose a root by canceling a factor common to both sides of an equation. (See I at the right.) You can avoid this mistake by remembering to add to your set of solutions all values of x that make the canceled factor zero. (See II below.) A simple way to avoid losing roots is to bring all terms to one side of the equation and then factor. (See III below.)

I. *Incorrect*

$$4x(\cancel{x^2 - 1}) = 3(\cancel{x^2 - 1})$$
$$4x = 3$$
$$x = \tfrac{3}{4}$$

II. *Correct*

$$4x(x^2 - 1) = 3(x^2 - 1)$$

If $x^2 - 1 = 0$, then both sides of the equation are equal to 0. Thus the roots include $x = \pm 1$.	If $x^2 - 1 \neq 0$, we can "cancel," getting $4x = 3$, or $x = \tfrac{3}{4}$.

III. *Correct*

$$4x(x^2 - 1) = 3(x^2 - 1)$$
$$4x(x^2 - 1) - 3(x^2 - 1) = 0$$
$$(x^2 - 1)(4x - 3) = 0$$
$$x = \pm 1, \tfrac{3}{4}$$

Polynomials   **53**

2. **Gaining a root.** Always check your solutions in the original equation to be sure you have not gained a root during the solution process. This often happens when both sides of an equation are squared (see Exercise 29) or when both sides of the equation are multiplied by an expression containing the variable, as in the following example.

EXAMPLE 4. Solve: $\dfrac{x+2}{x-2} + \dfrac{x-2}{x+2} = \dfrac{8-4x}{x^2-4}$

SOLUTION:

$$(x+2)^2 + (x-2)^2 = 8 - 4x \quad \text{Multiplying both sides}$$
$$(x^2 + 4x + 4) + (x^2 - 4x + 4) = 8 - 4x \quad \text{by } (x-2)(x+2)$$
$$2x^2 + 8 = 8 - 4x$$
$$2x^2 + 4x = 0$$
$$2x(x+2) = 0$$
$$x = 0, -2$$

Check:

$$x = 0$$

$$\frac{0+2}{0-2} + \frac{0-2}{0+2} \stackrel{?}{=} \frac{8-0}{0-4}$$

$$-1 + (-1) = -2$$

$$x = -2$$

$$\frac{-2+2}{-2-2} + \frac{-2-2}{-2+2} \stackrel{?}{=} \frac{8+8}{4-4}$$

Since two denominators are zero, the equation is meaningless. Thus, -2 is *not* a solution.

The solution is 0.

The previous examples have shown how to find the zeros of various polynomials. By reversing the process, we can find a polynomial that has a specified set of zeros. If n zeros are specified, a polynomial of degree n will be required. (We will prove this result later.)

EXAMPLE 5. Find a cubic polynomial with zeros 2, 0, and -2.

SOLUTION: If 2, 0, and -2 are zeros of the polynomial, then $x - 2$, $x - 0$, and $x - (-2)$, respectively, must be factors of the polynomial. We multiply these factors to find the required polynomial.

$$(x - 2)(x)(x + 2) = x^3 - 4x$$

You can easily verify that 2, 0, and -2 are zeros of $x^3 - 4x$.

ORAL EXERCISES 2-2

To which of the following categories does the given equation belong?
(1) a cubic equation with no constant term, which can therefore be factored
(2) a polynomial equation that can be solved by grouping terms
(3) a polynomial equation that has a quadratic form

1. $x^4 - 4x^2 - 12 = 0$

2. $x^3 + 6x^2 - 4x - 24 = 0$

3. $3x^3 + 7x^2 - 20x = 0$

4. $3x^3 - 16x^2 - 12x + 64 = 0$

5. $x^3 = 7x^2 - 6x$

6. $x^4 - 7x^2 - 8 = 0$

7. Comment on the following solution of the equation $9x^3 + 36x^2 = x + 4$:

$$9x^3 + 36x^2 = x + 4$$
$$9x^2(x + 4) = (x + 4)$$
$$9x^2 = 1$$
$$x = \pm\tfrac{1}{3}$$

8. Solve: $(2x - 5)(x + 6) = 7(x + 6)$

9. Solve: $(3x - 5)(4x - 1) = (4x - 1)$

10. Find a cubic polynomial with zeros 1, -2, and 0.

11. Find a cubic polynomial with zeros $\sqrt{2}$, $-\sqrt{2}$, and 0.

12. Find a cubic polynomial with zeros i, $-i$, and 2.

WRITTEN EXERCISES 2–2

A   **1–6.** Solve the equations given in Oral Exercises 1–6 for all real and imaginary roots.

In Exercises 7–16, solve the given equation for all real and imaginary roots.

7. $x^4 = 3x^2 + 28$

8. $x^3 + 2x^2 - 6x = 12$

9. $2x^3 - 3x^2 = 12 - 8x$

10. $4z^3 = 18z - z^2$

11. $2s^3 - 12s^2 + 18s = 0$

12. $2x^4 = -7x^2 + 15$

13. $2x^3 + 3x^2 - x = 0$

14. $10w^3 + 5w = 6w^2 + 3$

B   **15.** $2t^4 + 3t^2 - 1 = 0$

16. $a^2x^3 + abx^2 + acx = 0$
   (a, b, and c are constants.)

In Exercises 17–31, solve the given equation for all real roots.

17. $(4x + 7)(x - 1) = 2(x - 1)$

18. $(2x + 1)(4x^2 - 9) = 6(4x^2 - 9)$

19. $2w^2(4w - 1) = w(1 - 4w)$

20. $3(2x - 3)^2 = 4x(3 - 2x)$

21. $ax^2(b - x) = 2ax(x - b)^2$

22. $4x(x^2 - a^2) = 3x^2(a - x)$

(In Exercises 21 and 22, a and b represent constants.)

23. $\dfrac{x + 3}{x - 3} + \dfrac{x - 3}{x + 3} = \dfrac{18 - 6x}{x^2 - 9}$

24. $\dfrac{r}{r - 1} - \dfrac{r}{r + 1} = 1 + \dfrac{2}{r^2 - 1}$

25. $\dfrac{t^2 + 1}{t + 2} = \dfrac{t}{3} + \dfrac{5}{t + 2}$

26. $\dfrac{x + 2}{x^2 - x - 6} = 3 - \dfrac{4}{x - 3}$

27. $v - 2 = 3\sqrt{v - 4}$

28. $\sqrt{2x + 5} = x + 1$

C   **29.** $\sqrt{2x + 5} = 2\sqrt{2x} + 1$

30. $\sqrt{y - 3} = 1 - \sqrt{2y - 4}$

31. $\sqrt{\dfrac{x^2 + 1}{3}} + \sqrt{\dfrac{3}{x^2 + 1}} = \dfrac{3\sqrt{2}}{2}$

CALCULATOR EXERCISE

Approximate the real roots of $6x^8 - 19x^4 + 10 = 0$.

Graphs of polynomial functions

2-3/GRAPHING A QUADRATIC FUNCTION

The graph of the quadratic function $f(x) = ax^2 + bx + c$ is the set of points with coordinates (x, y) that satisfy the equation $y = ax^2 + bx + c$. The graph of a quadratic function is a **parabola,** a curve that can be seen in the cables of a suspension bridge and the path of a thrown ball. [See figures (a) and (b) below.] Parabolas can also be defined geometrically, as we shall see in Chapter 11. We shall also present certain geometric properties that make parabolas useful in a variety of fields, such as the design of telescopes, radar transmitters, and solar furnaces.

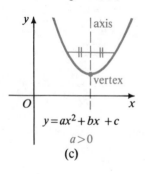

(a)

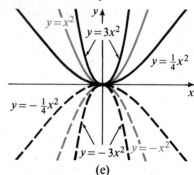

(b)

If a graph has line l as an axis of symmetry, each point of the graph can be paired with another point in such a way that l is the perpendicular bisector of the segment joining them. A parabola that is the graph of a quadratic function in x always has a vertical axis of symmetry. Points of the parabola on the same horizontal line are paired with one another. [See figure (c).] The **vertex** of the parabola is the point where the axis intersects the parabola. If $a > 0$, the parabola opens upward and y takes on its smallest value at the vertex of the parabola. [See figure (c).] Notice that the parabola slopes downward from left to right until it reaches a turning point at the vertex; then the parabola slopes upward. If $a < 0$, the parabola opens downward and y takes on its greatest value at the vertex of the parabola. [See figure (d).] Notice that the parabola slopes upward from left to right until it reaches a turning point at the vertex; then the parabola slopes downward. The bigger $|a|$ is, the narrower the parabola. [See figure (e).]

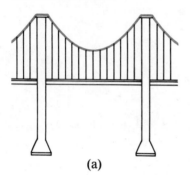

$y = ax^2 + bx + c$

$a > 0$

(c)

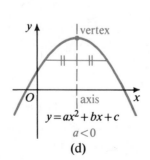

$y = ax^2 + bx + c$

$a < 0$

(d)

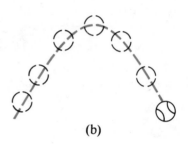

$y = x^2$ $y = 3x^2$ $y = \frac{1}{4}x^2$ $y = -\frac{1}{4}x^2$ $y = -3x^2$ $y = -x^2$

(e)

A parabola with equation $y = ax^2 + bx + c$ always intersects the y-axis at $(0, c)$ because $y = c$ when $x = 0$. The number c is called the **y-intercept** of the parabola. The **x-intercepts** of the parabola are the real-number solutions to the equation $ax^2 + bx + c = 0$. Since this quadratic equation may have two real solutions or just one or none at all, depending on the value of its **discriminant** $b^2 - 4ac$, we have the following three possibilities regarding the x-intercepts of a parabola:

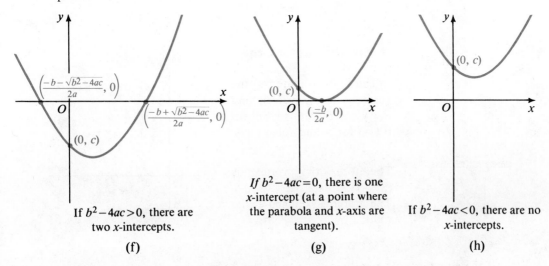

If $b^2 - 4ac > 0$, there are two x-intercepts.

(f)

If $b^2 - 4ac = 0$, there is one x-intercept (at a point where the parabola and x-axis are tangent).

(g)

If $b^2 - 4ac < 0$, there are no x-intercepts.

(h)

When a parabola has two x-intercepts, as in figure (f), you can use their values to find the equation of the axis of the parabola. If the x-intercepts are x_1 and x_2, then the midpoint of the (horizontal) segment joining them is $\left(\dfrac{x_1 + x_2}{2}, 0\right)$. Since the axis is the vertical line passing through the midpoint, its equation is $x = \dfrac{x_1 + x_2}{2}$. Example 1 shows this "averaging-zeros" method.

EXAMPLE 1. Given the parabola $y = x^2 - 6x + 8$ (that is, the parabola with equation $y = x^2 - 6x + 8$):

    a. Determine whether the parabola opens upward or downward.

    b. Does the parabola intersect the x-axis in two points, one point, or no point?

    c. What are the x- and y-intercepts of the parabola?

    d. Find the equation of the axis of symmetry and the coordinates of the vertex of the parabola.

    e. Sketch the parabola.

SOLUTION: **a.** Since the value of a is 1, which is greater than 0, the parabola opens upward.

    b. $b^2 - 4ac = (-6)^2 - 4(1)(8) = 4$

    Since the discriminant is positive, the parabola has two x-intercepts.

(*Solution continued on page 58*)

Polynomials   **57**

c. To find the x-intercepts, we set $y = 0$ and solve for x.

$$0 = x^2 - 6x + 8$$
$$0 = (x - 2)(x - 4)$$
$$x = 2 \text{ or } x = 4$$

The x-intercepts are 2 and 4. The y-intercept is always the constant term c. In this case, $c = 8$.

d. Since the average of the zeros is $\dfrac{2 + 4}{2}$, or 3, the axis of symmetry has the equation $x = 3$.

   The x-coordinate of the vertex is 3. (Why?) To find the y-coordinate, we substitute 3 for x and solve for y.

$$y = 3^2 - 6 \cdot 3 + 8 = -1$$

Therefore, the vertex is $(3, -1)$.

e. The graph is sketched at the right.

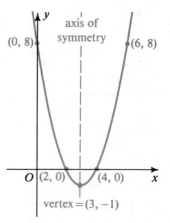

If the equation of a parabola is given in the form

$$y = a(x - h)^2 + k,$$

then the vertex of the parabola is (h, k). (See Exercise 51 for a proof.) Also, the equation of the axis of symmetry is $x = h$. We apply these results in Example 2.

EXAMPLE 2. Sketch the parabola $y = 2x^2 - 8x + 5$.

SOLUTION: 1. By completing the square,

$$y = 2x^2 - 8x + 5$$
$$= 2(x^2 - 4x \qquad) + 5$$
$$= 2(x^2 - 4x + 4) + 5 - 2 \cdot 4$$
$$= 2(x - 2)^2 - 3$$

The vertex is $(2, -3)$. The axis of symmetry is $x = 2$.

2. To find x-intercepts, set $y = 0$.

$$2(x - 2)^2 - 3 = 0$$
$$(x - 2)^2 = \tfrac{3}{2}$$
$$x - 2 = \pm\sqrt{\tfrac{3}{2}}$$
$$x = 2 \pm \sqrt{\tfrac{3}{2}}$$
$$x \approx 2 \pm 1.2 \approx 3.2, \ 0.8$$

3. The y-intercept is the constant term 5 of the original equation.

When the general quadratic equation $y = ax^2 + bx + c$ is rewritten as
$$y = a(x - h)^2 + k,$$
the value of h is $\dfrac{-b}{2a}$. (See Exercise 52 for a proof.) Therefore, the equation of the axis is $x = \dfrac{-b}{2a}$. This result can be used to prove that an equation of the axis is x = average of the roots (Exercise 53).

EXAMPLE 3. Find the vertex of the parabola $y = -3x^2 + 6x - 5$.

SOLUTION: The x-coordinate of the vertex is $\dfrac{-b}{2a} = \dfrac{-6}{2(-3)} = 1$. Substituting $x = 1$ in the equation of the parabola, we find that
$$y = -3 \cdot 1^2 + 6 \cdot 1 - 5 = -2.$$
Thus, the vertex is $(1, -2)$.

The methods used in Examples 1, 2, and 3 can be generalized as follows:

PROCEDURE FOR SKETCHING GRAPH OF $y = ax^2 + bx + c$, $a \neq 0$

1. Get a quick mental picture of the graph as follows:
   a. If $a > 0$, graph opens upward.
   If $a < 0$, graph opens downward.
   b. If $b^2 - 4ac > 0$, graph intersects x-axis in two points (two x-intercepts).
   If $b^2 - 4ac = 0$, graph is tangent to x-axis (one x-intercept).
   If $b^2 - 4ac < 0$, graph does not intersect x-axis (no x-intercepts).

2. Find x- and y-intercepts.
   a. y-intercept is always c.
   b. Solve $ax^2 + bx + c = 0$ to find x-intercepts. If $b^2 - 4ac$ is a perfect square, you can find the x-intercepts by factoring. Otherwise, use the quadratic formula.

3. Find the axis and vertex of the parabola by either of two methods.
   Method 1. The equation of the axis is always $x = \dfrac{-b}{2a}$. (This x-value is the average of the x-intercepts if there are any.) Substitute this value of x in $y = ax^2 + bx + c$ to find the y-coordinate of the vertex.
   Method 2. Rewrite the equation of the parabola in the form
   $$y = a(x - h)^2 + k.$$
   The vertex is then (h, k) and the equation of the axis is $x = h$.

ORAL EXERCISES 2–3

Find the vertex of the parabola whose equation is given, by substituting $x = \dfrac{-b}{2a}$ in the equation.

1. $y = x^2 - 2x + 7$ **2.** $y = 3x^2 + 6x$

For each of the following equations, tell whether its parabolic graph (**a**) opens upward or downward and (**b**) intersects the x-axis in two, one, or no points.

3. $y = x^2 - 2x + 1$ **4.** $y = x^2 - 2x + 3$

5. $y = -4x^2 + 8x$ **6.** $y = 9 - x^2$

7. $y = 5x^2 - 4x + 2$ **8.** $y = 3x^2 + 15x + 15$

By inspecting each of the following equations, find the vertex of its parabolic graph. (You do *not* need to substitute $x = \dfrac{-b}{2a}$ in the equation.)

9. $y = (x - 5)^2 + 4$ **10.** $y = -2(x - 7)^2 + 8$

11. $y = (x + 2)^2 - 3$ **12.** $y = a(x - h)^2 + k$

13. $y = x^2 + 5$ **14.** $y = -x^2 - 10$

Give the x-intercepts, if any, of each parabola whose equation is given.

15. $y = (x - 5)^2 - 1$ **16.** $y = (x - 2)^2$

17. $y = 2(x + 1)^2 - 6$ **18.** $y = (x + 7)^2 + 16$

WRITTEN EXERCISES 2-3

In Exercises 1–34, sketch the graph of each parabola whose equation is given. Show the coordinates of the vertex and any points where the parabola intersects the x- and y-axes. Also give the equation of the axis of symmetry.

In Exercises 1–12, use the method of Example 1.

A **1. a.** $y = x^2 - 6x$ **b.** $y = x^2 - 6x + 9$ **c.** $y = x^2 - 6x + 10$

2. a. $y = -x^2 + 4x$ **b.** $y = -x^2 + 4x - 4$ **c.** $y = -x^2 + 4x - 8$

3. $y = (x + 5)(x + 3)$ **4.** $y = (4 - x)(x + 2)$

5. $y = 9 - x^2$ **6.** $y = x^2 - 16$

7. $y = x^2 - 2x - 15$ **8.** $y = x^2 + 3x - 10$

9. $y = 4x^2 + 4x + 1$ **10.** $y = -x^2 - 6x - 9$

11. $y = 3 - 8x - 2x^2$ **12.** $y = 1 - 6x - 2x^2$

In Exercises 13–26, use the method of Example 2.

13. $y = (x - 4)^2 - 9$ **14.** $y = 4 - (x + 3)^2$

15. $y = 8 - 2(x + 1)^2$ **16.** $y = 6 - 3(x - 1)^2$

17. $y = x^2 - 2x - 7$ **18.** $y = x^2 + 4x + 9$

19. $y = x^2 + 6x + 1$ **20.** $y = x^2 - 10x + 20$

21. $y = 4x^2 - 8x + 2$ **22.** $y = -x^2 - 2x + 7$

23. $y = \frac{1}{2}x^2 + 4x + 8$ **24.** $y = -\frac{1}{3}x^2 + 2x + 1$

25. $y = -2x^2 - 4x + 3$ **26.** $y = -3x^2 - 12x - 3$

In Exercises 27–34, use the method that seems most appropriate to you.

27. $y = 4x^2 - 16x + 15$ **28.** $y = 12 - 3(x + 3)^2$

29. $y = x^2 + 2x + 17$

30. $y = 2x^2 - 4x + 11$

31. $y = 2(x + 3)^2 + 6$

32. $y = (2x - 7)(2x - 1)$

33. $y = -\frac{1}{4}x^2 + 6$

34. $y = \frac{1}{8}x^2 - x - 2$

In Exercises 35–38, you are given the equation of a line and the equation of a parabola. Sketch the graphs and solve the equations simultaneously to determine whether there are two, one, or no points of intersection. Give the coordinates of any points of intersection.

B **35.** $y = 4 - 2x$, $y = x^2 - 6x + 8$

36. $y = x + 3$, $y = 4x - x^2$

37. $y + x = -6$, $y = x^2 + 6x$

38. $2x + y = 10$, $y = 9 - x^2$

In Exercises 39–46, find an equation of the quadratic function. (*Suggestion:* Choose a form for the equation that uses the given data efficiently.)

EXAMPLE 4. Find an equation of the parabola with x-intercepts 3 and 6 and y-intercept -2.

SOLUTION: Any parabola with x-intercepts 3 and 6 has an equation of the form $y = a(x - 3)(x - 6)$ for some constant a. We use the fact that the parabola passes through $(0, -2)$ to find the value of a. We substitute 0 for x and -2 for y and then solve for a.

$$-2 = a(0 - 3)(0 - 6)$$
$$-\tfrac{1}{9} = a$$

Thus, an equation of the parabola is

$$y = -\tfrac{1}{9}(x - 3)(x - 6).$$

39. The parabola has x-intercepts 2 and -1 and y-intercept 6.

40. The parabola has x-intercepts 5 and 1 and y-intercept 1.

41. The parabola passes through the origin and has vertex $(4, 8)$. (*Hint:* $y = a(x - 4)^2 + 8$)

42. The parabola passes through the origin and has vertex $(3, -8)$.

43. The parabola is tangent to the x-axis at $(4, 0)$ and has y-intercept 6.

44. The parabola passes through $(-1, 0)$, $(3, 0)$, and $(0, 12)$.

45. The parabola passes through $(1, 2)$ and has vertex $(3, -5)$.

46. The parabola passes through $(-3, 4)$ and has vertex $(-1, 6)$.

47. a. If $f(x) = 2x^2 - 4x + 7$, show that $f(1 + k) = f(1 - k)$ for all real k.
b. How does part (a) show that the axis of symmetry is the line $x = 1$?

48. a. If $g(x) = 2(x - 3)^2 + 5$, show that $g(3 + k) = g(3 - k)$ for all real k.
b. Interpret the result of part (a) using the graph of $g(x)$.

49. a. If $h(x) = 2x^2 - 12x$, show that $h(3) \leq h(3 + k)$ for all real k.
b. Interpret the result of part (a) using the graph of $h(x)$.

50. a. If $f(x) = 9 + 8x - x^2$, show that $f(4) \geq f(4 + k)$ for all real k.
b. Interpret the result of part (a) using the graph of $f(x)$.

51. Show that if the equation of a parabola is written in the form

$$y = a(x - h)^2 + k$$

where $a \neq 0$, then the vertex of the parabola is (h, k). (*Hint:* Prove the theorem using two cases: (1) When $a > 0$ show that y takes on its smallest value at (h, k). (2) When $a < 0$, y takes on its greatest value at (h, k).)

52. Derive the following result: If the general quadratic equation

$$y = ax^2 + bx + c$$

is written as $\qquad y = a(x - h)^2 + k,$

then the value of h is $\dfrac{-b}{2a}$ and the value of k is $\dfrac{4ac - b^2}{4a}$.

53. Show that the average of the roots of $ax^2 + bx + c = 0$ is $\dfrac{-b}{2a}$.

54. A hauling company needs to determine whether a large house trailer can be moved along a highway that passes under a bridge with an opening in the shape of a parabolic arc, 12 m wide at the base and 6 m high in the center. If the trailer is 9 m wide and stands 3.2 m tall, measured from the ground to the top of the trailer, will it fit under the bridge?

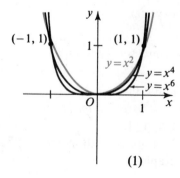

6 m

12 m

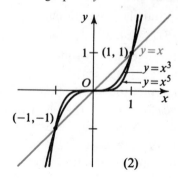

9 m

3.2 m

Rear view of trailer

2-4/GRAPHING POLYNOMIAL FUNCTIONS

You have already learned to graph the quadratic function $P(x) = ax^2 + bx + c$. In this section you will learn how to graph other polynomial functions. The simplest polynomial graphs are those of functions of the form $P(x) = x^n$, where n is a positive integer. The graphs for $n = 1$ through 6 are shown in figures (1) and (2) below. Notice that the graph of $y = x^n$ resembles that of $y = x^2$ if n is even, and that it resembles the graph of $y = x^3$ if n is odd.

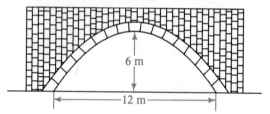

$(-1, 1)$

$(1, 1)$

$y = x^2$

$y = x^4$

$y = x^6$

(1)

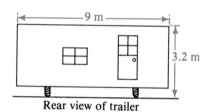

$(1, 1)$

$y = x$

$y = x^3$

$y = x^5$

$(-1, -1)$

(2)

If the polynomial defining a function can be factored into linear factors, the graph of the function is easy to sketch. Consider, for example, the graph of

$$y = (x - 1)(x - 2)(x - 3)$$

shown at the left below. You know immediately that:

(1) $y = 0$ when $x = 1$, 2, and 3.
(2) $y > 0$ when $x > 3$, since the factors $x - 1$, $x - 2$, and $x - 3$ are all positive.
(3) y changes sign whenever x passes a zero of the function. (Why?)

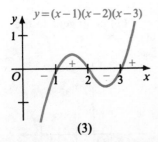

(3)

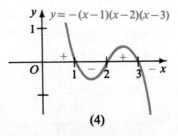

(4)

The graph of $y = -(x - 1)(x - 2)(x - 3)$, shown in figure (4) above, can be obtained using a similar analysis or by reflecting the graph in figure (3) in the x-axis.

In general, the graph of the **cubic function**

$$P(x) = ax^3 + bx^2 + cx + d$$

is shaped like a "sideways S," as shown below and on page 64. The graph of a cubic function can have at most two turning points, as shown below. If $a > 0$, the graph is in the first and third quadrants for large values of $|x|$. If $a < 0$, the graph is in the second and fourth quadrants for large values of $|x|$.

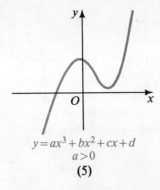

$y = ax^3 + bx^2 + cx + d$
$a > 0$
(5)

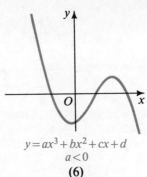

$y = ax^3 + bx^2 + cx + d$
$a < 0$
(6)

Generally, the graph of the **quartic function**

$$P(x) = ax^4 + bx^3 + cx^2 + dx + e$$

has a "W-shape" or "M-shape," as shown on the following page. The graph of a quartic function can have at most three turning points. If $a > 0$, the graph is in the first and second quadrants for large $|x|$; and if $a < 0$, the graph is in the third and fourth quadrants for large $|x|$.

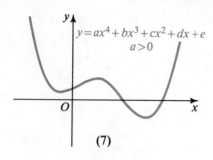

$$y = ax^4 + bx^3 + cx^2 + dx + e$$
$$a > 0$$

(7)

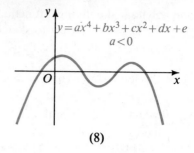

$$y = ax^4 + bx^3 + cx^2 + dx + e$$
$$a < 0$$

(8)

If a polynomial $P(x)$ has a squared factor such as $(x - c)^2$, then $x = c$ is a **double root** of $P(x) = 0$. In this case, the graph of $y = P(x)$ is tangent to the x-axis at $x = c$, as shown in figures (9), (10), and (11).

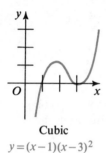

Cubic
$y = (x-1)(x-3)^2$
(9)

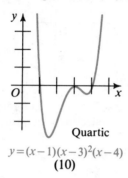

Quartic
$y = (x-1)(x-3)^2(x-4)$
(10)

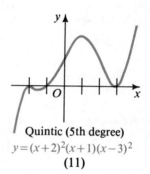

Quintic (5th degree)
$y = (x+2)^2(x+1)(x-3)^2$
(11)

If a polynomial $P(x)$ has a cubed factor such as $(x - c)^3$, then $x = c$ is called a **triple root** of $P(x) = 0$. In this case, the graph of $y = P(x)$ flattens out around $(c, 0)$ and crosses the x-axis at this point, as shown in the graphs below.

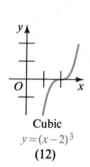

Cubic
$y = (x-2)^3$
(12)

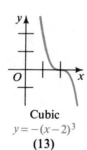

Cubic
$y = -(x-2)^3$
(13)

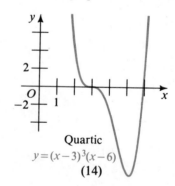

Quartic
$y = (x-3)^3(x-6)$
(14)

ORAL EXERCISES 2-4

1. Study the polynomial graphs in figures (9)–(14) above. In each case, consider the equation obtained by setting the polynomial equal to zero. Tell how many real roots this equation will have. Also indicate which real roots are double roots or triple roots.

2. Sketch the graphs of the following equations on the chalkboard.

    a. $y = (x + 2)(x - 2)(x - 4)$ **b.** $y = x(x + 2)(1 - x)$

    c. $y = -2x(x + 3)^2$ **d.** $y = x(3 - x)^2$

    e. $y = x^2(1 + x)(1 - x)$ **f.** $y = \frac{1}{4}x^2(x + 4)(x - 4)$

    g. $y = -x^3(x - 2)$ **h.** $y = \frac{1}{2}x^3(2 + x)$

3. In Exercise 2, you were given equations and asked to sketch their graphs. Can you reverse the process and give an equation for the following polynomial graphs? (Since the y-axis has no indicated scale, more than one equation can be correct.)

a.

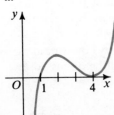

Cubic

b.

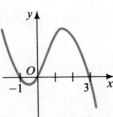

Cubic

c.

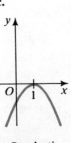

Quadratic

d.

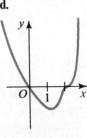

Quartic

WRITTEN EXERCISES 2-4

In Exercises 1–14, sketch the graph of the given equation.

A

  1. $y = (x + 1)(x - 2)(x - 4)$ **2.** $y = -(x + 3)(x + 2)(x - 1)$

  3. $y = -x(x + 5)(x + 3)$ **4.** $y = x^3 + x^2 - 4x - 4$ $x^2(x+1) - 4(x+1)$

  5. $y = x^2(x + 2)$ **6.** $y = (x - 2)^3$

  7. $y = x(x + 2)(x - 2)(x - 1)$ **8.** $y = x(1 - x)(1 + x)(2 + x)$

  9. $y = (x + 1)^3(x - 2)$ **10.** $y = x^4 - 5x^2 + 4$

 11. $y = -2x^3 - x^4$ **12.** $y = -x^3 - 3x^2 - 3x - 1$ $-x^2(x+3)$

 13. $y = x^2(x + 2)(x - 1)(x + 1)$ **14.** $y = x^2(1 - x)^2(2 + x)$

15. Study figure (1) on page 62; then make a sketch of $y = x^{100}$.

16. Study figure (2) on page 62; then make a sketch of $y = x^{101}$.

In Exercises 17–20, give an equation for each of the polynomial graphs shown. Since the y-axis has no indicated scale, more than one answer is possible.

17.

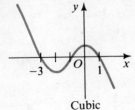

Cubic

18.

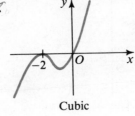

Cubic

Give an equation for the polynomial graph shown.

19.

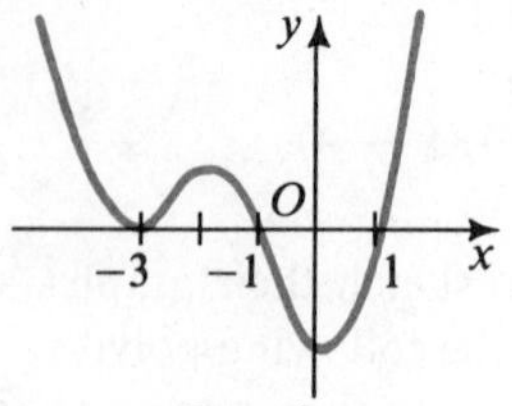

Quartic

20.

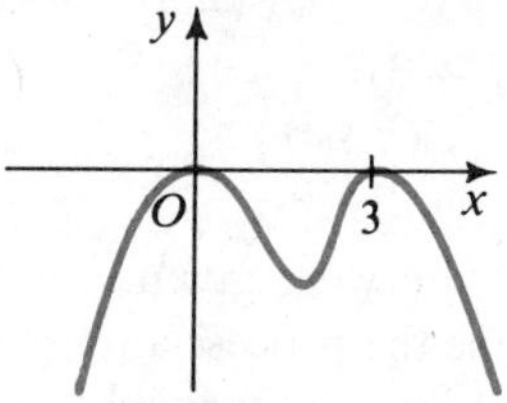

Quartic

21. **a.** Sketch the graph of $P(x) = 16x - x^3$.

b. Evaluate $P(2)$ and $P(2.1)$ and conclude that the highest point of the graph in the interval $0 \leq x \leq 4$ does not occur midway between the zeros of $P(x)$ at $x = 0$ and $x = 4$.

c. Show that $P(-x) = -P(x)$. (Because of this property, the origin is called a point of symmetry of the graph. This idea will be discussed in detail in Section 8-2.)

22. **a.** Sketch the graph of $P(x) = 2x^3 - 4x^2$.

b. Evaluate $P(1)$ and $P(1.1)$ and conclude that the lowest point of the graph in the interval $0 \leq x \leq 2$ does not occur midway between the zeros of $P(x)$ at $x = 0$ and $x = 2$.

In Exercises 23–26 you are given a linear equation and an equation of higher degree. Sketch their graphs on a set of axes and determine algebraically the coordinates of any points of intersection.

23. $y = x^3 - 4x$; $y = -3x$

24. $y = -x(x - 2)^2$; $y = -x$

25. $y = 4x^2 - x^4$; $y = 4$

26. $y = x^4 - 6x^2$; $y = -9$

In Exercises 27–30 give an equation for each of the polynomial graphs shown. Unlike Exercises 17–20, a scale on the y-axis is given.

B **27.**

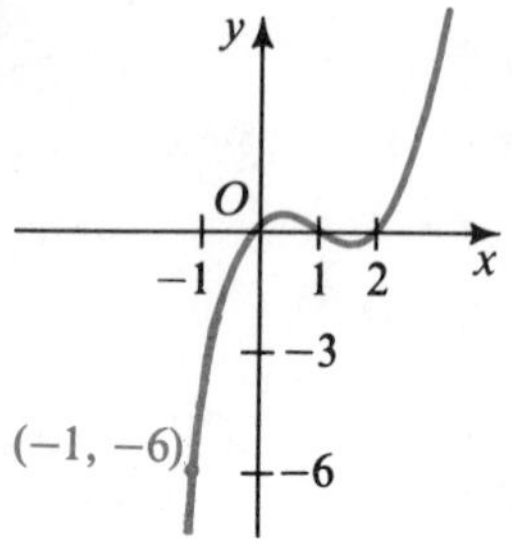

Cubic

28.

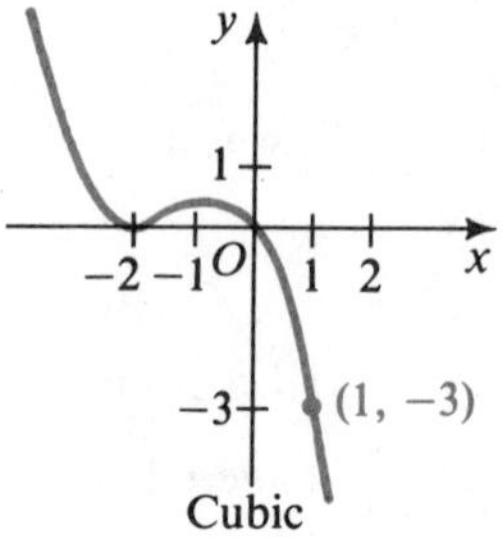

Cubic

29.

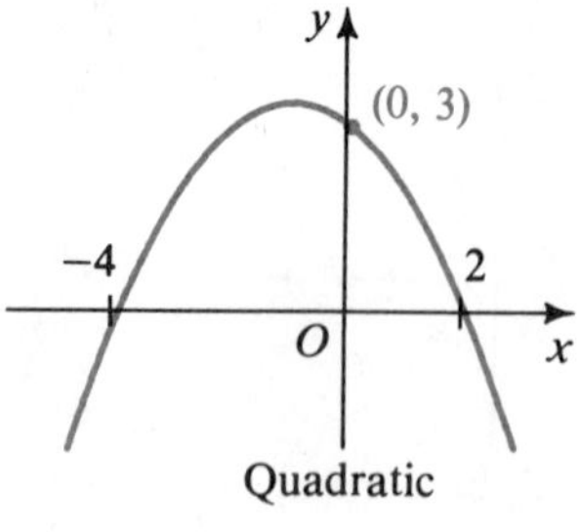

Quadratic

30.

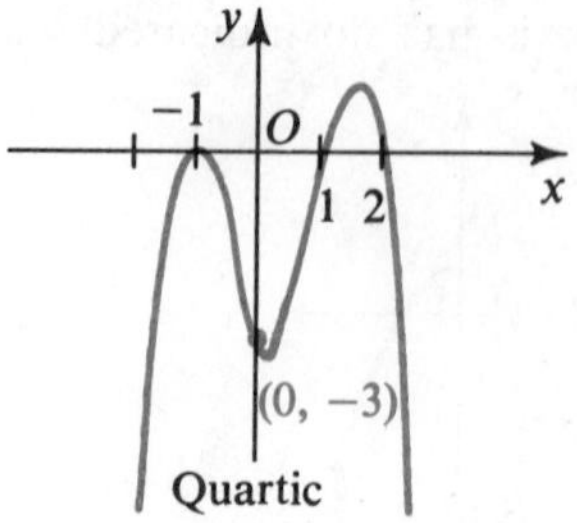

Quartic

31. Find an equation of the cubic function whose graph passes through the points $(3, 0)$ and $(1, 4)$, and is tangent to the x-axis at the origin.

32. Find an equation of the cubic function whose graph has y-intercept 8 and x-intercepts -4, -3, and 1.

33. Find an equation of the quartic function whose graph passes through $(0, -2)$ and is tangent to the x-axis at $(-1, 0)$ and $(2, 0)$.

34. Find an equation of the quadratic function whose graph passes through $(-1, 0)$ and has vertex $(1, 12)$.

35. Find an equation of the cubic polynomial $P(x)$ if $P(-3) = P(-1) = P(2) = 0$ and $P(0) = 6$.

36. Find an equation of the cubic polynomial $P(x)$ if $P(0) = 0$, $P(2) = -4$, and $P(x)$ is positive only when $x > 4$.

Roots of polynomial equations

2-5 / SYNTHETIC DIVISION; THE REMAINDER AND FACTOR THEOREMS

Synthetic division is a short-cut method of dividing a polynomial by a divisor of the form $x - a$. First let's review a long-division problem. We divide $3x^3 - 4x^2 - 9x - 1$ by $x - 2$ at the right.

$$\begin{array}{r} 3x^2 + 2x - 5 \\ x - 2 \overline{\smash{\big)}\ 3x^3 - 4x^2 - 9x - 1} \\ \underline{3x^3 - 6x^2} \\ 2x^2 - 9x \\ \underline{2x^2 - 4x} \\ -5x - 1 \\ \underline{-5x + 10} \\ -11 \end{array}$$

Next we will show how you can do this division more efficiently by using synthetic division.

Instead of writing $\quad x - 2 \overline{\smash{\big)}\ 3x^3 - 4x^2 - 9x - 1}$

write

$$\underline{2}\ |\ 3 \quad -4 \quad -9 \quad -1$$

Then bring down the first coefficient.
Write

$$3$$

Then multiply the first coefficient by 2 and add to the second coefficient as shown.

$$\begin{array}{r|rrrr} \underline{2} & 3 & -4 & -9 & -1 \\ & & 6 \\ \hline & 3 & 2 \end{array}$$

(*Division continued on page 68*)

Polynomials **67**

Continue the process of multiplying sums by 2 and adding, as shown below. (In general, you multiply by *a*.)

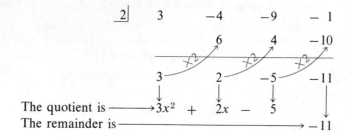

The quotient is ——→$3x^2$ + $2x$ − 5
The remainder is ——————————————→ −11

EXAMPLE 1. Divide $2x^3 + 7x^2 - 4$ by $x + 3$.

SOLUTION: Since the polynomial has no *x*-term, we must supply a zero coefficient for *x*. Also, because the divisor must have the form $x - a$, we write $x + 3$ as $x - (-3)$ and put -3 to the left of the coefficient row.

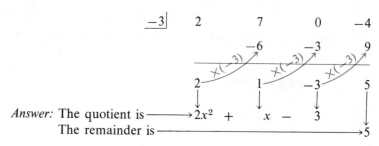

Answer: The quotient is ——→$2x^2$ + x − 3
The remainder is ——————————————→ 5

The Remainder and Factor Theorems

When a polynomial $P(x)$ is divided by $(x - a)$, the quotient $Q(x)$ is a polynomial in *x*, and the remainder R is a constant. These quantities are related as follows:

$$\text{Polynomial} = \text{Divisor} \times \text{Quotient} + \text{Remainder}$$
$$P(x) = (x - a) \cdot Q(x) + R$$

Substituting *a* for *x* in this equation, we get:

$$P(a) = (a - a) \cdot Q(a) + R$$
$$P(a) = (0) \cdot Q(a) + R$$
$$P(a) = R$$

This result is called the Remainder Theorem.

THE REMAINDER THEOREM. When a polynomial $P(x)$ is divided by $x - a$, the remainder $R = P(a)$.

EXAMPLE 2. Find the remainder when $P(x) = x^6 - 3x^5 + 6$ is divided by: **a.** $x - 1$ **b.** $x + 2$

SOLUTION: **a.** The remainder $R = P(1) = 1^6 - 3 \cdot 1^5 + 6 = 4$
      b. The remainder $R = P(-2) = (-2)^6 - 3(-2)^5 + 6 = 166$

An immediate consequence of the Remainder Theorem is the following Factor Theorem.

THE FACTOR THEOREM. $x - a$ is a factor of the polynomial $P(x)$ if and only if $P(a) = 0$.

EXAMPLE 3.   **a.** Is $x - 3$ a factor of $P(x) = x^4 - 6x^2 - 10x + 3$?
      b. Is $x + 2$ a factor?

SOLUTION: **a.** Since $P(3) = 3^4 - 6 \cdot 3^2 - 10 \cdot 3 + 3 = 0$, $x - 3$ is a factor.
      b. Since $P(-2) = (-2)^4 - 6(-2)^2 - 10(-2) + 3 = 15 \neq 0$, $x - (-2)$, or $x + 2$, is not a factor.

From the Remainder Theorem, we know that the remainder when a polynomial $P(x)$ is divided by $x - a$ is $P(a)$. Thus, we can evaluate $P(a)$ using synthetic division of $P(x)$ by $x - a$. This method is the one most commonly used with computers because it generally involves fewer steps than straightforward substitution of $x = a$ into $P(x)$.

EXAMPLE 4.   If $P(x) = 2x^3 - 7x^2 + 10x - 5$, evaluate $P(\frac{3}{2})$.

SOLUTION:

$$
\begin{array}{r|rrrr}
\tfrac{3}{2} & 2 & -7 & 10 & -5 \\
 & & 3 & -6 & 6 \\
\hline
 & 2 & -4 & 4 & \underline{|1} = \text{Remainder} = P(\tfrac{3}{2})
\end{array}
$$

ORAL EXERCISES 2-5

1. Is $x - 2$ a factor of $x^6 - 64$? Is $x + 2$? Explain.

2. Is $x - 2$ a factor of $x^5 + 32$? Is $x + 2$? Explain.

3. What is the remainder when $P(x) = x^{15} + 3x^{10} + 2$ is divided by $x - 1$? What is the remainder when $P(x)$ is divided by $x + 1$?

4. A division problem has been done by synthetic division below.

$$
\begin{array}{r|rrrrr}
-3 & 1 & 0 & -8 & 5 & -1 \\
 & & -3 & 9 & -3 & -6 \\
\hline
 & 1 & -3 & 1 & 2 & \underline{|-7}
\end{array}
$$

   a. What is the dividend? What is the divisor?
   b. What is the quotient? What is the remainder?
   c. What is the value of the polynomial when $x = -3$?

5. Repeat parts (a) and (b) of Exercise 4 for the synthetic division problem shown at the right. Then give the value of the polynomial when $x = -\frac{1}{2}$.

$$-\tfrac{1}{2} \begin{array}{|rrrrr} 2 & -3 & -2 & 1 \\ & -1 & 2 & 0 \\ \hline 2 & -4 & 0 & \underline{|1} \end{array}$$

6. a. Use synthetic division to divide $P(x) = x^3 + 3x^2 + x - 1$ by $x + 1$.
  b. Is $x + 1$ a factor of $P(x)$? If so, how would you find the other factors of $P(x)$?

WRITTEN EXERCISES 2-5

A

1. a. Use the Factor Theorem to decide whether $x - 1$ is a factor of $2x^{10} - x^7 - 1$.
  b. Is $x + 1$ a factor of $2x^{10} - x^7 - 1$?

2. a. Use the Factor Theorem to decide whether $x - 2$ is a factor of $x^5 - x^4 - 3x^2 - 4$.
  b. Is $x + 2$ a factor of $x^5 - x^4 - 3x^2 - 4$?

3. Use the Remainder Theorem to find the remainder when $x^5 - 2x^3 + x^2 - 4$ is divided by:
  a. $x - 1$ **b.** $x + 1$

4. Use the Remainder Theorem to find the remainder when $x^3 - 3x^2 + 5$ is divided by:
  a. $x - 2$ **b.** $x + 2$

In Exercises 5–12, find the quotient and the remainder when the first polynomial is divided by the second.

5. $x^3 - 2x^2 + 5x + 1$; $x - 1$ **6.** $2x^3 + x^2 + 3x + 7$; $x + 2$

7. $t^4 - 2t^3 + 5t + 2$; $t + 1$ **8.** $2x^4 - 3x^3 + 4x^2 - 5x + 2$; $x - 1$

9. $y^5 + y^3 + y$; $y - 3$ **10.** $k^2 - 3k^4$; $k + 2$

11. $3x^4 - 2x^3 + 5x^2 + x + 1$; $x^2 + 2x$ **12.** $x^5 + 3x^2 + 4$; $x^2 + 2x + 1$

13. If $P(x) = 3x^3 - 7x^2 + 2x + 3$, evaluate $P(\frac{1}{3})$ and $P(-\frac{2}{3})$ using the method of Example 4.

14. If $P(x) = 4x^4 - 3x^3 + 8x - 2$, evaluate $P(\frac{3}{4})$.

15. If $x + 2$ is a factor of $x^3 + 5x^2 + kx + 4$, find the value of k.

16. If $x + 1$ is a factor of $x^3 + kx - 3$, find the value of k.

17. Show that $x - a$ is a factor of $x^n - a^n$, for any positive integer n.

18. Show that $x + a$ is a factor of $x^n + a^n$, for any odd positive integer n.

19. When a polynomial $P(x)$ is divided by $2x + 1$, the quotient is $x^2 - x + 4$ and the remainder is 3. Find $P(x)$.

20. When a polynomial $P(x)$ is divided by $3x - 4$, the quotient is $x^3 + 2x + 2$ and the remainder is -1. Find $P(x)$.

21. Given that $x = 3$ is a root of the equation $2x^3 - 5x^2 - 4x + 3 = 0$, find the other roots.

22. Given that $x = -2$ is a root of the equation $6x^3 + 11x^2 - 4x - 4 = 0$, find the other roots.

B 23. Given that $x = 1$ and $x = -1$ are two roots of the equation $2x^4 - 9x^3 + 2x^2 + 9x - 4 = 0$, find the other roots.

24. Given that $x = -1$ and $x = 2$ are roots of the equation $x^4 - 2x^3 + x^2 - 4 = 0$, find the other roots.

25. Given that $x = 3$ is a double root of the equation $4x^4 - 24x^3 + 35x^2 + 6x - 9 = 0$, find the other roots.

26. Given that $x = 1$ is a double root of the equation $x^4 - 2x^3 + 2x^2 - 2x + 1 = 0$, find the other roots.

27. Use the Factor Theorem to show that $x - a$ is a factor of
$$x^2(a - b) + a^2(b - x) + b^2(x - a).$$

28. Use the Factor Theorem to show that $x - c$ is a factor of
$$(x - b)^3 + (b - c)^3 + (c - x)^3.$$

29. When $P(x) = x^3 + ax + b$ is divided by $x - 1$, the remainder is 2. When $P(x)$ is divided by $x - 2$, the remainder is 1. Find the values of a and b.

30. If $P(x) = x^3 + x^2 - x + 1$, find the remainder when $P(x)$ is divided by:
   a. $x - 1$ **b.** $x + 1$ **c.** $x^2 - 1$

C 31. Find all values of a for which $x^2 - ax^2 + (a + 1)x - 2$ is divisible by $x - a$.

32. Given $P(x) = ax^3 + bx^2 + cx + d$, find the relationship between a, b, c, and d if:
   a. $x - 1$ is a factor of $P(x)$.
   b. $x + 1$ is a factor of $P(x)$.
   c. $x^2 - 1$ is a factor of $P(x)$.

CALCULATOR EXERCISE

Given $f(x) = 8x^3 + 9x^2 - 6x - 5$, evaluate $f(7)$ by the two methods shown below and keep track of the number of times you press a calculator key for each method. Which method is more efficient?

Method 1. Substitute 7 for x and evaluate: $8 \cdot 7^3 + 9 \cdot 7^2 - 6 \cdot 7 - 5$

Method 2. Use synthetic division: $\underline{7 |}$ 8 9 -6 -5

2-6/RATIONAL ROOTS OF POLYNOMIAL EQUATIONS

If you were asked to solve $x^2 - 42x + 360 = 0$ by factoring, you might begin as follows:

$$x^2 - 42x + 360 = 0$$
$$(x - \quad)(x - \quad) = 0$$

You might try $x - 36$ or $x - 12$ as your first factor, but you would not try $x - 50$ since 50 is not a factor of 360. By considering only factors $x - p$ where

p is a factor of 360, you are, in essence, noticing that values of p are restricted to factors of the constant term 360.

In other words, we have concluded that any integral root of the given quadratic equation must be a factor of the constant term. Our reasoning depended upon certain facts about the equation: The coefficients were integers, the leading coefficient was 1, and the constant term was not 0. Whenever an equation satisfies these conditions, the *Integral Roots Theorem*, stated below, guarantees a similiar conclusion.

INTEGRAL ROOTS THEOREM. Let $P(x)$ be a polynomial of degree n such that its coefficients are integers, its leading coefficient is 1, and its constant term (k) is not 0:

$$P(x) = x^n + bx^{n-1} + cx^{n-2} + \cdots + k$$

If the integer p is a root of $P(x) = 0$, then p is a factor of k.

The Integral Roots Theorem is a special case of the *Rational Roots Theorem,* which is stated and proved below.

THE RATIONAL ROOTS THEOREM. Let $P(x)$ be a polynomial of degree n with integral coefficients:

$$P(x) = ax^n + bx^{n-1} + cx^{n-2} + \cdots + k$$

If $\dfrac{p}{q}$ is a root of $P(x) = 0$, where p and q are nonzero integers with no common factor other than 1, then p must be a factor of k, and q must be a factor of a.

Proof: We shall give a proof for a cubic polynomial. The reasoning used can be applied to construct a proof for a polynomial of any degree.

Given: (i) $P(x) = ax^3 + bx^2 + cx + k$ is a cubic polynomial with integral coefficients.

(ii) $\dfrac{p}{q}$ is a root of $P(x) = 0$.

(iii) p and q are nonzero integers with no common factors.

Prove: First we prove that p is a factor of k.

Since $\dfrac{p}{q}$ is a root of $P(x) = 0$:

$$a\left(\frac{p}{q}\right)^3 + b\left(\frac{p}{q}\right)^2 + c\left(\frac{p}{q}\right) + k = 0$$

Multiplying both sides of this equation by q^3 gives:

$$ap^3 + bqp^2 + cq^2p + q^3k = 0$$
$$ap^3 + bqp^2 + cq^2p = -q^3k$$
$$p(ap^2 + bqp + cq^2) = -q^3k$$

The last equation shows that p is a factor of $-q^3k$.

Since q and p have no common factors, p must be a factor of k.

Next we prove that q is a factor of a.

We begin by repeating the second equation from the first part of the proof:

$$ap^3 + bqp^2 + cq^2p + q^3k = 0$$
$$bqp^2 + cq^2p + q^3k = -ap^3$$
$$q(bp^2 + cqp + q^2k) = -ap^3$$

The last equation shows that q is a factor of $-ap^3$.

Since q and p have no common factors, q must be a factor of a.

EXAMPLE 1. Find all the rational roots of $P(x) = 2x^3 + x^2 + 5x - 3 = 0$.

SOLUTION: We apply the Rational Roots Theorem. The factors of the constant term are ± 1 and ± 3. The factors of the leading coefficient are ± 1 and ± 2. Therefore, the only possibilities for rational roots are ± 1, $\pm \frac{1}{2}$, $\pm \frac{3}{2}$, ± 3. Now we test these roots:

$$
\begin{array}{r|rrrl}
-1 & 2 & 1 & 5 & -3 \\
   &   & -2 & 1 & -6 \\
\hline
   & 2 & -1 & 6 & \underline{-9} = P(-1)
\end{array}
\qquad
\begin{array}{r|rrrl}
1 & 2 & 1 & 5 & -3 \\
  &   & 2 & 3 & 8 \\
\hline
  & 2 & 3 & 8 & \underline{5} = P(1)
\end{array}
$$

$$
\begin{array}{r|rrrl}
-\frac{1}{2} & 2 & 1 & 5 & -3 \\
             &   & -1 & 0 & -2\frac{1}{2} \\
\hline
             & 2 & 0 & 5 & \underline{-5\tfrac{1}{2}} = P(-\tfrac{1}{2})
\end{array}
\qquad
\begin{array}{r|rrrl}
\frac{1}{2} & 2 & 1 & 5 & -3 \\
            &   & 1 & 1 & 3 \\
\hline
            & 2 & 2 & 6 & \underline{0} = P(\tfrac{1}{2})
\end{array}
$$

The last synthetic division on the right shows that $\frac{1}{2}$ is a zero of $P(x)$. The division also shows that

$$2x^3 + x^2 + 5x - 3 = (x - \tfrac{1}{2})(2x^2 + 2x + 6).$$

Consequently, we can write $P(x) = 0$ as

$$(x - \tfrac{1}{2})(2x^2 + 2x + 6) = 0.$$

Any other roots of $P(x) = 0$ must be roots of

$$2x^2 + 2x + 6 = 0.$$

Since the discriminant of this equation is negative, there are only imaginary roots. Thus, $\frac{1}{2}$ is the only rational root of $P(x) = 0$.

When a is a root of the polynomial equation, the equation

$$P(x) \div (x - a) = 0$$

is called a **depressed equation**. In general, you can find the remaining roots of $P(x) = 0$ by finding solutions to the depressed equation.

EXAMPLE 2. Find all real and imaginary roots of $4x^3 + 5x + 3 = 0$.

SOLUTION: According to the rational roots theorem, the only possibilities for rational roots are ± 1, ± 3, $\pm\frac{1}{2}$, $\pm\frac{3}{2}$, $\pm\frac{1}{4}$, and $\pm\frac{3}{4}$. The process of checking these twelve numbers can be simplified by observing that there are no positive roots of the equation because a positive number when substituted into $4x^3 + 5x + 3$ will never give zero. Mental substitution in $P(x) = 4x^3 + 5x + 3$ shows that $P(0) = 3$ and $P(-1) = -6$. Since $P(x)$ changes signs between 0 and -1, there must be a real root between $x = 0$ and $x = -1$. Of course, this root might be irrational. If it is rational, however, it must be one of two possibilities: $-\frac{1}{2}$ or $-\frac{1}{4}$. The calculation below shows that $-\frac{1}{2}$ is a root.

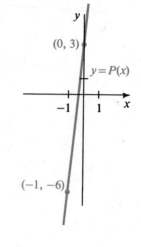

$$
\begin{array}{r|rrrr}
-\frac{1}{2} & 4 & 0 & 5 & 3 \\
 & & -2 & 1 & -3 \\
\hline
 & 4 & -2 & 6 & \underline{|\,0}
\end{array}
$$

Solving the depressed equation

$$4x^2 - 2x + 6 = 0$$

using the quadratic formula gives $x = \dfrac{1 \pm i\sqrt{23}}{4}$.

The solutions are $-\dfrac{1}{2}$, $\dfrac{1 - i\sqrt{23}}{4}$, and $\dfrac{1 + i\sqrt{23}}{4}$.

ORAL EXERCISES 2-6

Give the possible rational roots of each equation.

1. $x^3 - x^2 - x + 1 = 0$ **2.** $x^3 + 2x^2 - x - 2 = 0$

3. $x^4 - 10x^2 + 9 = 0$ **4.** $2x^3 - 9x^2 + 3x + 4 = 0$

5. $3x^3 - 4x^2 - 5x + 2 = 0$ **6.** $6x^3 + 7x^2 - 9x + 2 = 0$

7. $x^3 + 3x^2 - 13x - 15 = 0$ **8.** $2x^3 + 3x^2 - 23x - 12 = 0$

9. $3x^3 - 7x^2 - 2x + 8 = 0$ **10.** $8x^4 - 10x^3 - 14x^2 + 4x + 2 = 0$

11. If $P(x)$ is a polynomial such that $P(2) = -1$ and $P(3) = 2$, must the equation have a rational root between $x = 2$ and $x = 3$? Must the equation have a real root between $x = 2$ and $x = 3$?

12. a. Explain why the equation $x^3 + 2x^2 + x + 1 = 0$ has a real root between $x = -2$ and $x = -1$.
    b. Determine whether this real root is rational or irrational.
    c. Explain why this equation has no positive roots.

A **1-10.** Solve the equations in Oral Exercises 1–10, giving all rational roots.

In Exercises 11–18, solve each polynomial equation, giving all real and imaginary roots.

11. $2x^3 + 5x^2 - 1 = 0$ **12.** $4x^3 - 13x - 6 = 0$

13. $x^3 - 4x^2 + 4x - 16 = 0$ **14.** $x^3 + 9x^2 + 26x + 30 = 0$

15. $x^4 + 2x^3 - 2x^2 - 6x - 3 = 0$ **16.** $3x^3 - x^2 - 36x + 12 = 0$

17. $2x^3 - 7x + 2 = 0$ **18.** $2x^4 - x^3 - 7x^2 + x + 2 = 0$

19. Show that the equation $x^3 + x^2 - 3 = 0$ has no rational roots, but that it does have an irrational root between $x = 1$ and $x = 2$.

20. Show that the equation $x^4 - 4x + 2 = 0$ has no rational roots, but that it does have an irrational root between 0 and 1 and an irrational root between 1 and 2.

In Exercises 21–24 sketch the graphs of the two given equations on the same set of axes. Determine algebraically where the graphs intersect or are tangent.

21. $y = x^3 - 3x$ **22.** $y = 2x^3 + 3x^2$
    $y = 2$ $y = 1$

23. $y = x^3 - x$ **24.** $y = x^3 + 4x^2$
    $y = 3x$ $y = 3x + 18$

B **25. a.** The lateral height of a cone is 3 and its height is h. Show that its volume is
$$V(h) = \tfrac{1}{3}\pi(9 - h^2)h.$$

b. Find the two values of h that make $V(h) = \tfrac{10}{3}\pi$.

26. a. A cylinder is inscribed in a sphere with radius 4. Show that the volume of the cylinder is
$$V(x) = 2\pi x(16 - x^2).$$

b. Find the two values of x that make $V(x) = 42\pi$.

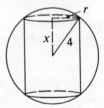

27. Show that the positive real cube root of 12 is irrational. (*Hint:* What equation must the cube root of 12 satisfy?)

28. Show that the positive fifth root of 100 is irrational.

C **29.** A wooden block is in the shape of a rectangular prism with dimensions n cm, $(n + 3)$ cm, and $(n + 9)$ cm, for some integer n. The surface of the block is painted and the block is then cut into 1 cm cubes by cuts parallel to the faces. If exactly half of these cubes have no paint on them, find the dimensions of the original block.

Consider the equation $P(x) = x^3 + x - 1 = 0$. A quick check shows that there are no rational roots, but that there is an irrational root between 0 and 1 because $P(0)$ is negative and $P(1)$ is positive.

a. Evaluate $P(0.5)$ and tell whether the root is greater or less than 0.5.

b. Evaluate $P(0.7)$ and tell whether the root is greater or less than 0.7.

c. Locate the root between two decimals in hundredths.

2-7/APPROXIMATING ROOTS OF POLYNOMIAL EQUATIONS (OPTIONAL)

In the preceding section, we were concerned with finding rational roots of polynomial equations. In this section, we shall discuss three methods of approximating roots of polynomial equations that apply to both rational and irrational roots.

The bracket-and-halving method

The *bracket-and-halving method* described below is particularly well suited to a calculator or computer, but it can also be used without either if the equation involved has simple coefficients. We illustrate the method with the equation

$$P(x) = x^3 + x - 1 = 0.$$

1. First, we mentally locate a root between two integers or have the computer do this. In this case, there is a root between $x = 0$ and $x = 1$ because $P(0)$ is negative and $P(1)$ is positive.

2. Now we divide the interval from 0 to 1 in half and determine in which half the root lies. Since $P(\frac{1}{2}) = \frac{1}{8} + \frac{1}{2} - 1 < 0$ and $P(1) > 0$, the root is between $\frac{1}{2}$ and 1.

3. Next, we divide the interval from $\frac{1}{2}$ to 1 in half and again find the half containing the root. Since $P(\frac{3}{4}) = \frac{27}{64} + \frac{3}{4} - 1 > 0$ and $P(\frac{1}{2}) < 0$, the root is between $\frac{1}{2}$ and $\frac{3}{4}$.

4. We continue to find intervals which bracket the root and then halve these intervals until we have the accuracy we desire.

The linear-interpolation method

The *linear-interpolation method* enables you to approximate roots by replacing part of the graph of a polynomial with a line segment. We shall use this method to approximate the positive root of

$$P(x) = x^3 + 2x - 5 = 0.$$

A quick check of the possible positive rational roots $(x = 1, 5)$ shows that neither is a root. Also, a mental check shows that $P(1) = -2 < 0$ and $P(2) = 7 > 0$, so there must be a root between $x = 1$ and $x = 2$. The linear-interpolation method assumes this root is approximately the point C where line segment $\overline{AB}$ crosses the x-axis. Consequently:

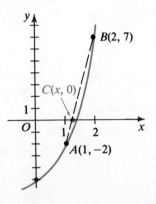

$$\text{Slope of } \overline{AC} = \text{Slope of } \overline{AB}$$
$$\frac{0 - (-2)}{x - 1} = \frac{7 - (-2)}{2 - 1}$$
$$\frac{2}{x - 1} = 9$$
$$x = 1\frac{2}{9}$$

From the graph we can see that the exact root lies between $x = 1\frac{2}{9}$ and $x = 2$. If we were not given the graph, we could observe that $P(1\frac{2}{9}) = -\frac{532}{729}$. Since $P(1\frac{2}{9})$ is negative, the exact root must lie between $x = 1\frac{2}{9}$ and $x = 2$. Then if we want greater accuracy, we can use linear interpolation again with $x = 1\frac{2}{9}$ and some greater value of x for which $P(x)$ is positive.

A computer search

A computer program can be written for almost any method of approximating roots of a polynomial equation. The method used in the following program is similar to the bracket-and-halving method, but the part of the x-axis containing the root is divided into tenths. The BASIC program below searches for a root of the equation $P(x) = 3x^4 - 5x^3 + 2x - 7 = 0$. First, we observe that $P(0) < 0$, $P(1) < 0$, and $P(2) > 0$. Thus, there is a root between 1 and 2. The first line of the program below tells the computer to consider $x = 1, 1.1, 1.2, 1.3, \ldots, 2.0$.

```
10  FOR X = 1 TO 2 STEP .1
20  LET Y = 3*X↑4 - 5*X↑3 + 2*X - 7
30  PRINT X, Y
40  NEXT X
50  END
```

The printout is shown on the next page.

```
1               -7
1.1             -7.0627
1.2             -7.0192
1.3             -6.8167
1.4             -6.3952
1.5             -5.6875
1.6             -4.6192
1.7             -3.1087
1.8             -1.0672
1.9              1.60131
2                5
```

Since the computer printout indicates that the value of *y* changes from negative to positive between 1.8 and 1.9, we change line 10 as shown below. The new printout follows.

```
10   FOR X = 1.8 TO 1.9 STEP .01
```

```
1.8             -1.0672
1.81            -.830208
1.82            -.586856
1.83            -.337038
1.84            -.0806561
1.85             .182395
1.86             .45222
1.87             .728919
1.88            1.01259
1.89            1.30335
1.9             1.6013
```

We can now modify line 10 again to consider values of X between 1.84 and 1.85. Continuing in this way, you can find the root to even more accuracy. If we had wanted to stop with an approximation to the nearest hundredth, we would choose X = 1.84 since $P(1.84) = 0.0806561$, which is nearer to 0 than $P(1.85)$.

The computer program we have used is not the most efficient one for finding roots. It is possible to write a more elaborate program that has the computer do all the work internally, instead of printing lists of values which require you to change line 10 to get more accuracy.

ORAL EXERCISES 2-7

Locate between two consecutive integers a real root of each equation.

1. $x^3 + x - 5 = 0$

2. $2x^3 - x - 3 = 0$

3. $x^4 - 2x^3 - 3x - 3 = 0$

4. $x^3 + 5x^2 - 3 = 0$

5. Study the table of values and give the approximate roots of $P(x) = 0$.

x	-3	-2	-1	0	1	2	3	4
$P(x)$	4	-1	-5	-2	3	10	-3	-16

Find all answers to the nearest tenth unless instructed otherwise.

Approximate the smallest positive real root of the given equation.

A **1-4.** Use the equations given in Oral Exercises 1–4.

   5. $x^3 - x^2 - 2 = 0$ **6.** $2x^3 + x - 4 = 0$

   7. $x^4 + 2x - 1 = 0$ **8.** $x^4 = 5x + 3$

B **9.** A box has length 4, width 2, and height 1. Another box with twice the volume has dimensions $4 + x$, $2 + x$, and $1 + x$. Find the approximate value of x.

   10. Suppose each face of a $3 \times 5 \times 4$ cm block of wood is shaved (or planed) by x cm. What is the approximate value of x that will give the shaved block half the original volume?

   11. a. Find in terms of x the volume of a regular square pyramid with base edges x and height $(x + 2)$.
   b. Find the approximate value of x for which the volume is 100.

   12. The radius of a cone is 2 more than the height of the cone. Find the approximate height if the volume of the cone is 40π.

   13. Show that $14x^4 - 9x^2 + 1 = 0$ has two real roots between 0 and 1.

C **14.** Given a polynomial $P(x)$ such that $P(a) < 0$ and $P(b) > 0$, show that the linear-interpolation estimate of the root of $P(x) = 0$ between a and b is

$$x = a - \frac{P(a) \cdot (b - a)}{P(b) - P(a)}.$$

COMPUTER EXERCISE *(The Crossed Ladders Problem)*

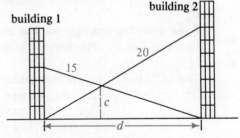

Two buildings are d units apart. A ladder 20 units long has its foot resting against building 1 and its top against the side of building 2. A second ladder 15 units long has its foot against building 2 and its top against the side of building 1. The ladders touch each other at a point c units above the ground.

1. Show that $\dfrac{1}{\sqrt{400 - d^2}} + \dfrac{1}{\sqrt{225 - d^2}} = \dfrac{1}{c}$. (Use similar triangles.)

2. a. Given $d = 12$, find the value of c using Exercise 1.
   b. If you are given $c = 8$, it is difficult to solve for d directly. It is easier to find an approximate value for d by using a computer to find the value of d for which the expression below changes sign. Do so.

$$\frac{1}{\sqrt{400 - d^2}} + \frac{1}{\sqrt{225 - d^2}} - \frac{1}{8}$$

2-8/FINDING MAXIMUMS AND MINIMUMS OF POLYNOMIAL FUNCTIONS

Quadratic functions

In the following example we apply the fact that the maximum or minimum value of a quadratic function occurs at the vertex of its graph.

EXAMPLE 1. A rectangular dog pen is constructed using a barn as one side and 60 m of fencing for the other three sides. Find the dimensions of the pen that give the greatest area.

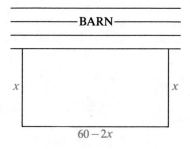

SOLUTION: Let $x =$ the length in meters of the side touching the barn. Then $(60 - 2x) =$ the length in meters of the side parallel to the barn. If $A(x)$ represents the area of the pen, then

$$A(x) = x(60 - 2x).$$

The zeros of $A(x)$ are at $x = 0$ and $x = 30$. Consequently, the maximum area will occur at the average of the zeros, or $x = 15$. Thus, the area will be a maximum when the dimensions are 15 m by 30 m, with the longer side being parallel to the barn. The maximum area is $A(15) = 450$ m².

Cubic functions (Optional)

In the following example we use a computer program to estimate the maximum value of a cubic function within a certain range of values.

EXAMPLE 2. Squares with sides of length x are cut from the corners of a rectangular piece of sheet metal with dimensions of 6 and 10. The metal is then folded to make an open-topped box. What is the maximum volume of such a box?

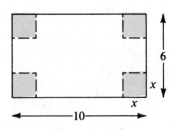

SOLUTION: The dimensions of the box are:

$$\text{height} = x$$
$$\text{length} = 10 - 2x$$
$$\text{width} = 6 - 2x$$

Thus, the volume of the box is given by the following function:

$$V = f(x) = x(10 - 2x)(6 - 2x)$$

The graph of the volume function at the right shows that V is positive when the value of x is between 0 and 3. This explains line 10 in the computer program below. (We do not consider values of x greater than 3 because they yield negative values for the width of the box.) Line 20 evaluates the volume V for values of x between 0 and 3. The variable M, introduced in line 5, will ultimately be the maximum value of V, but initially it is zero. Every time that a volume V is greater than the previous maximum M, line 40 replaces the previous value of M with the value of V. Line 50 replaces X1 with the corresponding value of X.

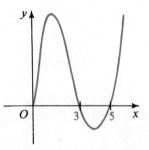

```
 5  LET M=0
10  FOR X=.01 TO 3 STEP .01
20  LET V=X*(6-2*X)*(10-2*X)
30  IF V <= M THEN 60
40  LET M=V
50  LET X1=X
60  NEXT X
70  PRINT "MAXIMUM VOLUME IS APPROXIMATELY";M;"AT X=";X1
80  END
```

The printout is shown below.

```
MAXIMUM VOLUME IS APPROXIMATELY 32.835 AT X = 1.21
```

ORAL EXERCISES 2-8

For each of the following quadratic functions, state **(a)** whether the function has a maximum or minimum value and **(b)** the value of x where the maximum or minimum occurs.

1. $f(x) = (x - 1)(x - 7)$ **2.** $g(x) = 8 - (x - 2)^2$ **3.** $h(x) = 2x^2 - 6x + 9$

4. (Optional) Study the graph of $V = f(x)$ shown in the solution to Example 2. You can see that V has minimum value when the value of x is between 3 and 5. Explain how you would modify the computer program to find this minimum value. (Ignore the fact that V represents a volume.)

WRITTEN EXERCISES 2-8

Quadratic functions

A   **1.** A gardener wants to make a rectangular enclosure using a wall as one side and 120 m of fencing for the other three sides. Express the area in terms of x, and find the value of x that gives the greatest area.

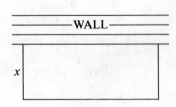

2. a. A rectangle has a perimeter of 80. If its width is x, express its length and its area in terms of x.

b. Make a graph showing the area for various values of x. What is the maximum area?

3. Suppose you had 102 m of fencing to make two side-by-side rectangular enclosures, as shown. What is the maximum area that you could enclose?

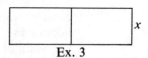

Ex. 3

4. Suppose you had to use exactly 200 m of fencing to make either one square enclosure or two separate square enclosures of any sizes you wished. What plan would give you the least area? What plan would give you the greatest?

5. Show that $(x - a)^2 + (x - b)^2$ has a minimum value when $x = \dfrac{a + b}{2}$.

6. Show that $(x - a)^2 + (x - b)^2 + (x - c)^2$ has a minimum value when $x = \dfrac{a + b + c}{3}$.

B **7.** Study Exercises 5 and 6 to make a generalization.

8. Show that of all rectangles having perimeter P, the square has the greatest area.

9. If a ball is thrown vertically upward at 30 m/s, its approximate height in meters t seconds later is given by the function: $f(t) = 30t - 5t^2$. How high does the ball go?

10. If a ball is thrown upward from a building 30 m tall, and the ball has a vertical velocity of 25 m/s, then its approximate height above the ground t seconds later is given by the function: $f(t) = 30 + 25t - 5t^2$.
a. How high does the ball go?
b. After how many seconds will the ball hit the ground?

11. The publisher of a magazine that has a circulation of 80,000 and sells for $1.60 a copy decides to raise the price of the magazine because of increased production and distribution costs. By surveying the readers of the magazine, the publisher finds that for each increase of 40¢ in price the magazine will lose 10,000 readers. What should the price per copy be to maximize the income?

12. An orange grower has 400 crates of fruit ready for market and will have 20 more for each day the grower waits. The present price is $60 per crate and will drop an estimated $2 per day for each day waited. In how many days should the grower ship the crop to maximize his income?

C **13.** It is possible to inscribe many rectangles in the isosceles triangle shown. What are the dimensions of the inscribed rectangle having the largest possible area?

14. Given the points $A(0, 5)$, $B(3, 7)$ and $C(6, 2)$, find the point P on the x-axis for which $(PA)^2 + (PB)^2 + (PC)^2$ is as small as possible.

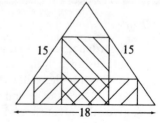

Cubic functions (*Optional*)

In Exercises 1–4, you will need to use a computer program similar to the one given in Example 2. Give the value of x or r to the nearest hundredth unless otherwise instructed.

A   **1.** Squares are cut from the corners of an 8 cm by 14 cm piece of sheet metal, as shown at the right. The metal is then folded to make a box.

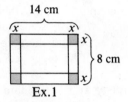

14 cm

8 cm

a. Show that the volume of this box is
$$V(x) = x(8 - 2x)(14 - 2x).$$

Ex.1

   b. Find the approximate value of x that makes the volume a maximum. Give the corresponding maximum volume.

2. A 10 cm by 20 cm piece of sheet metal is cut and folded as indicated in the diagram to make a box with a top.

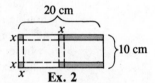

20 cm

10 cm

Ex. 2

   a. Show that the volume of the box is
$$V(x) = x(10 - x)(10 - 2x).$$

   b. Find the approximate value of x that makes the volume a maximum. Give the corresponding maximum volume.

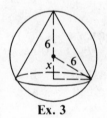

B   **3.** A cone is inscribed in a sphere of radius 6.

   a. Express its volume as a function of x. ($V = \frac{1}{3}\pi r^2 h$)

   b. Find the approximate value of x that makes the volume a maximum. Give the corresponding maximum volume.

Ex. 3

C   **4.** A cylinder is inscribed in a cone with height 10 and a base of radius 5, as shown in the diagram at the right. Find the approximate values of r and h for which the volume of the cylinder is a maximum. Give the corresponding maximum volume.

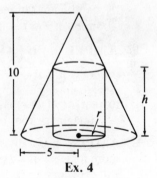

Ex. 4

CALCULATOR EXERCISES

A piece of wire 40 cm long is to be cut into two pieces. One piece will be bent to form a circle; the other will be bent to form a square.

1. Find the lengths of the two pieces that cause the sum of the areas of the circle and square to be a minimum.

2. How could you make the total area of the circle and the square a maximum?

2-9/GENERAL RESULTS FOR POLYNOMIAL EQUATIONS

In this section, we shall list five general theorems about polynomial equations. You may have discovered some of these results yourself as you were studying this chapter. The first theorem we shall state, known as the *Fundamental Theorem of Algebra*, is a cornerstone of much advanced work in mathematics. Its proof is beyond the scope of this book, but proofs of Theorems 2 through 5 below will be suggested in the exercises.

THEOREM 1. (Fundamental Theorem of Algebra) In the complex number system consisting of all real and imaginary numbers, a quadratic equation always has 2 roots and a cubic equation always has 3 roots (provided a double root is counted as 2 roots, a triple root is counted as 3 roots, and so on). In general, if $P(x)$ is an nth degree polynomial $(n > 0)$ whose coefficients are real or imaginary numbers, then $P(x) = 0$ has exactly n roots.

EXAMPLE 1. **a.** $3x^4 - 11x^3 + 19x^2 + 25x - 36 = 0$ has 4 roots.
b. $2ix^3 + \sqrt{5}x^2 + (3 + 2i)x + 7 = 0$ has 3 roots.

THEOREM 2. If $P(x)$ is a polynomial of odd degree with real coefficients, then $P(x) = 0$ has at least one real root.

EXAMPLE 2. Both of the equations $2x^3 + x^2 - 7 = 0$ and $x^9 - \sqrt{5}x^4 + \sqrt{6} = 0$ have at least one real root.

THEOREM 3. Suppose $P(x)$ is a polynomial with rational coefficients, and a and b are rational numbers such that $\sqrt{b}$ is irrational. If $a + \sqrt{b}$ is a root of $P(x) = 0$, then $a - \sqrt{b}$ is also a root of $P(x) = 0$.

EXAMPLE 3. If $\frac{3}{2} + \sqrt{5}$ is a root of $4x^3 - 16x^2 + x + 11 = 0$, then so is $\frac{3}{2} - \sqrt{5}$.

THEOREM 4. If $P(x)$ is a polynomial with real coefficients, and $a + bi$ is an imaginary root of $P(x) = 0$, then $a - bi$ is also a root.

EXAMPLE 4. If $1 + i\sqrt{2}$ is a root of $x^3 - x^2 + x + 3 = 0$, then so is $1 - i\sqrt{2}$.

THEOREM 5. For the equation $ax^n + bx^{n-1} + cx^{n-2} + \cdots + k = 0$, with $a \neq 0$:

the sum of the roots is $-\dfrac{b}{a}$.

the product of the roots is $\begin{cases} \dfrac{k}{a} & \text{if } n \text{ is even.} \\ -\dfrac{k}{a} & \text{if } n \text{ is odd.} \end{cases}$

EXAMPLE 5. **a.** For the equation $3x^2 + 5x + 1 = 0$, the sum of the roots is $-\dfrac{5}{3}$ and the product of the roots is $\dfrac{1}{3}$.

b. For the equation $2x^3 - 5x^2 - 3x + 9 = 0$, the sum of the roots is $\dfrac{5}{2}$ and the product of the roots is $-\dfrac{9}{2}$.

c. For the equation $x^4 + x^2 + x + 3 = 0$, the sum of the roots is 0 (because there is no x^3 term) and the product of the roots is 3.

Theorem 5 can be applied to the general quadratic equation

$$ax^2 + bx + c = 0$$

if we write it as

$$x^2 - \left(-\frac{b}{a}\right)x + \frac{c}{a} = 0.$$

Because $-\dfrac{b}{a}$ is the sum of the roots and $\dfrac{c}{a}$ is the product of the roots, the equation can be written as

$$x^2 - (\text{sum of roots})x + (\text{product of roots}) = 0.$$

EXAMPLE 6. Find a quadratic equation with roots $2 \pm 3i$.

SOLUTION: Find the sum of the roots: $(2 + 3i) + (2 - 3i) = 4$
Find the product of the roots: $(2 + 3i)(2 - 3i) = 13$
Write the equation: $x^2 - 4x + 13 = 0$

EXAMPLE 7. Find a cubic equation with integral coefficients that has no quadratic term, given that $3 + i\sqrt{2}$ is one of the roots.

SOLUTION: According to Theorem 4, if $3 + i\sqrt{2}$ is a root, then another root is $3 - i\sqrt{2}$. To find the third root, say r, we can make use of the fact that the sum of the roots is 0 since $b = 0$. Thus:

$$r + (3 - i\sqrt{2}) + (3 + i\sqrt{2}) = 0$$
$$r = -6$$

First we write a quadratic equation with roots $3 \pm i\sqrt{2}$:

$$x^2 - (\text{sum})x + \text{product} = 0$$
$$x^2 - 6x + 11 = 0$$

A cubic equation with roots -6 and $3 \pm i\sqrt{2}$ is:

$$(x + 6)(x^2 - 6x + 11) = 0$$
$$x^3 - 25x + 66 = 0$$

Another way to find such an equation is to multiply a group of factors in the form $(x - r)$, choosing one factor for each root:

$$[x - (3 + i\sqrt{2})][x - (3 - i\sqrt{2})][x - (-6)] = 0$$

Polynomials   **85**

The five theorems in this section have been proved in the last 300 years or so, but the study of polynomial equations has a long history, going back to the time of the ancient Greeks. The Greeks were able to solve quadratic equations by a geometric method, but an algebraic solution using a quadratic formula was not discovered until centuries later. The methods of equation solving of the Hindu Brahmagupta (about A.D. 628) or the Persian Omar Khayyám (about A.D. 1100) look complicated to us today because the notation they used was cumbersome. Even as late as the 1500's, mathematicians were using abbreviated Latin words in their equations. For example,

$$4 \text{ Se.} - 5 \text{ Pri.} - 7 \text{ N. aequatur } 0$$

was used instead of

$$4x^2 - 5x - 7 = 0$$

In a work published in 1545, Cardano stated a "cubic formula" and a "quartic formula"; these formulas gave the roots of cubic and quartic equations in terms of the coefficients of the equations and radicals. The discovery of the cubic formula has been attributed to the Italian mathematician Tartaglia, and the discovery of the quartic formula to Cardano's student Ferrari. The next goal of mathematicians was to discover a "quintic formula." In 1824, however, a young Norwegian, Niels Henrik Abel (1802–1829), proved that such a formula does not exist. That is, he proved that it is impossible, in general, to express the roots of a fifth-degree equation in terms of its coefficients and radicals.

The theory of polynomial equations has been used to answer a number of questions in geometry. For example, since the days of ancient Greece, people have tried to find a way to trisect an arbitrary angle with a straightedge and compass. Finally, in the 1800's, the theory of equations was used to prove that no such general method exists. As another example, in 1796 Gauss used the

Niels Henrik Abel

Omar Khayyám

equation $x^n = 1$ (along with a great deal of algebra) to determine which n-sided regular polygons could be constructed with straightedge and compass. In short, the study of polynomial equations has helped link together the once separate subjects of algebra and geometry.

ORAL EXERCISES 2-9

1. A cubic equation with integral coefficients has roots -1 and $\sqrt{3} + 2i$. What is the third root?

2. A quartic equation with integral coefficients has roots $-3 + \sqrt{2}$ and $-2i$. What are the other roots?

3. State the sum and product of the roots of the following equations:
   a. $2x^2 + 5x - 3 = 0$ **b.** $4x^3 - 2x^2 + 5x - 6 = 0$
   c. $x^4 + 2x^3 + 3x^2 = 0$ **d.** $x^5 + 32 = 0$

4. Use graphs to explain why a cubic equation with real coefficients must have at least one real root.

5. Use a graph to explain why a quartic equation with real coefficients does not necessarily have any real roots.

WRITTEN EXERCISES 2-9

Are the statements in Exercises 1–8 true or false? (All equations mentioned have real coefficients.)

A   1. Some cubic equations have no real roots.

2. Every polynomial equation has at least one real root.

3. The roots of a certain quartic equation are $x = \pm\frac{1}{2}$, 0, and $1 + i$.

4. The roots of a certain fifth-degree equation are $x = -3, 4, 1 - \sqrt{2}$, and $\pm i$.

5. The graph of a certain third-degree polynomial function is tangent to the x-axis at $x = -2, 1$, and 6.

6. No polynomial equation can have an odd number of imaginary roots.

7. Suppose that $P(x)$ is a polynomial with rational coefficients, a and b are rational, and $\sqrt{b}$ is irrational. If $a + \sqrt{b}$ is a root of the polynomial equation $P(x) = 0$, then $a - \sqrt{b}$ is also a root.

8. Suppose $P(x)$ is a polynomial with real coefficients. If $a + bi$ is an imaginary root of the polynomial equation $P(x) = 0$, then $P(x) = 0$ must have an even number of roots.

In Exercises 9–12, find the sum and product of the roots of the given equation.

9. $4x^2 - 3x + 6 = 0$ 10. $6x^3 - 9x^2 + x = 0$

11. $3x^3 + 5x^2 - x - 2 = 0$ 12. $x^4 - 4x^2 = 5$

Find a quadratic equation with integral coefficients that has the given roots.

13. $1 \pm i$　　　　**14.** $4 \pm \sqrt{3}$　　　　**15.** $3 \pm \sqrt{2}$　　　　**16.** $\dfrac{1 \pm i\sqrt{2}}{3}$

17. A cubic equation with integral coefficients has no quadratic term. If one root is $2 + i\sqrt{5}$, what are the other roots?

18. A quartic equation with integral coefficients has no cubic term and no constant term. If one root is $3 - i\sqrt{7}$, what are the other roots?

Find a cubic equation with integral coefficients having the given roots.

19. 2 and $4 + i$　　　　　　　　　　**20.** 3 and $7 - i$

21. $\dfrac{4 + i\sqrt{3}}{2}$ and -1　　　　　　　**22.** $i\sqrt{2}$ and 5

23. Find a quartic equation with integral coefficients and roots $5 - i\sqrt{3}$ and i.

24. Find a quartic equation with integral coefficients and roots $1 + i\sqrt{7}$ and $2i$.

25. Find integers c and d such that the equation $x^3 + cx + d = 0$ has one root $1 + \sqrt{3}$.

26. Find integers b and c such that the equation $x^3 + bx^2 + cx + 8 = 0$ has one root $2i$.

B　**27.** Find an integer c such that the equation $4x^3 + cx - 27 = 0$ has a double root.

28. Find an integer d such that the equation $x^3 + 4x^2 - 9x + d = 0$ has two roots that are additive inverses of each other.

29. Suppose $ax^3 + bx^2 + cx + d = 0$ has roots r_1, r_2, and r_3. Then

$$x^3 + \frac{b}{a}x^2 + \frac{c}{a}x + \frac{d}{a} = 0$$

has these same roots.

  a. Explain why $x^3 + \dfrac{b}{a}x^2 + \dfrac{c}{a}x + \dfrac{d}{a} = (x - r_1)(x - r_2)(x - r_3)$.

  b. By carrying out the multiplication of factors on the right side of the equation in part (a), show that:

$$\frac{-b}{a} = r_1 + r_2 + r_3 \quad \text{and} \quad \frac{-d}{a} = r_1 r_2 r_3$$

  c. Conclude from part (b) that the average of the roots of the cubic equation is $\dfrac{-b}{3a}$.

30. Use the technique of Exercise 29 to show that if

$$ax^4 + bx^3 + cx^2 + dx + e = 0$$

has roots r_1, r_2, r_3, and r_4, then:

$$\frac{-b}{a} = r_1 + r_2 + r_3 + r_4 \quad \text{and} \quad \frac{e}{a} = r_1 r_2 r_3 r_4$$

The purpose of Exercises 31–38 is to prove Theorem 4, page 84, for a cubic polynomial. (These exercises should be done sequentially.) Let $x = r + si$ and $z = u + vi$. The conjugates are then $\bar{x} = r - si$ and $\bar{z} = u - vi$.

C   **31.** Prove $\overline{x + z} = \bar{x} + \bar{z}$. (The conjugate of a sum is the sum of the conjugates.)

32. Prove $\overline{xz} = \bar{x} \cdot \bar{z}$. (The conjugate of a product is the product of the conjugates.)

33. Prove $\overline{x^2} = (\bar{x})^2$. (The conjugate of a square is the square of the conjugate.)

34. Prove $\overline{x^3} = (\bar{x})^3$. (The conjugate of a cube is the cube of the conjugate.)

35. a. Prove that the conjugate of a real number a is a.
   b. Prove $\overline{ax^3} = a(\bar{x})^3$.

36. Prove $\overline{bx^2} = b(\bar{x})^2$ and $\overline{cx} = c\bar{x}$.

37. Prove $\overline{ax^3 + bx^2 + cx + d} = a(\bar{x})^3 + b(\bar{x})^2 + c(\bar{x}) + d$.

38. If $P(x) = ax^3 + bx^2 + cx + d$ and $P(x) = 0$, then $P(\bar{x}) = 0$.

Note: The proof just given can be generalized by noting that $\overline{ax^n} = a(\bar{x})^n$. Moreover, a proof of Theorem 3, page 84, is similar to that just given if you interpret $\overline{a + \sqrt{b}}$ to mean $a - \sqrt{b}$, where a and b are rational numbers and $\sqrt{b}$ is an irrational number.

39. If $P(x) = x^3 - x$, the result of Exercise 37 tells you that $\overline{P(1 + i)} = P(1 - i)$. Verify this result by expressing $\overline{P(1 + i)}$ and $P(1 - i)$ in the form $a + bi$.

Chapter summary

1. A *polynomial* in x is an expression of the form
$$a_n x^n + a_{n-1} x^{n-1} + \cdots + a_2 x^2 + a_1 x + a_0,$$
where n is a nonnegative integer. In our study, the coefficients $a_0, a_1, \ldots, a_n$ were real numbers, although in general they may be either real or imaginary numbers. A *zero* of a polynomial is a number that makes the value of the related polynomial function equal zero.

2. If you divide a polynomial $P(x)$ by $x - a$, obtaining a quotient $Q(x)$ and a remainder R, then:
$$P(x) = (x - a)Q(x) + R$$
This equation leads to the Remainder and Factor Theorems:
Remainder Theorem: If $P(x)$ is divided by $x - a$, then the remainder $R = P(a)$.
Factor Theorem: $x - a$ is a factor of $P(x)$ if and only if $P(a) = 0$.

3. Several methods of solving polynomial equations were discussed.
   a. In Section 2-2 we solved polynomial equations by factoring. (See Examples 1, 2, and 3.)

b. In Section 2–7 the *bracket-and-halving* method and the *linear-interpola-tion method* were presented along with a third method used in a computer search.

c. In Section 2–6 polynomial equations were solved by using the *Rational Roots Theorem:* If $\dfrac{p}{q}$ is a root of the polynomial with integral coefficients $ax^n + \cdots + k = 0$, where p and q are nonzero integers with no common factor other than 1, then p must be a factor of k, and q must be a factor of a.

4. *Synthetic division* of polynomials is a compressed form of regular division. (See Example 1 on page 68.)

5. a. The steps in graphing a quadratic function are summarized on page 59.

b. The general shapes of the graphs of cubic and quartic functions are shown in figures (5), (6), (7), and (8) of Section 2–4.

c. If a polynomial $P(x)$ has a squared factor $(x - c)^2$, then the graph of $y = P(x)$ is tangent to the x-axis at $x = c$. (See figures (9)–(11) of Section 2–4.) If it has a cubed factor $(x - c)^3$, the graph of $y = P(x)$ flattens out at $x = c$ and crosses the x-axis at this point. (See figures (12)–(14) of Section 2–4.)

6. If (h, k) is the vertex of the graph of $P(x) = ax^2 + bx + c$, then the maximum or minimum value of $P(x)$ is located at (h, k).

7. To review the general results that hold for all polynomial equations, reread Theorems 1–5 of Section 2–9.

Chapter test

1. Given the polynomial $f(x) = x^3 - 4x$:
    a. Evaluate $f(2i)$.   **b.** Evaluate $f(x - 2)$.   **c.** Find the zeros of $f(x)$.

                2–1

2. Solve the equations for all real and imaginary roots.
    a. $x^4 - 7x^2 - 8 = 0$ **b.** $2x^3 - 7x^2 + 5x = 0$

                2–2

3. Sketch the graph of each parabola. Show the coordinates of the vertex of the parabola and the coordinates of any x-intercepts.
    a. $y = x^2 - 6x + 5$ **b.** $y = 8 - 2(x - 1)^2$

                2–3

4. Sketch the graph of each function. Show the coordinates of all points where the graph intersects or is tangent to the x-axis.
    a. $y = x^3 - 6x^2 + 9x$
    b. $y = 3x^3 - x^4$

                2–4

5. Sketch the graphs of $y = x^3 - 4x^2$ and $y = x - 4$ on the same set of axes. Determine algebraically where the graphs intersect.

6. Find an equation of the cubic graph at the right.

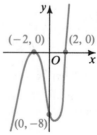

$(-2, 0)$ $(2, 0)$

$(0, -8)$

90   **Chapter 2**

7. Use the Remainder Theorem to find the remainder when $x^{101} - 7x^{94} + 6x^{41} + 10$ is divided by $x + 1$. 2–5

8. Use synthetic division to evaluate $P(\frac{3}{2})$ if
$$P(x) = 4x^4 - 3x^2 - 5x - 2.$$

9. Find all real and imaginary roots of the equation $2x^3 - x^2 + 2x - 1 = 0$. 2–6

10. Show that the equation $x^3 - 2x^2 + 4 = 0$ has no rational roots, but that it does have an irrational root between $x = -2$ and $x = -1$.

11. (Optional) Use linear interpolation to approximate to the nearest tenth the irrational root referred to in Exercise 10. 2–7

12. A rectangular enclosure is to be made using a barn as one side of the rectangle and 80 m of fencing for the other 3 sides. What is the maximum area of the enclosure? 2–8

13. Explain why it is impossible for a cubic equation with real coefficients to have as its roots -1, -3, and $2 - i$. 2–9

14. Find the sum and product of the roots of $2x^3 - 5x^2 - x + 8 = 0$.

CHAPTER THREE

Inequalities

OBJECTIVES

1. To solve and graph linear and polynomial inequalities in one and two variables.
2. To graph the solution set of a system of simultaneous inequalities.
3. (Optional) To solve certain applied problems involving systems of linear inequalities.
4. To use the discriminant to find an equation of a line tangent to a circle or parabola.

Linear and polynomial inequalities

3-1/LINEAR INEQUALITIES IN ONE VARIABLE; ABSOLUTE VALUE

In this chapter we shall discuss linear and quadratic inequalities and their graphs. Our study will fall into the following four categories:

Categories	*Examples*
1. Linear inequalities in one variable:	$2x - 5 > 1$; $|3x - 1| < 4$
2. Polynomial inequalities in one variable:	$2x^2 - x - 1 \geq 0$
3. Linear inequalities in two variables:	$2x + 3y \leq 6$
4. Polynomial inequalities in two variables:	$y > 3x^2 - 4x + 1$

The first of these categories is the subject of this section.

Industrial robots, such as those shown, are used by many industries. See pages 105–108 for an introduction to linear programming, a branch of mathematics that can be used to determine the most efficient use of resources.

The rules for solving linear inequalities are quite like the rules for solving linear equations.

Rule 1. You can add the same number to, or subtract the same number from, both sides of an inequality.

Rule 2. You can multiply or divide both sides of an inequality by the same positive number.

Rule 3. You can multiply or divide both sides of an inequality by the same negative number *if you reverse the inequality sign.*

EXAMPLE 1. Solve **(a)** $3x - 4 \leq 10 + x$; **(b)** $\dfrac{5 - 3x}{4} > 8.$

SOLUTION: **a.**

$$3x - 4 \leq 10 + x$$

Add 4 to both sides. $\qquad 3x \leq 14 + x$

Subtract x from both sides. $\qquad 2x \leq 14$

Divide both sides by 2. $\qquad x \leq 7$

b.

$$\frac{5 - 3x}{4} > 8$$

Multiply both sides by 4. $\qquad 5 - 3x > 32$

Subtract 5 from both sides. $\qquad -3x > 27$

Divide both sides by -3
and reverse inequality sign. $\qquad x < -9$

Linear inequalities in one variable are graphed by shading in certain regions of the number line. If an endpoint of such a region is to be included in the graph, a solid dot (●) is made. Otherwise, an open dot (o) indicates that the endpoint is not to be included in the graph. Some illustrations follow.

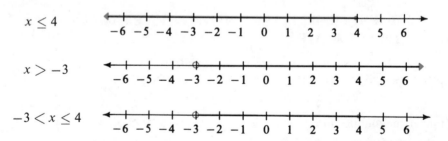

$x \leq 4$

$x > -3$

$-3 < x \leq 4$

The last inequality means $-3 < x$ and $x \leq 4$. The graph of this combined inequality is the intersection of the first two graphs shown above.

Warning. Care must be taken when inequalities are combined. For example, the statement $6 < x < 1$ says that $6 < x$ and $x < 1$. Since no real number x can satisfy both these inequalities, the combined inequality does not make any sense (except, perhaps, as a description of the empty set).

Absolute value

The absolute value of a number x, denoted $|x|$, can be interpreted as its distance from zero on the number line. Thinking in this way, you can solve linear equations and inequalities involving $|x|$.

EXAMPLES INVOLVING $|x|$

	1.	**2.**	**3.**
Sentence:	$|x| = c$	$|x| < c$	$|x| > c$
Meaning:	The distance from x to 0 is c.	The distance from x to 0 is less than c.	The distance from x to 0 is greater than c.
Graph:	(number line $-c$ 0 c)	(number line $-c$ 0 c)	(number line $-c$ 0 c)
Solution:	$x = \pm c$	$-c < x < c$	$x < -c$ or $x > c$

Equations and inequalities involving $|x - k|$, where k is a constant, can be solved by interpreting $|x - k|$ as the distance from x to k on the number line.

EXAMPLES INVOLVING $|x - k|$

	4.	**5.**	**6.**
Sentence:	$|x - 5| = 3$	$|x - 1| < 2$	$|x + 3| > 2$ (Think of $|x - (-3)| > 2$)
Meaning:	The distance from x to 5 is 3.	The distance from x to 1 is less than 2.	The distance from x to -3 is greater than 2.
Graph:	(number line 2 3 4 5 6 7 8)	(number line -1 0 1 2 3)	(number line -5 -4 -3 -2 -1 0)
Solution:	$x = 2$ or $x = 8$	$-1 < x < 3$	$x < -5$ or $x > -1$

Absolute-value sentences can be solved not only by the geometric method shown above but also by the algebraic method illustrated in the next example. This method is based on the three properties listed below. These properties are obtained by replacing x by $ax + b$ in Examples 1, 2, and 3 above.

	(1)	**(2)**	**(3)**
Sentence:	$|ax + b| = c$	$|ax + b| < c$	$|ax + b| > c$
Meaning:	$ax + b = \pm c$	$-c < ax + b < c$	$ax + b < -c$ or $ax + b > c$

EXAMPLE 2. Solve **(a)** $|3x - 9| > 4$; **(b)** $|2x + 5| \leq 7$.

SOLUTION: **(Algebraic Method)**

a. Using property (3), we see that $|3x - 9| > 4$ means
$$3x - 9 < -4 \text{ or } 3x - 9 > 4.$$
Thus: $3x < 5$ or $3x > 13$
$$x < \tfrac{5}{3} \text{ or } x > \tfrac{13}{3}$$

b. Using properties (1) and (2), we see that $|2x + 5| \leq 7$ means
$$-7 \leq 2x + 5 \leq 7.$$
Thus: $-7 \leq 2x + 5$ and $2x + 5 \leq 7$
$$-12 \leq 2x \text{ and } 2x \leq 2$$
$$-6 \leq x \text{ and } x \leq 1$$
That is: $-6 \leq x \leq 1$

SOLUTION: **(Geometric Method)** A geometric solution can also be given to these inequalities. To do so, we must use the fact that
$$|ab| = |a| \cdot |b|.$$
(See Exercise 37 on page 98.) This allows us to conclude, for example, that $|3x - 9| = |3| \cdot |x - 3| = 3|x - 3|$.

a. $|3x - 9| > 4$
$3|x - 3| > 4$
$|x - 3| > \tfrac{4}{3}$

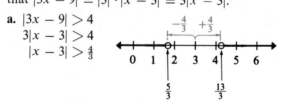

This means the distance from x to 3 is greater than $\tfrac{4}{3}$, which gives the picture above. Thus, $x < \tfrac{5}{3}$ or $x > \tfrac{13}{3}$.

b. $|2x + 5| \leq 7$
$2|x + \tfrac{5}{2}| \leq 7$
$|x - (-\tfrac{5}{2})| \leq \tfrac{7}{2}$

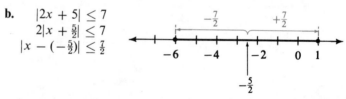

This means the distance from x to $-\tfrac{5}{2}$ is less than or equal to $\tfrac{7}{2}$, as illustrated above. Thus, $-6 \leq x \leq 1$.

ORAL EXERCISES 3-1

Solve for x.

1. $3x > -12$ **2.** $-3x > -12$ **3.** $1 - 2x < 11$

4. $|x| < 2$ **5.** $|x - 5| < 2$ **6.** $|x + 3| > 1$

In Exercises 7-12, match the inequalities with graphs a-f.

7. $1 < x \le 4$ **8.** $4 \le x$ **9.** $x < 1$

10. $x < 1$ or $x \ge 4$ **11.** $6 \ge x \ge 4$ **12.** $6 \le x \le 4$

a.
$$-4 \quad -2 \quad 0 \quad 2 \quad 4 \quad 6$$

b.
$$-4 \quad -2 \quad 0 \quad 2 \quad 4 \quad 6$$

c.
$$-4 \quad -2 \quad 0 \quad 2 \quad 4 \quad 6$$

d.
$$-4 \quad -2 \quad 0 \quad 2 \quad 4 \quad 6$$

e.
$$-4 \quad -2 \quad 0$$

f.
$$-4 \quad -2 \quad 0 \quad 2 \quad 4 \quad 6$$

WRITTEN EXERCISES 3-1

In Exercises 1-28, solve and graph the given equation or inequality. Give the solution as an interval. If there is no solution, so state.

A   **1.** $7x - 12 < 9$ **2.** $8x + 6 > 30$

3. $\dfrac{15 - 6x}{3} > 5$ **4.** $\dfrac{8 - 11x}{4} \le 13$

5. $\dfrac{1}{4}(x - 1) \le \dfrac{x + 4}{6}$ **6.** $\dfrac{2 - x}{3} < \dfrac{3 - 2x}{5}$

7. $2[5x - 3(x + 1)] > 3x - [(2x - 6) - x]$

8. $6 - [4(x - 3) - (3 - x)] \ge 3[2 - (x - 6)]$

9. $\dfrac{x + 2}{4} - \dfrac{2 - x}{3} + \dfrac{4x - 5}{6} < 4$ **10.** $\dfrac{4}{3}\left(x - \dfrac{1}{2}\right) + \dfrac{1}{2}x \ge \dfrac{2}{3}\left(2x - \dfrac{5}{2}\right)$

11. $|x| < 3$ **12.** $|x| \ge 5$

13. $|x - 4| < 3$ **14.** $|x - 7| > 3$

15. $|x + 7| \ge 3$ **16.** $|x + 2| < -1$

17. $|x - 8| = 4$ **18.** $|x + 3| = 7$

19. $|2x - 4| \le 5$ **20.** $|3x - 9| \ge 9$

21. $|4x + 8| \le 9$ **22.** $|6 - 3x| < 12$

23. $2 \le |x| \le 4$ **24.** $1 \le |x| < 5$

B   **25.** $1 < |x - 4| < 3$ **26.** $2 < |x - 6| < 5$

(*Hint for Ex. 25:* The distance from x to 4 is greater than 1 *and* less than 3.)

27. $0 < |x - 7| < 2$ **28.** $0 < |x + 3| < 1$

Solve for x. (Equations 29 and 30 have a quadratic form. See page 52.)

29. $|x|^2 - 6|x| + 8 = 0$ **30.** $|x|^2 - |x| - 2 = 0$

31. $|x|^3 - 2|x|^2 - 3|x| = 0$ **32.** $|x|^3 + 3|x|^2 + 4|x| = 0$

33. $\dfrac{1}{x} < 2$ (*Hint:* Consider the cases $x > 0$ and $x < 0$ separately.)

Inequalities   **97**

34. $\dfrac{1}{x-1} > 4$ (*Hint:* Consider the cases $x - 1 > 0$ and $x - 1 < 0$ separately.)

C   **35.** $|x| + |x - 2| = 2$ **36.** $|x| + |x - 2| > 5$

In Exercises 37–40, use the following definition of $|x|$: $|x| = \begin{cases} x \text{ if } x \geq 0. \\ -x \text{ if } x < 0. \end{cases}$

37. Show that $|ab| = |a| \cdot |b|$ when (1) $a > 0$ and $b > 0$; (2) $a < 0$ and $b > 0$; (3) $a > 0$ and $b < 0$; and (4) $a < 0$ and $b < 0$.

38. Show that $\left| \dfrac{a}{b} \right| = \dfrac{|a|}{|b|}$ either by considering the four cases of Exercise 37 or by using the result of Exercise 37.

39. a. Give three examples illustrating the triangle inequality:

$$|a + b| \leq |a| + |b|$$

   b. In Chapter 12 we shall define the absolute value of a complex number as follows: $|x + yi| = \sqrt{x^2 + y^2}$. Decide whether or not the triangle inequality holds if a and b are complex numbers.

40. Use the triangle inequality stated in Exercise 39(a) to prove:
   a. $|a - b| \leq |a| + |b|$ **b.** $|a| - |b| \leq |a - b|$

3-2 / POLYNOMIAL INEQUALITIES

If $P(x)$ is a polynomial, then $P(x) < 0$ and $P(x) > 0$ are called **polynomial inequalities.** A sketch of the graph of $y = P(x)$ is very useful in solving the related polynomial inequalities.

EXAMPLE 1.   Solve $x^3 - 2x^2 - 3x > 0$.

SOLUTION: 1. Find the zeros of $y = x^3 - 2x^2 - 3x$.

$$0 = x^3 - 2x^2 - 3x$$
$$0 = x(x + 1)(x - 3)$$

The zeros are $x = -1$, $x = 0$, and $x = 3$.

2. Plot the zeros and sketch in the rest of the graph.

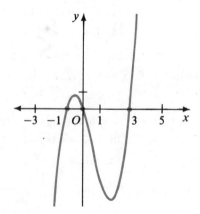

3. The sketch shows which points of the graph have positive y-coordinates. The corresponding x-coordinates make up the solution to the inequality. We show the signs of the y-coordinates on a **sign graph.**

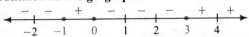

4. The sign graph shows that the y-coordinate is positive when x is between -1 and 0 or when x is greater than 3. Therefore, the solution is $-1 < x < 0$ or $x > 3$.

Although the polynomial in Example 1 changes sign at each of its zeros, this does not always happen. A polynomial $P(x)$ will not change sign at a zero c if c corresponds to the squared factor $(x - c)^2$.

EXAMPLE 2. Solve $(x - 4)^2(x^2 - 1) > 0$.

SOLUTION: 1. The zeros of $y = P(x) = (x - 4)^2(x^2 - 1)$ are $x = 4$, $x = -1$, and $x = 1$. Note the sign of $P(x)$ for some convenient value of x. For example, note that

$$P(0) = -16 < 0.$$

2. Determine the signs of $P(x)$ in the other intervals by marking a sign change at each zero except when the zero corresponds to a squared factor. We complete the sign graph by noting sign changes at -1 and 1 but none at 4 since it corresponds to a squared factor.

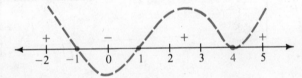

The solution is $x < -1$ or $1 < x < 4$ or $x > 4$.

The technique used in Example 2 can be used to solve inequalities involving functions of the form $\dfrac{P(x)}{Q(x)}$ where $P(x)$ and $Q(x)$ are polynomials or products of polynomials.

EXAMPLE 3. Solve $f(x) = \dfrac{(x + 2)(x - 5)^2}{x - 4} \le 0$.

SOLUTION: 1. Plot the zeros of all linear factors occurring in a numerator or a denominator. You can mark the zeros of the denominator to remind yourself to omit them from the solution. In this case there are zeros at $x = -2$, 4, and 5.

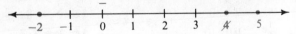

(*Solution continued on page 100*)

As you plot the zeros note the sign of $f(x)$ for some convenient value of x. In this case we note that $f(0) = -\dfrac{25}{2}$, a negative number.

2. Determine the signs of $f(x)$ in the other intervals by marking a sign change at each zero except when the zero corresponds to a squared factor. We complete the sign graph by noting sign changes at -2 and 4 but none at 5 since it corresponds to a squared factor.

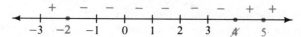

3. The solution is $-2 \le x < 4$ or $x = 5$. Notice that we omitted 4 from the solution since it is a zero of the denominator. We include 5, since $f(5) = 0$.

ORAL EXERCISES 3-2

In Exercises 1 and 2, use the given graph to solve the inequality for x.

1. $x^3 - 4x^2 - 4x + 16 > 0$ **2.** $x^3 + 5x^2 + 3x - 9 \le 0$

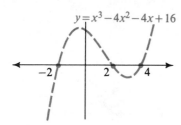

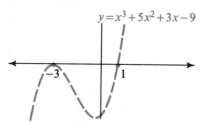

Solve for x.

3. $(x - 1)(x - 2)(x - 3)^2(x - 4) > 0$ **4.** $\dfrac{(x - 2)(x - 7)}{x - 4} < 0$

5. $(x^2 + 1)(x - 5) \le 0$ **6.** $|x^2 - 2x| < 0$

WRITTEN EXERCISES 3-2

Solve each inequality.

A   **1.** $(x - 3)(x + 4) > 0$ **2.** $(x + 7)(x + 9) > 0$

     3. $(x - 1)(x - 2)(x - 4) > 0$ **4.** $(1 - x)(x - 3)(x - 5) > 0$

     5. $(x - 4)(x - 2)^2(x - 3)^2 < 0$ **6.** $(2x - 5)^2(x + 3)(x + 2) < 0$

     7. $x^2 - 2x - 15 < 0$ **8.** $x^2 + 3x - 18 > 0$

     9. $2x^2 - x - 3 \ge 0$ **10.** $1 - 2x - 3x^2 < 0$

   11. $2x^2 + 5x - 7 \le 0$ **12.** $x^2 - 8x + 16 \le 0$

   13. $x^4 - 3x^2 - 10 > 0$ **14.** $3x^3 + 7x^2 - 6x \le 0$

15. $a^3 + 2a^2 - 4a - 8 > 0$

16. $b^4 - 16 < 0$

17. $n^3 - 7n + 6 < 0$

18. $2y^3 + 3y^2 - 1 \geq 0$

19. $2x^3 + x^2 - 5x < -2$

20. $r^3 - 9r > 8r^2$

21. $y^4 + y^3 < 4(y + 4)$

22. $x^4 + 6 < 5x^2$

23. $4x^4 - 4x^3 - 3x^2 + 4x - 1 > 0$

24. $2y^3 + y^2 - 12y + 9 < 0$

B

25. $\dfrac{(x - 3)(x - 4)}{(x - 5)(x - 6)^2} < 0$

26. $\dfrac{(x + 1)(x - 3)^2}{(x - 5)^2} > 0$

27. $\dfrac{(2x - 5)^3}{x^2 - 3x - 28} \geq 0$

28. $\dfrac{(3n - 12)^2}{3n^2 - 12} \leq 0$

29. $\dfrac{2x^2 + 7x + 8}{x^2 + 1} > 0$

30. $\dfrac{n^2 + 4n + 4}{n^2 + 4n} > 0$

31. The inequality $x^2 - 7x + 10 > 0$ is solved below by a purely algebraic method. Give the justification for each step.

   1. If $x^2 - 7x + 10 > 0$, then $(x - 2)(x - 5) > 0$.

   2. Therefore:

      $[x - 2 > 0$ and $x - 5 > 0]$ or $[x - 2 < 0$ and $x - 5 < 0]$

   3. $[x > 2$ and $x > 5]$ or $[x < 2$ and $x < 5]$

   4. $x > 5$ or $x < 2$

32. Follow the instructions of Exercise 31 for the inequality

$$12x^2 + 7x - 10 \leq 0.$$

   1. If $12x^2 + 7x - 10 \leq 0$, then $(3x - 2)(4x + 5) \leq 0$.

   2. Therefore:

      $[3x - 2 \geq 0$ and $4x + 5 \leq 0]$ or $[3x - 2 \leq 0$ and $4x + 5 \geq 0]$

   3. $[x \geq \frac{2}{3}$ and $x \leq -\frac{5}{4}]$ or $[x \leq \frac{2}{3}$ and $x \geq -\frac{5}{4}]$

   4.   [No solution] or $[-\frac{5}{4} \leq x \leq \frac{2}{3}]$

   5. Thus, $-\frac{5}{4} \leq x \leq \frac{2}{3}$.

In Exercises 33 and 34, solve the inequalities using the methods of Exercises 31 and 32.

33. $6x^2 + 7x - 20 > 0$

34. $4x^2 - 20x + 21 \leq 0$

Solve for x.

C

35. $\dfrac{|2x - 5|}{x^2 + 4} < 0$

36. $\dfrac{6x^2 + 13x - 5}{|5x^3 - 10|} < 0$

COMPUTER EXERCISE

Write a computer program that prints the solution of $ax^2 + bx + c > 0$ when you input values of a, b, and c.

3-3/INEQUALITIES IN TWO VARIABLES

Like solutions to equations in two variables, solutions to inequalities in two variables are ordered pairs of numbers. Thus, graphs of inequalities in two variables consist of points in the plane.

The graphs of linear and quadratic inequalities in two variables are closely related to the graphs of lines and parabolas. If a line has equation $y = mx + k$, any point (x, y) above the line satisfies the inequality $y > mx + k$, and any point below the line satisfies the inequality $y < mx + k$. A similar statement can be made for points above and below a parabola, as illustrated below.

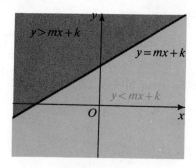
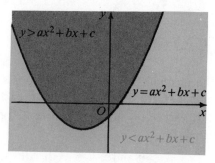

The method just given and another method of sketching graphs of inequalities are shown in the following example.

EXAMPLE 1. Sketch the graph of $y > x^2 - 2x - 8$.

SOLUTION: *Method* 1. Follow the approach suggested earlier. If a parabola has equation $y = x^2 - 2x - 8$, then any point (x, y) above the parabola satisfies the inequality. We show the parabola as a dashed curve since points on the parabola are not included in the solution.

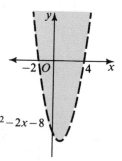

Method 2. Draw the graph of $y = x^2 - 2x - 8$, which separates the plane into two regions. Choose a point in one region and check to see whether or not it satisfies the inequality. For example, the point $(0, 0)$ satisfies the inequality. Thus the region containing $(0, 0)$ is shaded as the graph of the inequality.

EXAMPLE 2. Sketch the graph of
$$(x - 2)^2 + (y - 3)^2 \leq 10.$$

SOLUTION: The graph of the equation $(x - 2)^2 + (y - 3)^2 = 10$ is the circle with center $(2, 3)$ and radius $\sqrt{10}$. Since the point $(2, 3)$ obviously satisfies the inequality, we include $(2, 3)$ and all other points inside the circle in the shaded region.

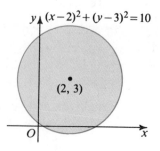

Another way to view this example is to rewrite the inequality as
$$\sqrt{(x-2)^2 + (y-3)^2} \le \sqrt{10},$$
which says that the distance between (x, y) and $(2, 3)$ is less than or equal to $\sqrt{10}$.

If you want to graph the solution of a system of simultaneous inequalities, you just take the intersection of the graphs of the individual inequalities. An example will illustrate this.

EXAMPLE 3.   Graph the solution set of the simultaneous inequalities $y \le 4 - x^2$ and $y > x + 2$.

SOLUTION: 1. The graph of $y \le 4 - x^2$ consists of points on or below the parabola $y = 4 - x^2$. This region is shaded in color.

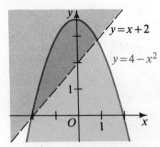

2. The graph of $y > x + 2$, shaded in gray, consists of points above the line $y = x + 2$. Because the points of the line are not included in the graph, we use a dashed line.

3. The intersection of the gray and colored regions is the graph of the solution set of the simultaneous inequalities.

If an inequality contains just one variable, you must be aware of whether you are working just on the number line or whether you are working with the whole coordinate plane. The two possibilities for the inequality $x \ge 3$ are shown below.

Problem.   Find all x such that $x \ge 3$.

Solution:

$$\xleftarrow{\quad\quad} \underset{-1\ \ 0\ \ 1\ \ 2\ \ 3\ \ 4\ \ 5}{+\ +\ +\ +\ +\ +\ +} \xrightarrow{\quad\quad}$$

Problem.   Find all (x, y) such that $x \ge 3$.

Solution:

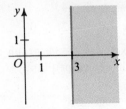

ORAL EXERCISES 3-3

In each of the following exercises, the coordinates of a point and an equation are given. Tell whether the point is on the graph of the equation, above it, or below it.

1. $(3, 4)$; $y = x$

2. $(3, 6)$; $y = 2x + 1$

3. $(1, -1)$; $y = 3x - 4$

4. $(0, 9)$; $y = x^2 - 8x + 9$

5. $(-1, 12)$; $y = 6 - 5x - x^2$

6. $(2, 10)$; $y = x^3 + x^2 + x + 1$

Inequalities   **103**

7. a. If a point (x, y) is in the first quadrant, what inequalities must be satisfied?

b. What if (x, y) is in the second quadrant? third quadrant? fourth quadrant?

Give a set of inequalities which define the shaded regions shown below.

8.
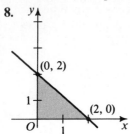

9.
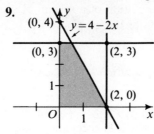

10.
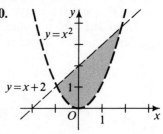

WRITTEN EXERCISES 3-3

In Exercises 1–20, graph the set of all (x, y) satisfying the given inequality.

A
1. $y \geq x$
2. $y \leq 2x - 1$
3. $3x + 4y < 12$
4. $4 - 2x < 3y$
5. $y \leq x^2$
6. $y \geq x^2 + 4x + 8$
7. $y < 2x^2 - 4x + 1$
8. $y > 3x^2 + 6x - 2$
9. $0 \leq x \leq 2$
10. $-1 \leq y \leq 3$
11. $|y| > 1$
12. $|x| \leq 2$
13. $|x - 3| < 2$
14. $2 < |x + 4| < 3$
15. $y > x^3 - 9x$
16. $y < x^4 - 5x^2 + 4$
17. $x^2 + y^2 \leq 4$
18. $x^2 + y^2 \geq 9$
19. $x^2 + y^2 - 2y \geq 0$
20. $x^2 + y^2 - 4x + 4y + 8 \leq 0$

In Exercises 21–34, graph all (x, y) that are solutions of the given system of simultaneous inequalities.

21. $x \geq 0$
$x + 2y \leq 4$

22. $y \leq 0$
$2x + y \leq 4$

23. $x < 0$
$3x - 2y \leq -6$

24. $y < 0$
$x - y > -1$

25. $y \geq x^2 - 2$
$y < x$

26. $y \leq 6 - x^2$
$2x - y \leq -3$

B
27. $0 \leq x \leq 3$
$0 \leq y \leq 2$
$y \leq 2 - x$

28. $-1 \leq x \leq 4$
$-2 \leq y \leq 5$
$5y \geq x + 11$

29. $|x| < 3$
$|y| < 1$

30. $|x| \geq 2$
$|y| \leq 4$

31. $x^2 + y^2 \le 36$
$|x| \le 4$

32. $x^2 + y^2 \le 12$
$|y| \ge 1$

33. $1 \le |x - 4| \le 3$
$1 \le |y - 4| \le 3$

34. $1 \le |x + 3| \le 3$
$1 \le |y + 4| \le 4$

35. On one set of axes, graph each of the following inequalities. Shade the region consisting of all points (x, y) that satisfy all the inequalities. Label each corner point of this region with its coordinates.

(1) $y \ge x$ (2) $x \ge 3$ (3) $y \le 8$ (4) $x + 2y \ge 16$

36. Repeat Exercise 35 for the following inequalities:

(1) $x \ge 0$ (2) $y \ge 3$ (3) $3 + y \le 12$ (4) $3y - x \le 24$

Sketch the graph of each equation or inequality.

C   **37.** $|x| + |y| = 2$ **38.** $|2x - 3y| < 6$

Applications of inequalities

3-4/LINEAR PROGRAMMING PROBLEMS (OPTIONAL)

A system of linear inequalities can sometimes be used to describe a group of alternatives from which a business must choose. The branch of mathematics called *linear programming* can be used to determine which of these alternatives is the most profitable or least expensive. Before we illustrate how this is done, we present a theorem we shall need. (The proof of this theorem is beyond the scope of this book.)

THEOREM   If a linear expression in two variables $P(x, y) = ax + by$ is to be evaluated for all points of a convex polygonal region, then the maximum value of P, if there is one, will occur at a vertex of the region. Also, the minimum value of P, if there is one, will occur at a vertex.

(*Note:* A polygonal region is *convex* if every line segment joining two points of the region is contained in the region. In general, a polygonal region is convex if a rubber band will fit snugly around it. Regions (a) and (b) below are convex, but (c) and (d) are not.

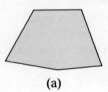   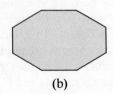

(a)   (b)   (c)   (d)

Illustration of Theorem: Suppose $P = 3x + 2y$ is to be evaluated at the points of the convex polygonal region with vertices $(0, 0)$, $(3, 6)$, $(6, 3)$, and $(6, 0)$, as shown at the right. As the diagram illustrates, the value of $P(x, y) = 3x + 2y$ is constant along any line with slope $-\frac{3}{2}$. Also, the values of $3x + 2y$ are greater as we move upward to the right. We know that the maximum value of $3x + 2y$ for points of the shaded region must occur at a vertex. By checking values we find that the maximum occurs at the vertex $(6, 3)$. Similarly, we find that the minimum value of $3x + 2y$ for points of the shaded region occurs at the vertex $(0, 0)$.

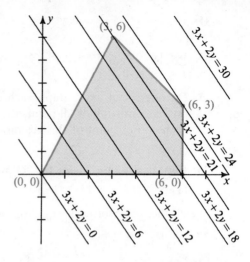

EXAMPLE 1. *Maximizing a Profit.* A small manufacturing company makes a $60 profit on each console TV it produces and a $40 profit on each portable TV. The table below shows that the production of a console TV requires 1 hour of Machine A, 1 hour of Machine B, and 4 hours of Machine C. The production of a portable TV requires 2 hours of A, 1 hour of B, and 1 hour of C. In a given day, Machines A, B, and C can work a maximum of 16, 9, and 24 hours, respectively. How many console TV's and how many portable TV's should be produced per day to maximize the profit?

	Console	Portable	Total hours available per day
Machine A	1 hour	2 hours	16
Machine B	1 hour	1 hour	9
Machine C	4 hours	1 hour	24
Profit per TV	$60	$40	

SOLUTION: Let $x =$ number of console TV's.
Let $y =$ number of portable TV's.

Step 1. Then we have the following inequalities:

1. $x \geq 0$
2. $y \geq 0$
3. $x + 2y \leq 16$ One hour of Machine A is needed for each console TV and 2 hours are needed for each portable. Thus, for x console TV's and y portable TV's, $x + 2y$ hours of Machine A are required. Since the machine is only available 16 hours a day, $x + 2y \leq 16$.
4. $x + y \leq 9$ Inequalities (4) and (5) are similar to (3) and express
5. $4x + y \leq 24$ the hour limitations for Machines B and C.

Step 2. The graphs below show the step-by-step procedure used to determine the region satisfying all five inequalities.

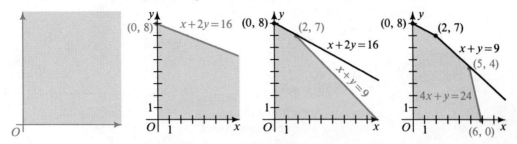

Step 3. The profit for x console TV's and y portable TV's is $P = 60x + 40y$ dollars. We want to know the maximum value of this polynomial for the points of the polygonal region shown at the far right above. According to the theorem, this maximum must occur at a vertex. So we try each vertex:

Vertex	Value of $P(x, y) = 60x + 40y$
$(0, 0)$	$P(0, 0) = 60 \cdot 0 + 40 \cdot 0 = \$0$
$(0, 8)$	$P(0, 8) = 60 \cdot 0 + 40 \cdot 8 = \$320$
$(2, 7)$	$P(2, 7) = 60 \cdot 2 + 40 \cdot 7 = \$400$
$(5, 4)$	$P(5, 4) = 60 \cdot 5 + 40 \cdot 4 = \$460$ ← Maximum profit
$(6, 0)$	$P(6, 0) = 60 \cdot 6 + 40 \cdot 0 = \$360$

The chart shows that the maximum profit ($460) occurs when 5 console and 4 portable TV's are produced per day.

EXAMPLE 2. *Minimizing a Cost.* Every day a certain person needs a dietary supplement of 4 milligrams of vitamin A, 11 milligrams of vitamin B, and 100 milligrams of vitamin C. Either of two brands of vitamin pills can be used: Brand X at 6¢ a pill or Brand Y at 8¢ a pill. The chart below shows that a Brand X pill supplies 2 milligrams of vitamin A, 3 milligrams of vitamin B, and 25 milligrams of vitamin C. Likewise, a Brand Y pill supplies 1, 4, and 50 milligrams of vitamins A, B, and C. How many pills of each brand should be taken each day in order to satisfy the minimum daily need most economically?

	Brand X	Brand Y	Minimum daily need
Vitamin A	2 mg	1 mg	4 mg
Vitamin B	3 mg	4 mg	11 mg
Vitamin C	25 mg	50 mg	100 mg
Cost per pill	6¢	8¢	

SOLUTION: Let x = number of Brand X pills.
Let y = number of Brand Y pills.

(Solution continued on page 108)

Step 1. You should be able to explain each of the following inequalities.

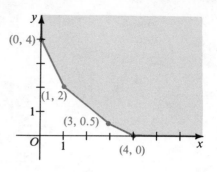

1. $x \geq 0$
2. $y \geq 0$
3. $2x + y \geq 4$
4. $3x + 4y \geq 11$
5. $x + 2y \geq 4$

Step 2. We determine the region satisfying all five inequalities.

Step 3. We evaluate the cost $C = 6x + 8y$ cents at each vertex.

At (x, y)	$C = 6x + 8y$
At $(0, 4)$	32¢
At $(1, 2)$	22¢ ⎱ Minimum cost
At $(3, 0.5)$	22¢ ⎰
At $(4, 0)$	24¢

The minimum cost occurs for both $(1, 2)$ and $(3, 0.5)$. It is, however, inconvenient to take half a pill. Therefore, 1 Brand X pill and 2 Brand Y pills is the best choice.

WRITTEN EXERCISES 3-4

A 1. **a.** Refer to Example 1 on page 106 and suppose each console TV manufactured produces an $80 profit and each portable TV manufactured produces a $55 profit. How many of each kind of TV should be manufactured to maximize the profit?
   b. If the profit on a console TV is $90 but is only $20 on a portable TV, show that the maximum profit is achieved by producing only console TV's.

2. **a.** Refer to Example 2 on page 107 and suppose each Brand X pill costs 3¢ and each Brand Y pill costs 9¢. How many of each type of pill should be taken if the cost is to be kept to the minimum?
   b. If each Brand X pill costs 10¢ and each Brand Y pill costs 4¢, show that the cost is minimized by taking only Brand Y pills.

3. Suppose the data of Example 1 on page 106 are changed as shown in the table below. How many of each kind of TV should be produced to maximize the profit?

	Console	Portable	Total hours available per day
Machine A	1	3	18
Machine B	1	1	8
Machine C	3	1	18
Profit per TV	$70	$40	

4. Suppose the data of Example 2 on page 107 are changed as shown in the table below. How many of each type of pill should be taken daily if the cost is to be kept to the minimum?

	Brand X	Brand Y	Minimum daily need
Vitamin A	4 mg	2 mg	10 mg
Vitamin B	6 mg	6 mg	24 mg
Vitamin C	25 mg	50 mg	125 mg
Cost per pill	12¢	15¢	

B 5. A grocer is preparing an order of dog food for a two-month period. The grocer normally stocks two brands: Pedigree Pellets and Mongrel Mash. These come in bags of the same size, and the grocer has display and storage space for a total of 800 bags. Because of the kinds of dogs that live near the store, the grocer expects to sell at least twice as much mash as pellets. The grocer's buying and selling prices are shown in the table below.

	(x)	(y)
	Pedigree Pellets	Mongrel Mash
Buy	$4.50	$1.50
Sell	7.00	3.00

If the grocer does not wish to order more than $1500 worth of dog food for the two-month period, how much of each kind should be ordered?
a. Let x = number of bags of Pedigree Pellets to be ordered.
   Let y = number of bags of Mongrel Mash to be ordered.
   Then explain why each of the following inequalities holds:

(1) $x \geq 0$ (4) $y \geq 2x$
(2) $y \geq 0$ (5) $4.50x + 1.50y \leq 1500$
(3) $x + y \leq 800$

b. What values of x and y satisfy the inequalities above and also yield the greatest profit if all the dog food is sold?

6. A sporting goods company manufactures two types of skis: a racing model and a free-style model. Each pair of racing skis requires 3 hours of labor, and the company can produce at most 20 pairs of racing skis per day. Each pair of free-style skis requires 2 hours of labor, and the company can produce at most 30 pairs of free-style skis per day. The maximum number of hours of labor available for ski production is 96. If the profit on each pair of racing skis is $30 and the profit on each pair of free-style skis is $40, find the number of each that should be manufactured in order to maximize profits.

C 7. Suppose in Exercise 6 that the profit on each pair of free-style skis is $20. Show that there are two different ways to maximize profits.

8. A smelting company receives a monthly order for at least 40 tons of iron, 60 tons of copper, and 40 tons of lead. It can fill this order by smelting either of two alloys, X and Y. Each railroad carload of alloy X will produce after smelting 1 ton of iron, 3 tons of copper, and 4 tons of lead. Each carload of alloy Y will produce after smelting 2 tons of iron, 2 tons of copper, and 1 ton of lead. If it costs $350 to smelt a carload of alloy X and $200 to smelt a carload of alloy Y, how many carloads of each alloy should be used to fill the order at the minimum cost?

3-5/USING THE DISCRIMINANT

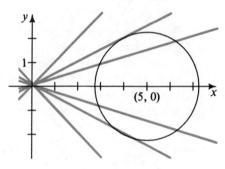

The diagram shows a circle with center $(5, 0)$ and radius $\sqrt{5}$, and also several lines through the origin. Some of these lines intersect the circle in two points, some in one point, and some in no points. To classify such lines by number of points of intersection, the discriminant (page 26) can be used. The following four steps show how.

1. The equation $y = mx$ represents a nonvertical line through the origin. Our goal is to determine which values of m give equations of lines that intersect the circle.

2. We start as if we were solving the equations of the line and circle simultaneously under the assumption that m is given.

$$\text{Line: } y = mx$$
$$\text{Circle: } (x - 5)^2 + y^2 = 5$$
$$\text{Substituting: } (x - 5)^2 + (mx)^2 = 5$$
$$x^2 - 10x + 25 + m^2x^2 = 5$$
$$(1 + m^2)x^2 - 10x + 20 = 0 \quad (1)$$

3. Instead of solving equation (1) we determine how many solutions it has. Equation (1) is a quadratic equation whose basic form is $ax^2 + bx + c = 0$. In this case, $a = 1 + m^2$, $b = -10$, and $c = 20$. The discriminant of this quadratic equation is:

$$D = b^2 - 4ac = (-10)^2 - 4(1 + m^2)(20) = 100 - 80 - 80m^2$$
$$= 20 - 80m^2$$

4. We note that D has zeros at $m = -\frac{1}{2}$ and $m = \frac{1}{2}$. A sign graph for D is shown below. From it we draw three conclusions.

a. If $-\frac{1}{2} < m < \frac{1}{2}$, $D > 0$. In this case, equation (1) has two distinct roots, indicating that the line intersects the circle in two points. Therefore, if $-\frac{1}{2} < m < \frac{1}{2}$, the line $y = mx$ intersects the circle in two points.

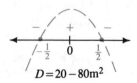

b. If $m = \pm\frac{1}{2}$, $D = 0$. In this case, equation (1) has a double root, indicating that the line is tangent to the circle. Therefore, if $m = -\frac{1}{2}$ or $m = \frac{1}{2}$, the line $y = mx$ is tangent to the circle.

c. If $m < -\frac{1}{2}$ or $m > \frac{1}{2}$, $D < 0$. In this case, equation (1) has no real roots, indicating that the line does not intersect the circle. Therefore, if $m < -\frac{1}{2}$ or $m > \frac{1}{2}$, the line $y = mx$ does not intersect the circle.

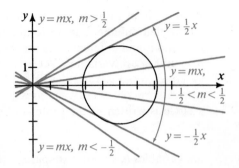

EXAMPLE 1. The diagram shows several lines with slope 3. One of these lines is tangent to the parabola

$$y = x^2.$$

Find an equation of this tangent line.

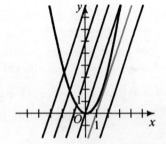

SOLUTION:
1. The equation $y = 3x + k$ represents a line with slope 3.
2. We start as if we were solving the equations of the line and the parabola simultaneously under the assumption that k is given.

$$\text{Line:} \quad y = 3x + k$$
$$\text{Parabola:} \quad y = x^2$$
$$\text{Substituting:} \quad x^2 = 3x + k$$
$$x^2 - 3x - k = 0 \quad (2)$$

3. To determine the number of roots equation (2) has, we find the discriminant.

$$D = b^2 - 4ac = (-3)^2 - 4(1)(-k) = 9 + 4k$$

4. If $D = 0$, equation (2) has a double root, indicating that the line intersects the parabola in a single point:

$$9 + 4k = 0$$
$$k = -\frac{9}{4}$$

Hence, the tangent line has the equation $y = 3x - \frac{9}{4}$.

Parameters

We often refer to the general form of an equation. For example,

$$ax^2 + bx + c = 0$$

is the general form of a quadratic equation, and

$$y = mx$$

is the general form of an equation of a nonvertical line through the origin. Notice that different quadratic equations are determined by different values of a, b, and c, and different lines are determined by different values of m. Thus, a, b, c, and m are variables. Yet as variables they play a different role than the variables x and y. Variables like a, b, c, and m, which are used to describe the general form of a set of equations, are called **parameters.** When the equation of a circle is written in the center-radius form,

$$(x - h)^2 + (y - k)^2 = r^2,$$

the variables h, k, and r are parameters. (The word *parameter* is also used in a slightly different sense which we shall consider in a later chapter.)

ORAL EXERCISES 3-5

If the given equation is written as an equivalent equation of the form $ax^2 + bx + c = 0$, tell the values of a, b, and c.

EXAMPLE 2. $kx^2 - x^2 + 2kx + 10 = k$

SOLUTION: We rewrite the equation as $(k - 1)x^2 + (2k)x + (10 - k) = 0$. Thus, $a = k - 1$, $b = 2k$, and $c = 10 - k$.

1. $kx^2 + x^2 + 3kx + k - 2 = 0$ **2.** $4x^2 + 3x + kx = 5$

3. $p^2x^2 + px - 2 = 2px$ **4.** $4y^2 + y + 1 = ry^2 + sy + t$

5. When the equation of a parabola is written in the form $y = (x - h)^2 + k$, which letters are used as parameters?

WRITTEN EXERCISES 3-5

In Exercises 1–4, find the values of k for which the given equation has a double root.

A   **1.** $2x^2 = 4x + 3 - k$

      2. $kx^2 - 4x + k = 0$

      3. $x^2 + (k - 1)x + k = 1$

      4. $kx^2 + x - 4kx - 2 = 0$

5. Find equations of the two lines through the origin that are tangent to the circle
$$(x - 12)^2 + y^2 = 36.$$

6. Find equations of the two lines through the origin that are tangent to the circle
$$x^2 + (y - 8)^2 = 16.$$

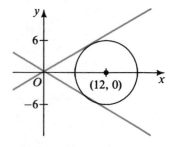

7. Find the values of k for which the line $y = 3x + k$ intersects the circle $x^2 + y^2 = 10$. Illustrate your solution with a sketch.

8. Find the values of k for which the line $x + y = k$ intersects the circle $x^2 + y^2 = 2$. Illustrate your solution with a sketch.

9. **a.** Sketch the circle $x^2 + y^2 - 8y + 8 = 0$.
   b. Determine which lines through the origin do *not* intersect the circle.

10. **a.** Sketch the parabola $y = x^2 + 9$.
    b. Determine which lines through the origin do *not* intersect the parabola.

11. **a.** Using k as a parameter, write an equation of a line with slope 4.
    b. Find the value of k for which the line will be tangent to the parabola $y = 6x - x^2$.
    c. Illustrate your solution with a sketch.

12. Find an equation of a line with slope 3 that is tangent to the parabola $y = x^2 + 5x$. Illustrate your solution with a sketch.

13. Find the values of r for which the circle $x^2 + y^2 = r^2$ intersects the line $2x + y = 5$.

B 14. A circle with center $(0, 5)$ and radius r intersects the line $3x - 4y = 5$. What are the possible values of r?

15. Show that the roots of $8x^2 + 6kx - 2 = -3k$ are real for all real values of k.

16. Show that the roots of $k^2(x^2 - 1) + 4kx + 20 = 0$ are real if $|k| \geq 4$.

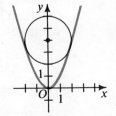

C 17. The diagram shows a circle with center $(0, 4)$ tangent to the parabola $y = x^2$. Find an equation of the circle.

Chapter summary

1. The rules for solving *linear inequalities* are quite like the rules for solving linear equations and are summarized on page 94. The graph of a linear inequality in one variable is part of the number line.

2. Simple equations and inequalities involving $|x - k|$ can be solved in two ways:
   a. By interpreting $|x - k|$ as the distance from x to k on the number line. (See Examples 1 through 6 on page 95.)
   b. By using the following properties where $c \geq 0$:

Statement	$\lvert ax + b \rvert = c$	$\lvert ax + b \rvert < c$	$\lvert ax + b \rvert > c$
Meaning	$ax + b = \pm c$	$-c < ax + b < c$	$ax + b < -c$ or $ax + b > c$

3. The solution to a *polynomial inequality* $P(x) > 0$ (or $P(x) < 0$) consists of those values of x for which the graph of the polynomial $y = P(x)$ is above (or below) the x-axis. (See Examples 1 through 3 of Section 3-2.)

4. Solutions to *inequalities in two variables* are ordered pairs of real numbers. Their graphs consist of points in the plane. (See Examples 1 and 2 of Section 3-3.) The graph of the solution of a system of inequalities is the intersection of the graphs of the individual inequalities. (This is illustrated in Example 3, page 103.)

5. A system of linear inequalities can sometimes be used to evaluate how profitable various business alternatives are. If a linear expression $P = ax + by$ is evaluated for all points of a convex polygonal region, the maximum and minimum values, if they exist, will each occur at a vertex.

6. The discriminant can be used to determine which of several lines intersect a circle or a parabola in two points, one point (tangent lines), or not at all. Simply solve the general equation of the line simultaneously with the equation of the circle or parabola, calculate the discriminant of the resulting quadratic equation, and then determine for which values of the variable the discriminant is zero, greater than zero, or less than zero.

7. Variables which are used to describe the general form of a set of equations are called *parameters*. For example, when the general quadratic equation is written as $ax^2 + bx + c = 0$, the variables a, b, and c are parameters.

Chapter test

1. Solve and graph each inequality on a number line.   3-1
   a. $8 - 4x < 6$ **b.** $|x - 3| < 5$ **c.** $-1 \leq |x + 1| \leq 4$

2. True or false for all a, b, and c?   3-2
   a. If $a < b$, then $ac < bc$.   **b.** If $a < b$, then $a^2 < b^2$.

3. Find the solution of each inequality.
   a. $4x^2 + 5x - 6 \leq 0$ **b.** $2x^3 - x^2 - x < 0$

Amalie Emmy Noether (1882–1935) received her degree from the University of Erlangen in 1916. She then taught at the University of Göttingen until 1933, when she and many other mathematicians of Jewish background were denied the right to teach. She came to the United States and taught at Bryn Mawr until her death.

Noether's greatest contribution was to the theory of rings. A ring is a certain type of abstract system with two binary operations.

4. Graph the solution set of the system given by: 3-3
   a. $x^2 + y^2 \le 36$ b. $y \ge x^2 - 2x$
       $|y| > 3$ $y \le 6 - x$

5. Give a set of inequalities that defines the shaded region shown.

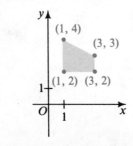

6. (Optional) An automobile manufacturer makes automobiles and trucks in a factory that is divided into two shops. Shop 1, which performs the basic assembly operation, must work 6 person-days on each truck but only 3 person-days on each automobile. Shop 2, which performs finishing operations, must work 4 person-days on each automobile or truck that it produces. Because of personnel and machine limitations, Shop 1 has 150 person-days per week available while Shop 2 has 120 person-days per week. If the manufacturer makes a profit of $500 on each truck and $350 on each automobile, how many of each should be produced each week to maximize the profit? 3-4

7. a. Write the general form of an equation that represents all lines with slope 2. Use k as a parameter. 3-5
   b. Find the values of k for which the line in part (a) intersects the circle $x^2 + y^2 = 5$.

CHAPTER FOUR

Functions

OBJECTIVES

1. To interpret the graph of a functional relationship and answer questions concerning the graph.
2. To find specified function values given an equation of the function.
3. To determine the domain, range, and zeros of a function and to sketch the graph of the function.
4. To write the equation and identify the domain of the composite of two functions.
5. To define functions from verbal descriptions.
6. To find an equation and sketch the graph of the inverse of a function.

Functions and their graphs

4-1/DEFINITION OF A FUNCTION

Examples of functions

When one quantity depends on another, the first is said to be a *function* of the second. For example, the value of the polynomial $2x^2 - 3x - 5$ depends on the value of x, so we say that the value of the polynomial is a function of x and write $f(x) = 2x^2 - 3x - 5$.* Although polynomial functions are quite important, they represent just one type of function. For example, the amount of time

---

*In this book, $f(x)$ will refer to a function of x and also to the value of the function at x. In some books, $f(x)$ is used only to represent the value of the function at x and the function itself is called f.

The Olympic rating of the sailing yacht shown is a function of its length, the area of its sails, and the volume of water it displaces. See the Calculator Exercise on page 124 for the formula used.

t required to travel 120 km on a superhighway is a function of the speed r. To keep this speed within ordinary ranges, let us assume that the car travels at least 20 km/h but no more than 120 km/h. Substituting in the formula $d = rt$ gives:

$$120 = rt$$

$$t = \frac{120}{r}$$

If we let $g(r)$ represent the time, then we can give the following formula for $g(r)$:

$$g(r) = \frac{120}{r} \qquad 20 \le r \le 120$$

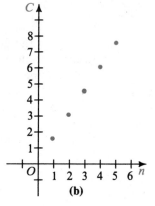

(a)

The graph of $t = g(r)$ is shown in figure (a).

An important aspect of $g(r)$ is that the values of r are restricted to real numbers greater than or equal to 20 and less than or equal to 120. We refer to this set of values as the *domain* of the function.

The domain of the next function we present is the set of integers from 1 through 5, inclusive. Suppose you plan to buy some mustard at a store that has 5 jars of mustard, each costing $1.50. The cost ($C$) of your purchase is a function of the number of jars you buy. We can write the following formula for $C = h(n)$ where n represents the number of jars:

$$h(n) = 1.50n \qquad n = 1, 2, 3, 4, 5$$

The graph of $C = h(n)$ is shown in figure (b). Notice that $h(n)$ takes on only the values 1.50, 3, 4.50, 6, and 7.50. We refer to the set of values just given as the *range* of the function.

(b)

The range of $g(r)$, as figure (a) suggests, consists of the real numbers between 1 and 6, inclusive. We can write this range as $\{t \mid 1 \le t \le 6\}$, read "the set of all t such that t is greater than or equal to 1 and less than or equal to 6."

Definition of a function

A **function** consists of the following:
1. A set of real numbers, called the **domain** of the function.
2. A rule that assigns to each element in the domain exactly one real number.

The set of real numbers assigned by the rule to the elements in the domain is called the **range** of the function. The graph of the function consists of all points (x, y) such that x is in the domain of the function and y is the corresponding element in the range.*

*Some people like to identify a function with its graph and so they consider a function as a set of ordered pairs. The set of first elements of these ordered pairs is the domain of the function and the set of second elements is the range.

The domain of a function

Some people like to think of the domain of a function as a set of input values to be substituted in the formula or rule for the function. The range can then be thought of as the corresponding set of output values. If the rule for a function is given without stating the domain, then the domain is understood to consist of all real numbers that can be substituted in the rule to produce real numbers.

EXAMPLE 1. State the domain of the function.

$$\text{a. } g(x) = \frac{1}{x-7} \qquad \text{b. } h(r) = \sqrt{1-r^2}$$

SOLUTION: **a.** The domain of $g(x)$ is the set of all real numbers except 7 since $\dfrac{1}{x-7}$ is not defined when the denominator is 0.

b. $\sqrt{1-r^2}$ is a real number only if $1-r^2 \geq 0$. Therefore, the domain of $h(r)$ is $\{r \mid -1 \leq r \leq 1\}$.

Functions of two variables

We can consider functions of two variables by permitting the domain to consist of sets of ordered pairs of real numbers.

The volume of a cylinder depends on both the radius (r) and the height (h). We can express the volume (V) as the function $f(r, h)$:

$$V = f(r, h) = \pi r^2 h \qquad r > 0, h > 0$$

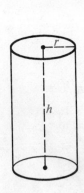 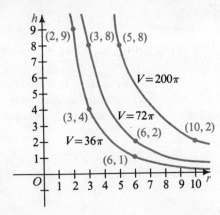

A function of two variables is more difficult to graph than a function of one variable. One common method is to plot curves along which the value of the function is constant. In the cylinder example the volume $\pi r^2 h$ is 36π when $(r, h) = (2, 9), (3, 4)$, and $(6, 1)$. These points and others for which $\pi r^2 h = 36\pi$ are shown in the graph above on the curve labeled $V = 36\pi$. Other curves of constant volume are also shown.

The following diagram shows curves of constant barometric pressure. Here the barometric pressure at a location is a function of the latitude and longitude.

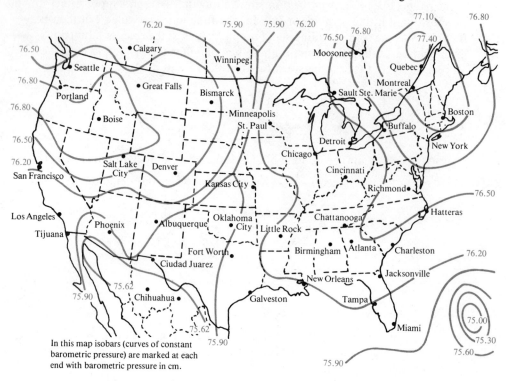

In this map isobars (curves of constant barometric pressure) are marked at each end with barometric pressure in cm.

ORAL EXERCISES 4-1

1. The graph at the left below shows the speed of a two-speed race car in a drag race.
   a. The graph illustrates that _?_ is a function of _?_.
   b. Explain why the graph has a little peak.
   c. How long does it take the car to reach a speed of 120 km/h?

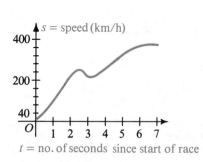

Ex. 1

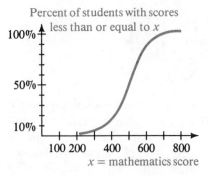

Ex. 2

2. The percentile rank of a student taking a mathematics aptitude test is a function of the student's score as shown on the graph at the right at the bottom of page 120.
   a. What is the domain of the function?
   b. Approximately what percent of the students have scores less than or equal to 600?
   c. Find the least score such that approximately 40% of the scores are less than or equal to it.

3. Describe in your own words the functional relationship given by each of the following formulas.
   a. $A = \pi r^2$
   b. $V = \frac{4}{3}\pi r^3$
   c. $A = \frac{1}{2}bh$
   d. $d = rt$
   e. $F = ma$
   f. $A = P(1 + r)^t$

4. If $f(x) = 3x - 1$, evaluate:
   a. $f(3)$
   b. $f(0)$
   c. $f(-1)$
   d. $f(-2)$

5. If $f(x) = x^2 - 5$, evaluate:
   a. $f(0)$
   b. $f(-2)$
   c. $f(\sqrt{5})$
   d. $f(-5)$

6. State the domain of the function.
   a. $f(x) = \dfrac{x}{x + 2}$
   b. $g(s) = \dfrac{1}{s - 1}$
   c. $h(y) = \dfrac{y}{4 - y^2}$

7. State the domain of the function.
   a. $g(t) = \sqrt{t}$
   b. $f(t) = \sqrt{t - 4}$
   c. $h(t) = \sqrt{t^2 - 4}$

8. Refer to the barometric pressure graph shown on page 120.
   a. Name two other cities with approximately the same pressure as San Francisco.
   b. Bad weather usually accompanies low pressure. In which cities is the weather apt to be poorest?

9. a. If $f(x, y) = \sqrt{x^2 + y^2}$, evaluate $f(3, 4)$, $f(-4, 3)$, and $f(0, 5)$.
   b. On the chalkboard sketch some constant curves of the function $f(x, y)$.

10. The function below is defined by different formulas for different intervals.

$$f(x) = \begin{cases} -x, & \text{if } x \leq 0 \\ \dfrac{1}{x}, & \text{if } 0 < x \leq 1 \\ 2x - x^2, & \text{if } x > 1 \end{cases}$$

   a. Evaluate $f(-2)$, $f(0.01)$, and $f(2)$.
   b. On the chalkboard sketch the graph of $f(x)$.

WRITTEN EXERCISES 4–1

A
1. The area of a square is a function of __?__.
2. The volume of a cube is a function of __?__.
3. The volume of a regular square pyramid is a function of __?__ and __?__.
4. The volume of a cone is a function of __?__ and __?__.

5. Find the domain of each function.

a. $f(x) = \dfrac{x + 2}{x - 9}$ **b.** $g(x) = \dfrac{1}{x}$ **c.** $h(x) = \dfrac{x + 2}{x^2 + 5x + 6}$

d. $f(x) = 3x - 2$ **e.** $f(x) = \sqrt{5 - x}$ **f.** $h(x) = \sqrt{x^2 - 25}$

6. For the linear function $f(x) = mx + k$, $f(1) = 4$ and $f(3) = 8$. Find the values of m and k.

7. If $x = 1, 2, 3$, then $y = 5, 7, 9$ respectively. Express y as a linear function of x.

8. Express the Celsius temperature C as a linear function of the equivalent Fahrenheit temperature F. (*Recall:* $0°C = 32°F$ and $100°C = 212°F$.)

9. In this exercise, we relate the number (n) of sides of a polygon and the sum (S) of interior angle measures.

a. Assuming that the points of the graph at the right are collinear, find an equation relating S and n.

b. What is the domain of the function S?

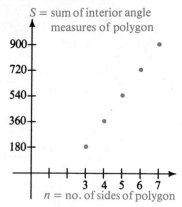

10. The graph at the right shows the result of investing $100, with interest compounded annually.

a. Approximately how many years does it take to double your money at 8% interest compounded annually? at 12% interest? at 16% interest?

b. The equation of the right-hand curve is $A = 100(1.08)^t$. Guess the equation of the other two curves.

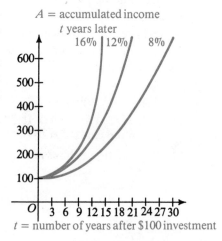

11. The graph at the right shows the fuel efficiency at various speeds for three different weights of cars. The fuel efficiency is given in kilometers per liter (km/L).

a. At what speed will a 2000 kg car have approximately the same fuel efficiency as a 1500 kg car traveling at 100 km/h?

b. A 1000 kg car has 3 L of gas in its tank and is 45 km from the nearest gas station. What is the fastest speed at which it should travel?

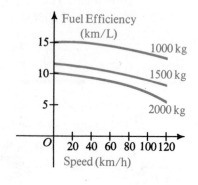

12. The graph at the right shows the wind-chill equivalent temperatures for $0°C$, $-12°C$, and $-24°C$.

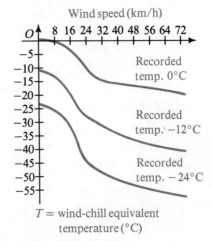

Wind speed (km/h)

8 16 24 32 40 48 56 64 72

Recorded temp. $0°C$

Recorded temp. $-12°C$

Recorded temp. $-24°C$

$T =$ wind-chill equivalent temperature $(°C)$

  a. If the recorded temperature is $-12°C$, approximately what wind speed will produce a wind-chill temperature of $-24°C$?

  b. In what wind speed range is the effect of the wind-chill temperature the greatest? the least?

  c. Would you rather be in a place where the recorded temperature is $0°C$ with a 48 km/h wind or $-12°C$ with an 8 km/h wind? Why?

13. The number of times (n) that a cricket chirps per minute is a linear function of the temperature C.

  a. Express n as a function of C, given that $(15°, 92)$ and $(20°, 132)$ are on the graph of the function.

  b. Sketch a graph of the function.

  c. What is the temperature when a cricket is chirping 116 times a minute?

Exercises 14–21 involve functions of two variables.

B   **14. a.** If $A(b, h) = \frac{1}{2}bh$, find $A(8, 3)$ and $A(16, 6)$.

  b. Give a geometric interpretation of the function $A(b, h)$.

15. a. If $V(r, h) = \frac{1}{3}\pi r^2 h$, find $V(2, 6)$ and $V(4, 12)$.

  b. Give a geometric interpretation of the function $V(r, h)$.

16. If $A(b, h) = \frac{1}{2}bh$, show that $A(3b, 3h) = 9 \cdot A(b, h)$.

17. If $V(r, h) = \pi r^2 h$, show that $V(2r, 2h) = 8 \cdot V(r, h)$.

18. a. If you travel at a constant rate r for t hours, the distance D that you travel is a function of r and t. Give a formula for this function.

  b. Draw a graph with r- and t-axes and show several points on the constant distance curve $D(r, t) = 200$.

  c. Draw the curve $D(r, t) = 400$.

19. a. The volume V of a square prism is a function of the side s of the square base and of the height h of the prism. Give an equation of this function.

  b. Draw a graph with s- and h-axes and show two curves of constant volume.

20. Look up "contour map" in a dictionary or encyclopedia. Draw a contour map of the mountainous region shown below.

21. Sketch a map of your state (or geographic region) and on it draw some curves of constant temperature that seem reasonable to you. (Assume the temperatures are for 12 noon.)

Refer to textbooks in the fields listed below to find several examples of functional relationships. Be prepared to discuss these relationships in class.

22. Physics, chemistry, or medicine

23. Biology, psychology, or economics

CALCULATOR EXERCISE

The Olympic rating R of a sailing yacht is a function of several variables.

$$R = 0.9\left(\frac{L\sqrt{A}}{12\sqrt[3]{V}} + \frac{L + \sqrt{A}}{4}\right)$$

L = length of yacht in meters
A = surface area of sails in square meters
V = volume of water yacht displaces in cubic meters

To be in the R-5.5 class at the Olympics, a yacht must have an Olympic rating less than 5.5. Does a yacht that is 10 m long, has 37 m² of sail, and displaces 8.5 m³ of water qualify?

4-2/GRAPHS OF FUNCTIONS

The domain of a function should be clear from the statement of the function. The range, however, is not always easily determined. The graph of the function can help you to visualize the range. For example, if you project lights horizontally toward the y-axis from a great distance away, the shadow of the graph on the y-axis gives you a picture of the range. Recall that a real number a is a zero of a function if $f(a) = 0$; also, the number a is an x-intercept of the graph since the graph passes through $(a, 0)$.

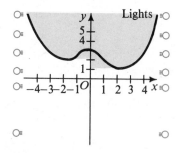

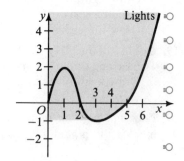

Domain: all real numbers

Range: $\{y \mid y \geq 1\}$

Zeros: The function has none.

Domain: $\{x \mid x \geq 0\}$

Range: $\{y \mid y \geq -1\}$

Zeros: The function has three zeros; they are 0, 2, and 5.

When you have an equation in x and y, that equation defines y as a function of x *provided* there is exactly one y-value for each x-value. For example, the equation $x^3 + y^3 = 1$ defines y as a function of x since to every x-value there corresponds exactly one y-value. On the other hand, the equation $y^2 = 4x$ does not define y as a function of x because when $x = 1$, y can be either 2 or -2.

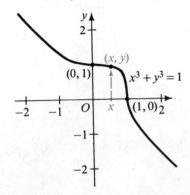
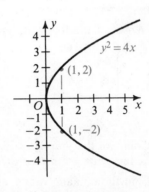

The graphs above illustrate the so-called vertical-line test for determining whether an equation defines a function.

THE VERTICAL-LINE TEST   If the graph of an equation is such that no vertical line intersects the graph in more than one point, then the graph is the graph of a function.

ORAL EXERCISES 4–2

1. State the domain, range, and zeros of the function.
   a. $f(x) = x$ **b.** $f(x) = x^2$ **c.** $f(x) = x^2 - 4$

2. The graph of $y = f(x)$ is shown. What are the domain and range of the function $f(x)$? What are the zeros of $f(x)$?

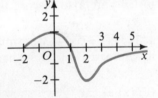

3. State the domain and range of $f(x) = (x - 3)^2 + 5$. How many zeros does $f(x)$ have?

4. The graphs of three equations in x and y are shown below. For each, tell whether or not it is the graph of a function. Explain your answer.

   a. 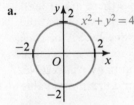   **b.** 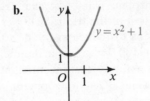   **c.**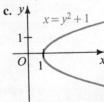

5. What is the domain of any equation of the form $f(x) = ax^2 + bx + c$ (the general equation of a parabola)?

6. Apply the vertical-line test to the line with equation $x = 4$. Is the graph the graph of a function? Explain your answer.

WRITTEN EXERCISES 4-2

A   **1.** Which of the following graphs are graphs of functions?

a. 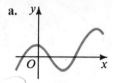   b. 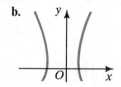   c.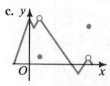

2. Compare the graphs of $y^2 = 9 - x^2$ and $y = \sqrt{9 - x^2}$. Which equation defines y as a function of x?

3. Find the domain, range, and zeros of the function.
   a. $f(x) = |x|$ **b.** $g(x) = |x - 2|$ **c.** $h(x) = |x| - 2$

In Exercises 4–11: **a.** State the domain and zeros (if any) of the function.
                    b. Sketch the graph of the function.
                    c. Determine the range of the function from the graph.

4. $f(x) = x^2 - 6x + 8$ **5.** $g(x) = 2 - (x - 3)^2$ **6.** $h(t) = \dfrac{1}{t}$

7. $p(x) = (x - 2)^3$ **8.** $g(x) = x^3 + 4x^2 - x - 4$ **9.** $f(x) = \sqrt{4 - x}$

10. $h(x) = \begin{cases} x^2, & \text{if } -2 \le x < 1 \\ 2 - x, & \text{if } 1 \le x < 4 \end{cases}$ **11.** $g(x) = \begin{cases} x - 1, & \text{if } x < 0 \\ x^2 - 2x - 3, & \text{if } 0 \le x \le 3 \\ 0, & \text{if } x > 3 \end{cases}$

12. a. If $f(x) = 3x - 15$, evaluate $f(x + 2) - f(x)$.
   b. Use part (a) to describe the change in the value of $f(x)$ when x increases by 2.

13. a. If $f(x) = 5x + 35$, evaluate $f(x - 6) - f(x)$.
   b. Use part (a) to describe the change in the value of $f(x)$ when x decreases by 6.

14. If $f(x) = 12x - 3x^2$, show that $f(2 + k) = f(2 - k)$ for all values of k. Interpret this result on a graph of $f(x)$.

B   **15.** An airplane with a 200-passenger capacity is to be chartered for a transatlantic flight. The airplane cannot be chartered unless there are at least 130 passengers. The cost of the trip is $240 per passenger, except that the airline company agrees to reduce everyone's ticket price by $1 for *each* ticket sold in excess of 130. Let x be the number of tickets sold in excess of 130 and let $f(x)$ be the total ticket income that the company will receive. Draw a graph of $f(x)$ and specify the domain, range, and maximum value of $f(x)$.

16. A company produces 100 radios each day and makes an $18 profit on each radio. By having the staff work overtime, the company can produce up to

150 radios per day, but the extra overtime pay makes the profit per radio go down. Suppose that when $100 + x$ radios per day are produced, the profit on each radio is $\$(18 - 0.10x)$. (This means, for example, that producing 110 radios cuts the profit per radio to \$17.) Let $f(x)$ be the total profit on $100 + x$ radios. Draw a graph of $f(x)$ and specify the domain, range, and maximum value of $f(x)$.

17. For which of the following functions, if any, does $f(a + b) = f(a) + f(b)$?

    a. $f(x) = x^2$ **b.** $f(x) = \dfrac{1}{x}$ **c.** $f(x) = 4x + 1$ **d.** $f(x) = 4x$

18. For which of the functions in Exercise 17 does $f(2x) = 2 \cdot f(x)$?

19. Suppose that all you know about a function $f(x)$ is that

$$f(a + b) = f(a) + f(b) \text{ for all } a \text{ and } b.$$

    a. Let $b = 0$ and prove $f(0) = 0$.
    b. Let $a = b = 1$ and prove $f(2) = 2f(1)$.
    c. Let $a = 1$ and $b = 2$ and prove $f(3) = 3f(1)$.
    d. Prove $f(4) = 4f(1)$.
    e. What generalization is suggested by parts (b), (c), and (d)?
    f. Let $a = b = \frac{1}{2}$ and prove $f(\frac{1}{2}) = \frac{1}{2}f(1)$.
    g. Prove $f(\frac{1}{3}) = \frac{1}{3}f(1)$.
    h. What generalization is suggested by parts (f) and (g)?
    i. Prove $f(-x) = -f(x)$ and $f(y - x) = f(y) - f(x)$.
    j. Can you think of any functions $f(x)$ that satisfy the original equation?

C   20. Suppose that all you know about a function $f(x)$ is that $f(ab) = f(a) \cdot f(b)$ for all a and b. Prove as much about $f(x)$ as you can and give some examples of functions that satisfy the original equation.

Relations among functions

4-3/THE COMPOSITION OF FUNCTIONS

The composite of two functions

Suppose $f(x) = x^2$ and $g(x) = 2x$. Then, as the diagram below suggests, $f(x)$ and $g(x)$ can be combined to produce a new function denoted $f(g(x))$, read "f of g of x." Since $g(x) = 2x$, $f(g(x)) = f(2x) = (2x)^2$. The function $f(g(x))$

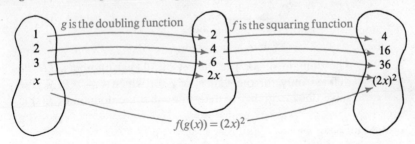

is called the **composite** of functions $f(x)$ and $g(x)$.* The operation that combines these functions to give $f(g(x))$ is called **composition.** To see that this operation is not commutative, note that $g(f(x)) = g(x^2) = 2x^2$, but we have seen that $f(g(x)) = (2x)^2 = 4x^2$.

EXAMPLE 1. Suppose $f(x) = \dfrac{1}{x}$ and $g(x) = 3x - 6$.

Find $f(g(x))$ and $g(f(x))$.

SOLUTION: $f(g(x)) = f(3x - 6) = \dfrac{1}{3x - 6}$

$$g(f(x)) = g\left(\frac{1}{x}\right) = 3\left(\frac{1}{x}\right) - 6 = \frac{3}{x} - 6$$

In the preceding example notice that the domain of $g(x)$ is the set of all real numbers, but the domain of $f(g(x))$ is the set of all real numbers except 2 since $\dfrac{1}{3x - 6}$ is not defined when $x = 2$. This illustrates that the domain of $f(g(x))$ does not always contain all the elements of the domain of $g(x)$.

In general, the domain of $f(g(x))$ consists of all elements of the domain of $g(x)$ for which $f(g(x))$ makes sense. In other words, a real number a is in the domain of $f(g(x))$ if a is in the domain of $g(x)$ and $g(a)$ is in the domain of $f(x)$.

EXAMPLE 2. If $f(x) = x^2$ and $g(x) = \sqrt{x - 2}$, find $f(g(x))$ and give its domain.

SOLUTION: $f(g(x)) = (\sqrt{x - 2})^2$
$= x - 2$

The domain of $g(x)$ is $\{x \mid x \geq 2\}$. Since the domain of $f(x)$ is all real numbers, all values of $g(x)$ are in the domain of $f(x)$. Therefore, the domain of $f(g(x))$ is $\{x \mid x \geq 2\}$. The domain of $f(g(x))$ can include only values in the domain of $g(x)$ even though the formula, $x - 2$, is defined for all real numbers.

EXAMPLE 3. If $L(t) = \sqrt{4 - t^2}$ and $M(t) = 2t$, find $L(M(t))$ and give its domain.

SOLUTION: $L(M(t)) = \sqrt{4 - (2t)^2}$
$= \sqrt{4(1 - t^2)}$
$= 2\sqrt{1 - t^2}$

The domain of $M(t)$ is the set of all real numbers. $L(t)$ is defined only for the values of t for which $4 - t^2 \geq 0$. We solve this inequality to determine that the domain of $L(t)$ is

*In some books the composite function $f(g(x))$ is denoted as $f \circ g$.

$\{t \mid -2 \le t \ge 2\}$. For any value of t, $M(t)$ will be in the domain of $L(t)$ only if:

$$-2 \le M(t) \le 2$$
$$-2 \le 2t \le 2$$
$$-1 \le t \le 1$$

Therefore, the domain of $L(M(t))$ is $\{t \mid -1 \le t \le 1\}$.

ORAL EXERCISES 4-3

In Exercises 1–12, let $f(x) = x^3$, $g(x) = \frac{x}{2}$, and $h(x) = x - 2$. Evaluate the following.

1. $f(g(4))$ **2.** $g(f(4))$ **3.** $f(h(2))$

4. $h(f(2))$ **5.** $g(h(0))$ **6.** $h(g(0))$

State the rule for each composite function.

7. $f(g(x))$ **8.** $g(f(x))$ **9.** $f(h(x))$

10. $h(f(x))$ **11.** $g(h(x))$ **12.** $h(g(x))$

WRITTEN EXERCISES 4-3

A   **1-6.** If $f(x) = \frac{x}{3}$, $g(x) = x^2$, and $h(x) = 3x + 2$, evaluate the expressions in Oral Exercises 1–6.

7-12. If $f(x) = \frac{4}{x}$, $g(x) = x^3$, and $h(x) = 4x - 1$, write the rule for each composite function in Oral Exercises 7–12.

In Exercises 13–20, let $f(x) = \sqrt{x}$, $g(x) = 6x - 3$, and $h(x) = \frac{x}{3}$. Evaluate the following.

B   **13.** $f(g(h(6)))$ **14.** $h(g(f(4)))$ **15.** $h(f(g(\frac{1}{2})))$ **16.** $g(h(f(9)))$

Write the rule for each composite function.

17. $f(g(h(x)))$ **18.** $h(g(f(x)))$ **19.** $h(f(g(x)))$ **20.** $g(h(f(x)))$

In Exercises 21–24, let $f(x) = 4x$, $g(x) = x^2 + 1$, and $h(x) = \frac{1}{x}$. Solve the equations.

21. $f(g(x)) = g(f(x))$ **22.** $f(h(x)) = h(f(x))$ **23.** $h(f(g(x))) = \frac{1}{4}$ **24.** $h(g(f(x))) = \frac{1}{2}$

In Exercises 25–28, let $f(x) = x^3$, $g(x) = \sqrt{x}$, $h(x) = x - 4$, and $k(x) = 2x$. Express each of the following as a composite of three of these functions.

25. $2(x - 4)^3$ **26.** $\sqrt{(x - 4)^3}$ **27.** $(2x - 8)^3$ **28.** $\sqrt{x^3 - 4}$

29. a. Express the radius r of a circle as a function of the circumference C.
    b. Express the area A of the circle as a function of C.

30. The volume and surface area of a sphere are given in terms of the radius by the following formulas: $V = \frac{4}{3}\pi r^3$ and $A = 4\pi r^2$.
    a. Express r as a function of A **b.** Express V as a function of A.

31. The speed of sound in air is $s = f(C) = 331 + 0.6C$ where C is the Celsius temperature and s is measured in meters per second (m/s). However, C is a function of the Fahrenheit temperature: $C = g(F) = \frac{5}{9}(F - 32)$. Express s as a function of F.

32. **a.** Express the area A and perimeter P of a semicircular region in terms of the radius r.
    b. Express A as a function of P.

33. **a.** If $y = 3x + 4$, $x = 2t - 7$, and $t = 3z + 1$, express y as a function of z.
    b. Express z as a function of y.

34. If $y = 4x - 7$, $x = \sqrt{2t - 6}$, and $t = 8z^2 + 3$, express y as a function of z.

Figure (a) below shows a swimmer's speed as a function of time. Figure (b) shows the swimmer's oxygen consumption as a function of the speed. The time is measured in seconds, the speed in meters per second, and the oxygen consumption in liters per minute (L/min). Use the graphs to answer Exercises 35 and 36.

35. Find the swimmer's speed and oxygen consumption after 20 seconds of swimming.

36. How many seconds have elapsed if the swimmer's oxygen consumption is 15 L/min?

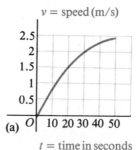

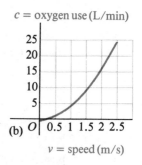

In Exercises 37–40, find $f(g(x))$ and $g(f(x))$ and give the domain of each.

37. $f(x) = 2x$, $g(x) = \sqrt{16 - x^2}$

38. $f(x) = \sqrt{x}$, $g(x) = \dfrac{1}{x - 4}$

39. $f(x) = x^2$, $g(x) = \sqrt{1 - x}$

40. $f(x) = x^2$, $g(x) = \sqrt{16 - x^2}$

41. If $g(x) = \dfrac{x + 3}{2}$, find $g(g(1))$, $g(g(g(1)))$, and $g(g(g(g(1))))$.

42. If $f(x) = 2x - 1$, show that $f(f(x)) = 4x - 3$. Find $f(f(f(x)))$.

C 43. The luminous intensity I of a light bulb is measured in candela (cd). A 100 watt bulb has an intensity of 130 cd. The law of illumination states that $E = \dfrac{I}{d^2}$, where E is illumination and d is the distance in meters to the light bulb. Suppose you hold a book 1 m away from a 100 watt bulb and begin walking away from the bulb at a rate of 1 m/s.
    a. Express E in terms of the time, t, in seconds after you begin walking.

130   Chapter 4

b. At what time will the illumination on the book be 1% of its original value?

44. The force F of repulsion between two electrical charges of magnitudes q_1 and q_2 is given by the formula

$$F = k\frac{q_1q_2}{r^2},$$

where k is a constant and r is the distance between the charges measured in meters (m). The magnitude of the charges is expressed in coulombs (C) and the force in newtons (N). (The actual value of k is 9×10^9 N $\cdot$ m^2/C^2.)

a. If two equal charges of magnitude 1 C and 10 cm apart move toward each other at a rate of 2 cm/s, then express F as a function of the time t after they begin moving. (Give the answer in terms of k.)

b. Find the amount of time t when the force is 100 times the original force.

CALCULATOR EXERCISE

The time T for one complete swing of a clock pendulum is $T = 2\pi\sqrt{\dfrac{L}{980}}$,

where L is the length of the pendulum in centimeters and T is in seconds. The earliest pendulum clocks became inaccurate with large temperature changes. A temperature change would cause a change in the length of a pendulum, which would in turn cause a change in the time of a swing. Suppose that a clock is calibrated to keep accurate time when the Celsius temperature c is 0°C. Suppose also that the length L of the pendulum varies with c, so that $L = 100 + 0.0004c$. How many seconds will the clock gain or lose in a day if the temperature is 25°C?

Historical Note: The problem of varying pendulum length was solved by placing a tube of mercury in the pendulum bob. As warm weather increased the length of the pendulum, it would also cause the mercury to rise, and the center of gravity of the pendulum bob would remain the same.

Pierre de Fermat (1601–1665) spent his adult life working as a civil servant, but his favorite pastime was the study of mathematics. He is best known for his work in number theory and his contributions to the development of coordinate geometry.

4-4/FORMING FUNCTIONS FROM VERBAL DESCRIPTIONS

An important problem of mathematics is finding the maximum or minimum value of a function. You have already seen examples involving minimizing costs and maximizing profits. Other such applications use mathematics to minimize the structural stress on a girder or to maximize the volume of a container made from a given amount of material.

Maximum and minimum values are often referred to as *extreme values*. Approximate extreme values of a function can be found using a computer. Exact extreme values are most often found using calculus. Whatever method is used, we almost always need to write a formula for the function. Building this skill is the goal of this section.

EXAMPLE 1. An open-topped box with a square base is to be constructed from sheet metal in such a way that the completed box is made of 2 m² of the sheet metal. Express the volume of the box as a function of the width of the base.

SOLUTION: 1. Sketch the box. Let h represent the height.
2. Find the volume as a function of the height and width.

$$V(w, h) = w^2 h$$

3. Express the area in terms of h and w. Then solve for h in terms of w.

$$\text{Area of sheeting used} = 2$$
$$\text{Area of the base} + 4 \cdot (\text{Area of a side}) = 2$$
$$w^2 \quad + \quad 4wh \quad = 2$$
$$h = \frac{2 - w^2}{4w}$$

4. Substitute to find V in terms of w alone.

$$V(w, h) = w^2 h$$
$$V(w) = w^2 \left(\frac{2 - w^2}{4w} \right)$$
$$= \frac{2w - w^3}{4}$$

EXAMPLE 2. A north-south bridle path intersects an east-west river at point O. At noon a horse and rider leave O traveling north at 12 km/h. At the same time, a boat is 25 km east of O traveling west at 16 km/h. Express the distance d between the horse and boat as a function of the number of hours t after noon.

SOLUTION: 1. Make a sketch showing the horse and boat at some time t. Let h be the horse's distance from O; let b be the boat's distance east of O.

2. By the Pythagorean Theorem:
$$d(h, b) = \sqrt{h^2 + b^2}$$

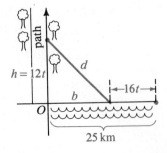

3. Next we use the formula $d = rt$ to find expressions for h and b in terms of t. Since the horse is traveling from O at 12 km/h, $h = 12t$. Since the boat is traveling toward O at 16 km/h from a distance of 25 km,

$$b = 25 - 16t.$$

4. Substitute to find d in terms of t alone.

$$d(h, b) = \sqrt{h^2 + b^2}$$
$$d(t) = \sqrt{(12t)^2 + (25 - 16t)^2}$$
$$= \sqrt{144t^2 + 625 - 800t + 256t^2}$$
$$= \sqrt{400t^2 - 800t + 625}$$

In Example 2, it is possible to determine when the distance between the horse and the boat is a minimum and what the minimum distance is. Since $d(t)$ is the square root of the function $f(t) = 400t^2 - 800t + 625$, $d(t)$ will be a minimum when $f(t)$ is a minimum. In Chapter 2 it was shown that a quadratic function will take on its maximum or minimum value at the vertex of its graph, which occurs when $t = \dfrac{-b}{2a}$. We substitute 400 for a and -800 for b:

$$t = \frac{-(-800)}{2(400)} = 1$$

Therefore, the horse and boat are closest 1 hour after noon. We find $d(1)$ to determine the minimum distance between them.

$$d(1) = \sqrt{400 \cdot 1^2 - 800 \cdot 1 + 625}$$
$$= \sqrt{225}$$
$$= 15$$

Therefore, the minimum distance between the horse and boat is 15 km.

In the discussion following Example 2 the function $d(t)$ was used to find the minimum distance between the boat and the horse. Had the boat or horse traveled at a different speed, the function $d(t)$ would have been different. Calculus can be used to show how the rate of the horse and the distance d are related. In fact, solving this type of problem is an important application of calculus. The first step in doing this is the same as the first step in finding the minimum of d. We must express d as a function of other variables, as was shown in Example 2. Example 3 also shows how to set up a function that would be useful when calculus is used to express the relationship among the rates.

EXAMPLE 3. Water is flowing into a conical tank 100 cm wide and 250 cm deep at a rate of 50 cm³/s. Find the volume V of the water in the tank as a function of the height h of the water. Then represent h as a function of the time t that the water has been flowing into the empty tank.

SOLUTION:

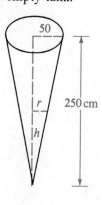

1. Let r represent the radius of the water surface and h the height of the water in the tank. By the formula, $V(r, h) = \frac{1}{3}\pi r^2 h$.

2. Find an expression for r in terms of h by using the similar triangles shown in a cross section of the tank. (See the second figure at the right.)

$$\frac{r}{h} = \frac{50}{250}$$

$$r = \frac{h}{5}$$

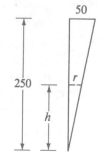

3. Substitute to find V in terms of h alone.

$$V(r, h) = \frac{1}{3}\pi r^2 h = \frac{1}{3}\pi\left(\frac{h}{5}\right)^2 h$$

$$V(h) = \frac{\pi}{75}h^3$$

4. To represent h as a function of the time t, notice that the volume of the water in the tank at t seconds is

$$V = 50t.$$

Substituting to find h in terms of t:

$$50t = \frac{\pi}{75}h^3$$

$$\sqrt[3]{\frac{3750t}{\pi}} = h$$

Therefore, $h(t) = \sqrt[3]{\dfrac{3750t}{\pi}}$

ORAL EXERCISES 4–4

1. **a.** Express the volume of a cube as a function of the edge e.
   b. Express e as a function of V.

2. Suppose a square has side s and diagonal d, as shown at the right.
   a. Express d as a function of s.
   b. Express s as a function of d.
   c. Express the area of the square as a function of d.

134   Chapter 4

3. Point A is 4 km north of point C, and point B is 8 km east of C. P is a point on $\overline{BC}$, at a distance of x km from C.

   a. Express $AP + PB$ as a function of x.

   b. What is the domain of this function?

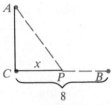
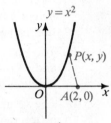

Ex. 3

4. $P(x, y)$ is a point on the parabola $y = x^2$.

   a. Express the distance from P to $A(2, 0)$ as a function of x and y.

   b. Express the distance as a function of x alone.

5. A runner starts north from point O at 6 m/s; at the same time a second runner sprints east from O at 8 m/s. Find the distance d between the runners t seconds later.

6. Answer Exercise 5 if the first runner had started 60 m north of O.

7. Answer the question of Example 1 on page 132 if the box had a square lid.

8. It costs $1 + \dfrac{1}{x}$ dollars per liter to manufacture x liters of a detergent.

   a. What is the total cost of manufacturing 100 liters? k liters?

   b. What is the profit in manufacturing k liters of detergent if it is sold at \$2.00 per liter? (profit = sales − cost.)

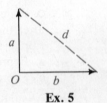

Ex. 4

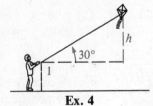

Ex. 5

WRITTEN EXERCISES 4-4

A

1. Express the area A of a $30°$–$60°$–$90°$ triangle as a function of the length h of the hypotenuse.

2. Express the area A of an equilateral triangle as a function of the perimeter.

3. A tourist walks n km at 4 km/h and then travels $2n$ km at 36 km/h by bus. Express the total traveling time as a function of n.

4. A student holds a kite and ball of string 1 m above the ground and begins to let the string out at a rate of 2 m/s. If the string is at a $30°$ angle to the horizontal, express the height of the kite as a function of the time t in seconds after it begins to fly.

Ex. 4

5. A store owner bought n dozen toy boats at a cost of \$3.00 per dozen, and sold them at \$.75 apiece. Express the profit as a function of n.

6. The cost of renting a large boat is 30 dollars per hour plus a usage fee roughly equivalent to x^3 cents per hour when the boat is operated at a speed of x km/h. Express the cost per kilometer as a function of x.

7. The height of a cylinder is twice the diameter. Express the total surface area as a function of the height h.

8. A pile of sand is in the shape of a cone with a diameter that is twice the height. Express the volume V of sand as a function of the height h.

9. A light 3 m above the ground causes a boy 180 cm tall to cast a shadow s cm long measured along the ground. Express s as a function of his distance d from the light.

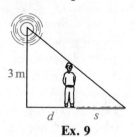

10. A girl 175 cm tall walks from a wall toward a light on the ground 15 m away. Express the height h of her shadow on the wall as a function of her distance d from the light.

Ex. 9

11. The surface area of a box with a square base and top is 3 m². Express the volume V as a function of the width w of the base.

12. The volume of a box with a square base and no top is 6 m³. Express the total surface area A as a function of the width x of the base.

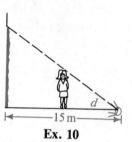

13. A stone is thrown into a lake, and t seconds after the splash the diameter of the circle of ripples is t meters.

Ex. 10

  a. Express the circumference of this circle as a function of t.
  b. Express the area of this circle as a function of t.

14. A balloon is blown up in such a way that its volume increases at a rate of 20 cm³/s.
  a. If the volume of the balloon was 100 cm³ when the process of inflation began, what will the volume be after t seconds of inflation?
  b. Express the radius of the balloon as a function of t, assuming that the balloon is spherical while it is being inflated.

15. A box with a square base and no top has volume 8 m³. The material for the base costs $8.00 per square meter and the material for the sides costs $6.00 per square meter. Express the cost of the materials used to make the box as a function of the width x of its base.

B 16. The volume of an open-topped box with a rectangular base of length 2 m and width x cm is 6 m³. Express the total surface area A as a function of the width x of the base.

17. A cylindrical can has a volume of 400π cm³. The material for the top and bottom costs 2¢ per square centimeter. The material for the vertical surface costs 1¢ per square centimeter. Express the cost of materials used to make the can as a function of the radius r.

18. At 2 P.M. bike A is 4 km north of point C and is traveling south at 16 km/h. At the same time, bike B is 2 km east of C and is traveling east at 12 km/h.
  a. Show that t hours after 2 P.M. the distance between the bikes is $\sqrt{400t^2 - 80t + 20}$.
  b. At what time is the distance between them the least?
  c. What is the distance between the bikes when they are closest?

19. A car leaves Oak Corners at 11:33 A.M. traveling south at 70 km/h. At the same time, another car is 65 km west of Oak Corners traveling east at 90 km/h.

    a. Express the distance d between the cars as a function of the time t after the first car left Oak Corners.

    b. Show that the cars are closest to each other at noon.

20. Water is flowing at the rate of 5 m³/s from the conical tank shown at the right.

    a. Find the volume V of the water as a function of the water level h.

    b. Find h as a function of the time t during which water has been flowing from the full tank.

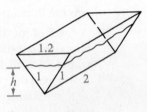

Ex. 20

21. $P(x, y)$ is an arbitrary point on the parabola $y = x^2$.

    a. Express the distance d from P to the point $(0, 1)$ as a function of the y-coordinate of P.

    b. What is the minimum distance d?

22. A trough is 2 m long, and its ends are triangles with sides of length 1 m, 1 m, and 1.2 m as shown in the diagram.

    a. Find the volume V of the water in the trough as a function of the water level h.

    b. If water is pumped into the empty trough at the rate of 6 L/min, find h as a function of the time t in minutes after the pumping begins. (1 m³ = 1000 L)

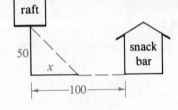

Ex. 22

23. $P(x, y)$ is an arbitrary point on the circle $x^2 + y^2 = 36$.

    a. Express the distance d from P to the point $A(8, 0)$ as a function of the x-coordinate of P.

    b. What are the domain and range of this function?

24. A lifeguard wants to swim to shore from a raft 50 m offshore and run to a snack bar 100 m down the beach, as shown in the diagram.

    a. If the lifeguard swims at 1 m/s and runs at 3 m/s, express the total swimming and running time as a function of x, where x represents the distance shown in the diagram.

    b. For which of the following values of x is the time $f(x)$ the least: 0, 50, or 100?

    c. Try other values of x to see whether there is a faster time.

Ex. 24

25. A rectangular area of 60 m² has a wall as one of its sides. The sides perpendicular to the wall are made of fencing that costs $2 per meter. The side parallel to the wall is made of decorative fencing that costs $3 per meter.

    a. Express the cost of the fencing as a function of the length x of a side perpendicular to the wall.

    b. Using calculus, one can determine that the cost is least when $x = 3\sqrt{5}$. What is the minimum cost?

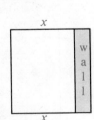

Ex. 25

26. A power station and a factory are on opposite sides of a river 60 m wide, as shown. A power line must be run from the station to the factory. It costs $25 per meter to run the cable in the river and $20 per meter on land.

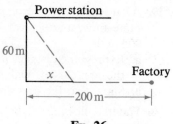

Ex. 26

a. Express the total cost as a function of x, where x represents the distance downstream from the power station to the point where the cable touches land.

b. Using calculus, one can determine that the cost is least when $x = 80$. What is the minimum cost?

27. A baseball diamond is a square with 90 foot sides. A runner (R in the diagram) has taken a 9 foot lead from first base. At the moment the ball is pitched, the runner runs toward second base at 27 ft/s. Express the runner's straight-line distance from home plate as a function of the time t after the ball is thrown. What are the domain and range of this function?

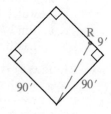

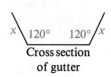

Ex. 27

28. A sheet of metal is 60 cm wide and 10 m long. It is bent along its width to form a gutter with a cross section that is an isosceles trapezoid with 120° angles, as shown in the diagram.

a. Express the volume V of the gutter as a function of x, where x represents the length in centimeters of one of the equal sides. (*Hint:* Volume = (area of base)(height))

b. For what value of x is the volume of the gutter a maximum?

Cross section
of gutter

Ex. 28

29. A window is in the shape of a rectangle with a semicircle on top, as shown in the diagram. The perimeter of the window is 300 cm.

a. Express the area as a function of the width w of the window.

b. For what value of w is the area a maximum?

30. Triangle OAB is an isosceles triangle with vertex O at the origin and vertices A and B on the parabola $y = 9 - x^2$, as shown in the diagram. Express the area of the triangle as a function of: **(a)** the x-coordinate of A; **(b)** the y-coordinate of B.

31. The isosceles trapezoid $ABCD$ is inscribed in a semicircle with radius 4 and center at the origin. Give the area of $ABCD$ as a function of the x-coordinate of C.

Ex. 29

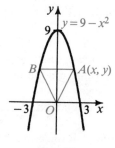

Ex. 30

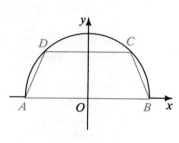

Ex. 31

32. A cylinder is inscribed in a sphere of radius 1. Find the volume of the cylinder as a function of the radius r of its base.

C **33.** A cone circumscribes a sphere with radius 1. Find the volume of the cone as a function of the radius r of the base.

34. A baseball player hits a ball into the farthest corner of the outfield and tries for an inside-the-park home run; that is, the player tries to run the bases and make it to home plate safely. Suppose that the player runs at 30 ft/s and stays strictly on the base lines, as shown in Exercise 27. Express the player's straight-line distance from home plate as a function $f(t)$ where t represents the time after the ball was hit. (You will need different formulas for different intervals of time.) Draw the graph of $f(t)$.

4-5 / INVERSE FUNCTIONS

If you choose a number, double it, and then take half of the result, the answer is the original number. You can also reverse the process by taking half of the number first and then doubling the result; again the answer is the original number. To say the same thing using functions, consider the doubling function

$f(x) = 2x$ and the halving function $g(x) = \dfrac{x}{2}$.

1. Double the number and halve the result:
$$g(f(x)) = \tfrac{1}{2}(2x) = x$$
2. Halve the number and double the result:
$$f(g(x)) = 2(\tfrac{1}{2}x) = x$$

The doubling and halving functions above are examples of *inverse functions*. The composite of two inverse functions is the function that always gives you the number you started with. This function $I(x) = x$ is called the *identity function* because it assigns each number to itself.

Another example of inverse functions is provided by the square root function $F(x) = \sqrt{x}$ and the squaring function $G(x) = x^2$ for $x \geq 0$.

1. Take the square root of x and square the result:
$$G(F(x)) = (\sqrt{x})^2 = x$$
2. Square x and take the square root of the result:
$$F(G(x)) = \sqrt{x^2} = x$$

The equation in statement 2 above is not true when $x < 0$. (For example, $F(G(-3)) = \sqrt{(-3)^2} = \sqrt{9} = 3$. So $F(G(-3)) \neq -3$.) That is why we did not allow the domain of $G(x)$ to include negative numbers. We restricted the domain of $G(x)$ so that condition 2 in the definition below would be satisfied.

In general, the functions $f(x)$ and $g(x)$ are called **inverse functions** if the following statements are true:

1. $g(f(x)) = x$ for all x in the domain of $f(x)$.
2. $f(g(x)) = x$ for all x in the domain of $g(x)$.

We denote the inverse of $f(x)$ by the symbol $f^{-1}(x)$, which is read "f inverse of x." Note that $f^{-1}(x)$ does *not* mean $\dfrac{1}{f(x)}$. The range of $f(x)$ is always the domain of $f^{-1}(x)$, and the range of $f^{-1}(x)$ is always the domain of $f(x)$.

EXAMPLE 1. If $f(x) = \dfrac{x-1}{2}$ and $g(x) = 2x + 1$, show that $f(x)$ and $g(x)$ are inverse functions.

SOLUTION: Since the domain and range of both $f(x)$ and $g(x)$ is the set of all real numbers, we proceed as follows:

1. For any real number x:

$$g(f(x)) = g\left(\frac{x-1}{2}\right) = 2\left(\frac{x-1}{2}\right) + 1 = x$$

2. For any real number x:

$$f(g(x)) = f(2x + 1) = \frac{(2x+1)-1}{2} = x$$

Thus, $f^{-1}(x) = g(x) = 2x + 1$.

The graphs of $f(x)$ and $f^{-1}(x)$ defined in Example 1 are shown at the right. Notice that for every point (a, b) on the graph of $y = f(x)$, there is a point (b, a) on the graph of $y = f^{-1}(x)$. This occurs because whenever $f(a) = b$, then $f^{-1}(b) = a$ by the definition of inverse functions. Notice also that if we fold the coordinate plane along the line $y = x$, then the graphs of $f(x)$ and $f^{-1}(x)$ will coincide. This occurs because the line $y = x$ is the

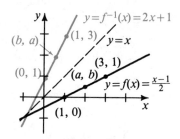

line of symmetry for any two points (a, b) and (b, a); that is, the line is the perpendicular bisector of the line segment joining the two points. In Chapter 8 we shall discuss symmetry in greater detail.

If the graph of $f^{-1}(x)$ can be obtained from the graph of $f(x)$ by changing every point (x, y) on the graph to the point (y, x), then the equation of $f^{-1}(x)$ can be obtained by interchanging the variables x and y in the equation of $f(x)$. The following example shows this process.

EXAMPLE 2. Let $f(x) = 4 - x^2$ for $x \geq 0$.
  a. Sketch the graph of $y = f^{-1}(x)$.
  b. Find $f^{-1}(x)$.

SOLUTION: **a.** We draw the graph of $y = f(x) = 4 - x^2$ and then use the symmetry about $y = x$ to sketch the graph of $y = f^{-1}(x)$.

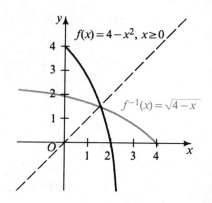

b. 1. Set $y = f(x)$. $y = 4 - x^2$, $x \geq 0$ [describes $y = f(x)$]

2. Switch x and y. $x = 4 - y^2$, $y \geq 0$ [describes $y = f^{-1}(x)$]

3. Solve for y. $y = \pm\sqrt{4 - x}$, $y \geq 0$

$$y = \sqrt{4 - x}$$

$$f^{-1}(x) = \sqrt{4 - x}$$

This gives us the formula $f^{-1}(x) = \sqrt{4 - x}$. The graph shows that the domain of $f^{-1}(x)$ is $\{x \mid x \leq 4\}$. Therefore, the inverse function is defined by $f^{-1}(x) = \sqrt{4 - x}$ for $x \leq 4$.

Not all functions have inverse functions. Functions that do have inverses are called **one-to-one functions.** This name is used because each x-value corresponds to exactly one y-value, and each y-value corresponds to exactly one x-value. One-to-one functions can be determined by the following graphical test:

THE HORIZONTAL-LINE TEST   If the graph of $y = f(x)$ is such that no horizontal line intersects the graph in more than one point, then $f(x)$ is one-to-one and has an inverse function.

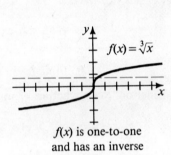

$f(x)$ is one-to-one
and has an inverse

$f(x)$ is not one-to-one and has no inverse
(compare with $f(x)$ in Example 2)

ORAL EXERCISES 4-5

1. Given that $f(2) = 3$, evaluate:
  a. $f^{-1}(3)$ **b.** $f(f^{-1}(3))$ **c.** $f^{-1}(f(2))$

2. Find $g^{-1}(x)$ if:
  a. $g(x) = 4x$ **b.** $g(x) = 3x + 2$ **c.** $g(x) = 2x - 1$

3. The figure shows the graph of $f(x) = x^3$.
  a. Name several points on the graph of $f^{-1}(x)$ and make a sketch of this graph on the chalkboard.
  b. Find $f^{-1}(x)$.

4. Is the function $f(x) = x^2$ a one-to-one function? Explain.

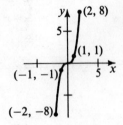

5. The graphs of $f(x)$, $g(x)$, and $h(x)$ are shown. Which functions are one-to-one? Which functions have inverses?

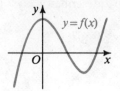
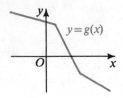
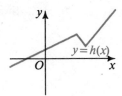

6. Which of the following functions have inverses?
 a. $f(x) = |x|$ **b.** $f(x) = x^3$
 c. $f(x) = x^2$ **d.** $f(x) = x^2$, $x \geq 0$

7. Explain why the Horizontal-Line Test (given on page 141) follows from the Vertical-Line Test (given on page 125).

WRITTEN EXERCISES 4-5

A
1. If $f(2) = 6$ and $f(3) = 7$, evaluate:
 a. $f^{-1}(6)$ **b.** $f^{-1}(f(3))$ **c.** $f(f^{-1}(7))$

2. If $g(3) = 5$ and $g(-1) = 5$, explain why $g(x)$ has no inverse.

3. Let $h(x) = 4x - 3$.
 a. Sketch the graphs of $h(x)$ and $h^{-1}(x)$. **b.** Find $h^{-1}(x)$.

4. Let $L(x) = \frac{1}{2}x - 4$.
 a. Sketch the graphs of $L(x)$ and $L^{-1}(x)$. **b.** Find $L^{-1}(x)$.

In Exercises 5–8, you are given the graph of a function. State whether or not the function has an inverse.

5. 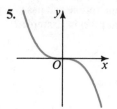 **6.** 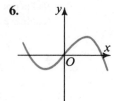 **7.** 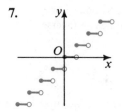 **8.**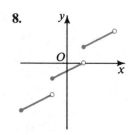

In Exercises 9–18 state whether the function $f(x)$ has an inverse. If $f^{-1}(x)$ exists, give an equation for $f^{-1}(x)$ in terms of x.

9. $f(x) = 3x - 5$ **10.** $f(x) = mx + k$, where $m \neq 0$

11. $f(x) = |x| - 2$ **12.** $f(x) = \sqrt[4]{x}$

13. $f(x) = \dfrac{1}{x}$ **14.** $f(x) = \dfrac{1}{x^2}$

15. $f(x) = \sqrt{5 - x}$ **16.** $f(x) = \sqrt{4 - x^2}$

17. $f(x) = \sqrt{5 - x^2}$ **18.** $f(x) = \sqrt[3]{1 + x^3}$

In Exercises 19–22, sketch the graphs of $g(x)$ and $g^{-1}(x)$ and find the formula for $g^{-1}(x)$.

B **19.** $g(x) = x^2 + 2,\ x \geq 0$ $\qquad\qquad$ **20.** $g(x) = 9 - x^2,\ x \leq 0$

21. $g(x) = (x - 1)^2 + 1,\ x \leq 1$ $\qquad$ **22.** $g(x) = (x - 4)^2 - 1,\ x \geq 4$

In Exercises 23–25, show that $h^{-1}(x) = h(x)$ and sketch the graph of $h(x)$.

23. $h(x) = \sqrt[3]{1 - x^3}$ $\qquad$ **24.** $h(x) = \dfrac{x}{x - 1}$ $\qquad$ **25.** $h(x) = \sqrt{1 - x^2},\ x \geq 0$

26. Using the results of Exercises 23–25, explain how the graph of $h(x)$ is related to the line $y = x$ when $h^{-1}(x) = h(x)$.

C **27. a.** Complete the following table.

$f(x)$	$g(x)$	$f(g(x))$	inverse of $f(g(x))$	$f^{-1}(x)$	$g^{-1}(x)$	$g^{-1}(f^{-1}(x))$
$\dfrac{x}{3}$	$x - 5$	?	?	?	?	?
$\dfrac{1}{x}$	$2x + 1$	?	?	?	?	?
x^3	$3x$	?	?	?	?	?

b. Using the results from the table above, make a conjecture about the inverse of $f(g(x))$. Can you prove your conjecture?

28. If $f(x)$ is a linear function such that $f(x + 2) - f(x) = 6$, find $f^{-1}(x + 2) - f^{-1}(x)$.

29. If $f(x)$ is a linear function such that $f(x + 1) - f(x) > 0$ for all x, what can you say about $f^{-1}(x + 1) - f^{-1}(x)$?

30. Suppose a, b, and c are constants such that $a \neq 0$. Let $P(x) = ax^2 + bx + c$ for $x \leq \dfrac{-b}{2a}$. Find $P^{-1}(x)$.

CALCULATOR EXERCISES

In the next chapter we shall define the functions $f(x) = e^x$ and $g(x) = \ln x$. You can use the calculator to learn something about these functions since you can evaluate them for various numbers in their domains.

1. Enter any number. Then press the buttons e^x and $\ln x$ alternately several times. What do you notice? Repeat this process for several other numbers. How would you describe the relationship between $f(x) = e^x$ and $g(x) = \ln x$?

2. By entering various numbers, determine whether e^x is defined for all real numbers.

3. By experimenting, determine the domain of $\ln x$.

Functions $\quad$ **143**

Chapter summary

1. When one quantity depends on another, the first quantity is said to be a *function* of the second. For example, the area of a circle is a function of the radius. The area of a rectangle is a function of two variables, the length and the width of the rectangle. One method of graphing a function of two variables is to plot curves along which the function is constant. (See examples on page 119.)

2. A function consists of a set of real numbers called the *domain* of the function and a rule that assigns to each element in the domain exactly one real number. The set of real numbers assigned by the rule is called the *range* of the function. If y is a function of x, we write $y = f(x)$. The graph of $f(x)$ consists of all points (x, y) such that x is in the domain of the function and y is the corresponding element in the range. Any root of the equation $f(x) = 0$ is called a *zero* of the function.

3. An equation in x and y will define y as a function of x provided there is exactly one y-value for each x-value. (See page 125 for examples.) A graphical means of determining whether y is a function of x is the Vertical-Line Test given on page 125.

4. The function $f(g(x))$ is called the *composite* of functions $f(x)$ and $g(x)$. The operation that combines these functions is called *composition*. For example, if $f(x) = x^3$ and $g(x) = 3x$, then

$$f(g(x)) = (3x)^3 \quad \text{and} \quad g(f(x)) = 3x^3.$$

(See pages 128 and 129 for other examples.)

5. An important problem of mathematics is finding the *extreme values* of a function. The first step in finding these values is to be able to write a formula for the function from a verbal description. (See Examples 1–3 of Section 4–4.)

6. The function $I(x) = x$ is called the *identity function*. The *inverse* of the function $f(x)$ is denoted $f^{-1}(x)$ and has the following properties:

   1. $f(f^{-1}(x)) = x$ and $f^{-1}(f(x)) = x$.
   2. The domain of $f^{-1}(x)$ is the range of $f(x)$.
   3. The range of $f^{-1}(x)$ is the domain of $f(x)$.

7. To obtain an equation which defines $f^{-1}(x)$, merely interchange the x's and y's in the equation which defines $f(x)$. To obtain the graph of $y = f^{-1}(x)$, reflect the graph of $y = f(x)$ in the line $y = x$.

8. A function $y = f(x)$ is called a *one-to-one function* if each x-value corresponds to exactly one y-value and each y-value corresponds to exactly one x-value. One-to-one functions are precisely those functions that have inverses. The *Horizontal-Line Test* (page 141) provides a method of looking at the graph of $f(x)$ to see whether $f(x)$ has an inverse.

Chapter test

1. Study the graph shown and answer the following questions.

    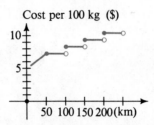
    Cost per 100 kg ($)

    a. What variable is a function of another variable?
    b. What is the cost of sending 400 kg of freight 170 km?

4-1

2. The area of a triangle is a function of the base b and the height h: $A = A(b, h) = \frac{1}{2}bh$. Sketch the graph of the curve of constant area $A = 50$.

3. If $f(x)$ is a linear function of x such that $f(2) = 1$ and $f(-2) = 3$, find an equation for $f(x)$.

4. Give the domain, range, and zeros of the following functions.
    a. $g(x) = x + 5$
    b. $f(x) = (x - 3)^2 + 2$
    c. the function $h(x)$ whose graph is shown

    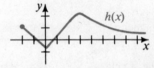
    $h(x)$

4-2

5. a. Sketch the graph of the function $f(x) = \sqrt{9 - x^2}$.
    b. Determine the domain, range, and zeros of $f(x)$.

6. If $f(x) = x^2 + 1$ and $g(x) = 2x - 1$, write the rule for the composite function.
    a. $f(g(x))$
    b. $g(f(x))$

4-3

7. Let $f(x) = x^3$, $g(x) = \sqrt{x}$, $h(x) = x + 2$, and $k(x) = 3x$. Express each of the following as a composite of three of these functions.
    a. $\sqrt{x^3 + 2}$
    b. $(3x + 6)^3$
    c. $3\sqrt{x + 2}$

8. A rectangle with length l and width w has diagonals that are 10 cm long. Express the area A of the rectangle as a function of w.

4-4

9. $P(x, y)$ is an arbitrary point on the circle $x^2 + y^2 = 1$.
    a. Express the distance from P to $(3, 0)$ as a function of the x-coordinate of P.
    b. Give the domain and range of the function.

    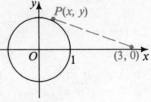

10. Let $f(x) = 2x - 3$.
    a. Sketch the graphs of $f(x)$ and $f^{-1}(x)$ on the same set of axes.
    b. Find the equation for $f^{-1}(x)$.

4-5

11. Let $f(x) = \sqrt{3 + x^2}$ and $g(x) = 3 + x$.
    a. Which of these functions has an inverse?
    b. Explain why the other function does not have an inverse.

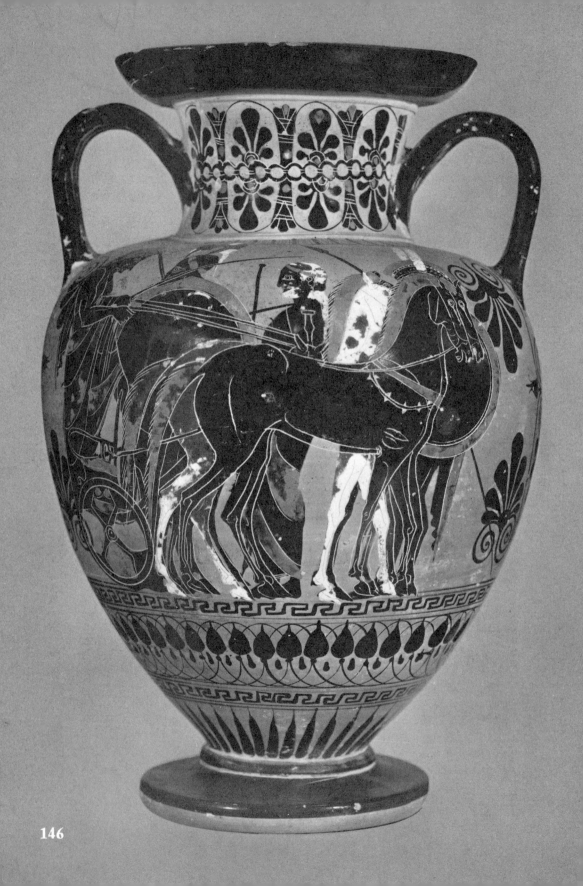

146

CHAPTER FIVE

Exponents and logarithms

OBJECTIVES

1. To use the laws of exponents and logarithms to simplify and evaluate expressions and solve equations.
2. To define and graph exponential functions and their inverse logarithmic functions.
3. To find logarithms of numbers to any base by applying a conversion formula to base 10 common logarithms.
4. To solve problems involving exponential growth and decay, including half-life problems.
5. To define the number e and the natural logarithm.

Exponents

5-1/INTEGRAL EXPONENTS

The earliest use of the symbol b^x was for positive integral values of x. By definition, b^x meant the product of x b's. The Laws of Exponents on the next page can be proved easily when x and y are positive integers.

The approximate ages of objects made from clay, such as the Greek vase shown, can be determined using the method of thermoluminescence. This method is based on the equation for exponential decay, which is given on page 174.

Same Bases

1. $b^x \cdot b^y = b^{x+y}$

2. If $x > y$ and $b \neq 0$, then $\dfrac{b^x}{b^y} = b^{x-y}$; if $x < y$ and $b \neq 0$, then $\dfrac{b^x}{b^y} = \dfrac{1}{b^{y-x}}$.

3. If $b \neq -1, 0,$ or 1, then $b^x = b^y$ if and only if $x = y$.

Same exponents

4. $(ab)^x = a^x b^x$

5. $\left(\dfrac{a}{b}\right)^x = \dfrac{a^x}{b^x} (b \neq 0)$

6. If a and b are positive, and $x \neq 0$, then $a^x = b^x$ if and only if $a = b$.

Power of a power

7. $(b^x)^y = b^{xy}$

As mathematics developed, the symbol b^x was also given meaning when x was zero or a negative integer. Later, b^x was defined when x was a fraction. Later still, x could be an irrational number. Finally, b^x was defined for an imaginary exponent x. At each stage of this development, b^x was deliberately defined so that the original laws for positive exponents would continue to hold in so far as possible for the new exponents.

Definition of b^0

If Law 1 is to hold for $y = 0$, then we must have $b^x \cdot b^0 = b^{x+0} = b^x$. Since b^0 behaves like 1, we define it to be 1:

$$b^0 = 1$$

Definition of b^{-x}

If Law 1 is to hold for $y = -x$ and $b \neq 0$, we must have $b^x \cdot b^{-x} = b^{x-x} = b^0 = 1$. Because $b^x \cdot b^{-x} = 1$, b^{-x} must be the reciprocal of b^x. Therefore we make the following definition for $x > 0$ and $b \neq 0$:

$$b^{-x} = \dfrac{1}{b^x}$$

EXAMPLE 1.   **a.** $3^{-2} = \dfrac{1}{3^2} = \dfrac{1}{9}$ **b.** $\left(\dfrac{2}{5}\right)^{-3} = \left(\dfrac{5}{2}\right)^3 = \dfrac{125}{8}$

The Laws of Exponents above can be shown to be true for negative integral and zero exponents using the laws as they apply to positive exponents. A proof of Law 1 for negative integral exponents follows. Similar proofs of some of the other laws are called for in Exercises 27–30 on page 151.

Proof of Law 1 for negative integral exponents.

Let $x > 0$, $y > 0$, and $b \neq 0$. Then:

$b^{-x} \cdot b^{-y} = \dfrac{1}{b^x} \cdot \dfrac{1}{b^y}$ Definition of negative exponent

$\qquad = \dfrac{1}{b^x \cdot b^y}$ Rule of algebra

$\qquad = \dfrac{1}{b^{x+y}}$ Law 1 for positive exponents

$\qquad = b^{-(x+y)}$ Definition of negative exponent

$\qquad = b^{(-x)+(-y)}$ Rule of algebra

Because negative and zero exponents are now defined, the restrictions in Law 2 are no longer necessary. For example,

$$\frac{4^2}{4^5} = 4^{2-5} = 4^{-3}.$$

Warning. Exponent Laws 4 and 5 apply only to powers of a product or quotient and not to powers of a sum or difference:

$$(a^{-2}b^{-2})^{-1} = a^2b^2$$

but
$$(a^{-2} + b^{-2})^{-1} \neq a^2 + b^2.$$

EXAMPLE 2. Write without negative exponents: $(a^{-2} + b^{-2})^{-1}$ where $a \neq 0$ and $b \neq 0$.

SOLUTION: $(a^{-2} + b^{-2})^{-1} = \left(\dfrac{1}{a^2} + \dfrac{1}{b^2}\right)^{-1}$

$\qquad\qquad\qquad\qquad = \left(\dfrac{b^2 + a^2}{a^2 b^2}\right)^{-1}$

$\qquad\qquad\qquad\qquad = \dfrac{a^2 b^2}{b^2 + a^2}$

A simplified expression should contain neither negative exponents nor powers of powers.

EXAMPLE 3. Simplify, where $x \neq 0$. **a.** $\dfrac{x^5 + x^{-2}}{x^{-3}}$ **b.** $\dfrac{x^5 \cdot x^{-2}}{x^{-3}}$

SOLUTION: **a.** $\dfrac{x^5 + x^{-2}}{x^{-3}} = \dfrac{x^5}{x^{-3}} + \dfrac{x^{-2}}{x^{-3}}$ **b.** $\dfrac{x^5 \cdot x^{-2}}{x^{-3}} = \dfrac{x^3}{x^{-3}}$

$\qquad\qquad\qquad\quad = x^8 + x \qquad\qquad\qquad\qquad\qquad = x^6$

We shall assume throughout the rest of this book without always stating it that variables are restricted so that there are no denominators of zero. In this section only, we assume that variables appearing as exponents represent integers.

ORAL EXERCISES 5-1

Simplify the expression.

1. a. 8^{-1} **b.** 8^{-2} **c.** $\left(\frac{2}{3}\right)^{-1}$ **d.** $\left(\frac{2}{3}\right)^{-2}$

2. a. 3^{-2} **b.** 2^{-3} **c.** $\left(\frac{2}{5}\right)^{-1}$ **d.** $\left(\frac{2}{5}\right)^{-3}$

3. a. $4 \cdot 3^{-2}$ **b.** $(4 \cdot 3)^{-2}$ **c.** $2 \cdot 5^0$ **d.** $(2 \cdot 5)^0$

4. a. $x^3 \cdot x^5$ **b.** $(x^3)^{-5}$ **c.** $(x^{-1})^{-2}$ **d.** $((a^{-1})^{-2})^{-2}$

5. a. $(3r^{-2})^{-1}$ **b.** $(5a^{-2})^2$ **c.** $((2b)^{-3})^{-1}$ **d.** $((-2x)^{-2})^{-2}$

6. a. $\dfrac{x^4}{x^7}$ **b.** $\dfrac{x^4}{x^{-7}}$ **c.** $\dfrac{x^2 \cdot x^{-3}}{2x^{-5}}$ **d.** $\dfrac{4r^{-3}}{(2r)^2}$

7. a. $\dfrac{12^3}{6^3}$ **b.** $\dfrac{8^n \cdot 3^n}{4^n}$ **c.** $\dfrac{(2n)^2}{2n^2}$ **d.** $\dfrac{(x^r)^2}{(x^2)^r}$

8. a. $(2^{-1} + 4^{-1})^{-1}$ **b.** $(2^{-1} \cdot 4^{-1})^{-1}$ **c.** $(3^{-1} - 3^{-2})^{-1}$ **d.** $(3^{-1} \cdot 3^{-2})^{-1}$

9. a. $x^2(x^3 + x^2)$ **b.** $2x^{-1}(x^2 + x)$ **c.** $x^{-2}(x^3 + 2x^2)$ **d.** $3x^{-3}(x^3 - 3x)$

10. a. $\dfrac{n^2 + n^{-2}}{n^{-2}}$ **b.** $\dfrac{n^2 \cdot n^{-2}}{n^{-2}}$ **c.** $\dfrac{3a^3 - a^6}{a^{-1}}$ **d.** $\dfrac{3a^3 \cdot a^6}{a^{-1}}$

WRITTEN EXERCISES 5-1

Simplify the expression.

A

1. a. $(-4)^{-2}$ **b.** -4^{-2} **c.** $(3^{-2})^{-2}$

2. a. $(-3)^{-4}$ **b.** -3^{-4} **c.** $(5^{-1})^{-2}$

3. a. $2 \cdot 8^{-2}$ **b.** $(2 \cdot 8)^{-2}$ **c.** $\left(\frac{6}{5}\right)^{-1}\left(\frac{10}{3}\right)^{-1}$

4. a. $5 \cdot 2^{-3}$ **b.** $(5 \cdot 2)^{-3}$ **c.** $6^{-2}\left(\frac{3}{2}\right)^{-2}$

5. a. $\dfrac{3^7 \cdot 3^{-4}}{3^4}$ **b.** $\dfrac{2^3 \cdot 2^{-5}}{2^2}$ **c.** $\dfrac{2^4 \cdot 3^4}{12^4}$

6. a. $\dfrac{n^3 \cdot n^{-2}}{n^5 \cdot n^{-1}}$ **b.** $\dfrac{2^{-9}}{2^{-7} \cdot 2^5}$ **c.** $\dfrac{8^{-2} \cdot 4^{-2}}{16^{-2}}$

7. a. $(m^{-1}n^{-2})^3$ **b.** $(-4a^{-5})^{-3}$ **c.** $(2x^{-1})^2 \cdot 2x^0$

8. a. $(a^{-2}b^3)^{-2}$ **b.** $(-3n^{-4})^3$ **c.** $(4r)^{-6} \cdot (4r)^6$

9. a. $(3a^{-2})^3 \cdot 3a^5$ **b.** $(-4x^3)^2 \cdot 3x^{-2}$ **c.** $(3n^2)^{-1} \cdot (3n^2)^7$

10. a. $(2r^{-1})^4(4r^2)^{-2}$ **b.** $(5na^{-2})^3 \cdot (n^3a)^{-2}$ **c.** $(3a^2b)^{-3}(3a^2b)^3$

11. a. $\dfrac{(2a^{-1})^2}{(2a^{-1})^{-2}}$ **b.** $\dfrac{(ab^2)^3}{(a^2b)^{-1}}$ **c.** $\dfrac{(-3n^{-3})^2}{-9n^{-4}}$

12. a. $\left(\dfrac{a}{b^2}\right)^{-2} \cdot \left(\dfrac{a}{b^2}\right)^{-3}$ **b.** $\dfrac{(-2r)^4}{(-2r)^{-2}}$ **c.** $\dfrac{5c^2d^3}{(5cd^{-2})^2}$

13. a. $2x^{-3}(x^5 - 2x^3)$ **b.** $\dfrac{3x^3 + 6x^2}{3x^{-2}}$ **c.** $\dfrac{8n^4 - 4n^{-2}}{2n^{-2}}$

14. a. $\dfrac{6a^{-2} + 9a^2}{3a^{-2}}$ **b.** $xy^{-2}(xy^2 - 3y^3)$ **c.** $2n^{-2}(n^4 - 2n^2)$

150   Chapter 5

15. a. $(2^{-2} + 2^{-3})^{-1}$ **b.** $(2^{-2} \cdot 2^{-3})^{-1}$ **c.** $(a^{-1} - b^{-1})^{-1}$

16. a. $(4^{-1} - 2^{-1})^2$ **b.** $(4^{-1} \div 2^{-1})^2$ **c.** $(a - a^{-1})^2$

Simplify by expressing all factors as powers of the same base.

17. a. $\dfrac{3^5 \cdot 9^4}{27^4}$ **b.** $\dfrac{125^{-3} \cdot 25}{5^{-8}}$ **c.** $\sqrt{\dfrac{8^n \cdot 2^7}{4^{-n}}}$

18. a. $\dfrac{4^9 \cdot 8^{-4}}{16^3}$ **b.** $\dfrac{3^7 \cdot 9^5}{\sqrt{27^{12}}}$ **c.** $\sqrt[3]{\dfrac{125^n \cdot 5^{4n}}{25^{-n}}}$

Write as a power of b.

19. a. $\dfrac{(b^n)^3}{b^n \cdot b^n}$ **b.** $\dfrac{(b^n)^2}{b^n \cdot b^{n+2}}$ **c.** $\dfrac{b \cdot b^n}{(b^3)^n}$

20. a. $\sqrt{\dfrac{b^{2n}}{b^{-2n}}}$ **b.** $\dfrac{(b \cdot b^n)^2}{(b^2)^n}$ **c.** $\sqrt{\dfrac{b^{1-n}}{b^{n-1}}}$

Simplify.

B   **21. a.** $\dfrac{2^{-1}}{2^{-2} + 2^{-3}}$ $\left(\textit{Hint:}\ \text{Multiply by}\ \dfrac{2^3}{2^3}.\right)$ **b.** $\dfrac{4^{-5}}{4^{-2} + 4^{-3}}$

22. a. $\dfrac{3^{-2}}{3^3 + 3^{-2}}$ **b.** $\dfrac{2^{-1} - 2^{-2}}{2^{-1} + 2^{-2}}$

23. a. $\dfrac{x^{-2} - y^{-2}}{x^{-1} - y^{-1}}$ **b.** $\dfrac{1 - y^{-1}}{y - y^{-1}}$

24. a. $\dfrac{x^{-1}}{x - x^{-1}}$ **b.** $\dfrac{4 - x^{-4}}{2 - x^{-2}}$

25. Prove Exponent Law 2 for positive integers using Law 1.

26. Prove Exponent Law 5 for positive integers using Law 4.

Use the laws of exponents for positive integers and the definitions of b^0 and b^{-x} to prove the following laws for negative exponents.

27. Law 4: $a^{-x} \cdot b^{-x} = (ab)^{-x},\ x > 0$

28. Law 7: $(b^{-x})^{-y} = b^{xy},\ x > 0,\ y > 0$

29. Law 3: $b^{-x} = b^{-y}$ if and only if $-x = -y,\ x > 0,\ y > 0$

30. Law 6: $a^{-x} = b^{-x}$ if and only if $a = b,\ x > 0$

Solve for x by rewriting the equation in quadratic form (see Chapter 2, page 52).

31. $2^{2x} - 3 \cdot 2^x + 2 = 0$ (Hint: $2^{2x} = (2^x)^2$.)

32. $3^{2x} - 10 \cdot 3^x + 9 = 0$ **33.** $4^{2x} - 3 \cdot 4^x - 4 = 0$

C   **34.** $2^x + 8 \cdot 2^{-x} = 9$ **35.** $2^x + 2^{-x} = \frac{5}{2}$

36. $2^{2x} - 3 \cdot 2^{x+1} + 8 = 0$ **37.** $3^{2x+1} - 10 \cdot 3^x + 3 = 0$

38. Simplify $\dfrac{a^{3x+7}b^2 - 2a^{3x+8}b + a^{3x+9}}{a^{3x+7} - b^2 a^{3x+5}}$.

In the previous section negative exponents were defined in such a way that the original laws for positive exponents continued to hold. In order to define rational exponents so that the original laws still hold, we must require that the base b be greater than 0.

Definition of $b^{\frac{p}{q}}$. (where p and q are positive integers, $b > 0$, and $b \neq 1$.)

First we consider the special cases $b^{\frac{1}{3}}$ and $b^{\frac{2}{3}}$.

Definition of $b^{\frac{1}{3}}$. If Law 7 is to hold for fractional exponents, we must have:

$$(b^{\frac{1}{3}})^3 = b$$

But if the cube of a number is b, then the number must be the cube root of b. Hence we define

$$b^{\frac{1}{3}} = \sqrt[3]{b}.$$

Definition of $b^{\frac{2}{3}}$. Again, if Law 7 is to hold for fractional exponents, then:

$$b^{\frac{2}{3}} = b^{(\frac{1}{3} \cdot 2)} = (b^{\frac{1}{3}})^2 = \left(\sqrt[3]{b}\right)^2$$

and

$$b^{\frac{2}{3}} = b^{(2 \cdot \frac{1}{3})} = (b^2)^{\frac{1}{3}} = \sqrt[3]{b^2}$$

The expressions $\left(\sqrt[3]{b}\right)^2$ and $\sqrt[3]{b^2}$ are equivalent according to Law 3 since each has the same third power. Hence we define $b^{\frac{2}{3}}$ to be either of the equivalent expressions

$$\left(\sqrt[3]{b}\right)^2 \quad \text{or} \quad \sqrt[3]{b^2}.$$

Definition of $b^{\frac{p}{q}}$. Because of reasoning similar to that above, we choose the following definitions.

$$b^{\frac{1}{q}} = \sqrt[q]{b} \quad \text{and} \quad b^{\frac{p}{q}} = \left(\sqrt[q]{b}\right)^p, \quad \text{or} \quad b^{\frac{p}{q}} = \sqrt[q]{b^p}$$

EXAMPLE 1. **a.** $16^{\frac{1}{4}} = \sqrt[4]{16} = 2$ **b.** $16^{-\frac{1}{4}} = (16^{\frac{1}{4}})^{-1} = 2^{-1} = \dfrac{1}{2}$

c. $8^{\frac{2}{3}} = \left(\sqrt[3]{8}\right)^2 = 4$ **d.** $8^{-\frac{2}{3}} = (8^{\frac{2}{3}})^{-1} = 4^{-1} = \dfrac{1}{4}$

All of the laws for integral exponents on page 148 continue to hold for rational exponents. Proofs of some of the laws for rational exponents are required in Exercises 45–49. Remember that when b is a negative number, $b^{\frac{p}{q}}$ is not defined because it is not possible to define such expressions consistently.

Laws 4 and 5 state that an exponent distributes over a product or a quotient. An exponent does not distribute over a sum or a difference, however. Part (a) of the next example shows the correct use of Law 4; part (b) shows that we must multiply out a power of a sum.

EXAMPLE 2. Simplify: **a.** $(a^{\frac{1}{2}} \cdot b^{\frac{1}{2}})^2$ **b.** $(a^{\frac{1}{2}} + b^{\frac{1}{2}})^2$

SOLUTION: **a.** $(a^{\frac{1}{2}} \cdot b^{\frac{1}{2}})^2 = (a^{\frac{1}{2}})^2 \cdot (b^{\frac{1}{2}})^2$ **b.** $(a^{\frac{1}{2}} + b^{\frac{1}{2}})^2 = (a^{\frac{1}{2}})^2 + 2a^{\frac{1}{2}}b^{\frac{1}{2}} + (b^{\frac{1}{2}})^2$

$\qquad\qquad\qquad = ab$ $\qquad\qquad\qquad\qquad\qquad = a + 2(ab)^{\frac{1}{2}} + b$

$\qquad\qquad\qquad\qquad\qquad\qquad\qquad\qquad\qquad\qquad\quad = a + 2\sqrt{ab} + b$

Examples 3 and 4 below show how the laws of exponents can be used to solve two kinds of equations containing exponents.

EXAMPLE 3.   (Variable in the exponent) Solve.

$\qquad\qquad$ **a.** $2^x = \frac{1}{8}$ $\qquad$ **b.** $9^{x+1} = \sqrt{27}$

SOLUTION: Express both sides of each equation as powers of the same base. Then apply Law 3.

$\qquad$ **a.** $2^x = \frac{1}{8}$ $\qquad\qquad$ **b.** $\quad 9^{x+1} = \sqrt{27}$

$\qquad\qquad 2^x = 2^{-3}$ $\qquad\qquad\qquad (3^2)^{x+1} = \sqrt{3^3}$

$\qquad\qquad\quad x = -3$ $\qquad\qquad\qquad\quad 3^{2x+2} = 3^{\frac{3}{2}}$

$\qquad\qquad\qquad\qquad\qquad\qquad\qquad 2x + 2 = \frac{3}{2}$

$\qquad\qquad\qquad\qquad\qquad\qquad\qquad\qquad\quad x = -\frac{1}{4}$

EXAMPLE 4.   (Variable in the base) Solve.

$\qquad\qquad$ **a.** $4x^{\frac{3}{2}} = 32$ $\qquad$ **b.** $(x - 1)^{-\frac{1}{4}} - 2 = 0$

SOLUTION: **a.** $4x^{\frac{3}{2}} = 32$ $\qquad\qquad\qquad\qquad$ **b.** $(x - 1)^{-\frac{1}{4}} - 2 = 0$

$\qquad\qquad\quad x^{\frac{3}{2}} = 8$ $\qquad\qquad\qquad\qquad\qquad\qquad (x - 1)^{-\frac{1}{4}} = 2$

Raise both sides of the $\qquad\qquad\qquad$ Raise both sides of the
equation to the $\frac{2}{3}$ power. $\qquad\qquad\quad$ equation to the -4 power.

$\qquad\qquad (x^{\frac{3}{2}})^{\frac{2}{3}} = 8^{\frac{2}{3}}$ $\qquad\qquad\qquad\qquad ((x - 1)^{-\frac{1}{4}})^{-4} = 2^{-4}$

$\qquad\qquad\qquad x = 4$ $\qquad\qquad\qquad\qquad\qquad\qquad\quad x - 1 = \frac{1}{16}$

$\qquad\qquad\qquad\qquad\qquad\qquad\qquad\qquad\qquad\qquad x = \frac{17}{16}$

Assume that variables represent real numbers throughout the rest of this section.

ORAL EXERCISES 5-2

Simplify each expression.

1. a. $4^{\frac{1}{2}}$ $\qquad\qquad$ **b.** $4^{-\frac{1}{2}}$ $\qquad\qquad$ **c.** $4^{\frac{3}{2}}$ $\qquad\qquad$ **d.** $4^{-\frac{3}{2}}$

2. a. $-9^{\frac{1}{2}}$ $\qquad\qquad$ **b.** $-9^{-\frac{1}{2}}$ $\qquad\qquad$ **c.** $8^{\frac{2}{3}}$ $\qquad\qquad$ **d.** $8^{-\frac{2}{3}}$

3. a. $100^{\frac{3}{2}}$ $\qquad\qquad$ **b.** $16^{-\frac{1}{2}}$ $\qquad\qquad$ **c.** $27^{-\frac{1}{3}}$ $\qquad\qquad$ **d.** $27^{-\frac{2}{3}}$

4. a. $(\frac{9}{25})^{-\frac{1}{2}}$ $\qquad$ **b.** $(\frac{4}{9})^{\frac{3}{2}}$ $\qquad\qquad$ **c.** $(36^{-1})^{-\frac{1}{2}}$ $\qquad$ **d.** $((\frac{1}{8})^{-2})^{\frac{1}{3}}$

5. a. $(16x^{16})^{\frac{1}{2}}$ $\qquad$ **b.** $(9n^{-4})^{\frac{1}{2}}$ $\qquad$ **c.** $(2x^{-\frac{1}{3}})^3$ $\qquad$ **d.** $(x^{\frac{1}{2}}y^{-\frac{3}{2}})^4$

6. a. $\left(\dfrac{125}{x^6}\right)^{\frac{1}{3}}$ $\qquad$ **b.** $\left(\dfrac{a^4}{a^{-6}}\right)^{\frac{1}{5}}$ $\qquad$ **c.** $(n^8 \cdot n^4)^{-\frac{1}{3}}$ $\qquad$ **d.** $\left(\dfrac{b^{\frac{1}{3}}}{b^{-\frac{1}{3}}}\right)^{\frac{1}{2}}$

7. a. $8^{\frac{2}{3}} \cdot 2^{\frac{2}{3}}$ $\qquad\quad$ **b.** $3^{\frac{1}{2}} \cdot 27^{\frac{1}{2}}$ $\qquad\quad$ **c.** $x^{\frac{1}{2}} \cdot x^{-\frac{1}{2}}$ $\qquad$ **d.** $2x^{\frac{3}{2}} \cdot 4x^{-\frac{1}{2}}$

8. a. $\dfrac{40^{\frac{3}{2}}}{10^{\frac{3}{2}}}$ **b.** $\dfrac{2^{\frac{1}{2}} \cdot 3^{\frac{1}{2}}}{24^{\frac{1}{2}}}$ **c.** $\dfrac{x^{\frac{1}{3}}}{2x^{-\frac{2}{3}}}$ **d.** $\dfrac{a^{\frac{3}{2}}}{(4a)^{\frac{3}{2}}}$

9. a. $x^{\frac{1}{2}}(2x^{\frac{1}{2}} - x^{-\frac{1}{2}})$ **b.** $\dfrac{a^{-\frac{1}{2}} + a^{\frac{1}{2}}}{a^{-\frac{1}{2}}}$ **c.** $\dfrac{x^{-\frac{2}{3}} + 2x^{\frac{1}{3}}}{x^{-\frac{5}{3}}}$

10. a. $x^n(x^n + x^{-n})$ **b.** $(9^{\frac{1}{2}} + 4^{\frac{1}{2}})^2$ **c.** $(64^{\frac{1}{3}} - 8^{\frac{1}{3}})^3$

Solve.

11. a. $(3^x)^2 = 3^{12}$ **b.** $9^x = 3^5$ **c.** $(5x)^2 = 125$

12. a. $x^{\frac{2}{3}} = 9$ **b.** $x^{-\frac{1}{2}} = 4$ **c.** $(2x)^{-\frac{1}{2}} = 4$

WRITTEN EXERCISES 5-2

Write the expression using positive rational exponents. Do not use radical signs.

A **1. a.** $\sqrt{x^5}$ **b.** $\sqrt[3]{y^2}$ **c.** $\left(\sqrt[6]{x}\right)^5$ **d.** $\sqrt{(2a)^3}$

 2. a. $\sqrt[3]{8x^7}$ **b.** $\left(\sqrt[4]{16x}\right)^3$ **c.** $\sqrt[3]{27x^{-6}y^2}$ **d.** $\left(\sqrt[3]{8n^8}\right)^2$

 3. a. $\sqrt[3]{125a^{-2}b}$ **b.** $\sqrt{36cd^3}$ **c.** $\sqrt[6]{8a^8b^{-4}}$ **d.** $\left(\sqrt[4]{81n}\right)^2$

 4. a. $\sqrt{\dfrac{2b^{-1}}{a^3}}$ **b.** $\sqrt[3]{\dfrac{n^2}{8n}}$ **c.** $\left(\sqrt[3]{a^2}\right)^5$ **d.** $\sqrt[5]{\dfrac{32x^{-1}}{y^2}}$

Write the expressions using radical signs and no negative exponents.

 5. a. $5^{\frac{1}{2}}x^{-\frac{1}{2}}$ **b.** $6^{\frac{1}{3}}x^{\frac{2}{3}}$ **c.** $x^{\frac{2}{5}}y^{\frac{3}{5}}$ **d.** $a^{\frac{4}{7}}b^{-\frac{4}{7}}$

 6. a. $3^{\frac{4}{3}}a^{-\frac{1}{3}}b^{\frac{2}{3}}$ **b.** $x^{\frac{1}{5}}y^{-\frac{2}{5}}z^{\frac{3}{5}}$ **c.** $(a^{\frac{1}{2}}b^{-\frac{1}{2}})^3$ **d.** $a^{\frac{1}{10}}b^{-\frac{1}{5}}$

Simplify.

 7. a. $\left(\dfrac{9}{25}\right)^{-\frac{1}{2}}$ **b.** $\left(\dfrac{9}{25}\right)^{-\frac{3}{2}}$ **c.** $\left(\dfrac{8}{b^3}\right)^{-\frac{1}{3}}$ **d.** $\left(\dfrac{a^6}{27}\right)^{-\frac{2}{3}}$

 8. a. $\left(\dfrac{27}{8}\right)^{-\frac{1}{3}}$ **b.** $\left(\dfrac{4}{9}\right)^{-\frac{3}{2}}$ **c.** $\left(\dfrac{125}{27}\right)^{\frac{2}{3}}$ **d.** $(32)^{-\frac{1}{5}}$

 9. a. $(16^{-\frac{3}{8}})^{\frac{5}{4}}$ **b.** $(25^{-\frac{1}{3}})^{-\frac{3}{2}}$ **c.** $(a^{-2}b)^{-\frac{1}{2}} \cdot ab^{\frac{1}{2}}$ **d.** $(4x^{-3})^{-\frac{1}{2}} \cdot 4x^{\frac{1}{2}}$

 10. a. $(1000a^{-6})^{-\frac{2}{3}}$ **b.** $((a^{\frac{2}{3}})^{-\frac{5}{4}})^{-6}$ **c.** $(3a^3)^{-\frac{1}{2}} \cdot (3a)^{\frac{1}{2}}$ **d.** $(25n^{-5})^{-\frac{3}{2}} \cdot \sqrt{n}$

 11. a. $(81^{\frac{1}{2}} - 9^{\frac{1}{2}})^2$ **b.** $(81^{\frac{1}{2}} \cdot 9^{\frac{1}{2}})^2$ **c.** $(3^{-2} + 4^{-2})^{-\frac{1}{2}}$ **d.** $(3^{-2} \cdot 4^{-2})^{-\frac{1}{2}}$

 12. a. $(6^{-2} + 8^{-2})^{\frac{1}{2}}$ **b.** $(6^{-2} \cdot 8^{-2})^{\frac{1}{2}}$ **c.** $(2^{\frac{1}{2}} - 2^{-\frac{1}{2}})^{-1}$ **d.** $(2^{\frac{1}{2}} \cdot 2^{-\frac{1}{2}})^{-1}$

 13. a. $\dfrac{(\frac{3}{2})^8}{(\frac{3}{2})^2}$ **b.** $\dfrac{8^{\frac{2}{3}}}{2^{\frac{2}{3}}}$ **c.** $\dfrac{9^{\frac{1}{6}} \cdot 9^{\frac{1}{4}}}{9^{-\frac{1}{12}}}$

 14. a. $\dfrac{1.5^4}{4.5^4}$ **b.** $\dfrac{4^{1.5}}{4^{4.5}}$ **c.** $\dfrac{8^{\frac{1}{2}} \cdot 8^{\frac{1}{4}}}{8^{\frac{1}{12}}}$

 15. a. $\sqrt{x}\sqrt[3]{x}\sqrt[6]{x}$ **b.** $\left(\dfrac{x^{\frac{1}{2}}}{x^{\frac{5}{2}}}\right)^2$ **c.** $(x^{\frac{1}{2}} + x^{\frac{5}{2}})^2$

 16. a. $\sqrt[4]{x}\sqrt[3]{x} \div \sqrt[6]{x}$ **b.** $\left(\dfrac{2x^{\frac{1}{2}}}{x^{-\frac{1}{2}}}\right)^2$ **c.** $(2x^{\frac{1}{2}} - x^{-\frac{1}{2}})^2$

154   Chapter 5

17. a. $a^{\frac{1}{2}}(a^{\frac{3}{2}} - 2a^{\frac{1}{2}})$ **b.** $2n^{\frac{1}{3}}(n^{\frac{2}{3}} + n^{-\frac{1}{3}})$ **c.** $x^{-\frac{1}{2}}(x^{\frac{5}{2}} - 2x^{\frac{3}{2}} + x^{\frac{1}{2}})$

18. a. $x^{\frac{1}{2}}(4x^{-\frac{1}{2}} + x^{\frac{3}{2}})$ **b.** $y^{-\frac{3}{2}}(y^{\frac{3}{2}} + 2y^{\frac{1}{2}})$ **c.** $2n^{-\frac{2}{3}}(n^{\frac{8}{3}} - 3n^{\frac{5}{3}} - n^{\frac{2}{3}})$

19. a. $\dfrac{x^{\frac{1}{2}} - 2x^{-\frac{1}{2}}}{x^{-\frac{1}{2}}}$ **b.** $\dfrac{y^{-\frac{1}{3}} - 3y^{\frac{2}{3}}}{y^{-\frac{4}{3}}}$ **c.** $\dfrac{4ab^{-\frac{1}{2}} - 2ab^{\frac{1}{2}}}{(a^2b)^{-\frac{1}{2}}}$

20. a. $\dfrac{2n^{\frac{1}{3}} - 4n^{-\frac{2}{3}}}{2n^{-\frac{2}{3}}}$ **b.** $\dfrac{x^{-\frac{1}{2}}(2x^{\frac{1}{2}} - x^{-\frac{1}{2}})}{x^{-1}}$ **c.** $\dfrac{2n^{\frac{1}{3}}(3n^{\frac{1}{3}} - 4n^{\frac{4}{3}})}{2n^{-\frac{1}{3}}}$

Solve.

21. a. $8^x = 2^6$ **b.** $9^{4x} = 81$ **c.** $8^{x-1} = 2^{x+1}$

22. a. $9^x = 3^{10}$ **b.** $8^x = 2^7 \cdot 4^9$ **c.** $27^{1-x} = (\frac{1}{9})^{2-x}$

23. a. $8^x = \sqrt[3]{16}$ **b.** $(\sqrt{2})^x = \sqrt[3]{2}$ **c.** $\sqrt[4]{9^x} = 27$

24. a. $\left(\dfrac{1}{3}\right)^x = \sqrt{27^8}$ **b.** $\sqrt{125^x} = \dfrac{5}{25^x}$ **c.** $\dfrac{(7^3)^x}{49} = 1$

25. a. $(8x)^{-3} = 64$ **b.** $8x^{-3} = 64$ **c.** $(8 + x)^{-3} = 64$

26. a. $(2x)^{-2} = 16$ **b.** $2x^{-2} = 16$ **c.** $4(x - 2)^{-2} = 16$

27. a. $5x^{\frac{1}{2}} = 20$ **b.** $(5x)^{-\frac{1}{2}} = 20$ **c.** $(5 - x)^{-\frac{1}{2}} = 20$

28. a. $8(x - 2)^3 = 27$ **b.** $[8(x - 2)]^3 = 27$ **c.** $2(1 - 4x)^{\frac{3}{2}} = 16$

Factor.

B **29. a.** $a^{\frac{3}{2}}b^{\frac{1}{2}} - a^{\frac{1}{2}}b^{\frac{3}{2}}$ (*Hint:* Factor out $a^{\frac{1}{2}}b^{\frac{1}{2}}$.) **b.** $a^{\frac{1}{2}}b^{-\frac{1}{2}} - a^{\frac{3}{2}}b^{\frac{1}{2}}$

30. a. $(x - 1)^{\frac{1}{2}} - x(x - 1)^{-\frac{1}{2}}$ **b.** $(x + 1)^{\frac{3}{2}} - 4(x + 1)^{\frac{1}{2}}$

31. a. $(x^2 + 1)^{\frac{3}{2}} - x^2(x^2 + 1)^{\frac{1}{2}}$ **b.** $(x^2 + 2)^{\frac{1}{2}} - x^2(x^2 + 2)^{-\frac{1}{2}}$

32. a. $(2x + 1)^{\frac{2}{3}} - 4(2x + 1)^{-\frac{1}{3}}$ **b.** $(1 + x^2)^{-\frac{3}{2}} - (1 + x^2)^{-\frac{1}{2}}$

33. a. $(2x - 1)^{\frac{1}{3}} - 2x(2x - 1)^{-\frac{2}{3}}$ **b.** $(x + 1)^{\frac{1}{2}}(x + 2)^{-\frac{1}{2}} - (x + 1)^{-\frac{1}{2}}(x + 2)^{\frac{1}{2}}$

Solve for x.

34. $\dfrac{2^{(x^2)}}{2^x} = 64$ **35.** $\dfrac{5^{(x^2)}}{(5^x)^2} = 125$

36. $\sqrt{\dfrac{9^{x+3}}{27^x}} = 81$ **37.** $\sqrt[3]{\dfrac{8^{x+1}}{16^x}} = 32$

Solve for x by rewriting the equation in quadratic form. (See page 52.)

38. $x^{\frac{2}{3}} - 7x^{\frac{1}{3}} + 12 = 0$ (*Hint:* $x^{\frac{2}{3}} - 7x^{\frac{1}{3}} + 12 = (x^{\frac{1}{3}})^2 - 7(x^{\frac{1}{3}}) + 12)$

39. $x^{\frac{2}{3}} - 4x^{\frac{1}{3}} - 5 = 0$ **40.** $x^{\frac{4}{3}} - 6x^{\frac{2}{3}} + 8 = 0$

41. $x - 5\sqrt{x} + 6 = 0$ **42.** $x^{\frac{1}{2}} + 2x^{-\frac{1}{2}} = 3$

43. $9^{2x} - 2 \cdot 9^x - 3 = 0$ **44.** $4^{2x} - 10 \cdot 4^x + 16 = 0$

45. Assume the Laws of Exponents on page 148 hold for all integral exponents. Use them and the definition of $b^{\frac{p}{q}}$ to prove the following lemmas.

Lemma 1: $(b^{\frac{p}{q}})^q = b^p$

Lemma 2: $(b^{\frac{p}{q}})^{qm} = b^{pm} = (b^{qm})^{\frac{p}{q}}$

Assume that the Laws of Exponents on page 148 hold for all integral exponents. Use these laws, the two lemmas proven in Exercise 45 and the definition of $b^{\frac{p}{q}}$ to prove the following laws for rational exponents.

C **46.** Prove Law 6: If a and b are positive, $a^{\frac{p}{q}} = b^{\frac{p}{q}}$ if and only if $a = b$. (p and q are integers, $p \neq 0$.) (*Hint*: $a^{\frac{p}{q}} = (a^{\frac{1}{q}})^p$. Now use Law 6 for integral exponents.)

47. Prove Law 1: $b^{\frac{p}{q}} \cdot b^{\frac{r}{s}} = b^{\frac{p}{q}+\frac{r}{s}}$ (*Hint*: Raise each side to the qs power.)

48. Prove Law 3: If $b > 0$ and $b \neq 1$, then $b^{\frac{p}{q}} = b^{\frac{r}{s}}$ if and only if $\dfrac{p}{q} = \dfrac{r}{s}$.

49. Prove Law 4: $(ab)^{\frac{p}{q}} = a^{\frac{p}{q}} \cdot b^{\frac{p}{q}}$.

50. Prove Law 7: $(b^{\frac{p}{q}})^{\frac{r}{s}} = b^{\frac{pr}{qs}}$.

CALCULATOR EXERCISES

If you play the A below middle C on a piano (see diagram), the piano string vibrates with a frequency of 220 vibrations per second, or 220 hertz (Hz). The frequencies for the notes above A are as follows:

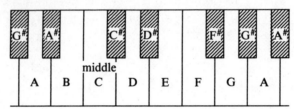

Note	$A^\sharp$ (A sharp)	B	Middle C	$C^\sharp$
Frequency (Hz)	$220 \cdot 2^{\frac{1}{12}}$ ≈ 233.1	$220 \cdot 2^{\frac{2}{12}}$ ≈ 246.9	$220 \cdot 2^{\frac{3}{12}}$ ≈ 261.6	$220 \cdot 2^{\frac{4}{12}}$ ≈ 277.2

1. Give the frequencies of the notes D through A above middle C that are shown on the keyboard above.

2. The frequency of the $G^\sharp$ to the left of A below middle C is $220 \cdot 2^{-\frac{1}{12}}$ Hz. Find the frequency to the nearest 0.1 Hz.

5-3 / EXPONENTIAL FUNCTIONS

We have defined b^x for all rational exponents x, but not for irrational exponents. For example, how is $3^{\sqrt{2}}$ defined? The irrational number $\sqrt{2}$ can be approximated by a sequence of rational numbers giving better and better accuracy.

$$\sqrt{2} \approx 1.4,\ 1.41,\ 1.414,\ 1.4142,\ 1.41421, \ldots$$

Thus, $3^{\sqrt{2}}$ can be approximated by a sequence of powers giving better and better accuracy.

$$3^{\sqrt{2}} \approx 3^{1.4},\ 3^{1.41},\ 3^{1.414},\ 3^{1.4142},\ 3^{1.41421}, \ldots$$

Notice that these powers of 3 all have rational exponents. It can be proved that if the sequence of exponents 1.4, 1.41, 1.414, . . . , has $\sqrt{2}$ as the limit, then the sequence of powers $3^{1.4}$, $3^{1.41}$, $3^{1.414}$, . . . also has a limit. This limit is defined to be $3^{\sqrt{2}}$.

In a similar way, b^x can be defined for any irrational exponent x if $b > 0$. In a more advanced course it can be proved that the laws of exponents will continue to hold for all real number exponents, that is, for both rational and irrational exponents.

Any function of the form $f(x) = ab^x$, $a > 0$, $b > 0$, $b \neq 1$, is called an **exponential function with base b.** You will see in later sections that these functions have many important applications. The domain of an exponential function is the set of all real numbers. The range is the set of positive real numbers. These facts are illustrated in the following graphs.

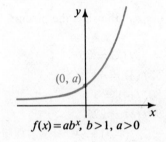

$f(x) = ab^x$, $b > 1$, $a > 0$

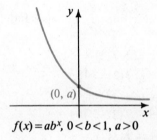

$f(x) = ab^x$, $0 < b < 1$, $a > 0$

EXAMPLE 1. Given that $f(x)$ is an exponential function, $f(0) = 3$, and $f(2) = 12$, evaluate $f(-2)$.

SOLUTION: Since $f(x)$ is an exponential function, $f(x) = ab^x$ for some constants a and b. We use the given values of $f(x)$ to solve for a and b.

1. Since $f(0) = 3$ 2. Since $f(2) = 12$
   $ab^0 = 3$ $3b^2 = 12$
   $a = 3$ $b = 2$

Therefore, $f(x) = 3 \cdot 2^x$ and $f(-2) = 3 \cdot 2^{-2} = \frac{3}{4}$.

If we apply the *Horizontal-Line Test* (page 141) to the graphs of exponential functions shown in this section, we see that each exponential function has an inverse. For example, the graphs of the function $f(x) = 3 \cdot 2^x$ and of $f(x)^{-1}$, the inverse of $f(x)$, are shown at the right. The inverses of exponential functions will be the subject of the next section.

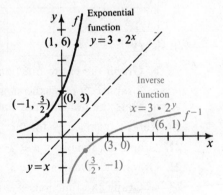

ORAL EXERCISES 5-3

1. If $f(x) = 2^x$, evaluate $f(-3)$ and $f^{-1}(16)$.
2. If $f(x) = 3^x$, evaluate $f(2)$ and $f^{-1}(\frac{1}{3})$.

Exponents and logarithms   **157**

3. If $f(x) = b^x$ and $f(3) = 1000$, find the value of b.

4. If $h(x) = ab^x$, $h(0) = 5$, and $h(1) = 15$, find the value of a and b.

5. The graph of $y = ab^x$ has y-intercept 7. Find the value of a.

6. If $g(x) = b^x$, and $g(2) > g(3)$, what can you say about b?

7. a. State the domain and range of $f(x) = 3 \cdot 4^x$.

    b. State the domain and range of $f^{-1}(x)$.

8. Is 200 or 600 a better approximation of 5^π?

WRITTEN EXERCISES 5-3

Write an equation of the exponential function $f(x) = ab^x$ that has the given values:

A

1. base $b = \frac{5}{3}$, $f(1) = 5$

2. base $b = 3$, $f(2) = 36$

3. $f(0) = \frac{1}{2}$, $f(2) = 18$

4. $f(0) = 5$, $f(3) = 40$

5. $f(0) = 64$, $f(2) = 4$

6. $f(0) = 80$, $f(4) = 5$

7. $f(1) = 8$, $f(2) = 16$

8. $f(1) = 9$, $f(3) = 225$

9. $f(-1) = \frac{3}{4}$, $f(2) = 48$

10. $f(-1) = 2$, $f(-3) = \frac{1}{32}$

11. $f(-1) = 12$, $f(-2) = 3$

12. $f(\frac{1}{2}) = 18$, $f(-\frac{1}{2}) = 2$

13. $f^{-1}(2) = 0$, $f^{-1}(6) = 1$

14. $f^{-1}(4) = 0$, $f^{-1}(100) = 2$

15. If $f(x) = 3 \cdot 4^x$, find $f^{-1}(12)$, $f^{-1}(\frac{3}{4})$, and $f^{-1}(\frac{3}{64})$.

16. If $f(x) = 5 \cdot 2^x$, find $f^{-1}(20)$, $f^{-1}(\frac{5}{4})$ and $f^{-1}(5)$.

17. If $f(x) = 3^x$, evaluate: **a.** $f(x + 1) \div f(x)$

                              b. $f(x + 2) \div f(x - 2)$

18. If $f(x) = 6 \cdot 2^x$, evaluate: **a.** $f(x + 2) \div f(x)$

                                    b. $f(x + 3) \div f(x - 3)$

19. a. Complete the table for the function $f(x) = 2^x$.

x	-2	-1	0	1	2
$f(x)$	?	?	?	?	?

    b. Graph the entire function.

    c. Graph $f^{-1}(x)$ on the same set of axes.

    d. State the domain and range of $f(x)$ and of $f^{-1}(x)$.

20. Repeat Exercise 19 for the function $f(x) = 3^x$.

21. Graph the functions $y = 2^x$ and $y = (\frac{1}{2})^x$ on the same set of axes.

22. Graph the functions $y = 3^x$ and $y = 3^{-x}$ on the same set of axes.

B

23. a. Graph the functions $y = 4^x$ and $y = 4^{|x|}$ on the same set of axes.

    b. How are the graphs related to each other?

24. a. Graph the functions $y = 2^x$ and $y = 2^{x-1}$ on the same set of axes.

    b. How are these graphs related to each other?

158   Chapter 5

25. If $f(x)$ is an exponential function with base b, show that $\dfrac{f(x + c)}{f(x)} = b^c$.

26. If $f(x) = 2^x$, show that $f(x + 2) - f(x) = 3 \cdot 2^x$.

27. If $f(x) = 10^{2x+1}$, show that $f(x + 1) - f(x)$ is divisible by 99 if x is a nonnegative integer.

28. If $f(x) = 5^{2x+1}$, show that $f(x + 1) - 5f(x)$ is divisible by 100 if x is a nonnegative integer.

C **29. a.** Graph the functions of $y = x^2$ and $y = 2^x$ on the same set of axes.
   b. How many solutions are there to the equation $x^2 = 2^x$?

30. a. Graph the functions $y = x^3$ and $y = 3^x$ on the same set of axes.
   b. How many solutions are there to the equation $x^3 = 3^x$?

CALCULATOR EXERCISES

1. Evaluate the following to find the approximate value of 2^π.
   a. $2^{3.1}$ **b.** $2^{3.14}$ **c.** $2^{3.141}$ **d.** $2^{3.1415}$ **e.** $2^{3.14159}$

In Exercises 2–4, solve for x to the nearest hundredth.

2. $1.5^x = 30$ **3.** $10^x = 152$ **4.** $8^x = .00236$

5. a. The bar graph below gives the U.S. population for each census since 1800. The tops of the bars lie approximately on the curve $y = ab^x$, where x is the number of years since 1800. Find the value of a; find the value of b to the nearest thousandth.
   b. Assuming that the U.S. population continues to grow as it has in the past, predict the U.S. population in the year 2000 and in the year 2050.

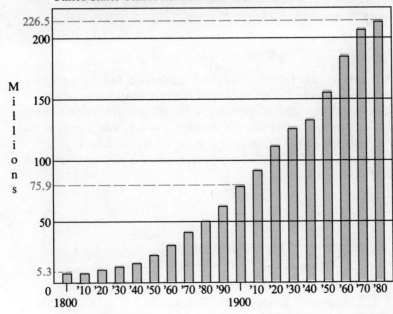

United States Census Results: 1800–1980

Logarithms

5-4/LOGARITHMIC FUNCTIONS

The inverse of the base 2 exponential function $f(x) = 2^x$ is called the base 2 logarithmic function $f^{-1}(x) = \log_2 x$, read "the logarithm of x to base 2" or "the logarithm to base 2 of x." The **base 2 logarithm** of a positive number N is the exponent k such that $2^k = N$. The graphs of $y = 2^x$ and $y = \log_2 x$, shown below, illustrate that $\log_2 3 \approx 1.58$ since $2^{1.58} \approx 3$.

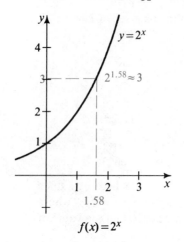

$$f(x) = 2^x$$

Domain: the real numbers
Range: the positive real numbers

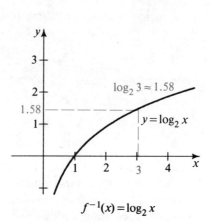

$$f^{-1}(x) = \log_2 x$$

Domain: the positive real numbers
Range: the real numbers

In general, the **base b exponential function** and the **base b logarithmic function** are inverses. In symbols, if $f(x) = b^x$, then $f^{-1}(x) = \log_b x$, and

$$f^{-1}(f(x)) = \log_b b^x = x,$$
$$f(f^{-1}(x)) = b^{\log_b x} = x \ (x > 0).$$

Recall that the exponential function $f(x) = b^x$ is defined only when b is a positive number not equal to 1. The domain of the exponential function is the set of all real numbers, and the range is the set of all positive real numbers. Consequently, the domain of the inverse function $y = \log_b x$ is the set of all positive real numbers, and the range is the set of all real numbers.

In general, if $b > 0$, and $b \neq 1$, the **base b logarithm** of a positive number N is the exponent k such that $b^k = N$:

$$\log_b N = k \quad \text{means} \quad N = b^k$$

Hence:
$$\log_2 8 = 3 \quad \text{because} \quad 8 = 2^3$$
$$\log_9 3 = \tfrac{1}{2} \quad \text{because} \quad 3 = 9^{\frac{1}{2}}$$
$$\log_5 \tfrac{1}{25} = -2 \quad \text{because} \quad \tfrac{1}{25} = 5^{-2}$$

EXAMPLE 1. Evaluate: **a.** $\log_3 81$ **b.** $\log_2 \tfrac{1}{8}$ **c.** $\log_{25} 125$

SOLUTION: **a.** Since $81 = 3^4$, $\log_3 81 = 4$

b. Since $\frac{1}{8} = 2^{-3}$, $\log_2 \frac{1}{8} = -3$

c. Since $125 = 25 \cdot 5 = 25 \cdot 25^{\frac{1}{2}} = 25^{\frac{3}{2}}$, $\log_{25} 125 = \frac{3}{2}$

Logarithms to base 10 are called **common logarithms** and were extremely important for doing heavy calculations with pencil and paper before the invention of computers and electronic calculators. Base 10 was used because the Hindu-Arabic number system has 10 as its base.

$$\log_{10} 10 = 1 \qquad \log_{10} 100 = 2 \qquad \log_{10} 1000 = 3$$

You can find the base 10 logarithm of a number by using a calculator or Table 2 on pages 626–627.

A logarithmic scale

Although logarithms are rarely used in calculations today, they have many applications related to sight, sound, and the other senses. For example, the decibel scale* for measuring the loudness of sound is a **logarithmic scale.** Let I_0 stand for the intensity of a sound that is just barely audible. This sound is assigned a decibel level of 0. If a sound has an intensity I, its decibel level is defined to be:

$$10 \log_{10} \left(\frac{I}{I_0} \right)$$

The table below illustrates that as sound intensities are multiplied by powers of 10^2, decibel levels increase by 20. This is typical of a logarithmic scale; a multiplication scale is converted into an addition scale.

Sound	I = sound intensity	decibel level $= 10 \log_{10} \left(\frac{I}{I_0} \right)$
Just barely audible	I_0	0
Leaves in a breeze	$10^2 I_0$	20
Soft recorded music	$10^4 I_0$	40
Two-person conversation	$10^6 I_0$	60
Loud stereo set	$10^8 I_0$	80
Subway train	$10^{10} I_0$	100
Jet at takeoff	$10^{12} I_0$	120
Pain at eardrum	$10^{13} I_0$	130

Scientists have discovered that people perceive a 40 decibel sound to be twice as loud as a 20 decibel sound even though the intensity ratio of the sounds is $10^4 \div 10^2$, or 100. Likewise, people perceive a 60 decibel sound to be twice as loud as a 30 decibel sound even though the intensity ratio is $10^6 \div 10^3$, or 1000. However, the actual decibel increase may be much less than expected when an intensity level is doubled, as the next example illustrates.

*Although the *decibel* is used almost exclusively today, the original unit is *bel,* named in honor of Alexander Graham Bell (1847–1922). The bel turned out to be too large a unit.

EXAMPLE 2. Two loud stereos are playing the same music simultaneously at 80 decibels each. What is the decibel level of the two sets together?

SOLUTION: Since one stereo at 80 decibels has an intensity $10^8 I_0$, two stereos will have an intensity I that is twice that amount.

$$I = 2 \times (10^8 I_0)$$
$$= 10^{0.3} \times (10^8 I_0) \text{ because } \log_{10} 2 \approx 0.3$$

Therefore, $I = 10^{8.3} I_0$, so the decibel level corresponding to I is 83.

ORAL EXERCISES 5-4

Write the equation in exponential form.

1. $\log_5 25 = 2$ **2.** $\log_5 125 = 3$ **3.** $\log_5 \frac{1}{5} = -1$

4. $\log_5 \sqrt{5} = \frac{1}{2}$ **5.** $\log_4 8 = \frac{3}{2}$ **6.** $\log_8 4 = \frac{2}{3}$

Find each logarithm.

7. $\log_7 49$ **8.** $\log_2 16$ **9.** $\log_2 \frac{1}{4}$

10. $\log_7 \frac{1}{7}$ **11.** $\log_7 \sqrt{\frac{1}{7}}$ **12.** $\log_{10} 10\sqrt{10}$

Solve each equation for x.

13. $\log_5 x = 4$ **14.** $\log_5 x = 0$ **15.** $\log_5 x = -2$

16. $\log_x 121 = 2$ **17.** $\log_x 64 = 3$ **18.** $\log_x 4 = \frac{1}{2}$

Find the decibel level of a sound with the given intensity I using the fact that $\log_{10} 2 \approx 0.3$.

19. $I = 10^{3.2} I_0$ **20.** $I = 10^{11.2} I_0$ **21.** $I = 2 \times 10^{11.2} I_0$

WRITTEN EXERCISES 5-4

Express each equation in exponential form.

A   **1.** $\log_8 \frac{1}{2} = -\dfrac{1}{3}$ **2.** $\log_5 \frac{1}{125} = -3$ **3.** $\log_6 1 = 0$

     4. $\log_{\sqrt{2}} 8 = 6$ **5.** $\log_{\sqrt{3}} 9 = 4$ **6.** $\log_\pi \pi = 1$

Evaluate.

7. a. $\log_2 4$ **b.** $\log_2 32$ **c.** $\log_2 64$ **d.** $\log_2 (2^{10})$

8. a. $\log_3 9$ **b.** $\log_3 27$ **c.** $\log_3 243$ **d.** $\log_3 (3^8)$

9. a. $\log_5 0.2$ **b.** $\log_5 \frac{1}{125}$ **c.** $\log_5 \sqrt[3]{5}$ **d.** $\log_5 1$

10. a. $\log_4 64$ **b.** $\log_4 \frac{1}{64}$ **c.** $\log_4 \sqrt[4]{4}$ **d.** $\log_4 1$

11. a. $\log_6 36$ **b.** $\log_{36} 6$ **c.** $\log_6 6\sqrt{6}$ **d.** $\log_6 \sqrt[3]{\frac{1}{6}}$

12. a. $\log_9 27$ **b.** $\log_4 8$ **c.** $\log_8 16$ **d.** $\log_{81} 243$

13. a. $\log_5 25\sqrt{5}$ **b.** $\log_{10} 0.001$ **c.** $\log_{\frac{1}{3}} 9$ **d.** $\log_{\sqrt{2}} 4$

14. a. $\log_4 8\sqrt{2}$ **b.** $\log_b b^5$ **c.** $\log_{\frac{1}{2}} 8$ **d.** $\log_{\sqrt{5}} 125$

Solve for x.

15. a. $\log_{10} x = 3$ **b.** $\log_{10} |x| = 3$

16. a. $\log_6 x = 2$ **b.** $\log_6 (x + 2) = 2$

B   **17. a.** $\log_4 x = \frac{3}{2}$ **b.** $\log_4 (4x) = \frac{3}{2}$

18. a. $\log_3 |x| = 2$ **b.** $\log_3 |x - 3| = 2$

19. a. $\log_x 64 = 3$ **b.** $\log_x 27 = \frac{3}{2}$

20. a. $\log_x x = 1$ **b.** $\log_x 1 = 0$

21. a. $\log_3 (x^2 - 7) = 2$ **b.** $\log_4 (x^2 - 3x) = 1$

22. a. $\log_5 \sqrt{2x^2 - 3} = \frac{3}{2}$ **b.** $\log_{16} \sqrt[3]{100 - x^2} = \frac{1}{2}$

23. a. $\log_5 (\log_3 x) = 0$ **b.** $\log_6 (\log_4 (\log_2 x)) = 0$

24. a. $\log_2 (\log_x 64) = 1$ **b.** $\log_7 (\log_5 (\log_3 |x|)) = 0$

25. a. Graph the functions $f(x) = \log_3 x$ and $g(x) = 3^x$.

     b. State the domain, range, and zeros of each function.

26. For the given functions, state the domain and range of $f(g(x))$ and $g(f(x))$.

     a. $f(x) = \log_2 x$, $g(x) = |x|$ **b.** $f(x) = \log_2 x$, $g(x) = x - 2$

Find the decibel level for each sound with the given intensity I.

27. a. Average car at 70 km/h, $I = 10^{6.8} I_0$ **b.** Whisper, $I = 10^3 I_0$

28. a. Softly played flute, $I = 10^{4.1} I_0$ **b.** Vacuum cleaner, $I = 10^{7.5} I_0$

29. a. Find the decibel level of two stereos, each playing the same music simultaneously at 62 decibels. (See Example 2.)

     b. What would be the decibel level if there were three stereos instead of two? (*Hint:* $3 \approx 10^{0.5}$)

30. Four new identical sports cars are involved in an acceleration test. If the decibel level of one car accelerating from rest to 50 km/h is 80 decibels, what would be the decibel level for all four cars accelerating simultaneously?

31. The *pH* of a solution measures how acidic or alkaline it is. By definition,

$$pH = -\log_{10} (\text{Hydrogen Ion Concentration}).$$

Pure water, which has a *pH* of 7, is considered neutral. A *pH* less than 7 is acidic; a *pH* more than 7 is alkaline. Find the *pH* of the following solutions and classify them as acidic, neutral, or alkaline, given that $3 \approx 10^{0.5}$ and $5 \approx 10^{0.7}$.

Solution	Hydrogen ion concentration (moles per liter)
Human gastric juices	10^{-2}
Acid rain	3×10^{-5}
Pure water	10^{-7}
Good soil for potatoes	5×10^{-8}
Sea water	10^{-9}

32. The observed brightness of stars is classified by magnitude. Two stars can be compared by giving their magnitude difference d or their brightness ratio r. The numbers d and r are related by the equation $d \approx 2.5 \log_{10} r$. Comparing a first magnitude star with a sixth magnitude star, $d = 6 - 1 = 5$. Find the value of r.

33. **a.** Evaluate $\log_4 16$ and $\log_{16} 4$.   **b.** Evaluate $\log_9 27$ and $\log_{27} 9$.

    c. Prove: $\log_b a = \dfrac{1}{\log_a b}$.

34. **a.** Show that $\log_2 4 + \log_2 8 = \log_2 32$ by evaluating the 3 logarithms.

    b. Show that $\log_9 3 + \log_9 27 = \log_9 81$ by evaluating the 3 logarithms.

    c. Make a generalization based on parts (a) and (b). Prove the generalization.

C   35. Prove that $\log_{10} 2$ is irrational. (*Hint:* Use an indirect proof and assume that $\log_{10} 2$ is rational; that is, assume $\log_{10} 2 = \frac{p}{q}$, where p and q are integers and $\frac{p}{q}$ is in lowest terms.)

CALCULATOR EXERCISES

1. **a.** Is $\log_2 25$ more than 3? more than 4? more than 5?

   b. Find $\log_2 25$ to the nearest hundredth.

2. Find $\log_\pi 1000$ to the nearest hundredth.

5-5/PROPERTIES OF LOGARITHMS

Since the logarithmic function $y = \log_b x$ is the inverse of the exponential function $y = b^x$, it is not surprising that the laws of logarithms given below are very closely related to the laws of exponents given on page 148.

LAWS OF LOGARITHMS (M, N, $b > 0$, $b \neq 1$)

1. $\log_b MN = \log_b M + \log_b N$ 2. $\log_b \dfrac{M}{N} = \log_b M - \log_b N$

3. $\log_b M = \log_b N$ if and only if $M = N$.

4. $\log_b M^k = k \log_b M$, for any real number k.

Proofs: Suppose $\log_b M = x$ and $\log_b N = y$, so that $M = b^x$ and $N = b^y$.

Law 1	Law 2	Law 4
$MN = b^x \cdot b^y = b^{x+y}$	$\dfrac{M}{N} = \dfrac{b^x}{b^y} = b^{x-y}$	$M^k = (b^x)^k = b^{kx}$
$\therefore \log_b MN = x + y$	$\therefore \log_b \dfrac{M}{N} = x - y$	$\therefore \log_b M^k = kx$
$\log_b MN = \log_b M + \log_b N$	$\log_b \dfrac{M}{N} = \log_b M - \log_b N$	$\log_b M^k = k \log_b M$

Law 3 is simply a restatement, in terms of logarithms, of Law 3 of exponents on page 148.

If you know the logarithms of M and N, then you can use the laws of logarithms to find the logarithm of a more complicated expression in M and N, as illustrated in the following examples. In these examples, we shall write "log M" instead of "$\log_b M$" if the result does not depend on a particular choice of b.

EXAMPLE 1.   Express $\log MN^2$ in terms of $\log M$ and $\log N$.

SOLUTION: $\log MN^2 = \log M + \log N^2$ (Law 1)
$= \log M + 2 \log N$ (Law 4)

EXAMPLE 2.   Express $\log \sqrt{\dfrac{M^3}{N}}$ in terms of $\log M$ and $\log N$.

SOLUTION: $\log \sqrt{\dfrac{M^3}{N}} = \log \left(\dfrac{M^3}{N}\right)^{\frac{1}{2}} = \frac{1}{2} \log \left(\dfrac{M^3}{N}\right)$ (Law 4)

$= \frac{1}{2}(\log M^3 - \log N)$ (Law 2)

$= \frac{1}{2}(3 \log M - \log N)$ (Law 4)

In Examples 1 and 2, the logarithm of an expression was broken up into separate logarithms. In Examples 3 and 4, which follow, the separate logarithms are reassembled to give a single logarithm.

EXAMPLE 3.   Simplify: $\log_2 100 - 2 \log_2 5$

SOLUTION: $\log_2 100 - 2 \log_2 5 = \log_2 100 - \log_2 5^2$ (Law 4)

$= \log_2 100 - \log_2 25$

$= \log_2 \frac{100}{25}$ (Law 2)

$= \log_2 4 = 2$

Note that we did not need to know the values of $\log_2 100$ and $\log_2 5$ to do this problem.

EXAMPLE 4.   Express y in terms of x if $\log y = \frac{1}{3} \log x + \log 4$.

SOLUTION: $\log y = \frac{1}{3} \log x + \log 4$

$\log y = \log x^{\frac{1}{3}} + \log 4$ (Law 4)

$\log y = \log 4x^{\frac{1}{3}}$ (Law 1)

$\therefore y = 4x^{\frac{1}{3}}$ (Law 3)

In the preceding section we presented several scientific applications of *common logarithms*. It is customary to omit the subscript 10 when writing common logarithms: that is, $\log x$ denotes the common logarithm of x. Table 2 on pages 626–627 gives the common logarithms of numbers expressed in hundredths from 1.00 to 9.99. The table gives approximate values of $\log N$ accurate to four decimal places. It is understood that each entry in the body of the table is preceded by a decimal point. To find $\log 4.36$, read down the N column to 4.3 then across to the column headed 6. You should find that $\log 4.36 = 0.6395$. We shall use the equal sign for convenience.

EXAMPLE 5. Evaluate $\log 5 \sqrt{6}$.

SOLUTION: $\log 5 \sqrt{6} = \log 5 + \log \sqrt{6}$

$= \log 5 + \log 6^{\frac{1}{2}}$

$= \log 5 + \frac{1}{2} \log 6$

$= 0.6990 + \frac{1}{2}(0.7782)$

$= 0.6990 + 0.3891$

$= 1.0881$

Example 6 shows how to find the common logarithm of any number with 3 significant digits using entries from Table 2. If you have a calculator, however, you can evaluate the logarithms directly.

EXAMPLE 6. Evaluate. **a.** $\log 67{,}400$ **b.** $\log 0.00017$

SOLUTION: **a.** $\log 67{,}400 = \log (6.74 \times 10^4)$ **b.** $\log 0.00017 = \log (1.7 \times 10^{-4})$

$= \log 10^4 + \log 6.74$ $\qquad = \log 10^{-4} + \log 1.7$

$= 4 + 0.8287$ $\qquad\qquad = -4 + 0.2304$

$= 4.8287$ $\qquad\qquad$ (If $0 < N < 1$, $\log N$ is often left in this form.)

EXAMPLE 7. Find the value of N. **a.** $\log N = 2.9763$ **b.** $\log N = 0.4014 - 3$

SOLUTION: **a.** $\log N = 2.9763$ $\qquad\qquad$ **b.** $\log N = 0.4014 - 3$

$N = 10^{2.9763}$ $\qquad\qquad\qquad\qquad N = 10^{0.4014-3}$

$N = 10^2 \times 10^{0.9763}$ $\qquad\qquad\qquad N = 10^{-3} \times 10^{0.4014}$

Locate .9763 in $\qquad\qquad\qquad$ Locate .4014 in
Table 2, page 627. $\qquad\qquad\qquad$ Table 2, page 626.

$N = 10^2 \times 9.47$ $\qquad\qquad\qquad\quad N = 10^{-3} \times 2.52$

$= 947$ $\qquad\qquad\qquad\qquad\qquad = .00252$

The number N such that $\log N = x$ is called the **antilogarithm** of x. Part (a) of Example 7 shows that the antilogarithm of 2.9763 is 947. When tables are used to find logarithms and antilogarithms, interpolation can be used to achieve greater accuracy. (See *Using Tables* on page 624.)

ORAL EXERCISES 5-5

Write the base b logarithm of the given expression in terms of $\log M$ and $\log N$.

1. M^2N $\quad$ **2.** $\dfrac{M^2}{N}$ $\quad$ **3.** $\sqrt{\dfrac{M}{N}}$ $\quad$ **4.** $\sqrt[3]{MN}$ $\quad$ **5.** $M\sqrt{N}$ $\quad$ **6.** $\dfrac{M^2}{N^3}$

Use the laws of logarithms to express each of the following as a single logarithm or as a rational number.

7. $\log_5 2 + \log_5 3$ $\qquad\qquad$ **8.** $\log_3 5 + \log_3 4$

9. $\log_4 12 - \log_4 3$ $\qquad\qquad$ **10.** $\log_7 3 + \log_7 6 - \log_7 2$

11. $\log_6 4 + 2 \log_6 3$ $\qquad\qquad$ **12.** $\frac{1}{2} \log_9 36 - \log_9 2$

13. $\log M + 2 \log N$

14. $2 \log P - \log Q$

15. $\log M + \log N + \log P$

16. $\log M + \log N - 3 \log P$

17. $\frac{1}{2} \log M - \frac{1}{2} \log N$

18. $\log R + \frac{1}{3} \log S$

Is the statement true for all positive values of M and N?

19. $\log (M + N) = \log M + \log N$

20. $\log \left(\dfrac{M}{N}\right) = \dfrac{\log M}{\log N}$

WRITTEN EXERCISES 5-5

Write the base b logarithm of the given expression in terms of $\log M$ and $\log N$.

A **1.** $(MN)^2$

2. $\dfrac{M}{N^2}$

3. $\sqrt[3]{\dfrac{M}{N}}$

4. $M\sqrt[4]{N}$

5. $M^2 \sqrt{N}$

6. $\dfrac{1}{M}$

Write the given expression as a rational number, if possible, or as a single logarithm.

7. $\log_5 2 + \log_5 3 + \log_5 4$

8. $\log 8 + \log 5 - \log 4$

9. $\frac{1}{2} \log_6 9 + \log_6 5$

10. $\log_2 48 - \frac{1}{3} \log_2 27$

11. $2 \log_3 6 - 2 \log_3 2$

12. $\frac{1}{2} \log_7 5 + 3 \log_7 2$

13. $\log M - 3 \log N$

14. $4 \log M + \frac{1}{2} \log N$

15. $\log A + 2 \log B - 3 \log C$

16. $\frac{1}{2}(\log M + \log N - \log P)$

17. $\frac{1}{3}(2 \log M - \log N - \log P)$

18. $5(\log A + \log B) - 2 \log C$

19. $\log_2 \pi + 2 \log_2 r$

20. $\log_5 4 - \log_5 3 + \log_5 \pi + 3 \log_5 r$

21. $2 \log 5 + \log 4$

22. $\log_8 \sqrt{80} - \log_8 \sqrt{5}$

Evaluate the given common logarithm.

23. $\log 275$

24. $\log 0.00386$

25. $\log \sqrt{3}$

26. $\log \sqrt[3]{62}$

Find the value of N.

27. $\log N = 5.7419$

28. $\log N = 8.5378$

29. $\log N = 0.7332 - 2$

30. $\log N = 0.2672 - 6$

31. Express y in terms of x if:

  a. $\log y = 2 \log x$

  b. $\log y = 3 \log x + \log 5$

  c. $\log y - \log x = 2 \log 7$

  d. $\log y = \log 4 - 2 \log x$

32. Express y in terms of x if:

  a. $\log y = -\log x$

  b. $\log y = 2 \log x + \log 2$

  c. $\log y + \frac{1}{2} \log x = \log 3$

  d. $\log y = \frac{1}{3}(\log 4 + \log x)$

B **33.** If $\log_8 3 = r$ and $\log_8 5 = s$, express the given logarithm in terms of r and s.

  a. $\log_8 75$ **b.** $\log_8 225$ **c.** $\log_8 0.12$ **d.** $\log_8 \frac{3}{64}$

34. If $\log_9 5 = x$ and $\log_9 4 = y$, express the given logarithm in terms of x and y.

  a. $\log_9 100$ **b.** $\log_9 36$ **c.** $\log_9 6\frac{1}{4}$ **d.** $\log_9 3.2$

Solve.

35. $\log_2 (x + 2) + \log_2 5 = 4$ **36.** $\log_4 (2x + 1) - \log_4 (x - 2) = 1$

37. $\log_6 (x + 1) + \log_6 x = 1$ **38.** $\log_3 x + \log_3 (x - 2) = 1$

39. $\log_4 (x - 4) + \log_4 x = \log_4 5$ **40.** $\log_2 (x^2 + 8) = \log_2 x + \log_2 6$

C

41. Suppose that all you know about a function f is that $f(ab) = f(a) + f(b)$ for all positive numbers a and b.

   a. Evaluate $f(1)$.

   b. Prove $f(a^2) = 2f(a)$ and $f(a^3) = 3f(a)$. What generalization does this suggest?

   c. Prove $f(\sqrt{a}) = \frac{1}{2}f(a)$ and $f(\sqrt[3]{a}) = \frac{1}{3}f(a)$. What generalization does this suggest?

   d. Prove $f\left(\dfrac{1}{b}\right) = -f(b)$.

   e. Prove $f\left(\dfrac{a}{b}\right) = f(a) - f(b)$.

   f. Can you think of a function $f(x)$ which satisfies the original equation?

   g. If $f(10) = 1$, find the values of x for which $f(x) = 2$ and $f(x) = 3$.

42. The astronomer Johannes Kepler (1571–1630) spent years discovering a relationship between the time T for a planet to revolve about the sun and the average distance d of the planet from the sun. In his lifetime, Kepler had data for only the six planets then known. In the graph at the left below, T, in days, and d, in millions of kilometers, are plotted for several planets.

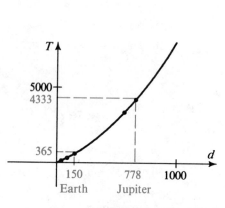
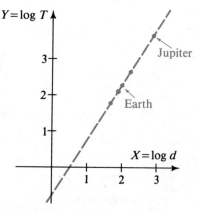

If common logarithms of the data are plotted, the relationship between T and d is found more easily because the new graph is a straight line (see right graph above) whose equation is, approximately,

$$Y = \tfrac{3}{2}X - 0.7$$

$$\text{or } \log T = \tfrac{3}{2}\log d - 0.7$$

Use this equation to find Kepler's equation giving T in terms of d. This equation is customarily stated without logarithms and with integral exponents.

168   Chapter 5

43. Suppose the number $9^{(9^9)}$ were typed on a long strip of paper, 3 digits per centimeter. How many kilometers long would the paper have to be? Give your answer to three significant digits. (*Hint:* The base 10 logarithm of a number can tell you how many digits the number has.)

44. The *Richter scale* is a system for rating the severity of earthquakes. The scale was revised in 1979, so that the Richter magnitude, R, of an earthquake is defined by the formula

$$R = 0.67 \log{(0.37E)} + 1.46$$

where E is the energy, in kilowatt-hours, released by the earthquake. The 1906 San Francisco earthquake is rated 7.9 on the revised Richter scale. How much energy was released by this earthquake?

5-6 / EXPONENTIAL EQUATIONS; CHANGING BASES

Equations in which the variable occurs as an exponent are called **exponential equations.** Here are two examples of exponential equations that you already can solve.

$$2^{t-3} = 8 \qquad\qquad 9^{2t} = 3\sqrt[3]{3}$$
$$2^{t-3} = 2^3 \qquad\qquad 3^{4t} = 3^{\frac{4}{3}}$$
$$t - 3 = 3 \qquad\qquad 4t = \frac{4}{3}$$
$$t = 6 \qquad\qquad t = \frac{1}{3}$$

These exponential equations are special because both sides of each equation can easily be expressed as powers of the same number. This is not the case for some exponential equations.

EXAMPLE 1. Solve: **a.** $2^t = 7$ **b.** $(0.4)^x = 5$

SOLUTION: Take logarithms of both sides of the equation. You can use any base logarithm, but base 10 is particularly useful because these are available on your calculator and in tables.

a. $2^t = 7$

$\log 2^t = \log 7$

$t \log 2 = \log 7$

$t = \dfrac{\log 7}{\log 2}$

$= \dfrac{0.8451}{0.3010}$

$= 2.81$

b. $(0.4)^x = 5$

$x \log 0.4 = \log 5$

$x = \dfrac{\log 5}{\log 0.4}$

$= \dfrac{0.6990}{0.6021 - 1}$

$= \dfrac{0.6990}{-0.3979}$

$= -1.76$

The technique used in Example 1 can be used to find logarithms to any base.

EXAMPLE 2. Evaluate $\log_3 52$.

SOLUTION: Let $\log_3 52 = x$.

$$\text{Then} \quad 3^x = 52$$
$$\log 3^x = \log 52$$
$$x \log 3 = \log 52$$
$$x = \frac{\log 52}{\log 3} = \frac{1.7160}{0.4771} = 3.60$$

There is a remarkably simple formula that can be used to find the logarithm of a number in one base if logarithms in another base are known.

THEOREM:
$$\log_b c = \frac{\log_a c}{\log_a b}$$

Proof:
$$\text{Let} \quad \log_b c = x$$
$$\text{Then} \quad b^x = c$$
$$\log_a b^x = \log_a c$$
$$x \log_a b = \log_a c$$
$$x = \frac{\log_a c}{\log_a b}$$
$$\log_b c = \frac{\log_a c}{\log_a b}$$

In practice, this formula is often applied to find logarithms in another base by using common logarithms.

EXAMPLE 3. $\log_6 88 = \dfrac{\log 88}{\log 6} = \dfrac{1.9445}{0.7782} \approx 2.50$

When $a = c$ in the theorem, the following corollary is obtained.

COROLLARY: $\log_b c = \dfrac{1}{\log_c b}$

EXAMPLE 4. $\log_8 2 = \dfrac{1}{\log_2 8} = \dfrac{1}{3}$

ORAL EXERCISES 5-6

Solve the equation. Leave answers in radical form when appropriate.

1. $x^3 = 81$ **2.** $3^x = 81$ **3.** $4^x = 8$ **4.** $x^4 = 8$

Solve the equation. Use a calculator or the table on pages 626–627.

5. $10^x = 3$ **6.** $10^x = 8.1$ **7.** $10^x = 256$ **8.** $100^x = 302$

Solve. Express answers to three significant digits.

A
1. $3^x = 12$
2. $2^x = 100$
3. $(1.06)^x = 3$
4. $(0.98)^x = 0.5$

5. $2^{-x} = 18$
6. $3^{-x} = 0.01$
7. $\sqrt{3^x} = 50$
8. $\sqrt[3]{7^x} = 4$

For each pair of equations, solve one of them by taking logs of both sides. Solve the other by expressing both sides as a power of the same number.

9. **a.** $4^x = 16\sqrt{2}$
10. **a.** $9^x = 4$
11. **a.** $25^x = \sqrt[5]{5^x}$
12. **a.** $8^x = \sqrt[3]{\dfrac{2}{4^x}}$

b. $4^x = 20$
b. $9^x = \dfrac{3}{3^x}$
b. $25^x = 2$
b. $8^x = \sqrt[3]{5}$

Find each logarithm to the nearest hundredth.

13. $\log_5 9$
14. $\log_3 20$
15. $\log_9 0.5$
16. $\log_4 0.01$

B
17. Prove the corollary on page 170: $\log_b c = \dfrac{1}{\log_c b}$.

18. Prove: $\log_a b \log_b c = \log_a c$.

Evaluate the expression. (Use the results of Exercises 17 and 18.)

19. $\log_3 4 \cdot \log_4 3$
20. $\log_3 2 \cdot \log_2 27$
21. $\log_{25} 8 \cdot \log_8 5$

22. $\dfrac{1}{\log_2 6} + \dfrac{1}{\log_3 6}$
23. $\dfrac{1}{\log_4 6} + \dfrac{1}{\log_9 6}$
24. $\log_3 8 \cdot \log_8 9$

If $\log_b M = x$ and $\log_b N = y$, express the following in terms of x and y.

25. **a.** $\log_M b$ **b.** $\log_M b^2$ **c.** $\log_{MN} b$
26. **a.** $\log_N b$ **b.** $\log_N \sqrt{b}$ **c.** $\log_{\frac{M}{N}} b$

Solve. Express x as a logarithm if necessary. For example, if $5^x = 2$, then $x = \log_5 2$.

27. $2^{2x} - 2^x - 6 = 0$ (*Hint:* $2^{2x} - 2^x - 6 = (2^x)^2 - (2^x) - 6$)

28. $3^{2x} - 5 \cdot 3^x + 4 = 0$
29. $6^{x+1} - 6^{-x} = 5$

30. $3^{2x+1} - 7 \cdot 3^x + 2 = 0$
31. $2^{4x} - 5 \cdot 2^{2x+1} + 16 = 0$

C
32. Prove: $a^{\log b} = b^{\log a}$
33. Prove: $\dfrac{1}{\log_a ab} + \dfrac{1}{\log_b ab} = 1$

34. Solve: $4 \log_x 2 + \log_2 x = 5$

35. The intensity I of a beam of light after passing through a material is given by the formula

$$I = (4^{-ct})I_0$$

where I_0 is the initial intensity, t is the thickness, in centimeters, of the material, and c is a constant called the absorption factor. Ocean water absorbs light with an absorption factor of $c = 0.0101$.

a. At what depth will a vertical beam of light be reduced to 50% of its initial intensity? (*Hint:* Let $I = 0.5I_0$.)

b. At what depth will a vertical beam of light be reduced to 2% of its initial intensity?

1. Solve the equation $x^x = \pi$. Give the solution to the nearest hundredth.

2. Evaluate $\left(1 + \dfrac{1}{n}\right)^n$ for the given value of x. Give results to five decimal places.

   a. $n = 1000$ **b.** $n = 10{,}000$ **c.** $n = 100{,}000$

   (Note: These values approach an important constant, the number e.)

5-7 / EXPONENTIAL GROWTH

When a quantity is growing by a fixed percent per year, the growth is said to be exponential. For example, if world population is now P_0, read "P sub zero," and is growing at the rate of 2% per year, then next year the population will be $P_0 + 0.02P_0$ or $P_0(1.02)$. The population two years from now will be $P_0(1.02) + 0.02(P_0(1.02))$ or $P_0(1.02)^2$. The table shows the population P at a time t years from now.

$$P = P_0(1.02)^t$$

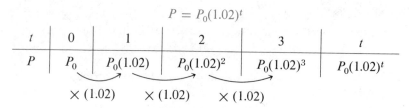

t	0	1	2	3	t
P	P_0	$P_0(1.02)$	$P_0(1.02)^2$	$P_0(1.02)^3$	$P_0(1.02)^t$

All quantities that **grow exponentially** have equations that can be written in the form

$$N = N_0 b^t$$

where N_0 is the original amount, N is the amount at time t, and b is 1 plus the growth rate. If the growth rate is given as a percent, be sure to express it as a decimal.

EXAMPLE 1. Suppose that the cost of a jumbo jar of peanut butter has been increasing at the rate of 7% per year. If the cost of the peanut butter is now $2.00, what will it be in 10 years? What was the cost 5 years ago?

SOLUTION: Cost in t years = original cost $\times (1.07)^t$
$$C = 2(1.07)^t$$

In 10 years, $C = 2(1.07)^{10}$
$$\approx 2(1.967) \approx 3.93$$
The cost in 10 years will be $3.93.

5 years ago, $C = 2(1.07)^{-5}$
$$= \frac{2}{(1.07)^5} \approx \frac{2}{1.403} \approx 1.43$$

The cost 5 years ago was $1.43.

EXAMPLE 2. During one year the world consumption of crude oil increased at the rate of 5%. Suppose the world consumption of crude oil continued to increase at the rate of 5% per year. How long would it take to double the world's consumption?

SOLUTION: Let N_0 be the present consumption and N the consumption in t years. Then $N = N_0(1.05)^t$
Since consumption is to double, we have $N = 2N_0$.

$$2N_0 = N_0(1.05)^t$$
$$2 = (1.05)^t$$
$$\log 2 = t \log 1.05$$
$$t = \frac{\log 2}{\log 1.05}$$
$$t = \frac{0.3010}{0.0212} \approx 14.2$$

Consumption will double in approximately 14 years at this rate.

The time required for a quantity that is growing exponentially to double is called the **doubling time.** In Example 2, the doubling time at a 5% growth rate was 14 years. In Example 1, the cost nearly doubled in 10 years at a 7% growth rate.

The **rule of 70** provides an approximation of the doubling times for exponential growth. If a quantity is growing at r% per year, then

$$\text{doubling time} \approx 70 \div r.$$

For example, when the growth rate is 3.5% per year, the doubling time is $70 \div 3.5 \approx 20$ years. The rule of 70 works best for small values of r. A derivation is given in the exercises of Section 5–8.

EXAMPLE 3. During the last several years the cost of tuition at a state university has been increasing by 14% per year. At this rate how long will it take for the tuition to:
a. double? **b.** quadruple? **c.** increase by a factor of 8?

SOLUTION: By the rule of 70, the doubling time is $70 \div 14 = 5$ years.
a. 5 years
b. Since $4 = 2^2$, the tuition will quadruple in $2 \cdot 5 = 10$ years.
c. Since $8 = 2^3$, the tuition will increase by a factor of 8 in $3 \cdot 5 = 15$ years.

EXAMPLE 4. A person's annual income increased by a fixed percentage each year from $13,000 in 1974 to $22,300 in 1982. Find the annual rate of increase to the nearest percent.

(Solution continued on page 174.)

Exponents and logarithms **173**

Income in t years = 1974 income $\times b^t$

$$22,300 = 13,000\, b^8$$
$$b^8 = 1.715$$

Using logarithms, $b = 1.07$

Since $b = 1 +$ growth rate, the annual rate of increase was 0.07, or 7%.

If a quantity decreases by a fixed percent each time period, it is said to **decay exponentially** (instead of grow exponentially). For example, if a quantity N_0 decreases by 5% each year, then after a year there is 95% of N_0, or $N_0(0.95)$, remaining. The amount remaining after 2 years is $N_0(0.95)^2$. In general, the amount remaining after t years is

$$N = N_0 b^t$$

where b is 1 minus the rate of decrease. Again, be sure that the rate is expressed as a decimal.

EXAMPLE 5. Suppose that a radioactive substance decays so that the amount present decreases by 11% per day. How much will be present after 6 days if the original amount weighs 40 kg?

SOLUTION: The rate of decrease is $b = 1 - 0.11 = 0.89$

$$N = N_0 b^t = 40(0.89)^6 \approx 40(0.5) \approx 20 \text{ kg}$$

In the last example, the 40 kg will decay to 20 kg in 6 days. In another 6 days, this 20 kg will decay to just 10 kg. This 6 day period during which half of the quantity decays is called its **half-life.** Some radioactive isotopes have half-lives measured in seconds, while others (such as uranium 235) have half-lives measured in millions of years.

Consider again the substance with a half-life of 6 days referred to in Example 5. To find how much of the original 40 kg will remain after 30 days, you could reason that 30 days is 5 half-lives. Thus the original 40 kg is multiplied by $\frac{1}{2}$ five times, and the amount remaining is

$$N = 40\left(\frac{1}{2}\right)^5 = \frac{40}{32} = 1.25 \text{ kg}$$

In general, if a substance has a half-life h, then in time t, there will be $\frac{t}{h}$ half-lives and the original amount N_0 is multiplied by one-half $\frac{t}{h}$ times.

$$\text{Amount at time } t = \text{Original Amount} \times \left(\frac{1}{2}\right)^{\frac{t}{\text{half-life}}}$$

$$N = N_0 (\tfrac{1}{2})^{\frac{t}{h}}$$

The half-life of a substance decaying exponentially is similar to the doubling time of a substance growing exponentially. These and other comparisons are summarized on the next page.

	Exponential decay	Exponential growth

Exponential decay

r = rate of decrease per time period

$b = 1 - r$ (r expressed as a decimal)

h = half-life
= time required for quantity to be halved

$N = N_0 b^t$

$N = N_0(\frac{1}{2})^{\frac{t}{h}}$

Exponential growth

r = rate of increase per time period

$b = 1 + r$ (r expressed as a decimal)

d = doubling time
= time required for quantity to be doubled

$N = N_0 b^t$

$N = N_0(2)^{\frac{t}{d}}$

ORAL EXERCISES 5-7

Complete each table.

1.

a.	If N increases by	3%	15%	4.6%	120%	?
b.	Then multiply N_0 by	?	?	?	?	1.105

2.

a.	If N decreases by	12%	7.5%	80%	?	100%
b.	Then multiply N_0 by	?	?	?	.68	

3.

	Item	Annual rate of increase	Cost now	Cost in t years
a.	Bike	5%	$200.00	?
b.	Jeans	8%	$20.00	?
c.	Movie ticket	10.5%	$4.50	?
d.	Loaf of bread	?	$1.25	$1.25(1.06)^t$

4.

	Item	Annual rate of decrease	Value now	Value in t years
a.	Car	20%	$9800	?
b.	Boat	15%	$2200	?
c.	Skates	?	$100	$100(0.75)^t$

5. Use the *Rule of 70* to complete this table.

	Item	Annual rate of increase	Approximate doubling time	Quadrupling time
a.	Population	3%	?	?
b.	Money in the bank	5%	?	?
c.	Cost of ice cream cone	8%	?	?

Complete each table.

	Item	Annual rate of increase	Cost now	Cost in 10 years	Cost in 20 years
A 1.	Airplane ticket	15%	$300	?	?
2.	Swim suit	8%	$35	?	?
3.	Jar of mustard	7%	$1	?	?
4.	College tuition	10%	$6000	?	?

	Item	Annual rate of decrease	Value now	Value in 3 years	Value in 6 years
5.	Farm tractor	25%	$35,000	?	?
6.	Industrial equipment	10%	$200,000	?	?
7.	Value of the dollar	6%	$1	?	?
8.	Value of the dollar	8%	$1	?	?

9. The half-life of a radioactive isotope is 4 days. If 3200 grams are present now, how much will be present in:
   a. 4 days   b. 8 days?   c. 20 days?   d. t days?

10. The half-life of radium is 1600 years. If you have 1000 grams now, how much will there be in:
    a. 3200 years?   b. 16,000 years?   c. 800 years?   d. t years?

11. The value of a new car decreases 20% each year. Complete the table. V is the value of the car, and t is the age in years. Give each value to the nearest $100. Make a graph to show the relationship between V and t.

t	0	1	2	3	4	5
V	$10,000	?	?	?	?	?

Use the values in your table. Let the horizontal axis be the t-axis and the vertical axis the V-axis.

12. Suppose the population of a nation is growing at 3% per year. If the population was 30,000,000 in 1975, what will the population be in 1995, to the nearest million?

13. If grocery prices increase 1% a month for a whole year, how much would groceries that cost $100 at the beginning of the year cost at the end of the year?

14. The cost of goods and services in an urban area increased 1.5% last month. If this rate continues, what will be the annual rate of increase?

15. According to legend, in 1626 Manhattan Island was purchased for trinkets worth approximately $24. If the $24 had been invested instead at a rate of 6% interest per year, what would be its value in 1996? Compare this with a recent total of $34,000,000,000 in assessed values for Manhattan.

16. If $1000 is invested so that it grows at the rate of 10% per year, what will the investment be worth in 20 years?

17. A population is growing at a rate of 3% per year. According to the Rule of 70, in approximately how many years will the population double?

18. Repeat Exercise 17 for a growth rate of 6%.

19. If the price of sneakers increases at 6% per year, about how long will it take to double the price?

20. If the price of theater tickets increases at 8% per year, about how long will it take to double the price?

21. The price of firewood 4 years ago was $140 per cord. Today, a cord of wood costs $182. What has been the annual rate of increase in the cost, to the nearest percent?

22. A house bought for $41,000 in 1978 was sold for $58,000 in 1983. What was the annual rate of appreciation (increase) in the value of the house, to the nearest percent?

23. The *consumer price index* (CPI) is a measure of the average cost of goods and services. The U.S. government set the index at 100 for 1967, the base year. In 1977, the index was 181.5. What was the average annual rate of increase over the decade from 1967 to 1977?

24. Look in an almanac and find the current consumer price index. Then determine the average annual rate of increase for this index since 1967.

B
25. When a certain drug enters the blood stream, it gradually dilutes, decreasing exponentially with a half-life of 3 days. If the initial amount of the drug in the blood stream is A_0, what will the amount be 30 days later?

26. Radioactive iodine with a half-life of 8.1 days is used to determine if people have a thyroid deficiency. An amount, N_0, of the iodine is injected into the blood stream and is absorbed by a healthy thyroid gland. By measuring the thyroid's radioactivity at various later times, it is possible to tell whether or not the thyroid is functioning normally. Express in terms of N_0 the amount of radioactive iodine which should be present in a healthy thyroid gland 5 days after it was injected into the blood stream.

27. Assume that half of the representatives serving in Congress at any given time will not be serving in 8 years; that is, the half-life of the number of original representatives is 8 years. Of the 435 representatives, about how many will be serving in 4 years? 12 years?

28. All living organisms contain a small amount of the radioactive isotope C^{14} called carbon 14. When an organism dies, the amount of C^{14} present decays exponentially. By measuring the radioactivity N of, say, an ancient skeleton of an animal and comparing it with the radioactivity N_0 of living animals, archaeologists can tell the approximate time of death of the animal.
a. Given the half-life of C^{14} is 5700 years, write an equation relating N, N_0, and the time t since the animal's death.
b. Suppose it is found that, for a certain animal, $N = \frac{1}{10} N_0$. About how long ago did the animal die to the nearest 100 years?

29. An archaeologist unearths a piece of wood that may have come from King Nebuchadnezzar's Hanging Garden of Babylon, about 600 B.C. The archaeologist measures the radioactivity of the C^{14} of this wood and finds $N = 0.8 N_0$. Is it possible that the wood could be from the Hanging Garden? (The half-life of C^{14} is 5700 years.)

30. A culture of yeast doubles in size every 3 hours. If the yeast population is now N_0, what will it be 24 hours from now?

31. A certain bond investment will double your money every 8 years. If you invest $1000 now, what will it be worth in 20 years?

32. A house purchased four years ago for $40,000 was sold for $60,000. If the value of the house increases exponentially, how much would it be worth next year?

33. The population of a country increased from 100 million to 200 million in 60 years. How long will it take, to the nearest year, for the population to increase from 200 million to 300 million at the same rate?

34. Suppose a certain population increases by 30% every 10 years. By what percent does it increase each year?

35. During one year food prices increased 16%. What was the monthly rate of increase to the nearest tenth of a percent?

36. *Newton's Law of Cooling* states that the difference in the temperatures of a warm body and its cooler surroundings decreases exponentially. Suppose you pour a cup of coffee at 100°C in a room at 20°C. Then the initial difference in temperatures is 80°C, and the temperature difference D at time t in minutes is $D = 80b^t$. If you find the temperature is 90°C at time $t = 1$, how long must you wait before the coffee cools to 45°C? Assume that the room remains at 20°C.

37. Apply Newton's Law of Cooling (see Exercise 36) to a murder mystery: At precisely 3:48 P.M., our hero, Inspector Khan discovers the murdered victim in an abandoned warehouse in which the temperature is 21°C. The inspector immediately determines that the body temperature of the victim is 31°C. One hour later the body temperature is 29°C.
   a. If T is the temperature of the body t minutes after the inspector discovers it, then $T - 21 = ab^t$. Find the values of a and b.
   b. Since at death the body temperature is 37°C(98.6°F), set $T = 37$ in your equation from part (a). Then tell the inspector the time of death.

CALCULATOR EXERCISES

1. Suppose you have a large sheet of ordinary paper, 0.015 cm thick, and suppose you tear the paper in half and put the pieces on top of each other. Then you tear the two pieces in half and stack the four pieces on top of each other. Suppose you were able to keep tearing and stacking in this manner a total of 50 times, so that there would be 2^{50} pieces of paper in the pile. How high, in kilometers, would the pile be? The average distance to the sun is about 150 million km.

2. Suppose a bacteria population doubles its size every 45 minutes. Thus, if the initial population is P, then it is $2P$ after 45 minutes, $4P$ after 90 minutes, and $8P$ after 135 minutes. Find the approximate population after 24 hours.

3. Many scientists believe that there were equal amounts of Uranium 235 and Uranium 238 when Earth was formed. Because Uranium 235 decays faster than Uranium 238, the ratio of their present amounts is about $1 : 139$. That is, in the table below, $\dfrac{A}{B} = \dfrac{1}{139}$. Use this information to calculate the approximate age of Earth, according to this theory.

	Half-life (years)	Amount present when Earth was formed	Amount present now (t years later)
Uranium 235	$h_1 = 7.13 \times 10^8$	N_0	$A = N_0(\tfrac{1}{2})^{\frac{t}{h_1}}$
Uranium 238	$h_2 = 4.51 \times 10^9$	N_0	$B = N_0(\tfrac{1}{2})^{\frac{t}{h_2}}$

5-8 / THE NUMBER e AND THE NATURAL LOGARITHM

You have now studied many exponential functions of the form $f(x) = b^x$, $b > 0$, $b \neq 1$. In advanced mathematics, the most important base is an irrational number called e, and the most important exponential function is $f(x) = e^x$. The number e is defined as follows:

$$e = \lim_{n \to \infty} \left(1 + \frac{1}{n}\right)^n$$

This equation is read "e is the limit, as n approaches infinity, of $\left(1 + \frac{1}{n}\right)^n$."

Although we will not study limits until a later chapter, you can get an idea of the value of e by substituting larger and larger values of n into the expression $\left(1 + \frac{1}{n}\right)^n$. In the table below, values of $\left(1 + \frac{1}{n}\right)^n$ are rounded to five decimal places.

n	100	1000	10,000	100,000
$\left(1 + \frac{1}{n}\right)^n$	2.70481	2.71692	2.71815	2.71827

You can see that as n increases the value of $\left(1 + \frac{1}{n}\right)^n$ appears to get closer and closer to some number. The Swiss mathematician Euler (1707–1783) proved that this actually happens, and this limiting number is called e in his honor.

$$e \approx 2.718281828459045$$

Values of the function $y = e^x$ can be obtained using a calculator or a table of values, such as Table 6 on page 644. The graph of the exponential function $y = e^x$ is shown at the right. The graph of the inverse function $y = \log_e x$ is also shown; $\log_e x$ is called the **natural logarithm** of x. The symbol $\ln x$ is usually used instead of $\log_e x$.

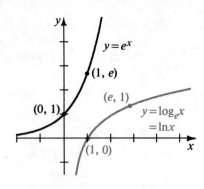

If necessary, common logarithms can be used to evaluate natural logarithms. Substituting e for b and 10 for a in the change of base formula on page 170 gives the following useful result:

$$\ln c = \frac{\log c}{\log e}$$

EXAMPLE 1. $\ln 5 = \dfrac{\log 5}{\log e} = \dfrac{0.6990}{0.4343} = 1.6095$

Interest and the number e

The table below shows how to calculate the effective annual yield on an investment at 12% annual interest when the interest is compounded at various time periods.

Interest period	% Growth each period	Growth factor during period	Growth factor for year	Effective annual yield
Annually	12%	1.12	$(1.12)^1 = 1.12$	12%
Semiannually	6%	1.06	$(1.06)^2 = 1.1236$	12.36%
Quarterly	3%	1.03	$(1.03)^4 = 1.1255$	12.55%
Monthly	1%	1.01	$(1.01)^{12} = 1.1268$	12.68%
Daily (1 yr = 360 da)	$\dfrac{12}{360}\%$	$1 + \dfrac{0.12}{360}$	$\left(1 + \dfrac{0.12}{360}\right)^{360} = 1.1275$	12.75%
k times per year	$\dfrac{12}{k}\%$	$1 + \dfrac{0.12}{k}$	$\left(1 + \dfrac{0.12}{k}\right)^{k}$	?

You might wonder if you could dramatically increase the effective annual yield by investing in a plan that compounded every hour or even every second. In other words, what happens to the effective annual yield as k gets larger?

$$\left(1 + \frac{0.12}{k}\right)^k = \left(1 + \frac{1}{\frac{k}{0.12}}\right)^k$$

$$= \left[\left(1 + \frac{1}{\frac{k}{0.12}}\right)^{\frac{k}{0.12}}\right]^{0.12}$$

$$= \left[\left(1 + \frac{1}{n}\right)^n\right]^{0.12}, \text{ if we let } n = \frac{k}{0.12}.$$

Since the limit of $\left(1 + \frac{1}{n}\right)^n$ as n approaches infinity is e, then the limit of the above expression is $e^{0.12}$, which is approximately 1.1275. Thus you can see that if the 12% annual rate were compounded every second, or even "continuously," the best you could achieve would be an annual yield of approximately 12.75%.

In general, if you invest P dollars at an $r\%$ annual rate that is compounded continuously, then t years later your investment will be worth Pe^{rt} dollars. See Exercises 40 and 41 on page 182 for proofs.

ORAL EXERCISES 5-8

Give each statement in exponential form.

1. $\log_e 6 = 1.8$ **2.** $\ln 3 = 1.1$ **3.** $\ln 100 = 4.6$

Evaluate.

4. $\log_2 2$ **5.** $\log_3 9$ **6.** $\ln e$

7. $\log 100$ **8.** $\ln e^5$ **9.** $\ln \frac{1}{e^2}$

Give each statement in logarithmic form.

10. $e^{3.9} = 50$ **11.** $e = 2.718$ **12.** $e^{-0.7} = \frac{1}{2}$

13. Give the approximate value of $\left(1 + \frac{1}{n}\right)^n$ when $n = 1,000,000$.

14. If money is invested at 8% compounded semiannually, then each year the investment is multiplied by $(1.04)^2$. What is it multiplied by if the compounding is done   **a.** quarterly?   **b.** 8 times a year?   **c.** continuously?

WRITTEN EXERCISES 5-8

Give each statement in exponential form.

A   **1. a.** $\ln 20 \approx 3$ **b.** $\ln 10,000 \approx 9.2$ **c.** $\ln 0.3 \approx -1.2$ **d.** $\ln e = 1$

Give each statement in logarithmic form.

2. a. $e^2 \approx 7.389$ **b.** $e^{5.298} \approx 200$ **c.** $e^0 = 1$ **d.** $\sqrt{e} \approx 1.65$

Evaluate.

3. a. $\ln e^2$ **b.** $\ln e^3$ **c.** $\ln \frac{1}{e}$ **d.** $\ln \sqrt{e}$

4. a. $\ln e^4$ **b.** $\ln \frac{1}{e^3}$ **c.** $\ln \sqrt[3]{e}$ **d.** $\ln 1$

Write as a single logarithm.

5. $\ln 12 - \ln 3 - \ln 2$ **6.** $\ln 6 + \ln 8 - \ln 2$

7. $\frac{1}{2} \ln 25 + \ln 2$ **8.** $2 \ln 8 - \ln 16$

9. $\frac{3}{2} \ln 25 + 3 \ln 2$ **10.** $\frac{2}{3} \ln 8 + 2 \ln 5$

11. $\ln 3e^2 + \ln \frac{6}{e^2}$ **12.** $\ln \sqrt{2e} - \ln \sqrt{8e}$

Simplify the given expression.

13. a. $\ln e^x$ **b.** $e^{\ln x}$ **c.** $e^{2\ln x}$ **d.** $e^{-\ln x}$

14. a. $\ln e^{3x}$ **b.** $e^{3\ln x}$ **c.** $e^{\ln \sqrt{x}}$ **d.** $e^{-\frac{1}{2}\ln x}$

In Exercises 15–26, solve for x. (Answers can be left in terms of e if necessary.)

15. $\ln(x-2)=3$ **16.** $\ln(x+1)=2$

17. $2\ln x = 8$ **18.** $\ln x^3 = 18$

B   **19.** $\ln\dfrac{x}{e}=2$ **20.** $\ln\dfrac{1}{x}=-5$

21. $\ln x - \ln(x-1)=\ln 2$ **22.** $(\ln x)^2 = 16$

23. $\ln(\ln x)=0$ **24.** $\ln|\ln x|=0$

25. $\ln x + \ln(x+3)=\ln 10$ **26.** $(\ln x)^2 - 6(\ln x)+9=0$

Solve each exponential equation by taking natural logarithms of both sides.

27. a. $e^x = 3$ **b.** $e^{2x}=3$ **c.** $e^{x-1}=3$

28. a. $e^{2x}=5$ **b.** $\sqrt{e^x}=9$ **c.** $e^{\sqrt{x}}=9$

29. Find the effective annual yield of an investment that pays 10% annual interest compounded
a. semiannually   **b.** quarterly   **c.** continuously

30. Repeat Exercise 29 if the annual rate is 16% instead of 10%.

31. Sketch the graph of $f(x)=\ln\left(\frac{1}{x}\right)$.

32. Sketch the graph of $f(x)=e^{-x^2}$.

Solve each equation by writing it in a quadratic form. Express solutions as natural logarithms.

33. $e^{2x}-5e^x+6=0$ **34.** $e^{2x}-e^x-6=0$

35. $e^{2x}-2e^x-10=0$ **36.** $e^{2x}-6e^x+9=0$

37. $e^{2x+2}+e^{x+2}=6e^2$ **38.** $e^{3x}+10e^{-x}=7e^x$

39. Prove that \$1.00 invested at a rate r (r a decimal) compounded continuously will grow to e^r dollars in one year. (*Hint:* See pages 180–181.)

40. Prove that P dollars invested at a rate r (r a decimal) compounded continuously will grow to Pe^{rt} dollars in t years. (See Exercise 39.)

41. Derive the rule of 70 as follows: If compounded continuously at a rate of $r\%$ the quantity P will grow to an amount $A = Pe^{0.01r}$ in one year and to $A = Pe^{0.01rt}$ in t years. Let $A = 2P$, take the natural logarithm of both sides, and solve for t.

C   **42.** It can be proved that e equals the limit of the sum of the infinite series

$$1 + \frac{1}{1!} + \frac{1}{2!} + \frac{1}{3!} + \cdots + \frac{1}{n!}$$

Approximate e by using the first 5 terms of this series. (*Note:* The symbol $n!$, read "n factorial," denotes the product $n\cdot(n-1)\cdot(n-2)\cdots 2\cdot 1$. For example, $4! = 4\cdot 3\cdot 2\cdot 1 = 24$.)

43. The points $A(0, f(0))$ and $B(h, f(h))$ are on the graph of the function $f(x) = e^x$. Find the slope of $\overline{AB}$ if h is
  a. 1   **b.** 0.1   **c.** 0.01

44. It can be proved that the graph of $y = e^x$ is tangent to the line $y = 1 + x$. Thus, $e^x \approx 1 + x$ when $|x|$ is small, and also $e \approx (1 + x)^{\frac{1}{x}}$ when $|x|$ is small. Show that this approximation for e agrees with the definition of e.

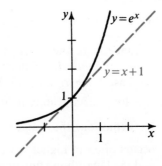

CALCULATOR EXERCISES

How long will the world's supply of crude oil last?
How long will the U.S.'s supply of coal last?

The expiration time T, in years, of a natural resource is the time remaining until it is all used up. If one assumes that the current growth rate of consumption remains constant, then:

$$T = \frac{1}{r} \ln \left(\frac{rR}{C} + 1 \right) \text{ where}$$

C = current consumption
r = current growth rate of consumption
R = resource size

Suppose that the world consumption of oil is growing at the rate of 7% per year ($r = 0.07$) and current consumption is approximately 17×10^9 barrels per year. Find the expiration time for these estimates of R.

1. $R = 1691 \times 10^9$ barrels. (estimate of remaining crude oil)

2. $R = 1881 \times 10^9$ barrels. (estimate of remaining crude oil plus shale oil)

An approximation of $n!$ for large n is given by *Stirling's Formula* below. (Exercise 42 on page 182 reviews the meaning of $n!$.)

$$n! \approx \sqrt{2\pi n} \cdot n^n \cdot e^{-n}$$

This approximation is useful for larger values of n.

3. Calculate 50!.

4. Use Stirling's Formula to approximate 50!.

5. Which is larger, e^π or π^e?

6. a. Evaluate $\left(1 - \frac{1}{n}\right)^n$ for $n = 100$, 10,000, and 1,000,000.

  b. Compare your answers in (a) with a decimal approximation for e^{-1}.

  c. What appears to be $\lim\limits_{n \to \infty} \left(1 - \frac{1}{n}\right)^n$?

Chapter summary

1. *Zero, negative,* and *fractional exponents* are defined as follows:

   a. $b^0 = 1$ **b.** $b^{-x} = \dfrac{1}{b^x}$ **c.** $b^{\frac{p}{q}} = \sqrt[q]{b^p}$ or $\left(\sqrt[q]{b}\right)^p$

2. **a.** A function of the form $f(x) = ab^x$ $(a > 0,\ b > 0,\ b \neq 1)$ is called an *exponential function* with base b. If $a = 1$, then the inverse of $f(x) = b^x$ is the base b *logarithmic function* $f^{-1}(x) = \log_b x$.

   b. $y = \log_b x$ if and only if $b^y = x$.   **c.** $\log_b b^x = x$ and $b^{\log_b x} = x$.

   d. By definition, the number $e = \underset{n \to \infty}{\text{limit}}\left(1 + \dfrac{1}{n}\right)^n \approx 2.7183$. This number occurs often in higher mathematics. The exponential function is $y = e^x$ and the logarithmic function is $y = \log_e x$, usually written as $y = \ln x$ for the *natural logarithm* of x.

3. Since the exponential and logarithmic functions are inverses, their laws are closely related.

Laws of Exponents	*Corresponding Laws of Logarithms*
1. $b^x \cdot b^y = b^{x+y}$	1. $\log_b MN = \log_b M + \log_b N$
2. $\dfrac{b^x}{b^y} = b^{x-y}$	2. $\log_b \dfrac{M}{N} = \log_b M - \log_b N$
3. $b^x = b^y$ if and only if $x = y$. $(b \neq 0, 1, -1)$	3. $\log_b M = \log_b N$ if and only if $M = N$.
4. $(ab)^x = a^x b^x$	
5. $\left(\dfrac{a}{b}\right)^x = \dfrac{a^x}{b^x}$	
6. $a^x = b^x$ if and only if $a = b$ $(a > 0, b > 0, x \neq 0)$	
7. $(b^x)^y = b^{xy}$	4. $\log_b M^k = k \log_b M$

4. To solve an *exponential equation,* take the logarithm of both sides of the equation (see Example 1, Section 5–6). In the special case where both sides of the equation can be expressed as powers of the same number, logarithms are not necessary. (See page 169.)

5. To find the logarithm of a number to any base use the method of Example 2, page 170, or apply the theorem on page 170. As a corollary to the theorem we have $\log_b c = \dfrac{1}{\log_c b}$.

6. A quantity that increases (or decreases) by the same percent each year is said to *grow* (or *decay*) *exponentially.* The exponential growth (or decay) is given by the equation

$$N = N_0 b^t,$$

where N_0 is the original amount, N is the amount at time t, and b is 1 plus (or minus) the growth (or decay) rate, expressed as a decimal. (See Examples 1–5 of Section 5–7.)

7. The *rule of 70* provides an approximation of the doubling times for exponential growth. If a quantity is growing at $r\%$ per year, then

$$\text{doubling time} \approx 72 \div r.$$

8. The amount of time in which a quantity decays or depreciates to half of its original value is called its *half-life*.

Chapter test

(No tables or calculators are required for this test.)

1. Evaluate the following: 5-1

    a. $(6^{-1})^{-2}$ **b.** $\dfrac{3^5 \cdot 3^{-4}}{3^{-2}}$ **c.** $(3^{-2} + 3^0)^{-1}$ **d.** $\sqrt{\dfrac{4^6}{2^{-4}}}$

2. Simplify the following: 5-2

    a. $(8a^3)^{\frac{2}{3}}$ **b.** $(9x^{\frac{2}{3}})^{-\frac{1}{2}}$ **c.** $\dfrac{2x^{\frac{1}{2}} \cdot x^{-\frac{1}{2}}}{2x^{-\frac{1}{2}}}$ **d.** $\dfrac{2x^{\frac{1}{2}} + x^{-\frac{1}{2}}}{2x^{-\frac{1}{2}}}$

3. Solve.

    a. $(2x)^{-2} = 16$ **b.** $5y^{-\frac{1}{2}} = 20$ **c.** $8^y = \dfrac{2^y}{2}$ **d.** $x^{\frac{2}{3}} - 4x^{\frac{1}{3}} + 3 = 0$

4. If $f(0) = \frac{1}{3}$ and $f(2) = 12$, write an equation of the exponential function $f(x) = ab^x$.   5-3

5. Evaluate.   5-4

    a. $\log_2 16$ **b.** $\log_5 125\sqrt{5}$ **c.** $\log_{10} 0.001$ **d.** $\log_6 \sqrt[3]{\frac{1}{6}}$

6. a. Sketch the graphs of $y = 2^x$ and $y = \log_2 x$ on the same set of axes.

    b. State the domain and range of each function.

7. Express each of the following base b logarithms in terms of $\log M$ and $\log N$.   5-5

    a. $\log (MN)^2$ **b.** $\log \dfrac{\sqrt[3]{M}}{N}$

8. If $\log y = 3 \log x - \log 3$, express y in terms of x.

9. Solve.

    a. $\log_5 |x| = -2$ **b.** $\log_4 (x + 3) + \log_4 (x - 3) = 2$

10. Solve the equation $5^x = 8$ using $\log_{10} 5 \approx 0.7$ and $\log_{10} 2 \approx 0.3$.   5-6

11. The value of a new \$10,000 car depreciates 30% per year. What is its value after 2 years?   5-7

12. A radioactive element has a half-life of 6 years. If you have 10 kg of the element now, how much will be left after 3 years?

13. Evaluate.   5-8

    a. $\ln e$ **b.** $\ln e^4$ **c.** $e^{\ln 4}$

14. a. If you invest \$1000 at 12% interest, compounded quarterly, how much money will you have after 1 year?

    b. Approximately how long will it take to double your money?

Cumulative review (Chapters 1–5)

1. If $A = (0, -3)$ and $B = (-8, 1)$, find:
   a. the midpoint and length of $\overline{AB}$
   b. an equation in slope-intercept form for $\overline{AB}$
   c. an equation for the perpendicular bisector of $\overline{AB}$

2. Solve simultaneously and illustrate with a sketch:
$$4x + 3y = 4$$
$$3x + y = -2$$

3. Line j has equation $2x - 5y = -3$. Write an equation in general form for the line through $(-1, 4)$ that is (a) parallel to j; (b) perpendicular to j.

4. Simplify: a. $\dfrac{1}{\sqrt{-32}} + \sqrt{-8}$ b. $(-3 + 2i)(7 - i)$ c. $\dfrac{i}{1 - 2i}$

5. Use the discriminant of the equation $9x^2 - 18x - 16 = 0$ to show that the roots are rational. Then solve the equation by factoring.

6. Solve by completing the square: $x^2 - 8x - 12 = 0$.

7. Solve: $\dfrac{t}{t - 2} = \dfrac{5}{t + 7}$

8. Find the radius of the circle through $(-2, -10)$ and with center $(4, -2)$. Write an equation for the circle.

9. Find the center and radius of the circle with equation
$$x^2 + y^2 - 6x + 12y - 4 = 0.$$

10. $ABCD$ is the parallelogram shown.
    a. What are the coordinates of C? of M?
    b. Write an equation expressing the fact that $\overline{AC} \perp \overline{BD}$.
    c. Simplify the equation in part (b) to show that $a^2 = b^2 + c^2$.
    d. What kind of quadrilateral must $ABCD$ be? Why?

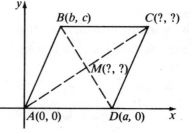

11. If $f(x) = x^2 - 3x + 1$, find the zeros of $f(x)$. Then evaluate $f(x + 1)$ and $f(-i)$.

12. Solve: a. $2x^3 - 7x^2 + 4x - 14 = 0$ b. $z^4 - 12z^2 + 32 = 0$

13. Sketch the graph of the parabola with equation $y = -2x^2 - 4x - 2$. Label the vertex and the axis of symmetry.

14. Sketch the graphs of $y = -3x$ and $y = x^3 - 4x^2$ on a set of axes. Determine algebraically the coordinates of each point of intersection.

15. Show that $x = \frac{2}{3}$ is a root of $3x^3 - 8x^2 + 19x - 10 = 0$.

16. When a polynomial $P(x)$ is divided by $x - 2$, the quotient is $3x^2 - x - 5$ and the remainder is -1. Find $P(x)$ and $P(-1)$.

17. a. Show that $x^3 + x^2 + 5 = 0$ has no rational root but does have an irrational root between -2 and -3.

   b. (Optional) Approximate the root in part (a) to the nearest tenth.

18. A ball is thrown upward with a speed of 20 m/s from a hill 32 m high. The approximate height of the ball t seconds later is given by $h(t) = 32 + 20t - 5t^2$. What is the maximum height attained by the ball?

19. A quartic equation $x^4 + ax^3 + bx + c = 0$ has integral coefficients. Two of its roots are i and $2 - \sqrt{5}$. Find the sum and product of the roots and use these to find the values of a, b, and c.

20. Solve and graph on a number line:

   a. $|6 - 4x| > 2$ **b.** $x^4 - 8x^2 - 9 < 0$

21. Find the values of k for which the line $x - y = k$ **(a)** is tangent to the circle $x^2 + y^2 = 4$; **(b)** intersects the circle in two points.

22. For the linear function $f(x) = mx + k$, $f(1) = -2$ and $f(-2) = 11$.

   a. Find the values of m and k.

   b. Find the change in the value of $f(x)$ when x increases by 2.

23. If $f(x) = 3x - 1$ and $g(x) = \frac{1}{2}x^2$, find $f(g(4))$ and $g(f(4))$.

24. What must be true of a function $f(x)$ for $f^{-1}(x)$ to exist?

25. If $h(x) = \dfrac{1}{x - 1}$, find $h^{-1}(x)$. Then give the domain and the range of $h(x)$ and $h^{-1}(x)$.

26. Evaluate: **a.** $(2^{-1} \cdot 2^{-3})^{-2}$ **b.** $(2^{-1} - 2^{-3})^{-2}$

27. Solve: **a.** $27^{-\frac{2}{3}} = 3^0 \cdot 9^{\frac{3x}{2}}$ **b.** $(1 + x)^{-\frac{1}{4}} = \dfrac{1}{2^{-1}}$

28. If $f(x)$ is an exponential function with $f(0) = 16$ and $f(3) = \frac{27}{4}$, evaluate $f(-1)$.

29. Evaluate: **a.** $\log_8 32$ **b.** $\log_8 \sqrt[6]{\dfrac{1}{64}}$ **c.** $\log_{10} (0.01)^{-3}$

30. If $g(x) = \log_4 x$, graph $g(x)$ and $g^{-1}(x)$. State a formula for $g^{-1}(x)$.

31. Simplify: $\log_{15} 10 + 2 \log_{15} 3 - \frac{1}{2} \log_{15} 36$

32. Between what two integers does $\log 486$ lie?

33. If $\log 3.09 \approx 0.4900$, solve to the nearest hundredth:

   a. $(3.09)^x = 1000$ **b.** $\log_{0.1} 3.09 = x$

34. Carbon-11 has a half-life of 20 min. If 500 g are present now, how much will remain after one hour?

35. Express $\ln \dfrac{5}{e^2}$ in terms of common logarithms.

36. Suppose you invest \$1000 at 9% interest for 4 years. Write an expression that represents the value of the investment if the interest is compounded **(a)** monthly; **(b)** continuously. About how long will it take for the value of the investment to double?

CHAPTER SIX

Trigonometric functions

OBJECTIVES

1. To convert between degree and radian measure.
2. To find the area and arclength of a sector.
3. To define angles in terms of positive and negative rotations.
4. To graph, and find values of, the trigonometric functions.
5. To find the inclination of a line.
6. To simplify trigonometric expressions, prove trigonometric identities, and solve trigonometric equations.

Angles, arcs, and sectors

6-1/MEASUREMENT OF ANGLES

Perhaps the most natural unit in which to measure very large angles is the **revolution.** It is this unit that is used when people talk about racing an automobile engine at 5000 revolutions per minute. And it is common to refer to a phonograph record as a "45 RPM" or a "$33\frac{1}{3}$ RPM," where RPM stands for revolutions per minute.

In the photograph of San Francisco at the left, the frame houses appear larger than the skyscrapers in the background. The mathematical concept of apparent size, discussed on page 196, provides an explanation of such misleading visual relationships.

189

To measure smaller angles, we need a smaller unit. The most common unit for measuring smaller angles is the **degree.** The convention of having 360 degrees in a complete revolution can be traced to the fact that the Babylonian numeration system was based on the number 60. One theory is that mathematicians using this system naturally subdivided the angles of an equilateral triangle into 60 equal pieces. Each of these little pieces came to be regarded as a unit and eventually was called a degree. Since six equilateral triangles can be arranged within a circle as shown, $6 \times 60 = 360$ degrees was called the measure of a complete revolution.

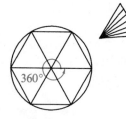

Each degree can be subdivided into decimal parts or, from the Babylonian base-60 numeration system, into 60 parts called minutes. Each minute in turn can be subdivided into 60 parts called seconds.* The measure of, say, 34 degrees, 15 minutes, and 20 seconds is written $34°15'20''$. To convert between decimal degrees and degrees, minutes, and seconds, you can use a calculator or reason as follows.

$$12.3° = 12° + 0.3(60)' = 12°18'$$
$$25°20'6'' = 25° + \tfrac{20}{60}° + \tfrac{6}{3600}° = 25.335°$$

Relatively recently in mathematical history, another unit of angle measurement, the **radian,** has come into widespread use. An angle of 1 radian is shown in the diagram at the left below. When an arc of a circle has the same length as the radius of the circle, the measure of its central angle, $\angle AOB$, is by definition 1 radian. As shown at the right, the central angle will have measure 1.5 radians

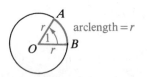

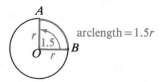

when the arclength is 1.5 times the radius. In general, the radian measure of the central angle, $\angle AOB$, is the number of radius units in the length of arc AB. This accounts for the name radian. In the diagram at the right, the measure θ (Greek theta) of the central angle is given by

arclength $= s = r\theta$

$$\theta = \frac{s}{r}.$$

Let us use this equation to see how many radians correspond to a complete revolution of 360°. In this case, the arclength of a complete revolution is the circumference of the circle, $2\pi r$, so that

$$\theta = \frac{s}{r} = \frac{2\pi r}{r} = 2\pi.$$

*The first subdivision of a degree was called in Latin *pars minuta prima* (the first small part), from which we get "minute." The second subdivision was called *pars minuta secunda* (the second small part), from which we get "second."

Hence, a complete revolution measured in radians is 2π and measured in degrees is 360. Consequently,

$$2\pi \text{ radians} = 360 \text{ degrees}, \quad \text{and}$$
$$\pi \text{ radians} = 180 \text{ degrees}.$$

This gives us the following two conversion formulas:

$$1 \text{ radian} = \frac{180}{\pi} \text{ degrees} \approx 57.2958 \text{ degrees}$$

$$1 \text{ degree} = \frac{\pi}{180} \text{ radians} \approx 0.0174533 \text{ radians}$$

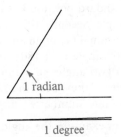

Angle measures that come out evenly in degrees do not come out evenly in radians, and vice versa. That is why angles measured in radians are frequently given as fractional multiples of π. Angles whose measures are multiples of $\frac{\pi}{4}$, $\frac{\pi}{3}$, and $\frac{\pi}{6}$ appear often in trigonometry. The following diagrams will help you keep the degree conversions for these angles in mind. Note that degree measure, such as 90°, is usually written with the degree symbol (°), while radian measure, such as $\frac{\pi}{2}$, is usually written without any special symbol.

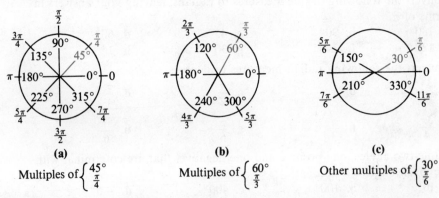

In trigonometry, an angle often represents a rotation about a point. Thus in each diagram below you can think of rotating the inital ray of an angle (the positive x-axis) to its terminal ray. Counterclockwise rotations are positive, and clockwise rotations are negative.

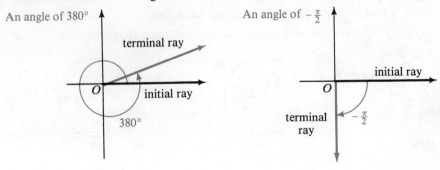

An angle whose vertex is at the origin and whose initial ray is the positive x-axis is said to be in **standard position.** If the terminal ray of an angle in standard position lies in the first quadrant, the angle is called a first-quadrant angle. Second-, third-, and fourth-quadrant angles are similarly defined. If the terminal ray of an angle in standard position lies on either axis—for example, a $90°$ angle or a $180°$ angle—the angle is called a **quadrantal angle.**

Two angles in standard position are called **coterminal angles** if they have the same terminal ray. In fact there are an infinite number of angles coterminal with any given angle.

For example, the angles $\frac{\pi}{4} \pm 2\pi, \frac{\pi}{4} \pm 4\pi, \ldots, \frac{\pi}{4} \pm 2n\pi$ (n an

integer) are all coterminal with the angle $\frac{\pi}{4}$.

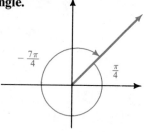

Coterminal angles

ORAL EXERCISES 6-1

1. Convert the following degree measures to radians, leaving your answers in terms of π.
   a. $180°$ **b.** $90°$ **c.** $45°$ **d.** $60°$
   e. $120°$ **f.** $240°$ **g.** $30°$ **h.** $1°$

2. Convert the following radian measures to degrees.
   a. 2π **b.** π **c.** $\frac{\pi}{2}$ **d.** $\frac{\pi}{4}$

   e. $\frac{3\pi}{4}$ **f.** $\frac{\pi}{3}$ **g.** $\frac{\pi}{6}$ **h.** $\frac{5\pi}{6}$

3. Name two angles, one positive and one negative, that are coterminal with the given angle.
   a. $10°$ **b.** $100°$ **c.** $400°$ **d.** $-5°$

   e. π **f.** $\frac{\pi}{2}$ **g.** 4π **h.** $-\frac{\pi}{3}$

4. **a.** The equation $\theta = (60 + 360n)°$, where n is an integer, represents all angles θ coterminal with an angle of _?_ °.
   b. What would be the equivalent equation in radians?

5. Give the radian measure of θ in each of the following diagrams.

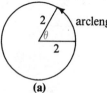

(a)

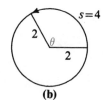

(b)

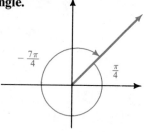

(c)

A   **1.** Study figures (a), (b), and (c) on page 191. Then, without looking at these figures, make a copy of each. Time yourself!

2. Give the degree measure and radian measure of the angle between the hour hand and the minute hand of a clock at
   a. 1 o'clock   **b.** 2 o'clock   **c.** 4 o'clock
   d. 5 o'clock   **e.** 5:30 P.M.

Express in radians, leaving answers in terms of π.

3. a. 315°   **b.** 225°   **c.** 15°   **d.** −45°

4. a. −90°   **b.** 135°   **c.** −180°   **d.** −225°

5. a. −120°   **b.** −240°   **c.** 300°   **d.** 360°

6. a. 210°   **b.** −135°   **c.** −210°   **d.** −315°

Convert the following radian measures to degrees.

7. a. $-\dfrac{\pi}{2}$ **b.** $\dfrac{4\pi}{3}$ **c.** $-\dfrac{3\pi}{4}$ **d.** $-\dfrac{\pi}{6}$

8. a. $-\dfrac{5\pi}{6}$ **b.** -2π **c.** $\dfrac{5\pi}{4}$ **d.** $-\dfrac{\pi}{3}$

9. a. π **b.** $-\dfrac{3\pi}{2}$ **c.** $\dfrac{2\pi}{3}$ **d.** $\dfrac{7\pi}{6}$

10. a. $-\dfrac{\pi}{4}$ **b.** $\dfrac{7\pi}{4}$ **c.** 4π **d.** $\dfrac{11\pi}{6}$

11. Give the radian measure of θ if
   a. $r = 5$ and $s = 6$.
   b. $r = 8$ and $s = 6$.

12. Give the radian measure of θ if
   a. $r = 4$ and $s = 5$.
   b. $r = 6$ and $s = 15$.

Ex. 11, 12

Besides using a calculator or the conversion formulas given on page 191, you can convert from degrees to radians or from radians to degrees and minutes by using Table 3, 4, or 5 at the back of the book.

EXAMPLE.   Convert **(a)** 196° to radians and **(b)** 1.65 radians to degrees and minutes.

SOLUTION: **a.** 196° = 180° + 16°
   $\approx \pi + 0.28$ (from Table 3)
   $\approx 3.14 + 0.28 = 3.42$

   b. 1.65 = 1 + 0.65
   $\approx 57°18' + 37°15'$ (from Table 5)
   $= 94°33'$

Express in radians, giving answers to the nearest hundredth.

13. a. 95°   **b.** 110°   **c.** 95°10′   **d.** 119.2°
14. a. 212°   **b.** 365°   **c.** 200°40′   **d.** 240.8°

Convert the following radian measures to degrees. Give your answers to the nearest minute or tenth of a degree.

15. a. 1.6   **b.** 1.7   **c.** 1.21   **d.** 1.32
16. a. 2.2   **b.** 3.7   **c.** 2.82   **d.** 3.41

Estimate (by sight) the size in radians of the angles shown below. Then measure the angles with a protractor and convert from degrees to radians to find their actual size.

17.   **18.**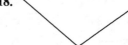

Give the measures of two angles, one of which has a positive measure and one of which has a negative measure, that are coterminal with the given angle.

19. a. 500°   **b.** −60°   **c.** $\dfrac{\pi}{4}$ **d.** $-\dfrac{2\pi}{3}$

20. a. 1000°   **b.** −100°   **c.** $\dfrac{4\pi}{3}$ **d.** $-\dfrac{\pi}{6}$

B **21. a.** 360°30′   **b.** −90°40′   **c.** 3°21′   **d.** 115°15′
       22. a. 180°20′   **b.** −270°30′   **c.** 11°44′   **d.** 172°11′
       23. a. 28.5°   **b.** 116.3°   **c.** −60.4°   **d.** −315.3°
       24. a. 38.4°   **b.** 127.6°   **c.** −50.8°   **d.** −320.7°

25. Let n be any integer. Give an expression in terms of n for the measure of all angles that are coterminal with an angle of 29.7°.

26. Let n be any integer. Give an expression in terms of n for the measure of all angles that are coterminal with an angle of −116°10′.

Each of the following exercises gives the speed of a revolving gear. Find **(a)** the number of degrees per minute through which each gear turns and **(b)** the number of radians per minute. Give your answers to the nearest hundredth.

27. 35 RPM   **28.** 27 RPM   **29.** 2.5 RPM
30. 6.5 RPM   **31.** 14.6 RPM   **32.** 19.8 RPM

CALCULATOR EXERCISES

Convert the following from decimal degrees to degrees, minutes, and seconds.

1. 12.8°   **2.** 74.38°   **3.** −18.44°

Convert the following to decimal degrees.

4. 56°12′   **5.** 80°25′48″   **6.** 17°32′30″

7. Find the smallest positive angle coterminal with an angle of one million degrees.

8. Convert **(a)** 83.7° and **(b)** 271.4° to radians.

9. Convert **(a)** 3 radians and **(b)** 1.23 radians to decimal degrees.

10. Here is an alternate method for converting from decimal degree measure to radian measure. The method uses the functions sine and inverse sine that will be discussed in later sections. First, put the calculator into degree mode. Enter the degree measure to be converted and press the key for the sine function. Next, switch the calculator to radian mode. Finally, press the key for the inverse sine function (INV SIN, ARC SIN or SIN⁻¹). This method works for angles between −90° and +90°. Can you figure out why it works?

11. Invent a method like that in Exercise 10 for switching from radians to degrees.

6-2 / SECTORS OF CIRCLES

Pictured in color at the right is a **sector** of a circle. Your geometrical intuition should tell you that, like the arclength s, the area K of the sector can be expressed in terms of the radius r and the central angle θ.* Take a simple example. Suppose the central angle is 60° and the radius is 12. Then the arclength of the sector is $\frac{60}{360} = \frac{1}{6}$ of the whole circumference, or $\frac{1}{6}(2\pi r) = \frac{1}{6}(2\pi \cdot 12) = 4\pi$. Similarly, the area of the sector is $\frac{1}{6}$ of the area of the whole circle, or $\frac{1}{6}\pi r^2 = \frac{1}{6}\pi \cdot 12^2 = 24\pi$.

In general, if θ is expressed in degrees, then the arclength s and the area K are given by the following formulas:

$$(1) \quad \text{Arclength } s = \frac{\theta}{360} \cdot 2\pi r \quad (\theta \text{ in degrees})$$

$$(2) \quad \text{Area } K = \frac{\theta}{360} \cdot \pi r^2 \quad (\theta \text{ in degrees})$$

If θ is given in radians, we have:

$$(1a) \quad s = \frac{\theta}{2\pi} \cdot 2\pi r, \text{ or } s = r\theta \quad (\theta \text{ in radians})$$

$$(2a) \quad K = \frac{\theta}{2\pi} \cdot \pi r^2, \text{ or } K = \frac{1}{2}r^2\theta \quad (\theta \text{ in radians})$$

Note that equation (1a), when written in the form $\theta = \frac{s}{r}$, is the equation that defines the radian measure of an angle (page 190). By combining equations (1a) and (2a), we can obtain still another area formula:

$$K = \tfrac{1}{2}r^2\theta = \tfrac{1}{2}r(r\theta) = \tfrac{1}{2}rs$$

$$(3a) \quad K = \tfrac{1}{2}rs$$

*In this text, θ (or a similar symbol) will sometimes represent an angle and other times represent its measure. The meaning will be clear from context.

Notice that formulas (1a) and (2a) on the previous page are more straightforward than formulas (1) and (2). In fact, one reason for using radian measure is that many formulas in calculus are expressed more simply in radians than in degrees.

EXAMPLE 1. A sector of a circle has arclength 6 cm and area 75 cm². Find its radius and the measure of its central angle.

SOLUTION: Let r be the unknown radius and θ the measure of the central angle in radians. Then, substituting in formula (3a),

$$K = \tfrac{1}{2}rs; \quad 75 = \tfrac{1}{2}r(6); \quad r = 25 \text{ cm}$$

From the arclength formula (1a), $\theta = \dfrac{s}{r} = \dfrac{6}{25} = 0.24$ radians, or approximately 14°.

Apparent size

It is a commonplace observation that objects close to you look big, whereas the same objects far away look small. If you think about it, you will realize that how big something looks is determined not only by its size but also by the angle it subtends at your eye. The measure of this angle is called the **apparent size** of the object. For example, the planet Jupiter looks small to people on Earth not because it is small but because it subtends a small angle at our eyes. (See diagram below.)

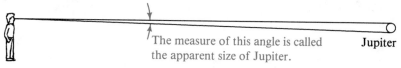

The measure of this angle is called the apparent size of Jupiter.   Jupiter

Let us suppose that Jupiter has an apparent size of $0.01° \approx 0.0001745$ radians when it is 8×10^8 kilometers away from Earth. The exaggerated figure below shows that the diameter of Jupiter is approximately the arclength of a sector. Thus we can find the diameter by using formula (1a):

diameter $\approx s = r\theta$
$\approx (8 \times 10^8)(0.0001745)$
$\approx 140,000$ km

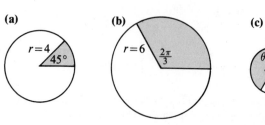

$r = 8 \times 10^8$ km
$\theta = 0.0001745$
diameter $\approx$ arc length of sector

ORAL EXERCISES 6-2

1. Find the arclength and area of each of the sectors shown at the right.

(a) $r = 4$, $45°$

(b) $r = 6$, $\dfrac{2\pi}{3}$

(c) $\theta = 4$, $r = 2$

2. a. The apparent size of a tall building 2 km away is 0.05 radians. What is the approximate height of the building?

b. Why can you apply a sector formula to triangle ABC and get a good approximation to the distance BC?

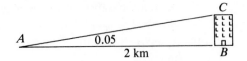

REVIEW EXERCISES

(More practice in converting between degrees and radians)

1. Convert the following degree measures to radian measures. Give answers as fractions of π.

a. 60°	**b.** 120°	**c.** 150°	**d.** −90°
e. 270°	**f.** 45°	**g.** 135°	**h.** −225°
i. 315°	**j.** 210°	**k.** 330°	**l.** −300°

2. Convert the following radian measures to degree measures.

a. $\dfrac{\pi}{6}$	**b.** $\dfrac{5\pi}{6}$	**c.** $\dfrac{\pi}{2}$	**d.** $\dfrac{2\pi}{3}$
e. $-\dfrac{\pi}{4}$	**f.** $\dfrac{3\pi}{4}$	**g.** $-\dfrac{\pi}{3}$	**h.** $\dfrac{5\pi}{3}$
i. $-\dfrac{3\pi}{2}$	**j.** $\dfrac{7\pi}{4}$	**k.** $\dfrac{7\pi}{6}$	**l.** π

WRITTEN EXERCISES 6–2

A

1. A circular sector is drawn with radius 6 cm and central angle 0.5 radians.
 a. Draw the sector with ruler and protractor.
 b. Find its arclength and area.

2. A circular sector is drawn with radius 5 cm and central angle 3 radians.
 a. Draw the sector.
 b. Find its arclength and area.

3. A circular sector has arclength 11 cm and central angle 2.2 radians. Find its radius and area.

4. A circular sector has arclength 2.0 cm and central angle 0.4 radians. Find its radius and area.

5. A sector of a circle has area 25 cm² and central angle 0.5 radians. What are its radius and arclength?

6. A sector of a circle has area 90 cm² and central angle 0.2 radians. What are its radius and arclength?

7. A sector has central angle 30° and arclength 3.5 cm. What is its area to the nearest square centimeter?

8. A sector has central angle 24° and arclength 8.4 cm. What is its area to the nearest square centimeter?

9. A sector has perimeter 7 cm and area 3 cm². What is its radius? (Two answers)

10. A sector has perimeter 12 cm and area 8 cm². What is its radius? (Two answers)

In Exercises 11 and 12 leave answers in terms of π.

11. A phonograph record turns at 45 RPM (revolutions per minute).
    a. Through how many degrees does it turn in a minute?
    b. Through how many radians does it turn in a minute?
    c. If the diameter of the record is 18 cm, find the distance that a point on the rim travels in 1 minute.

12. Find the distance traveled in 1 minute by a point on the rim of a $33\frac{1}{3}$ RPM phonograph record with a 30-centimeter diameter.

13. The diameter of the moon is about 3500 km. Its apparent size is about 0.0087 radians. About how far is it from Earth?

14. At its closest approach, Mars is about 56 million km from Earth. At this time, its apparent size is 0.00012 radians. What is the diameter of Mars?

B 15. The moon and the sun have approximately the same apparent size for viewers on Earth (but not for viewers on other planets). The distances from Earth to the moon and the sun are approximately 4×10^5 km and 1.5×10^8 km, respectively. The diameter of the moon is approximately 3500 km. What is the diameter of the sun? (*Hint:* Use the fact that the sun and moon have the same apparent size.)

16. What is the apparent size of an object 1 cm long held at arm's length?

17. A cow is tethered to a post alongside a barn 10 m wide and 30 m long. If the post is 10 m from a corner of the barn and if the rope is 30 m long, find the cow's total grazing area.

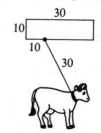

18. This exercise will show how Eratosthenes (about 276 B.C.–194 B.C.) determined the circumference of Earth. It was reported to Eratosthenes that at noon on the first day of summer the sun was directly overhead in the city of Syene because there was no shadow in a deep well. Eratosthenes observed at this same time in the city of Alexandria that the sun's rays made an angle $\theta = 7\frac{1}{5}°$ with a vertical pole.

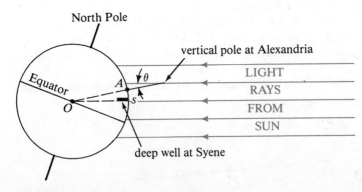

a. How did Eratosthenes conclude that the measure of $\angle AOS = \theta = 7\frac{1}{5}°$?

b. If Alexandria was known to be 5000 stadia due north of Syene, show how Eratosthenes could conclude that the circumference of Earth was about 250,000 stadia (1 stadium $\approx$ 0.168 km).

c. What is the percentage difference between Eratosthenes' value for the circumference of Earth and the modern value of approximately 40,067 km?

$$\left(\text{Percentage difference} = \frac{\text{old value} - \text{modern value}}{\text{modern value}} \times 100.\right)$$

19. You are traveling in a car toward a certain mountain at a speed of 80 km/h. The apparent size of the mountain is $\frac{1}{2}°$. Fifteen minutes later the same mountain has an apparent size of $1°$. Approximately how tall is the mountain?

20. A ship is approaching a lighthouse known to be 20 m high. The apparent size of the lighthouse is 0.005 radians. Ten minutes later the lighthouse has an apparent size of 0.010 radians. What is the speed of the ship in km/h?

21. The sector shown has perimeter 20 cm.

a. Show that $\theta = \dfrac{20}{r} - 2$ and that the area of the sector is $K = 10r - r^2$.

b. What value of r gives the maximum possible area of a sector of perimeter 20 cm? (*Suggestion:* It might help to make a graph of all the possibilities, showing the radii r on the horizontal axis and the areas K on the vertical axis.)

c. What is the measure of the central angle of the sector of maximum area?

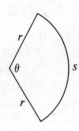

22. The purpose of this exercise is to derive the formula for the area of a circle by first deriving the formula for the area of a sector. Consider the sector with radius r and arclength s shown in the diagram. Inscribed in the sector are n congruent isosceles triangles, with heights h and bases b ($n = 5$ in the diagram).

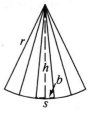

  a. Show that the total area of the inscribed triangles is $\frac{1}{2}nbh$.
  b. As n gets larger and larger,
     (1) the value of h gets closer and closer to the value of _?_.
     (2) the value of nb gets closer and closer to the value of _?_.
  c. Use parts (a) and (b) to explain how to derive the formula $K = \frac{1}{2}rs$.
  d. Derive the formula for the area of a circle from $K = \frac{1}{2}rs$.

CALCULATOR EXERCISES

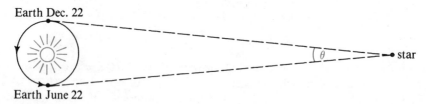

Earth Dec. 22

Earth June 22

star

Some stars are so far away that they do not appear to shift as Earth orbits the sun. Suppose, however, that the "nearby" star shown above appears to shift through an arc of $\theta = 0°0'2.2''$ when viewed on the first day of winter and the first day of summer. We determine θ by measuring the angle between the star's former position and its current position.

1. If the distance from Earth to the sun is about 1.5×10^8 km, find the approximate distance from Earth to the star.

2. Give the distance found in Exercise 1 in light years. (A light year is the distance light travels in one year. Use the fact that light travels 3.00×10^8 m/s.)

The trigonometric functions

6-3/THE SINE AND COSINE FUNCTIONS

If $P(x, y)$ is a point on the circle $x^2 + y^2 = r^2$ and if θ is an angle in standard position with terminal ray OP, then the **sine** and **cosine** of θ are defined as follows in terms of the coordinates of P.

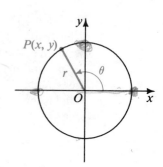

$$\text{sine of } \theta = \sin \theta = \frac{y}{r} \qquad\qquad \text{cosine of } \theta = \cos \theta = \frac{x}{r}$$

EXAMPLE 1. If the terminal ray of θ passes through $(-3, 2)$, find $\sin \theta$ and $\cos \theta$.

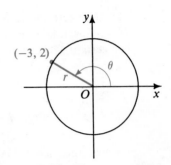

SOLUTION: Since $r^2 = (-3)^2 + 2^2 = 13$,
$$r = \sqrt{13}.$$

Thus, $\sin \theta = \dfrac{y}{r} = \dfrac{2}{\sqrt{13}}$ and

$$\cos \theta = \dfrac{x}{r} = \dfrac{-3}{\sqrt{13}}.$$

EXAMPLE 2. If θ is a fourth-quadrant angle and $\sin \theta = -\frac{5}{13}$, find $\cos \theta$.

SOLUTION: Make a diagram of a circle with radius 13 as shown. Then use the equation of the circle to find x. Notice that since $\sin \theta = \dfrac{y}{r} = -\dfrac{5}{13}$ and r is always positive, $y = -5$.

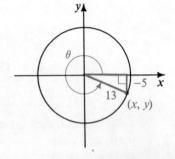

$$x^2 + y^2 = r^2$$
$$x^2 + (-5)^2 = 13^2$$
$$x^2 + 25 = 169$$
$$x^2 = 144$$
$$x = \pm 12$$

Since θ is a fourth-quadrant angle, x is positive. Therefore, $x = 12$ and

$$\cos \theta = \dfrac{x}{r} = \dfrac{12}{13}.$$

Although the definitions of $\sin \theta$ and $\cos \theta$ refer to the radius r of a circle, the values of $\sin \theta$ and $\cos \theta$ depend only on the size of θ and not on the size of r. You can see this by referring to the circles shown in the diagram below. In the small circle,

$$\sin \theta = \frac{y\text{-coordinate of } P}{\text{radius}} = \frac{AP}{OP}.$$

In the large circle,

$$\sin \theta = \frac{y\text{-coordinate of } Q}{\text{radius}} = \frac{BQ}{NQ}.$$

Since $\triangle OAP$ and $\triangle NBQ$ are similar (AA similarity), the two sine ratios above are the same even though radius OP is smaller than radius NQ. Given a particular value for θ, then, we can choose any value for r to find $\sin \theta$ and $\cos \theta$.

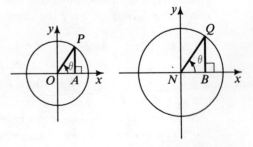

The easiest radius with which to work is $r = 1$. In this case the circle, $x^2 + y^2 = 1$, is called the **unit circle.** Using the unit circle can help us remember certain facts about the sine and cosine functions.

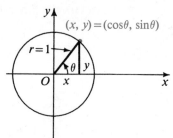

$$\sin \theta = \frac{y}{r} = \frac{y}{1} = y$$

$$\cos \theta = \frac{x}{r} = \frac{x}{1} = x$$

For example, since $\sin \theta = y$, $\sin \theta$ is positive when y is positive (first and second quadrants). Likewise, the sign of x determines the sign of $\cos \theta$.

Function value	Quadrant of the terminal ray of θ			
	First	Second	Third	Fourth
$\sin \theta$	+	+	−	−
$\cos \theta$	+	−	−	+

We can also remember the domain and range of $\sin \theta$ and $\cos \theta$ by referring to the unit circle. Since $\sin \theta$ and $\cos \theta$ are defined for any angle θ, θ can be any real number. Since the values of $\sin \theta$ and $\cos \theta$ are coordinates of points on the unit circle, the range of both functions is the set of real numbers between -1 and 1, inclusive.

EXAMPLE 3. Find **(a)** $\sin 90°$, **(b)** $\sin 450°$, and **(c)** $\cos (-\pi)$.

SOLUTION:

a.
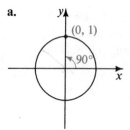

b.
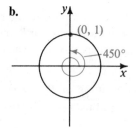

c.
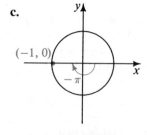

$\sin 90° = y\text{-coordinate} = 1$ $\sin 450° = y\text{-coordinate} = 1$ $\cos(-\pi) = x\text{-coordinate} = -1$

Figures (a) and (b) in Example 3 show coterminal angles whose sines are equal to 1. Example 4 shows that there are many other such angles. All coterminal angles, in fact, have the same sine.

EXAMPLE 4. Solve $\sin \theta = 1$ for θ in degrees.

SOLUTION: $\theta = 90°, \ 90° + 360°, \ 90° + 2 \cdot 360°, \ 90° + 3 \cdot 360°, \ldots,$
$90° - 360°, \ 90° - 2 \cdot 360°, \ 90° - 3 \cdot 360°, \ldots.$

Note that every solution in Example 4 equals $90°$ plus or minus a multiple of $360°$. This can be written more conveniently as $\theta = 90° + n \cdot 360°$, where n is any integer (positive, negative, or zero). If the angle were to be found in radians, the corresponding statement would be $\theta = \frac{\pi}{2} + n \cdot 2\pi$ or $\theta = \frac{\pi}{2} + 2\pi n$.

You can see from the previous example and from the given definitions that $\sin\theta$ and $\cos\theta$ repeat their values every 360° or 2π radians. Formally this means that

$$\sin(\theta + 360°) = \sin\theta \qquad \cos(\theta + 360°) = \cos\theta$$
$$\sin(\theta + 2\pi) = \sin\theta \qquad \cos(\theta + 2\pi) = \cos\theta$$

These facts are usually summarized by saying that the sine and cosine functions are periodic and that they have a **fundamental period** of 360°, or 2π radians. It is the periodic nature of these functions that makes them useful in describing many repetitive phenomena such as tides, sound waves, and the orbital paths of satellites.

ORAL EXERCISES 6-3

Find $\sin\theta$ and $\cos\theta$.

1.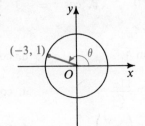

2.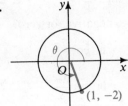

3.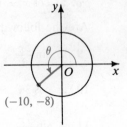

4. What is the largest value of the sine function? What is the smallest value?

5. There are many values of θ for which $\cos\theta = 0$. Name several.

6. a. Explain the meaning of $\theta = 45° + n \cdot 360°$.
   b. What is the equivalent statement if θ is expressed in radians instead of in degrees?

7. For what values of θ in degrees does $\sin\theta = -1$? $\sin\theta = 0$?

8. For what values of θ in degrees does $\cos\theta = 1$? $\cos\theta = -1$?

9. Does $\cos\theta$ increase or decrease as
   a. θ increases from 0° to 90°?
   b. θ increases from 90° to 180°?
   c. θ increases from 180° to 270°?
   d. θ increases from 270° to 360°?

10. Answer Exercise 9 for $\sin\theta$.

11. Are the following positive or negative?
   a. $\sin 165°$ **b.** $\sin 265°$ **c.** $\cos 210°$ **d.** $\cos 310°$
   e. $\sin\frac{5\pi}{6}$ **f.** $\cos\frac{5\pi}{6}$ **g.** $\sin\frac{4\pi}{3}$ **h.** $\cos\frac{5\pi}{3}$
   i. $\sin 2$ **j.** $\cos 2$ **k.** $\sin 4$ **l.** $\cos 4$

Trigonometric functions **203**

Evaluate without using a table or a calculator.

A **1. a.** $\sin 180°$ **b.** $\cos 180°$ **c.** $\sin 270°$ **d.** $\cos 270°$

2. a. $\sin(-90°)$ **b.** $\cos(-90°)$ **c.** $\sin 360°$ **d.** $\cos 360°$

3. a. $\sin \pi$ **b.** $\cos(-\pi)$ **c.** $\sin \dfrac{3\pi}{2}$ **d.** $\cos \dfrac{\pi}{2}$

4. a. $\cos 2\pi$ **b.** $\sin\left(-\dfrac{\pi}{2}\right)$ **c.** $\sin 3\pi$ **d.** $\cos\left(-\dfrac{3\pi}{2}\right)$

Name the quadrant described.

5. a. $\sin \theta > 0$ and $\cos \theta < 0$ **b.** $\sin \theta < 0$ and $\cos \theta < 0$

6. a. $\sin \theta < 0$ and $\cos \theta > 0$ **b.** $\sin \theta > 0$ and $\sin(90° + \theta) > 0$

Answer Exercises 7–12 without using a table or a calculator.
Solve each equation for *all* x in radians.

7. a. $\sin x = 1$ **b.** $\cos x = -1$ **c.** $\sin x = 0$ **d.** $\sin x = 2$

8. a. $\cos x = 1$ **b.** $\sin x = -1$ **c.** $\cos x = 0$ **d.** $\cos x = -3$

Are the following positive, negative, or zero?

9. a. $\sin \dfrac{5\pi}{3}$ **b.** $\cos \dfrac{7\pi}{6}$ **c.** $\sin \dfrac{\pi}{4}$ **d.** $\cos \dfrac{3\pi}{4}$

10. a. $\cos 3\pi$ **b.** $\sin \dfrac{2\pi}{3}$ **c.** $\sin \dfrac{11\pi}{6}$ **d.** $\cos\left(-\dfrac{\pi}{2}\right)$

11. a. $\sin \dfrac{7\pi}{4}$ **b.** $\cos \dfrac{3\pi}{2}$ **c.** $\sin\left(-\dfrac{\pi}{6}\right)$ **d.** $\sin \dfrac{\pi}{3}$

12. a. $\cos\left(-\dfrac{\pi}{3}\right)$ **b.** $\cos \dfrac{\pi}{6}$ **c.** $\sin \dfrac{5\pi}{4}$ **d.** $\cos \dfrac{7\pi}{4}$

Find $\sin \theta$ and $\cos \theta$.

13. 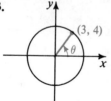   **14.** 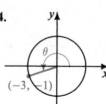   **15.** 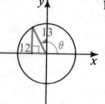   **16.**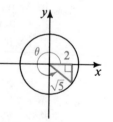

Complete the table. (A sketch such as that in Example 2 on page 201 may be helpful.)

	17.	**18.**	**19.**	**20.**	**21.**	**22.**	**23.**	**24.**
Quadrant	I	II	III	IV	II	III	IV	II
$\sin \theta$	$\frac{4}{5}$	$\frac{12}{13}$	?	?	$\frac{1}{5}$	$-\frac{3}{7}$	?	$\frac{1}{9}$
$\cos \theta$	?	?	$-\frac{24}{25}$	$\frac{15}{17}$	?	?	$\frac{3}{4}$	?

Complete the statement by replacing the question mark with one of the symbols $<$, $>$, or $=$.

B **25.** $\sin 40°$? $\sin 30°$ **26.** $\cos 40°$? $\cos 30°$ **27.** $\sin 172°$? $\sin 8°$

28. $\sin 310°$? $\sin 230°$ **29.** $\sin 130°$? $\sin 50°$ **30.** $\cos 50°$? $\cos(-50°)$

31. $\sin 169°$? $\sin 168°$ **32.** $\cos \dfrac{\pi}{2}$? $\cos \dfrac{\pi}{4}$ **33.** $\sin 3$? $\sin(-3)$

34. List in order of increasing size: $\sin 1$, $\sin 2$, $\sin 3$, $\sin 4$.

35. List in order of increasing size: $\cos 1$, $\cos 2$, $\cos 3$, $\cos 4$.

36. a. What are the rectangular coordinates of the points P and Q where the line $y = \frac{1}{2}$ intersects the unit circle?

 b. Explain how part (a) shows that if $\sin \theta = \dfrac{1}{2}$, then $\cos \theta = \dfrac{\pm\sqrt{3}}{2}$.

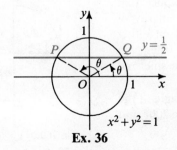

Ex. 36

Ex. 37

37. a. What are the rectangular coordinates of the points P and Q where the line $x = -\frac{1}{2}$ intersects the unit circle?

 b. Explain how part (a) shows that if $\cos \theta = -\dfrac{1}{2}$, then $\sin \theta = \dfrac{\pm\sqrt{3}}{2}$.

38. In the diagram of the unit circle at the right, z is measured in radians.

 a. Show that the length of arc $PQ = z$.

 b. Show that the length of $\overline{PA}$ is $\sin z$.

 c. Parts (a) and (b) show that $\sin z \approx z$ for a small angle z. Use Table 5 on pages 641–643 or a calculator to see if $\sin z \approx z$ when z is a very small number of radians.

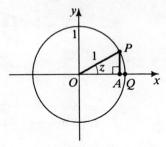

COMPUTER EXERCISES

1. Use a computer to approximate the value of the infinite series

$$x - \frac{x^3}{3!} + \frac{x^5}{5!} - \frac{x^7}{7!} + \frac{x^9}{9!} - \cdots$$

to five decimal places when $x = 1$, 2, and $\dfrac{\pi}{2} \approx \dfrac{1}{2}(3.14159) \approx 1.57080$.

Compare your answers with the values of SIN 1, SIN 2, and SIN 1.57080.

2. Use a computer to approximate the value of the infinite series

$$1 - \frac{x^2}{2!} + \frac{x^4}{4!} - \frac{x^6}{6!} + \frac{x^8}{8!} - \cdots$$

to five decimal places when $x = 1, 2,$ and $\pi \approx 3.14159$. Compare your answers with the values of COS 1, COS 2, and COS 3.14159.

6–4/EVALUATING SINES AND COSINES

Reference angles

Let α (Greek alpha) be an acute angle in standard position. Suppose, for example, that $\alpha = 20°$. Notice that the terminal ray of $\alpha = 20°$ and the terminal ray of $180° - \alpha = 160°$ are symmetrically located on either side of the y-axis. The sine and cosine of $160°$, then, can be deduced as shown in the diagram, if the sine and cosine of $20°$ are known.

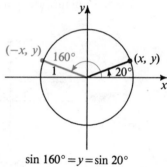

$$\sin 160° = y = \sin 20°$$
$$\cos 160° = -x = -\cos 20°$$

The angle $\alpha = 20°$ is called the **reference angle** of the $160°$ angle. It is also the reference angle for the $200°$ and $340°$ angles shown below.

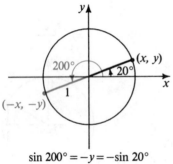

$$\sin 200° = -y = -\sin 20°$$
$$\cos 200° = -x = -\cos 20°$$

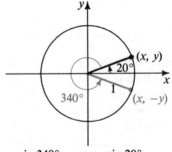

$$\sin 340° = -y = -\sin 20°$$
$$\cos 340° = x = \cos 20°$$

In general, the acute angle α is the reference angle for the angles $\theta = 180° - \alpha$, $180° + \alpha$, and $360° - \alpha = -\alpha$. You can find the reference angle of α by finding the acute positive angle that the terminal ray of α makes with the positive or negative part of the x-axis.

Using calculators or tables

The easiest way to find the sine or cosine of most angles is to use a calculator that has the sine and cosine functions. Often such calculators allow you to enter angles in degrees, radians, or grads. (A grad is another unit for measuring angles; 100 grads is the measure of a right angle.)

If you do not have access to a calculator, there are tables at the back of the book that evaluate $\sin \theta$ and $\cos \theta$ for first-quadrant values of θ. Check to see if you can find the following entries.

In Table 4: $\sin 20° = 0.3420$ $\cos 65.3° = 0.4179$

In Table 5: $\sin 0.55 = 0.5227$ $\cos 1.20 = 0.3624$

If θ is not a first-quadrant angle, then you must use reference angles to find $\sin \theta$ and $\cos \theta$.

To find $\sin \theta$ [figure (a)]:

1. Look up the sine of the reference angle α.
2. If θ is in Quadrant II, $\sin \theta = \sin \alpha$, but if θ is in Quadrant III or IV, $\sin \theta = -\sin \alpha$.

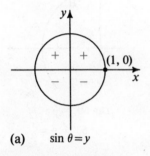

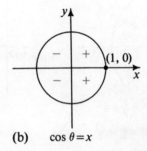

(a) $\sin \theta = y$ (b) $\cos \theta = x$

To find $\cos \theta$ [figure (b)]:

1. Look up the cosine of the reference angle α.
2. If θ is in Quadrant IV, $\cos \theta = \cos \alpha$, but if θ is in Quadrant II or III, $\cos \theta = -\cos \alpha$.

EXAMPLE 1. Find $\sin 130°$ and $\cos 130°$ to four decimal places.

SOLUTION: The reference angle of $130°$ is $180° - 130° = 50°$. Since $130°$ is a Quadrant II angle,

$\sin 130° = \sin 50° \approx 0.7660$ and
$\cos 130° = -\cos 50° \approx -0.6428$.

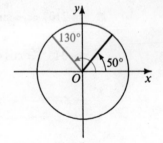

EXAMPLE 2. Find $\sin 4$ and $\cos 4$ to the nearest hundredth.

SOLUTION: The reference angle of 4 radians is $4 - \pi \approx 4 - 3.14 = 0.86$. Since 4 radians is a Quadrant III angle, we have

$\sin 4 \approx -\sin 0.86 \approx -0.76$ and
$\cos 4 \approx -\cos 0.86 \approx -0.65$.

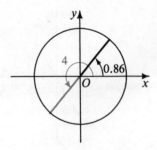

Trigonometric functions   **207**

Finding sines and cosines of special angles

Because angles that are multiples of 30° and 45° occur so often in mathematics, it can be useful to know their sines and cosines without having to consult a calculator or table. To do this, you need these facts.

1. In a 30–60–90° triangle, the sides are in the ratio $1:\sqrt{3}:2$.
2. In a 45–45–90° triangle, the sides are in the ratio $\sqrt{2}:\sqrt{2}:2$.

EXAMPLES.

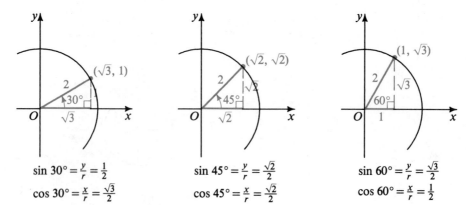

$$\sin 30° = \frac{y}{r} = \frac{1}{2}$$
$$\cos 30° = \frac{x}{r} = \frac{\sqrt{3}}{2}$$

$$\sin 45° = \frac{y}{r} = \frac{\sqrt{2}}{2}$$
$$\cos 45° = \frac{x}{r} = \frac{\sqrt{2}}{2}$$

$$\sin 60° = \frac{y}{r} = \frac{\sqrt{3}}{2}$$
$$\cos 60° = \frac{x}{r} = \frac{1}{2}$$

The information obtained from the preceding diagrams is summarized in the table at the right. To find the sines and cosines of other multiples of 30° or 45°, use reference angles. For example,

$$\sin 210° = -\sin 30° = -\frac{1}{2};$$

$$\cos 315° = \cos 45° = \frac{\sqrt{2}}{2}.$$

θ (degrees)	(radians)	$\sin \theta$	$\cos \theta$
0°	0	0	1
30°	$\dfrac{\pi}{6}$	$\dfrac{1}{2}$	$\dfrac{\sqrt{3}}{2}$
45°	$\dfrac{\pi}{4}$	$\dfrac{\sqrt{2}}{2}$	$\dfrac{\sqrt{2}}{2}$
60°	$\dfrac{\pi}{3}$	$\dfrac{\sqrt{3}}{2}$	$\dfrac{1}{2}$
90°	$\dfrac{\pi}{2}$	1	0

The values of $\sin \theta$ and $\cos \theta$ are shown graphically in the following diagrams.

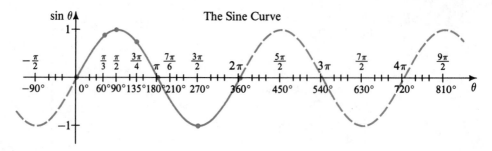

The Sine Curve

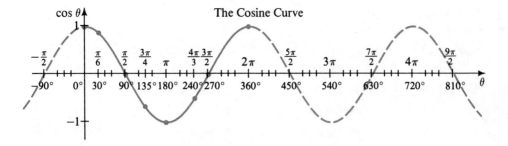

The Cosine Curve

Notice how the graphs show that both functions have a fundamental period of 360°, or 2π. Notice also that the graph of the cosine function is the graph of the sine function shifted 90° to the left.

ORAL EXERCISES 6-4

Use a calculator or table to evaluate the following.

1. sin 10° **2.** sin 28.6° **3.** sin 54° **4.** cos 32°

5. cos 73° **6.** sin 0.3 **7.** cos 1.4 **8.** cos 0.39

9. What is the reference angle for θ if
  a. $\theta = 170°$ **b.** $\theta = 205.1°$ **c.** $\theta = 310°$ **d.** $\theta = 3$?

10. Name another angle with the same sine as: **a.** 70° **b.** 200°

11. Name another angle with the same cosine as: **a.** 40° **b.** 165°

12. Give each sine in terms of the sine of its reference angle.
  (*Example:* sin 190° = −sin 10°)
  a. sin 170° **b.** sin 330° **c.** sin 220° **d.** sin 400° **e.** sin (−15°)

13. Give each cosine in terms of the cosine of its reference angle.
  a. cos 160° **b.** cos 182° **c.** cos 300° **d.** cos 365° **e.** cos (−100°)

14. Evaluate the following with Table 3 or a calculator.
  a. sin 188° **b.** sin 110° **c.** cos 355° **d.** cos 365°

15. Evaluate the following with Table 5 or a calculator.
  a. sin 0.43 **b.** cos 1.09 **c.** sin 3 **d.** cos 4

Study the sines and cosines of 30°, 45°, and 60°. Then give the values of the following in simplest radical form, without using a table or a calculator.

16. a. sin 45° **b.** sin 135° **c.** sin 225° **d.** sin 315°

17. a. cos 60° **b.** cos 120° **c.** cos 240° **d.** sin 300°

18. a. sin 30° **b.** sin (−30°) **c.** cos 30° **d.** cos (−30°)

19. a. sin 120° **b.** cos 120° **c.** sin $\dfrac{7\pi}{6}$ **d.** cos $\dfrac{7\pi}{6}$

20. a. cos $\dfrac{\pi}{4}$ **b.** sin $\left(-\dfrac{\pi}{3}\right)$ **c.** cos $\dfrac{5\pi}{6}$ **d.** sin (−300°)

WRITTEN EXERCISES 6-4

Evaluate to the nearest hundredth with tables or a calculator.

A **1. a.** $\sin 28°$ **b.** $\sin 46°$ **c.** $\cos 65.3°$ **d.** $\sin 0.49$

2. a. $\sin 81°$ **b.** $\cos 0.3°$ **c.** $\cos 1.17$ **d.** $\sin 1.11$

3. a. $\sin 160°$ **b.** $\cos 205.3°$ **c.** $\cos 302.1°$ **d.** $\sin 207°18'$

4. a. $\cos 238°$ **b.** $\sin 403°$ **c.** $\sin 285.7°$ **d.** $\cos 132°6'$

5. a. $\sin 1.2$ **b.** $\sin 2.2$ **c.** $\cos 3.5$ **d.** $\cos 5$

6. a. $\sin 3.54$ **b.** $\cos 6$ **c.** $\sin (-2)$ **d.** $\cos 3$

Study the sines and cosines of 30°, 45°, and 60°. Then give the values of the following in simplest radical form, without using a table or a calculator.

7. a. $\cos 45°$ **b.** $\cos 135°$ **c.** $\sin 210°$ **d.** $\cos 330°$

8. a. $\sin (-60°)$ **b.** $\cos 300°$ **c.** $\cos 225°$ **d.** $\sin (-225°)$

9. a. $\sin 150°$ **b.** $\cos (-240°)$ **c.** $\sin (-135°)$ **d.** $\cos \left(-\dfrac{\pi}{6}\right)$

10. a. $\cos 210°$ **b.** $\cos 90°$ **c.** $\sin (-120°)$ **d.** $\sin (-315°)$

11. a. $\sin \dfrac{\pi}{6}$ **b.** $\cos \dfrac{\pi}{3}$ **c.** $\cos \dfrac{2\pi}{3}$ **d.** $\sin \dfrac{3\pi}{4}$

12. a. $\cos \dfrac{\pi}{4}$ **b.** $\sin \left(-\dfrac{\pi}{4}\right)$ **c.** $\sin \dfrac{5\pi}{3}$ **d.** $\cos \left(-\dfrac{7\pi}{6}\right)$

13. a. $\cos 2\pi$ **b.** $\sin \dfrac{11\pi}{6}$ **c.** $\cos \dfrac{5\pi}{6}$ **d.** $\cos \dfrac{3\pi}{4}$

14. a. $\cos \left(-\dfrac{\pi}{3}\right)$ **b.** $\sin \pi$ **c.** $\sin \dfrac{5\pi}{4}$ **d.** $\sin \left(-\dfrac{\pi}{6}\right)$

15. The "natural" way to graph $y = \sin x$ or $y = \cos x$, is to measure x in radians and use the same real-number scale on both axes. Sketch the graph of $y = \sin x$ using this method, letting $\dfrac{\sqrt{2}}{2} \approx 0.71$ and $\dfrac{\sqrt{3}}{2} \approx 0.87$.

16. Sketch the graph of $y = \cos x$ using the method described in Exercise 15.

Evaluate without using tables or a calculator.

B **17. a.** $\log_2 \sin \dfrac{\pi}{6}$ **b.** $\log_2 \cos 2\pi$ **c.** $\log_2 \sin \dfrac{\pi}{4}$

18. a. $\log_2 \cos \left(-\dfrac{\pi}{3}\right)$ **b.** $\log_2 \sin \dfrac{3\pi}{4}$ **c.** $\log_2 \sin \dfrac{\pi}{2}$

19. The latitude of a point on Earth is the degree measure of the smallest arc from the point to the equator. The latitude of the point P in the diagram, for example, equals the degree measure of arc PE. At what latitude is the circumference of the circle of latitude at P half the distance around the equator?

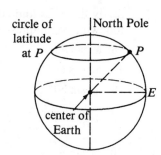

20. The radius of Earth is approximately 6400 km, and the latitude of Durham, North Carolina, is 36°N (north of the equator). About how far from Durham is the North Pole?

21. Peking, China, is due north of Perth, Australia. The latitude of Peking is 39°55′N and the latitude of Perth is 31°58′S. How far apart are the two cities? (Use 6400 km for the radius of Earth.)

22. Memphis, Tennessee, is due north of New Orleans, Louisiana. The latitude of Memphis is 36°6′N, and the latitude of New Orleans is 30°N. How far apart are the two cities? (Use 6400 km for the radius of Earth.)

23. Earth's rotational speed at the equator can be found by dividing the circumference of the equator by 24 hours: 40,074 km ÷ 24 h ≈ 1670 km/h. What is Earth's rotational speed at (a) Bangor, Maine (latitude 45°N), and (b) Esquina, Argentina (latitude 30°S)?

24. What is Earth's rotational speed at (a) Anchorage, Alaska (latitude 60°N), and (b) Rio de Janeiro, Brazil (latitude 23°S)?

25. a. Prove that Earth's rotational speed at latitude L is 1670 cos L.
    b. Find Earth's rotational speed at the North Pole.

26. How is sin $(-\theta)$ related to sin θ? (Hint: See diagram at right.)

27. How is cos $(-\theta)$ related to cos θ?

28. Rome, Italy, and Boston, Massachusetts, have approximately the same latitude (42°N). A plane flying from Rome due west to Boston is able to "stay with the sun," leaving Rome with the sun overhead and landing in Boston with the sun overhead. How fast is the plane flying? (Hint: See Exercise 25.)

C 29. A piston rod PQ, 4 units long, is connected to the rim of a wheel at point P, and to a piston at point Q. As P moves counterclockwise around the wheel at 1 radian per second, Q slides left and right in the piston. What are the coordinates of P and Q in terms of time t in seconds? Assume that P is at (1, 0) when $t = 0$.

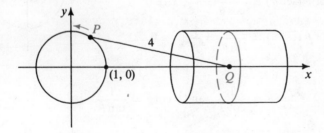

30. Sketch the graph of $y = \log_{10} \sin x$.
31. Sketch the graph of $y = \log_2 \cos x$.

A great circle on Earth's surface is a circle whose center is at Earth's center. The shortest distance measured in degrees from one point to another on Earth's surface is along the arc of a great circle. To find the shortest distance between A and B along Earth's surface, we can work with the spherical triangle ABN shown in the diagram, whose sides are arcs of great circles. The important measurements in this triangle are:

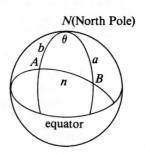

N(North Pole)

great circle arc $AB = n°$
great circle arc $AN = b° = 90° -$ latitude of A
great circle arc $BN = a° = 90° -$ latitude of B
$\theta =$ difference in longitudes of A and B

Use the formula $\cos n = \cos a \cos b + \sin a \sin b \cos \theta$ to find the great circle distance between Rome and Boston given the following data.

	latitude	longitude
Rome	41.9°N	12.5°E (East)
Boston	42.3°N	71.1°W (West)

6-5/OTHER TRIGONOMETRIC FUNCTIONS

We can define four other trigonometric functions of θ in terms of $\sin \theta$ and $\cos \theta$:

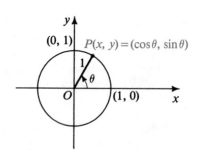

$$\text{tangent of } \theta \ (\tan \theta) = \frac{\sin \theta}{\cos \theta} = \frac{y}{x}, \ x \neq 0$$

$$\text{cotangent of } \theta \ (\cot \theta) = \frac{\cos \theta}{\sin \theta} = \frac{x}{y}, \ y \neq 0$$

$$\text{secant of } \theta \ (\sec \theta) = \frac{1}{\cos \theta} = \frac{1}{x}, \ x \neq 0$$

$$\text{cosecant of } \theta \ (\csc \theta) = \frac{1}{\sin \theta} = \frac{1}{y}, \ y \neq 0$$

$P(x, y) = (\cos \theta, \sin \theta)$
$(0, 1)$
$(1, 0)$

Notice that

$$\tan \theta = \frac{1}{\cot \theta} \quad \text{and} \quad \cot \theta = \frac{1}{\tan \theta}.$$

The signs of the six trigonometric functions in the various quadrants are summarized in the table at the right.

	I	II	III	IV
$\sin \theta$ and $\csc \theta$	+	+	−	−
$\cos \theta$ and $\sec \theta$	+	−	−	+
$\tan \theta$ and $\cot \theta$	+	−	+	−

Now let us find some values of these new trigonometric functions and study their graphs.

EXAMPLE 1. Use a calculator or tables to evaluate the following to the nearest hundredth.

    a. cot 165°   **b.** csc 203°   **c.** sec 6

SOLUTION:

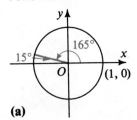 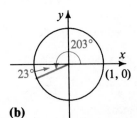 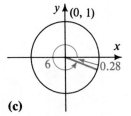

(a)   **(b)**   **(c)**

  a. Calculator: $\cot 165° = \dfrac{1}{\tan 165°} \approx -3.73$

      Table: The reference angle of 165° is $180° - 165° = 15°$. Since the cotangent is negative in Quadrant II,

$$\cot 165° = -\cot 15° \approx -3.73 \text{ (Table 3)}.$$

  b. Calculator: $\csc 203° = \dfrac{1}{\sin 203°} \approx -2.56$

      Table: The reference angle of 203° is $203° - 180° = 23°$. Since the cosecant is negative in Quadrant III,

$$\csc 203° = -\csc 23° \approx -2.56 \text{ (Table 3)}.$$

  c. Calculator: Set the calculator in radian mode.

$$\sec 6 = \frac{1}{\cos 6} \approx 1.04$$

      Table: The reference angle of 6 radians is $2\pi - 6 \approx 2(3.14) - 6 = 0.28$. Since the secant is positive in Quadrant IV,

$$\sec 6 \approx \sec 0.28 \approx 1.04 \text{ (Table 5)}.$$

EXAMPLE 2. Without using a calculator or tables, evaluate the following.

    a. tan 135°   **b.** cot 240°   **c.** $\csc \dfrac{11\pi}{6}$

SOLUTION: **a.** The reference angle of 135° is $180° - 135° = 45°$. Since the tangent is negative in Quadrant II,

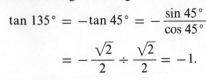 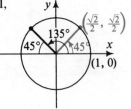

$$\tan 135° = -\tan 45° = -\frac{\sin 45°}{\cos 45°}$$

$$= -\frac{\sqrt{2}}{2} \div \frac{\sqrt{2}}{2} = -1.$$

(*Solution continued on page 214*)

                                      Trigonometric functions

b. The reference angle of $240°$ is $240° - 180°$
$= 60°$. Since the cotangent is positive in
Quadrant III,

$$\cot 240° = \cot 60° = \frac{\cos 60°}{\sin 60°}$$

$$= \frac{1}{2} \div \frac{\sqrt{3}}{2} = \frac{1}{\sqrt{3}}, \text{ or } \frac{\sqrt{3}}{3}.$$

c. The reference angle of $\frac{11\pi}{6}$ is $\frac{\pi}{6}$. Since
the cosecant is negative in Quadrant IV,

$$\csc \frac{11\pi}{6} = -\csc \frac{\pi}{6} = -\frac{1}{\sin \frac{\pi}{6}} = \frac{-1}{\frac{1}{2}}$$

$$= -2.$$

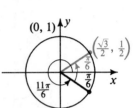

The tangent graph

Picture the point $P(x, y)$ traveling along the circumference of the unit circle.
When $\theta = 0°$, P is at $(1, 0)$ and $\tan \theta = \frac{y}{x} = \frac{0}{1} = 0$. As θ increases to $\frac{\pi}{2}$,

y increases and x decreases, so that $\tan \theta = \frac{y}{x}$ gets bigger.

When $\theta = \frac{\pi}{4}$, $y = x$, so that $\tan \frac{\pi}{4} = \frac{y}{x} = 1$. When $\theta = \frac{\pi}{2}$, P

is at $(0, 1)$, so that $\tan \theta = \frac{1}{0}$, which is undefined.

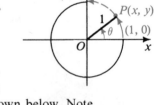

By considering $P(x, y)$ as it travels farther around the circumference of the unit circle, the other values of $\tan \theta$ can be
analyzed in a similar way. The graph of $\tan \theta$ versus θ is shown below. Note
that the tangent has period π (or $180°$). The graph of the cotangent function is
similar and is left for you to draw as an exercise (Exercise 18 on page 217).

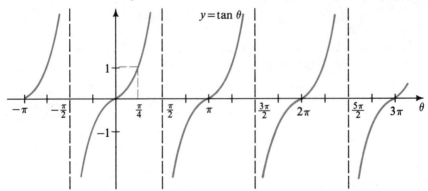

The secant graph

Since $\sec \theta = \dfrac{1}{\cos \theta}$, to draw the secant graph it is helpful to draw the cosine graph first, as shown in black in the following diagram. The secant graph shown in red can be drawn using the cosine graph and these four facts:

1. $\sec \theta = 1$ when $\cos \theta = 1$; at $\theta = 0, \pm 2\pi, \pm 4\pi, \ldots$

2. $\sec \theta = -1$ when $\cos \theta = -1$; at $\theta = \pm \pi, \pm 3\pi, \pm 5\pi, \ldots$

3. $\sec \theta$ is not defined when $\cos \theta = 0$; at $\theta = \pm \dfrac{\pi}{2}, \pm \dfrac{3\pi}{2}, \pm \dfrac{5\pi}{2}, \ldots$

4. $|\sec \theta|$ gets larger as $|\cos \theta|$ gets smaller.

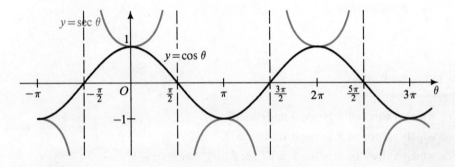

The graph shows that the secant function, like the cosine function, has period 2π (or $360°$). This means that values of the secant function repeat every 2π radians. The cosecant graph is similar to the secant graph and can be found by a similar analysis. (See Exercise 17 on page 217.)

Inclination and slope

The inclination of a line is the measure of the angle from the positive direction of the x-axis to the line. The diagrams below show lines with inclinations $35°$ and $155°$. The theorem that follows states that the tangents of these angles are the slopes of the lines.

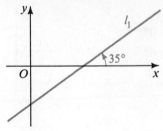

slope of $l_1 = \tan 35° \approx 0.7002$

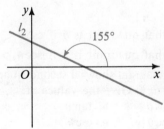

slope of $l_2 = \tan 155° \approx -0.4663$

Trigonometric functions   **215**

THEOREM The slope of a line equals the tangent of its inclination.

Proof Suppose the line l_1 (see diagram) passes through the origin and has inclination θ. Then $P = (\cos\theta, \sin\theta)$ is a point where l_1 and the unit circle intersect. The slope of l_1 is

$$\frac{\Delta y}{\Delta x} = \frac{\sin\theta - 0}{\cos\theta - 0} = \tan\theta.$$

If the line l_2 has inclination θ and does not pass through the origin, it is parallel to l_1. (Why?) Thus l_2 must have the same slope as l_1. (Why?) Therefore, the slope of l_2 also equals $\tan\theta$.

ORAL EXERCISES 6-5

1. For which values of θ is $\sec\theta$ undefined?

2. For which values of θ is $\tan\theta$ undefined?

3. For which values of θ is $\sec\theta = 0$? $\sec\theta = 1$? $\sec\theta = -1$?

4. For which values of θ is $\tan\theta = 0$? $\tan\theta = 1$? $\tan\theta = -1$?

5. What is the period of the tangent function?

6. What is the period of the secant function?

7. **a.** What is the reference angle for an angle of $147°$?
   b. If $\tan 33° \approx 0.6494$, what is $\tan 147°$? $\tan 213°$? $\tan 327°$?

8. If $\sec 15° \approx 1.0353$, find: **(a)** $\sec 165°$; **(b)** $\sec 195°$; **(c)** $\sec 345°$.

9. Use a calculator or Table 3 to find:
   a. $\cot 185°$ **b.** $\csc 310°$

10. Lines l_1 and l_2, shown below, have inclinations of $45°$ and $60°$, respectively. What are the slopes of these lines?

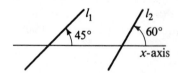

11. In what quadrant is θ if $\csc\theta < 0$ and $\tan\theta > 0$?

12. In what quadrant is θ if $\sec\theta > 0$ and $\cot\theta < 0$?

13. The diagram shows a second-quadrant angle θ with $\sin\theta = \frac{4}{5}$. Give the values of:
    a. $\cos\theta$ **b.** $\tan\theta$ **c.** $\cot\theta$
    d. $\sec\theta$ **e.** $\csc\theta$

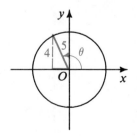

In Exercises 1-4, express the given quantity as the same trigonometric function of its reference angle.

EXAMPLE. $\sec 220° = -\sec 40°$

A
1. **a.** $\sin 195°$ **b.** $\sec 280°$ **c.** $\tan(-140°)$ **d.** $\sec 2$

2. **a.** $\cot 285°$ **b.** $\sec(-105°)$ **c.** $\csc 600°$ **d.** $\tan 3$

3. **a.** $\tan 820°$ **b.** $\sec 290°$ **c.** $\cot 185°$ **d.** $\csc 4$

4. **a.** $\tan 160°$ **b.** $\csc 115°$ **c.** $\sec 235°$ **d.** $\cot 5$

5. For which values of x (in radians) is $\csc x$
   a. undefined?   **b.** 0?   **c.** 1?   **d.** -1?

6. For which values of x (in radians) is $\cot x$
   a. undefined?   **b.** 0?   **c.** 1?   **d.** -1?

In Exercises 7-14, find the exact value or state that the value is undefined.

EXAMPLE. $\tan 120° = -\tan 60° = -\dfrac{\sin 60°}{\cos 60°}$

$$= -\frac{\sqrt{3}}{2} \div \frac{1}{2} = -\sqrt{3}$$

7. **a.** $\cos 120°$ **b.** $\sec 120°$ **c.** $\sin 120°$ **d.** $\tan 120°$

8. **a.** $\sin 225°$ **b.** $\csc 225°$ **c.** $\tan 225°$ **d.** $\sec 225°$

9. **a.** $\csc 90°$ **b.** $\sec 180°$ **c.** $\tan 240°$ **d.** $\cot 0°$

10. **a.** $\csc 150°$ **b.** $\csc 0°$ **c.** $\tan 315°$ **d.** $\sec 315°$

11. **a.** $\csc \pi$ **b.** $\tan \dfrac{2\pi}{3}$ **c.** $\cot \dfrac{\pi}{2}$ **d.** $\sec \dfrac{5\pi}{6}$

12. **a.** $\tan \dfrac{\pi}{2}$ **b.** $\cot \dfrac{7\pi}{4}$ **c.** $\sec(-3\pi)$ **d.** $\csc \dfrac{7\pi}{6}$

13. **a.** $\log_2\left(\sec \dfrac{\pi}{3}\right)$ **b.** $\log_2\left(\tan \dfrac{\pi}{4}\right)$ **c.** $\log_2\left(\csc \dfrac{3\pi}{4}\right)$

14. **a.** $\log_2\left(\cos \dfrac{5\pi}{3}\right)$ **b.** $\log_2\left(\sec \dfrac{\pi}{4}\right)$ **c.** $\log_3\left(\tan \dfrac{\pi}{3}\right)$

In Exercises 15 and 16, use a calculator or tables to evaluate the given quantities.

15. **a.** $\tan 100°$ **b.** $\cot 276°$ **c.** $\csc 5$ **d.** $\sec 2.14$

16. **a.** $\sec(-11°)$ **b.** $\csc 233°$ **c.** $\tan 3$ **d.** $\cot 7.28$

17. Make a graph of $\csc \theta$ versus θ for θ between $-180°$ and $540°$. Compare your graph with the graph of $\sec \theta$ versus θ shown on page 215.

18. Make a graph of $\cot \theta$ versus θ for θ between $-180°$ and $540°$. Compare your graph with the graph of $\tan \theta$ versus θ shown on page 214.

In Exercises 19–24, find the other five trigonometric functions of x.

19. $\sin x = \dfrac{5}{13}, \dfrac{\pi}{2} < x < \pi$　　　　　**20.** $\cos x = \dfrac{24}{25}, -\dfrac{\pi}{2} < x < 0$

21. $\tan x = \dfrac{3}{4}, \pi < x < 2\pi$　　　　　**22.** $\cot x = -\dfrac{12}{5}, 0 < x < \pi$

23. $\sec x = -3, 0 < x < \pi$　　　　　**24.** $\csc x = -5, \dfrac{\pi}{2} < x < \dfrac{3\pi}{2}$

25. a. Verify that $\sin^2 \dfrac{\pi}{3} + \cos^2 \dfrac{\pi}{3} = 1$. $\left[\textit{Note: } \sin^2 \dfrac{\pi}{3} \text{ means } \left(\sin \dfrac{\pi}{3}\right)^2.\right]$

　　b. Can you find any other values of x for which $\sin^2 x + \cos^2 x = 1$?

26. For which of the following values of x is it true that $\csc^2 x - \cot^2 x = 1$?

　　a. $x = \dfrac{3\pi}{4}$　　　　**b.** $x = \dfrac{5\pi}{6}$　　　　**c.** $x = \dfrac{\pi}{2}$

In Exercises 27–30, find the inclinations of the given lines. Give your answers in degrees to the nearest 10 minutes, using Table 3.

27. a. the line $y = x$　　　　　**b.** the line joining $(-1, 2)$ and $(4, 1)$

28. a. the line $y = -x$　　　　　**b.** the line joining $(-1, 1)$ and $(4, 2)$

29. a. the line $3x + 5y = 8$　　　　　**b.** a line perpendicular to $4x + 3y = 12$

30. a. the line $x + 4y = 7$　　　　　**b.** a line parallel to $2x + 3y = -6$

31. Give the domain, range, and period of **(a)** the tangent function and **(b)** the secant function.

32. Give the domain, range, and period of **(a)** the cotangent function and **(b)** the cosecant function.

33. What is the domain of $f(x) = \log \tan x$?

34. What is the domain of $g(x) = \log \cot x$?

B　**35. a.** Complete the statement: $\tan (x + \pi) = \underline{\ ?\ }$.

　　b. Make an analogous statement for a function $f(x)$ with period p.

36. True or false? A periodic function cannot be one-to-one.

37. a. Show that if two lines have inclinations α and β (beta), respectively, then one of the angles between the lines is $\theta = \alpha - \beta$.

　　b. Let $A = (1, 1)$, $B = (5, 2)$, and $C = (2, 6)$. Find the inclinations of the three sides of $\triangle ABC$ and use these to find the angles of the triangle.

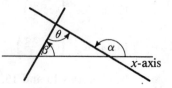

CALCULATOR EXERCISE

Consider the equation $\tan x = x$ (x in radians).

a. One solution to this equation is easy to guess. What is it?

b. Estimate to the nearest hundredth the solution of this equation that is between 1 and 5 by using the trial and error method with your calculator. (*Hint:* Consider the graphs of $y = \tan x$ and $y = x$ on the same axes.)

Identities and equations

6-6/RELATIONSHIPS AMONG THE FUNCTIONS

In this section we shall investigate some of the various relationships among the trigonometric functions; namely, the reciprocal relationships, the cofunction relationships, and the Pythagorean relationships. All of these relationships are derived from the definitions of the six trigonometric functions.

Reciprocal relationships

$$\csc \theta = \frac{1}{\sin \theta}, \quad \text{and} \quad \sin \theta = \frac{1}{\csc \theta}, \quad \sin \theta \neq 0$$

$$\sec \theta = \frac{1}{\cos \theta}, \quad \text{and} \quad \cos \theta = \frac{1}{\sec \theta}, \quad \cos \theta \neq 0$$

$$\cot \theta = \frac{\cos \theta}{\sin \theta} = \frac{1}{\tan \theta}, \quad \text{and} \quad \tan \theta = \frac{\sin \theta}{\cos \theta} = \frac{1}{\cot \theta},$$

$$\sin \theta \neq 0, \tan \theta \neq 0 \qquad\qquad \cos \theta \neq 0, \cot \theta \neq 0$$

Negatives

In Exercises 26 and 27 on page 211 we saw that:

$$\sin (-\theta) = -\sin \theta \quad \text{and} \quad \cos (-\theta) = \cos \theta$$

Using the reciprocal relationships it is easy to show that the following are true:

$$\csc (-\theta) = -\csc \theta \qquad \sec (-\theta) = \sec \theta$$
$$\tan (-\theta) = -\tan \theta \qquad \cot (-\theta) = -\cot \theta$$

Cofunction relationships

The sine and cosine are called **cofunctions,** as are the tangent and cotangent, and the secant and cosecant. The cofunctions are related as follows:

$$\sin \theta = \cos (90° - \theta) \quad \text{and} \quad \cos \theta = \sin (90° - \theta)$$
$$\tan \theta = \cot (90° - \theta) \quad \text{and} \quad \cot \theta = \tan (90° - \theta)$$
$$\sec \theta = \csc (90° - \theta) \quad \text{and} \quad \csc \theta = \sec (90° - \theta)$$

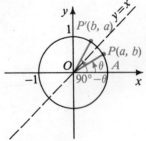

The reason for the cofunction relationship can be seen from the diagram at the right. If the sum of the measures of $\angle POA$ and $\angle P'OA$ totals $90°$, then points P and P' are symmetric with respect to the line $y = x$. Hence, if $P = (a, b)$, then $P' = (b, a)$. Consequently,

$\sin \theta = y$-coordinate of P
$\quad = x$-coordinate of P'
$\quad = \cos (90° - \theta)$

$\cos \theta = x$-coordinate of P
$\quad = y$-coordinate of P'
$\quad = \sin (90° - \theta)$

The other cofunction relationships follow from those stated on the preceding page. For example,

$$\tan \theta = \frac{\sin \theta}{\cos \theta} = \frac{\cos (90° - \theta)}{\sin (90° - \theta)} = \cot (90° - \theta)$$

$$\sec \theta = (you \; supply \; the \; steps) = \csc (90° - \theta)$$

You should convince yourself that the preceding argument remains the same if the diagram is changed so that θ is not in Quadrant I. In general,

$$\text{function of } \theta = \text{cofunction of the complement of } \theta$$

Pythagorean relationships

Recall that if P is a point on the unit circle shown, then by definition, $\cos \theta = x$ and $\sin \theta = y$. Substituting into the unit circle's equation $x^2 + y^2 = 1$, we have $(\cos \theta)^2 + (\sin \theta)^2 = 1$, or, more simply, $\sin^2 \theta + \cos^2 \theta = 1$. This gives the first of the three Pythagorean relationships that follow. (See Exercises 40 and 41 for suggestions on how to prove the other two.)

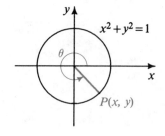

$$\sin^2 \theta + \cos^2 \theta = 1$$
$$1 + \tan^2 \theta = \sec^2 \theta$$
$$1 + \cot^2 \theta = \csc^2 \theta$$

Identities

The third Pythagorean relationship, $1 + \cot^2 \theta = \csc^2 \theta$, holds for all values of θ except $\theta = n\pi$, n an integer. For these values, $\cot \theta$ and $\csc \theta$ are undefined. Likewise, every trigonometric relationship discussed in this section holds for all values of θ except those for which either side of the equation is undefined. Such relationships are often called **trigonometric identities**, just as $(a + b)^2 = a^2 + 2ab + b^2$ is called an *algebraic identity*. In the following examples we shall use these basic trigonometric identities to simplify other expressions and prove other identities.

EXAMPLE 1.　Simplify: $\sec \theta - \sin \theta \cdot \tan \theta$

SOLUTION:　We use the definitions of $\tan \theta$ and $\sec \theta$ to express these functions in terms of $\sin \theta$ and $\cos \theta$. This yields

$$\sec \theta - \sin \theta \cdot \tan \theta$$

$$= \frac{1}{\cos \theta} - \sin \theta \cdot \frac{\sin \theta}{\cos \theta}$$

$$= \frac{1 - \sin^2 \theta}{\cos \theta} = \frac{\cos^2 \theta}{\cos \theta} = \cos \theta$$

EXAMPLE 2. Prove: $\dfrac{\cot A(1 + \tan^2 A)}{\tan A} = \csc^2 A$

SOLUTION: It is tempting to want to multiply both sides of this equation by $\tan A$, but the difficulty in doing this is that we must be sure that each step in the proof is reversible. Consequently, we follow the usual practice in proving trigonometric identities; that is, we simplify each side separately until we reach a recognizable identity.

Method 1.

$$\frac{\cot A(1 + \tan^2 A)}{\tan A} \overset{?}{=} \csc^2 A$$

Since $1 + \tan^2 A = \sec^2 A,$ $\qquad \dfrac{\cot A \cdot \sec^2 A}{\tan A} \quad\Big|\quad \csc^2 A$

Since $\dfrac{1}{\tan A} = \cot A,$ $\qquad \cot^2 A \cdot \sec^2 A \quad\Big|\quad \csc^2 A$

$$\frac{\cos^2 A}{\sin^2 A} \cdot \frac{1}{\cos^2 A} \quad\Big|\quad \csc^2 A$$

$$\frac{1}{\sin^2 A} \quad\Big|\quad \csc^2 A$$

$$\csc^2 A = \csc^2 A$$

Method 2. Sometimes you may not see which relationship to apply in proving an identity. If this happens, try expressing all of the trigonometric functions in terms of sines and cosines. Usually this method takes longer, but it can be effective if all else fails.

$$\frac{\cot A(1 + \tan^2 A)}{\tan A} \overset{?}{=} \csc^2 A$$

By using definitions, $\qquad \dfrac{\dfrac{\cos A}{\sin A}\left(1 + \dfrac{\sin^2 A}{\cos^2 A}\right)}{\dfrac{\sin A}{\cos A}} \quad\Big|\quad \dfrac{1}{\sin^2 A}$

By multiplying the numerator and denominator by $\dfrac{\cos A}{\sin A}$,

$$\frac{\cos A}{\sin A} \cdot \frac{\cos A}{\sin A}\left(1 + \frac{\sin^2 A}{\cos^2 A}\right) \quad\Big|\quad \frac{1}{\sin^2 A}$$

$$\frac{\cos^2 A}{\sin^2 A}\left(1 + \frac{\sin^2 A}{\cos^2 A}\right) \quad\Big|\quad \frac{1}{\sin^2 A}$$

$$\frac{\cos^2 A}{\sin^2 A} + 1 \quad\Big|\quad \frac{1}{\sin^2 A}$$

$$\frac{\cos^2 A + \sin^2 A}{\sin^2 A} \quad\Big|\quad \frac{1}{\sin^2 A}$$

Since $\cos^2 A + \sin^2 A = 1,$ $\qquad \dfrac{1}{\sin^2 A} = \dfrac{1}{\sin^2 A}$

1. Since $\sin^2 \theta + \cos^2 \theta = 1$, then

    a. $1 - \sin^2 \theta = \underline{\ ?\ }$ **b.** $1 - \cos^2 \theta = \underline{\ ?\ }$

2. Since $1 + \tan^2 \theta = \sec^2 \theta$, then

    a. $\sec^2 \theta - 1 = \underline{\ ?\ }$ **b.** $\sec^2 \theta - \tan^2 \theta = \underline{\ ?\ }$

3. Since $1 + \cot^2 \theta = \csc^2 \theta$, then

    a. $\csc^2 \theta - \cot^2 \theta = \underline{\ ?\ }$ **b.** $\csc^2 \theta - 1 = \underline{\ ?\ }$

Simplify the following.

4. a. $\tan \theta \cdot \cos \theta$ **b.** $\tan (90° - A)$ **c.** $\cos \left(\dfrac{\pi}{2} - x \right)$

5. a. $\dfrac{1 - \cos^2 x}{\sin x}$ **b.** $\cos y \cdot \sec y$ **c.** $\sin B \cdot \csc B$

6. a. $(1 - \sin x)(1 + \sin x)$ **b.** $\sin^2 x - 1$ **c.** $(\sec x - 1)(\sec x + 1)$

7. a. $\dfrac{1}{\tan A}$ **b.** $\tan A \cdot \cot A$ **c.** $\cot y \cdot \sin y$

8. a. $\cos B \cdot \csc B$ **b.** $\dfrac{1}{\sin \left(\dfrac{\pi}{2} - A \right)}$ **c.** $\cot^2 x - \csc^2 x$

9. Evaluate: **a.** $\sin^2 \dfrac{5\pi}{6} + \cos^2 \dfrac{5\pi}{6}$ **b.** $\sec^2 \pi - \tan^2 \pi$

10. True or false?

    a. $\sin (90° - \theta) = \cos \theta$ **b.** $\sin 70° = \cos 20°$

    c. $\sin 100° = \cos (-10°)$ **d.** $\tan (90° - A) = \cot A$

    e. $\tan 89° = \cot 1°$ **f.** $\tan 190° = \cot (-100°)$

Tell how you would simplify each complex fraction.

11. a. $\dfrac{t + \dfrac{1}{t}}{t}$ **b.** $\dfrac{\tan A + \dfrac{1}{\tan A}}{\tan A}$

12. a. $\dfrac{\dfrac{1}{x} - x}{\dfrac{1}{x}}$ **b.** $\dfrac{\dfrac{1}{\cos B} - \cos B}{\dfrac{1}{\cos B}}$

13. a. $\dfrac{y + \dfrac{1}{y}}{\dfrac{1}{y}}$ **b.** $\dfrac{\cot \alpha + \dfrac{1}{\cot \alpha}}{\dfrac{1}{\cot \alpha}}$

14. a. $\dfrac{\dfrac{y}{x} + \dfrac{x}{y}}{\dfrac{1}{xy}}$ **b.** $\dfrac{\dfrac{\sin \theta}{\cos \theta} + \dfrac{\cos \theta}{\sin \theta}}{\dfrac{1}{\cos \theta \sin \theta}}$

Simplify the given expression.

A 1. **a.** $\cos^2 \theta + \sin^2 \theta$ **b.** $(1 - \cos \theta)(1 + \cos \theta)$ **c.** $(\sin \theta - 1)(\sin \theta + 1)$

 2. **a.** $1 + \tan^2 x$ **b.** $(\sec x - 1)(\sec x + 1)$ **c.** $\tan^2 x - \sec^2 x$

 3. **a.** $1 + \cot^2 A$ **b.** $(\csc A - 1)(\csc A + 1)$ **c.** $\dfrac{1}{\sin^2 A} - \dfrac{1}{\tan^2 A}$

 4. **a.** $\dfrac{1}{\cos (90° - \theta)}$ **b.** $1 - \dfrac{\sin^2 \theta}{\tan^2 \theta}$ **c.** $\dfrac{1}{\cos^2 \theta} - \dfrac{1}{\cot^2 \theta}$

 5. **a.** $\dfrac{1}{\sin \left(\dfrac{\pi}{2} - x\right)}$ **b.** $\sec^2 x(1 - \sin^2 x)$ **c.** $\dfrac{\sin x}{\cot \left(\dfrac{\pi}{2} - x\right)}$

 6. **a.** $\cos \theta \cdot \cot (90° - \theta)$ **b.** $\csc^2 x(1 - \cos^2 x)$ **c.** $\cos \theta (\sec \theta - \cos \theta)$

 7. **a.** $\cot A \cdot \sec A \cdot \sin A$ **b.** $\cos^2 A (\sec^2 A - 1)$ **c.** $\sin \theta (\csc \theta - \sin \theta)$

 8. **a.** $\dfrac{\tan x}{\csc \left(\dfrac{\pi}{2} - x\right)}$ **b.** $\dfrac{\sec^2 x - \tan^2 x}{\csc \left(\dfrac{\pi}{2} - x\right)}$ **c.** $\cos x \cdot \csc x \cdot \tan x$

In Exercises 9–14, the results of part (a) should help you simplify the expressions in part (b).

 9. **a.** $\dfrac{\dfrac{y}{1} + \dfrac{x}{1}}{y \quad x}$ **b.** $\dfrac{\sin \theta}{\csc \theta} + \dfrac{\cos \theta}{\sec \theta}$

 10. **a.** $\dfrac{\dfrac{s}{1} - 1}{s}$ **b.** $\dfrac{\sec \theta}{\cos \theta} - 1$

 11. **a.** $\dfrac{\dfrac{a}{1} - \dfrac{b}{1}}{a \quad b}$ **b.** $\dfrac{\sec \theta}{\cos \theta} - \dfrac{\tan \theta}{\cot \theta}$

 12. **a.** $\dfrac{1}{x + \dfrac{y^2}{x}}$ **b.** $\dfrac{1}{\cos \theta + \dfrac{\sin^2 \theta}{\cos \theta}}$

 13. **a.** $\dfrac{1}{t + \dfrac{1}{t}}$ **b.** $\dfrac{1}{\cot (90° - \theta) + \dfrac{1}{\tan \theta}}$

 14. **a.** $\dfrac{\dfrac{1}{st}}{\dfrac{s}{t} - \dfrac{t}{s}}$ **b.** $\dfrac{\dfrac{1}{\sec A \tan A}}{\dfrac{\sec A}{\tan A} - \dfrac{\tan A}{\sec A}}$

 15. $\sin A \tan A + \sin (90° - A)$ 16. $\csc A - \cos A \cot A$

 17. $(\sec B - \tan B)(\sec B + \tan B)$ 18. $(1 - \cos B)(\csc B + \cot B)$

 19. $(\csc x - \cot x)(\sec x + 1)$ 20. $(1 - \cos x)(1 + \sec x) \cos x$

Simplify the given expression.

21. $\dfrac{1}{1 + \tan^2 \theta} + \dfrac{1}{1 + \cot^2 \theta}$

22. $\dfrac{1 + \tan^2 \theta}{1 + \cot^2 \theta}$

23. $\dfrac{\tan x + \cot x}{\sec^2 x}$

24. $\dfrac{\sin x \cos x}{1 - \cos^2 x}$

25. $(\sin x + \cos x)^2 + (\sin x - \cos x)^2$

26. $(\csc y - \cot y)(1 + \cos y)$

27. $\dfrac{\cot^2 \theta}{1 + \csc \theta} + \sin \theta \cdot \csc \theta$

28. $\dfrac{\tan^2 \theta}{\sec \theta + 1} + 1$

29. $\dfrac{\sec y + \csc y}{1 + \tan y}$

30. $\cos^3 y + \cos y \sin^2 y$

31. $\sin^4 x + 2 \sin^2 x \cos^2 x + \cos^4 x$

32. $(\sec^2 \theta - 1)(\csc^2 \theta - 1)$

33. $\dfrac{\sin \theta}{1 + \cos \theta} + \dfrac{1 + \cos \theta}{\sin \theta}$

34. $\dfrac{\sin \theta \cot \theta + \cos \theta}{2 \tan (90° - \theta)}$

35. $\dfrac{\csc \theta - \sin \theta}{\cot^2 \theta}$

36. $\dfrac{\sin^4 \theta - \cos^4 \theta}{\sin^2 \theta - \cos^2 \theta}$

B 37. Use the Pythagorean Theorem to explain the Pythagorean relationships.

38. Given that $x = a \cos \theta - b \sin \theta$ and $y = a \sin \theta + b \cos \theta$, show that $x^2 + y^2 = a^2 + b^2$.

39. If $x = a (1 - \cos \theta)$ and $y = a \sin \theta$, show that $x^2 + y^2 = 2ax$.

40. Use the equation $\sin^2 \theta + \cos^2 \theta = 1$ to prove that $\tan^2 \theta + 1 = \sec^2 \theta$.

41. Use the equation $\sin^2 \theta + \cos^2 \theta = 1$ to prove that $\cot^2 \theta + 1 = \csc^2 \theta$.

In Exercises 42–59, prove the given identity.

42. $\cot^2 \theta + \cos^2 \theta + \sin^2 \theta = \csc^2 \theta$

43. $\dfrac{\cot \theta - \tan \theta}{\sin \theta \cos \theta} = \csc^2 \theta - \sec^2 \theta$

44. $\dfrac{\sin \theta}{\csc \theta} + \dfrac{\cos \theta}{\sec \theta} = \sin \theta \csc \theta$

45. $\dfrac{1 - \sin^2 \theta}{1 + \cot^2 \theta} = \sin^2 \theta \cos^2 \theta$

46. $\tan^2 x - \sin^2 x = \tan^2 x \sin^2 x$

47. $\dfrac{\tan^2 x}{1 + \tan^2 x} = \sin^2 x$

48. $\dfrac{\sin \theta}{\sin \theta + \cos \theta} = \dfrac{\tan \theta}{1 + \tan \theta}$

49. $\dfrac{\cos \theta}{1 + \sin \theta} + \dfrac{1 + \sin \theta}{\cos \theta} = 2 \csc (90° - \theta)$

50. $\dfrac{\sec^4 \theta - 1}{\tan^2 \theta} = \tan^2 \theta + 2$

51. $(\cot A + \tan A)^2 = \csc^2 A \sec^2 A$

52. $\dfrac{\tan x}{1 + \sec x} + \dfrac{1 + \sec x}{\tan x} = 2 \csc x$

53. $\dfrac{\tan x + \tan y}{\cot x + \cot y} = \tan x \tan y$

54. $\cos^4 \theta - \sin^4 \theta = 1 - 2 \sin^2 \theta$

55. $\sec A + \csc A = \dfrac{1 + \tan A}{\sin A}$

56. $\sec^2 A + \csc^2 A = \sec^2 A \csc^2 A$

57. $\cot^2 x - \cos^2 x = \cot^2 x \cdot \sin^2 \left(\dfrac{\pi}{2} - x\right)$

58. $\log (1 - \sin \theta) + \log (1 + \sin \theta) = 2 \log \cos \theta$

59. $\log (\sec A - \tan A) + \log (\sec A + \tan A) = 0$

224   Chapter 6

60. $\overline{AB}$ is tangent to the unit circle at $B = (1, 0)$.

  a. Why is $\triangle OPQ \sim \triangle OAB$?

  b. Use part (a) to explain why

$$(1) \; \frac{PQ}{OQ} = \frac{AB}{OB} \quad \text{and} \quad (2) \; \frac{OP}{OQ} = \frac{AO}{BO}.$$

  c. Use part (b) to show that (1) $AB = \tan \theta$ and (2) $AO = \sec \theta$. (This should help explain why certain trigonometric functions are called *tangent* and *secant*. Note that segments AB and AO lie on the lines tangent and secant, respectively, to the circle from A.)

  d. Now use right triangle OAB to prove $\sec^2 \theta = 1 + \tan^2 \theta$.

  e. Extend $\overline{AO}$ to meet the circle at C. A theorem from geometry states that $(AB)^2 = AP \cdot AC$. Use this fact to prove $\tan^2 \theta = (\sec \theta - 1)(\sec \theta + 1)$.

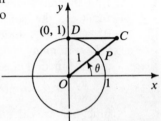

61. Segment CD is tangent to the unit circle at $(0, 1)$. Show that

$$(1) \; CD = \cot \theta \quad \text{and} \quad (2) \; CO = \csc \theta.$$

(You should first understand Exercise 60 before doing this exercise.)

Exercises 62–65 refer to the following example.

EXAMPLE. Express $\tan \theta$ in terms of $\sin \theta$.

  SOLUTION: $\tan \theta = \dfrac{\sin \theta}{\cos \theta} \leftarrow$

$\cos \theta = \pm \sqrt{1 - \sin^2 \theta}$, because $\cos^2 \theta = 1 - \sin^2 \theta$.

$$= \frac{\sin \theta}{\pm \sqrt{1 - \sin^2 \theta}}$$

The $\pm$ sign is used in the answer because we do not know the quadrant of θ. If θ is in the second or third quadrant, we use the $-$ sign, because $\cos \theta < 0$. If θ is in the first or fourth quadrant, we use the $+$ sign.

62. Express $\tan \theta$ in terms of $\cos \theta$.
    63. Express $\sec \theta$ in terms of $\sin \theta$.

64. Express $\cot \theta$ in terms of $\sin \theta$.
    65. Express $\csc \theta$ in terms of $\cos \theta$.

In Exercises 66–69, simplify the given expression.

C **66.** $\dfrac{(\cot A - \cos A)(1 + \sin A)}{\cos^3 A}$

67. $\left(\dfrac{2 - \sec^2 \theta}{\sec^2 \theta} \right)\left(\dfrac{\cos \theta + \sin \theta}{\cos \theta - \sin \theta} \right) - \dfrac{2 \tan \theta}{1 + \tan^2 \theta}$

68. $(1 + \cot \theta - \csc \theta)(1 + \tan \theta + \sec \theta)$

69. $(3 \sin x - 4 \sin^3 x)^2 + (4 \cos^3 x - 3 \cos x)^2$

70. If $x = a(\sec \theta + \tan \theta)$, show that $\cos \theta = \dfrac{2ax}{a^2 + x^2}$.

71. If you replace all the functions in an identity by their corresponding cofunctions, another identity will result. Here are two examples.

Original Identity: (1) $\sec^2 \theta = 1 + \tan^2 \theta$ (2) $\tan \theta = \sin \theta / \cos \theta$

                 $\downarrow$ $\downarrow$ $\downarrow$ $\downarrow$ $\downarrow$

     New Identity: $\csc^2 \theta = 1 + \cot^2 \theta$ $\cot \theta = \cos \theta / \sin \theta$

Explain why this works.

72. Prove $\sqrt{\dfrac{1 - \sin x}{1 + \sin x}} = |\sec x - \tan x|$.

COMPUTER EXERCISES

1. Imagine that your computer is capable of calculating only one trigonometric function, the sine. Write a program for which the input is any θ, $0° \leq \theta \leq 90°$, and the outputs are the six trigonometric functions of θ. [*Note:* In many computer languages the angles in the domain of a trigonometric function must be in radians. In this case, θ degrees must be converted to $\dfrac{\pi}{180} \cdot x$ (or approximately $0.017453x$) radians before the trigonometric functions are calculated.]

2. Repeat Exercise 1, allowing the input to be any θ, $0° \leq \theta < 360°$.

6-7 / TRIGONOMETRIC EQUATIONS I

In Section 6-6 we discussed identities such as $\sin^2 \theta + \cos^2 \theta = 1$, which are true for all values of θ (except, of course, those values for which either side of the equation is undefined). In this section, we shall be concerned with equations, such as $\sin \theta = 2 \cos \theta$, that are true for *some* but *not all* values of θ. An equation of this sort is called a **trigonometric equation,** and the **solutions** of the equation are the numbers θ that make it a true statement.

The sine graph at the right illustrates that there are many solutions to the trigonometric equation $\sin x = 0.5$. We know that $x = \dfrac{\pi}{6}$ and $x = \dfrac{5\pi}{6}$, for example, are two solutions. Since we know that the period of $\sin x$ is 2π, by using $\dfrac{\pi}{6}$ as the reference angle we can determine that the rest of the solutions are $x = \dfrac{\pi}{6} + 2n\pi$ and $x = \dfrac{5\pi}{6} + 2n\pi$, where n is any integer.

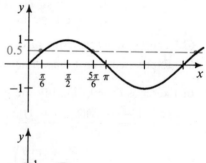

The same idea applies when solving an equation such as $\sin x = 0.6$, except that we must use a table or calculator to find the reference angle as shown in Example 1.

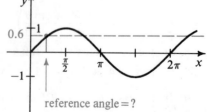

reference angle = ?

EXAMPLE 1. Find all x to the nearest thousandth between 0 and 2π for which $\sin x = 0.6$.

SOLUTION: **Step 1:** Use Table 5 to find a sine value near 0.6000. The corresponding value of $x \approx 0.644$. Or enter 0.6 on a calculator in radian mode and then find the INV SIN (or ARC SIN) of 0.6. You should get $x \approx 0.644$.

Step 2: 0.644 is the measure of a reference angle for other solutions. Since $\sin x$ is positive, we have another solution in Quadrant II, namely, $x = \pi - 0.644 \approx 2.498$. Therefore, $x \approx 0.644, 2.498$.

Note that if in Example 1 you had been asked to find all values of x for which $\sin x = 0.6$, and not just those between 0 and 2π, your answer would be $x = 0.644 + n \cdot 2\pi$ and $x = 2.498 + n \cdot 2\pi$, for any integer n.

Throughout the rest of the book, we will use roman letters such as x and y to refer to angle measures in radians; we will use Greek letters such as α and θ to refer to angle measures in degrees.

EXAMPLE 2. Solve $\cos x = -0.25$ for $0 \le x < 2\pi$.

SOLUTION: **Step 1:** Temporarily ignore the negative sign, and consider $\cos x = 0.25$. With a table or a calculator, find the reference angle $x \approx 1.32$ radians. If you use a calculator, be sure it is in radian mode.

Step 2: Since $\cos x$ is negative, solutions are in Quadrants II and III.
Quadrant II solution: $x \approx \pi - 1.32 \approx 3.14 - 1.32 = 1.82$
Quadrant III solution: $x \approx \pi + 1.32 \approx 3.14 + 1.32 = 4.46$

EXAMPLE 3. Solve $\csc^2 x = 5$ for $0 \le x < 2\pi$.

SOLUTION: **Preliminary Step:** $\csc^2 x = 5$ implies that $\csc x = \pm\sqrt{5} \approx \pm 2.2361$,

Step 1: Find the reference angle with a table or a calculator in radian mode. If you use a calculator, you must first rewrite $\csc x \approx \pm 2.2361$ as $\sin x \approx \pm\dfrac{1}{2.2361} \approx \pm 0.4472$ and then find the INV SIN (ARC SIN or SIN^{-1}). The reference angle is $x \approx 0.46$.

Step 2: $\csc x = +\sqrt{5}$ will give solutions in Quadrants I and II; $\csc x = -\sqrt{5}$ will give solutions in Quadrants III and IV. The four solutions are $x \approx 0.46$, $\pi - 0.46 \approx 2.68$, $\pi + 0.46 \approx 3.61$, and $2\pi - 0.46 \approx 5.82$.

EXAMPLE 4. Solve $2 \cot \theta + 5 = 0$ for $0° \le \theta < 360°$.

SOLUTION: **Preliminary Step:**
Solving $2 \cot \theta + 5 = 0$ is like solving $2x + 5 = 0$ for x.

$$2 \cot \theta = -5 \qquad\qquad 2x = -5$$
$$\cot \theta = -\tfrac{5}{2} \qquad\qquad x = -\tfrac{5}{2}$$

(Solution continued on page 228)

Trigonometric functions   **227**

Step 1: Temporarily ignore the negative sign and consider $\cot \theta = \frac{5}{2}$. Find the reference angle with a table or a calculator in degree mode. If you use a calculator, rewrite $\cot \theta = \frac{5}{2}$ as $\tan \theta = \frac{2}{5}$, enter $\frac{2}{5}$, and then find INV TAN (ARC TAN or TAN^{-1}). The reference angle is $\theta \approx 21.8°$.

Step 2: Since $\cot \theta$ is negative, the solutions are in Quadrants II and IV.
Quadrant II solution: $\theta = 180° - 21.8° = 158.2°$
Quadrant IV solution: $\theta = 360° - 21.8° = 338.2°$

ORAL EXERCISES 6-7

Give solutions for $0° \le \theta < 360°$ to the nearest degree.

1. $\cos \theta = 0.4$ 2. $\cos \theta = -0.4$ 3. $\sin \theta = 0.72$ 4. $\sin \theta = -0.72$
5. $\tan \theta = 1.9$ 6. $\tan \theta = -1.9$ 7. $\csc \theta = 10$ 8. $\cot \theta = -5$

Give solutions for $0 \le x < 2\pi$ to the nearest hundredth of a radian.

9. $\sin x = 0.6210$ 10. $\sin x = -0.6210$ 11. $\sec x = 4$ 12. $\sec x = -4$

Solve for $0° \le \theta < 360°$ without using tables or a calculator.

13. $\cos \theta = \dfrac{1}{2}$ 14. $\sin \theta = -\dfrac{\sqrt{2}}{2}$ 15. $\csc \theta = 2$ 16. $\cot \theta = -1$

Solve, giving all solutions (not just those between $0°$ and $360°$).

17. $\cos \theta = -1$ 18. $\tan \theta = 1$

WRITTEN EXERCISES 6-7

Solve for $0° \le \theta < 360°$. Give solutions to the nearest tenth of a degree.

A   1. a. $\sin \theta = 0.7$ b. $\sin \theta = -0.7$
    2. a. $\cos \theta = 0.42$ b. $\cos \theta = -0.42$
    3. a. $\tan \theta = 1.2$ b. $\tan \theta = -1.2$
    4. a. $\cot \theta = 3$ b. $\cot \theta = -3$
    5. a. $\sec \theta = 5$ b. $\sec \theta = -5$
    6. a. $\csc \theta = 10$ b. $\csc \theta = -10$
    7. $3 \cos \theta = 1$ 8. $4 \sin \theta = 3$
    9. $5 \sec \theta + 6 = 0$ 10. $2 \tan \theta + 1 = 0$
    11. $6 \csc \theta - 9 = 0$ 12. $4 \cot \theta - 5 = 0$

Solve for $0 \le x < 2\pi$. Give solutions to the nearest hundredth of a radian.

13. $3 \sin x + 2 = 4$ 14. $8 = 9 \cos x + 2$
15. $\sec^2 x = 9$ 16. $\csc^2 x = 16$
17. $\dfrac{5 \csc x}{3} = \dfrac{9}{4}$ 18. $\dfrac{3 \cot x}{4} + 1 = 0$

Solve for $0° \leq \theta < 360°$ without using tables or a calculator.

19. $\sin \theta = -\frac{1}{2}$

20. $\cos \theta = \frac{1}{2}$

21. $\cos \theta = -\frac{\sqrt{3}}{2}$

22. $\sin \theta = \frac{\sqrt{3}}{2}$

23. $\tan \theta = 1$

24. $\cot \theta = -1$

25. $\sec \theta = 2$

26. $\csc \theta = -\frac{2}{\sqrt{3}}$

27. $\csc \theta = 0$

28. $2 \sec \theta + 1 = 0$

Solve for $0 \leq x < 2\pi$ without using tables or a calculator.

29. $\cot x = 0$

30. $\tan x = 0$

31. $2 \cos x + 1 = 0$

32. $2 \sin x - 1 = 0$

33. $\tan x = \sqrt{3}$

34. $\sin x = -\frac{\sqrt{2}}{2}$

35. $\csc^2 x = 4$

36. $\tan^2 x = 1$

37. $|\cos x| = 1$

38. $|\sec x| = \sqrt{2}$

B 39. $\log_2 (\sin x) = 0$

40. $\log_2 (\cos x) = -1$

41. $\log_3 (\tan x) = \frac{1}{2}$

42. $\log_2 |\csc x| = 1$

43. $\log_{\sqrt{3}} (\cot x) = 1$

44. $\log_4 (\sec x) = 0$

Solve for x in radians, giving all solutions (not just those between 0 and 2π).

45. $\tan x = \cot x$

46. $\sec x = \cos x$

47. $2 \sin x = \csc x$

48. $3 \cot x = \tan x$

6-8/TRIGONOMETRIC EQUATIONS II

As we saw in Example 4 on pages 227–228, trigonometric equations are solved in the same way that algebraic equations are solved. For example, compare the following solutions of the quadratic equations $x^2 - 3x - 4 = 0$ and $\cos^2 \theta - 3 \cos \theta - 4 = 0$.

$$x^2 - 3x - 4 = 0 \qquad \cos^2 \theta - 3 \cos \theta - 4 = 0$$
$$(x + 1)(x - 4) = 0 \qquad (\cos \theta + 1)(\cos \theta - 4) = 0$$
$$x = -1, 4 \qquad \cos \theta = -1 \text{ or } \cos \theta = 4$$
$$\theta = 180° + n \cdot 360°$$

(Note that $\cos \theta = 4$ has no solution.)

Some trigonometric equations that are not quadratic can be transformed into equations that have the quadratic form.

EXAMPLE 1. Solve $\sin^2 x - \sin x = \cos^2 x$ for $0 \leq x < 2\pi$.

SOLUTION: To get an equation involving only $\sin x$, substitute $1 - \sin^2 x$ for $\cos^2 x$.

$$\sin^2 x - \sin x = \cos^2 x$$
$$\sin^2 x - \sin x = 1 - \sin^2 x$$
$$2 \sin^2 x - \sin x - 1 = 0 \quad \text{(This is like } 2y^2 - y - 1 = 0.\text{)}$$
$$(2 \sin x + 1)(\sin x - 1) = 0$$
$$\sin x = -\tfrac{1}{2} \quad \text{or} \quad \sin x = 1$$
$$x = \frac{7\pi}{6}, \frac{11\pi}{6}, \frac{\pi}{2}$$

EXAMPLE 2. Solve $\sin x \tan x = \sin x$ for $0 \leq x < 2\pi$.

SOLUTION:

$$\sin x \tan x = \sin x$$
$$\sin x \tan x - \sin x = 0$$
$$\sin x (\tan x - 1) = 0$$

$$\sin x = 0 \qquad \text{or} \qquad \tan x - 1 = 0$$
$$x = 0, \pi \qquad\qquad \tan x = 1$$
$$x = \frac{\pi}{4}, \frac{5\pi}{4}$$

The solutions are $0, \frac{\pi}{4}, \pi, \frac{5\pi}{4}$.

Notice in Example 2 that we did not cancel the factor $\sin x$ from both sides of the equation. Doing so would have caused us to lose a root. (See the warning concerning losing roots on page 53.) In the next example, there is no common factor for both sides of the equation. In this case, then, there is no difficulty in dividing both sides by $\sin \theta$, as long as $\sin \theta \neq 0$, of course. We do not lose a root since values of θ for which $\sin \theta = 0$ are clearly not solutions to the original equation.

EXAMPLE 3. Solve $2 \sin \theta = \cos \theta$ for $0° \leq \theta < 360°$.

SOLUTION: $2 \sin \theta = \cos \theta$

$$2 = \frac{\cos \theta}{\sin \theta} \quad \text{(Dividing both sides by } \sin \theta\text{)}$$
$$2 = \cot \theta$$
$$\theta = 26.6°, 206.6°$$

At first glance, Example 4 looks much like Example 3. However, they are very different. If you divide both sides of the equation in Example 4 by $\sin \theta$, you get an even more difficult equation instead of a simpler one. The strategy in solving Example 4, then, is to square both sides of the equation and then replace $\sin^2 \theta$ by $1 - \cos^2 \theta$, to give a quadratic equation.

230   Chapter 6

EXAMPLE 4. Solve $2 \sin \theta = \cos \theta + 2$ for $0° \leq \theta < 360°$.

SOLUTION:
$$2 \sin \theta = \cos \theta + 2$$
$$4 \sin^2 \theta = 4 + 4 \cos \theta + \cos^2 \theta \quad \text{(Squaring both sides)}$$
$$4(1 - \cos^2 \theta) = 4 + 4 \cos \theta + \cos^2 \theta$$
$$4 - 4 \cos^2 \theta = 4 + 4 \cos \theta + \cos^2 \theta$$
$$0 = 4 \cos \theta + 5 \cos^2 \theta$$
$$0 = \cos \theta (4 + 5 \cos \theta)$$

$$\cos \theta = 0 \qquad \text{or} \quad 4 + 5 \cos \theta = 0$$
$$\theta = 90°, 270° \qquad \cos \theta = -\tfrac{4}{5}$$
$$\theta \approx 180° - 36.9° = 143.1°, \text{ or}$$
$$\theta \approx 180° + 36.9° = 216.9°$$

The values $90°$, $270°$, $143.2°$, and $216.8°$ are all solutions of the equation $0 = \cos \theta (4 + 5 \cos \theta)$. Because we squared the original equation, they may not all be solutions of the original equation. Any solution of the original equation, however, must be one of these four values. Therefore, we must check each of these solutions in the original equation.

Check:

$\theta = 90°$

$2 \sin 90° - \cos 90° \overset{?}{=} 2$
$2(1) - 0 \overset{?}{=} 2$
$2 = 2$

Hence, $\theta = 90°$ is a solution.

$\theta = 270°$

$2 \sin 270° - \cos 270° \overset{?}{=} 2$
$2(-1) - 0 \overset{?}{=} 2$
$-2 \neq 2$

Hence, $\theta = 270°$ is not a solution.

The diagram at the right illustrates that since $\cos 143.2°$ and $\cos 216.8° \approx -\tfrac{4}{5}$, $\sin 143.2° \approx \tfrac{3}{5}$ and $\sin 216.8° \approx -\tfrac{3}{5}$. Thus, the remaining checks are as follows.

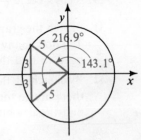

$\theta = 143.2°$

$2 \sin 143.2° - \cos 143.2° \overset{?}{=} 2$
$2(\tfrac{3}{5}) - (-\tfrac{4}{5}) \overset{?}{=} 2$
$\tfrac{10}{5} = 2$

Hence, $\theta = 143.2°$ is a solution.

$\theta = 216.8°$

$2 \sin 216.8° - \cos 216.8° \overset{?}{=} 2$
$2(-\tfrac{3}{5}) - (-\tfrac{4}{5}) \overset{?}{=} 2$
$-\tfrac{2}{5} \neq 2$

Hence, $\theta = 216.8°$ is not a solution.

WRITTEN EXERCISES 6-8

Solve for θ, $0° \leq \theta < 360°$. Give solutions to the nearest tenth of a degree.

A
1. $2 \cos^2 \theta - 3 \cos \theta + 1 = 0$
2. $\cos^2 \theta - 4 \cos \theta = 5$
3. $6 \sin^2 \theta - 7 \sin \theta + 2 = 0$
4. $2 \tan^2 \theta = 3 \tan \theta - 1$

Solve for θ, $0° \le \theta < 360°$. Give solutions to the nearest tenth of a degree.

5. $6 \sin^2 \theta = 7 - 5 \cos \theta$

6. $8 \cos^2 \theta = 9 - 6 \sin \theta$

7. $2 \cos^2 \theta + \sin \theta + 1 = 0$

8. $\cos^2 \theta - 3 \sin \theta = 3$

Solve for x, $0 \le x < 2\pi$. Give solutions to the nearest hundredth of a radian.

9. $\cos x \tan x = \cos x$

10. $\sec x \sin x = 2 \sin x$

11. $\sin^2 x = \sin x$

12. $\tan^2 x = \tan x$

13. $2 \cos^2 x = \cos x$

14. $\csc^3 x = 4 \csc x$

15. $\sin x = \cos x$

16. $\sin x + \cos x = 0$

17. $3 \sin x = \cos x$

18. $\sec x = 2 \csc x$

B

19. $\cot x = 2 \cos x$

20. $3 \sin^2 x = 2 \cos x \sin x$

21. $\tan^2 x = 2 \tan x \sin x$

22. $2 \sin x \cos x = \tan x$

23. $\sec^2 x = 3 \tan x - 1$

24. $2 \csc^2 x = 3 \cot^2 x - 1$

25. $\cot^2 x = 1 + \csc x$

26. $2 \sec^2 x + \tan x = 5$

27. $\cos^2 x - 2 \cos x - 1 = 0$ (*Hint:* Use the quadratic formula to solve for $\cos x$.)

28. $\sin^2 x + \sin x - 1 = 0$ (*Hint:* Use the quadratic formula to solve for $\sin x$.)

29. $3 \cos x \cot x + 7 = 5 \csc x$

30. $4 \cos x + \sec x = 8$

31. Sketch the graphs of $y = \frac{1}{2}x$ and $y = \sin x$ (x in radians) on the same axes. Tell how many solutions there are to the equation $\frac{1}{2}x = \sin x$. (You do not need to find these solutions.)

32. By sketching the graphs of $y = \frac{1}{10}x$ and $y = \sin x$, tell how many solutions there are to the equation $\sin x = \frac{1}{10}x$ (x in radians). (You do not need to find these solutions.)

Solve for x, $0 \le x < 2\pi$. Give solutions to the nearest hundredth of a radian.

33. $2 \sin^3 x - \sin^2 x - 2 \sin x + 1 = 0$
(*Hint:* First solve $2x^3 - x^2 - 2x + 1 = 0$.)

34. $4 \sin^3 x + 4 \sin^2 x - 3 \sin x - 3 = 0$

35. $\sec^3 x + \sec^2 x - 4 \sec x - 4 = 0$

C

36. $2 \cos^3 x + \cos^2 x - 5 \cos x + 2 = 0$

37. $\tan^3 x + 4 \tan^2 x + 4 \tan x + 3 = 0$

38. $2 \cos^2 x - \cos x = 2 - \sec x$

39. $\csc^2 x - 2 \csc x = 2 - 4 \sin x$

40. $2 \cos x = 2 + \sin x$

41. $1 - \sin x = 3 \cos x$

42. $(9^{\tan x})^{\cos x} = \dfrac{1}{3}$

43. $\dfrac{16^{\sin x}}{2^{\cos^2 x}} = 2 \cdot 2^{\sin^2 x}$

Chapter summary

1. Angles can be measured in *revolutions, gradients* (grads), *degrees,* or *radians.* The special conversion formulas between degrees and radians are shown on page 191. Two angles in standard position are called *coterminal angles* if they have the same terminal ray.

2. The number of radians in $\angle AOB$ shown at the right is defined by the formula

$$\theta = \frac{s}{r}.$$

   Thus, the length of arc AB is $s = r\theta$. The area of sector AOB is

$$K = \tfrac{1}{2}r^2\theta \ (\theta \text{ in radians})$$

   or

$$K = \tfrac{1}{2}rs.$$

3. If $P(x, y)$ is a point on the unit circle $x^2 + y^2 = 1$, and if θ is an angle in standard position with terminal ray OP, then the six *trigonometric functions* are defined as follows:

			$\tan\theta = \dfrac{\sin\theta}{\cos\theta}$	$\cot\theta = \dfrac{\cos\theta}{\sin\theta}$	$\sec\theta = \dfrac{1}{\cos\theta}$	$\csc\theta = \dfrac{1}{\sin\theta}$
Function	$\sin\theta = y$	$\cos\theta = x$				
Domain	all θ	all θ	$\theta \neq \dfrac{\pi}{2} + n\pi$	$\theta \neq n\pi$	$\theta \neq \dfrac{\pi}{2} + n\pi$	$\theta \neq n\pi$
Range	$\lvert\sin\theta\rvert \leq 1$	$\lvert\cos\theta\rvert \leq 1$	all reals	all reals	$\lvert\sec\theta\rvert \geq 1$	$\lvert\csc\theta\rvert \geq 1$
Period	2π	2π	π	π	2π	2π

   Graphs of the functions are shown on pages 208, 209, 214, and 215.

4. Values of the six trigonometric functions for the special angles $30°, 45°,$ and $60°$ may be found without using tables or a calculator by knowing the following relationships:

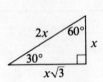

5. The acute angle α is called the *reference angle* of $\theta = 180° \pm \alpha$ and $\theta = 360° - \alpha$. To evaluate a trigonometric function of θ, first find the function of α. Then affix a plus sign or minus sign depending on the function and the quadrant involved. (See the diagram at the right.)

6. A *trigonometric identity* is an equation that is true for all values of θ for which both sides of the equation are defined. In proving such identities, the following relationships are helpful:

PYTHAGOREAN RELATIONSHIPS

$$\sin^2 \theta + \cos^2 \theta = 1$$

$$1 + \tan^2 \theta = \sec^2 \theta$$

$$1 + \cot^2 \theta = \csc^2 \theta$$

RECIPROCAL RELATIONSHIPS

$$\csc \theta = \frac{1}{\sin \theta} \quad \text{and} \quad \sin \theta = \frac{1}{\csc \theta}$$

$$\sec \theta = \frac{1}{\cos \theta} \quad \text{and} \quad \cos \theta = \frac{1}{\sec \theta}$$

$$\cot \theta = \frac{1}{\tan \theta} \quad \text{and} \quad \tan \theta = \frac{1}{\cot \theta}$$

COFUNCTION RELATIONSHIPS

$$\sin\left(\frac{\pi}{2} - \theta\right) = \cos \theta \quad \text{and} \quad \cos\left(\frac{\pi}{2} - \theta\right) = \sin \theta$$

$$\tan\left(\frac{\pi}{2} - \theta\right) = \cot \theta \quad \text{and} \quad \cot\left(\frac{\pi}{2} - \theta\right) = \tan \theta$$

$$\sec\left(\frac{\pi}{2} - \theta\right) = \csc \theta \quad \text{and} \quad \csc\left(\frac{\pi}{2} - \theta\right) = \sec \theta$$

7. Techniques for solving elementary *trigonometric equations* are illustrated in Examples 1–4 of Section 6–7. More advanced techniques are shown in Examples 1–4 of Section 6-8.

8. If a line has *inclination* θ and slope m, then $\tan \theta = m$.

Chapter test

1. Convert the following radian measures to degrees. 6–1

   a. $\frac{\pi}{6}$ **b.** $\frac{3\pi}{4}$ **c.** $-\frac{7\pi}{3}$ **d.** 2

2. Convert the following degree measures to radians.
   a. 45°   **b.** 150°   **c.** −225°   **d.** 300°

3. A circular sector has central angle 1.4 radians and arclength 21 cm. Find the radius and area. 6–2

4. Find $\sin \theta$ and $\cos \theta$. 6–3

   a.

   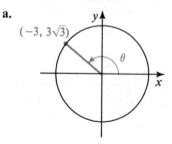

   b.

   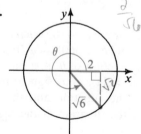

234   Chapter 6

5. Evaluate without using a table or a calculator.

   a. $\cos 180°$ **b.** $\sin 270°$ **c.** $\cos\left(-\dfrac{\pi}{2}\right)$ **d.** $\sin\left(-\dfrac{3\pi}{2}\right)$

6. Give the values of the following in simplest radical form without using a   6-4
table or a calculator.

   a. $\sin 135°$ **b.** $\cos\left(-\dfrac{5\pi}{3}\right)$ **c.** $\sin\left(\dfrac{5\pi}{4}\right)$ **d.** $\cos 150°$

7. a. Sketch the graph of $y = \sin x$ for $-\pi \le x \le 3\pi$.
   b. Give the range and the period of the sine function.

8. Without using a table or a calculator, solve for $0 \le x < 2\pi$.   6-5

   a. $\sin x = -\dfrac{\sqrt{3}}{2}$ **b.** $\tan x = -1$ **c.** $\sec x = \sqrt{2}$

9. Give the domain, range, and period of the tangent function.

10. Simplify the following: 6-6

   a. $\cot A\,(\sec A - \cos A)$ **b.** $\dfrac{\cot\theta}{\sin\left(\dfrac{\pi}{2} - \theta\right)}$

   c. $(\sec x + \tan x)(1 - \sin x)$ **d.** $\dfrac{\cot\alpha + \tan\alpha}{\csc^2\alpha}$

11. Solve the following equations for $0 \le x < 2\pi$.   6-7
   a. $2\cos x = \sin x$ **b.** $\sin x = \csc x$

12. Solve the following equations for $0° \le \theta < 360°$.   6-8
   a. $2\cos^2\theta + 3\sin\theta - 3 = 0$ **b.** $\cos\theta\cot\theta = \cos\theta$

CHAPTER SEVEN

Triangle trigonometry

OBJECTIVES

1. To use trigonometry to solve right triangles.
2. To find the area of a triangle given two sides and the included angle.
3. To use the Law of Sines and the Law of Cosines to solve triangles.
4. To graph, and find values of, the inverse trigonometric functions.

Solving triangles and finding their areas

7-1/RIGHT TRIANGLE TRIGONOMETRY

In the last chapter we defined the trigonometric functions in terms of coordinates of points on a circle. In this chapter our emphasis shifts from circles to triangles. You will see that trigonometric relationships can be used to find unknown parts of triangles given certain known parts. For example, if you know the lengths of the sides of a triangle, then you can find the measures of its angles. In this section, we shall consider how trigonometry can be applied to right triangles.

Pictured at the left is the Köhlbrand Bridge in Hamburg, West Germany. The structure's triangular supports offer a dramatic example of the use of triangular shapes in architecture.

The two right triangles shown in the diagrams both have an acute angle θ and are, therefore, similar. Consequently, lengths of corresponding sides are proportional, and we have the following equations for an acute angle θ in any right triangle:

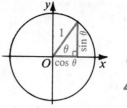 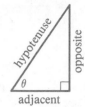

$$\frac{\sin \theta}{1} = \frac{\text{length of opposite side}}{\text{length of hypotenuse}} \qquad \frac{\cos \theta}{1} = \frac{\text{length of adjacent side}}{\text{length of hypotenuse}}$$

$$\tan \theta = \frac{\sin \theta}{\cos \theta} = \frac{\text{length of opposite side}}{\text{length of adjacent side}}$$

By the reciprocal relationships, we also have:

$$\csc \theta = \frac{\text{length of hypotenuse}}{\text{length of opposite side}} \qquad \sec \theta = \frac{\text{length of hypotenuse}}{\text{length of adjacent side}}$$

$$\cot \theta = \frac{\text{length of adjacent side}}{\text{length of opposite side}}$$

Applications of these equations are given in the following examples. Notice, in these examples, the convention of using a capital letter to denote an angle and the corresponding lower-case letter to denote the length of the side opposite that angle.

EXAMPLE 1. For right triangle ABC, use a calculator or tables to find the values of b and c to the nearest tenth.

SOLUTION: To find the value of b, use either $\tan 28°$ or $\cot 28°$.

Calculator: $\tan 28° = \dfrac{\text{opposite}}{\text{adjacent}} = \dfrac{40}{b}$

$$b = \frac{40}{\tan 28°} \approx 75.2$$

Table: $\cot 28° = \dfrac{\text{adjacent}}{\text{opposite}} = \dfrac{b}{40}$

$b = 40 \cdot \cot 28° \approx 40(1.881) \approx 75.2$

To find the value of c, use either $\sin 28°$ or $\csc 28°$.

Calculator: $\sin 28° = \dfrac{\text{opposite}}{\text{hypotenuse}} = \dfrac{40}{c}$

$$c = \frac{40}{\sin 28°} \approx 85.2$$

Table: $\csc 28° = \dfrac{\text{hypotenuse}}{\text{opposite}} = \dfrac{c}{40}$

$c = 40 \cdot \csc 28° \approx 40(2.130) = 85.2$

EXAMPLE 2. A triangle has sides of lengths 8, 8, and 4. Find the measures of the angles of the triangle to the nearest tenth of a degree.

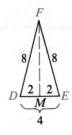

SOLUTION: If the altitude to the base of isosceles triangle *DEF* is drawn, two congruent right triangles are formed. In $\triangle DMF$, we have

$$\cos D = \frac{\text{adjacent}}{\text{hypotenuse}} = \frac{2}{8} = 0.2500$$

$$\angle D \approx 75.5°$$

Therefore, $\angle E \approx 75.5°$.

$$\angle F \approx 180° - 2(75.5°) = 29.0°$$

Notice, in Example 2, that we have written cos *D* as an abbreviation for the cosine of the measure of $\angle D$.

Usually a measurement that is approximated by using a calculator is more accurate than one obtained by using trigonometric tables. When tables with four significant digits are used, angle measures are generally accurate to tenths of a degree and lengths are generally accurate to three significant digits. Throughout this chapter, round your answers in this way unless you are directed otherwise.

In Exercises 19–24 of Section 6–5, you learned how to find the value of any trigonometric function of an angle given the quadrant of the terminal ray of the angle and the value of one trigonometric function. Our final example illustrates this idea for an acute angle of a right triangle.

EXAMPLE 3. If $\angle R$ is acute and $\sin R = x$, find the other trigonometric functions of $\angle R$ in terms of x.

SOLUTION: We draw a right triangle *RST* with hypotenuse of length 1 unit and the side opposite $\angle R$ of length x units. Then, by the Pythagorean Theorem, $RT = \sqrt{1 - x^2}$. Hence, using the definitions on page 238,

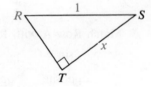

$$\cos R = \frac{\sqrt{1 - x^2}}{1} = \sqrt{1 - x^2}$$

$$\tan R = \frac{x}{\sqrt{1 - x^2}}$$

$$\csc R = \frac{1}{x}$$

$$\sec R = \frac{1}{\sqrt{1 - x^2}}.$$

$$\cot R = \frac{\sqrt{1 - x^2}}{x}$$

1. **a.** Referring to the given diagram, express the sine, cosine, and tangent of α in terms of x, y, and z.
   b. What are the sine, cosine, and tangent of β?
   c. True or false? $\sin \alpha = \cos \beta$
   $$\tan \alpha = \cot \beta$$
   $$\sec \alpha = \csc \beta$$

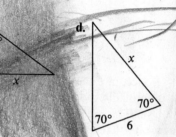

2. Name two reciprocal trigonometric functions that could be used to find the value of x.

   a.  **b.** 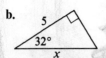 **c.**   **d.**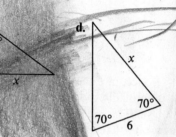

3. In Exercise 2(a), one student found the value of x using $\sin 50° = \dfrac{x}{10}$ while another found the value of x using $\cos 40° = \dfrac{x}{10}$. Who is correct?

4. State two equations, using different trigonometric functions, that can be used to find the value of θ.

   a.   **b.** 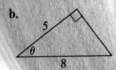 **c.** 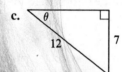 **d.**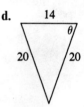

5. Match Row A with Row B.

A:	sin	cos	tan	cot	sec	csc
B:	$\dfrac{\text{opposite}}{\text{adjacent}}$	$\dfrac{\text{opposite}}{\text{hypotenuse}}$	$\dfrac{\text{adjacent}}{\text{opposite}}$	$\dfrac{\text{hypotenuse}}{\text{opposite}}$	$\dfrac{\text{adjacent}}{\text{hypotenuse}}$	$\dfrac{\text{hypotenuse}}{\text{adjacent}}$

WRITTEN EXERCISES 7-1

Throughout the exercises, give angle measures to the nearest tenth of a degree and lengths to three significant digits.

A 1. In $\triangle ABC$, $\angle A = 90°$, $\angle B = 25°$, and $a = 18$. Find b and c.
   2. In $\triangle PQR$, $\angle P = 90°$, $\angle Q = 64°$, and $p = 27$. Find q and r.
   3. In $\triangle DEF$, $\angle D = 90°$, $\angle E = 12°$, and $e = 9$. Find d and f.
   4. In $\triangle XYZ$, $\angle X = 90°$, $\angle Y = 37°$, and $z = 25$. Find x and y.

5. Use the diagram at the right to evaluate:
   a. $\sin A$ b. $\cos B$ c. $\tan A$
   d. $\cot B$ e. $\sec A$ f. $\csc B$
   g. the measure of $\angle A$
   h. the measure of $\angle B$

6. Find the measures of the acute angles of a right triangle whose legs are 9 cm and 16 cm long.

7. Find the measures of the angles of an isosceles triangle whose sides are 6, 6, and 8. Also find the area of the triangle.

8. Find the measures of the angles of an isosceles triangle whose sides are 15, 15, and 18. Also find the area of the triangle.

9. The legs of an isosceles triangle are each 21 cm long and the angle between them has measure 52°. What is the length of the third side?

10. The angle between the two congruent sides of an isosceles triangle has measure 112° and the length of the base is 20 m. What is the length of the altitude drawn to the base?

11. An isosceles trapezoid has sides whose lengths are in the ratio $5:8:5:14$. How large is the angle between one of the legs and the shorter base? (*Hint:* Draw the altitudes to the longer base.)

12. In an isosceles trapezoid one of the legs makes an angle of 35° with the longer base. The bases have lengths 12 cm and 20 cm. What are the lengths of the equal legs? (See *Hint* for Exercise 11.)

13. A rectangle is 14 cm wide and 48 cm long. Find the measure of the acute angle between its diagonals.

14. In the figure, $\overline{PA}$ and $\overline{PB}$ are tangents to a circle with radius $OA = 6$. If the measure of $\angle APB$ is 42°, find PA and PB.

$\angle A$ is acute. Use the given information to express the other trigonometric functions of $\angle A$ in terms of t.

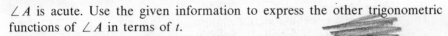

15. $\cos A = \dfrac{2}{t}$ 16. $\csc A = t$

17. $\cot A = \dfrac{2t}{3}$ 18. $\tan A = \dfrac{1}{4}t$

19. The diagram below shows that from a point 250 m from a building, the *angle of elevation* of the top of the building is 5°; that is, an observer's line of sight must be elevated 5° from the horizontal to point C. In the last chapter, we said that $\triangle ABC$ is approximately the same as a sector with central angle A.

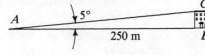

   a. Use the sector formula $s = r\theta$ to find BC approximately. (Remember to express θ in radians.)
   b. Use right triangle trigonometry to find BC more accurately. Compare your answers.

20. Refer to Exercise 19 and suppose the angle of elevation is 10° instead of 5°. Find BC approximately by using the sector arclength formula. Then find BC by using right triangle trigonometry. Compare your answers.

21. From the top of a lighthouse the *angle of depression* of a buoy is 30°; that is, the lighthouse attendant's line of sight must be depressed 30° from the horizontal to the location of the buoy. If the buoy is 75 m from the base of the lighthouse, how tall is the lighthouse?

22. Suppose in Exercise 21 the angle of depression is 15° instead of 30°. What is the height of the lighthouse? Is your answer half that of Exercise 21?

Exercises 23–26 refer to $\triangle ABC$.

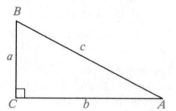

23. Express c in terms of a and a trigonometric function of $\angle A$.

24. Express c in terms of a and $\angle B$.

25. Express b in terms of a and $\angle B$.

26. Express a in terms of b and $\angle B$.

B **27.** From A and B, 10 m apart, the angles of elevation of the top of a tower are 40° and 54°, respectively. Find the height of the tower.

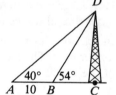

28. In $\triangle XYZ$, $XY = 100$, $\angle X = 51°$, and $\angle Y = 28°$. Find the length of the altitude from Z.

29. A line tangent to the circle $x^2 + y^2 = 1$ at P intersects the axes at T and S, as shown in the diagram.

a. Show that $PT = \tan\theta$ and $OT = \sec\theta$. (This may explain the use of the names *tangent* and *secant* functions. Another interpretation is given in Exercise 60 of Section 6–6.)

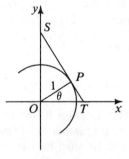

b. Use the diagram and part (a) to show that $\sec^2\theta = 1 + \tan^2\theta$.

c. Describe what happens to point T as θ increases from 0° to 90°, from 90° to 180°, from 180° to 270°, and from 270° to 360°. Your answers should agree with the graphs of $\tan\theta$ and $\sec\theta$.

30. a. Use the diagram in Exercise 29 to show that $PS = \cot\theta$ and $OS = \csc\theta$.

b. Show that $\csc^2\theta = 1 + \cot^2\theta$.

c. Describe what happens to point S as θ increases from 0° to 90°, from 90° to 180°, from 180° to 270°, and from 270° to 360°. Your answers should agree with the graphs of $\cot\theta$ and $\csc\theta$.

Express x and y in terms of the given angles α and β and the given side a.

31.

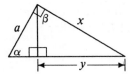

32.

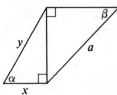

33.

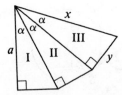

34.

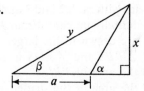

Use the figure at the right to do Exercises 35–38.

35. Show that $c = b \cos A + a \cos B$.

36. Show that $a \sin B = b \sin A$.

37. Show that the area of $\triangle ABC$ equals $\frac{1}{2}ac \sin B$.

38. Show that the area of $\triangle ABC$ equals $\frac{1}{2}bc \sin A$.

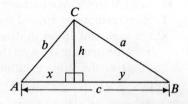

39. In the diagram, $\overline{AC}$ is tangent to a circle with radius OA of 12 cm. If the measure of $\angle AOB$ is $28°$, find the area to the nearest square centimeter of the region that is inside $\triangle OAC$ and outside circle O.

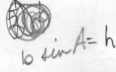

40. A circle has radius 20 cm. Find to the nearest cm² the area of the smaller of the two pieces into which it is cut by a chord of length 16 cm.

41. Find the measure of the angle θ formed by a diagonal of a cube and a diagonal of one of the faces of the cube.

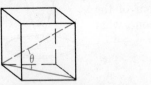

Ex. 41

Ex. 42

42. Find the measure of the angle α formed by the two diagonals of a cube.

C **43.** Derive a formula for the area A of a regular polygon of n sides which is inscribed in a circle of radius r. Your answer should give A in terms of n, r, and trigonometric functions of θ, where $\theta = \dfrac{180°}{n} = \dfrac{\pi}{n}$. Make sure you see why $\theta = \dfrac{\pi}{n}$.

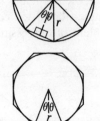

44. Derive a formula for the area K of a regular polygon of n sides which is circumscribed about a circle of radius r.

Use your answers from Exercises 43 and 44 on the preceding page to write a computer program which will print the values of A, K, and $K - A$ when $r = 1$ and $n = 10, 20, 30, \ldots , 100$. Interpret your answer.

CALCULATOR EXERCISES

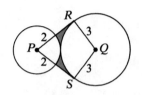

1. Two circles with radii 2 and 3 and centers P and Q are externally tangent. From P, tangents $\overline{PR}$ and $\overline{PS}$ are drawn to the larger circle.
   a. Find $\angle RPS$ and $\angle RQS$ in radians.
   b. Find the area of $PRQS$.
   c. Find the area of the shaded region.

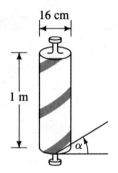

2. The red stripe on a barber pole makes two complete revolutions around the pole. If the pole is 1 m high and 16 cm in diameter, what angle α does the stripe make with the horizontal? (*Hint:* Think of the stripe as a line.)

3. In $\triangle ABC$, $\angle C = 90°$, $\angle A = 15°$, and $BC = 1$. E is a point on $\overline{AC}$ such that $\angle EBC = 60°$.
   a. Draw a diagram and use it to show that
   $$\tan 15° = \frac{1}{2 + \sqrt{3}}.$$
   b. Use a calculator to evaluate $\tan 15°$ and $\frac{1}{2 + \sqrt{3}}$. Compare your results.

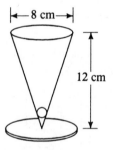

4. An olive is dropped into a cone-shaped glass with dimensions shown. If the bottom of the olive is 2 cm above the vertex of the cone, what is the radius of the olive?

7-2 / AREA OF A TRIANGLE

When the lengths of two sides of a triangle and the measure of the included angle are given, the triangle is uniquely determined. This fact is a consequence of the side-angle-side (SAS) congruence theorem in geometry. We now consider the problem of expressing the area of a triangle in terms of this information.

Suppose we are given the lengths a and b in triangle ABC and the measure of the included angle C. (See diagrams below.) Let the area of the triangle be denoted by K. We want to express K in terms of a, b, and $\angle C$.

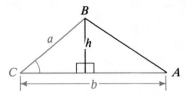 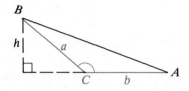

If the length of the altitude from B is h, then the area of the triangle is given by

$$K = \tfrac{1}{2}bh.$$

By right-triangle trigonometry, we know that $\dfrac{h}{a} = \sin C$, or

$$h = a \sin C.$$

Substituting this expression for h, we find that

$$K = \tfrac{1}{2}ab \sin C.$$

By drawing other altitudes in $\triangle ABC$, we could show that

$$K = \tfrac{1}{2}bc \sin A \quad \text{and} \quad K = \tfrac{1}{2}ac \sin B.$$

Instead of remembering a specific formula, you may find it easier to remember the basic pattern of these area formulas:

$$\text{Area} = \tfrac{1}{2} \cdot (\text{one side}) \cdot (\text{another side}) \cdot (\text{sine of included angle})$$

EXAMPLE 1. Two sides of a triangle have lengths 7 cm and 4 cm, respectively. The included angle has measure 150°. Find the area of the triangle.

SOLUTION: $K = \tfrac{1}{2} \cdot 7 \cdot 4 \cdot \sin 150° = \tfrac{1}{2} \cdot 7 \cdot 4 \cdot \tfrac{1}{2} = 7 \text{ cm}^2$

EXAMPLE 2. The area of $\triangle PQR$ is 15. If $p = 5$ and $q = 10$, find the measure of $\angle R$.

SOLUTION: $K = \tfrac{1}{2}pq \sin R$

$15 = \tfrac{1}{2} \cdot 5 \cdot 10 \cdot \sin R = 25 \sin R$

$\sin R = \tfrac{15}{25} = \tfrac{3}{5} = 0.6$

$\angle R \approx 36.9°$ or $\angle R \approx 180° - 36.9° = 143.1°$

ORAL EXERCISES 7-2

1. Two adjacent sides of a triangle have lengths of 5 and 8. If these sides form a 30° angle, what is the area of the triangle?

Find the area of each triangle.

2.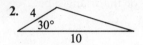

3.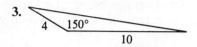

4.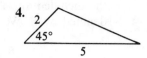

5. A triangle with area 5 has two sides of lengths 4 and 5.
   a. Find the sine of the angle included between these sides.
   b. Find the included angle. (Two answers are possible.)

WRITTEN EXERCISES 7–2

Throughout the exercises, give areas in radical form or rounded to three significant digits. Give lengths to three significant digits and angle measures to the nearest tenth of a degree.

In Exercises 1–4, find the area of $\triangle ABC$.

A **1. a.** $a = 4$, $b = 5$, $\angle C = 30°$ **2. a.** $b = 3$, $c = 8$, $\angle A = 120°$
   b. $a = 4$, $b = 5$, $\angle C = 150°$ **b.** $b = 3$, $c = 8$, $\angle A = 60°$

  3. a. $a = 6$, $c = 2$, $\angle B = 45°$ **4. a.** $a = 10$, $b = 20$, $\angle C = 70°$
   b. $a = 6$, $c = 2$, $\angle B = 135°$ **b.** $a = 10$, $b = 20$, $\angle C = 110°$

5. Suppose that $\angle C$ in $\triangle ABC$ is a right angle. Explain why the formula $K = \frac{1}{2}ab \sin C$ is valid.

6. Suppose that $\angle C$ in $\triangle ABC$ is obtuse, as in the diagram at the right on page 245. Show that $h = a \sin C$ and so the formula $K = \frac{1}{2}ab \sin C$ is valid.

7. Find the area of $\triangle XYZ$ if $x = 16$, $y = 25$, and $\angle Z = 52°$.

8. Find the area of $\triangle RST$ if $\angle S = 125°$, $r = 6$, and $t = 15$.

9. The area of $\triangle ABC$ is $18\sqrt{3}$. If $\angle B = 60°$ and $c = 8$, find the value of a.

10. The area of $\triangle DEF$ is $15\sqrt{2}$. If $\angle D = 135°$ and $e = 10$, find the value of f.

11. The area of $\triangle ABC$ is 15. If $a = 12$ and $b = 5$, find the two possible values for the measure of $\angle C$.

12. The area of $\triangle PQR$ is 9. If $q = 4$ and $r = 9$, find the two possible values for the measure of $\angle P$.

13. Find the area of a regular octagon inscribed in a circle of radius 40 cm.

14. Find the area of a regular 12-sided polygon inscribed in a circle of radius 8 cm.

15. Two adjacent sides of a parallelogram have lengths 6 cm and 7 cm, and the measure of the included angle is 30°. Find the area of the parallelogram.

16. Sketch a parallelogram with sides of lengths a and b and with an acute angle θ. Express the area of the parallelogram in terms of a, b, and θ.

17. The area of a parallelogram is 24 cm². If it has a 60° angle and one of the sides is 8 cm long, find the length of one of the adjacent sides.

18. The area of a rhombus is 18 cm², and one of the angles is 30°. What is the length of a side?

B 19. Suppose a triangle has two sides of lengths 3 cm and 4 cm and the included angle has measure θ. Express the area of the triangle as a function of θ. State the domain and range of the function and sketch its graph.

20. Suppose a triangle has two sides of lengths a and b. If the angle between these sides varies, what is the maximum possible area that the triangle can attain? Explain.

In Exercises 21–24, use the diagram given.

21. Find the area of a segment of a circle of radius $r = 5$ if the central angle $\theta = 2$ radians.

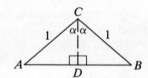

22. Derive a formula for the area of the shaded segment in terms of θ radians and r.

23. If $\theta = \dfrac{\pi}{6}$ and the area of the segment is 36, find the value of r.

24. Find the area of the segment formed by a chord 24 cm long in a circle with radius 13 cm.

Graph the region satisfying both inequalities and find its area.

25. $x^2 + y^2 \le 36,\ y \ge 3$ 26. $x^2 + y^2 \le 9,\ x \ge 1$

C 27. $x^2 + y^2 \le 9,\ x^2 + y^2 - 10x + 9 \le 0$

28. For the isosceles triangle shown, calculate the area of the triangle in two different ways. Then derive the formula

$$\sin 2\alpha = 2 \sin \alpha \cos \alpha.$$

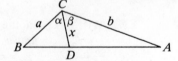

29. For the figure, $a = 1$, $b = 2$, $\alpha = 60°$, and $\beta = 60°$.
   a. Express the areas of $\triangle BCD$ and $\triangle ACD$ in terms of x.
   b. Find the area of $\triangle ABC$.
   c. Derive and solve an equation for x.

30. Generalize the method of Exercise 29 to express x in terms of a, b, α, and β.

CALCULATOR EXERCISE

In a circle of radius 10, there is a segment with area 95. Find the value of θ to the nearest tenth of a radian. (You will need to solve an equation by trial and error.)

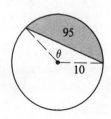

7-3/THE LAW OF SINES

When there are several methods for solving a problem, useful results can sometimes be obtained by comparing the solutions. From Section 7–2 we know three ways to calculate the area of the triangle ABC shown at the right, depending on which pair of sides is viewed as given. The three equal expressions for *the area* are

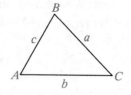

$$\tfrac{1}{2}bc \sin A = \tfrac{1}{2}ac \sin B = \tfrac{1}{2}ab \sin C.$$

If each of these expressions is divided by $\frac{1}{2}abc$, we obtain the *Law of Sines*.

LAW OF SINES: In $\triangle ABC$, $\dfrac{\sin A}{a} = \dfrac{\sin B}{b} = \dfrac{\sin C}{c}.$

That is, the sines of the angles of a triangle are proportional to the lengths of the opposite sides. For example, in a $30°$–$60°$–$90°$ triangle,

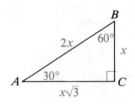

$$\sin 30° : \sin 60° : \sin 90° = a:b:c$$

$$\frac{1}{2} : \frac{\sqrt{3}}{2} : 1 = a:b:c$$

If we know the measures of two angles and a side of a triangle, we can use the Law of Sines to **solve the triangle** (that is, to find the measures of the remaining sides and angles). For example, suppose a civil engineer wants to determine the distances from points A and B to an inaccessible point C, as shown. Since the distance c and angles A and B can be measured, the distances a and b can be found with the Law of Sines.

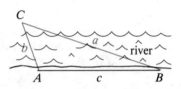

EXAMPLE 1. In the diagram above, suppose $\angle A = 110°$, $\angle B = 40°$, and $c = 25$ m. Solve $\triangle ABC$.

SOLUTION: $\angle C = 180° - 110° - 40° = 30°$. By the Law of Sines,

$$\frac{\sin 110°}{a} = \frac{\sin 40°}{b} = \frac{\sin 30°}{25}.$$

Since $\dfrac{\sin 30°}{25} = \dfrac{\frac{1}{2}}{25} = \dfrac{1}{50}$,

$$\frac{\sin 110°}{a} = \frac{\sin 40°}{b} = \frac{1}{50},$$

and we have

$a = 50 \sin 110° \approx 50(0.9397) \approx 47.0$ m
$b = 50 \sin 40° \approx 50(0.6428) \approx 32.1$ m

Suppose you are asked to draw a triangle ABC when given the measure of $\angle A$ and the lengths a and b. Depending on the values of a and b, and on the measure of $\angle A$, you will be able to draw no, one, or two triangles. These possibilities are illustrated in the following examples.

EXAMPLE 2. Draw $\triangle ABC$ if $\angle A = 30°$, $a = 4$, and $b = 10$.

SOLUTION:
1. Draw $\angle A$ with measure $30°$.
2. Locate point C so that $AC = 10$.
3. With center C, draw an arc with radius $a = 4$.
4. Notice that this arc fails to intersect ray AB, so that no triangle is formed.

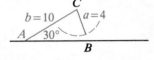

Law of Sines Verification of No Triangle

$$\frac{\sin B}{b} = \frac{\sin A}{a}$$

$$\sin B = \frac{b \sin A}{a} = \frac{10 \sin 30°}{4} = 1.25$$

Since $\sin B > 1$, there is no solution.

EXAMPLE 3. Draw $\triangle ABC$ if $\angle A = 30°$, $a = 5$, and $b = 10$.

SOLUTION: Follow the procedure outlined in Example 2.

Law of Sines Verification of One Triangle

$$\frac{\sin B}{b} = \frac{\sin A}{a}$$

$$\sin B = \frac{b \sin A}{a} = \frac{10 \sin 30°}{5} = 1$$

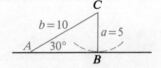

Hence, $\angle B = 90°$. Note that this agrees with the construction since segment AB is tangent to the dotted circular arc and is, therefore, perpendicular to the radius CB. Also, $\angle C = 60°$ and $c = 5\sqrt{3}$.

EXAMPLE 4. Draw $\triangle ABC$ if $\angle A = 30°$, $a = 5$, and $b = 8$.

SOLUTION: Follow the procedure outlined in Example 2. Note that $\triangle AB_1C$ and $\triangle AB_2C$ satisfy the given information.

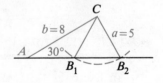

Law of Sines Verification of Two Triangles

$$\frac{\sin B}{b} = \frac{\sin A}{a}$$

$$\sin B = \frac{b \sin A}{a} = \frac{8 \sin 30°}{5} = 0.8$$

$$\angle B \approx 53.1° \quad \text{or} \quad \angle B \approx 180° - 53.1° = 126.9°$$

Example 4 shows that if the sine of an unknown angle is found to be less than one, then there may be two possible measures for that angle. On the other hand, the next example shows that other information may eliminate one of these possibilities.

EXAMPLE 5. In $\triangle RST$, $\angle S = 126°$, $s = 12$, and $t = 7$. Find the measure of $\angle T$.

SOLUTION:
$$\frac{\sin T}{t} = \frac{\sin S}{s}$$

$$\sin T = \frac{t \sin S}{s} = \frac{7 \sin 126°}{12} \approx 0.4719$$

$$\angle T \approx 28.2° \quad \text{or} \quad 151.8°$$

Since $\angle S = 126°$ and $\angle R + \angle S + \angle T = 180°$, $\angle T$ cannot have measure 151.8°. Thus, $\angle T = 28.2°$.

ORAL EXERCISES 7-3

Consider a triangle ABC.

1. If $\angle A \geq 90°$, what can you conclude about the measure of $\angle B$? Explain.
2. If $\angle B$ has a greater measure than $\angle C$, what must be true of b and c? Why?

In Exercises 3–5, state an equation you can use to solve for x.

3.

4.

5.

6. Suppose that $a = 8$ and $b = 6$. Draw a diagram to show that $\triangle ABC$ is uniquely determined for each of the following measures of $\angle A$:
   a. $\angle A = 45°$ b. $\angle A = 90°$ c. $\angle A = 120°$
7. Use the Law of Sines to show that $\triangle ABC$ in Exercise 6(a) is unique.

WRITTEN EXERCISES 7-3

In Exercises 1–10, solve each $\triangle ABC$, giving angle measures to the nearest tenth of a degree and lengths in simplest radical form or to three significant digits. Be alert to problems with no solution or with two solutions.

A
1. $\angle A = 45°$, $\angle B = 60°$, $a = 14$ 2. $\angle B = 30°$, $\angle C = 45°$, $b = 9$
3. $\angle B = 30°$, $\angle A = 135°$, $b = 4$ 4. $\angle A = 60°$, $\angle B = 75°$, $c = 10$
5. $\angle A = 25°$, $b = 3$, $a = 2$ 6. $\angle B = 36°$, $a = 10$, $b = 8$
7. $\angle A = 76°$, $a = 12$, $b = 4$ 8. $\angle B = 130°$, $b = 15$, $c = 11$
9. $\angle B = 40°$, $a = 12$, $b = 6$ 10. $\angle C = 112°$, $c = 5$, $a = 7$

11. In $\triangle RST$, $\angle R = 140°$ and $s = \frac{3}{4}r$. Find the measures of $\angle S$ and $\angle T$.

12. In $\triangle DEF$, $\angle F = 120°$ and $f = \frac{4}{5}e$. Find the measures of $\angle D$ and $\angle E$.

13. A fire tower at point A is 30 km north of a fire tower at point B. A fire at point F is observed from both towers. If $\angle FAB = 54°$ and $\angle ABF = 31°$, find AF.

14. From lighthouses P and Q, 16 km apart, a disabled ship S is sighted. If $\angle SPQ = 44°$ and $\angle SQP = 66°$, find the distance from S to the nearer lighthouse.

15. In $\triangle XYZ$, $\tan X = \frac{12}{5}$, $\tan Y = 1$, and $x = 24$. Find the value of y in simplest radical form. (*Hint:* Find $\sin X$ and $\sin Y$.)

16. In $\triangle ABC$, $\cot B = \dfrac{3}{4}$, $\cot C = \dfrac{\sqrt{5}}{2}$, and $b = 6$. Find the value of c in simplest radical form.

17. In $\triangle PQR$, $\cos Q = \dfrac{\sqrt{3}}{2}$, $\cos R = -\dfrac{2}{3}$, and $q = 6$. Find the value of r in simplest radical form.

18. In $\triangle LMN$, $\sec N = \dfrac{29}{21}$, $\sec M = -\dfrac{5}{3}$, and $n = 25$. Find the value of m in simplest radical form.

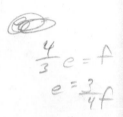

B 19. In this exercise, we shall prove that in $\triangle ABC$ the three equal ratios $\dfrac{a}{\sin A}$, $\dfrac{b}{\sin B}$, $\dfrac{c}{\sin C}$ are each equal to the diameter of its circumcircle.

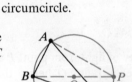

    a. Let circle O be the circumcircle and let $\overline{BP}$ be the diameter through B. Show that $\angle P$ and $\angle C$ have the same measure.

    b. Show that $\dfrac{AB}{BP} = \sin P$.

    c. Use parts (a) and (b) to show that
$$\frac{c}{\sin C} = \text{diameter} = \frac{b}{\sin B} = \frac{a}{\sin A}.$$

20. A triangle with angles $50°$, $60°$, and $70°$ has all three vertices on a circle of radius 8 cm. Find the lengths of the three sides. (*Hint:* See Exercise 19.)

21. Prove that the area of $\triangle ABC$ is given by
$$K = \frac{1}{2}\left(\frac{\sin B \sin C}{\sin A}\right)a^2.$$
State two other expressions for K, one in terms of b and the other in terms of c.

22. Use the results of Exercises 19 and 21 to show that the ratio of the area of $\triangle ABC$ to the area of its circumcircle is $\dfrac{2}{\pi}\sin A \sin B \sin C$.

23. The purpose of this exercise is to use the Law
of Sines to prove that the angle bisector of a
triangle divides the opposite side in the ratio
of the two adjacent sides; that is, in the dia-
gram at the right, to prove that $\dfrac{x}{y} = \dfrac{a}{b}$.

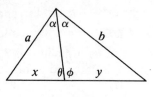

a. Prove that $\sin \theta = \sin \phi$. (ϕ is the Greek letter *phi*.)

b. Prove that $\dfrac{a}{x} = \dfrac{\sin \theta}{\sin \alpha}$ and $\dfrac{b}{y} = \dfrac{\sin \phi}{\sin \alpha}$.

c. Prove the stated theorem.

C **24.** Suppose that two circles with centers P and Q have radii 1 and 2, respectively. P is the midpoint of $\overline{OQ}$. An arbitrary line through O intersects the
circles at A, B, C, and D.
a. Prove: A is the midpoint of $\overline{OC}$.
b. What special property do you think B has?

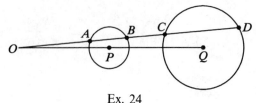

Ex. 24

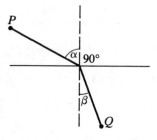

Ex. 25

25. Given: $\angle DAC = \angle BAE = 90°$
Prove: $DE = BC \tan x \tan y$

CALCULATOR EXERCISE

A ray of light from point P above the water is bent
to reach point Q below the water. According to
Snell's Law,

$$\frac{\text{speed of light in air}}{\text{speed of light in water}} = \frac{\sin \alpha}{\sin \beta}$$

If $\alpha = 45°$, find β. Use the fact that the speed of
light in air is 3.00×10^8 km/s and the speed of
light in water is 2.25×10^8 km/s.

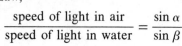

7-4/THE LAW OF COSINES

In Section 7-2 we mentioned that by the SAS congruence theorem a triangle is
uniquely determined if the lengths of two sides and the measure of the included angle are given. By the SSS (side-side-side) congruence theorem, a
triangle is also uniquely determined if the lengths of three sides are given. The
Law of Cosines can be used to solve a triangle in either of these cases.

252   Chapter 7

LAW OF COSINES: In $\triangle ABC$, $c^2 = a^2 + b^2 - 2ab \cos C$.

Proof: You can see from the diagram that $B = (a, 0)$ and $A = (b \cos C, b \sin C)$. Then use the distance formula to find the square of the distance c from A to B.

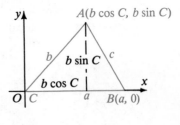

$$c^2 = (b \cos C - a)^2 + (b \sin C - 0)^2$$
$$= b^2 \cos^2 C - 2ab \cos C + a^2 + b^2 \sin^2 C$$
$$= b^2 (\cos^2 C + \sin^2 C) - 2ab \cos C + a^2$$
$$= b^2 - 2ab \cos C + a^2$$
$$c^2 = a^2 + b^2 - 2ab \cos C$$

When using the Law of Cosines, it helps to remember it in the following schematic form:

$$\left(\begin{matrix} \text{side} \\ \text{opposite} \\ \text{angle} \end{matrix}\right)^2 = \left(\begin{matrix} \text{side} \\ \text{adjacent} \\ \text{to angle} \end{matrix}\right)^2 + \left(\begin{matrix} \text{other side} \\ \text{adjacent} \\ \text{to angle} \end{matrix}\right)^2 - 2\left(\begin{matrix} \text{one} \\ \text{adjacent} \\ \text{side} \end{matrix}\right) \cdot \left(\begin{matrix} \text{other} \\ \text{adjacent} \\ \text{side} \end{matrix}\right) \cdot \cos \text{(angle)}$$

Note that when $\angle C = 90°$ the Law of Cosines reduces to $c^2 = a^2 + b^2$. Therefore the Law of Cosines includes the Pythagorean Theorem as a special case and is, consequently, more flexible and useful than the Pythagorean Theorem. When $\angle C$ is acute, c^2 is less than $a^2 + b^2$ by the amount $2ab \cos C$; when $\angle C$ is obtuse, $\cos C$ is negative and so c^2 is greater than $a^2 + b^2$.

EXAMPLE 1. Suppose two sides of a triangle have lengths 3 cm and 7 cm, respectively, and the included angle has measure 120°. Find the length of the third side.

SOLUTION: The triangle is determined by SAS. Let c be the length of the third side. Then:

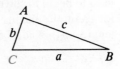

$$c^2 = 3^2 + 7^2 - 2 \cdot 3 \cdot 7 \cdot \cos 120°$$
$$= 9 + 49 - 2 \cdot 3 \cdot 7 \cdot (-0.5)$$
$$= 58 + 21 = 79$$
$$c = \sqrt{79} \approx 8.89 \text{ cm (to three significant digits)}$$

If we solve the Law of Cosines for $\cos C$, we obtain

$$\cos C = \frac{a^2 + b^2 - c^2}{2ab}.$$

This form of the Law of Cosines can be used to find the measures of the angles of a triangle given the lengths of its three sides. You can remember this formula in the schematic form:

$$\cos \text{(angle)} = \frac{(\text{adjacent})^2 + (\text{adjacent})^2 - (\text{opposite})^2}{2(\text{adjacent}) \cdot (\text{adjacent})}$$

Notice in Example 1 that you can now apply either the Law of Sines or the Law of Cosines to determine that the measures of the acute angles are 43.0° and 17.0° to the nearest tenth of a degree.

Triangle trigonometry   **253**

EXAMPLE 2. The lengths of three sides of a triangle are 5, 10, and 12. Solve the triangle.

SOLUTION: Recall that the largest angle is opposite the longest side.

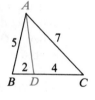

$$\cos \alpha = \frac{5^2 + 10^2 - 12^2}{2(5 \cdot 10)} = -0.19$$

$$\alpha \approx 101.0°$$

$$\cos \beta = \frac{12^2 + 10^2 - 5^2}{2(12 \cdot 10)} = 0.9125$$

$$\beta \approx 24.1°$$

$$\theta \approx 180° - (101.0° + 24.1°) = 54.9°$$

This example shows that the Law of Cosines is more effective than the Law of Sines in determining angles, since the cosine of an acute angle is positive and the cosine of an obtuse angle is negative, whereas both of these angles have positive sines.

EXAMPLE 3. In the diagram at the right, $AB = 5$, $BD = 2$, $DC = 4$, and $CA = 7$. Find AD using neither trigonometric tables nor a calculator.

SOLUTION: First we apply the Law of Cosines in $\triangle ABC$:

$$\cos B = \frac{5^2 + 6^2 - 7^2}{2(5 \cdot 6)} = 0.2$$

Then we apply the Law of Cosines in $\triangle ABD$:

$$(AD)^2 = 2^2 + 5^2 - 2(2 \cdot 5) \cos B$$
$$= 29 - 20(0.2) = 25$$
$$AD = 5$$

In this section and the last, we have seen various applications of the Law of Sines and the Law of Cosines. The situations in which each of these laws can be used to begin to solve a triangle are summarized below. The abbreviation ASA, for example, means that you are given the measures of two angles and the length of the side between them.

Law of Cosines

SAS   To find third side
SSS   To find any angle

Law of Sines

ASA ⎫
AAS ⎭   To find remaining sides
         (Note that third angle can be computed directly.)
SSA   To find the angle opposite the given side
         (Two noncongruent triangles can result from this case, as shown on page 249.)

254   Chapter 7

Once you have computed an additional side or angle, you can often use whichever law you prefer to find the remaining measures. Whenever you use the Law of Sines, remember that for every real t between 0 and 1, there is an acute $\angle A$ and a supplementary obtuse $\angle B$ such that $\sin A = \sin B = t$. Oral Exercises 4–6 show how you can tell which angle is correct.

ORAL EXERCISES 7–4

In Exercises 1–3, state an equation you can use to solve for x.

1.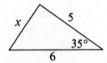

2.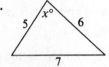

3.

You know that a triangle is completely determined by the lengths of two of its sides and the measure of the angle included by these sides. In $\triangle XYZ$, $x = 4$, $y = 8$, and $\angle Z = 50°$.

4. Use the Law of Cosines to find z to the nearest hundredth.

5. Use the Law of Sines to find the measure of $\angle Y$ to the nearest tenth of a degree. Then compute the measure of $\angle X$.

6. Note that $x < z < y$.
  a. What does this imply about the measures of $\angle X$, $\angle Y$, and $\angle Z$?
  b. Do your answers to Exercises 5 and 6(a) agree? If not, explain where an error in reasoning occurred.

WRITTEN EXERCISES 7–4

Solve each triangle. Give lengths to three significant digits and angle measures to the nearest tenth of a degree.

A **1.** $a = 8$, $b = 5$, $\angle C = 60°$ **2.** $t = 16$, $s = 14$, $\angle R = 120°$

3. $x = 9$, $y = 40$, $z = 41$ **4.** $a = 6$, $b = 10$, $c = 7$

5. $p = 3$, $q = 8$, $\angle R = 50°$ **6.** $d = 5$, $e = 9$, $\angle F = 115°$

7. $a = 8$, $b = 7$, $c = 13$ **8.** $x = 10$, $y = 11$, $z = 12$

9. Find the measure of the largest angle of a triangle with sides 24 cm, 7 cm, and 25 cm long.

10. Use the Law of Cosines to verify that $\triangle ABC$ is a 30°–60°–90° triangle when $a = 3\sqrt{3}$, $b = 6\sqrt{3}$, $c = 9$.

11. In $\triangle ABC$, $a = 7$, $b = 5$, $c = 4\sqrt{2}$.
  a. Find $\cos C$.
  b. Find $\sin C$ by using the identity $\sin^2 C + \cos^2 C = 1$.
  c. Find the area of $\triangle ABC$ by using the formula $K = \frac{1}{2}ab \sin C$.

12. Repeat Exercise 11 if $a = 10$, $b = 11$, $c = 3\sqrt{5}$.

In Exercises 13 and 14, use the method given in Example 3, page 254, to find *AD*.

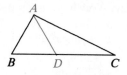

13. $AB = 8$ cm, $BD = 7$ cm, $DC = 5$ cm, $AC = 10$ cm

14. $AB = 5$ cm, $BD = 5$ cm, $DC = 3$ cm, $AC = 7$ cm

In Exercises 15 and 16, find the length of the median from A in the given $\triangle ABC$. Leave answers in simplest radical form.

15. $a = 8$ cm, $b = 4$ cm, $c = 6$ cm   **16.** $a = 12$ cm, $b = 13$ cm, $c = 5$ cm

17. A parallelogram has a 70° angle and sides 6 cm and 10 cm long. How long are its diagonals?

18. A parallelogram has a 140° angle and sides 3 cm and 7 cm long. How long are its diagonals?

B   **19.** In $\triangle ABC$, $AB = 20$, $BC = 15$, and $AC = 10$. $\overline{CD}$ bisects $\angle ACB$.

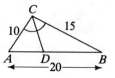

    a. Use the results of Exercise 23, page 252, to find *AD* and *DB*.

    b. Use the method of Example 3, page 254, to find *CD*.

20. Repeat Exercise 19 if $AB = 10$, $BC = 12$, and $AC = 8$.

21. The area of $\triangle PQR$ is 84. If $r = 14$ and $q = 13$, find sin *P*. Use a trigonometric identity to find two possible values of cos *P*. Find two possible values for *p*.

22. Repeat Exercise 21 if $\triangle PQR$ has area 12, $r = 8$, and $q = 5$.

23. Give a geometric interpretation of the Law of Cosines for the "triangle" with two sides *a* and *b* and "included angle" $C = 0°$.

24. Give a geometric interpretation of the Law of Cosines for the "triangle" with two sides *a* and *b* and "included angle" $C = 180°$.

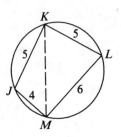

25. Quadrilateral *JKLM* is inscribed in a circle as shown. Find the length of each diagonal. (*Hint:* Opposite angles of an inscribed quadrilateral are supplementary. Express cos *L* in terms of cos *J* and cos *K* in terms of cos *M*.)

C   **26. a.** Prove that in $\triangle ABC$ the length *x* of the median from *C* is given by
$$x = \tfrac{1}{2}\sqrt{2a^2 + 2b^2 - c^2}.$$

    b. What does this formula reduce to when $\angle C = 90°$?

27. Segment *CD* bisects $\angle ACB$ in the diagram.

    a. Use the method of Exercise 19(a) to show that

$$p = \frac{bc}{a + b} \text{ and } q = \frac{ac}{a + b}.$$

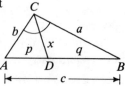

    b. Prove that $x^2 = ab - pq$.

    c. Use part (b) to find *CD* in Exercise 19.

28. Suppose a parallelogram has two adjacent sides of lengths a and b and diagonals of lengths x and y. Show that $x^2 + y^2 = 2a^2 + 2b^2$.

29. The purpose of this exercise is to derive *Hero's Formula* for the area, K, of $\triangle ABC$:

$$K = \sqrt{s(s-a)(s-b)(s-c)} \quad \text{where } s = \frac{a+b+c}{2}.$$

   a. Using the formula $\cos C = \dfrac{a^2 + b^2 - c^2}{2ab}$, show that:

$$
\begin{aligned}
\sin^2 C &= 1 - \cos^2 C \\
&= (1 + \cos C)(1 - \cos C) \\
&= \frac{(a+b)^2 - c^2}{2ab} \cdot \frac{c^2 - (a-b)^2}{2ab} \\
&= \frac{(a+b+c)(a+b-c)(c+a-b)(c-a+b)}{4a^2b^2} \\
&= \frac{(2s)(2s-2c)(2s-2b)(2s-2a)}{4a^2b^2}
\end{aligned}
$$

   b. Use part (a) and the formula $K = \frac{1}{2}ab \sin C$ to derive Hero's Formula.

CALCULATOR EXERCISES

In surveying work, compass readings are usually given as the acute angle from the north-south line toward the east or west. Some examples are shown below.

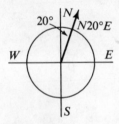
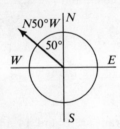
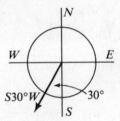

Give each answer to four significant digits.

1. Find the area of a piece of land described as follows: From the iron post at the old mill, proceed 85 rods south, then 67 rods in the direction N80°E, then 120 rods N10°W to the Mill Road, then along the Mill Road to the starting point. (A rod is approximately 5.03 m.)

2. Two ships leave San Francisco Bay at 12 noon. One travels in the direction N52°W at a speed of 20 knots. (A knot is a nautical mile per hour; a nautical mile is about 1.15 miles.) The other ship travels in the direction S80°W at 15 knots. How far apart (in nautical miles) are the ships at 4 P.M.?

3. Fire tower A is 8 km southwest of fire tower B. A fire is spotted from A in the direction N32°E. From B, the fire is in the direction N63°W. Find the distance from the fire to the nearer tower.

Where appropriate, give angle measures to the nearest tenth of a degree and express lengths of sides in simplest radical form or to three significant digits.

1. **a.** Find the area of $\triangle PQR$ if $q = 6$, $r = 7$, and $\angle P = 50°$. Also find p.
   b. Find the area of $\triangle PQR$ if $p = 7$, $q = 10$, and $\angle R = 130°$. Also find r.

2. Find the largest angle in a triangle with sides 3, 6, and 7.

3. Find the largest angle in a triangle with sides 7, 12, and 13.

4. In $\triangle RST$, $\angle R = 75°$, $\angle S = 45°$, and $t = 3$. Find r and s.

5. Three measurements in $\triangle ABC$ are given as $\angle A = 60°$, $a = 4$, and $b = 5$. Show that at least one of the measurements is incorrect.

6. A polygon with 180 sides is inscribed in a circle with radius 1. Find its area. Compare your answer with π.

7. In $\triangle ABC$, $\cos A = -0.6$. Find $\sin A$ and $\tan A$ using neither tables nor a calculator.

8. In $\triangle ABC$, $a = 17$, $b = 10$, and $c = 21$. Find:
   a. $\cos A$ **b.** $\sin A$ **c.** the area of $\triangle ABC$.

9. In $\triangle XYZ$, $\angle X = 21.1°$, $x = 6$, and $y = 9$. Find the possible measures of $\angle Y$.

10. In parallelogram $ABCD$, $\angle A = 60°$, $AB = 5$, and $AD = 8$.
    a. Find its area.   **b.** Find the lengths of both diagonals.

11. A triangle has area 21 cm² and two of its sides are 9 cm and 14 cm long. Find the possible measures of the angle formed by these sides.

12. In $\triangle DEF$, $\angle D = 36°$, $\angle E = 64°$, and $f = 8$. Find d and e.

13. $\triangle ABC$ is similar to $\triangle DEF$; $\angle A = 120°$.
    a. Find the lengths a, e, and f.
    b. Find the ratio of the areas of the triangles.

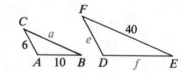

14. The diagonals of a parallelogram are 8 and 14 and they meet at a 60° angle. Find the area and the perimeter of the parallelogram.

15. An obtuse triangle with area 12 has two sides of lengths 4 and 10. Find the length of the third side.

16. The perimeter of a regular decagon (10 sides) is 240. Find its area.

17. A triangle has sides 6, 9, and 10. How long is the segment that bisects the largest angle? (*Hint:* Use the result of Exercise 23 on page 252.)

18. In $\triangle ABC$, $\tan A = 1$, $\tan B = \frac{3}{4}$, and $b = 18$. Find a.

19. If $180° < x < 360°$ and $\tan x = -\frac{1}{5}$, find $\sin x$ and $\cos x$.

20. If $180° < x < 360°$ and $\sec x = \frac{3}{2}$, find $\sin x$ and $\cot x$.

21. In $\triangle PQR$, $\tan P = -\sqrt{3}$. Find $\sin P$ and $\cos P$.

22. In $\triangle DEF$, $\sec F = -\sqrt{2}$. Find $\angle F$ and $\tan F$.

In geometry you can prove two triangles congruent by SSS. This means that when the lengths of the three sides of a triangle are given, its shape is completely determined. Exercises 23 and 24 below illustrate this principle. Exercises 25 and 26 illustrate the principle for SAS and ASA.

B   **23.** In $\triangle ABC$, $a = 5$, $b = 8$, and $c = 7$. (Given SSS)
  a. Solve $\triangle ABC$.
  b. Find the area. Use $K = \frac{1}{2}ab \sin C$ or Hero's Formula (Ex. 29, page 257).
  c. Find the length of the altitude to $\overline{AC}$.
  d. Find the length of the median from B to $\overline{AC}$ (Ex. 26, page 256).
  e. Find the length of the angle bisector from B to $\overline{AC}$ (Ex. 27, page 256).
  f. Find the radius R of the circumscribed circle (Ex. 19, page 251).
  g. Find the radius r of the inscribed circle. ($r = 2 \times$ area/perimeter)

24. Repeat Exercise 23 if $a = 13$, $b = 14$, and $c = 15$. (Given SSS)

25. Repeat Exercise 23 if $a = 9$, $b = 10$, and $\cos C = -\frac{3}{5}$. (Given SAS)

26. Repeat Exercise 23 if $\sin A = \frac{8}{17}$, $c = 21$, and $\sin B = \frac{\sqrt{2}}{2}$. (Given ASA) Assume that $\angle B$ is obtuse.

27. a. The consecutive sides of a quadrilateral inscribed in a circle have lengths 1, 4, 3, and 2. Find the length of each diagonal, using the fact that opposite angles are supplementary.

  b. Check your answer using Ptolemy's Theorem: If $ABCD$ is inscribed in a circle, then $AC \cdot BD = AB \cdot DC + AD \cdot BC$.

28. A triangle DEF is inscribed in a circle of radius 3. $\angle D = 120°$ and $\angle E = 15°$. Find the lengths d, e, and f. (*Hint:* Find the measures of arc DE and arc DF.)

29. Prove that the altitude from A in $\triangle ABC$ has length $p = \dfrac{a}{\cot B + \cot C}$.

C   **30.** Two circles are tangent externally. Common tangents to the circles form an angle of measure $2x$. Prove that the ratio of the radii of the circles is $\dfrac{1 - \sin x}{1 + \sin x}$. (*Hint:* Express $\sin x$ in terms of the two radii.)

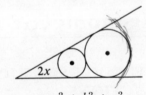

31. Prove that in any $\triangle ABC$, $\cot A + \cot B + \cot C = \dfrac{a^2 + b^2 + c^2}{4(\text{area of } \triangle ABC)}$.

32. Given $\triangle ABC$ with $c^2 = \dfrac{a^3 + b^3 + c^3}{a + b + c}$. Find the measure of $\angle C$.

33. $\overline{AC}$ is a ladder leaning against the side of a house at an angle α with the ground. If the foot of the ladder slides y units from A to B, the top of the ladder slides x units from C to D and the ladder makes an angle β with the ground.

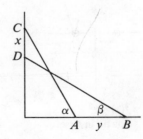

Prove: $x = \dfrac{y(\sin \alpha - \sin \beta)}{\cos \beta - \cos \alpha}$.

34. A lifeguard at the seashore is watching a ship with a smokestack 30 m above water level as the ship steams out to sea. The lifeguard's eye level is 4 m above water level. About how far will the ship be from shore when the stack disappears from view? Assume that the radius of the earth is about 6400 km.

35. In $\triangle XYZ$, $\overline{YR}$ and $\overline{ZS}$ are altitudes. Prove that $RS = x \cos X$.

36. **a.** $\overline{AD}$ is a diameter of the circle and is tangent to line l at D. If $AD = 6$ and $\overline{BC} \perp l$, express AB and BC in terms of θ.
    b. What value of θ makes the sum $AB + BC$ a maximum?

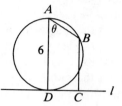

37. Each circle in the diagram is tangent to the other three circles. The largest three circles have radii of 1, 2, and 3, respectively. Find the radius of the smallest circle.

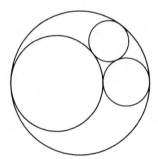

Trigonometric inverses

7-5/THE INVERSE TRIGONOMETRIC FUNCTIONS

Since the squaring function $f(x) = x^2$ is not one-to-one, it does not have an inverse. (See the graph at the left below.) However, if you consider only $x \geq 0$, the restricted function does have an inverse, namely, the square root function.

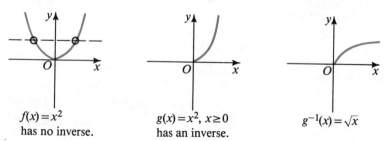

$f(x) = x^2$
has no inverse.

$g(x) = x^2$, $x \geq 0$
has an inverse.

$g^{-1}(x) = \sqrt{x}$

A similar situation exists for the tangent function $f(x) = \tan x$. The graph at the left below shows that $\tan x$ is not one-to-one and thus has no inverse. However, if you restrict the values of x to those between $-\frac{\pi}{2}$ and $\frac{\pi}{2}$, the restricted function, which we shall call $F(x) = \text{Tan } x$, is one-to-one. Its inverse is denoted $\text{Tan}^{-1} x$ or $\text{Arctan } x$ and read "the inverse tangent of x." (Some calculators use Inv tan x.)

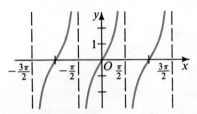

$f(x) = \tan x$
has no inverse.
Domain: $\{x: x \neq \frac{\pi}{2} + n\pi\}$
Range: Real numbers

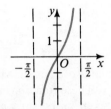

$F(x) = \text{Tan } (x)$
Domain: $\{x: -\frac{\pi}{2} < x < \frac{\pi}{2}\}$
Range: Real numbers

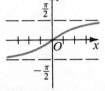

$F^{-1}(x) = \text{Tan}^{-1}(x)$
Domain: Real numbers
Range: $\{y: -\frac{\pi}{2} < y < \frac{\pi}{2}\}$

Notice that Tan x takes on every value that tan x takes on, but because the domain of Tan x is smaller, it does so exactly once. For example, Tan $x = 1$ only when $x = \frac{\pi}{4}$, whereas tan $x = 1$ when $x = \frac{\pi}{4}, \frac{5\pi}{4}, \frac{9\pi}{4}$, and so forth. You may think of $\text{Tan}^{-1} x$ as the unique real number y between $-\frac{\pi}{2}$ and $\frac{\pi}{2}$ whose tangent is x.

EXAMPLE 1. Evaluate: **a.** $\text{Tan}^{-1}(-\sqrt{3})$ **b.** $\text{Arctan}(-1)$ **c.** $\text{Tan}^{-1} 2$

SOLUTION: **a.** $\text{Tan}^{-1}(-\sqrt{3})$ is the real number between $-\frac{\pi}{2}$ and $\frac{\pi}{2}$ whose tangent is $-\sqrt{3}$. Therefore, $\text{Tan}^{-1}(-\sqrt{3}) = -\frac{\pi}{3}$.

b. $\text{Arctan}(-1)$ is the real number between $-\frac{\pi}{2}$ and $\frac{\pi}{2}$ whose tangent is -1. Therefore, $\text{Arctan}(-1) = -\frac{\pi}{4}$.

c. $\text{Tan}^{-1} 2$ represents the real number between $-\frac{\pi}{2}$ and $\frac{\pi}{2}$ whose tangent is 2. Using a calculator or trigonometric tables, you find $\text{Tan}^{-1} 2 \approx 1.11$.

It is possible to define the other inverse trigonometric functions in a manner similar to that used for the inverse tangent function. We shall carry out this procedure for the inverse sine and inverse cosine functions. In Exercises 43–45 you will define and graph the inverse cotangent function, the inverse cosecant function, and the inverse secant function.

Triangle trigonometry   **261**

1. Note that the sine and cosine functions are not one-to-one, so neither has an inverse.

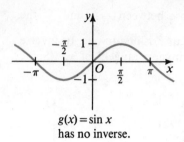

$g(x) = \sin x$
has no inverse.

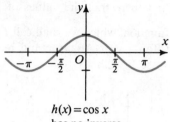

$h(x) = \cos x$
has no inverse.

2. We shall restrict the domains of x, as shown below, to create new functions, $y = \text{Sin } x$ and $y = \text{Cos } x$. These functions are one-to-one and they take on all the values that are taken on by $\sin x$ and $\cos x$.

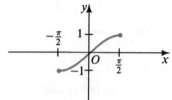

$G(x) = \text{Sin } x$
Domain: $\{x: -\frac{\pi}{2} \le x \le \frac{\pi}{2}\}$
Range: $\{y: -1 \le y \le 1\}$

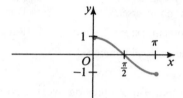

$H(x) = \text{Cos } x$
Domain: $\{x: 0 \le x \le \pi\}$
Range: $\{y: -1 \le y \le 1\}$

3. When we reflect the graphs of $y = \text{Sin } x$ and $y = \text{Cos } x$ in the line $y = x$, we obtain the graphs of their inverse functions, $\text{Sin}^{-1} x$ and $\text{Cos}^{-1} x$.

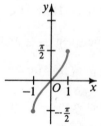

$G^{-1}(x) = \text{Sin}^{-1}x = \text{Arcsin } x$
Domain: $\{x: -1 \le x \le 1\}$
Range: $\{y: -\frac{\pi}{2} \le y \le \frac{\pi}{2}\}$

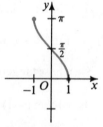

$H^{-1}(x) = \text{Cos}^{-1}x = \text{Arccos } x$
Domain: $\{x: -1 \le x \le 1\}$
Range: $\{y: 0 \le y \le \pi\}$

Notice that $\text{Cos}^{-1} x$ takes on values between 0 and π, whereas $\text{Tan}^{-1} x$ and $\text{Sin}^{-1} x$ take on values between $-\frac{\pi}{2}$ and $\frac{\pi}{2}$. Also, remember that unlike $\text{Cos}^{-1} x$ and $\text{Sin}^{-1} x$, $\text{Tan}^{-1} x$ cannot assume its boundary values, $-\frac{\pi}{2}$ and $\frac{\pi}{2}$.

EXAMPLE 2. Evaluate: **a.** $\mathrm{Cos}^{-1}\left(-\dfrac{1}{2}\right)$ **b.** Arcsin 0.6   **c.** $\mathrm{Sin}^{-1}(-0.6)$

SOLUTION: **a.** $\mathrm{Cos}^{-1}\left(-\dfrac{1}{2}\right)$ is the real number between 0 and π, inclusive, whose cosine is $-\dfrac{1}{2}$. Therefore, $\mathrm{Cos}^{-1}\left(-\dfrac{1}{2}\right) = \dfrac{2\pi}{3}$.

b. Arcsin 0.6 is the real number between $-\dfrac{\pi}{2}$ and $\dfrac{\pi}{2}$, inclusive, whose sine is 0.6. Using a calculator or tables, Arcsin $0.6 \approx 0.64$.

c. Using the results of part (b), we have:
$\mathrm{Sin}^{-1}(-0.6) = -\mathrm{Sin}^{-1}\,0.6 \approx -0.64$.

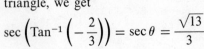

EXAMPLE 3. Evaluate: **a.** $\tan\left(\mathrm{Sin}^{-1}\dfrac{3}{5}\right)$ **b.** $\sec\left(\mathrm{Tan}^{-1}\left(-\dfrac{2}{3}\right)\right)$

SOLUTION: **a.** This is a familiar problem in disguise. If we let $\theta = \mathrm{Sin}^{-1}\dfrac{3}{5}$, then we are asked to find $\tan\theta$, given that $\sin\theta = \dfrac{3}{5}$ and $0 < \theta < \dfrac{\pi}{2}$. Drawing the familiar diagram, we get

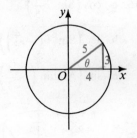

$$\tan\left(\mathrm{Sin}^{-1}\dfrac{3}{5}\right) = \tan\theta = \dfrac{3}{4}.$$

b. Let $\theta = \mathrm{Tan}^{-1}\left(-\dfrac{2}{3}\right)$, so that θ is a fourth-quadrant angle such that $\tan\theta = -\dfrac{2}{3}$. Drawing a reference triangle, we get

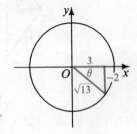

$$\sec\left(\mathrm{Tan}^{-1}\left(-\dfrac{2}{3}\right)\right) = \sec\theta = \dfrac{\sqrt{13}}{3}$$

Of course, if you use a calculator to solve Example 3, you find directly that $\tan(\mathrm{Sin}^{-1}\frac{3}{5}) = 0.75$ and that $\sec(\mathrm{Tan}^{-1}(-\frac{2}{3})) \approx 1.202$. Nonetheless, study the diagrams above so that you understand the reasoning behind these answers.

In our work with the inverse trigonometric functions we shall express angle measures in radians exclusively. For example, we consider $\mathrm{Sin}^{-1}1$ to be $\dfrac{\pi}{2}$ rather than $90°$. The use of radians is good preparation for the applications of trigonometry, many of which have nothing to do with angles. We shall explore some of these applications in the next chapter.

Evaluate each expression using neither tables nor a calculator.

1. $\text{Sin}^{-1} 0$

2. $\text{Cos}^{-1} \dfrac{1}{2}$

3. $\text{Arcsin} \dfrac{\sqrt{3}}{2}$

4. $\text{Arccos}\,(-1)$

5. $\text{Tan}^{-1} 0$

6. $\text{Arctan}\,1$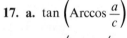

7. $\cos\left(\text{Sin}^{-1} 1\right)$

8. $\sin\left(\text{Cos}^{-1}\left(-\dfrac{\sqrt{3}}{2}\right)\right)$

9. $\tan\left(\text{Arccos}\dfrac{\sqrt{2}}{2}\right)$

10. $\text{Arcsin}\left(\sin\dfrac{\pi}{2}\right)$

11. $\text{Arctan}\left(\tan\dfrac{5\pi}{6}\right)$

12. $\cos\left(\text{Tan}^{-1}\left(-\dfrac{\sqrt{3}}{3}\right)\right)$

Use Table 5 to evaluate each expression to the nearest hundredth of a radian.

13. $\text{Arccos}\,0.7$

14. $\text{Tan}^{-1}\,(-1.3)$

15. $\text{Sin}^{-1}\,(-0.25)$

Use the diagrams to evaluate each expression.

16. **a.** $\cos\left(\text{Sin}^{-1}\dfrac{4}{5}\right)$

    b. $\sec\left(\text{Sin}^{-1}\left(-\dfrac{4}{5}\right)\right)$

    c. $\sin\left(\text{Arctan}\left(-\dfrac{4}{3}\right)\right)$

17. **a.** $\tan\left(\text{Arccos}\dfrac{a}{c}\right)$

    b. $\tan\left(\text{Cos}^{-1}\left(-\dfrac{a}{c}\right)\right)$

    c. $\cos\left(\text{Arctan}\left(-\dfrac{b}{a}\right)\right)$

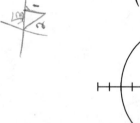

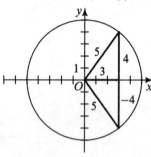

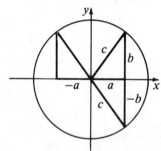

Find the exact value of each expression in terms of π.

A

1. $\text{Arcsin}\,(-1)$

2. $\text{Cos}^{-1} 0$

3. $\text{Sin}^{-1}\dfrac{1}{2}$

4. $\text{Arcsin}\left(-\dfrac{1}{2}\right)$

5. $\text{Arccos}\left(-\dfrac{\sqrt{2}}{2}\right)$

6. $\text{Arccos}\dfrac{\sqrt{3}}{2}$

7. $\text{Tan}^{-1}\,(-1)$

8. $\text{Arctan}\dfrac{\sqrt{3}}{3}$

9. $\text{Arcsin}\left(-\dfrac{\sqrt{3}}{2}\right)$

10. $\text{Cos}^{-1} 1$

11. $\text{Tan}^{-1}\sqrt{3}$

12. $\text{Sin}^{-1} 0$

13. $\text{Arccos}\left(-\dfrac{1}{2}\right)$

14. $\text{Arctan}\,0$

15. $\text{Sin}^{-1}\left(-\dfrac{\sqrt{2}}{2}\right)$

Use tables or a calculator to evaluate each expression to the nearest hundredth of a radian.

16. a. $\text{Sin}^{-1} 0.4$ **b.** $\text{Sin}^{-1}(-0.4)$ **c.** $\text{Cos}^{-1}(-0.4)$

17. a. $\text{Arcsin}\left(-\frac{2}{3}\right)$ **b.** $\text{Arccos}\left(-\frac{3}{4}\right)$ **c.** $\text{Arctan } 6$

18. a. $\text{Sin}^{-1} 0.8$ **b.** $\text{Cos}^{-1} 0.8$ **c.** $\text{Sin}^{-1} 0.8 + \text{Cos}^{-1} 0.8$

19. a. $\text{Tan}^{-1}(-2)$ **b.** $\text{Tan}^{-1}\left(-\frac{1}{2}\right)$ **c.** $\text{Tan}^{-1}(-2) + \text{Tan}^{-1}\left(-\frac{1}{2}\right)$

Use the method of Example 3 to evaluate the expression.

20. a. $\tan\left(\text{Cos}^{-1}\frac{12}{13}\right)$ **b.** $\tan\left(\text{Cos}^{-1}\left(-\frac{12}{13}\right)\right)$

21. a. $\sin\left(\text{Arccos}\frac{1}{5}\right)$ **b.** $\sin\left(\text{Arccos}\left(-\frac{1}{5}\right)\right)$

22. a. $\csc\left(\text{Tan}^{-1} 1.05\right)$ **b.** $\sec\left(\text{Arcsin}(-0.5)\right)$

23. a. $\cos\left(\text{Arctan}(-0.2)\right)$ **b.** $\sin\left(\text{Tan}^{-1}(-0.2)\right)$

Evaluate without a calculator or tables.

24. a. $\text{Arcsin}\left(\sin\frac{\pi}{3}\right)$ **b.** $\text{Arcsin}\left(\sin\frac{2\pi}{3}\right)$

25. a. $\text{Cos}^{-1}\left(\cos\frac{\pi}{4}\right)$ **b.** $\text{Cos}^{-1}\left(\cos\frac{5\pi}{4}\right)$

26. a. $\text{Tan}^{-1}\left(\tan\frac{2\pi}{3}\right)$ **b.** $\text{Tan}^{-1}\left(\tan\left(-\frac{\pi}{4}\right)\right)$

27. a. $\text{Sin}^{-1}\left(\sin\left(-\frac{2\pi}{3}\right)\right)$ **b.** $\text{Cos}^{-1}\left(\cos\left(-\frac{2\pi}{3}\right)\right)$

28. What is wrong with the expression $\text{Cos}^{-1} 3$? What happens when you try to evaluate the expression on a calculator or a computer?

29. Repeat Exercise 28, using the expression $\text{Arcsin}\left(-\sqrt{3}\right)$.

B   **30.** Draw the graph of $y = -\text{Tan}^{-1} x$.

31. Draw the graphs of $y = \text{Sin}^{-1} x$ and $y = \text{Arccos } x$ from memory.

32. Express $\sin\left(\text{Cos}^{-1} x\right)$, $-1 \le x \le 1$, in terms of x.

33. Express $\tan\left(\text{Arcsin } y\right)$, $-1 < y < 1$, in terms of y.

34. a. Evaluate $\sin\left(\text{Sin}^{-1}\frac{1}{2}\right)$, $\sin\left(\text{Sin}^{-1} 1\right)$, and $\sin\left(\text{Sin}^{-1}(-2)\right)$.

     b. Do you think that $\sin\left(\text{Sin}^{-1} x\right) = x$ for *all* x?

     c. Evaluate $\text{Sin}^{-1}\left(\sin\frac{\pi}{4}\right)$ and $\text{Sin}^{-1}\left(\sin\left(-\frac{\pi}{3}\right)\right)$.

     d. Find a value of x for which $\text{Sin}^{-1}(\sin x) \ne x$.

     e. Does part (d) contradict the property of inverse functions which states that $f^{-1}(f(x)) = x$ for all x in the domain of $f(x)$? Explain.

True or false? If false, give a counterexample.

35. a. $\text{Tan}\left(\text{Tan}^{-1} x\right) = x$ **b.** $\text{Tan}^{-1}(\tan x) = x$

36. a. $\text{Sin}^{-1}(-x) = -\text{Sin}^{-1} x$, $-1 \le x \le 1$

     b. $\text{Cos}^{-1}(-x) = -\text{Cos}^{-1} x$, $-1 \le x \le 1$

37. $\text{Tan}^{-1} x = \dfrac{1}{\text{Tan } x}$

The expression always has the same value. Find this value and explain why the expression has a constant value.

38. $\text{Tan}^{-1} x + \text{Tan}^{-1}\left(\dfrac{1}{x}\right)$, $x > 0$

39. $\text{Sin}^{-1} x + \text{Cos}^{-1} x$, $-1 \le x \le 1$

40. $\text{Cos}^{-1} x + \text{Cos}^{-1}(-x)$, $-1 \le x \le 1$

41. Sketch the graph of $y = \text{Arcsin}(\sin x)$.

42. Sketch the graph of $y = \text{Arccos}(\cos x)$.

43. Let $y = \text{Cot } x$ be a function identical to the cotangent function except that its domain is $0 < x < \pi$. State the domain and range of $y = \text{Cot}^{-1} x$ and draw its graph.

44. Let $y = \text{Sec } x$ be a function identical to the secant function except that its domain is $0 \le x \le \pi$, $x \ne \dfrac{\pi}{2}$. State the domain and range of $y = \text{Sec}^{-1} x$ and draw its graph.

45. Let $y = \text{Csc } x$ be a function identical to the cosecant function except that its domain is $-\dfrac{\pi}{2} \le x \le \dfrac{\pi}{2}$, $x \ne 0$. State the domain and range of $y = \text{Arccsc } x$ and draw its graph.

46. a. Evaluate $\text{Arcsec } 2$ and $\text{Arccsc } \sqrt{2}$.
   b. Define $\text{Arcsec } x$ in terms of $\text{Arccos } x$ and $\text{Arccsc } x$ in terms of $\text{Arcsin } x$.

47. Using calculus one can prove that

$$\text{Tan}^{-1} x = x - \frac{x^3}{3} + \frac{x^5}{5} - \frac{x^7}{7} + \cdots, \text{ for } |x| \le 1.$$

Use this relationship to show that

$$\pi = 4\left(1 - \frac{1}{3} + \frac{1}{5} - \frac{1}{7} + \cdots\right).$$

C 48. If $\text{Sin}^{-1} x + \text{Sin}^{-1} y = \dfrac{\pi}{2}$, show that $x^2 + y^2 = 1$.

CALCULATOR EXERCISES

1. Try to evaluate $\text{Sin}^{-1} \pi$. Explain the result.

Solve each equation to four significant digits.

2. $\text{Arctan } x = \dfrac{\pi}{8}$ (*Hint:* Take the tangent of both sides.)

3. $2 \text{Sin}^{-1} x = \dfrac{5\pi}{36}$　　　　　　**4.** $5 \text{Arccos}(1.2x) = 7$

5. $3 \text{Arcsin}\left(\dfrac{2x - 7}{13}\right) = \pi$　　　**6.** $\text{Tan}^{-1} \sqrt{x} = 0.4$

7-6/USING COMPUTERS TO SOLVE TRIGONOMETRIC EQUATIONS

Some trigonometry problems lead to equations which can be solved most easily by a computer or calculator search. The computer is generally more efficient.

EXAMPLE 1. The diagram shows an end view of a cylindrical oil drum with radius 1. Through a hole in the top, a vertical rod is lowered to touch the bottom of the tank. When the rod is removed, the oil level in the tank can be read from the oil mark on the rod. Where on the rod should markings be put to show the tank is $\frac{3}{4}$ full, $\frac{1}{2}$ full, and $\frac{1}{4}$ full?

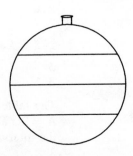

SOLUTION: Consider the circle $x^2 + y^2 = 1$, shown in the diagram below. To find where to put the "$\frac{3}{4}$ full" mark, we need to find the horizontal line $y = k$ which cuts off a segment equal to $\frac{1}{4}$ of the circle's area, or $\frac{\pi}{4}$.

$$\frac{\pi}{4} = \frac{1}{2} \cdot 1^2 \cdot \alpha - \frac{1}{2} \cdot 1^2 \cdot \sin \alpha$$

$$\frac{\pi}{4} = \frac{\alpha - \sin \alpha}{2}$$

$$\frac{\pi}{2} = \alpha - \sin \alpha$$

$$0 = \alpha - \sin \alpha - \frac{\pi}{2}$$

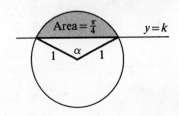

It is obvious that the solution for α is between 0 and π. In fact, using trial and error, you can determine that α is between 2 and π. Accordingly, line 30 of the program below considers values of α between 2 and π. The letter A denotes angle alpha. Also, Arctan $1 = \frac{\pi}{4}$, so 4 ATN (1) equals π. (The BASIC symbol for Arctan x is ATN (x).)

```
10  LET P=4*ATN(1)
20  PRINT "A", "A-SIN(A)-P/2"
30  FOR A=2 TO P STEP .1
40  PRINT A, A-SIN(A)-P/2
50  NEXT A
60  END
```

When this program is run we get the table of values shown at the top of the following page.

```
A                    A-SIN(A)-P/2
2                    -.480094
2.1                  -.334006
2.2                  -.179293
2.3                  -.165021E-1
2.4                   .15374
2.5                   .33073
2.6                   .513701
2.7                   .701822
2.8                   .894214
2.9                  1.08995
3                    1.28808
3.1                  1.48762
```

The table shows that the sign of A − sin(A) − P/2 changes between A = 2.3 and A = 2.4. Thus we modify line 30 to take a closer look at the interval 2.3 ≤ A ≤ 2.4.

```
30  FOR A=2.30 TO 2.40 STEP .01
```

When the program is run again we get the following table.

```
A                    A-SIN(A)-P/2
2.3                  -.165017E-1
2.31                  .198126E-3
2.32                  .169719E-1
2.33                  .033819
2.34                  .507386E-1
2.35                  .06773
2.36                  .847926E-1
2.37                  .101925
2.38                  .119128
2.39                  .1364
2.4                   .15374
```

This table shows there is a root in the interval 2.30 < A < 2.31. If we required even more precision, we could further modify line 30 as follows:

```
30  FOR A=2.300 TO 2.310 STEP .001
```

To the nearest hundredth, the solution is A = 2.31. Consequently, the required value of k is

$$k = \cos \frac{A}{2}$$

$$= \cos \frac{2.31}{2} \approx .4039$$

Thus the desired lines which give the four equal areas are $y \approx 0.4039$, $y = 0$, and $y \approx -0.4039$.

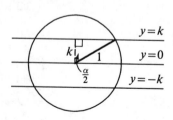

Solve the given equation using a computer. Either use a program like that given in the text or write one of your own.

A   **1.** $\cos x = x$

3. $\sin x = 2x - 3$

5. $\cos x = 2x$

2. $\sin x = x^2$

4. $\cos x = 3x - 2$

6. $\tan x = 2x$
   (Change P to P/2 in line 30)

B   **7.** $\text{Tan}^{-1} x = x - 1$

8. $\text{Tan}^{-1} x = \dfrac{1}{x}$

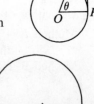

9. $\overline{RP}$ is a tangent to a circle with radius 1. If the area of sector OQP equals half the area of $\triangle ORP$, find θ in radians $(-2\pi \leq \theta \leq 2\pi)$.

10. Use the diagram for Exercise 9 to find the value of θ for which segment QR and arc QP have the same length.

11. The circle $x^2 + y^2 = 1$ is cut into 3 equal areas by two horizontal lines. What are the equations of these lines?

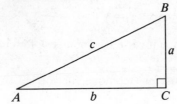

C   **12.** A goat is tethered to a stake at the edge of a circular field with radius 1 unit. How long should the rope be so that the goat can graze over half of the field?

Chapter summary

1. If you know the lengths of two sides of a right triangle, or the length of one side and the measure of one acute angle, then you can find the measures of the remaining sides and angles using the trigonometric functions. In the diagram below:

$$\sin A = \frac{\text{opposite}}{\text{hypotenuse}} = \frac{a}{c} \qquad \csc A = \frac{1}{\sin A} = \frac{c}{a}$$

$$\cos A = \frac{\text{adjacent}}{\text{hypotenuse}} = \frac{b}{c} \qquad \sec A = \frac{1}{\cos A} = \frac{c}{b}$$

$$\tan A = \frac{\text{opposite}}{\text{adjacent}} = \frac{a}{b} \qquad \cot A = \frac{1}{\tan A} = \frac{b}{a}$$

2. In any $\triangle ABC$ the following relationships hold:

Law of Sines: $\dfrac{\sin A}{a} = \dfrac{\sin B}{b} = \dfrac{\sin C}{c}$

Law of Cosines: $c^2 = a^2 + b^2 - 2ab \cos C$.

Another form of this law is $\cos C = \dfrac{a^2 + b^2 - c^2}{2ab}$.

To use these laws you need to know three of the six measures of a triangle, and at least one must be the length of a side. The summary on page 254 shows the conditions for which each law is used. Remember that the SSA (side-side-nonincluded angle) case may result in 0, 1, or 2 triangles.

Area Formulas. Use the formula below to find the area given SAS:

$$K = \tfrac{1}{2}ab \sin C$$

Use either of the following methods to find the area given SSS:

Method A: Use the Law of Cosines to find $\cos C$. Then find $\sin C$. Then use the formula above.

Method B: (Optional) Use Hero's Formula (derived in Exercise 29, page 257):

$$K = \sqrt{s(s-a)(s-b)(s-c)} \quad \text{where } s = \frac{a+b+c}{2}$$

3. a. None of the trigonometric functions has an inverse because none is a one-to-one function. However, by restricting the domains of the trigonometric functions, we can define new functions which do have inverses.

b. The domains and ranges of three of these *inverse trigonometric functions* are as follows:

Equation	*Domain*	*Range*
$y = \mathrm{Sin}^{-1} x = \mathrm{Arcsin}\, x$	$-1 \le x \le 1$	$-\dfrac{\pi}{2} \le y \le \dfrac{\pi}{2}$
$y = \mathrm{Cos}^{-1} x = \mathrm{Arccos}\, x$	$-1 \le x \le 1$	$0 \le y \le \pi$
$y = \mathrm{Tan}^{-1} x = \mathrm{Arctan}\, x$	all real x	$-\dfrac{\pi}{2} < y < \dfrac{\pi}{2}$

c. The equation $y = \mathrm{Sin}^{-1} x$ is read

"y equals the inverse sine of x"

or "y is the number between $-\dfrac{\pi}{2}$ and $\dfrac{\pi}{2}$ whose sine is x."

Hence $y = \mathrm{Sin}^{-1} x$ means $\sin y = x$ and $-\dfrac{\pi}{2} \le y \le \dfrac{\pi}{2}$.

d. The graphs of these three inverse trigonometric functions are shown on pages 261 and 262.

4. A computer search is a useful technique for solving some trigonometric equations.

Chapter test

1. A rhombus has diagonals of lengths 8 and 12. Find the measure to the nearest tenth of a degree of an acute angle of the rhombus. (*Hint:* The diagonals are perpendicular.)   7-1

2. The angle formed by the legs of an isosceles triangle has measure 70°. If the base of the triangle is 26 cm long, find the length of each leg to the nearest tenth.

3. The area of $\triangle XYZ$ is 42 cm². If $x = 15$ and $y = 10$, find the possible measures of $\angle Z$.   7-2

4. A regular polygon with 20 sides is inscribed in a circle with radius 7 cm. Find the area of the polygon.

5. How many different triangles can be constructed using the given information?   7-3
   a. $p = 5$, $q = 4$, $\angle Q = 74°$ **b.** $p = 9$, $q = 8$, $\angle P = 23°$

6. Observers at J and K, 30 km apart, sight an airplane at angles of elevation of 40° and 75°, respectively, as shown in the diagram. How far is the plane from each observer?

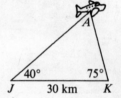

7. A triangle has sides of lengths 5, 6, and 7. Find the measure of the smallest angle to the nearest tenth of a degree.   7-4

8. In $\triangle ABC$, $a = 3$, $b = 6$, and $\angle C = 120°$. Find the exact value of c and the exact area of the triangle.

9. State the domain and the range of the inverse cosine function.   7-5

10. Evaluate without tables and without a calculator:
    a. $\text{Arccos}(-1)$ **b.** $\text{Sin}^{-1} 0$
    c. $\cos\left(\text{Tan}^{-1}\sqrt{3}\right)$ **d.** $\sin\left(\text{Arctan}\left(-\frac{15}{8}\right)\right)$

11. (Optional) Briefly describe how you could use a computer to find a positive real solution to the equation $3\cos x = x$.   7-6

CHAPTER EIGHT

Advanced graphing

OBJECTIVES

1. To determine whether a function is periodic and to find the period and amplitude of a periodic function.
2. To sketch graphs of equations of the form $y = cf(x), y = f(cx), y = |f(x)|,$ $y = f(-x), y = f^{-1}(x),$ and $y - k = f(x - h)$ given the graph of $y = f(x).$
3. To determine whether a graph is symmetric with respect to the coordinate axes, the origin, or the line $y = x.$
4. To sketch the graph of an equation of the form $y = \dfrac{1}{f(x)}$ and find its asymptotes.
5. To study various applications of the sine and cosine functions.
6. To approximate the area under a curve.

8-1/STRETCHING GRAPHS; PERIODIC FUNCTIONS

Period and amplitude

In this section we shall see that simple changes in a function's equation will stretch or shrink its graph. If the function is *periodic*, these stretches and shrinks will affect its *period* and *amplitude*. These terms are defined on the following page.

Tidal waters, such as those pictured at the left, rise and fall in a pattern that repeats itself every 12.4 hours. See pages 296 and 297 for a mathematical description of the tides.

If $f(x)$ has domain D and there is a number $p > 0$ such that $f(x + p) = f(x)$ for all x in D, then $f(x)$ is a *periodic function*. If p is the smallest such positive number, then p is called the **period** of the function. If a periodic function has a maximum value M and a minimum value m, the **amplitude** of the function is $\dfrac{M - m}{2}$.

EXAMPLE 1. The graph of a periodic function $f(x)$ is shown below. Find **(a)** its period; **(b)** its amplitude; **(c)** $f(1000)$.

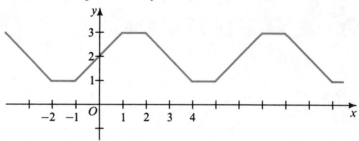

SOLUTION: **a.** $f(x + 6) = f(x)$ for all x. The period is 6.

b. Maximum $M = 3$; minimum $m = 1$

$$\text{Amplitude} = \frac{M - m}{2} = \frac{3 - 1}{2} = 1$$

c. First we observe that $1000 \div 6$ is 166 with remainder 4. Thus, we can write 1000 as $4 + 6 \cdot 166$ and proceed as follows:

$$f(1000) = f(4 + 6 \cdot 166) = f(4) = 1$$

The graph of $y = cf(x)$

If c is positive (and unequal to 1), then the graph of $y = cf(x)$ looks like the graph of $y = f(x)$ except that it has been *vertically* stretched or shrunk. Some examples will illustrate the possibilities.

EXAMPLE 2. If $f(x) = \sin x$, sketch the graphs of $y = 2f(x)$ and $y = \frac{1}{2}f(x)$.

SOLUTION: First we sketch the graph of $y = \sin x$. Then we can obtain the graph of

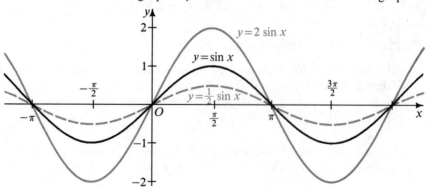

$y = 2 \sin x$ from this graph by noting that the y-coordinate of each point on the graph of $y = 2 \sin x$ is twice the y-coordinate of the corresponding point on the graph of $y = \sin x$. Similarly, the y-coordinate of each point on the graph of $y = \frac{1}{2} \sin x$ is one half the y-coordinate of the corresponding point on the graph of $y = \sin x$.

In general, if $f(x)$ is a periodic function with period p and amplitude A, then $cf(x)$ is a periodic function with period p and amplitude cA.

The graph of $y = f(cx)$

Students sometimes think the functions $f(cx)$ and $cf(x)$ are the same, but they are usually different. For example, $\sin 2x \neq 2 \sin x$, $\log 3x \neq 3 \log x$, and $\sqrt{4x} \neq 4\sqrt{x}$ in general. In Example 2 you saw that the constant c in $y = cf(x)$ has the effect of stretching or shrinking the graph of $y = f(x)$ in the vertical direction, while leaving it unchanged in the horizontal direction. In the following example you will see that the constant c in $y = f(cx)$ has the effect of stretching or shrinking the graph in the horizontal direction, while leaving it unchanged in the vertical direction.

EXAMPLE 3. If $f(x) = \sin x$, sketch the graphs of $y = f(2x)$ and $y = f(\frac{1}{2}x)$.

SOLUTION:

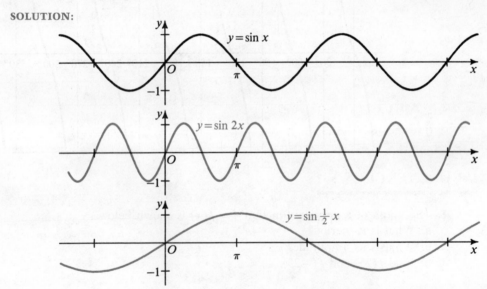

If you compare the first two graphs just shown, you can see that the function $y = \sin 2x$ completes its cycle of sine values twice as fast as $y = \sin x$. Thus its period is only half as long, or $\frac{1}{2}(2\pi) = \pi$. On the other hand, by comparing the first and third graphs just shown, you can see that $y = \sin \frac{1}{2}x$ has period 4π.

In general, if a function $f(x)$ is periodic with period p, then $f(cx)$ is periodic with period $\frac{p}{c}$ $(c \neq 0)$.

A comparison of the graphs of $y = \sin 2x$ and $y = \sin x$ also shows that the equation $\sin 2x = 1$ has twice as many solutions between 0 and 2π as the equation $\sin x = 1$. Algebraic solutions to these equations are given below. Notice that solving for $0 \le x < 2\pi$ is equivalent to solving for $0 \le 2x < 4\pi$.

$$\sin x = 1 \quad (0 \le x < 2\pi) \qquad\qquad \sin 2x = 1 \quad (0 \le 2x < 4\pi)$$

$$x = \frac{\pi}{2} \qquad\qquad\qquad 2x = \frac{\pi}{2}, \frac{5\pi}{2}$$

$$x = \frac{\pi}{4}, \frac{5\pi}{4}$$

EXAMPLE 4. Solve $\tan 3\theta = 1$ for $0° \le \theta < 360°$.

SOLUTION: Solving for $0° \le \theta < 360°$ is equivalent to solving for $0° \le 3\theta < 1080°$. Our strategy is to solve for 3θ and then for θ.

$$\tan 3\theta = 1 \quad (0 \le 3\theta < 1080°)$$
$$3\theta = 45°, 225°, 405°, 585°, 765°, 945°$$
$$\theta = 15°, 75°, 135°, 195°, 255°, 315°$$

These six solutions are shown on the graph of $y = \tan 3x$ below.

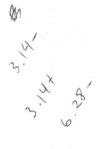

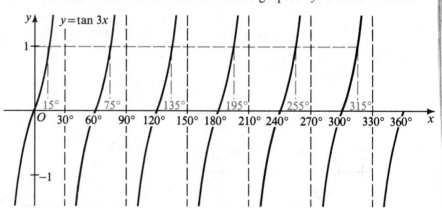

ORAL EXERCISES 8-1

1. The graph of a periodic function $y = f(x)$ is shown below.
   a. What is its period?
   b. What is its amplitude?
   c. Find $f(25)$ and $f(-25)$.

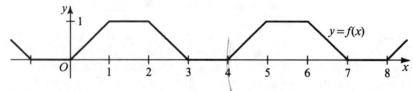

$y = f(x)$

2. Given the graph of $y = f(x)$ shown above, sketch the graphs of $y = 2f(x)$ and $y = \frac{1}{2}f(x)$ on the chalkboard. Give the period and amplitude of each.

3. Given the graph of $y = f(x)$ shown on page 276, sketch the graphs of $y = f(2x)$ and $y = f(\frac{1}{2}x)$. Give the period and amplitude of each.

4. Give the period and amplitude of each function.
   a. $y = 4 \cos 2x$ **b.** $y = 3 \sin \frac{1}{2}x$

Tell how many solutions each equation has between 0 and 2π.

5. **a.** $\sin x = \frac{1}{2}$ **b.** $\sin 2x = \frac{1}{2}$ **c.** $\sin 3x = \frac{1}{2}$

6. **a.** $\cos x = -1$ **b.** $\cos 2x = -1$ **c.** $\cos 3x = -1$

7. **a.** $\tan x = 1$ **b.** $\tan 2x = 1$ **c.** $\tan \dfrac{x}{2} = 1$

WRITTEN EXERCISES 8-1

The graph of a function $f(x)$ is given. Tell whether $f(x)$ is periodic. If so, give its period and amplitude, then find $f(1000)$ and $f(-1000)$.

A   **1.**

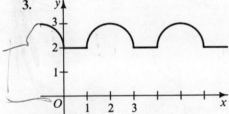

2.

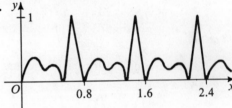

3.

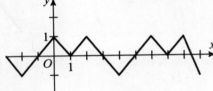

4.

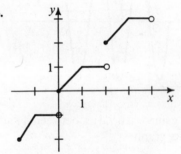

In Exercises 5 and 6 you are given the graph of $y = f(x)$. Sketch the graphs of:
a. $y = 3f(x)$ **b.** $y = \frac{1}{2}f(x)$ **c.** $y = f(2x)$ **d.** $y = f(\frac{1}{2}x)$

5.

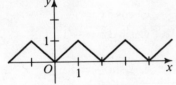

6.

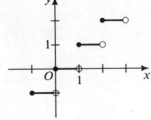

7. For each function, sketch the graph and give the period and amplitude.
   a. $y = 2 \cos x$ **b.** $y = \frac{1}{2} \cos x$ **c.** $y = \cos 2x$ **d.** $y = \cos \frac{1}{2} x$

8. Give the period of each function.

   a. $y = \tan x$ **b.** $y = \tan 3x$ **c.** $y = \tan \frac{x}{3}$

   d. Is the amplitude of $f(x) = \tan x$ defined? Explain your answer.

Sketch the graph of each function. Give its period and amplitude if any.

9. **a.** $y = \sin 3x$ **b.** $y = 2 \sin 3x$ **c.** $y = \sin \frac{1}{3} x$

10. **a.** $y = 2 \sec x$ **b.** $y = 2 \sec 2x$
    (Refer to the graph of the secant function on page 215.)

11. **a.** $y = \tan 2x$ **b.** $y = \tan \frac{1}{2} x$
    (Refer to the graph of the tangent function on page 214.)

12. **a.** $y = \frac{1}{2} \csc x$ **b.** $y = \frac{1}{2} \csc \frac{1}{2} x$

13. **a.** $y = 3 \cos 2\pi t$ **b.** $y = 5 \sin \frac{2\pi}{365} t$

14. **a.** $y = 4 \sin \pi t$ **b.** $y = \tan \frac{\pi}{2} t$

Give an equation for each sine and cosine *wave* (curve).

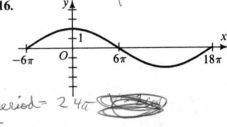

15.

16.

17. A sine wave varies between 4 and -4 with period 12. Find an equation for its graph.

18. A cosine wave varies between -9 and 9 with period 5. Find an equation for its graph.

Solve for $0 \le x < 2\pi$.

19. **a.** $\cos x = 1$ **b.** $\cos 2x = 1$ **c.** $\cos 3x = 1$

20. **a.** $\sin x = -1$ **b.** $\sin 2x = -1$ **c.** $\sin 3x = -1$

B 21. **a.** $2 \sin x = 1$ **b.** $2 \sin 2x = 1$ **c.** $2 \sin \frac{x}{2} = 1$

22. **a.** $\tan x = -1$ **b.** $\tan 2x = -1$ **c.** $\tan \frac{x}{2} = -1$

Solve for $0° \leq \theta < 360°$.

23. $\cos 3\theta = \dfrac{\sqrt{3}}{2}$

24. $\sin \dfrac{\theta}{4} = \dfrac{\sqrt{2}}{2}$

25. $\cot 2\theta = 1$

26. $\sec^2 \dfrac{\theta}{2} = 4$

27. Use the graph of $y = 2^x$ shown on page 160 to sketch the graphs of the following functions.

   a. $y = \dfrac{1}{2} \cdot 2^x$

   b. $y = 2^{\left(\frac{x}{2}\right)}$

28. Use the graph of $y = \log_2 x$ shown on page 160 to sketch the graphs of the following functions.

   a. $y = 2 \log_2 x$

   b. $y = \log_2 2x$

29. a. Sketch the graph of (i) $y = x^2(6 - x)$ and (ii) $y = (2x)^2(6 - 2x)$.

   b. If graph (i) has a high point at $(4, 32)$, what is the high point on graph (ii)?

   c. What would be the high point on the graph of $y = 3x^2(6 - x)$?

In Exercises 30–32, sketch the graph of each equation.

30. a. $y = x^3 - 3x^2 + 2x$

   b. $y = (\tfrac{1}{2}x)^3 - 3(\tfrac{1}{2}x)^2 + 2(\tfrac{1}{2}x)$

31. a. $x^2 + y^2 = 4$

   b. $(2x)^2 + y^2 = 4$

32. a. $x^2 + y^2 = 36$

   b. $(2x)^2 + (3y)^2 = 36$

33. Sketch the graphs of $y = \dfrac{x}{4}$ and $y = \cos 2x$ on the same set of axes (x in radians). State how many solutions there are for the equation $\dfrac{x}{4} = \cos 2x$.

34. Sketch the graphs of $y = 4 - x^2$ and $y = 4 \sin x$ on the same set of axes (x in radians). State how many solutions there are for the equation $4 \sin x = 4 - x^2$.

35. If $f(x)$ and $g(x)$ both have period p and $h(x) = f(x) + g(x)$, show that
$$h(x + p) = h(x)$$
is an identity.

36. What is the period of the function $f(x) = \sin \dfrac{x}{2} + \cos \dfrac{x}{3}$?

C 37. A function $f(x)$ is defined for all real numbers as follows:
$$f(x) = 1 \text{ if } x \text{ is rational.}$$
$$f(x) = 0 \text{ if } x \text{ is irrational.}$$

   a. Show that if p is a rational number, $f(x + p) = f(x)$ for all x.

   b. Does this function have a period p?

   c. If p is an irrational number, can you find a value for x such that $f(x + p) \neq f(x)$?

   d. If p is an irrational number, can you find a value for x such that $f(x + p) = f(x)$?

8-2/REFLECTING GRAPHS; SYMMETRY

Reflection in the x-axis

The graph of $y = -f(x)$. The graph of $y = -f(x)$ is obtained by reflecting the graph of $y = f(x)$ in the x-axis.

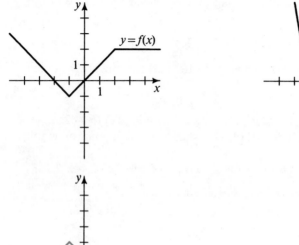

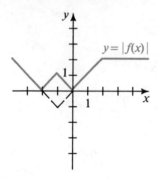

The graph of $y = |f(x)|$. The graph of $y = |f(x)|$ is the same as the graph of $y = f(x)$ when $f(x) \geq 0$ and is the same as the graph of $y = -f(x)$ when $f(x) < 0$. This principle is applied to the graphs just shown to produce the following graphs.

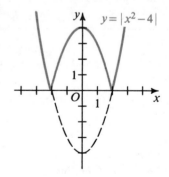

Reflection in the y-axis

The graph of $y = f(-x)$. The graph of $y = f(-x)$ is obtained by reflecting the graph of $y = f(x)$ in the y-axis, as shown on the next page.

280   Chapter 8

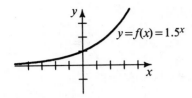

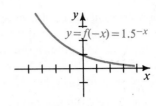

Reflection in the line y = x

Reflecting the graph of an equation in the line $y = x$ is equivalent to interchanging the x's and the y's in the equation.

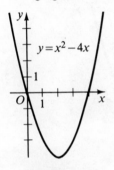

original graph
and equation

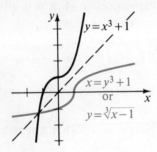

reflection in $y=x$

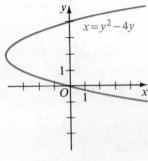

reflected graph
and altered equation

Reflection in the line $y = x$ was used in Section 4–5 to obtain the graph of $y = f^{-1}(x)$, the inverse of the one-to-one function $y = f(x)$.

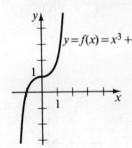

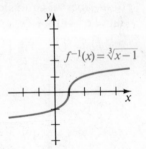

Symmetry

A line l is called an **axis of symmetry** of a graph if it is possible to pair the points of the graph in such a way that l is the perpendicular bisector of the segment joining each pair. (See the figure at the right.)

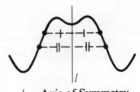

l = Axis of Symmetry

A point O is called a **point of symmetry** of a graph if it is possible to pair the points of the graph in such a way that O is the midpoint of the segment joining each pair. (See the figure at the right.)

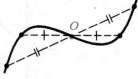

You have seen that the graph of $y = ax^2 + bx + c$ has an axis of symmetry with equation $x = \dfrac{-b}{2a}$. In Exercise 30 of

O = Point of Symmetry

Section 8–3 we shall show that the graph of $y = ax^3 + bx^2 + cx + d$ has a point of symmetry at the point where $x = \dfrac{-b}{3a}$.

Every quadratic graph has a line of symmetry.

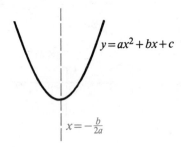

$y = ax^2 + bx + c$

$x = -\dfrac{b}{2a}$

Every cubic graph has a point of symmetry.

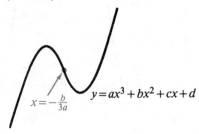

$x = -\dfrac{b}{3a}$

$y = ax^3 + bx^2 + cx + d$

SPECIAL TESTS FOR SYMMETRY

1. Symmetry in the x-axis: $(x, -y)$ is on the graph whenever (x, y) is.

   Test: Does an equivalent equation result when you substitute $(x, -y)$ for (x, y)?

   Example: $(x - 1)^2 + y^2 = 1$ ⟵ These equations have
   $(x - 1)^2 + (-y)^2 = 1$ ⟵ same solutions.

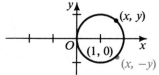

2. Symmetry in the y-axis: $(-x, y)$ is on the graph whenever (x, y) is.

   Test: Does an equivalent equation result when you substitute $(-x, y)$ for (x, y)?

   Example: $y = x^2$ ⟵ These equations have
   $y = (-x)^2$ ⟵ same solutions.

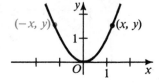

3. Symmetry in the line $y = x$: (y, x) is on the graph whenever (x, y) is.

   Test: Does an equivalent equation result when you substitute (y, x) for (x, y)?

   Example: $xy = \frac{1}{6}$ ⟵ These equations have
   $yx = \frac{1}{6}$ ⟵ same solutions.

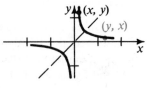

4. Symmetry in the origin: $(-x, -y)$ is on the graph whenever (x, y) is.

   Test: Does an equivalent equation result when you substitute $(-x, -y)$ for (x, y)?

   Example: $y = x^3$ ⟵ These equations have
   $(-y) = (-x)^3$ ⟵ same solutions.

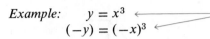

ORAL EXERCISES 8-2

1. The graph of $y = f(x)$ is shown at the left below. On the chalkboard sketch the following graphs:

   a. $y = -f(x)$ **b.** $y = |f(x)|$ **c.** $y = f(-x)$

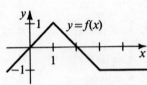

Ex. 1

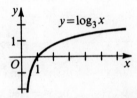

Ex. 2

2. The graph of $y = \log_3 x$ is shown at the right above. On the chalkboard sketch the graph of:

   a. $y = -\log_3 x$ **b.** $y = \log_3 (-x)$
   c. $y = |\log_3 x|$ **d.** $x = \log_3 y$

3. For each of the following graphs, give the equation(s) of its line(s) of symmetry (if any) and give the coordinates of its point of symmetry (if any).

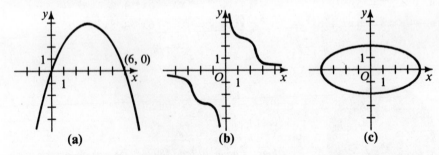

(a)   (b)   (c)

4. Test each of the following equations to see whether its graph has symmetry with respect to: (i) the x-axis; (ii) the y-axis; (iii) the line $y = x$; and (iv) the origin.

   a. $x^4 + y^4 = 1$ **b.** $xy^2 = 1$ **c.** $y = \cos x$

5. Give the equation of the axis of symmetry of each parabola.
   a. $y = x^2 - 8x - 7$ **b.** $y = 8x - 4x^2$ **c.** $y = x^2 + 3$

6. Give the coordinates of the point of symmetry of each cubic graph.
   a. $y = x^3 - 6x^2 + 5x + 7$ **b.** $y = 9x + 6x^2 + 2x^3$ **c.** $y = 3x^3 - 3x + 7$

Given the graph of $f(x)$, sketch the graphs of:

a. $y = -f(x)$ **b.** $y = |-f(x)|$ **c.** $y = f(-x)$

A   **1.** 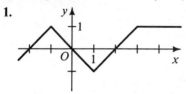   **2.**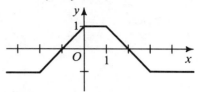

3. Use the graph of Exercise 1 to sketch the graph of $y = -2f(x)$.

4. Use the graph of Exercise 2 to sketch the graph of $y = -3f(x)$.

Sketch the graphs of the following:

5. a. $y = -\sin x$ **b.** $y = |\sin x|$ **c.** $y = |2 \sin x|$ **d.** $x = \sin y$

6. a. $y = -\cos x$ **b.** $y = |\cos x|$ **c.** $y = \left|\cos \dfrac{x}{2}\right|$ **d.** $x = \cos y$

7. a. $y = -\tan x$ **b.** $y = \tan(-x)$ **c.** $y = |\tan x|$ **d.** $x = \tan y$

8. a. $y = -\sec x$ **b.** $y = |\sec x|$ **c.** $y = \sec(-x)$ **d.** $y = \sec(-2x)$

9. a. $y = \log_2 x$ **b.** $y = \log_2(-x)$ **c.** $y = |\log_2 x|$ **d.** $x = \log_2 y$

10. a. $y = 2^x$ **b.** $y = 2^{-x}$ **c.** $y = -2^x$ **d.** $x = 2^y$

11. a. $y = |2 - 2x|$ **b.** $y = -|2 - 2x|$

12. a. $y = |3x - 6|$ **b.** $y = -|3x - 6|$

13. a. $y = x^3 - 6x^2 + 9x$ **b.** $y = |x^3 - 6x^2 + 9x|$

14. a. $y = 4x + x^2$ **b.** $y = |4x + x^2|$

15. a. $y = x^4 - 5x^2 + 4$ **b.** $y = |x^4 - 5x^2 + 4|$

16. a. $y = x^4 - 6x^3 + 8x^2$ **b.** $y = |x^4 - 6x^3 + 8x^2|$

Percy Lavon Julian (1899–1975) was an American chemist who received international recognition. He synthesized several important drugs, including cortisone, widely used in the treatment of arthritis, and physostingine, used in the treatment of glaucoma.

17. Test each equation to see if its graph has symmetry with respect to: (i) the x-axis; (ii) the y-axis; (iii) the line $y = x$, and (iv) the origin.

   a. $y^2x = 2$ **b.** $x^3 + y^3 = 1$ **c.** $y = \tan x$

18. Repeat Exercise 17 for each of the following equations:

   a. $x^2 + xy = 4$ **b.** $|x| + |y| = 1$ **c.** $y = \sec x$

19. The graph of a cubic function has a low point at $(5, -3)$ and a symmetry point at $(0, 4)$. What is a high point of the graph?

20. **a.** Find the symmetry point of the graph of $y = -x^3 + 15x^2 - 48x + 45$.
   b. $(2, 1)$ is a low point of the graph. What is a high point?

21. **a.** Sketch the graph of $y = 2x^2 - x^3$. What is a low point of the graph?
   b. Find the symmetry point of the graph and thus deduce the coordinates of a high point.

22. **a.** Sketch the graph of $y = -x^3 - 6x^2 - 9x$. What is a low point?
   b. Find the symmetry point and thus deduce the coordinates of a high point.

Sketch the graph of $y = f(x)$. Then give the equation and graph of $y = f^{-1}(x)$.

23. $f(x) = 3x - 3$ 24. $f(x) = x^2$, $x \geq 0$ 25. $f(x) = x^3$ 26. $f(x) = x^3 - 1$

B   27. **a.** How is $\log \dfrac{1}{x}$ related to $\log x$?

    b. On the same set of axes, sketch the graphs of $y = \log_2 x$ and $y = \log_2 \dfrac{1}{x}$.

28. **a.** How is $\log x^2$ related to $\log x$?
    b. On the same set of axes, sketch the graphs of $y = \log_2 x$ and $y = \log_2 x^2$, $x > 0$.

Sketch the graph of each equation. (*Hint:* Sketch the part of the graph in the first quadrant; then use symmetry.)

29. $|x| + |y| = 2$ 30. $|x|^{\frac{1}{2}} + |y|^{\frac{1}{2}} = 2$

By sketching graphs, determine the number of solutions of the given equation.

31. $|\sin x| = x^2$ 32. $\log_{10} x = \sin x$ 33. $2^x = \cos x$ 34. $-\cos x = x^2$

Use the following definitions to do Exercises 35–40:

$$f(x) \text{ is an } even \text{ } function \text{ if } f(-x) = f(x)$$
$$f(x) \text{ is an } odd \text{ } function \text{ if } f(-x) = -f(x)$$

35. Classify each of the six trigonometric functions as odd or even.

36. Classify each function as odd or even:
   a. $f(x) = x^2$ **b.** $f(x) = x^3$ **c.** $f(x) = x^4 - 2x^2$ **d.** $f(x) = x^5 - 4x^3$
   e. Use your answers to parts (a) through (d) to guess the reasons for using the terms "odd" and "even" as they are applied to functions.

37. What kind of symmetry does the graph of an even function have?

38. What kind of symmetry does the graph of an odd function have?

39. Suppose $f(x)$ and $g(x)$ are both odd functions and $h(x) = f(x) \cdot g(x)$. Prove that $h(x)$ is even.

40. Suppose $f(x)$ is odd and $g(x)$ is even. Prove that $h(x) = f(x) \cdot g(x)$ is odd.

41. Sketch the graphs of $y = \text{Sin}^{-1}(\sin x)$ and $y = \text{Cos}^{-1}(\cos x)$ on the same axes. Find the area of the region enclosed between the graphs for $\frac{\pi}{2} \leq x \leq 2\pi$.

CALCULATOR EXERCISE

Sketch the graph of $x^{\frac{2}{3}} + y^{\frac{2}{3}} = 1$. Symmetry will be helpful.

8-3/TRANSLATING GRAPHS

The graph of $y - k = f(x - h)$

The diagrams below show the circle $x^2 + y^2 = r^2$ and the parabola $y = x^2$. Notice how the equations of the circle and parabola change when these figures are translated h units in the x-direction and k units in the y-direction. Each x in the original equation is replaced by $(x - h)$ and each y is replaced by $(y - k)$.

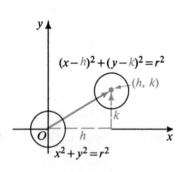

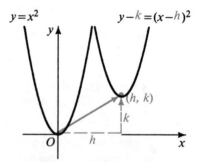

The situation illustrated in the preceding diagrams holds for the graph of any equation. Thus, in the diagram at the right, if the graph of $y = f(x)$ is translated h units in the x-direction and k units in the y-direction, then the equation of the new graph is

$$y - k = f(x - h).$$

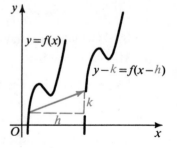

EXAMPLE 1. Sketch the graphs of $y = |x|$, $y - 2 = |x - 3|$, and $y = |x + 5|$.

SOLUTION:

Notice that the equation $y = |x + 5|$ is equivalent to
$$y - 0 = |x - (-5)|,$$
so that the original graph of $y = |x|$ is translated -5 units in the x-direction and 0 units in the y-direction.

The graph of $y = f(x) + k$

The graph of $y = f(x) + k$ is the graph of $y = f(x)$ translated k units in the y-direction. Rewriting the equation $y = f(x) + k$ as $y - k = f(x - 0)$ makes it easy to see how this agrees with the previous formula.

EXAMPLE 2. Sketch the graphs of $y = x^2 + 2$ and $y = x^2 - 3$.

SOLUTION: The graph of $y = x^2 + 2$ is the graph of $y = x^2$ translated 2 units in the y-direction. The graph of $y = x^2 - 3$ is the graph of $y = x^2$ translated -3 units in the y-direction.

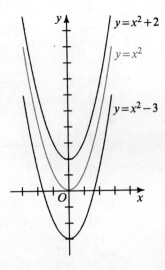

SUMMARY OF GRAPHICAL METHODS

So far in this chapter we have showed how certain simple changes in the equation of a curve can stretch the curve, reflect it, or translate it. These results are summarized below.

Change in equation $(c > 1)$	*Change in graph*
$y = f(x)$ changed to $y = cf(x)$	Vertical stretch by a factor of c
$y = f(x)$ changed to $y = \dfrac{1}{c}f(x)$	Vertical shrink by a factor of $\dfrac{1}{c}$
$y = f(x)$ changed to $y = f(cx)$	Horizontal shrink by a factor of c
$y = f(x)$ changed to $y = f\left(\dfrac{x}{c}\right)$	Horizontal stretch by a factor of c
$y = f(x)$ changed to $y = -f(x)$	Reflection in x-axis
$y = f(x)$ changed to $y = f(-x)$	Reflection in y-axis
$y = f(x)$ changed to $x = f(y)$	Reflection in the line $y = x$
$y = f(x)$ changed to $y = f(x - h)$	Horizontal translation h units
$y = f(x)$ changed to $y - k = f(x)$	Vertical translation k units

EXAMPLE 3. Sketch the graph of $y = -3 \sin 2\left(x - \dfrac{\pi}{6}\right)$.

SOLUTION: This graph is obtained from the graph of $y = \sin x$ by a series of four transformations.

Original graph

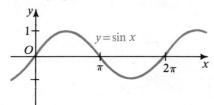

1. A horizontal shrink by a factor of 2.

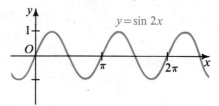

2. A vertical stretch by a factor of 3.

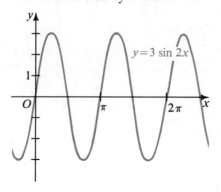

3. A reflection in the x-axis

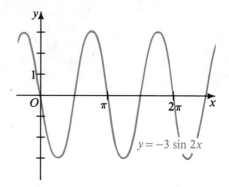

4. A translation of $\dfrac{\pi}{6}$ units right

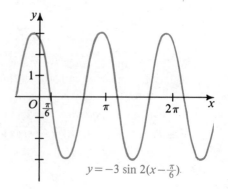

EXAMPLE 4. Let the roots of the equation $4x^2 - 9x + 5 = 0$ be x_1 and x_2. Give a related equation having roots

a. $x_1 + 2$ and $x_2 + 2$
b. $x_1 - 3$ and $x_2 - 3$
c. $2x_1$ and $2x_2$

SOLUTION: **a.** If the roots are increased by 2, the graph of $y = 4x^2 - 9x + 5$ must be translated 2 units right. Therefore replace x by $x - 2$.

$$4(x - 2)^2 - 9(x - 2) + 5 = 0$$

b. To decrease roots by 3, translate the graph 3 units left and replace x by $x + 3$.

$$4(x + 3)^2 - 9(x + 3) + 5 = 0$$

c. If the roots are multiplied by 2, the graph of $y = 4x^2 - 9x + 5$ must be horizontally stretched. Therefore replace x by $\frac{x}{2}$.

$$4\left(\frac{x}{2}\right)^2 - 9\left(\frac{x}{2}\right) + 5 = 0$$

ORAL EXERCISES 8-3

1. Given the graph of $y = f(x)$ shown at the right, sketch on the chalkboard the graph of:
   a. $y - 2 = f(x - 3)$
   b. $y + 1 = f(x + 4)$
   c. $y = f(x) + 1$
   d. $y = f(x) - 1$

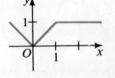

2. Describe how you would transform the graph of $y = \cos x$ to obtain the graph of each.

   a. $y = \cos\left(x + \frac{\pi}{4}\right)$ **b.** $y = 2\cos\left(x - \frac{\pi}{4}\right)$

   c. $y = -\cos 2x$ **d.** $y - 3 = \cos(x + 2)$

   e. $y = 4 + \cos\frac{1}{2}x$ **f.** $y - 4 = 3\cos\frac{1}{2}\left(x - \frac{\pi}{2}\right)$

3. Suppose you know that 1, 3, and 5 are the roots of $x^3 - 9x^2 + 23x - 15 = 0$. Find the roots of each equation.
   a. $(x - 1)^3 - 9(x - 1)^2 + 23(x - 1) - 15 = 0$
   b. $(x + 1)^3 - 9(x + 1)^2 + 23(x + 1) - 15 = 0$
   c. $\left(\frac{x}{10}\right)^3 - 9\left(\frac{x}{10}\right)^2 - 23\left(\frac{x}{10}\right) - 15 = 0$
   d. $(10x)^3 - 9(10x)^2 + 23(10x) - 15 = 0$

WRITTEN EXERCISES 8-3

Sketch the graph of each equation.

A   1. **a.** $y = |x|$ **b.** $y = |x + 2|$
         c. $y - 3 = |x - 4|$ **d.** $y + 1 = |x + 3|$
      2. **a.** $y = |x - 3|$ **b.** $y = |x| - 3$
         c. $y = -|x|$ **d.** $y = |-x|$

Sketch the graph of each equation.

3. a. $y = \sqrt{x}$ **b.** $y = \sqrt{x - 4}$
   c. $y - 2 = \sqrt{x + 1}$ **d.** $y + 5 = \sqrt{x - 3}$

4. a. $y = \sqrt{x - 1}$ **b.** $y = \sqrt{x} - 1$
   c. $y = 2\sqrt{x}$ **d.** $y = \sqrt{2x}$

5. a. $y = 3^x$ **b.** $y = -3^{x-2}$
   c. $y = 3^x - 2$ **d.** $y - 1 = 3^{x+1}$

6. a. $y = -3^x$ **b.** $y = -3^{-x}$
   c. $y - 4 = -3^{-(x+1)}$ **d.** $y = 3^{-x+2} - 2$

7. a. $y = \log_2 x$ **b.** $y = \log_2 (x - 2)$
   c. $y = \log_2 x - 2$ **d.** $y = 2 - \log_2 x$

8. a. $y = \log_2 2x$ **b.** $y = 2\log_2 x$
   c. $y = \log_2 \left(\dfrac{1}{x}\right)$ **d.** $y = \log_2 (-x)$

9. a. $y = \cos \left(x - \dfrac{\pi}{6}\right)$ **b.** $y = 3\cos \left(x - \dfrac{\pi}{6}\right)$
   c. $y = 3\cos 2x$ **d.** $y = 3\cos 2\left(x - \dfrac{\pi}{6}\right)$

10. a. $y = \sin \left(x - \dfrac{\pi}{2}\right)$ **b.** $y = \sin x - 1$
   c. $y = 2 + \sin \dfrac{1}{2}x$ **d.** $y = 2 + \sin \dfrac{1}{2}\left(x - \dfrac{\pi}{2}\right)$

11. a. $y = x^2 - 4$ **b.** $y = (-x)^2 - 4$
   c. $y = -2(x^2 - 4)$ **d.** $y = (x + 2)^2 - 4$

12. a. $y = x^3$ **b.** $y = -(x - 2)^3$
   c. $y = 1 + x^3$ **d.** $y - 5 = \frac{1}{2}(x + 1)^3$

13. Sketch the graph of each equation:

  a. $y = \tan x$ **b.** $y = \tan \left(x - \dfrac{\pi}{2}\right)$ **c.** $y = \tan \left(\dfrac{\pi}{2} - x\right)$

  d. The graph in part (c) should be the same as the graph of the trigonometric function $y = \underline{\text{?}}$.

14. Sketch the graph of each equation:

  a. $y = \cos x$ **b.** $y = \cos \left(x - \dfrac{\pi}{2}\right)$ **c.** $y = \cos \left(\dfrac{\pi}{2} - x\right)$

  d. The graph in part (c) should be the same as the graph of the trigonometric function $y = \underline{\text{?}}$.

Sketch the graph of each equation.

B   **15.** $y = 3 - \sin \pi x$ **16.** $y = 2 + 3\sin \left(2x - \dfrac{\pi}{2}\right)$

    17. $y = 4\cos (2x - \pi)$ **18.** $y - 2 = 2\cos (2\pi x)$

    19. $y = 4 + 3\sin 2(x - 1)$ **20.** $y = 1 + \tan (-\frac{1}{2}x)$

Give an equation of each sine or cosine wave.

21.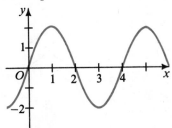

22.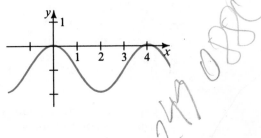

23.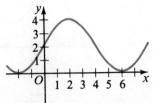

24.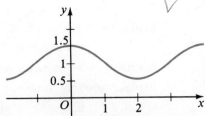

25. The equation $2x^3 - 23x^2 + 58x + 35 = 0$ has roots 5, 7, and $-\frac{1}{2}$. Give a related equation having the indicated roots.

    a. 6, 8, and $\frac{1}{2}$ **b.** 4, 6, and $-\frac{3}{2}$ **c.** 10, 14, and -1

26. The equation $x^3 - 8x^2 + 32x - 64 = 0$ has roots 4, $2 \pm 2i\sqrt{3}$. Give a related equation having the indicated roots.

    a. 2, $1 \pm i\sqrt{3}$ **b.** 12, $6 \pm 6i\sqrt{3}$ **c.** 0, $-2 \pm 2i\sqrt{3}$

The graphs below show the potential energy in a mechanical system as a function of the time t. If the sum of the potential energy and the kinetic energy of the system is a constant 400 joules, sketch the graph of the kinetic energy as a function of the time t.

27. Potential Energy (joules)

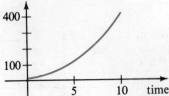

28. Potential Energy (joules)

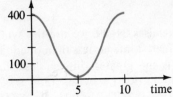

29. a. Show that the graph of $y = x^3 - 4x$ has symmetry in the origin by using the symmetry test on page 283 of Section 8-2.

    b. Use part (a) to deduce the symmetry point of the graph of $y = x^3 - 4x + 7$.

    c. Use your answer to part (b) to deduce the symmetry point of the graph of $y = (x - 1)^3 - 4(x - 1) + 7$.

    d. As a check on your answer to part (c), rewrite the equation in part (c) in the form $y = ax^3 + bx^2 + cx + d$ and use the formula $x = \dfrac{-b}{3a}$.

Advanced graphing   **291**

30. The purpose of this exercise is to show that the graph of
$$y = ax^3 + bx^2 + cx + d$$
has a point of symmetry at $x = \dfrac{-b}{3a}$.

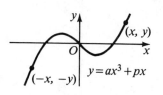

a. If $y = f(x) = ax^3 + px$, show that
$$f(-x) = -f(x).$$

b. Explain how part (a) shows that the origin is a symmetry point of the graph of
$$y = ax^3 + px.$$

c. Explain why the graph of
$$y = ax^3 + px + q$$
has $(0, q)$ as a symmetry point.

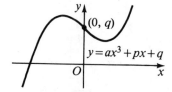

d. Explain why the graph of
$$y = a(x - h)^3 + p(x - h) + q$$
has (h, q) as a symmetry point.

e. Suppose the equation
$$y = ax^3 + bx^2 + cx + d$$
is rewritten in the equivalent form
$$y = a(x - h)^3 + p(x - h) + q.$$

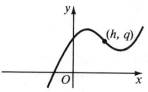

Conclude that: (1) $ax^3 + bx^2 + cx + d = a(x - h)^3 + p(x - h) + q$

(2) $ax^3 + bx^2 + cx + d = ax^3 - 3ahx^2 + \text{(other terms)}$

(3) $b = -3ah$ and $h = \dfrac{-b}{3a}$

(4) $y = ax^3 + bx^2 + cx + d$ has a point of symmetry

at $x = \dfrac{-b}{3a}$.

In Exercises 31–39, we use the symbol $[x]$ to represent the greatest integer less than or equal to x. For example, $[4.3]$ is the greatest integer less than or equal to 4.3, which is 4. Likewise, $[5.7] = 5$, $[-1] = -1$, and $[-1.1] = -2$. The graph of $y = f(x) = [x]$ is shown at the right.

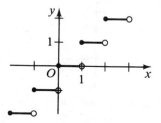

31. Sketch the graphs of $y = f(2x)$ and $y = 2f(x)$.

32. Sketch the graphs of $y = [\frac{1}{2}x]$ and $y = \frac{1}{2}[x]$.

33. Compare the graphs of $y = [x + 1]$ and $y = [x] + 1$.

34. The cost of parking a car in a parking lot is $2.00 for each hour or fraction of an hour. (Thus, parking for $2\frac{1}{4}$ hours costs $6.00.) Draw a graph showing cost as a function of parking time. Give an equation of this function using the greatest integer function.

35. The meter in a taxi reads 70 cents when you start out. It then jumps by 30 cents when you reach $\frac{1}{5}$ mile, $\frac{2}{5}$ mile, $\frac{3}{5}$ mile, and so on. Draw a graph showing cost as a function of distance traveled. Give an equation of this function using the greatest integer function.

Sketch the graph of each equation or inequality.

36. a. $|y| = [x]$ **b.** $|y| \le [x]$

37. a. $[y] = |x|$ **b.** $[y] \le |x|$

C   **38.** $[x]^2 + [y]^2 = 4$ **39.** $[x]^2 + [y]^2 \le 4$

      40. $y = |4 - |x||$ **41.** $y = |4 - |4 - |x|||$

8-4/GRAPH OF $y = \dfrac{1}{f(x)}$; HORIZONTAL AND VERTICAL ASYMPTOTES

In this section we shall compare the graph of $y = f(x)$ with the graph of

$$y = \frac{1}{f(x)}.$$

Two general observations will be helpful before we consider specific examples:

1. $\dfrac{1}{f(x)}$ is $\begin{cases} \text{positive when } f(x) \text{ is positive.} \\ \text{negative when } f(x) \text{ is negative.} \\ \text{undefined when } f(x) = 0. \end{cases}$

2. $\left| \dfrac{1}{f(x)} \right|$ is $\begin{cases} 1 \text{ when } |f(x)| = 1. \\ \text{near zero when } |f(x)| \text{ is large.} \\ \text{large when } |f(x)| \text{ is near zero.} \end{cases}$

EXAMPLE 1.   Sketch the graphs of $y = x - 2$ and $y = \dfrac{1}{x - 2}$.

SOLUTION: The graph of $y = x - 2$ is a straight line with slope 1 and y-intercept -2.

Since $x - 2$ and $\dfrac{1}{x - 2}$ have the same sign, we can see that $\dfrac{1}{x - 2}$ is positive when $x > 2$ and negative when $x < 2$. When $x = 2$, $\dfrac{1}{x - 2}$ is undefined. Moreover, when x is near 2, $x - 2$ is near 0, so that $\left| \dfrac{1}{x - 2} \right|$ is very large. When $|x|$ is large,

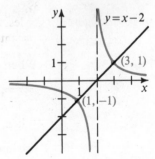

so is $|x - 2|$, so that $\left| \dfrac{1}{x - 2} \right|$ is near 0.

Asymptotes

An **asymptote** is a line that a curve approaches more and more closely until the distance between the curve and line almost vanishes. In this section we shall be concerned with asymptotes that are horizontal or vertical.

Imagine a point $P = (x, y)$ moving along the graph of $y = \dfrac{1}{x - 2}$ shown in Example 1. Notice that as x approaches 2, the value of $|y|$ becomes arbitrarily large. In this case, the vertical line $x = 2$ is called a *vertical asymptote* of the graph. The graph also has a *horizontal asymptote,* namely the line $y = 0$ (the x-axis). Notice that y gets nearer and nearer zero as $|x|$ gets larger and larger. In general:

(1)   The line $x = a$ is a **vertical asymptote** of the graph of $y = f(x)$ if $|f(x)|$ becomes arbitrarily large as x approaches a either from the left or from the right. (See diagram at the left below.)

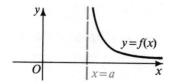
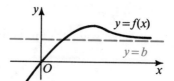

(2)   The line $y = b$ is a **horizontal asymptote** of the graph of $y = f(x)$ if $f(x)$ approaches b as x becomes arbitrarily large in either the positive or the negative direction. (See diagram at the right above.)

EXAMPLE 2.   Sketch the graphs of $y = x^2 - 1$ and $y = \dfrac{1}{x^2 - 1}$.

SOLUTION: The graph of $y = x^2 - 1$ is a parabola with x-intercepts ± 1.

When $|x| > 1$, $x^2 - 1$ and

$\dfrac{1}{x^2 - 1}$ are positive.

When $|x| < 1$, $x^2 - 1$ and

$\dfrac{1}{x^2 - 1}$ are negative.

When $x = \pm 1$, $x^2 - 1 = 0$ and

$\dfrac{1}{x^2 - 1}$ is undefined.

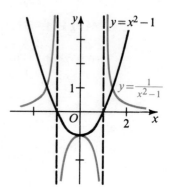

When x approaches ± 1, $\left| \dfrac{1}{x^2 - 1} \right|$

becomes arbitrarily large.

Consequently, the lines $x = 1$ and $x = -1$ are vertical asymptotes of

the graph of $y = \dfrac{1}{x^2 - 1}$. The line $y = 0$ (the x-axis) is a horizontal

asymptote of the graph because $\dfrac{1}{x^2 - 1}$ approaches zero as $|x|$ becomes arbitrarily large.

1. Sketch the graphs of $y = x + 2$ and $y = \dfrac{1}{x + 2}$ on the chalkboard.

2. Sketch the graphs of $y = x^2 - 4$ and $y = \dfrac{1}{x^2 - 4}$ on the chalkboard.

WRITTEN EXERCISES 8-4

In Exercises 1–12, sketch the graphs of the two given equations on the same set of axes. State the vertical and horizontal asymptotes, if any, for the second equation.

A **1.** $y = x + 3, \quad y = \dfrac{1}{x + 3}$

2. $y = x - 4, \quad y = \dfrac{1}{x - 4}$

3. $y = x^2, \quad y = \dfrac{1}{x^2}$

4. $y = x^2 + 1, \quad y = \dfrac{1}{x^2 + 1}$

5. $y = 9 - x^2, \quad y = \dfrac{1}{9 - x^2}$

6. $y = x^2 - 4, \quad y = \dfrac{1}{x^2 - 4}$

7. $y = \sqrt{x}, \quad y = \dfrac{1}{\sqrt{x}}$

8. $y = |x|, \quad y = \dfrac{1}{|x|}$

9. $y = \sin x, \quad y = \csc x$

10. $y = 3 \cos x, \quad y = \tfrac{1}{3} \sec x$

11. $y = 2^x, \quad y = 2^{-x}$

12. $y = \log_2 x, \quad y = \dfrac{1}{\log_2 x}$

B **13.** Given the graph of $y = f(x)$ shown in Oral Exercise 1 of Section 8–1, sketch the graph of $y = \dfrac{1}{f(x)}$.

14. Given the graph of $y = f(x)$ shown below, sketch the graph of $y = \dfrac{1}{f(x)}$.

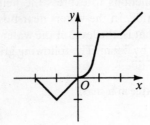

15. Give an example of an equation whose graph has the lines $x = -4$ and $x = 3$ as vertical asymptotes.

16. Give an example of an equation whose graph has $x = -2$, $x = 1$, and $x = 2$ as vertical asymptotes.

17. Consider the equation $y = 1 + \dfrac{1}{x + 1}$.

    a. What line is a vertical asymptote of the graph of $y = 1 + \dfrac{1}{x + 1}$?

    b. What value is $\left| \dfrac{1}{x + 1} \right|$ near when $|x|$ is large?

    c. Since $y = 1 + \dfrac{1}{x + 1}$, what value is y near when $|x|$ is large?

    d. What line is a horizontal asymptote of the graph of $y = 1 + \dfrac{1}{x + 1}$?

    e. Sketch the graph.

18. Give the vertical and horizontal asymptotes, if any, of the graph of $y = \dfrac{2x - 5}{x - 3}$. (*Hint:* See Exercise 17.)

8-5 / APPLICATIONS OF SINES AND COSINES

Many applications of sines and cosines make use of the periodic nature of these functions but have nothing to do with angles. A typical application might involve sin t or cos t where t represents time. If $t = 1.5$ hours, for example, you evaluate sin t as if t were in radians. Table 5 or a calculator gives sin $1.5 \approx 0.9975$.

Tides

The tide in a large body of water is the periodic variation in the depth of the water caused by the gravitational effects of the moon and the sun. *Low tide* occurs when the water level is lowest; *high tide* occurs when the water level is highest. We call the difference between the water level at a given time and the level at low tide the *height* of the tide. The tide comes in and goes out every 12.4 hours.

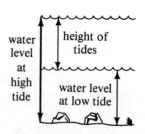

    We can use trigonometric functions to express the height of the tide as a function of the time since high tide. In the waters near Boston, Massachusetts, the average high tide is 2.9 m; that is, the level of the water at high tide exceeds the level of the water at low tide by 2.9 m. The following graph shows the depth of the water t hours after high tide.

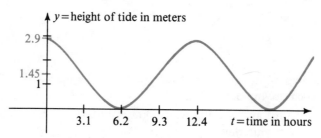

The tidal curve shown in the diagram is approximately a cosine wave with equation

$$y = 1.45 + 1.45 \cos \frac{2\pi}{12.4} t$$

Translates cosine graph by 1.45 units vertically

Amplitude of 1.45

Period of 12.4

The average height of the tide is $\frac{2.9 + 0}{2}$, or 1.45, meters. The graph is translated vertically by this average height and also has this average height as its amplitude.

EXAMPLE 1. Find the height of the tide near Boston 2 hours after high tide.

SOLUTION: Substitute 2 for t in the equation.

$$y = 1.45 + 1.45 \cos \frac{2\pi}{12.4}(2)$$

$$= 1.45 + 1.45 \cos 1.01$$

$$\approx 1.45 + 1.45(0.5319) \approx 2.22$$

The height of the tide 2 hours after high tide is about 2.22 m.

Electricity

Ordinary household electrical circuits are 60 cycle alternating current (AC) circuits. This means that the voltage oscillates like the sine wave in the following diagram at a *frequency* of 60 cycles per second, or 1 cycle every $\frac{1}{60}$ second. The period of a sine wave is always the reciprocal of the frequency. In this example, the period is $\frac{1}{60}$ and the frequency is 60. Most household circuits are called 110-volt circuits because they deliver energy at the same rate as a direct current of 110 volts. In actuality, maximum voltage in a 110-volt circuit is $110 \sqrt{2}$ volts.

The voltage equation is

$$V(t) = 110 \sqrt{2} \sin \frac{2\pi}{\frac{1}{60}} t = 110 \sqrt{2} \sin 60 \cdot 2\pi t$$

Amplitude of $110 \sqrt{2}$ Period of $\frac{1}{60}$ Frequency of 60

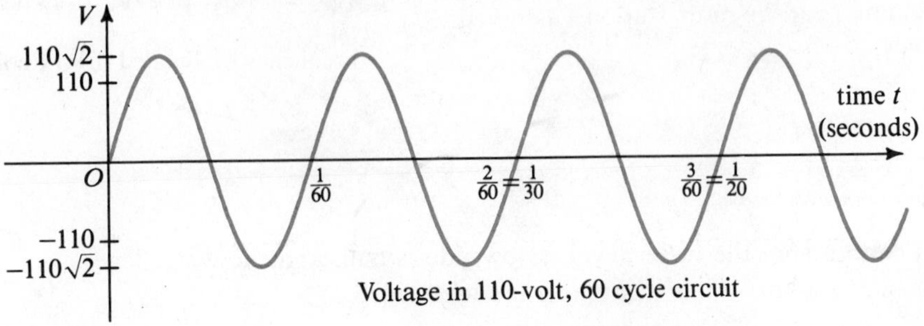

Voltage in 110-volt, 60 cycle circuit

Radio

The metric unit for cycles per second is the **hertz** (Hz). Household appliances operate at a frequency of 60 Hz. In many other applications, the frequencies are much higher and are measured in greater units such as the kilohertz (kHz) or megahertz (MHz); $1 \text{ kHz} = 10^3 \text{ Hz}$, and $1 \text{ MHz} = 10^6 \text{ Hz}$. For example, when an AM radio is tuned to the number 800 on the dial it will amplify signals broadcast at a frequency of 800 kHz, or 800,000 cycles per second. Similarly, the number 95 on the FM radio dial stands for 95 MHz, or 95,000,000 cycles per second.

The initials AM stand for **amplitude modulation.** When an AM radio is tuned to 800 on its dial it filters out all signals except those at 800 kHz. The sound you hear is communicated by varying, or "modulating," the amplitude. This means that the constant amplitude A is replaced by a variable amplitude $A(t)$ which is a function of time, and thus the signal has the form

$$A(t) \sin 2\pi ft,$$

where f is a constant that determines the frequency. The function $A(t)$ carries the information about the sound to be reproduced. An exaggerated graph of the situation follows.

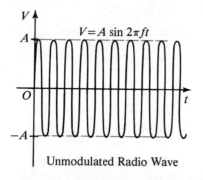

Unmodulated Radio Wave

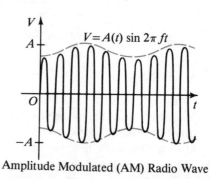

Amplitude Modulated (AM) Radio Wave

The initials FM stand for **frequency modulation.** In this method of broadcasting, the information is communicated by varying the frequency; that is, the constant frequency f is replaced by a variable frequency $f(t)$, which is a function of time. An exaggerated graph of the situation appears at the right. A complication of FM broadcasting is that $f(t)$ must remain near the radio station's assigned frequency.

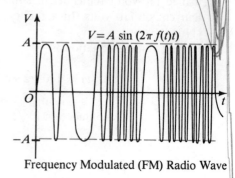

Frequency Modulated (FM) Radio Wave

ORAL EXERCISES 8–5

1. At a certain spot the water level at low tide is 6 m, and the water level at high tide is 14 m. What is the average high tide?

2. Refer to the equation and graph for the height of the tide on pages 296–297. How would you modify the equation for waters in which the high tide is 3.2 m?

3. Refer to the equation and graph describing a 110-volt circuit on page 297. How would you modify the equation to describe a 220-volt circuit?

4. Give the amplitude, frequency, and period of each function.
   a. $V = 400 \sin 1000\pi t$ b. $V = 50 \sin 200{,}000\pi t$

$$y = 4 + 4 \cos \frac{2\pi}{12.4} t$$

WRITTEN EXERCISES 8–5

$$y = 4 + 4 \cos .5t$$

$$\text{for } t = 3 \longrightarrow y = 4 + 4 \cos(1.5) =$$

In Exercises 1–6, which refer to tides, use the equations and graph for tides in the Boston area as a model when necessary. In Exercises 1–4, give an equation and graph for the tides at each of the given locations.

$$4.28$$

$$t = 2 \longrightarrow y = 4 + 4 \cos(1) = 6.16$$

	Location	Average high tide	Period
A	1. Savannah, Georgia	2.7 m	12.4 h
	2. Eastport, Maine	5.5 m	12.4 h
	3. Vancouver, British Columbia	3.2 m	12.4 h
	4. Newport, Rhode Island	1.1 m	12.4 h

then
$$6.16 - 4.28$$

5. The Bay of Fundy is famous for its average high tide of about 8 m. Calculate the change in water level from time $t = 2$ to $t = 3$ hours after high tide.

6. Show that the change in water level at the Bay of Fundy between $t = 0$ and $t = 2.1$ is approximately the same as the change between $t = 2.1$ and $t = 3.1$.

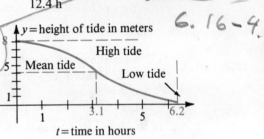

y=height of tide in meters

High tide

Mean tide

Low tide

t=time in hours

Exercises 7–12 concern electricity, radio, television, and light waves.

7. What are the amplitude, frequency, and period for each of the following functions of the time t?
   a. $V = 220 \sin 100\pi t$ b. $V = 100 \sin 50\pi t$

8. Write an equation that gives the voltage as a function of time if the voltage has amplitude 150 volts and frequency 60 hertz.

9. In a reference book find the frequency range for VHF (very high frequency) and UHF (ultra high frequency) television.

10. In a reference book find the following:
    a. the frequency of blue light
    b. the range of frequencies for infra-red light

B 11. Which of the following functions could represent a typical FM signal and which could represent a typical AM signal (t is in seconds)?
    a. $A(1 + \sin 500\pi t) \sin (2{,}000{,}000\pi t)$
    b. $A \sin (200{,}000{,}000 + 10{,}000 \sin 500\pi t)t$

12. Express in kilohertz the broadcast frequency of an AM station whose equation is $V = A(t) \sin 2\pi \cdot 950{,}000t$.

Exercises 13 and 14 refer to a satellite that is launched from Cape Canaveral and begins to orbit the earth along a path that, viewed from Earth's surface, approximates the cosine wave shown below. (Viewed from space, the satellite's path is in a plane passing through the center of Earth.)

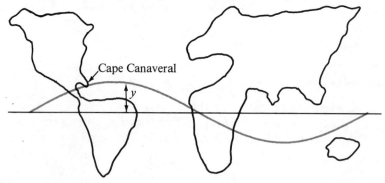

We can write an equation that gives the satellite's distance from the equator. Let t = time in minutes after blastoff, and let y = the distance in kilometers from the equator north to a point below the satellite's orbit. (When y is negative, the satellite is south of the equator.) This satellite travels along a path with equation $y = 4600 \cos \dfrac{\pi}{60}(t - 15)$.

13. After the given number of minutes have elapsed, what is the satellite's distance north or south of the equator?
 a. $t = 30$ **b.** $t = 45$ **c.** $t = 90$
 d. Find the distance from the equator to Cape Canaveral. (*Hint:* Set t = ?)

14. a. What is the satellite's maximum distance north of the equator?
 b. How many minutes after blastoff does it take the satellite to reach (for the first time) its maximum distance north of the equator? its maximum distance south of the equator?
 c. How long does it take for the satellite to orbit Earth?

Exercises 15–19 concern estimating the hours of daylight at various locations throughout the world. The graph and equation on the next page show the approximate number of hours between sunrise and sunset at various times of the year in Denver, Colorado.*

15. a. Find the number of hours between sunrise and sunset on April 1.
 b. At what times of the year is the amount of daylight changing most rapidly?
 c. How would you modify the equation if 16 and 8 were the maximum and minimum hours of daylight?

---

*The daylight equation can be made more precise by replacing the 12 by 12.17. The extra 0.17 hour is due to (1) refraction, which allows us to see the sun when it is below the horizon plane, and (2) the recording of sunrises and sunsets when the edge, rather than the middle, of the sun appears or disappears.

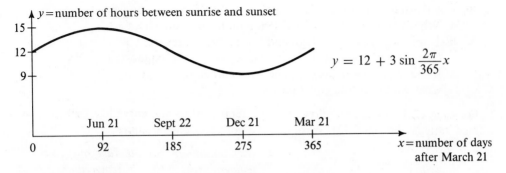

y = number of hours between sunrise and sunset

$$y = 12 + 3 \sin \frac{2\pi}{365} x$$

x = number of days after March 21

16. The equation and graph given above apply to Denver and all other locations at latitude 39°44′N. Modify the equation and graph for a location at latitude 39°44′S.

In Exercises 17 and 18, modify the equation and graph given above so that they apply to the given location. Use the modified equation to find the number of hours between sunrise and sunset on New Year's Day.

Location	Maximum hours of daylight	Minimum hours of daylight
17. Seattle, Washington	16.0 (June 21)	8.5 (December 21)
18. Chicago, Illinois	15.2 (June 21)	9.2 (December 21)

19. Consult a local almanac or newspaper to find the maximum and minimum hours of daylight where you live. Then modify the equation and graph given above so that they apply. Use the modified equation to find the number of hours between sunset and sunrise on New Year's Day.

Exercises 20–25 concern average daily temperatures. In the following graph the average monthly temperatures for New Orleans, Louisiana, are plotted for the middle of each month. These points can be connected to give a smooth curve that gives an approximation of the average daily temperatures. The curve has as its equation

$$T = 20 + 8 \sin \frac{2\pi}{365}(x - 105),$$

where T represents the average daily Celsius temperature and x represents the number of the day in the year.

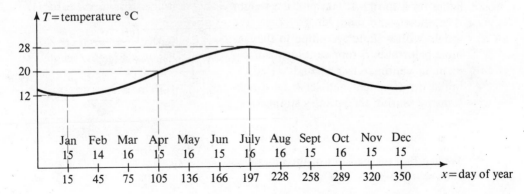

T = temperature °C

x = day of year

Advanced graphing **301**

20. Use the temperature equation and graph on the preceding page.
  a. What is the average annual temperature in New Orleans?
  b. If the average annual high and low temperatures were 30°C and 10°C (instead of 28°C and 12°C), what would be the equation for T?
  c. If the average annual high and low temperatures were 30°C and 6°C, what would be the equation for T?

For each of the locations below, sketch a temperature sine curve like that for New Orleans on page 301. Also give the equation of the curve.

Location	Average maximum temperature	Average minimum temperature
21. New York, New York	28°C (July 15)	3°C (January 15)
22. Winnipeg, Manitoba, Canada	26°C (July 15)	−14°C (January 15)
23. Rio de Janeiro, Brazil	28°C (January 15)	22°C (July 15)
24. Sidney, Australia	26°C (January 15)	15°C (July 15)

25. Consult a local almanac or newspaper to find the average maximum and minimum temperatures for the area where you live. Sketch a temperature sine curve and give the equation of the curve.

26. When the note called *concert A* is sounded on a piano, the piano string vibrates with a frequency of 440 Hz. The equation that gives the displacement of a point on the vibrating piano string is $D = A \sin 880\pi t$ where t is the number of seconds after the string is struck and A is a constant that depends on how hard the string is hit and the point's position on the string. (This formula applies only for the first few seconds.)

  a. What will be the displacement when $t = \dfrac{1}{440}$ second?

  b. What is the earliest time t for which the displacement is maximum?
  c. The piano string for A below middle C has a frequency of 220 Hz. Write an equation for its displacement.

CALCULATOR EXERCISES

Applications to springs and pendulums

1. Suppose a weight with mass m grams hangs on a spring. If you pull the weight A cm downward and let go of it, the weight will oscillate according to the formula below. Let D represent the displacement in centimeters t seconds after the initial displacement and let k be a constant measuring the spring's stiffness.

$$D = -A \cos \sqrt{\frac{k}{m}}\, t$$

We can find the period t of the motion as shown at the right.

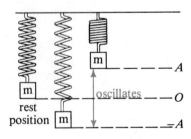

$$t = \frac{2\pi}{\sqrt{\dfrac{k}{m}}} = 2\pi \sqrt{\frac{m}{k}}$$

a. Suppose you put a weight with known mass $m = 100$ g on the spring and you time its period to be $t = 1.1$ seconds. Find the value of the spring's constant k.

b. Suppose you put another weight with unknown mass m on the spring and observe that its period is $t = 1.4$ seconds. Find m. (Use the value of k found in part (a).)

2. When a pendulum swings back and forth through a small arc, its horizontal displacement is given by the formula below. Let D be the horizontal displacement in centimeters of the pendulum t seconds after passing through 0, let A represent the maximum displacement, and let l represent the length of the pendulum.

$$D \approx A \sin\left(\sqrt{\frac{980}{l}}\right) t$$

a. If the length l of the pendulum is 100 cm, find the earliest time t for which the displacement is maximum.

b. How long is a clock pendulum that has a period of 1 second?

8-6/THE AREA UNDER A CURVE (Optional)

In geometry you studied areas of triangles, parallelograms, circles, and other special figures; but you can find areas of still other figures. One such figure is the region R under the curve $y = \sin x$ and above the x-axis.

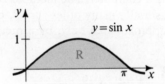

The area of region R can be approximated by calculating the areas of many rectangles having base vertices at $0, 0.1, 0.2, \ldots, 3.1$, as shown in the following diagram. We chose 3.1 as the last vertex because π equals 3.1 to the nearest tenth. You can see that the sum of the areas of the first few rectangles is greater than the corresponding area under the curve, while the sum of the areas of the last few rectangles is less than the corresponding area under the curve.

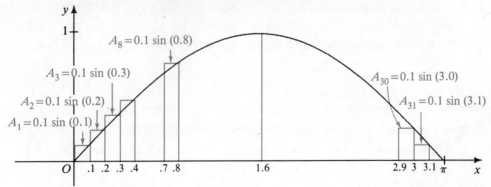

The calculations involved in approximating this area are ideally suited for the computer. Here is a simple program in the BASIC language for calculating the total area of all the rectangles we've discussed.

```
20  FOR X = .1 TO 3.1 STEP .1
30  LET Y = SIN(X)
40  LET A = A + .1*Y
50  NEXT X
60  PRINT "AREA IS APPROXIMATELY"; A
70  END
```

When the program is run, the following printout results:

```
AREA IS APPROXIMATELY 1.99955
```

A better approximation can be found if we use rectangles placed at 0, 0.01, 0.02, ..., 3.14, instead of 0.0, 0.1, ..., 3.1. All that is necessary is to change lines 20 and 40 as shown below. Notice that we use 3.14, which is correct to the nearest hundredth, for π.

```
20  FOR X = .01 TO 3.14 STEP .01
40  LET A = A + .01*Y
```

These changes give the following printout:

```
AREA IS APPROXIMATELY 1.99999
```

In calculus it is possible to prove that the area of region R is *exactly* 2 without using a computer. The methods of calculus can also give exact answers for the areas under many other curves, but not all. The areas under some curves can be approximated only, by a method like that we used for the sine curve.

ORAL EXERCISES 8-6

Tell how you would modify the computer program above to approximate the area of the region described.

1. The region under the curve $y = \sin x$, above the x-axis and between $x = 1$ and $x = 2$.

2. The region under the curve $y = x^2$, above the x-axis and between $x = 0$ and $x = 4$.

3. One of the regions enclosed between the curves $y = \sin x$ and $y = -\sin x$.

WRITTEN EXERCISES 8-6

Run a computer program to find the area of each region described.

A 1. The region under the curve $y = \tan x$, above the x-axis and between $x = 0$ and $x = \dfrac{\pi}{4}$

2. The region under the curve $y = \cos x$, above the x-axis and between $x = 0$ and $x = \dfrac{\pi}{2}$

3. The region under the curve $y = x^2$, above the x-axis and between 0 and 1

4. The region under the curve $y = x^3$, above the x-axis and between 0 and 1

5. Consider the area under the curve $y = x^n$, above the x-axis and between 0 and 1.
   a. When $n = 1$, you know from geometry that the area is _?_ .
   b. When $n = 2$, Exercise 3 shows the area is about $\frac{1}{3}$.
   c. When $n = 3$, Exercise 4 shows the area is about $\frac{1}{4}$.
   d. Make a conjecture for $n = 4$. Use the computer to check your guess.

6. a. Find the area of region A under the curve $y = \dfrac{1}{x}$, above the x-axis and between $x = 1$ and $x = 2$. Compare your result with the natural logarithm of 2. (The BASIC function LOG(X) gives the natural logarithm of X.)

   b. Find the area of region B under the curve $y = \dfrac{1}{x}$, above the x-axis and between $x = 1$ and $x = 3$. Compare your result with the natural logarithm of 3.

   c. Find the area of region C under the curve $y = \dfrac{1}{x}$, above the x-axis and between $x = 1$ and $x = 6$. Compare your result with the natural logarithm of 6.

   d. From the results of parts (a)–(c), we can observe that area of A + area of B = area of C. Also, $\ln 2 + \ln 3 = \ln \underline{\ ?\ }$. Thus, areas under the curve $y = \dfrac{1}{x}$ have logarithmic properties.

B   7. In Exercise 6 you found that the area under the curve $y = \dfrac{1}{x}$ between $x = 1$ and $x = 2$ was less than 1, and the area between $x = 1$ and $x = 3$ was greater than 1. Thus, there must be a number x between 2 and 3 so that the area between 1 and x is exactly 1 (see diagram). Write a computer program to find the approximate value of this number x.

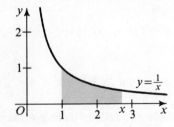

8. The area under the curve $y = \dfrac{4}{1 + x^2}$, above the x-axis and between 0 and 1, is a special number. Write a computer program to find it.

Chapter summary

1. $f(x)$ is a *periodic function* if there is a number $p > 0$ such that for all x in the domain D of $f(x)$:

$$f(x + p) = f(x)$$

If p is the smallest such positive number, then p is called the *period* of the function. The *amplitude* of the function is $\dfrac{M - m}{2}$, where M and m are the maximum and minimum values, respectively, of the function.

2. Tests for *symmetry* are summarized on pages 282–283.

3. Certain changes in the equation of a curve will stretch, shrink, reflect, or translate the graph of the curve. These changes are summarized on page 287.

4. To obtain the graph of $y = \dfrac{1}{f(x)}$ from the graph of $y = f(x)$, keep in mind the following facts (and see Examples 1 and 2, pages 293 and 294).

   a. $y = \dfrac{1}{f(x)}$ is $\begin{cases} \text{undefined when } f(x) = 0. \\ \text{positive (or negative) when } f(x) \text{ is positive (or negative).} \end{cases}$

   b. $|y| = \left| \dfrac{1}{f(x)} \right|$ is $\begin{cases} 1 \text{ when } |f(x)| = 1. \\ \text{large when } |f(x)| \text{ is near zero.} \\ \text{near zero when } |f(x)| \text{ is large.} \end{cases}$

5. The line $x = a$ is a *vertical asymptote* of the graph of $y = f(x)$ if $|f(x)|$ becomes arbitrarily large as x approaches a either from the left or from the right. The line $y = b$ is a *horizontal asymptote* of the graph of $y = f(x)$ if $f(x)$ approaches b as x becomes arbitrarily large in either the positive or the negative direction.

6. The graphs of the sine and cosine functions have many applications, including the description of tides, electricity, radio waves, satellite orbits, sunrise and sunset, weather, musical tones, and the motion of springs and pendulums.

7. A computer or programmable calculator can be used to approximate the area under a specified curve.

Chapter test

1. a. Sketch the graph of $y = -\frac{1}{2}\sin 2x$ for $-\pi \le x \le 2\pi$.   8-1
   b. What is the period of this function?

2. Do the graphs of the following equations have symmetry with respect to   8-2
   (i) the x-axis, (ii) the y-axis, (iii) the line $y = x$, and (iv) the origin?

   a. $x^2 y = 2$ b. $y = \sin x$

For Exercises 3–9, use the graph of $y = f(x)$ shown.

3. Sketch the graph of $y = 2f(x)$.

4. Sketch the graph of $y = f(2x)$.

5. Sketch the graph of $y = |f(x)|$.

6. Sketch the graph of $y = f(-x)$.

7. Sketch the graph of $y = f(x) + 2$.

8. Sketch the graph of $y = f(x + 2)$.

9. Sketch the graph of $y - 3 = f(x - 1)$.

10. Sketch the graph of each equation.

    a. $y = \log(x - 1)$ **b.** $y = \log x - 1$

11. Give the equation of the cosine curve shown.

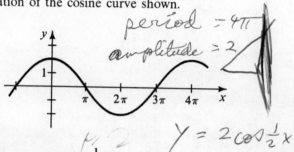

12. a. Sketch the graph of $y = \dfrac{1}{x - 5}$.

    b. Give equations of its asymptotes.

13. At a certain location, the longest amount of time between sunrise and sunset is about 16 hours on June 21, and the shortest amount of time is about 8 hours on December 21. If x is the number of days past March 21 (the date midway between December 21 and June 21), express the amount of time between sunrise and sunset as a function of x and sketch a graph of the function.

14. (Optional, computer program required) Find the area of the region under the curve $y = x^2$, above the x-axis and between $x = 1$ and $x = 2$.

308

CHAPTER NINE

Trigonometric addition formulas

OBJECTIVES

1. To use the trigonometric addition, double-angle, and half-angle formulas to simplify and evaluate expressions, prove identities, and solve equations.
2. To find the angle between two intersecting lines.

Deriving formulas

Trigonometric functions are used in science and engineering to describe light and sound waves. For example, the figure below shows the wave pattern of a note sounded by a violin. The equation of this wave involves sines and cosines of 2α, 3α, and higher multiples of α. Our goal in this chapter is to gain experience working with expressions like $\sin 2\alpha$ and $\cos 3\alpha$. We shall derive formulas showing, for example, how $\cos 2\alpha$ is related to $\cos \alpha$ and how $\cos(\alpha + \beta)$ is related to $\cos \alpha$ and $\cos \beta$.

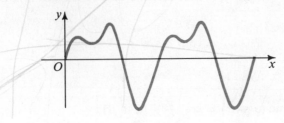

Any note sounded by a violin has a wave pattern similar to the graph above. The higher the note, the more quickly the wave pattern repeats itself.

Formulas for cos (α − β) and cos (α + β)

Let A and B be points on the unit circle with coordinates as shown. The measure of $\angle AOB$, then, is $\alpha - \beta$. Recall that the distance AB can be found by using the Law of Cosines or by using the Distance Formula.

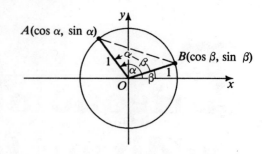

Using the Law of Cosines:

$$(AB)^2 = 1^2 + 1^2 - 2 \cdot 1 \cdot 1 \cdot \cos(\alpha - \beta)$$
$$= 2 - 2\cos(\alpha - \beta)$$

Using the Distance Formula:

$$(AB)^2 = (\cos\alpha - \cos\beta)^2 + (\sin\alpha - \sin\beta)^2$$
$$= \cos^2\alpha - 2\cos\alpha\cos\beta + \cos^2\beta + \sin^2\alpha - 2\sin\alpha\sin\beta + \sin^2\beta$$
$$= (\cos^2\alpha + \sin^2\alpha) + (\cos^2\beta + \sin^2\beta) - 2(\sin\alpha\sin\beta + \cos\alpha\cos\beta)$$
$$= 2 - 2(\cos\alpha\cos\beta + \sin\alpha\sin\beta)$$

If we equate the two expressions for $(AB)^2$, we get:

$$2 - 2\cos(\alpha - \beta) = 2 - 2(\cos\alpha\cos\beta + \sin\alpha\sin\beta)$$

and

$$\cos(\alpha - \beta) = \cos\alpha\cos\beta + \sin\alpha\sin\beta \qquad (1)$$

If we use the formula for cos (α − β) and replace β by −β, we get:

$$\cos(\alpha - (-\beta)) = \cos\alpha\cos(-\beta) + \sin\alpha\sin(-\beta)$$

so that

$$\cos(\alpha + \beta) = \cos\alpha\cos\beta - \sin\alpha\sin\beta \qquad (2)$$

In the last step, we substituted $\cos(-\beta) = \cos\beta$ and $\sin(-\beta) = -\sin\beta$.

Formulas for sin (α + β) and sin (α − β)

In this derivation, we use the fact that the sine and cosine are *cofunctions;* that is,

$$\sin\theta = \cos\left(\frac{\pi}{2} - \theta\right).$$

Let $\theta = \alpha + \beta$. Then,

$$\sin(\alpha + \beta) = \cos\left[\frac{\pi}{2} - (\alpha + \beta)\right]$$
$$= \cos\left[\left(\frac{\pi}{2} - \alpha\right) - \beta\right].$$

Use formula (1) to get:

$$\cos\left[\left(\frac{\pi}{2} - \alpha\right) - \beta\right] = \cos\left(\frac{\pi}{2} - \alpha\right)\cos\beta + \sin\left(\frac{\pi}{2} - \alpha\right)\sin\beta$$

so that

$$\sin(\alpha + \beta) = \sin\alpha\cos\beta + \cos\alpha\sin\beta \qquad (3)$$

If we use the formula for $\sin(\alpha + \beta)$ and replace β by $-\beta$ we get:

$$\sin(\alpha + (-\beta)) = \sin\alpha\cos(-\beta) + \cos\alpha\sin(-\beta)$$

so that

$$\sin(\alpha - \beta) = \sin\alpha\cos\beta - \cos\alpha\sin\beta \qquad (4)$$

SUMMARY OF FORMULAS

$$\sin(\alpha \pm \beta) = \sin\alpha\cos\beta \pm \cos\alpha\sin\beta$$
$$\cos(\alpha \pm \beta) = \cos\alpha\cos\beta \mp \sin\alpha\sin\beta$$

EXAMPLE 1. Show that $\sin\left(\frac{3\pi}{2} - \theta\right) = -\cos\theta$.

SOLUTION: We use formula (4) with $\alpha = \frac{3\pi}{2}$ and $\beta = \theta$.

$$\sin\left(\frac{3\pi}{2} - \theta\right) = \sin\frac{3\pi}{2}\cos\theta - \cos\frac{3\pi}{2}\sin\theta$$
$$= (-1)\cos\theta - (0)\sin\theta$$
$$= -\cos\theta$$

EXAMPLE 2. Simplify: **a.** $\cos 50° \cos 10° - \sin 50° \sin 10°$

b. $\sin\frac{5\pi}{12}\cos\frac{\pi}{12} + \cos\frac{5\pi}{12}\sin\frac{\pi}{12}$

SOLUTION: The two given expressions have the patterns shown in formulas (2) and (3), respectively.

a. $\cos\alpha\cos\beta - \sin\alpha\sin\beta = \cos(\alpha + \beta)$

$$\cos 50° \cos 10° - \sin 50° \sin 10° = \cos(50° + 10°)$$
$$= \cos 60°$$
$$= \tfrac{1}{2}$$

b. $\sin\alpha\cos\beta + \cos\alpha\sin\beta = \sin(\alpha + \beta)$

$$\sin\frac{5\pi}{12}\cos\frac{\pi}{12} + \cos\frac{5\pi}{12}\sin\frac{\pi}{12} = \sin\left(\frac{5\pi}{12} + \frac{\pi}{12}\right)$$
$$= \sin\frac{\pi}{2}$$
$$= 1$$

Trigonometric addition formulas   **311**

EXAMPLE 3. Evaluate sin 105° without a calculator or tables.

SOLUTION: We would like to express 105° as the sum or difference of two numbers whose sines and cosines we know. Since 105° = 60° + 45°, we have:

$$\sin 105° = \sin (60° + 45°)$$
$$= \sin 60° \cos 45° + \cos 60° \sin 45°$$
$$= \left(\frac{\sqrt{3}}{2}\right)\left(\frac{\sqrt{2}}{2}\right) + \left(\frac{1}{2}\right)\left(\frac{\sqrt{2}}{2}\right)$$
$$= \frac{\sqrt{6} + \sqrt{2}}{4}$$

Therefore, $\sin 105° = \dfrac{\sqrt{6} + \sqrt{2}}{4}$.

EXAMPLE 4. If $\sin \alpha = \dfrac{4}{5}$ and $\sin \beta = \dfrac{5}{13}$, and if $0 < \alpha < \dfrac{\pi}{2} < \beta < \pi$, find $\cos (\alpha + \beta)$.

SOLUTION: If $\sin \alpha = \dfrac{4}{5}$ and $0 < \alpha < \dfrac{\pi}{2}$, then

$$\cos \alpha = \frac{3}{5}.$$

If $\sin \beta = \dfrac{5}{13}$ and $\dfrac{\pi}{2} < \beta < \pi$, then

$$\cos \beta = -\frac{12}{13}.$$

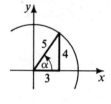

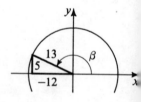

Thus,

$$\cos (\alpha + \beta) = \cos \alpha \cos \beta - \sin \alpha \sin \beta$$
$$= \left(\frac{3}{5}\right)\left(-\frac{12}{13}\right) - \left(\frac{4}{5}\right)\left(\frac{5}{13}\right)$$
$$= -\frac{56}{65}$$

Optional formulas

The following formulas are derived in Exercises 42–44.

$$\sin x + \sin y = 2 \sin \frac{x + y}{2} \cos \frac{x - y}{2}$$

$$\sin x - \sin y = 2 \cos \frac{x + y}{2} \sin \frac{x - y}{2}$$

$$\cos x + \cos y = 2 \cos \frac{x + y}{2} \cos \frac{x - y}{2}$$

$$\cos x - \cos y = -2 \sin \frac{x + y}{2} \sin \frac{x - y}{2}$$

Make up a single example which shows that, in general:

1. $\sin 2x \neq 2 \sin x$ **2.** $\cos (a - b) \neq \cos a - \cos b$

Are there any values of a and b for which the following are true? If so, give an example.

3. $\sin (a - b) = \sin a - \sin b$ **4.** $\sin (a + b) = \sin a + \sin b$

5. Simplify each of the following:

    a. $\sin 10° \cos 80° + \cos 10° \sin 80°$ **b.** $\sin 45° \cos 15° - \cos 45° \sin 15°$

    c. $\cos \dfrac{7\pi}{12} \cos \dfrac{5\pi}{12} - \sin \dfrac{7\pi}{12} \sin \dfrac{5\pi}{12}$ **d.** $\cos 220° \cos 10° + \sin 220° \sin 10°$

6. Explain how to evaluate $\sin 15°$ without a calculator or tables.

WRITTEN EXERCISES 9-1

In Exercises 1–6, simplify the given expression.

A   **1.** $\sin 75° \cos 15° + \cos 75° \sin 15°$

    2. $\cos 105° \cos 15° + \sin 105° \sin 15°$

    3. $\cos \dfrac{5\pi}{12} \cos \dfrac{\pi}{12} - \sin \dfrac{5\pi}{12} \sin \dfrac{\pi}{12}$

    4. $\sin \dfrac{4\pi}{3} \cos \dfrac{\pi}{3} - \cos \dfrac{4\pi}{3} \sin \dfrac{\pi}{3}$

    5. $\sin 3x \cos 2x - \cos 3x \sin 2x$

    6. $\cos 2x \cos x + \sin 2x \sin x$

In Exercises 7–10, prove the given identity.

    7. $\cos \left(\dfrac{\pi}{2} + \theta \right) = -\sin \theta$ **8.** $\cos \left(\dfrac{\pi}{2} - \theta \right) = \sin \theta$

    9. $\sin (\pi + x) = -\sin x$ **10.** $\cos (\pi + x) = -\cos x$

In Exercises 11–16, evaluate the given expression without using a calculator or tables.

    11. $\cos 75°$ **12.** $\sin 15°$

    13. $\cos \dfrac{7\pi}{12}$ **14.** $\sin \dfrac{11\pi}{12}$

    15. $\sin (-15°)$ **16.** $\cos (-165°)$

In Exercises 17–22, simplify the given expression.

    17. $\sin (30° + x) + \sin (30° - x)$ **18.** $\cos (30° + x) + \cos (30° - x)$

    19. $\cos \left(\dfrac{\pi}{3} + \theta \right) + \cos \left(\dfrac{\pi}{3} - \theta \right)$ **20.** $\sin \left(\dfrac{\pi}{4} + \theta \right) + \sin \left(\dfrac{\pi}{4} - \theta \right)$

    21. $\cos \left(\dfrac{3\pi}{2} + \theta \right) + \cos \left(\dfrac{3\pi}{2} - \theta \right)$ **22.** $\sin (\pi + \theta) + \sin (\pi - \theta)$

Trigonometric addition formulas   **313**

B 23. $\triangle ABC$ is an acute triangle with $\cos A = \frac{1}{3}$ and $\cos B = \frac{1}{4}$.
   a. Find $\cos (A + B)$.   **b.** Find $\cos C$.

24. Show that in $\triangle ABC$, $\sin C = \sin A \cos B + \cos A \sin B$.

25. If $\sin \alpha = \dfrac{3}{5}$ and $\sin \beta = \dfrac{24}{25}$, and if $0 < \alpha < \dfrac{\pi}{2} < \beta < \pi$, find $\sin (\alpha + \beta)$.

26. If $\sin \alpha = \dfrac{4}{5}$ and $\sin \beta = \dfrac{1}{2}$, and if $\dfrac{\pi}{2} < \alpha < \beta < \pi$, find $\sin (\alpha - \beta)$.

27. If $\tan \alpha = \dfrac{4}{3}$ and $\tan \beta = \dfrac{12}{5}$, and if $0 < \alpha < \beta < \dfrac{\pi}{2}$, find $\sin (\alpha + \beta)$.

28. If $0 < \alpha < \dfrac{\pi}{2} < \beta < \pi$, and if $\sec \alpha = \dfrac{5}{4}$ and $\tan \beta = -1$, find $\cos (\alpha + \beta)$.

In Exercises 29–36, simplify the given expression.

29. $\sin (x + y) \cos y - \cos (x + y) \sin y$. (*Hint:* It is easier *not* to expand $\sin (x + y)$ and $\cos (x + y)$ but rather to let $\alpha = x + y$ and $\beta = y$ and recognize the resulting expression.)

30. $\cos (x + y) \cos x + \sin (x + y) \sin x$

31. $\dfrac{\sin (\alpha + \beta) + \sin (\alpha - \beta)}{\cos \alpha \cos \beta}$

32. $\dfrac{\cos (\alpha + \beta) + \cos (\alpha - \beta)}{\cos \alpha \cos \beta}$

33. $\cos x \cos y (\tan x + \tan y)$

34. $\sin x \sin y (\cot x \cot y - 1)$

35. $\sin (x + y) \sec x \sec y$

36. $\cos \left(x + \dfrac{\pi}{3} \right) + \sin \left(x - \dfrac{\pi}{6} \right)$

37. Use formula (3) to derive a formula for $\sin 2x$.

38. Derive a formula for $\cos 2x$.

39. Using formulas (1) and (4), verify that $\sin^2 (\alpha - \beta) + \cos^2 (\alpha - \beta) = 1$.

C 40. Evaluate $\sin (\text{Tan}^{-1} \frac{1}{2} + \text{Tan}^{-1} \frac{1}{3})$ without a calculator or tables.

41. Evaluate $\cos (\text{Tan}^{-1} \frac{1}{2} + \text{Tan}^{-1} 2)$ without a calculator or tables.

42. **a.** Derive the formula

$$\sin x + \sin y = 2 \sin \frac{x + y}{2} \cos \frac{x - y}{2}.$$

   (*Hint:* Show that $\sin (\alpha + \beta) + \sin (\alpha - \beta) = 2 \sin \alpha \cos \beta$, and then substitute $\alpha = \dfrac{x + y}{2}$ and $\beta = \dfrac{x - y}{2}$.)

   b. Use the formula in part (a) to show that $\sin 40° + \sin 20° = \cos 10°$.
   c. Use the formula in part (a) to derive a formula for $\sin x - \sin y$.

43. Derive the formula

$$\cos x + \cos y = 2 \cos \frac{x + y}{2} \cos \frac{x - y}{2}$$

   by simplifying $\cos (\alpha + \beta) + \cos (\alpha - \beta)$ and then substituting as in Exercise 42.

314

44. Derive the formula

$$\cos x - \cos y = -2 \sin \frac{x+y}{2} \sin \frac{x-y}{2}$$

by simplifying $\cos(\alpha + \beta) - \cos(\alpha - \beta)$ and then substituting as in Exercise 42.

45. The purpose of this exercise is to use the figure to the right to derive the formula for $\sin(\alpha + \beta)$ when α, β, and $\alpha + \beta$ are acute. Show the following:

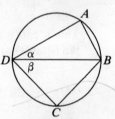

Ex. 45, 46

a. $OR = \cos \beta$
b. $SP = RQ = \sin \alpha \cos \beta$
c. $TR = \sin \beta$
d. measure of $\angle NTR = \alpha$
e. $TS = \sin \beta \cos \alpha$
f. $TP = \sin(\alpha + \beta)$
g. Use the results of parts (b) and (e) to show that $TP = \sin \alpha \cos \beta + \cos \alpha \sin \beta$.
h. Conclude that $\sin(\alpha + \beta) = \sin \alpha \cos \beta + \cos \alpha \sin \beta$.

46. Use the diagram of Exercise 45 to derive the formula for $\cos(\alpha + \beta)$. (*Hint:* $\cos(\alpha + \beta) = OP = OQ - PQ = OQ - RS$.)

47. Ptolemy's Theorem states that if $ABCD$ is inscribed in a circle, then

$$AB \cdot CD + AD \cdot BC = AC \cdot BD.$$

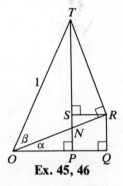

Ex. 47

Consider a special case of this theorem in which BD is a diameter with length 1.

a. Show that $AB = \sin \alpha$ and $AD = \cos \alpha$.
b. Find BC and CD in terms of β.
c. Show that $AC = \sin(\alpha + \beta)$.
d. Use parts (a), (b), (c), and Ptolemy's Theorem to derive the formula for $\sin(\alpha + \beta)$.

COMPUTER EXERCISE

Write a computer program that will print a table of sines and cosines for $1°$, $2°, \ldots, 90°$ given only that $\sin 1° = .0174524$. *Hint:* Here is how you could begin:

1. Since $\sin^2 1° + \cos^2 1° = 1$,

$$\cos 1° = \sqrt{1 - \sin^2 1°} = \sqrt{1 - (.0174524)^2},$$

which the computer will evaluate as .999848.

2. $\sin 2° = \sin(1° + 1°) = \sin 1° \cos 1° + \cos 1° \sin 1° = .0348995$.
   $\cos 2° = \sqrt{1 - \sin^2 2°} = .999391$.

3. $\sin 3° = \sin(2° + 1°) = \sin 2° \cos 1° + \cos 2° \sin 1°$, and so on.

Trigonometric addition formulas   **315**

Formula for tan (α + β)

To derive a formula expressing $\tan(\alpha + \beta)$ in terms of $\tan \alpha$ and $\tan \beta$, we use the formulas for $\sin(\alpha + \beta)$ and $\cos(\alpha + \beta)$ given on page 311 and the fact that $\tan(\alpha + \beta) = \dfrac{\sin(\alpha + \beta)}{\cos(\alpha + \beta)}$.

$$\tan(\alpha + \beta) = \frac{\sin(\alpha + \beta)}{\cos(\alpha + \beta)} \qquad (\cos(\alpha + \beta) \neq 0)$$

$$= \frac{\sin \alpha \cos \beta + \cos \alpha \sin \beta}{\cos \alpha \cos \beta - \sin \alpha \sin \beta}$$

Dividing numerator and denominator by $\cos \alpha \cos \beta$ gives:

$$\tan(\alpha + \beta) = \frac{\dfrac{\sin \alpha \cos \beta}{\cos \alpha \cos \beta} + \dfrac{\cos \alpha \sin \beta}{\cos \alpha \cos \beta}}{\dfrac{\cos \alpha \cos \beta}{\cos \alpha \cos \beta} - \dfrac{\sin \alpha \sin \beta}{\cos \alpha \cos \beta}} \qquad (\cos \alpha \cos \beta \neq 0)$$

$$\tan(\alpha + \beta) = \frac{\tan \alpha + \tan \beta}{1 - \tan \alpha \tan \beta} \tag{5}$$

This identity is valid for all values of α and β for which $\tan \alpha$, $\tan \beta$, and $\tan(\alpha + \beta)$ are defined.

EXAMPLE 1. If $\tan \alpha = \frac{1}{3}$ and $\tan \beta = \frac{1}{2}$:
  a. find $\tan(\alpha + \beta)$;
  b. show that $\tan^{-1}\dfrac{1}{3} + \tan^{-1}\dfrac{1}{2} = \dfrac{\pi}{4}$.

SOLUTION: **a.** $\tan(\alpha + \beta) = \dfrac{\tan \alpha + \tan \beta}{1 - \tan \alpha \tan \beta}$

$$\tan(\alpha + \beta) = \frac{\frac{1}{3} + \frac{1}{2}}{1 - (\frac{1}{3})(\frac{1}{2})} = \frac{\frac{5}{6}}{\frac{5}{6}} = 1$$

b. Let $\alpha = \mathrm{Tan}^{-1}\dfrac{1}{3}$ and $\beta = \mathrm{Tan}^{-1}\dfrac{1}{2}$. Since α and β are between 0 and $\dfrac{\pi}{2}$, $\alpha + \beta$ is between 0 and π. From part (a), we know that $\tan(\alpha + \beta) = 1$. Thus, since $\dfrac{\pi}{4}$ is the only angle between 0 and π whose tangent is 1,

$$\alpha + \beta = \frac{\pi}{4}.$$

That is,

$$\mathrm{Tan}^{-1}\frac{1}{3} + \mathrm{Tan}^{-1}\frac{1}{2} = \frac{\pi}{4}.$$

Formula for tan $(\alpha - \beta)$

To derive a formula for tan $(\alpha - \beta)$, we simply replace β with $-\beta$ in formula (5), page 316, and use the fact that $\tan(-\beta) = -\tan\beta$. (Can you prove this?)

$$\tan(\alpha - \beta) = \tan(\alpha + (-\beta)) \qquad \tan(\alpha - \beta) = \frac{\tan\alpha + \tan(-\beta)}{1 - \tan\alpha\tan(-\beta)}$$

$$\tan(\alpha - \beta) = \frac{\tan\alpha - \tan\beta}{1 + \tan\alpha\tan\beta} \tag{6}$$

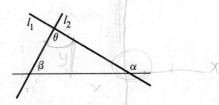

The formula for tan $(\alpha - \beta)$ can be used to find the angle between two intersecting lines. For example, consider the angle θ between lines l_1 and l_2 with slopes m_1 and m_2. Suppose the inclinations of these lines are α and β, respectively. Then:

1. $\tan\alpha = $ slope of $l_1 = m_1$ and $\tan\beta = $ slope of $l_2 = m_2$.

2. $\alpha = \beta + \theta$, because the measure of an exterior angle of a triangle equals the sum of the measures of the two opposite interior angles. Therefore, $\theta = \alpha - \beta$ and $\tan\theta = \tan(\alpha - \beta)$.

3. Thus, substituting in formula (6) gives,

$$\tan\theta = \frac{m_1 - m_2}{1 + m_1 m_2} \tag{7}$$

EXAMPLE 2. Find the angle between the lines $y = 3x$ and $y = 5 - 2x$.

SOLUTION: We first note that we can call $m_1 = 3$ and $m_2 = -2$, or $m_1 = -2$ and $m_2 = 3$. These two possibilities give us the two supplementary angles, θ_1 and θ_2, between the lines.

(1) Letting $m_1 = -2$ and $m_2 = 3$, and using formula (7), gives

$$\tan\theta_1 = \frac{-2 - 3}{1 + (-2)(3)} = \frac{-5}{-5} = 1.$$

Thus $\theta_1 = 45°$.

(2) Letting $m_1 = 3$ and $m_2 = -2$ gives

$$\tan\theta_2 = \frac{3 - (-2)}{1 + (3)(-2)} = \frac{5}{-5} = -1.$$

Thus $\theta_2 = 135°$.

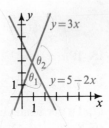

ORAL EXERCISES 9-2

1. If $\tan\alpha = 2$ and $\tan\beta = 3$, find **(a)** $\tan(\alpha + \beta)$ and **(b)** $\tan(\alpha - \beta)$.

2. Simplify: **a.** $\dfrac{\tan 15° + \tan 30°}{1 - \tan 15°\tan 30°}$ **b.** $\dfrac{\tan 150° - \tan 15°}{1 + \tan 150°\tan 15°}$

3. Interpret formula (7) when $1 + m_1 m_2 = 0$.

A   **1.** If $\tan \alpha = \frac{2}{3}$ and $\tan \beta = \frac{1}{2}$, find **(a)** $\tan (\alpha + \beta)$ and **(b)** $\tan (\alpha - \beta)$.

2. If $\tan \alpha = 2$ and $\tan \beta = -\frac{1}{3}$, find **(a)** $\tan (\alpha + \beta)$ and **(b)** $\tan (\alpha - \beta)$.

In Exercises 3–8, evaluate the given expression.

3. $\dfrac{\tan 75° - \tan 30°}{1 + \tan 75° \tan 30°}$

4. $\dfrac{\tan 100° + \tan 20°}{1 - \tan 100° \tan 20°}$

5. $\dfrac{\tan \dfrac{2\pi}{3} + \tan \dfrac{\pi}{12}}{1 - \tan \dfrac{2\pi}{3} \tan \dfrac{\pi}{12}}$

6. $\dfrac{\tan \dfrac{4\pi}{3} - \tan \dfrac{\pi}{12}}{1 + \tan \dfrac{4\pi}{3} \tan \dfrac{\pi}{12}}$

7. $\dfrac{\tan \left(\dfrac{\pi}{3} - \theta\right) + \tan \theta}{1 - \tan \left(\dfrac{\pi}{3} - \theta\right) \tan \theta}$

8. $\dfrac{\tan \left(\dfrac{\pi}{8} - \theta\right) + \tan \left(\dfrac{\pi}{8} + \theta\right)}{1 - \tan \left(\dfrac{\pi}{8} - \theta\right) \tan \left(\dfrac{\pi}{8} + \theta\right)}$

9. Evaluate $\tan \left(\dfrac{\pi}{4} + \theta\right)$ if $\tan \theta = \dfrac{1}{2}$.

10. Evaluate $\tan \left(\dfrac{3\pi}{4} - \theta\right)$ if $\tan \theta = \dfrac{1}{3}$.

11. Evaluate without tables: $\tan 75°$ and $\cot 75°$.

12. Evaluate without tables: $\tan 165°$ and $\cot 165°$.

B   **13.** If $0 < \alpha < \beta < \dfrac{\pi}{2}$ and $\tan \alpha = \dfrac{1}{4}$ and $\tan \beta = \dfrac{3}{5}$, find $\tan (\alpha + \beta)$. Then

show that $\mathrm{Tan}^{-1} \dfrac{1}{4} + \mathrm{Tan}^{-1} \dfrac{3}{5} = \dfrac{\pi}{4}$.

14. If $0 < \beta < \alpha < \dfrac{\pi}{2}$ and $\tan \alpha = 3$ and $\tan \beta = \dfrac{1}{2}$, find $\tan (\alpha - \beta)$. Then

show that $\mathrm{Tan}^{-1} 3 - \mathrm{Tan}^{-1} \dfrac{1}{2} = \dfrac{\pi}{4}$.

15. If $\alpha = \mathrm{Tan}^{-1} 2$ and $\beta = \mathrm{Tan}^{-1} 3$, show that $\tan (\alpha + \beta) = -1$.

16. If $\alpha = \mathrm{Tan}^{-1} 5$ and $\beta = \mathrm{Tan}^{-1} \dfrac{2}{3}$, show that $\alpha - \beta = \dfrac{\pi}{4}$.

17. Under what conditions does $\tan (\alpha + \beta) = \tan \alpha + \tan \beta$?

18. Under what conditions does $\tan (\alpha - \beta) = \tan \alpha - \tan \beta$?

19. Find the two supplementary angles formed by the lines $y = 3x - 5$ and $y = x + 4$.

20. Find the two supplementary angles formed by the lines $3x + 2y = 5$ and $4x - 3y = 1$.

21. Given $A = (0, 0)$, $B = (2, 3)$, and $C = (1, -5)$, find the measure of $\angle BAC$ by using the two different methods outlined in parts (a) and (b) below.
   a. Find the slopes of lines AB and AC and use formula (7).
   b. Use the Law of Cosines to find $\cos \angle BAC$.

22. Given $A = (3, 1)$, $B = (14, -1)$, and $C = (5, 5)$, find the measure of $\angle BAC$. (See Exercise 21.)

23. If $\cot \alpha = 2$ and $\cot \beta = \frac{2}{3}$, find $\cot (\alpha + \beta)$.

24. If $\cot \alpha = \frac{3}{2}$ and $\cot \beta = \frac{1}{2}$, find $\cot (\alpha - \beta)$.

25. If $0 < \alpha < \beta < \frac{\pi}{2}$, and $\sin \alpha = \frac{3}{5}$ and $\cos \beta = \frac{5}{13}$, find without tables:

    a. $\sin (\alpha + \beta)$ **b.** $\cos (\alpha + \beta)$ **c.** $\tan (\alpha + \beta)$

26. If $\frac{\pi}{2} < \alpha < \beta < \pi$, and $\sin \alpha = \frac{4}{5}$ and $\tan \beta = -\frac{3}{4}$, find without tables:

    a. $\sin (\alpha + \beta)$ **b.** $\cos (\alpha + \beta)$ **c.** $\tan (\alpha + \beta)$

27. Derive a formula for $\tan 2x$ in terms of $\tan x$.

C   28. **a.** Line l bisects the indicated angle formed by lines l_1 and l_2. If the slopes of these three lines are m, m_1, and m_2, respectively, show that

$$\frac{m_1 - m}{1 + m_1 m} = \frac{m - m_2}{1 + m_2 m}.$$

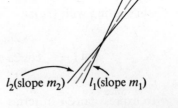

    b. If l_1 and l_2 have equations $y = 2x$ and $y = x$, find an equation of l.

29. Prove that in any $\triangle ABC$, $\tan A + \tan B + \tan C = \tan A \tan B \tan C$.

30. Verify that $\frac{\pi}{4} = 4 \operatorname{Tan}^{-1} \frac{1}{5} - \operatorname{Tan}^{-1} \frac{1}{239}$. (*Hint:* Let $\alpha = \operatorname{Tan}^{-1} \frac{1}{5}$ and $\beta = \operatorname{Tan}^{-1} \frac{1}{239}$, and find $\tan (4\alpha - \beta)$. In your work you will need to know that if $\tan \alpha = \frac{1}{5}$, then $\tan 2\alpha = \underline{\ ?\ }$ and $\tan 4\alpha = \underline{\ ?\ }$.)

CALCULATOR EXERCISE

Given $A(4, 7)$, $B(9, 1)$ and $C(-3, -5)$. Find $\angle BAC$ by using each of the methods in Exercise 21.

9–3 / DOUBLE-ANGLE AND HALF-ANGLE FORMULAS

If you know the value of $\sin \alpha$, you do *not* double it to find $\sin 2\alpha$. Nor do you halve it to find $\sin \frac{1}{2}\alpha$. Our goal in this section is to derive formulas that correctly express the sine, cosine, and tangent of 2α and $\frac{1}{2}\alpha$ in terms of functions of α. These formulas are called **double-angle** and **half-angle** formulas.

Double-angle formulas

The following double-angle formulas are merely special cases of the formulas for $\sin (\alpha + \beta)$, $\cos (\alpha + \beta)$, and $\tan (\alpha + \beta)$. If you set $\beta = \alpha$ in each of these, you obtain the corresponding double-angle formula.

Since $\sin(\alpha + \beta) = \sin \alpha \cos \beta + \cos \alpha \sin \beta$:

$$\sin(\alpha + \alpha) = \sin \alpha \cos \alpha + \cos \alpha \sin \alpha$$
$$\sin 2\alpha = 2 \sin \alpha \cos \alpha \qquad (8)$$

Since $\cos(\alpha + \beta) = \cos \alpha \cos \beta - \sin \alpha \sin \beta$:

$$\cos(\alpha + \alpha) = \cos \alpha \cos \alpha - \sin \alpha \sin \alpha$$
$$\cos 2\alpha = \cos^2 \alpha - \sin^2 \alpha \qquad (9a)$$

Using the fact that $\sin^2 \alpha + \cos^2 \alpha = 1$, we can transform (9a) into alternative formulas for $\cos 2\alpha$. Starting with (9a):

$$\cos 2\alpha = \cos^2 \alpha - \sin^2 \alpha$$
$$\cos 2\alpha = (1 - \sin^2 \alpha) - \sin^2 \alpha$$
$$\cos 2\alpha = 1 - 2 \sin^2 \alpha \qquad (9b)$$

Again starting with (9a):

$$\cos 2\alpha = \cos^2 \alpha - \sin^2 \alpha$$
$$\cos 2\alpha = \cos^2 \alpha - (1 - \cos^2 \alpha)$$
$$\cos 2\alpha = 2 \cos^2 \alpha - 1 \qquad (9c)$$

To express $\tan 2\alpha$ in terms of $\tan \alpha$, we proceed as follows:

$$\tan(\alpha + \beta) = \frac{\tan \alpha + \tan \beta}{1 - \tan \alpha \tan \beta}$$
$$\tan(\alpha + \alpha) = \frac{\tan \alpha + \tan \alpha}{1 - \tan \alpha \tan \alpha}$$
$$\tan 2\alpha = \frac{2 \tan \alpha}{1 - \tan^2 \alpha} \qquad (10)$$

EXAMPLE 1. If $\sin \alpha = \dfrac{4}{5}$ and $0 < \alpha < \dfrac{\pi}{2}$, find $\sin 2\alpha$, $\cos 2\alpha$, and $\tan 2\alpha$.

SOLUTION: From the given information, we know that $\cos \alpha = \frac{3}{5}$ and $\tan \alpha = \frac{4}{3}$. (See diagram.)

$$\sin 2\alpha = 2 \sin \alpha \cos \alpha$$
$$= 2(\tfrac{4}{5})(\tfrac{3}{5}) = \tfrac{24}{25}$$
$$\cos 2\alpha = \cos^2 \alpha - \sin^2 \alpha$$
$$= (\tfrac{3}{5})^2 - (\tfrac{4}{5})^2 = -\tfrac{7}{25}$$
$$\tan 2\alpha = \frac{2 \tan \alpha}{1 - \tan^2 \alpha}$$
$$= \frac{2(\tfrac{4}{3})}{1 - (\tfrac{4}{3})^2} = (\tfrac{8}{3})(-\tfrac{9}{7}) = -\tfrac{24}{7}$$

You can check this result by noting that $\tan 2\alpha = \dfrac{\sin 2\alpha}{\cos 2\alpha} = -\dfrac{24}{7}$.

EXAMPLE 2. Derive a formula for $\sin 4x$ in terms of functions of x.

SOLUTION: We use formula (8), replacing α with $2x$.

$$\sin 2\alpha = 2 \sin \alpha \cos \alpha$$

$$\sin 4x = 2 \underline{\sin 2x} \underline{\cos 2x}$$

$$= 2(2 \sin x \cos x)(\cos^2 x - \sin^2 x)$$

$$= 4 \sin x \cos^3 x - 4 \sin^3 x \cos x$$

A different form of $\cos 2x$ will yield different but equivalent results.

Half-angle formulas

To obtain the sine and cosine half-angle formulas, we use formulas 9(b) and 9(c), replacing α with $\dfrac{x}{2}$.

$$\cos 2\alpha = 1 - 2 \sin^2 \alpha$$

$$\cos 2\left(\frac{x}{2}\right) = 1 - 2 \sin^2 \left(\frac{x}{2}\right)$$

$$\cos x = 1 - 2 \sin^2 \left(\frac{x}{2}\right)$$

$$2 \sin^2 \left(\frac{x}{2}\right) = 1 - \cos x$$

$$\sin \frac{x}{2} = \pm \sqrt{\frac{1 - \cos x}{2}} \quad (11)$$

$$\cos 2\alpha = 2 \cos^2 \alpha - 1$$

$$\cos 2\left(\frac{x}{2}\right) = 2 \cos^2 \left(\frac{x}{2}\right) - 1$$

$$\cos x = 2 \cos^2 \left(\frac{x}{2}\right) - 1$$

$$2 \cos^2 \left(\frac{x}{2}\right) = 1 + \cos x$$

$$\cos \frac{x}{2} = \pm \sqrt{\frac{1 + \cos x}{2}} \quad (12)$$

When you use the sine and cosine half-angle formulas, you determine whether to use the plus sign or the minus sign by noting in which quadrant $\dfrac{x}{2}$ lies. The following example illustrates this.

EXAMPLE 3. Find **(a)** $\sin 105°$ and **(b)** $\cos \dfrac{5\pi}{8}$ without using tables.

SOLUTION: **a.** Since $105° = \dfrac{1}{2}(210°)$, we can use formula (11) with $x = 210°$.

Moreover, because $105°$ is in the second quadrant, $\sin 105°$ is positive, and we choose the plus sign in formula (11).

$$\sin 105° = \sin \frac{210°}{2} = +\sqrt{\frac{1 - \cos 210°}{2}} = \sqrt{\frac{1 - \left(-\dfrac{\sqrt{3}}{2}\right)}{2}}$$

$$= \sqrt{\frac{2 + \sqrt{3}}{4}} = \frac{\sqrt{2 + \sqrt{3}}}{2}$$

(*Solution continued on page 322*)

Trigonometric addition formulas   **321**

In Example 3 on page 312, $\sin 105° = \dfrac{\sqrt{6} + \sqrt{2}}{4}$, and in this example, $\sin 105° = \dfrac{\sqrt{2 + \sqrt{3}}}{2}$. You might like to reconcile these answers by squaring both numbers.

b. Since $\dfrac{5\pi}{8} = \dfrac{1}{2}\left(\dfrac{5\pi}{4}\right)$, we can use formula (12) with $x = \dfrac{5\pi}{4}$. Because $\dfrac{5\pi}{8}$ is in the second quadrant, $\cos \dfrac{5\pi}{8}$ is negative, and we choose the minus sign in the formula.

$$\cos \frac{5\pi}{8} = \cos \frac{1}{2}\left(\frac{5\pi}{4}\right)$$

$$= -\sqrt{\frac{1 + \cos \dfrac{5\pi}{4}}{2}}$$

$$= -\sqrt{\frac{1 + \left(-\dfrac{\sqrt{2}}{2}\right)}{2}}$$

$$= -\sqrt{\frac{2 - \sqrt{2}}{4}} = -\frac{\sqrt{2 - \sqrt{2}}}{2}$$

To derive a formula for $\tan \dfrac{x}{2}$, we can divide equation (11) by equation (12), obtaining

$$\tan \frac{x}{2} = \pm\sqrt{\frac{1 - \cos x}{1 + \cos x}}.$$

However, the following formulas are more useful:

$$\tan \frac{x}{2} = \frac{\sin x}{1 + \cos x} = \frac{1 - \cos x}{\sin x} \tag{13}$$

These formulas can be verified by replacing each x in "$\sin x$" and "$\cos x$" by $2\left(\dfrac{x}{2}\right)$ and using the double-angle formulas. (See Exercises 43 and 44, page 324.)

ORAL EXERCISES 9-3

In Exercises 1–10, simplify the given expression.

1. $2 \sin 10° \cos 10°$

2. $\cos^2 15° - \sin^2 15°$

3. $1 - 2 \sin^2 35°$

4. $2 \cos^2 25° - 1$

5. $2 \sin 3x \cos 3x$

6. $\cos^2 5x - \sin^2 5x$

7. $\dfrac{2 \tan 50°}{1 - \tan^2 50°}$ **8.** $\dfrac{2 \tan 40°}{1 - \tan^2 40°}$

9. $1 - \sin^2 x$ **10.** $1 - 2 \sin^2 x$

11. If $\cos 70° \approx 0.342$, explain how you can find $\cos 35°$ without tables.

12. If $\sin A = \dfrac{4}{5}$, $0° < A < 90°$, find **(a)** $\cos A$, **(b)** $\sin 2A$, **(c)** $\sin \dfrac{A}{2}$.

WRITTEN EXERCISES 9–3

In Exercises 1–4, $\angle A$ is an acute angle. Use the given information to find $\sin 2A$, $\cos 2A$, and $\tan 2A$.

A **1.** $\sin A = \frac{5}{13}$ **2.** $\cos A = \frac{4}{5}$ **3.** $\cos A = \frac{24}{25}$ **4.** $\tan A = \frac{1}{2}$

In Exercises 5–8, $\angle A$ is an acute angle.

5. If $\sin A = \frac{3}{5}$, find $\sin 2A$ and $\sin 4A$.

6. If $\cos A = \frac{1}{3}$, find $\cos 2A$ and $\cos 4A$.

7. If $\cos A = \dfrac{1}{5}$, find **(a)** $\cos 2A$, **(b)** $\cos \dfrac{A}{2}$.

8. If $\cos A = \dfrac{1}{4}$, find **(a)** $\sin 2A$, **(b)** $\sin \dfrac{A}{2}$.

In Exercises 9–20, simplify the given expression.

9. $2 \cos^2 10° - 1$ **10.** $2 \sin \dfrac{x}{2} \cos \dfrac{x}{2}$

11. $1 - 2 \sin^2 20°$ **12.** $\dfrac{4 \tan x}{1 - \tan^2 x}$

13. $2 \sin 35° \cos 35°$ **14.** $\cos^2 4A - \sin^2 4A$

15. $\dfrac{2 \tan 25°}{1 - \tan^2 25°}$ **16.** $2 \cos^2 3x - 1$

17. $\cos^2 40° - \sin^2 40°$ **18.** $1 - 2 \sin^2 \dfrac{x}{2}$

19. $\sqrt{\dfrac{1 - \cos 80°}{2}}$ **20.** $\sqrt{\dfrac{1 + \cos 70°}{2}}$

In Exercises 21–26, evaluate the given expression.

21. $2 \cos^2 \dfrac{\pi}{8} - 1$ **22.** $\dfrac{2 \tan \dfrac{\pi}{8}}{1 - \tan^2 \dfrac{\pi}{8}}$

23. $\sin 15° \cos 15°$ **24.** $1 - 2 \sin^2 \dfrac{\pi}{12}$

25. $\cos^2 \dfrac{\pi}{12} - \sin^2 \dfrac{\pi}{12}$ **26.** $4 \sin \dfrac{\pi}{8} \cos \dfrac{\pi}{8}$

Trigonometric addition formulas   **323**

27. Find $\cos 105°$ using **(a)** the $\cos(\alpha + \beta)$ formula and **(b)** the $\cos \dfrac{x}{2}$ formula.

28. Find $\sin 75°$ using **(a)** the $\sin(\alpha + \beta)$ formula and **(b)** the $\sin \dfrac{x}{2}$ formula.

B   **29.** Evaluate $\sin 2(\operatorname{Cos}^{-1} \tfrac{3}{5})$. (*Hint:* Let $\alpha = \operatorname{Cos}^{-1} \tfrac{3}{5}$.)

      30. Evaluate: **a.** $\cos 2(\operatorname{Sin}^{-1} \tfrac{5}{13})$ **b.** $\tan 2(\operatorname{Tan}^{-1} \tfrac{1}{3})$

In Exercises 31–44, prove the given identity.

31. $\dfrac{\sin 2A}{1 - \cos 2A} = \cot A$ **32.** $\dfrac{1 - \cos 2A}{1 + \cos 2A} = \tan^2 A$

33. $\left(\sin \dfrac{x}{2} + \cos \dfrac{x}{2}\right)^2 = 1 + \sin x$ **34.** $\tan x(1 + \cos 2x) = \sin 2x$

35. $\cot x - 2 \cot 2x = \tan x$ **36.** $\dfrac{1 - \tan^2 x}{1 + \tan^2 x} = \cos 2x$

37. $\sin 4x = 4 \sin x \cos x \cos 2x$ **38.** $\dfrac{1 + \sin A - \cos 2A}{\cos A + \sin 2A} = \tan A$

39. $\csc 2A + \cot 2A = \cot A$ **40.** $\dfrac{\tan 2A \tan A}{\tan 2A - \tan A} = \sin 2A$

41. $2 \cos^2 \left(\dfrac{\pi}{4} - \dfrac{x}{2}\right) - 1 = \sin x$

42. $2 \sin \left(\dfrac{\pi}{3} - \theta\right) \cos \left(\dfrac{\pi}{3} - \theta\right) = \sin \left(\dfrac{\pi}{3} + 2\theta\right)$

(*Hint for Exs. 43 and 44:* On the right side of each equation, replace x with $2\left(\dfrac{x}{2}\right)$ and then use the double-angle formulas.) Another proof of Exercise 43 is given in Exercise 45.

43. $\tan \dfrac{x}{2} = \dfrac{\sin x}{1 + \cos x}$ **44.** $\tan \dfrac{x}{2} = \dfrac{1 - \cos x}{\sin x}$

45. The diagram shows a unit circle, with the measure of $\angle BOC = \theta$.

     a. Explain why the measure of $\angle BAO = \dfrac{\theta}{2}$.

     b. Explain why $\tan \dfrac{\theta}{2} = \dfrac{\sin \theta}{1 + \cos \theta}$.

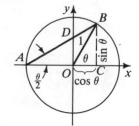

46. Use the rhombus and the steps below to derive the formula for $\sin 2\theta$.

     a. Show that the area of the rhombus is $a^2 \sin 2\theta$.

     b. Use the fact that the diagonals of a rhombus are perpendicular to show that its area is

$$2a^2 \sin \theta \cos \theta.$$

     c. Use your answers from parts (a) and (b) to get a formula for $\sin 2\theta$.

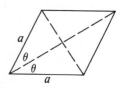

47. What is the maximum value of $4 \sin x \cos x$?

48. What is the minimum value of $\cos^2 x - \sin^2 x$?

Christine

Simplify.

49. $\dfrac{1 + \cos 2x}{\cot x}$

50. $(1 - \sin^2 x)(1 - \tan^2 x)$

51. $\dfrac{2 \tan x}{1 + \tan^2 x}$

52. $\csc 2x - \cot 2x$

53. $\dfrac{(1 + \tan^2 x)(1 - \cos 2x)}{2}$

54. $\sin x \tan x + \cos 2x \sec x$

55. $\dfrac{\sin 3x}{\sin x} - \dfrac{\cos 3x}{\cos x}$ (*Hint:* Combine terms over a common denominator.)

56. $\cos^2 \left(\dfrac{\pi}{4} - \dfrac{x}{2}\right) - \sin^2 \left(\dfrac{\pi}{4} - \dfrac{x}{2}\right)$

57. Evaluate $\log_2 (\sin x) + \log_2 (\cos 2x)$ when $x = \dfrac{\pi}{6}$.

58. Evaluate $\log 2 + \log \tan x - \log (1 - \tan^2 x)$ when $x = \dfrac{\pi}{8}$.

59. Evaluate $\dfrac{9^{\cos^2 \theta}}{9^{\sin^2 \theta}}$ when $\theta = \dfrac{\pi}{3}$.

60. Evaluate $\dfrac{4^{2 \cos^2 \theta}}{4}$ when $\theta = \dfrac{\pi}{3}$.

Prove each identity.

61. $\sin 3x = 3 \sin x - 4 \sin^3 x$

62. $\cos 3x = 4 \cos^3 x - 3 \cos x$

C

63. $\log (1 - \cos 2x) - \log (1 + \cos 2x) = 2 \log \tan x$

64. $\log \sec 2\theta = 2 \log \sec \theta - \log (2 - \sec^2 \theta)$

65. Given: $AB = AC = 1$, $\angle A = 36°$
$\overrightarrow{BD}$ bisects $\angle B$.

   a. Prove that $\triangle ABC \sim \triangle BCD$.

   b. Use similar triangles to show that $\dfrac{x}{1 - x} = \dfrac{1}{x}$.

   c. Show that $x = \dfrac{\sqrt{5} - 1}{2}$.

   d. Draw the bisector of $\angle A$ and show that $\sin 18° = \dfrac{\sqrt{5} - 1}{4} = \cos 72°$.

   e. Draw the perpendicular from B to $\overline{AC}$ and show that $\cos 36° = \dfrac{\sqrt{5} + 1}{4} = \sin 54°$.

66. In $\triangle ABC$, the measure of $\angle B$ is twice that of $\angle C$.
a. Use the Law of Sines to show that $b = 2c \cos C$.
b. Use the Law of Cosines to show that $b^2 = c(a + c)$.

67. The lengths of the sides of a triangle are consecutive integers n, $n + 1$, $n + 2$, and the largest angle is twice the smallest angle θ.

a. Use the Law of Sines to show that $\cos \theta = \dfrac{n + 2}{2n}$.

b. Use the Law of Cosines to show that $\cos \theta = \dfrac{n + 5}{2(n + 2)}$.

c. Use (a) and (b) to find n.

CALCULATOR EXERCISES

1. Given that $\sin 10° \approx 0.173648$, find **(a)** the sine and **(b)** the cosine of $20°$, $30°$, $40°$, $50°$, $60°$, $70°$, and $80°$ without using the sine or cosine keys.

2. Given that $\cos 12° \approx 0.978148$, find **(a)** the sine and **(b)** the cosine of $24°$, $36°$, $48°$, $72°$, $84°$, $96°$, $108°$, and $120°$ without using the sine or cosine keys.

Applying formulas

9-4/SOLVING EQUATIONS

In the last three sections, we have derived formulas for trigonometric functions of $\alpha \pm \beta$ and 2α. In this section, we shall use these formulas to solve equations. The following rules of thumb will be helpful in our solutions.

Guidelines for solving trigonometric equations

RULE 1. If the equation involves functions of x and $2x$, use the double-angle formulas to replace the functions of $2x$ by functions of x. (See Examples 1 and 2 that follow.)

RULE 2. If the equation involves only functions of $2x$, it is usually better to solve for $2x$ rather than using the double-angle formulas. (See Method 1 of Example 3.)

RULE 3. Be careful not to lose roots when you divide both sides of an equation by a function of the variable. [See Examples 1 and 2(a).]

EXAMPLE 1. Solve $\sin 2x = \sin x$ for $0 \leq x < 2\pi$.

SOLUTION:

$$\sin 2x = \sin x$$
$$2 \sin x \cos x = \sin x \longleftarrow$$

Warning: If you divide by $\sin x$ be sure to follow Rule 3.

$\sin x = 0$	$2 \cos x = 1$
$x = 0, \pi$	$\cos x = \frac{1}{2}$
	$x = \frac{\pi}{3}, \frac{5\pi}{3}$

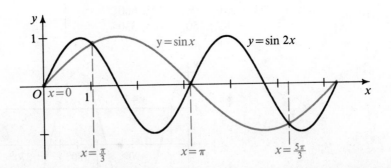

EXAMPLE 2. Solve each equation for $0 \le x < 2\pi$.

  a. $\cos 2x = 1 - \sin x$
  b. $3 \cos 2x + \cos x = 2$

SOLUTION: For equation (a), we use the formula $\cos 2x = 1 - 2 \sin^2 x$ to obtain a quadratic equation in $\sin x$. For equation (b), we use $\cos 2x = 2 \cos^2 x - 1$ to obtain a quadratic equation in $\cos x$.

a.
$$\cos 2x = 1 - \sin x$$
$$1 - 2 \sin^2 x = 1 - \sin x$$
$$2 \sin^2 x = \sin x$$

$\sin x = 0$	$2 \sin x = 1$
$x = 0, \pi$	$\sin x = \frac{1}{2}$
	$x = \frac{\pi}{6}, \frac{5\pi}{6}$

b.
$$3 \cos 2x + \cos x = 2$$
$$3(2 \cos^2 x - 1) + \cos x = 2$$
$$6 \cos^2 x + \cos x - 5 = 0$$
$$(6 \cos x - 5)(\cos x + 1) = 0$$

$\cos x = \frac{5}{6}$	$\cos x = -1$
$x \approx 0.59, 5.70$	$x = \pi$

EXAMPLE 3. Solve $2 \cos 2x = 1$ for $0° \le x < 360°$.

SOLUTION: Two methods of solution are shown. In Method 1, we solve for $\cos 2x$, and in Method 2 we solve for $\cos x$. Method 1 is recommended because it is somewhat easier and it can also be used to solve equations like $2 \cos 3x = 1$ or $2 \cos \frac{x}{4} = 1$.

(*Solution continued on page 328*)

Method 1	Method 2

Method 1

$$2 \cos 2x = 1$$
$$\cos 2x = \tfrac{1}{2}$$

Since $0° \leq x < 360°$, then $0° \leq 2x < 720°$. Thus we must find values of $2x$ between $0°$ and $720°$ for which $\cos 2x = \tfrac{1}{2}$:

$$2x = 60°, 300°, 420°, 660°$$
$$x = 30°, 150°, 210°, 330°$$

Method 2

$$2 \cos 2x = 1$$
$$2(2 \cos^2 x - 1) = 1$$
$$4 \cos^2 x - 2 = 1$$
$$4 \cos^2 x = 3$$
$$\cos x = \pm \frac{\sqrt{3}}{2}$$
$$x = 30°, 150°, 210°, 330°$$

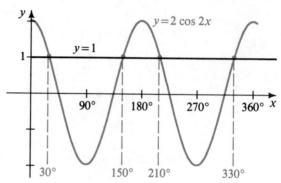

ORAL EXERCISES 9-4

1. When solving the equation $2 \sin 2x = 1$, would you use the formula for $\sin 2x$ or would you solve for $\sin 2x$?

2. When solving the equation $\cos 2x = \sin x$, which formula for $\cos 2x$ would you use?

Describe the method you would use to solve the following equations for $0° \leq x < 360°$.

3. $\cos 2x = \cos x$

4. $\sin^2 x = \sin x$

5. $\sin x = \cos x$

6. $\sin 2x = \cos 2x$

7. $\sin 3x = \cos 3x$

8. $\tan (x - 10°) = 1$

9. When solving the equation $\sin 4x = \sin 2x$, would you solve for $\sin x$ or $\sin 2x$?

10. When solving the equation $\cos 4x = 1 - 3 \cos 2x$, would you solve for $\cos x$ or $\cos 2x$?

11. In Exercise 9, which double-angle formula would you use to simplify $\sin 4x$?

12. In Exercise 10, which double-angle formula would you use to simplify $\cos 4x$?

A **1-10.** Solve the equations given in Oral Exercises 1–10 for $0° \leq x < 360°$.

11. a. On the same set of axes, sketch the graphs of $y = \sin 2x$ and $y = \tan x$ for $0 \leq x < 2\pi$.

    b. Determine where the graphs intersect by solving $\sin 2x = \tan x$.

12. a. On the same set of axes, sketch the graphs of $y = \cos 2x$ and $y = \cos x$ for $0 \leq x < 2\pi$.

    b. Determine where the graphs intersect by solving $\cos 2x = \cos x$.

In Exercises 13–16, solve the given pair of equations for $0 \leq x < 2\pi$. Then sketch their graphs on the same set of axes and show their intersection points.

13. $y = \cos 2x$ **14.** $y = \cos x$
     $y = \sin x$ $y = \sin 2x$

15. $y = \cot x$ **16.** $y = \cot x$
     $y = 2 \sin 2x$ $y = \frac{1}{2} \sin 2x$

Solve for $0° \leq x < 360°$.

B **17.** $2 \cos (x + 45°) = 1$ **18.** $\cot (x - 20°) = 1$

19. $\sin (180° - 2x) = \sin (90° - x)$ **20.** $\cos (2x - 180°) - \sin (x - 90°) = 0$

21. $\sin (60° - x) = 2 \sin x$ **22.** $2 \sin (30° + x) = 3 \cos x$

23. $\sin 2x = \sqrt{2} \sin (270° - x)$ **24.** $\sin (x + 45°) + \cos (x - 45°) = \sqrt{3}$

25. $\tan (x + 45°) + \tan (x - 45°) = \frac{3}{2}$ **26.** $\tan (45° - x) + \cot (45° - x) = 4$

Solve for $0 \leq x < 2\pi$.

27. $\sin x \cos x = \frac{1}{2}$ **28.** $\cos (3x + \pi) = \dfrac{\sqrt{2}}{2}$

29. $\tan 2x = 3 \tan x$ **30.** $\tan 2x + \tan x = 0$

31. $\cos 2x = 5 \sin^2 x - \cos^2 x$ **32.** $\sin 2x \sec x + 2 \cos x = 0$

33. $\cos 2x = \sec x$ **34.** $\sin x \cos 2x = 1$

35. Sketch on the same axes the graphs of $y = \sin 2x$ and $y = \ln (x + 1)$. Tell how *many* solutions there are for the equation $\sin 2x = \ln (x + 1)$.

36. How *many* solutions are there for the equation $\sec 2x = 2^{-x}$?

Solve for $0 \leq x < 2\pi$.

C **37.** $\sin 3x = \sin 5x + \sin x$ [Use one of the formulas on page 312.]

38. $\cos 3x + \cos x = \cos 2x$ **39.** $\sin 3x - \sin x = 2 \cos 2x$

Solve.

40. $\text{Cos}^{-1} 2x = \text{Sin}^{-1} x$ **41.** $\text{Tan}^{-1} 2x = \text{Sin}^{-1} x$

42. $\text{Sin}^{-1} 2x = \text{Cos}^{-1} x$ **43.** $\text{Sin}^{-1} 2x = \text{Tan}^{-1} x$

44. $\text{Cos}^{-1} 2x = \text{Cot}^{-1} x$ **45.** $\text{Sin}^{-1} x = \text{Csc}^{-1} 2x$

Chapter summary

1. The six *trigonometric addition formulas* as derived in Sections 9–1 and 9–2 are as follows:

$$\sin(\alpha + \beta) = \sin\alpha\cos\beta + \cos\alpha\sin\beta$$
$$\sin(\alpha - \beta) = \sin\alpha\cos\beta - \cos\alpha\sin\beta$$
$$\cos(\alpha + \beta) = \cos\alpha\cos\beta - \sin\alpha\sin\beta$$
$$\cos(\alpha - \beta) = \cos\alpha\cos\beta + \sin\alpha\sin\beta$$
$$\tan(\alpha + \beta) = \frac{\tan\alpha + \tan\beta}{1 - \tan\alpha\tan\beta}$$
$$\tan(\alpha - \beta) = \frac{\tan\alpha - \tan\beta}{1 + \tan\alpha\tan\beta}$$

2. The *double-angle formulas* below are derived by setting $\beta = \alpha$ in the formulas for $\sin(\alpha + \beta)$, $\cos(\alpha + \beta)$, and $\tan(\alpha + \beta)$.

$$\sin 2\alpha = 2\sin\alpha\cos\alpha$$
$$\cos 2\alpha = \cos^2\alpha - \sin^2\alpha$$
$$\text{(or)} \quad = 1 - 2\sin^2\alpha$$
$$\text{(or)} \quad = 2\cos^2\alpha - 1$$
$$\tan 2\alpha = \frac{2\tan\alpha}{1 - \tan^2\alpha}$$

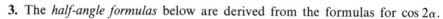

3. The *half-angle formulas* below are derived from the formulas for $\cos 2\alpha$.

$$\sin\frac{x}{2} = \pm\sqrt{\frac{1 - \cos x}{2}}$$
$$\cos\frac{x}{2} = \pm\sqrt{\frac{1 + \cos x}{2}}$$
$$\tan\frac{x}{2} = \frac{\sin x}{1 + \cos x} = \frac{1 - \cos x}{\sin x}$$

Mary Fairfax Somerville (1780–1872) was a self-taught Scottish mathematician who began her studies with Euclid's geometry and then went on to master a variety of mathematical subjects, including trigonometry, conic sections, and Newton's calculus. She wrote several books designed to make the current scientific theories, and the mathematical basis for them, more familiar to the general public.

4. (Optional) The formulas for converting trigonometric sums to products are given on page 312.

5. General guidelines for solving trigonometric equations are suggested in Rules 1–3 of Section 9–4.

6. The angle θ between lines l_1 and l_2 with slopes m_1 and m_2 is given by the formula

$$\tan \theta = \frac{m_1 - m_2}{1 + m_1 m_2}.$$

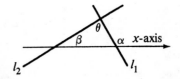

Chapter test

No tables or calculators are required for this test.

9–1

1. Simplify the following:
   a. $\cos(30° + x) + \cos(30° - x)$
   b. $\cos 75° \cos 15° + \sin 75° \sin 15°$

2. Evaluate $\cos 15°$ without using tables or a calculator.

3. Evaluate $\tan\left(\frac{5\pi}{4} - \theta\right)$ if $\tan \theta = -\frac{1}{3}$. 9–2

4. If $\tan \alpha = \frac{4}{3}$ and $\tan \beta = -\frac{1}{2}$, show that $\tan(\alpha + \beta) = \tan(\pi - \beta)$.

5. If $\angle A$ is acute and $\cos A = \frac{4}{5}$, evaluate (a) $\sin A$, (b) $\cos 2A$, (c) $\sin 2A$, 9–3
   (d) $\sin 4A$.

6. Simplify the following:

   a. $\dfrac{\sin 2x}{1 - \cos 2x}$ b. $(1 + \tan^2 y)(\cos 2y - 1)$

   c. $\dfrac{\tan t}{\sec t + 1}$ d. $\cos^2 \dfrac{x}{2} - \sin^2 \dfrac{x}{2}$ $= \cos x$

7. Evaluate the given expression.

   a. $2 \cos^2 \dfrac{\pi}{12} - 1$ b. $4 \sin \dfrac{\pi}{6} \cos \dfrac{\pi}{6}$

8. Prove the identity.

   a. $(1 + \cot^2 x)(1 - \cos 2x) = 2$ b. $\dfrac{\sin \theta \sec \theta}{\tan \theta + \cot \theta} = \cos^2 \theta - \cos 2\theta$

9. a. On the same set of axes, sketch the graphs of $y = \cos 2\theta$ and $y = \sin \theta$ 9–4
      for $0° \le \theta < 360°$.
   b. Determine where the graphs intersect by solving $\cos 2\theta = \sin \theta$.

10. a. Sketch the graph of $y = \cos 2x$ for $0 \le x < 2\pi$.
    b. For which of these values of x does $\cos 2x = \frac{1}{2}$?

Trigonometric addition formulas   **331**

CHAPTER TEN

Polar coordinates and complex numbers

OBJECTIVES

1. To convert from polar coordinates to rectangular coordinates and vice versa.
2. To graph equations using polar coordinates.
3. To represent complex numbers in the complex plane.
4. To find powers of complex numbers.

10-1/POLAR COORDINATES AND GRAPHS

The position of a point P in the plane can be described by giving its distance r from a point O, called the **pole**, and the measure of an angle formed by ray OP and a reference ray, called the **polar axis**. Although the choice of the pole and

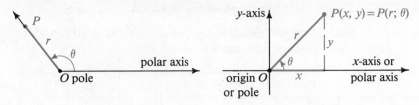

polar axis is arbitrary, it is customary to let the pole coincide with the origin in a rectangular coordinate system and to let the polar axis coincide with the

Spiral shapes, such as that seen in the X-ray of the seashell shown at the left, abound in nature. Equations for spirals are most easily written using polar coordinates.

positive x-axis, as shown on page 333. It is also customary to measure angles counterclockwise from the x-axis. (In navigational applications, however, angles are usually measured toward the east from a ray directed to the north.)

The numbers r and θ are called the **polar coordinates** of P, and we write $P = (r; \theta)$. (See the diagram at the right on the previous page.) The semicolon is used here to avoid confusion with the usual rectangular coordinates. The polar coordinates r and θ are called, respectively, the **polar distance** and the **polar angle** of point P; θ can be measured in degrees or radians. See figures (a)–(c). Figures (b) and (c) show that the same point can be represented by more than one pair of polar coordinates.

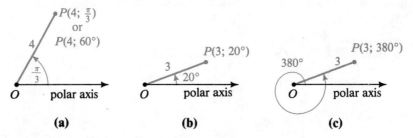

(a)   **(b)**   **(c)**

When an equation of a curve is given in terms of r and θ, it is called a **polar equation.** When drawing graphs of polar equations, we allow a negative value of r to represent the negative of the distance r. When $r < 0$, then, the point $(r; \theta)$ is plotted as $(|r|; \theta + \pi)$. This is shown below.

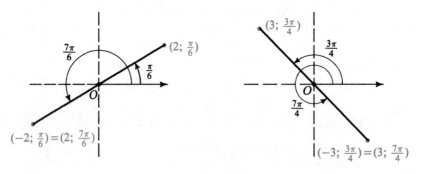

EXAMPLE 1. Sketch the polar graph of the equation $r = 2 \cos \theta$.

SOLUTION: Since the cosine function has period 360°, we need consider only values of θ between 0° and 360°.

θ	0°	30°	45°	60°	90°	120°	135°	150°	180°
$r = 2 \cos \theta$	2	$\sqrt{3}$	$\sqrt{2}$	1	0	-1	$-\sqrt{2}$	$-\sqrt{3}$	-2

θ	180°	210°	225°	240°	270°	300°	315°	330°	360°
$r = 2 \cos \theta$	-2	$-\sqrt{3}$	$-\sqrt{2}$	-1	0	1	$\sqrt{2}$	$\sqrt{3}$	2

At the top of the next page the complete graph is given for

values of θ between $0°$ and $180°$. For values of θ between $180°$ and $360°$, the same graph is traced again.

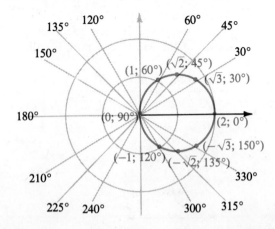

The relationship between polar and rectangular coordinates is illustrated by the diagram and given in the equations below.

(1) $x = r \cos \theta$

(2) $y = r \sin \theta$

From these equations, we also have:

(3) $r = \pm \sqrt{x^2 + y^2}$

(4) $\tan \theta = \dfrac{y}{x}$

(5) $\cos \theta = \dfrac{x}{r} = \dfrac{x}{\pm \sqrt{x^2 + y^2}}$

(6) $\sin \theta = \dfrac{y}{r} = \dfrac{y}{\pm \sqrt{x^2 + y^2}}$

These equations can be used to transform polar equations into rectangular equations, and vice versa.

EXAMPLE 2. Example 1 shows the graph of the polar equation $r = 2 \cos \theta$. What is the rectangular equation of this curve?

SOLUTION: Use equations (3) and (5).

$$r = 2 \cos \theta$$

$$\pm \sqrt{x^2 + y^2} = 2\left(\frac{x}{\pm \sqrt{x^2 + y^2}} \right)$$

$$x^2 + y^2 = 2x$$

The last rectangular equation can be rewritten equivalently as

$$(x - 1)^2 + y^2 = 1$$

to confirm that the graph is a circle with center $(1, 0)$ and radius 1.

Polar coordinates and complex numbers **335**

EXAMPLE 3. Sketch the polar graph of the equation $r = 2 \sin 2\theta$ and find a rectangular equation of the graph.

SOLUTION: We make a table of values for $0° \leq \theta \leq 360°$ and then sketch the portion of the curve in each quadrant, as follows. The completed graph, shown in the fourth diagram, is called a four-leaved rose.

θ	0°	30°	45°	60°	90°
$r = 2 \sin 2\theta$	0	$\sqrt{3}$	2	$\sqrt{3}$	0

θ	120°	135°	150°	180°
r	$-\sqrt{3}$	-2	$-\sqrt{3}$	0

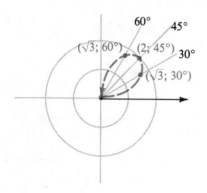

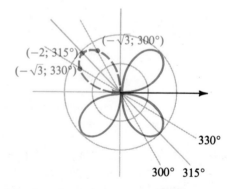

θ	210°	225°	240°	270°
r	$\sqrt{3}$	2	$\sqrt{3}$	0

θ	300°	315°	330°	360°
r	$-\sqrt{3}$	-2	$-\sqrt{3}$	0

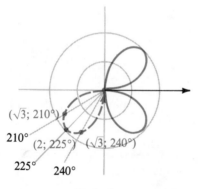

To find a rectangular equation of the four-leaved rose, we use equations (3), (6), and (5) on page 335:

$$r = 2 \sin 2\theta$$
$$r = 4 \sin \theta \cdot \cos \theta$$
$$\pm\sqrt{x^2 + y^2} = 4\left(\frac{y}{\pm\sqrt{x^2 + y^2}}\right)\left(\frac{x}{\pm\sqrt{x^2 + y^2}}\right)$$
$$\pm(x^2 + y^2)^{\frac{3}{2}} = 4yx$$
$$(x^2 + y^2)^3 = 16y^2x^2$$

Example 3 illustrates that the polar equation of a curve can be much simpler than its rectangular equation. The polar equations of the following curves are all rather simple, but the corresponding rectangular equations are extremely complicated.

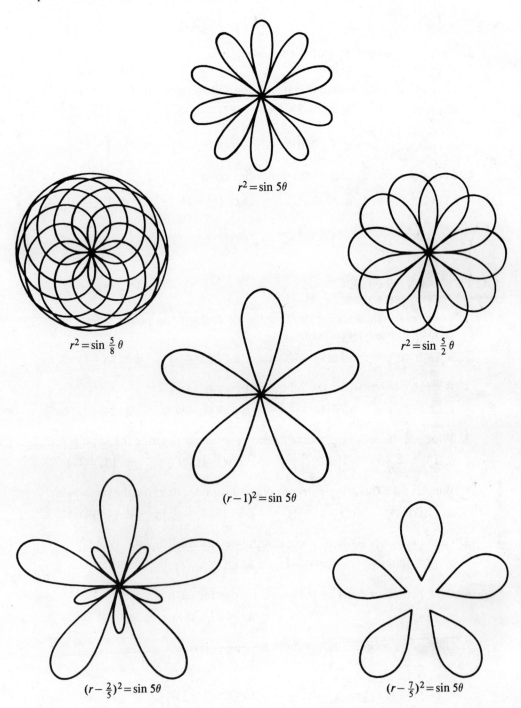

$r^2 = \sin 5\theta$

$r^2 = \sin \frac{5}{8} \theta$

$r^2 = \sin \frac{5}{2} \theta$

$(r-1)^2 = \sin 5\theta$

$(r - \frac{2}{5})^2 = \sin 5\theta$

$(r - \frac{7}{5})^2 = \sin 5\theta$

1. Give the rectangular coordinates of the following points.
   a. $(6; 90°)$ **b.** $(3; -90°)$ **c.** $(2; 540°)$ **d.** $(-4; 0°)$
   e. $\left(-1; \dfrac{\pi}{2}\right)$ **f.** $(5; \pi)$ **g.** $\left(3; -\dfrac{3\pi}{2}\right)$ **h.** $\left(\dfrac{1}{2}; 4\pi\right)$

2. Give polar coordinates of the following points.
   a. $(3, 0)$ **b.** $(0, -8)$ **c.** $(-4, 0)$ **d.** $(0, 1)$

3. Give the rectangular coordinates of the following points.
   a. $(8; 45°)$ **b.** $(3; -270°)$ **c.** $\left(-4; \dfrac{3\pi}{4}\right)$ **d.** $\left(2; -\dfrac{\pi}{4}\right)$
   e. $(-2; 60°)$ **f.** $(1; 390°)$ **g.** $(10; -\pi)$ **h.** $\left(2; \dfrac{5\pi}{6}\right)$

4. Give polar coordinates of the following points.
   a. $(3, 3)$ **b.** $(2, 2\sqrt{3})$ **c.** $(1, -1)$ **d.** $\left(\dfrac{\sqrt{3}}{2}, -\dfrac{1}{2}\right)$

WRITTEN EXERCISES 10-1

A   1. Make a sketch showing the location of each of the points whose polar coordinates are given below.
   a. $A(2; 40°)$ **b.** $B(3; -100°)$ **c.** $C\left(5; \dfrac{3\pi}{2}\right)$ **d.** $D(-4; 10°)$

2. Repeat Exercise 1 for the following points.
   a. $A(1; 80°)$ **b.** $B\left(6; -\dfrac{\pi}{6}\right)$ **c.** $C(2; 490°)$ **d.** $D(-3; 110°)$

3. Which of the following pairs of polar coordinates describe the same point?
   a. $\left(3; -\dfrac{\pi}{4}\right)$ **b.** $\left(-3; \dfrac{\pi}{4}\right)$ **c.** $(3; 315°)$ **d.** $\left(3; \dfrac{15\pi}{4}\right)$

4. Which of the following pairs of polar coordinates describe the same point?
   a. $(4; 270°)$ **b.** $(-4; 90°)$ **c.** $\left(-4; -\dfrac{\pi}{2}\right)$ **d.** $(4; 990°)$

5. Give two other pairs of polar coordinates for each point.
   a. $P(3; 180°)$ **b.** $Q\left(2; -\dfrac{\pi}{3}\right)$

6. Give two other pairs of polar coordinates for each point.
   a. $P\left(4; \dfrac{3\pi}{4}\right)$ **b.** $Q(1; -90°)$

7. Give polar coordinates for the following points.
   a. $(-2, 2)$ **b.** $(5, 0)$
   c. $(\sqrt{2}, -\sqrt{2})$ **d.** $(-\sqrt{3}, 1)$
   e. $(-4, -4)$ **f.** $(3, -3\sqrt{3})$

8. Repeat Exercise 7 for the following points.

a. $(-1, -1)$ **b.** $(0, 12)$ **c.** $\left(\dfrac{1}{2}, \dfrac{\sqrt{3}}{2}\right)$

d. $(-2, 0)$ **e.** $\left(\dfrac{\sqrt{2}}{2}, \dfrac{\sqrt{2}}{2}\right)$ **f.** $(-\sqrt{3}, 3)$

9. Give the rectangular coordinates of the following points.

a. $(4; 120°)$ **b.** $(-3; 90°)$ **c.** $\left(1; \dfrac{5\pi}{6}\right)$ **d.** $\left(2; \dfrac{3\pi}{4}\right)$

10. Repeat Exercise 9 for the following points.

a. $(2; 225°)$ **b.** $(6; -30°)$ **c.** $\left(10; -\dfrac{3\pi}{2}\right)$ **d.** $\left(-4; \dfrac{\pi}{3}\right)$

11. Use a calculator or Table 3 to find the rectangular coordinates of the following points to three decimal places.
     a. $(1; 20°)$ **b.** $(2; 20°)$ **c.** $(1; 200°)$ **d.** $(1; 2)$

12. Repeat Exercise 11 for the following points.
     a. $(1; 65°)$ **b.** $(4; 65°)$ **c.** $(1; 350°)$ **d.** $(1; 3)$

13. Use a calculator or Table 4 to find polar coordinates for the following points. Give θ to the nearest tenth of a degree.
     a. $(3, 4)$ **b.** $(1, 2)$ **c.** $(-2, 3)$

14. Repeat Exercise 13 for the following points.
     a. $(5, 2)$ **b.** $(8, -6)$ **c.** $(-1, -4)$

In Exercises 15–22, sketch the polar graph of the given polar equation. Also, give a rectangular equation of each graph.

15. $r = \sin \theta$ **16.** $r = \cos \theta$

17. $r = 1 - \sin \theta$ (cardioid)   **18.** $r = 1 + \cos \theta$ (cardioid)

19. $r = 1 + 2 \sin \theta$ (limaçon)   **20.** $r = 1 - 2 \cos \theta$ (limaçon)

21. $r = \sec \theta$ **22.** $r = 2 \csc \theta$

23. a. The rectangular equations $x = 4$ and $y = 2$ have very simple graphs. Sketch them.

     b. Do the polar equations $r = 4$ and $\theta = 2$ (radians) have simple graphs? Sketch them.

24. a. Show that the line with rectangular equation $x = 3$ has polar equation $r = 3 \sec \theta$.

     b. Find a polar equation of the line with rectangular equation $y = 3$.

In Exercises 25–34, sketch the polar graph of the given polar equation. Use radian measure for θ.

B   **25.** $r = \cos 2\theta$ (4-leaved rose)   **26.** $r = \cos 4\theta$ (8-leaved rose)

     27. $r = \sin 3\theta$ (3-leaved rose)   **28.** $r = \sin 5\theta$ (5-leaved rose)

     29. $r = 2\theta$ (spiral of Archimedes)   **30.** $r = \dfrac{2}{\theta}$ (hyperbolic spiral)

     31. $r = e^{\theta}$ (logarithmic spiral)   **32.** $r^2\theta = 1$ (lituus)
        (*Hint* for Ex. 31: Use a calculator or Table 6 on page 644.)

33. $r^2 = 4 \sin 2\theta$ (lemniscate)   **34.** $r = \cos \frac{1}{2}\theta$

35. Match each statement in column (A) with the correct statement in column (B) about the graph of $r = f(\theta)$.

(A)	(B)

a. $f(-\theta) = f(\theta)$ **a.** Graph has symmetry in the pole.

b. $f\left(\dfrac{\pi}{2} + \theta\right) = f\left(\dfrac{\pi}{2} - \theta\right)$ **b.** Graph has a horizontal axis of symmetry.

c. $f(\pi + \theta) = f(\theta)$ **c.** Graph has a vertical axis of symmetry.

C **36. a.** Determine a common point of the graphs of $r = 1 + \sin\theta$ and $r = 2 \sin\theta$ by solving the equations simultaneously.

    b. Now sketch the two graphs and note that they have another common point whose polar coordinates do not satisfy both equations. Explain why this happens.

37. A circle has center $C = (a; \alpha)$ and radius R. Use the Law of Cosines to show that the polar equation of the circle is:

$$r^2 - 2ar \cos(\theta - \alpha) + a^2 = R^2$$

CALCULATOR EXERCISES

Many calculators have keys for converting between rectangular and polar coordinates. If your calculator has this feature, use it to make these conversions. Give θ to the nearest tenth of a degree and all other coordinates to four decimal places.

Convert the given rectangular coordinates to polar coordinates.

1. $(2, 3)$ **2.** $(5, -7)$ **3.** $(-4, 3)$ **4.** $(-2, -5)$

Convert the given polar coordinates to rectangular coordinates.

5. $(3; 40°)$ **6.** $(5; 100°)$ **7.** $(4; 1.2)$ **8.** $(-3; 17.9)$

10-2 / A GEOMETRIC REPRESENTATION OF COMPLEX NUMBERS

In 1806 an obscure Parisian bookkeeper published an essay in which complex numbers were represented geometrically. The bookkeeper was Jean Robert Argand (1768–1822), and diagrams such as the one at the right are called **Argand diagrams** in his honor. As shown in this diagram, the complex number $3 + 4i$ can be represented by the point $(3, 4)$ or by an arrow from the origin to $(3, 4)$. Likewise, the complex number $5 - 2i$ can be represented by the point $(5, -2)$ or by an arrow from the origin to $(5, -2)$. When points of the coordinate plane represent complex numbers, this plane is called the **complex number plane** or, simply, the

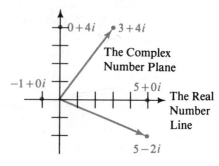

complex plane. An important subset of this plane is the real number line. A typical point on the real number line is $(a, 0)$, which represents the complex number $a + 0i$.

The point representing the complex number $z = a + bi$ can be given either in rectangular coordinates (a, b) or in polar coordinates $(r; \theta)$. (See the diagram below.) Since $a = r \cos \theta$ and $b = r \sin \theta$, we have the following two ways of specifying z:

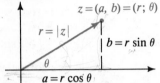

Rectangular form: $z = (a, b) = \underbrace{a}\ +\ \underbrace{b}\ i$

Polar form: $\quad\quad z = (r; \theta) = r \cos \theta + r \sin \theta\ \ i$
$$= r(\cos \theta + i \sin \theta)$$

The length of the arrow representing z is called the **absolute value** of z and is denoted by $|z|$. The diagram shows that
$$|z| = \sqrt{a^2 + b^2}.$$

Converting from rectangular to polar form

EXAMPLE 1. Express the complex numbers **(a)** $z_1 = -1 - i$ and **(b)** $z_2 = 3 + 4i$ in polar form.

SOLUTION: **a.** $|z_1| = |-1 - i| = \sqrt{(-1)^2 + (-1)^2} = \sqrt{2}$

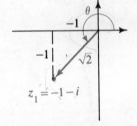

As the diagram shows, θ is a third-quadrant angle whose tangent is 1, so $\theta = 225°$. The polar coordinates for z_1 are $(\sqrt{2}; 225°)$; in polar form,
$$z_1 = \sqrt{2}\,(\cos 225° + i \sin 225°).$$
Since $225°$ and $-135°$ are coterminal, we also have
$$z_1 = \sqrt{2}\,(\cos(-135°) + i \sin(-135°)).$$

b. $|z_2| = |3 + 4i| = \sqrt{3^2 + 4^2} = 5$

θ is a first-quadrant angle whose tangent is $\frac{4}{3}$, or approximately 1.3333. By Table 3, $\theta \approx 53.1°$. Therefore, $z_2 = (5; 53.1°)$, or
$$z_2 = 5(\cos 53.1° + i \sin 53.1°).$$

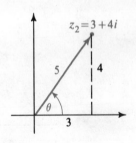

Converting from polar to rectangular form

EXAMPLE 2. Express the complex number $\left(2; \frac{\pi}{6}\right)$ in rectangular form.

SOLUTION: $\left(2; \frac{\pi}{6}\right) = 2\left(\cos \frac{\pi}{6} + i \sin \frac{\pi}{6}\right)$
$$= 2\left(\frac{\sqrt{3}}{2} + \frac{i}{2}\right)$$
$$= \sqrt{3} + i$$

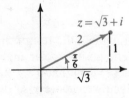

Polar coordinates and complex numbers   **341**

The polar form of complex numbers is used mainly in applications of the following theorem and De Moivre's Theorem given in Section 10–3.

THEOREM. If $z_1 = r(\cos \alpha + i \sin \alpha)$

and $z_2 = s(\cos \beta + i \sin \beta)$,

then $z_1 z_2 = rs[\cos (\alpha + \beta) + i \sin (\alpha + \beta)]$.

Proof: $z_1 z_2 = r(\cos \alpha + i \sin \alpha) \cdot s(\cos \beta + i \sin \beta)$
$$= rs[(\cos \alpha \cos \beta - \sin \alpha \sin \beta)$$
$$+ i(\sin \alpha \cos \beta + \cos \alpha \sin \beta)]$$

Using formulas (2) and (3), pages 310 and 311,

$$z_1 z_2 = rs[\cos (\alpha + \beta) + i \sin (\alpha + \beta)].$$

This theorem can be paraphrased as follows:

To multiply two complex numbers in polar form,
1. multiply their absolute values,
2. add their polar angles.

$$(r; \alpha) \cdot (s; \beta) = (rs; \alpha + \beta)$$

EXAMPLE 3. Express in polar form the numbers $z_1 = 3 + 3i$, $z_2 = 1 + i\sqrt{3}$, and their product $z_1 z_2$.

SOLUTION: z_1 and z_2 are represented by points which have polar coordinates $(3\sqrt{2}; 45°)$ and $(2; 60°)$, respectively. According to the previous theorem, the point representing $z_1 z_2$ has polar coordinates

$$(3\sqrt{2} \cdot 2; 45° + 60°) = (6\sqrt{2}; 105°).$$

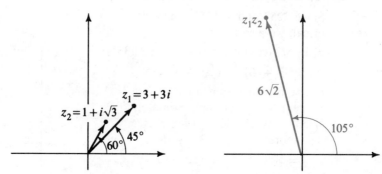

Thus, polar forms of the three complex numbers are as follows:

$$z_1 = 3\sqrt{2}(\cos 45° + i \sin 45°)$$
$$z_2 = 2(\cos 60° + i \sin 60°)$$
$$z_1 z_2 = 6\sqrt{2}(\cos 105° + i \sin 105°)$$

Throughout the rest of this book, we shall give final answers using positive radii and positive polar angles. If the polar angle θ is measured in degrees, then we choose θ so that $0° < \theta < 360°$; if the polar angle x is measured in radians, then we choose x so that $0 \le x < 2\pi$.

Give the polar form of each complex number.

1. $1 + i$ **2.** i **3.** -3 **4.** $\sqrt{3} - i$

Give the rectangular form of each complex number.

5. $(5; 90°)$ **6.** $(3; \pi)$ **7.** $(4; 45°)$ **8.** $(6; 30°)$

Give the polar form of each product.

9. $(2; 20°)(3; 30°)$ **10.** $\left(5; \dfrac{\pi}{4}\right)\left(2; \dfrac{3\pi}{4}\right)$

WRITTEN EXERCISES 10-2

Give the polar form of each complex number. Give angle measures in degrees, to the nearest tenth when necessary.

A **1.** $-1 + i$ **2.** $-3i$ **3.** $1 + i\sqrt{3}$ **4.** $-2 - 2i$
  5. -7 **6.** $2\sqrt{3} + 2i$ **7.** $3 - 4i$ **8.** $5 + 12i$

Give the rectangular form of each complex number.

9. $\left(6; \dfrac{4\pi}{3}\right)$ **10.** $\left(8; \dfrac{3\pi}{4}\right)$ **11.** $(9; 2\pi)$ **12.** $(2; \pi)$

Give the polar and rectangular form of each product.

13. $(5; 30°)(2; 60°)$ **14.** $(2; 115°)(3; 65°)$

15. $\left(8; \dfrac{\pi}{3}\right)\left(\dfrac{1}{2}; -\dfrac{2\pi}{3}\right)$ **16.** $\left(4; \dfrac{\pi}{4}\right)\left(3; \dfrac{\pi}{2}\right)$

In Exercises 17-20, **(a)** express z_1, z_2, and $z_1 z_2$ in polar form, **(b)** convert $z_1 z_2$ to rectangular form, **(c)** multiply z_1 and z_2 in rectangular form and see if your answer agrees with **(b)**, and **(d)** show z_1, z_2, and $z_1 z_2$ in an Argand diagram.

17. $z_1 = 2 + 2i\sqrt{3}$, $z_2 = \sqrt{3} - i$ **18.** $z_1 = 3 + 3i$, $z_2 = -2i$
19. $z_1 = 2 + 2i$, $z_2 = 2 - 2i$ **20.** $z_1 = -1 + i\sqrt{3}$, $z_2 = -1 - i\sqrt{3}$

21. Suppose $a + bi$ is multiplied by $1 + i$. By how many degrees must the arrow from $(0, 0)$ to (a, b) be rotated to coincide with the arrow from $(0, 0)$ to the product?

22. Suppose $a + bi$ is multiplied by $\sqrt{3} + i$. By how many degrees must the arrow from $(0, 0)$ to (a, b) be rotated to coincide with the arrow from $(0, 0)$ to the product?

Note: Exercises 23-28 preview the next section and should be worked in order (rather than odds and then evens).

23. Restate the theorem on page 342 for the special case when $z_2 = z_1$.

24. Square the number $\cos \theta + i \sin \theta$ and show that your result equals $\cos 2\theta + i \sin 2\theta$.

25. From Exercise 24, we have that $(\cos \theta + i \sin \theta)^2 = \cos 2\theta + i \sin 2\theta$. Make a conjecture about $(\cos \theta + i \sin \theta)^3$.

26. The diagram shows the complex number $z = \cos 30° + i \sin 30°$. Copy this diagram and on it show z^2, z^3, and z^4. (*Hint:* Use the results of Exercises 24 and 25.)

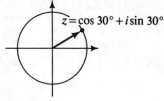

Ex. 26

B **27. a.** If $z = \cos \theta + i \sin \theta$, represent z and $\bar{z}$ in an Argand diagram.

   b. Explain why $\bar{z} = \cos(-\theta) + i \sin(-\theta)$
   $= \cos \theta - i \sin \theta$.

28. If $z = \cos \theta + i \sin \theta$, show that $\dfrac{1}{z} = \cos \theta - i \sin \theta$.

29. Let $z_1 = r(\cos \alpha + i \sin \alpha)$ and $z_2 = s(\cos \beta + i \sin \beta)$. Prove that

$$\frac{z_1}{z_2} = \frac{r}{s}(\cos(\alpha - \beta) + i \sin(\alpha - \beta))$$

by multiplying numerator and denominator of $\dfrac{z_1}{z_2}$ by $(\cos \beta - i \sin \beta)$. (This result shows that to divide two complex numbers, you divide their absolute values and subtract their polar angles.)

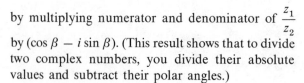

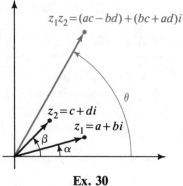

Ex. 30

C **30.** (An alternate proof of the theorem on page 342) Recall that if $z_1 = a + bi$ and $z_2 = c + di$, then $z_1 z_2 = (ac - bd) + (bc + ad)i$.

   a. From the diagram, $\tan \alpha = \dfrac{b}{a}$. Similarly, $\tan \beta = \underline{?}$ and $\tan \theta = \underline{?}$.

   b. Use the values of $\tan \alpha$ and $\tan \beta$ and the formula for $\tan(\alpha + \beta)$ to show that $\tan(\alpha + \beta) = \tan \theta$. Then conclude that $\alpha + \beta = \theta$.

   c. By definition, $|z_1| = \sqrt{a^2 + b^2}$, $|z_2| = \underline{?}$, and $|z_1 z_2| = \underline{?}$. Show that $|z_1 z_2| = |z_1| \cdot |z_2|$.

31. Why are these two triangles similar?

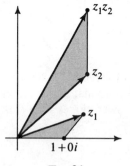

Ex. 31

CALCULATOR EXERCISES

Give **(a)** the polar form of z_1, z_2, and $z_1 z_2$, and **(b)** the rectangular form of $z_1 z_2$.

1. $z_1 = 19 - 27i$, $z_2 = 31 + 13i$ **2.** $z_1 = -43 + 25i$, $z_2 = 57 + 31i$

10-3/POWERS OF COMPLEX NUMBERS

The chief advantage of using the polar form of a complex number is that it makes it easy to find the powers and roots of that number. To find the square of a complex number, we use the theorem about products given on page 342:

$$(r; \alpha)(s; \beta) = (rs; \alpha + \beta)$$

Therefore, $(r; \alpha)^2 = (r; \alpha)(r; \alpha) = (r^2; 2\alpha)$

To find $(r; \alpha)^3$, we let $(r; \alpha)^3 = (r; \alpha)(r^2; 2\alpha)$ and use the same theorem:

$$(r; \alpha)^3 = (r; \alpha)(r^2; 2\alpha) = (r^3; 3\alpha)$$

The pattern can be generalized for any positive integer n:

$$(r; \alpha)^n = (r^n; n\alpha)$$

Exercises 9 and 10 on page 348 provide hints for proving that this equation holds for zero and negative integral values of n as well. This is our next theorem, named after Abraham De Moivre.

DE MOIVRE'S THEOREM. If $z = r(\cos \theta + i \sin \theta)$, then for any integer n:
$$z^n = r^n(\cos n\theta + i \sin n\theta)$$

Restatement: If $z = (r; \theta)$, then $z^n = (r^n; n\theta)$.

EXAMPLE 1. If $z = \dfrac{1}{2} + \dfrac{i\sqrt{3}}{2}$, calculate z^2, z^3, z^4, z^5, and z^6. Show your answers in an Argand diagram.

SOLUTION: First we put z in polar form and then we use De Moivre's Theorem.

$$z = \frac{1}{2} + \frac{i\sqrt{3}}{2} = 1\left(\cos\frac{\pi}{3} + i \sin\frac{\pi}{3}\right)$$

Thus:
$$z^2 = 1^2\left(\cos\frac{2\pi}{3} + i \sin\frac{2\pi}{3}\right) = -\frac{1}{2} + \frac{i\sqrt{3}}{2}$$

$$z^3 = 1^3\left(\cos\frac{3\pi}{3} + i \sin\frac{3\pi}{3}\right) = -1$$

$$z^4 = 1^4\left(\cos\frac{4\pi}{3} + i \sin\frac{4\pi}{3}\right) = -\frac{1}{2} - \frac{i\sqrt{3}}{2}$$

$$z^5 = 1^5\left(\cos\frac{5\pi}{3} + i \sin\frac{5\pi}{3}\right) = \frac{1}{2} - \frac{i\sqrt{3}}{2}$$

$$z^6 = 1^6\left(\cos\frac{6\pi}{3} + i \sin\frac{6\pi}{3}\right) = 1$$

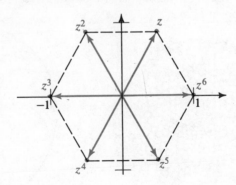

The Argand diagram above shows that these six powers of z are closely related to the construction of a regular hexagon.

EXAMPLE 2. **a.** If $z = 1 + i$, use De Moivre's Theorem to calculate z^{10}.

b. Show several powers of z in an Argand diagram.

SOLUTION: **a.** $z = 1 + i = \sqrt{2}\,(\cos 45° + i \sin 45°)$. Thus:

$$z^{10} = (\sqrt{2})^{10}(\cos 450° + i \sin 450°)$$
$$= 32(0 + i \cdot 1) = 32i$$

b. A number of powers of z are shown in the following diagram.

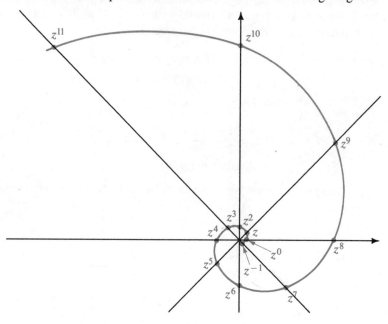

In the diagram above, a smooth spiral has been drawn through the points z, z^2, z^3, and so on. This spiral is called an **equiangular spiral** because it crosses all lines through the origin at the same angle.

Spirals can be found throughout nature. For example, the horns and claws of some animals are equiangular spirals, and the seeds of a sunflower form spirals, as illustrated at the left below. The spiral of the chambered nautilus shell shown at the right below bears a striking resemblance to the "powers of z spiral" shown above.

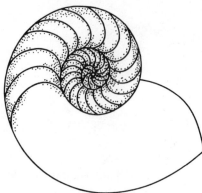

Give the polar form of each of the following.

1. $(2; 45°)^2$ **2.** $(2; 45°)^3$

3. $(\sqrt{2}; -18°)^4$ **4.** $(1; 36°)^{10}$

5. $\left(4; \dfrac{\pi}{6}\right)^3$ **6.** $\left(\sqrt{3}; \dfrac{5\pi}{6}\right)^6$

7. Let $z = -1 + i$.
  a. Express z in polar form.
  b. Express z^6 in polar form.
  c. Express z^6 in rectangular form.

8. Let $z = 1 - i$.
  a. Evaluate z^{10} by writing z as $\sqrt{2}(\cos 315° + i \sin 315°)$, applying De Moivre's Theorem, and simplifying the result.
  b. Evaluate z^{10} by writing z as $\sqrt{2}[\cos(-45°) + i \sin(-45°)]$, applying De Moivre's Theorem, and simplifying the result.
  c. Which calculation was easier? Why?

A   **1.** If $z = \dfrac{\sqrt{3}}{2} + \dfrac{i}{2}$, express z in polar form. Calculate z^2 and z^3 using De Moivre's Theorem, and show z, z^2, and z^3 in an Argand diagram.

     2. If $z = i$, show z, z^2, z^3, z^4, and z^{10} in an Argand diagram.

     3. a. If $z = 1 - i$, express z in polar form and show z^{-1}, z^0, z, z^2, z^3, z^4, z^5, z^6, z^7, and z^8 in an Argand diagram.
  b. Show that $z^{12} = -64$.

The German mathematician Carl Friedrich Gauss (1777–1855) determined with the help of complex numbers just which regular polygons could be constructed with straightedge and compass. One of the polygons Gauss discovered could be constructed was the 17-sided regular polygon. He was so pleased with this discovery that he requested that such a polygon be inscribed on his tombstone. Although this request was never fulfilled, a 17-sided star was inscribed on a monument to Gauss in his birthplace, Brunswick, Germany.

4. a. If $z = 1 + i\sqrt{3}$, express z in polar form and show z^{-2}, z^{-1}, z^0, z, z^2, z^3, and z^4 in an Argand diagram.

b. Show that $z^{18} = 2^{18}$.

5. Compute $(1 - i\sqrt{3})^3$ by two methods:

a. Carry out the multiplication $(1 - i\sqrt{3})(1 - i\sqrt{3})(1 - i\sqrt{3})$ and simplify.

b. Apply De Moivre's Theorem and then express your answer in rectangular form.

6. Compute $(-1 - i)^3$ by two methods:

a. Carry out the multiplication $(-1 - i)(-1 - i)(-1 - i)$ and simplify.

b. Apply De Moivre's Theorem and then express your answer in rectangular form.

7. Refer to the figure on page 345 showing the regular hexagon and powers of z. Do you see that z^7 simplifies to z? Simplify z^{14}, z^{40}, z^{-1}, and z^{-11}.

8. Refer to the figure on page 346 showing the powers of $1 + i$. Simplify z^{16} and z^{-2}.

B   **9. a.** Verify that De Moivre's Theorem is true when $n = 0$.

b. Prove De Moivre's Theorem for the case when $n = -1$ by writing

$$z^{-1} = \frac{1}{z} = \frac{1}{r(\cos\theta + i\sin\theta)}$$

and simplifying.

10. In Exercise 9, you proved De Moivre's Theorem for $n = -1$. Now prove it for $n = -2$ by writing $z^{-2} = (z^{-1})^2$. (A similar proof of De Moivre's Theorem can be made for any negative integer n.)

11. Make an Argand diagram showing the number i. On the same diagram, show a number z whose cube is i.

12. Make an Argand diagram showing the number $-i$. On the same diagram, show a number z whose cube is $-i$.

In Exercises 13–15, use the following identity:

$$re^{i\theta} = r(\cos\theta + i\sin\theta)$$

13. Prove De Moivre's Theorem for all real numbers n. Assume that the laws of exponents (page 148) hold for imaginary exponents.

14. Show that $e^{i\pi} = -1$ and $e^{\frac{i\pi}{2}} = i$.

15. Is i^i a real number or an imaginary number? (*Hint:* Let $r = 1$ and $\theta = \frac{\pi}{2}$ in the identity $re^{i\theta} = r(\cos\theta + i\sin\theta)$, simplify the result, and then raise both sides of the equation to the power i. A calculator will be helpful to evaluate your answer.)

C   **16.** An equiangular spiral is often called a logarithmic spiral because if the point $(r; \theta)$ is on the spiral, then $\ln r = k\theta$, for some constant k. Suppose

348   Chapter 10

the point $z = (r; \theta)$ is on such a spiral, so that $\ln r = k\theta$. Prove that the points z^2, z^3, and z^4 also satisfy this equation and hence are on the spiral.

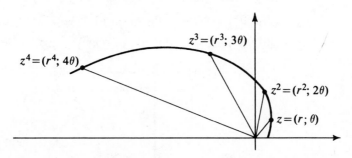

$z^3 = (r^3; 3\theta)$

$z^4 = (r^4; 4\theta)$

$z^2 = (r^2; 2\theta)$

$z = (r; \theta)$

10-4 / ROOTS OF COMPLEX NUMBERS

In this section, we use De Moivre's Theorem to find roots of complex numbers. In general, a nonzero complex number will have two square roots, three cube roots, four fourth roots, and k kth roots. Here are two examples:

EXAMPLE 1. Find the three cube roots of $8i$.

SOLUTION: This problem can be solved by thinking of $8i$ and its cube root z as points in the plane. The polar coordinates of these points are shown in the Argand diagram. Since z is a cube root of $8i$, we have:

$$z^3 = 8i$$

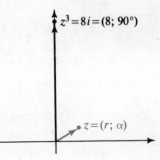

$z^3 = 8i = (8; 90°)$

$z = (r; \alpha)$

By replacing z and $8i$ with their polar coordinates, we have

$(r; \alpha)^3 = (8; 90°)$,

or, by using De Moivre's Theorem,

$(r^3; 3\alpha) = (8; 90°)$.

Thus: $r^3 = 8$ and $3\alpha = 90°$ or $(90° + \text{a multiple of } 360°)$

$r = 2$ $\alpha = 30°$ or $(30° + \text{a multiple of } 120°)$

$\alpha = 30°,\quad 150°,\quad 270°,\quad 390°,\quad 510°$

$+120° \quad +120° \quad +120° \quad +120°$

Note that there are only three distinct values of α. The fourth value, $390°$, is $360°$ more than the first value, $30°$; the fifth value is $360°$ more than the second value, and so on. Consequently, we can get our answers by considering only the values $30°$, $150°$, and $270°$.

(*Solution continued on page 350*)

Therefore, the three cube roots of $8i$ are:

$$z_1 = (2; 30°) = 2(\cos 30° + i \sin 30°) = \sqrt{3} + i$$
$$z_2 = (2; 150°) = 2(\cos 150° + i \sin 150°) = -\sqrt{3} + i$$
$$z_3 = (2; 270°) = 2(\cos 270° + i \sin 270°) = -2i$$

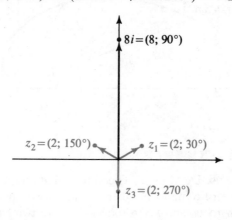

EXAMPLE 2. Find the four fourth roots of -16.

SOLUTION: If z is a fourth root of -16, then:

$$z^4 = -16$$

$(r; \alpha)^4 = (16; \pi)$, by replacing z and -16 with their polar coordinates

$(r^4; 4\alpha) = (16; \pi)$, by using De Moivre's Theorem

Thus: $r^4 = 16$ and $4\alpha = \pi$ or $(\pi + $ a multiple of $2\pi)$

$r = 2$ $\qquad$ $\alpha = \dfrac{\pi}{4}$ or $\left(\dfrac{\pi}{4} + \text{a multiple of } \dfrac{\pi}{2}\right)$

Therefore, the four fourth roots of -16 are:

$$z_1 = \left(2; \frac{\pi}{4}\right) = 2\left(\cos \frac{\pi}{4} + i \sin \frac{\pi}{4}\right) = \sqrt{2} + i\sqrt{2}$$
$$z_2 = \left(2; \frac{3\pi}{4}\right) = 2\left(\cos \frac{3\pi}{4} + i \sin \frac{3\pi}{4}\right) = -\sqrt{2} + i\sqrt{2}$$
$$z_3 = \left(2; \frac{5\pi}{4}\right) = 2\left(\cos \frac{5\pi}{4} + i \sin \frac{5\pi}{4}\right) = -\sqrt{2} - i\sqrt{2}$$
$$z_4 = \left(2; \frac{7\pi}{4}\right) = 2\left(\cos \frac{7\pi}{4} + i \sin \frac{7\pi}{4}\right) = \sqrt{2} - i\sqrt{2}$$

The method given in these examples can be generalized as follows:

FINDING THE NTH ROOTS OF A COMPLEX NUMBER

1. Plot the complex number as a point in the plane and find its polar coordinates $(r; \theta)$.

2. The n nth roots are then $(r^{\frac{1}{n}}; \alpha)$,

   where $\alpha = \dfrac{\theta}{n}$ or $\left(\dfrac{\theta}{n} + \text{a multiple of } \dfrac{2\pi}{n}\right)$.

(*Note:* This two-step procedure is the same as using De Moivre's Theorem with n replaced by $\dfrac{1}{n}$.)

WRITTEN EXERCISES 10-4

A

1. Find the three cube roots of i.
2. Find the three cube roots of $-i$.
3. Find the three cube roots of 8.
4. Find the three cube roots of -8.
5. Find the fourth roots of 16.
6. Find the fourth roots of $-\dfrac{1}{2} - \dfrac{i\sqrt{3}}{2}$.

7. **a.** Refer to the answers in Example 1, pages 349–350, and show that $z_1 + z_2 + z_3 = 0$ and $z_1 z_2 z_3 = 8i$.
   b. In Example 1 any cube root of $8i$ must satisfy the equation $z^3 = 8i$. Use Theorem 5 on page 84 to find the sum and product of the roots of this equation. Your answers should agree with part (a).

8. **a.** Refer to the answers in Example 2, page 350, and show that $z_1 + z_2 + z_3 + z_4 = 0$ and $z_1 z_2 z_3 z_4 = 16$.
   b. In Example 2 any fourth root of -16 must satisfy the equation $z^4 = -16$. Use Theorem 5 on page 84 to find the sum and product of the roots of this equation. Your answers should agree with part (a).

In Exercises 9–11 use the following formulas for factoring a sum and difference of two cubes.

$$(1) \quad a^3 + b^3 = (a + b)(a^2 - ab + b^2)$$

$$(2) \quad a^3 - b^3 = (a - b)(a^2 + ab + b^2)$$

9. The three cube roots of 8 should satisfy the equation $z^3 - 8 = 0$. Use formula (2) to solve this equation. Your answers should agree with those of Exercise 3.

10. The three cube roots of -8 should satisfy the equation $z^3 + 8 = 0$. Use formula (1) to solve this equation, and check your answers with those of Exercise 4.

B   **11. a.** The six sixth roots of 1 should satisfy the equation $z^6 - 1 = 0$. By factoring $z^6 - 1$ as a difference of two squares and then using formulas (1) and (2), show that

$$z^6 - 1 = (z - 1)(z + 1)(z^2 + z + 1)(z^2 - z + 1)$$

b. Use part (a) to find the six sixth roots of 1 and plot your answers on an Argand diagram.

12. a. Factor $z^4 + 4$ by writing it as $(z^4 + 4z^2 + 4) - 4z^2$.
b. Use part (a) to solve $z^4 = -4$ and plot your answers on an Argand diagram.
c. Use De Moivre's Theorem to verify your answer to (b).

13. Find the fifth roots of 32. (Use a calculator or trigonometric tables.)

14. Find the cube roots of $3\sqrt{3} + 3i$. (Use a calculator or trigonometric tables.)

15. A prime number of the form $N = 2^{2^n} + 1$ is called a Fermat prime.
a. Find the first four Fermat primes by substituting $n = 0$, 1, 2, and 3.
b. Gauss stated that a regular polygon of N sides could be constructed with a straightedge and compass if and only if $N = 2^j$ ($j \geq 2$) or $N = 2^k$ ($k \geq 0$) times a product of one or more different Fermat primes. For example, since 3 and 5 are Fermat primes, then a polygon of $2^2 \cdot 3 \cdot 5 = 60$ sides is constructible. Decide whether a polygon is constructible if it has 120 sides, 17 sides, 7 sides, 9 sides, 10 sides, 255 sides.

16. Explain what the solutions to the equation $z^{17} = 1$ have to do with a regular polygon of 17 sides.

C   **17.** Let G be the set of all n nth roots of 1.
a. Show that if z_1 and $z_2 \in G$, then $z_1 z_2 \in G$.
b. Show that if $z_1 \in G$, then $(z_1)^{-1} \in G$.
(G is an example of a type of algebraic system called a group.)

CALCULATOR EXERCISE

Find the three cube roots of $7 + 9i$ in rectangular form.

Chapter summary

1. When we write $P = (r; \theta)$, we are describing the point P by its polar coordinates r and θ. The polar distance r is the distance from the pole to P and the polar angle θ is an angle measured from the polar axis to the ray OP. Although a point has only one pair of rectangular coordinates, it has many pairs of polar coordinates. For example,

$$(-2; 20°) = (2; 200°) = (2; -160°).$$

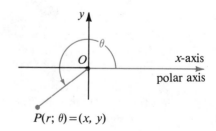

2. Conversion formulas:

polar to rectangular	rectangular to polar
$x = r \cos \theta$	$r = \sqrt{x^2 + y^2}$
$y = r \sin \theta$	$\theta = \pm \mathrm{Tan}^{-1} \dfrac{y}{x}$

3. In an Argand diagram, the complex number $x + yi$ is represented by the point (x, y) or by an arrow from the origin to (x, y). When the points of the plane correspond to complex numbers in this way, the plane is called the complex plane.

4. The absolute value of a complex number $z = x + yi$ is

$$|z| = \sqrt{x^2 + y^2},$$

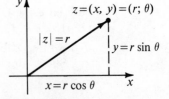

the distance from the origin to (x, y). The diagram at the right illustrates two ways of expressing z:

rectangular form: $z = (x, y) = x + yi$
polar form: $\quad z = (r; \theta) = r(\cos \theta + i \sin \theta)$

5. To multiply two complex numbers in polar form, multiply their absolute values and add their polar angles. That is:

$$(r; \alpha)(s; \beta) = (rs; \alpha + \beta)$$

Powers of complex numbers can be found using De Moivre's Theorem:

$$\text{If } z = (r; \theta), \text{ then } z^n = (r^n; n\theta)$$

6. The n nth roots of a complex number $z = (r; \theta)$ are $(r^{\frac{1}{n}}; \alpha)$, where

$$\alpha = \frac{\theta}{n} \text{ or } \alpha = \frac{\theta}{n} + \text{ a multiple of } \frac{2\pi}{n}.$$

See Example 1, page 349.

Chapter test

1. Complete the table:

rectangular coordinates (x, y)	$(3, 3)$	$(0, -2)$	$(-1, \sqrt{3})$	?	?
polar coordinates $(r; \theta)$	?	?	?	$\left(6, -\dfrac{\pi}{2}\right)$	$(8, \pi)$

2. a. Sketch the polar graph of the equation $r = 1 + 2 \cos \theta$.
b. Find a rectangular equation of this graph.

3. Sketch the graph of each polar equation.
a. $r = 3$ **b.** $\theta = 3$ **c.** $r = \theta$

4. a. If $z_1 = 4 + 3i$ and $z_2 = 3 - 2i$, represent z_1, z_2, and $z_1 + z_2$ in an Argand diagram.

    b. Evaluate $|z_1|$, $|z_2|$, and $|z_1 + z_2|$.

5. a. If $z_1 = 4(\cos 120° + i \sin 120°)$ and $z_2 = 2(\cos 30° + i \sin 30°)$, represent z_1, z_2, and $z_1 z_2$ in an Argand diagram.

    b. Express $z_1 z_2$ in rectangular form.

6. a. Express $z = -1 - i$ in polar form.

    b. Show z, z^2, z^3, and z^4 in an Argand diagram.

    c. Evaluate z^{10}.

7. a. If $z = r(\cos\theta + i \sin\theta)$, then according to De Moivre's Theorem, $z^2 = \underline{\ ?\ }$.

    b. Without using De Moivre's Theorem, square both sides of the equation $z = r(\cos\theta + i \sin\theta)$ and prove that your answer to part (a) is correct.

8. Find the three cube roots of $-8i$, expressing each answer in the form $x + yi$. Show your three answers in an Argand diagram.

10-2

10-3

10-4

Cumulative review (Chapters 6–10)

1. A sector of a circle of radius 4 cm has arclength 5 cm. Find:
   a. the measure of the central angle in radians
   b. the area of the sector

2. Evaluate without using a table or calculator.
   a. $\sin 120°$ **b.** $\sec \dfrac{5\pi}{4}$ **c.** $\cos\left(-\dfrac{3\pi}{2}\right)$ **d.** $\tan\left(-\dfrac{7\pi}{6}\right)$

3. Sketch the graph of $y = |\sin x|$. What is the domain of this function? What is the range?

4. Complete the table.

	a.	**b.**	**c.**	**d.**
Quadrant	I	II	III	IV
$\cot \theta$	?	$-\dfrac{12}{5}$	$\dfrac{3}{4}$	?
$\sin \theta$	$\dfrac{24}{25}$	?	?	$-\dfrac{21}{29}$

5. If $\sec x = \sqrt{5}$ and $\pi \le x \le 2\pi$, find the other five trigonometric functions of x without using a table or calculator.

6. Prove the following identities:
   a. $\dfrac{\cot x - \tan x}{\cot x + \tan x} = 1 - 2\sin^2 x$
   b. $(\tan x + \cot x)^2 = \sec^2 x + \csc^2 x$

7. Find the angle of inclination of the line $x + 2y = 7$.

8. Simplify: $\dfrac{\cos \theta}{1 - \sin \theta} - \dfrac{\cos \theta}{1 + \sin \theta}$

9. Solve for x, $0 \le x < 2\pi$:
   a. $2 \sec x + 3 = 0$ **b.** $\sin x \cos x = 2 \cot x$

10. Solve to the nearest tenth of a degree, $0° \le \theta < 360°$:
    $$4 \sin^2 \theta + \cos^2 \theta = -4 \sin \theta \cos \theta$$

11. Find the lengths of the diagonals of a rhombus with side 11.5 cm and acute angles 70°.

12. If $\angle A$ is acute and $\sec A = y$, express $\cot A$ in terms of y.

13. Two people are standing 100 m apart on the bank of a river. If a rock on the opposite bank is N25°E from one person and N15°W from the other, how wide is the river?

14. Show that no triangle with sides 5 cm and 8 cm can have an area of 24 cm².

15. A parallelogram has adjacent sides 10 and 12. Its area is 96 sq. units. Find:
    a. the angles of the parallelogram
    b. the length of each diagonal

16. $\triangle ABC$, shown at the right, is equilateral and $AB = 15$. If $\overline{AD}$ and $\overline{AE}$ trisect $\angle BAC$, find BD, DE, and EC.

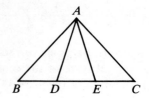

17. Evaluate without using tables or calculators.

   a. $\text{Cos}^{-1}\left(-\dfrac{\sqrt{3}}{2}\right)$

   b. $\cos(\text{Arctan }\sqrt{3})$

   c. $\sin\left(\text{Sin}^{-1}\left(-\dfrac{1}{2}\right)\right)$

   d. $\text{Arcsin}\left(\sin\dfrac{3\pi}{4}\right)$

   e. $\csc\left(\text{Tan}^{-1}\left(-\dfrac{1}{2}\right)\right)$

18. Sketch the graph of $f(t) = 3\sin\dfrac{\pi t}{2}$. State the period and the amplitude of this function.

19. Solve for x, $0 \le x < 2\pi$: $2\cos 3x + 1 = 0$

20. Test each of the following equations to see whether its graph has symmetry with respect to: (i) the x-axis; (ii) the y-axis; (iii) the line $y = x$; (iv) the origin.

   a. $y^2 = \dfrac{2}{x^2 + 1}$

   b. $xy - x = y$

   c. $y = 2^{\sin x}$

21. Sketch the graph of $y = -\dfrac{1}{2}\cos\left(x - \dfrac{\pi}{4}\right)$.

22. Sketch the graphs of $y = |x^2 - 1|$ and $y = \dfrac{1}{|x^2 - 1|}$ on the same axes.

23. Find the amplitude, frequency, and period of the function
$$V = 150\sin 72\pi t.$$

24. Write an equation for a sine wave that has amplitude 3 and period 8.

25. Prove the following identities:

   a. $\tan x \sin 2x = 2\sin^2 x$

   b. $\dfrac{1 - \sin 2x}{\cos 2x} = \dfrac{1 - \tan x}{1 + \tan x}$

26. Evaluate without a calculator or tables.

   a. $\cos 20° \cos 115° - \sin 20° \sin 115°$

   b. $\sin\dfrac{7\pi}{12}\cos\dfrac{7\pi}{12}$

   c. $\dfrac{\tan 105° - \tan 75°}{1 + \tan 105° \tan 75°}$

27. Evaluate $\sin 255°$ without a calculator or tables.

28. Find the measure of the acute angle formed by the lines $y = 2x + 5$ and $y = -x - 1$, to the nearest tenth of a degree.

29. Given that $0 < \alpha < \dfrac{\pi}{2} < \beta < \pi$, $\cos \alpha = \dfrac{\sqrt{3}}{2}$, and $\sin \beta = \dfrac{5}{13}$. Find, without a calculator or tables:

    a. $\cos(\alpha - \beta)$ **b.** $\cos 2\beta$ **c.** $\sin \dfrac{\beta}{2}$

30. Simplify: $\dfrac{\tan x}{1 + \tan x} + \dfrac{\tan x}{1 - \tan x}$

31. Solve for x, $0 \le x < 360°$: $\tan(6x + 5°) = 1$

32. Solve for x, $0 \le x < 2\pi$: $\sin 2x = \cot x$

33. Sketch the graphs of $y = \cos 2x$ and $y = 2x^2$ on the same axes. Then state the *number* of solutions to the equation $\cos 2x = 2x^2$.

34. a. Give polar coordinates for the point $(-3, 3\sqrt{3})$.

    b. Give rectangular coordinates for the point $\left(8; \dfrac{7\pi}{4}\right)$.

35. Sketch the polar graph for $r = 2 \cos 3\theta$.

36. Write the rectangular equation for the given polar equation and draw its graph.

    a. $r = \cos 2\theta$ **b.** $r = 3 \cos \theta$

Given $z_1 = 1 - i$ and $z_2 = -\sqrt{3} + i$:

37. Express z_1 and z_2 in polar form.

38. Express $z_1 z_2$ in polar form.

39. a. Express $(z_2)^{10}$ in polar form.
    b. Express $(z_2)^{10}$ in rectangular form.

40. Find all the cube roots of $-27i$, in polar form.

358

CHAPTER ELEVEN

Conic sections

OBJECTIVES

1. To identify the type of conic section an equation represents, and to locate its vertex (vertices), and focus (foci).
2. To graph and find the equations of parabolas, ellipses, and hyperbolas.
3. To measure the eccentricity of ellipses and hyperbolas.
4. To find the intersection points of a system of conic equations.
5. To solve applied problems involving conic sections.

Introduction

In Section 1–6 you studied equations of the form $(x - h)^2 + (y - k)^2 = r^2$, $r > 0$, whose graph is, of course, the circle with center (h, k) and radius r. Then, in Section 2–3, you learned to sketch parabolas with equations of the form $y = a(x - h)^2 + k$. Notice that the equations of circles and parabolas are quadratic equations in two variables. In this chapter we will continue our examination of the various graphs of the **general second-degree equation:**

$$Ax^2 + Bxy + Cy^2 + Dx + Ey + F = 0$$

This group of equations has many interesting properties in common. One important property is geometric.

The picture of Saturn shown at the left was assembled from images transmitted by the Voyager 2 satellite. Saturn's elliptical rings and several of Saturn's moons can be seen.

Imagine that the double cones shown below were extended indefinitely up and down. If these cones are sliced by a plane tilted at various angles, the resulting cross sections are called **conic sections.** As shown below, a circle, an ellipse, a parabola, and a hyperbola are conic sections.

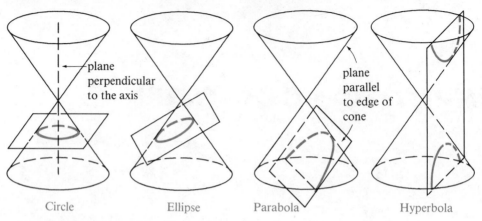

Circle   Ellipse   Parabola   Hyperbola

It is also possible to slice the double cone to obtain a single point, a line, or a pair of lines. Do you see how? These extreme cases are called degenerate conic sections.

The graph of every general second-degree equation in two variables is a conic section or possibly a degenerate conic section, and every conic section has an equation of the form given above. Here are some examples of conic sections and their equations:

Circle: $x^2 + y^2 = 9$

Ellipse: $x^2 + 4y^2 = 9$

Hyperbola: $x^2 - 4y^2 = 9$, $xy = 6$

Parabola: $x^2 + 2x + 5 = y$

Degenerate Circle: $x^2 + y^2 = 0$

Degenerate Ellipse: $25x^2 + 16y^2 = 0$

11-1/ELLIPSES

The diagrams at the top of page 361 illustrate the curve called an **ellipse.** All ellipses have two axes of symmetry; in the examples that follow, the x- and y-axes are the axes of symmetry. The portions of the axes of symmetry that are on or within the ellipse are called the **major axis** and the **minor axis** of the ellipse. It is customary to let $2a$ represent the length of the major, or longer, axis and to let $2b$ ($b < a$) represent the length of the minor axis. If the major axis of an ellipse is horizontal and the **center** of the ellipse is at the origin, then the equation and graph of the ellipse are as shown at the left on page 361. The situation when the major axis is vertical is shown at the right. In each case, the **vertices** of the ellipse are the endpoints of its major axis.

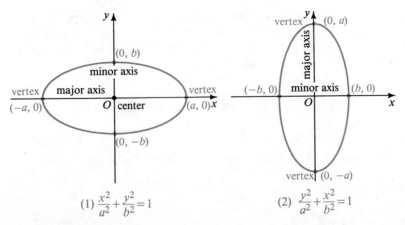

$$(1) \ \frac{x^2}{a^2} + \frac{y^2}{b^2} = 1 \qquad\qquad (2) \ \frac{y^2}{a^2} + \frac{x^2}{b^2} = 1$$

Note that equations (1) and (2) above have the general form

$$(3) \ \frac{x^2}{(x\text{-intercept})^2} + \frac{y^2}{(y\text{-intercept})^2} = 1.$$

EXAMPLE 1. Sketch the graph of $\dfrac{x^2}{9} + \dfrac{y^2}{4} = 1$.

SOLUTION: If we set $y = 0$ and solve for x, we can see that the x-intercepts are $\pm\sqrt{9} = \pm 3$. If we set $x = 0$ and solve for y, we see that the y-intercepts are $\pm\sqrt{4} = \pm 2$. The equation $y = \pm\dfrac{2}{3}\sqrt{9 - x^2}$ can be used

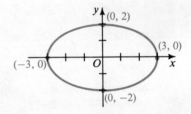

to compute the coordinates of other points on the ellipse. When $x = 2$, $y \approx \pm 1.5$. Thus, using the symmetry of the ellipse, $(2, 1.5)$, $(-2, 1.5)$, $(-2, -1.5)$, and $(2, -1.5)$ are all on the ellipse.

The equation $\dfrac{x^2}{a^2} + \dfrac{y^2}{b^2} = 1$ can be considered as an *algebraic definition* of an ellipse with center at the origin and horizontal and vertical axes. An ellipse can also be defined *geometrically* as follows.

Geometric definition of an ellipse

Consider two fixed points, $F_1 = (c, 0)$ and $F_2 = (-c, 0)$ and a constant a, $0 < c < a$. The set of all points P in the plane such that

$$PF_1 + PF_2 = 2a$$

is an ellipse. Points F_1 and F_2 are called the **foci** of the ellipse. (*Foci* is the plural of *focus*.)

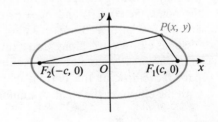

Now let us show that this geometric definition of an ellipse is compatible with the algebraic one. To do this, we shall use the distance formula to express PF_1 and PF_2 in terms of x, y, and c. Since $PF_1 + PF_2 = 2a$, we have:

$$\sqrt{(x - c)^2 + y^2} + \sqrt{(x + c)^2 + y^2} = 2a$$

$$\sqrt{(x + c)^2 + y^2} = 2a - \sqrt{(x - c)^2 + y^2}$$

Squaring: $\quad (x + c)^2 + y^2 = 4a^2 - 4a\sqrt{(x - c)^2 + y^2} + [(x - c)^2 + y^2]$

$\qquad x^2 + 2cx + c^2 + y^2 = 4a^2 - 4a\sqrt{(x - c)^2 + y^2} + x^2 - 2cx + c^2 + y^2$

Simplifying: $\qquad 4cx = 4a^2 - 4a\sqrt{(x - c)^2 + y^2}$

Dividing by 4 and
rearranging terms: $\quad cx - a^2 = -a\sqrt{(x - c)^2 + y^2}$

Squaring again:

$$c^2x^2 - 2ca^2x + a^4 = a^2[(x - c)^2 + y^2]$$
$$= a^2x^2 - 2ca^2x + a^2c^2 + a^2y^2$$

Simplifying: $\qquad c^2x^2 + a^4 = a^2x^2 + a^2c^2 + a^2y^2$

Rearranging $\qquad a^4 - a^2c^2 = a^2x^2 - c^2x^2 + a^2y^2$

terms: $\qquad a^2(a^2 - c^2) = (a^2 - c^2)x^2 + a^2y^2$

We recall that $a > c$ and
substitute $b^2 = a^2 - c^2$:

$$a^2b^2 = b^2x^2 + a^2y^2$$

Dividing by a^2b^2: $\qquad 1 = \dfrac{x^2}{a^2} + \dfrac{y^2}{b^2}$

The steps in the derivation of this equation can be reversed to show that any point satisfying the equation must also satisfy the condition $PF_1 + PF_2 = 2a$. Thus the algebraic and geometric definitions are equivalent.

In the preceding derivation, we substituted $b^2 = a^2 - c^2$. This relationship between a, b, and c is illustrated for another ellipse in the diagram at the right. The diagram also shows the points $F_1 = (c, 0)$ and $F_2 = (-c, 0)$ that are the foci of the ellipse.

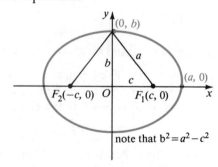

Some ellipses are "long and skinny" and others are nearly circular. For example, compare the ellipse at the bottom of page 361 with the ellipse at the right. A number called the **eccentricity** of an ellipse is used to measure the amount of its elongation. It is defined as follows:

$$e = \text{eccentricity} = \frac{\text{distance from center to focus}}{\text{distance from center to vertex}}$$

$$= \frac{c}{a}, \quad 0 < c < a.$$

If the eccentricity of an ellipse is near 0, then the ellipse is almost circular. For this reason, a circle is sometimes considered to be a "special ellipse" with eccentricity 0. It is clear that the eccentricity of an ellipse is a positive number less than one.

EXAMPLE 2. Find an equation of an ellipse with center at the origin, eccentricity 0.4, and vertex (0, 5).

SOLUTION: Since $a = 5$ and $e = \frac{c}{a} = 0.4$,

we have $c = 5(0.4) = 2$ and $b^2 = a^2 - c^2 = 5^2 - 2^2 = 21$. We substitute to write an equation for the ellipse:

$$\frac{y^2}{a^2} + \frac{x^2}{b^2} = 1$$

$$\frac{y^2}{25} + \frac{x^2}{21} = 1$$

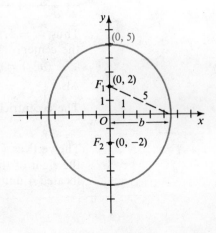

If an ellipse centered at the origin is translated h units along the x-axis and k units along the y-axis, its equation is changed by replacing x by $x - h$ and y by $y - k$. Notice that the translated ellipse has center (h, k).

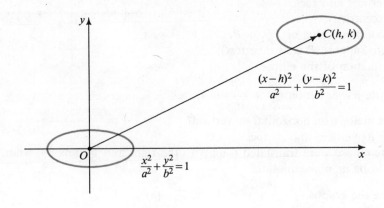

EXAMPLE 3. Find the center, vertices, foci, and eccentricity of the ellipse with equation

$$9x^2 + 25y^2 - 108x + 150y + 324 = 0$$

SOLUTION: Complete the square:

$$9x^2 - 108x + 25y^2 + 150y = -324$$
$$9(x^2 - 12x \qquad) + 25(y^2 + 6y \qquad) = -324$$
$$9(x^2 - 12x + 36) + 25(y^2 + 6y + 9) = -324 + 9 \cdot 36 + 25 \cdot 9$$
$$9(x - 6)^2 + 25(y + 3)^2 = 225$$
$$\frac{(x - 6)^2}{25} + \frac{(y + 3)^2}{9} = 1$$

This equation has the standard form

$$\frac{(x - h)^2}{a^2} + \frac{(y - k)^2}{b^2} = 1.$$

Thus, $a = 5$, $b = 3$, and the center

$$(h, k) = (6, -3).$$

Also, $\quad c = \sqrt{a^2 - b^2} = \sqrt{5^2 - 3^2} = 4.$

The eccentricity $\quad e = \dfrac{c}{a} = \dfrac{4}{5}.$

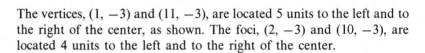

The vertices, $(1, -3)$ and $(11, -3)$, are located 5 units to the left and to the right of the center, as shown. The foci, $(2, -3)$ and $(10, -3)$, are located 4 units to the left and to the right of the center.

ORAL EXERCISES 11-1

1. Study the ellipse shown and state:
    a. whether its major axis is horizontal or vertical.
    b. its vertices and foci.
    c. its eccentricity.
    d. the constant value of $PF_1 + PF_2$. (*Hint:* Suppose P is at a vertex.)
    e. an equation of the ellipse.

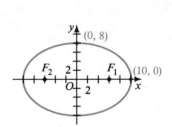

2. An ellipse has equation $\dfrac{x^2}{25} + \dfrac{y^2}{169} = 1.$

    a. Is its major axis horizontal or vertical?
    b. Find its vertices and its foci.
    c. If the ellipse were translated 6 units to the right and 7 units up, what would be its new equation?

3. Describe the graphs of $\dfrac{x^2}{9} + y^2 < 1$ and $\dfrac{x^2}{9} + y^2 > 1$.

4. If F_1 and F_2 are two fixed points in three-dimensional space, describe the set of points P for which **(a)** $PF_1 + PF_2 = 8$, **(b)** $PF_1 + PF_2 < 8$.

5. The equation of an ellipse is $\dfrac{(x + 5)^2}{4} + \dfrac{(y - 8)^2}{3} = 1$.

    a. Find the coordinates of its center.

    b. Find the values of a, b, c, and e.

WRITTEN EXERCISES 11-1

Sketch the ellipse whose equation is given. Find the coordinates of its vertices and foci and find the eccentricity.

A   **1.** $\dfrac{x^2}{36} + \dfrac{y^2}{16} = 1$ **2.** $\dfrac{x^2}{4} + \dfrac{y^2}{9} = 1$ **3.** $\dfrac{x^2}{16} + \dfrac{y^2}{25} = 1$

    4. $4x^2 + 25y^2 = 100$ **5.** $9x^2 + 25y^2 = 225$ **6.** $6.25x^2 + 4y^2 = 25$

Find an equation of the ellipse, with center at the origin, which satisfies the given conditions.

    7. A vertex at $(7, 0)$ and minor axis 2 units long

    8. A vertex at $(0, -9)$ and minor axis 6 units long

    9. A vertex at $(0, -13)$ and a focus at $(0, -5)$

    10. A vertex at $(17, 0)$ and a focus at $(8, 0)$

Sketch the graphs of the given equations on the same set of axes. Then determine algebraically where the graphs intersect. (*Hint:* You can use the procedure outlined on page 34.)

    11. $9x^2 + 2y^2 = 18$ **12.** $x^2 + 4y^2 = 400$
         $3x + y = -3$ $x - 2y = 28$

    13. $2x^2 + y^2 = 9$ **14.** $x^2 + 4y^2 = 16$
         $y - 4x = 9$ $|x| = 2$

    15. The ellipses $4x^2 + 5y^2 = 81$ and $5x^2 + 4y^2 = 81$ intersect in four points, all of which are on a circle.

    a. Illustrate this graphically.

    b. Add the equations of the ellipses to find an equation of the circle.

    c. Explain why the intersecting points of the ellipses must lie on this circle.

    16. Two ellipses have equations $5x^2 + 3y^2 = 64$ and $3x^2 + 5y^2 = 64$. Show algebraically that their common points are also on the circle with equation $x^2 + y^2 = 16$.

Sketch each semi-ellipse. Label each x- and y-intercept.

    17. a. $y = \sqrt{16 - 4x^2}$ **18. a.** $x = \dfrac{\sqrt{16 - y^2}}{2}$

       b. $y = -\sqrt{16 - 4x^2}$ **b.** $x = \dfrac{-\sqrt{16 - y^2}}{2}$

Sketch the ellipse whose equation is given. Find the coordinates of its vertices and foci and find the eccentricity.

19. $\dfrac{(x-5)^2}{25} + \dfrac{(y+3)^2}{9} = 1$

20. $\dfrac{(x+6)^2}{12} + \dfrac{(y-4)^2}{16} = 1$

21. $9(x-3)^2 + 4(y+5)^2 = 36$

22. $(x+1)^2 + 4(y+3)^2 = 9$

B 23. $x^2 + 25y^2 - 6x - 100y + 84 = 0$

24. $9x^2 + y^2 + 18x - 6y + 9 = 0$

25. $25x^2 + 4y^2 + 200x = 0$

26. $9x^2 + 16y^2 - 18x - 64y - 71 = 0$

Find an equation of the ellipse described.

27. Center is $(3, 7)$; one focus is $(6, 7)$; one vertex is $(8, 7)$.

28. Center is $(4, -1)$; one vertex is $(4, -5)$; eccentricity $= \dfrac{3}{4}$.

29. Vertices are $(5, 9)$ and $(5, 1)$; eccentricity $= 0.5$.

30. Foci are at $(0, -3)$ and $(10, -3)$; major axis is 26 units long.

31. Center is $(5, 6)$; the ellipse is tangent to both axes.

32. Vertices are $(-8, 3)$ and $(12, 3)$; one focus is $(8, 3)$.

33. **a.** If $F_1 = (3, 0)$, $F_2 = (-3, 0)$, and $P = (x, y)$, write an equation that expresses the fact that $PF_1 + PF_2 = 10$.

   b. Simplify this equation to one of the form $\dfrac{x^2}{a^2} + \dfrac{y^2}{b^2} = 1$.

34. **a.** If $F_1 = (0, 15)$, $F_2 = (0, -15)$, and $P = (x, y)$, write an equation that expresses the fact that $PF_1 + PF_2 = 34$.

   b. Simplify this equation to one of the form $\dfrac{y^2}{a^2} + \dfrac{x^2}{b^2} = 1$.

Sketch the graph of each inequality.

35. $\dfrac{x^2}{25} + \dfrac{y^2}{9} \le 1$

36. $\dfrac{x^2}{4} + \dfrac{y^2}{16} \ge 1$

37. $(x-3)^2 + \dfrac{(y+4)^2}{4} \ge 1$

38. $x^2 + y^2 - 4x + 12y + 40 \le 0$

C 39. The equations $x = a \cos \theta$ and $y = b \sin \theta$ are called the *parametric equations* of the ellipse $\dfrac{x^2}{a^2} + \dfrac{y^2}{b^2} = 1$.

   a. Show by substitution that the parametric equations satisfy the rectangular equation.

   b. What are the parametric equations of the circle $x^2 + y^2 = r^2$?

   c. What is the eccentricity of the ellipse with parametric equations $x = 4 \cos \theta$, $y = 2 \sin \theta$?

40. Guess the parametric equations of the ellipse with equation $\dfrac{(x-7)^2}{25} + \dfrac{(y-3)^2}{16} = 1$. (*Hint:* See Exercise 39.)

41. *P* is a point 1 m from the bottom of a ladder that is 3 m long. Suppose the foot of the ladder slides along the ground as the top slides down the wall. Show that *P* follows an elliptical path. (*Hint:* Express the coordinates of *P* in terms of θ. Then use the results of Exercise 39.)

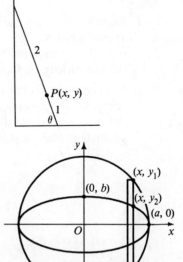

42. The diagram at right below shows the graphs of $x^2 + y^2 = a^2$ and $\dfrac{x^2}{a^2} + \dfrac{y^2}{b^2} = 1$ and also a thin rectangle through the points (x, y_1) on the circle and (x, y_2) on the ellipse.

a. Solve both equations for *y*, and show that
$$y_2 = \frac{b}{a}y_1.$$

b. If the rectangle's area is *K*, explain why the area of that portion of the rectangle within the ellipse is approximately $\dfrac{b}{a}K$.

c. Derive the formula $A = \pi ab$ for the area of the ellipse.

COMPUTER EXERCISES

1. The area of the ellipse $\dfrac{x^2}{25} + \dfrac{y^2}{1} = 1$, shown below, can be approximated by adding the areas of ten rectangles in the first quadrant and then multiplying this sum by 4. The area of the leftmost rectangle is $\frac{1}{2} \cdot 1$; the area of the next is $\frac{1}{2}y_1$; and of the next is $\frac{1}{2}y_2$. How do you find y_1 and y_2?

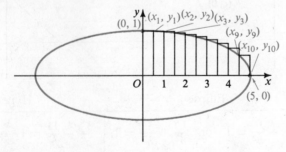

Write a computer program that will print the sum of these 10 rectangles and also the approximate area of the ellipse.

2. Modify your program of Exercise 1 to print a more accurate approximation of the area of the ellipse.

1. Earth's orbit is approximately an ellipse with the sun at one focus. During January, Earth is closest to the sun, a distance of 9.14×10^7 km, while during July, Earth is farthest from the sun, a distance of 9.44×10^7 km. Find the eccentricity of Earth's orbit.

2. Most comets travel in elliptical orbits with the sun at a focus. Halley's comet reappears about every 76 years, and it passes within 88 million kilometers of the sun. Its greatest distance from the sun is approximately 5282 million kilometers. Make a sketch of the orbit of the comet and give its eccentricity.

11-2/HYPERBOLAS

The equation of a hyperbola with center at the origin and horizontal and vertical axes of symmetry is very similar to the equation of an ellipse in the same position. Moreover, the definitions and terminology for ellipses and hyperbolas are similar.

Let us consider an ellipse and a hyperbola with horizontal major axes, as shown. Both graphs have two foci $F_1(c, 0)$ and $F_2(-c, 0)$, two vertices $V_1(a, 0)$ and $V_2(-a, 0)$, and a major axis $\overline{V_1 V_2}$. The dotted lines in the right-hand diagram are called *asymptotes*. They are helpful guides for drawing a hyperbola and will be explained shortly.

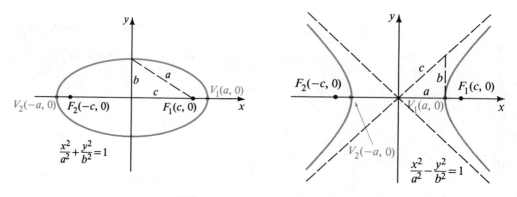

Ellipse   Hyperbola

	Ellipse	Hyperbola
Equation	$\dfrac{x^2}{a^2} + \dfrac{y^2}{b^2} = 1$ or $\dfrac{y^2}{a^2} + \dfrac{x^2}{b^2} = 1$	$\dfrac{x^2}{a^2} - \dfrac{y^2}{b^2} = 1$ or $\dfrac{y^2}{a^2} - \dfrac{x^2}{b^2} = 1$
Geometric Definition	$PF_1 + PF_2 = 2a \ (0 < c < a)$	$|PF_1 - PF_2| = 2a \ (0 < a < c)$
Eccentricity	$0 < e = \dfrac{c}{a} < 1 \ (c^2 = a^2 - b^2)$	$e = \dfrac{c}{a} > 1 \ (c^2 = a^2 + b^2)$

The eccentricity e of a hyperbola with center at the origin and horizontal and vertical axes of symmetry is defined as it was for the ellipse:

$$e = \frac{\text{distance from center to focus}}{\text{distance from center to vertex}} = \frac{c}{a}.$$

For any hyperbola $e > 1$.

The preceding chart shows that for every point P on a hyperbola, $|PF_1 - PF_2| = 2a$, $0 < a < c$. In order to derive the equation

$$\frac{x^2}{a^2} - \frac{y^2}{b^2} = 1$$

from this equation, we begin as follows:

$$|PF_1 - PF_2| = 2a$$
$$\sqrt{(x - c)^2 + y^2} - \sqrt{(x + c)^2 + y^2} = \pm 2a$$

The rest of the derivation is similar to that for the ellipse on page 362 except that where we substituted $b^2 = a^2 - c^2$ in the equation of the ellipse, we must now substitute $b^2 = c^2 - a^2$ in the equation of the hyperbola. This relationship between a, b, and c is illustrated in the diagram at the right.

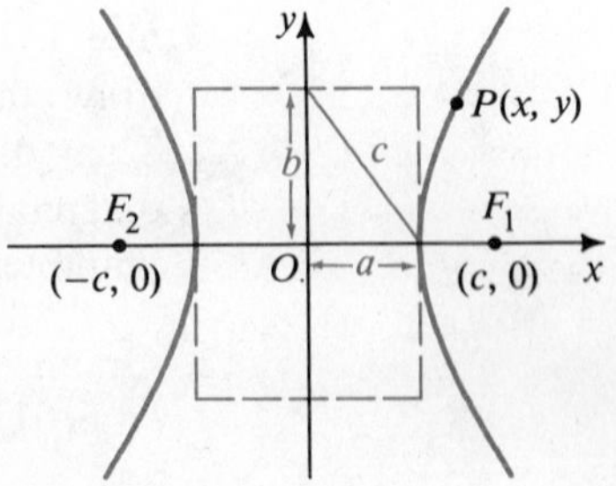

The equation

$$\frac{x^2}{a^2} - \frac{y^2}{b^2} = 1$$

can be solved for y as follows:

$$\frac{y^2}{b^2} = \frac{x^2}{a^2} - 1 = \frac{x^2 - a^2}{a^2}$$

$$y^2 = \frac{b^2}{a^2}(x^2 - a^2)$$

$$y = \pm\frac{b}{a}\sqrt{x^2 - a^2}$$

When the equation of the hyperbola is written in this form, we can deduce two things about its graph:

(1) Since we must take the square root of $x^2 - a^2$, we cannot have $x^2 < a^2$. That is, there are no points (x, y) on the hyperbola in the interval $-a < x < a$.

(2) When $|x|$ is very large, $\sqrt{x^2 - a^2}$ is approximately the same as $\sqrt{x^2} = |x|$. Therefore, when $|x|$ is large,

$$y = \pm\frac{b}{a}\sqrt{x^2 - a^2} \approx \pm\frac{b}{a}x.$$

The lines $y = \pm\frac{b}{a}x$ are called the **asymptotes** of the hyperbola and are illustrated in the diagram following. Notice that the hyperbola gets closer and closer to its asymptotes as $|x|$ gets larger. Other curves can also have asymptotes. In general, an asymptote of a curve is a line that the curve approaches more and more closely until they touch or almost touch.

The asymptotes for the hyperbola $\frac{x^2}{a^2} - \frac{y^2}{b^2} = 1$ are the diagonals of a rectangle with dimensions $2a$ and $2b$. Drawing the rectangle and its diagonals is a good first step in drawing the hyperbola. The next example illustrates this.

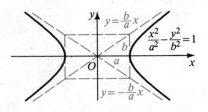

EXAMPLE 1. Sketch the hyperbola with equation $\frac{x^2}{36} - \frac{y^2}{9} = 1$. Find its eccentricity.

SOLUTION: 1. Comparing the given equation with the standard equation shows that $a = 6$ and $b = 3$. Use these values to draw the rectangle shown. Then draw and extend the diagonals, which are the asymptotes of the hyperbola.

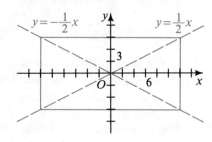

2. Graph the vertices, $(\pm 6, 0)$. Solve for y:

$$y = \pm \frac{1}{2}\sqrt{x^2 - 36}.$$

Use this equation and the symmetry of the hyperbola to graph additional points. The asymptotes will help you check your work.

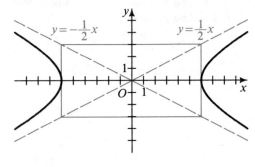

3. $c^2 = a^2 + b^2 = 36 + 9 = 45$
$c = \sqrt{45} = 3\sqrt{5}$

$e = \frac{c}{a} = \frac{3\sqrt{5}}{6} = \frac{\sqrt{5}}{2}$

So far, we have been concerned with hyperbolas whose major axes are horizontal. The graph of

$$\frac{y^2}{a^2} - \frac{x^2}{b^2} = 1$$

is also a hyperbola, but with a vertical major axis. Its asymptotes have equations $y = \pm \frac{a}{b}x$. Contrast the hyperbola in Example 2 on the next page with the one in Example 1.

EXAMPLE 2. Sketch the hyperbola with equation $\dfrac{y^2}{9} - \dfrac{x^2}{36} = 1$. Find its foci and its eccentricity.

SOLUTION:

1. Begin by drawing the same rectangle as in Example 1. Notice that the vertices $(0, 3)$ and $(0, -3)$ lie on the y-axis. Solve for y:

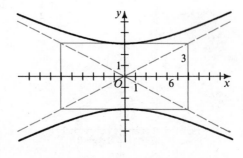

$$y = \pm \frac{1}{2}\sqrt{x^2 + 36}.$$

Compute additional points to complete the graph as shown.

2. From the standard equation, $a = 3$ and $b = 6$. Thus,

$$c^2 = a^2 + b^2$$
$$= 9 + 36 = 45$$

and

$$c = 3\sqrt{5}.$$
$$F_1(0, c) = (0, 3\sqrt{5})$$

and

$$F_2(0, -c) = (0, -3\sqrt{5}).$$

$$e = \frac{c}{a} = \frac{3\sqrt{5}}{3}$$
$$= \sqrt{5}$$

The eccentricity of a hyperbola is a measure of its "wideness." The eccentricity of the hyperbola in Example 2 is $\sqrt{5} \approx 2.2$, and the eccentricity of the hyperbola in Example 1 is $\dfrac{\sqrt{5}}{2} \approx 1.1$. As you would expect, the hyperbola with the greater eccentricity is more open than the one with the smaller value of e.

The diagrams below show hyperbolas with simple equations, that is, equations of the form $xy = c$ where c is a nonzero constant. As we shall see in Example 3, the coordinate axes are the asymptotes for these hyperbolas.

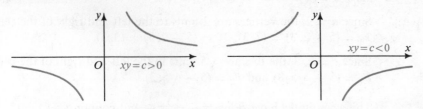

If a hyperbola is translated h units along the x-axis and k units along the y-axis, its equation is obtained by replacing x by $x - h$ and y by $y - k$. Its new center is the point (h, k).

EXAMPLE 3. Sketch the graphs of $xy = 6$ and $(x - 2)(y + 3) = 6$.

SOLUTION: The graph of the hyperbola with equation $xy = 6$ is shown at the left below. If we translate this graph 2 units to the right and 3 units down, we get the graph of $(x - 2)(y + 3) = 6$, as shown at the right below.

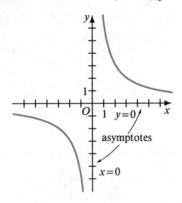
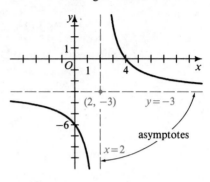

EXAMPLE 4. Find the center, vertices, foci, and asymptotes of the hyperbola with equation $x^2 - 4y^2 - 10x + 24y - 15 = 0$. Sketch its graph.

SOLUTION: Complete the squares:

$$x^2 - 4y^2 - 10x + 24y - 15 = 0$$
$$(x^2 - 10x \quad) - 4(y^2 - 6y \quad) = 15$$
$$(x^2 - 10x + 25) - 4(y^2 - 6y + 9) = 15 + 25 - 4 \cdot 9$$
$$(x - 5)^2 - 4(y - 3)^2 = 4$$
$$\frac{(x - 5)^2}{4} - \frac{(y - 3)^2}{1} = 1$$

This equation has the standard form

$$\frac{(x - h)^2}{a^2} - \frac{(y - k)^2}{b^2} = 1.$$

Thus $a = 2$, $b = 1$, and the center $(h, k) = (5, 3)$. Also, $c = \sqrt{a^2 + b^2} = \sqrt{4 + 1} = \sqrt{5}$.

Since $a = 2$, the vertices are 2 units to the left and right of the center. $V_1 = (5 + 2, 3) = (7, 3)$; $V_2 = (5 - 2, 3) = (3, 3)$.

Since $c = \sqrt{5}$, the foci are $\sqrt{5}$ units to the left and right of the center. $F_1 = (5 + \sqrt{5}, 3)$ and $F_2 = (5 - \sqrt{5}, 3)$.

The asymptotes have slope $\pm \dfrac{b}{a} = \pm \dfrac{1}{2}$ and contain $(h, k) = (5, 3)$.

Thus, their equations are $\dfrac{y - 3}{x - 5} = \dfrac{1}{2}$ and $\dfrac{y - 3}{x - 5} = -\dfrac{1}{2}$.

You may find it helpful to graph the hyperbola by considering points on the graph of $\dfrac{x^2}{4} - \dfrac{y^2}{1} = 1$ and translating each point 5 units to the right and 3 units up.

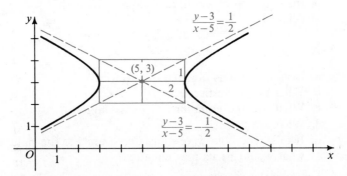

One use of hyperbolas is illustrated in the diagram below. Points A and B represent two *long range navigational* stations (abbreviated LORAN). These stations simultaneously send electronic signals to ships at sea. A receiver aboard the ship converts the time difference in receiving these signals into a distance difference, $PA - PB$. This distance difference locates the ship on a hyperbola with foci A and B. By receiving signals from stations B and C, the ship can be located on a second hyperbola, and therefore its position can be determined.

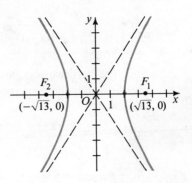

ORAL EXERCISES 11-2

1. Study the hyperbola shown and tell:
   a. its vertices.
   b. its foci.
   c. its eccentricity.
   d. whether its major axis is horizontal or vertical.
   e. its equation.
   f. the equations of its asymptotes.
   g. Of the hyperbolas shown in Examples 1 and 2, which has the greater eccentricity? Why?

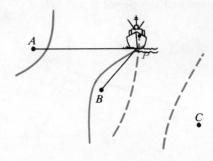

2. Study the hyperbola shown and tell:

    a. its equation.

    b. the equations of its asymptotes.

    c. its vertices.

    d. the equation of a congruent hyperbola with center $(-2, 1)$.

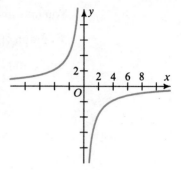

3. A hyperbola has equation $\dfrac{y^2}{25} - \dfrac{x^2}{1} = 1$.

    a. Is its major axis horizontal or vertical?

    b. What are its vertices?

    c. What are the equations of its asymptotes?

    d. What are its foci?

    e. If this hyperbola were translated 6 units to the right and 5 units down, what would be its new equation?

4. Describe the graphs of $x^2 - y^2 < 4$ and $x^2 - y^2 > 4$.

5. In figure (a) below, a piece of string is tied to two thumbtacks at A and B. Then a pencil traces along the paper keeping the string taut. What sort of curve will the pencil trace? Why?

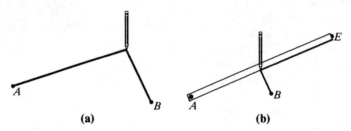

       (a)   **(b)**

6. In figure (b) above, a ruler is pivoted at A. One end of a piece of string is thumbtacked to the end of the ruler and the other end of the string is thumbtacked to the paper at B. If the string is held taut against the ruler by a pencil, and if the ruler is rotated about A, what sort of curve will the pencil trace? Why?

WRITTEN EXERCISES 11-2

Find the vertices, foci, and eccentricity of the hyperbola whose equation is given. Sketch the hyperbola. Include the asymptotes and their equations in your sketch.

A   **1.** $x^2 - y^2 = 1$ **2.** $y^2 - x^2 = 1$ **3.** $\dfrac{y^2}{4} - \dfrac{x^2}{1} = 1$

        4. $\dfrac{x^2}{16} - \dfrac{y^2}{9} = 1$ **5.** $4y^2 - 9x^2 = 36$ **6.** $25x^2 - 9y^2 = 225$

Find an equation of the hyperbola, with center at the origin, that satisfies the given conditions.

7. A vertex at $(6, 0)$ and a focus at $(10, 0)$

8. A vertex at $(0, -12)$ and a focus at $(0, -13)$

9. A vertex at $(0, -2)$ and an asymptote with equation $y = -x$

10. A vertex at $(8, 0)$ and an asymptote with equation $y = \frac{1}{2}x$

Sketch each hyperbola and its asymptotes.

11. **a.** $xy = 12$ **b.** $xy = -12$ **c.** $(x + 3)(y + 4) = 12$

12. **a.** $xy = 8$ **b.** $xy = -8$ **c.** $(x - 2)(y + 4) = -8$

13. On one set of axes, sketch the graphs of $x^2 - y^2 = k$ where $k = 4, 1, 0,$ $-1,$ and -4.

14. On one set of axes, sketch the graphs of $xy = k$ where $k = 12, 8, 4, 1,$ and 0.

Sketch the graphs of the given equations on the same set of axes. Then determine algebraically where the graphs intersect.

15. $x + y = 5$
    $x^2 - y^2 = 16$

16. $y^2 - x^2 = 9$
    $4x - y = 3$

17. $xy = -20$
    $x + 2y = 6$

18. $xy = 24$
    $2x + y = -14$

Sketch each half-hyperbola. Draw and label the asymptotes.

19. **a.** $y = \sqrt{16 + 4x^2}$

    b. $y = -\sqrt{16 + 4x^2}$

20. **a.** $x = \dfrac{\sqrt{16 + y^2}}{2}$

    b. $x = -\dfrac{\sqrt{16 + y^2}}{2}$

Sketch the hyperbola whose equation is given. Find the coordinates of the vertices and the foci, and the equations for the asymptotes. Give the eccentricity of each hyperbola.

21. $\dfrac{(x - 6)^2}{36} - \dfrac{(y - 8)^2}{64} = 1$

22. $\dfrac{(y + 5)^2}{16} - \dfrac{x^2}{9} = 1$

B 23. $y^2 - x^2 - 2y + 4x - 4 = 0$ 24. $x^2 - 4y^2 - 2x + 16y - 19 = 0$

25. $4x^2 - y^2 + 32x - 10y + 35 = 0$ 26. $y^2 - 9x^2 - 6y - 18x - 9 = 0$

Find an equation of the hyperbola described.

27. Center is $(5, 0)$; one vertex is $(9, 0)$; eccentricity $= \frac{5}{4}$.

28. Vertices are $(4, 0)$ and $(4, 8)$; asymptotes have slopes ± 1.

29. Foci are $(-4, -3)$ and $(-4, 7)$; one vertex is $(-4, 5)$.

30. Hyperbola passes through $(0, 0)$; asymptotes are $x = 5$ and $y = 4$.

31. **a.** If $F_1 = (10, 0)$, $F_2 = (-10, 0)$, and $P = (x, y)$, write an equation that expresses the fact that $|PF_1 - PF_2| = 16$.

    b. Simplify this equation to one of the form $\dfrac{x^2}{a^2} - \dfrac{y^2}{b^2} = 1$.

32. a. If $F_1 = (0, 13)$, $F_2 = (0, -13)$, and $P = (x, y)$, write an equation that expresses the fact that $|PF_1 - PF_2| = 10$.

b. Simplify this equation to one of the form $\dfrac{y^2}{a^2} - \dfrac{x^2}{b^2} = 1$.

Sketch the graph of each inequality.

33. a. $xy \geq 8$ **b.** $xy \geq 1$ **c.** $xy \geq 0$

34. a. $xy \leq -4$ **b.** $xy \leq -1$ **c.** $xy \leq 0$

35. $x^2 - 4y^2 < 4$ **36.** $y^2 - x^2 > 1$

37. $(x - 5)^2 - (y - 4)^2 \geq 1$ **38.** $(x + 4)(y + 3) \leq 12$

39. The equations $x = a \sec \theta$ and $y = b \tan \theta$ are called the *parametric equations* of the hyperbola $\dfrac{x^2}{a^2} - \dfrac{y^2}{b^2} = 1$.

  a. Show by substitution that the parametric equations satisfy the rectangular equation.

  b. What is the eccentricity of the hyperbola with parametric equations $x = 5 \sec \theta$ and $y = 12 \tan \theta$?

C   **40.** Complete the derivation of the equation $\dfrac{x^2}{a^2} - \dfrac{y^2}{b^2} = 1$ that was begun on page 369.

  41. Let $F_1 = (c, c)$, $F_2 = (-c, -c)$, and $P = (x, y)$. Show that the condition $|PF_1 - PF_2| = 2c$, $c > 0$, simplifies to the equation $xy = \frac{1}{2}c^2$.

  42. Let $x = \dfrac{e^\theta + e^{-\theta}}{2}$ and $y = \dfrac{e^\theta - e^{-\theta}}{2}$. Show that for all values of θ, the point (x, y) is on the hyperbola $x^2 - y^2 = 1$.

  Note: x and y are called the hyperbolic cosine and hyperbolic sine of θ, respectively. (See the following Calculator Exercise.)

CALCULATOR EXERCISE

Sin θ and cos θ are often called "circular functions" in contrast with the "hyperbolic functions" sinh θ and cosh θ (sinh θ is pronounced "sinch θ" and cosh θ sounds like "gosh θ"). For every value of θ, the point $(\cos \theta, \sin \theta)$ lies on the "unit circle" $x^2 + y^2 = 1$ and the point $(\cosh \theta, \sinh \theta)$ lies on the "unit hyperbola" $x^2 - y^2 = 1$.

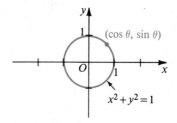
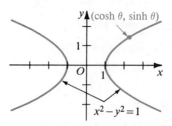

If your calculator has hyperbolic functions, use it to evaluate $(\cosh \theta, \sinh \theta)$ for $\theta = 0, \pm 1, \pm 2, \pm \pi$. Plot these points and sketch the curve that contains them. For each of these values of θ, verify that $\cosh^2 \theta - \sinh^2 \theta = 1$.

A particle moves so that its position (x, y) at time t is given by the parametric equations $x = \sec t$ and $y = \tan t$.

a. If your computer has a visual display, write a program that will show the path of the particle for $0 \le t \le 2\pi$.

b. If your computer does not have a visual display, write a program that will print a table giving the x- and y-coordinates of the particle for $0 \le t \le 2\pi$. Then sketch the path of the particle.

11-3 / PARABOLAS

When we studied quadratic equations in Section 2–3, we stated that their graphs were parabolas without actually defining a parabola. In this section we shall define a parabola and present certain general results concerning parabolas.

Definition of parabola

A parabola is the set of all points P in a plane that are equidistant from a fixed point F, called the focus, and a fixed line D (not containing F), called the directrix. That is, if F and D are fixed, then the set of all points P such that $PF = PN$ is a parabola (see diagram).

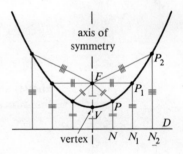

The line through the focus F perpendicular to the directrix D is a line of symmetry for the parabola. This line of symmetry is sometimes called the axis of symmetry. It intersects the parabola at its vertex V. Notice that the vertex, like every other point on the parabola, is equidistant from F and D.

Equation of a parabola with vertex at (0, 0)

If the line $y = -p, p \ne 0$, is the directrix of a parabola with focus $F(0, p)$, then by the definition above:

$$PF = PN$$
$$\sqrt{x^2 + (y - p)^2} = |y + p|$$

Squaring and simplifying, we get:

$$x^2 + (y - p)^2 = (y + p)^2$$
$$x^2 + y^2 - 2yp + p^2 = y^2 + 2yp + p^2$$
$$x^2 = 4py$$

$$(1) \quad y = \frac{1}{4p}x^2$$

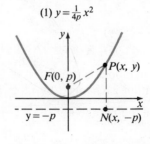

The graphs and equations below show other important parabolas with a vertex at $(0, 0)$. The derivations of their equations are analogous to that shown on the previous page. (See Exercise 30.)

(2) $y = -\frac{1}{4p}x^2$

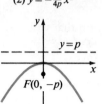

(3) $x = \frac{1}{4p}y^2$

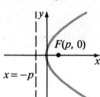

(4) $x = -\frac{1}{4p}y^2$

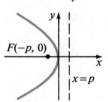

EXAMPLE 1. Find the focus and directrix of each parabola whose equation is given.

     a. $y = 2x^2$ **b.** $x = \frac{1}{20}y^2$

SOLUTION: **a.** Comparing $y = 2x^2$ and $y = \frac{1}{4p}x^2$ (equation (1) on page 377) shows that $p = \frac{1}{8}$. Thus the focus is $F\left(0, \frac{1}{8}\right)$ and the directrix is $y = -\frac{1}{8}$.

     b. Comparing $x = \frac{1}{20}y^2$ with $x = \frac{1}{4p}y^2$ (equation (3) above) shows that $p = 5$. Thus the focus is $F(5, 0)$ and the directrix is $x = -5$.

EXAMPLE 2. Find an equation of the parabola with vertex $(0, 0)$ and directrix $x = 2$.

SOLUTION: A sketch shows that the parabola has an equation of type (4), $x = -\frac{1}{4p}y^2$. Since p is the distance from the vertex to the directrix, $p = 2$. The equation is $x = -\frac{1}{4 \cdot 2}y^2$, or $x = -\frac{1}{8}y^2$.

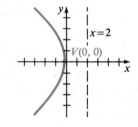

Equation of a parabola with vertex at (h, k)

If a parabola is translated so that its vertex is (h, k) instead of $(0, 0)$, then its equation is changed by replacing x by $x - h$ and y by $y - k$. For example, when the parabola $y = \frac{1}{4p}x^2$ is translated so its vertex is (h, k), its new equation is $y - k = \frac{1}{4p}(x - h)^2$. This observation is used in the next examples.

EXAMPLE 3. Find the vertex, axis, focus, and directrix of the parabola with equation $y = 2x^2 - 12x + 7$.

SOLUTION: Complete the square.
$$y = 2x^2 - 12x + 7$$
$$y - 7 = 2(x^2 - 6x \qquad)$$
$$y - 7 + 2 \cdot 9 = 2(x^2 - 6x + 9)$$
$$y + 11 = 2(x - 3)^2$$

Compare:
$$y - k = \frac{1}{4p}(x - h)^2$$

Thus vertex $= (h, k) = (3, -11)$, and $p = \frac{1}{8}$.
The axis has equation $x = 3$.
The focus is $(3, -10\frac{7}{8})$ and the directrix has equation $y = -11\frac{1}{8}$.

EXAMPLE 4. Find an equation of the parabola with focus $F(1, -1)$ and directrix $y = -7$.

SOLUTION: *Method 1.* Since the vertex of the parabola is half-way between the focus and the directrix, the vertex is $V(1, -4)$. Also $p = 3$, the distance from the focus to the vertex. This parabola is related to the type (1) parabola $y = \frac{1}{4 \cdot 3}x^2$ by a translation 1 unit right and 4 units down.

Hence its equation is $y + 4 = \frac{1}{4 \cdot 3}(x - 1)^2$.

Method 2. Use the definition of a parabola as the set of points $P(x, y)$ equidistant from the focus $F(1, -1)$ and the directrix $y = -7$.

$$PF = PN$$
$$\sqrt{(x - 1)^2 + (y + 1)^2} = |y - (-7)|$$
$$x^2 - 2x + 1 + y^2 + 2y + 1 = y^2 + 14y + 49$$
$$x^2 - 2x + 1 = 12y + 48$$
$$(x - 1)^2 = 12(y + 4)$$
$$\frac{1}{12}(x - 1)^2 = y + 4$$

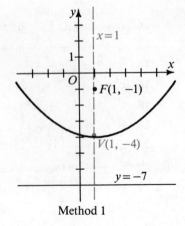

Method 1

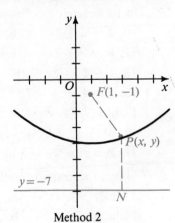

Method 2

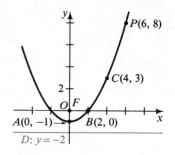

1. The points A, B, C, and P all lie on the parabola $y = \frac{1}{4}x^2 - 1$, whose focus is $F = (0, 0)$ and whose directrix D is the line $y = -2$.
   a. Evaluate the distances from A to F and A to D.
   b. Evaluate the distances from B to F and B to D.
   c. Evaluate the distances from C to F and C to D.
   d. Evaluate the distances from P to F and P to D.

2. The point $P = (x, y)$ is on the parabola whose focus F is $(-4, 0)$ and whose directrix is the y-axis.
   a. Express in terms of x and y
   (1) the distance PF;
   (2) the distance PN, from P to the directrix.
   b. Explain how to *derive* an equation of the parabola.

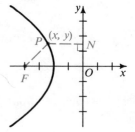

Give the focus and the directrix of each parabola.

3. $y = x^2$
4. $y = -2x^2$
5. $y - 3 = \frac{1}{4}(x + 6)^2$
6. $x = y^2$
7. $x = (y - 2)^2$
8. $x + 5 = -3(y - 1)^2$

WRITTEN EXERCISES 11-3

Find an equation of the parabola specified and sketch its graph.

A
1. Focus, $(-1, 0)$; directrix, $x = 1$
2. Focus, $(0, \frac{1}{4})$; directrix, $y = -\frac{1}{4}$
3. Focus, $(2, \frac{5}{2})$; directrix, $y = \frac{3}{2}$
4. Focus, $(1, 2)$; directrix, $x = 5$
5. Vertex, $(0, 0)$; focus, $(0, -\frac{1}{4})$
6. Vertex, $(0, 0)$; focus, $(\frac{3}{2}, 0)$
7. Vertex, $(-2, -1)$; directrix, $x = -4$
8. Vertex, $(-3, 2)$; directrix, $y = \frac{9}{2}$

For each parabola give the coordinates of its vertex and focus and the equation of its directrix.

9. a. $y = \frac{1}{8}x^2$ b. $x = \frac{1}{8}y^2$
10. a. $y = -\frac{1}{12}x^2$ b. $x = -\frac{1}{12}y^2$
11. a. $y = -2x^2$ b. $x = -2y^2$
12. a. $y = x^2 + 1$ b. $x = y^2 + 1$
13. $y - 1 = \frac{1}{4}(x - 2)^2$
14. $y + 3 = \frac{1}{8}(x - 5)^2$
15. $x - 4 = (y - 7)^2$
16. $x + 2 = -2(y - 3)^2$
17. $y = 2x^2 - 8x + 3$
18. $y = 5 - 6x - 3x^2$
19. $6x - x^2 = 8y + 1$
20. $4y = x^2 - 8x + 12$
21. $x = y^2 - 2y - 5$
22. $y^2 - 6y + 16x + 25 = 0$

Sketch the graph of each "half-parabola."

23. a. $y = \sqrt{x}$ b. $y = -\sqrt{x}$ c. $y = -\sqrt{x - 3}$ d. $y = \sqrt{x + 2} - 1$
24. a. $x = \sqrt{y}$ b. $x = -\sqrt{y}$ c. $x = \sqrt{y - 4}$ d. $x = -\sqrt{y + 3} + 1$

25. The diagram at the right shows many points $P(x, y)$ that are equidistant from $F(0, 6)$ and the x-axis.

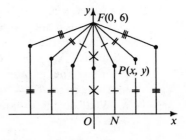

   a. Write an equation that states that $PF = PN$.

   b. Simplify the equation and show that it has the form

$$y = ax^2 + bx + c.$$

In Exercises 26–29 the coordinates of point F and an equation of line D are given. Write and simplify an equation that specifies the set of points $P(x, y)$ that are equidistant from F and D. Give your answer in the form $y = ax^2 + bx + c$ or $x = dy^2 + ey + f$.

26. $F = (0, 2)$
   $D: y = -2$

27. $F = (0, -3)$
   $D: x = -2$

28. $F = (-1, 3)$
   $D: x = 0$

29. $F(-7, -5)$
   $D: y = -7$

B 30. Derive Equation (3), page 378, by showing that a parabola with focus $(p, 0)$ and directrix $x = -p$ has equation $x = \dfrac{1}{4p}y^2$.

31. The diagram at the right shows a 45 m tower whose top is represented by the point $(0, 45)$. If a ball is thrown horizontally from the top of the tower at 30 m/s, its position (x, y), t seconds later, is given by the parametric equations $x = 30t$ and $y = 45 - 5t^2$.

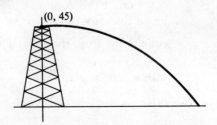

   a. How far from the base of the tower does the ball land?

   b. Find a single equation in x and y that describes the path of the ball.

32. A football is kicked from a point 30 yards directly in front of the goal posts. If a coordinate system is set up with the ball at $(0, 0)$ as shown, the position (x, y) of the ball t seconds after it is kicked is given by the parametric equations $x = 40t$ and $y = 40t - 16t^2$ where x and y are in feet.

   a. Find a single equation in x and y that describes the path of the ball.

   b. Will the ball pass over the goal post crossbar that is 10 feet above the ground?

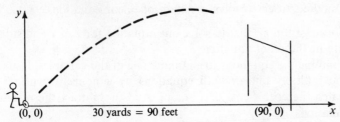

33. A particle moves so that its position at time t is given by the parametric equations $y = \cos 2t$ and $x = \cos t$. Show that the path of the particle is parabolic.

34. A projectile is fired from a cannon whose angle of elevation is θ and whose muzzle velocity is v. If the muzzle is at the origin of a coordinate system, the position $P(x, y)$ of the particle t seconds later is given by the parametric equations $x = (v \cos \theta)t$ and $y = (v \sin \theta)t - 5t^2$.

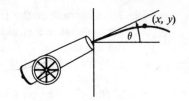

a. What do these parametric equations give when $\theta = 90°$?

b. Show that the projectile will hit the ground ($y = 0$) when $t = \dfrac{v \sin \theta}{5}$.

Also show that it will hit the ground at $x = \dfrac{v^2 \sin 2\theta}{10}$.

c. Use part (b) to conclude that the cannon can fire the longest horizontal distance when $\theta = 45°$.

d. Derive an equation of the path of the projectile. This equation will show that the path is a parabola.

C **35. a.** Suppose a parabola has focus $F(p, q)$ and directrix $y = k$. Prove that the equation of the parabola has the form $y = ax^2 + bx + c$, $a \neq 0$. (This proves that any parabola with a vertical axis of symmetry has an equation of this form.)

b. Prove that any point $P = (x, y)$ on the graph of $y = ax^2 + bx + c$ is equidistant from the point $F = \left(\dfrac{-b}{2a}, \dfrac{4ac - b^2 + 1}{4a}\right)$ and the line D with equation $y = \dfrac{4ac - b^2 - 1}{4a}$.

(This proves that the graph of $y = ax^2 + bx + c$ is a parabola.)

CALCULATOR EXERCISE

Refer to Exercise 34 and suppose that the cannon is fired at an angle $\theta = 28°$ with the ground, and with a velocity $v = 98$. Where and when will the projectile hit the ground?

11-4/SYSTEMS OF SECOND-DEGREE EQUATIONS

The methods for solving two second-degree equations in two unknowns are similar to the methods used to solve a system of linear equations.

Method 1: Use substitution; that is, solve one equation for x or y and substitute in the other equation.

Method 2: Combine the equations to eliminate one of the variables. Usually, you multiply one or both equations by a nonzero constant to match the coefficients of x^2 or y^2. Add or subtract the resulting equations.

EXAMPLE 1. Solve the system: $xy = 6$
$x^2 + 4y^2 = 64$

SOLUTION: We use Method 1 because the first equation is easy to solve for x.

Step 1. *Solve one equation for x or y.* In this case we solve $xy = 6$ for x:

$$x = \frac{6}{y}$$

Step 2. *Substitute in the other equation and solve.* In this case we substitute $\frac{6}{y}$ for x in the second equation.

$$\left(\frac{6}{y}\right)^2 + 4y^2 = \frac{36}{y^2} + 4y^2 = 64$$

$$4y^4 - 64y^2 + 36 = y^4 - 16y^2 + 9 = 0$$

$$y^2 = \frac{16 \pm \sqrt{16^2 - 4 \cdot 9}}{2} = \frac{16 \pm \sqrt{220}}{2} = 8 \pm \sqrt{55}$$

$$y^2 \approx 8 \pm 7.416 = 15.416 \text{ or } 0.584$$

$$y \approx \pm 3.926 \text{ or } \pm 0.764$$

Step 3. *Find the corresponding values of the other variable.* In this case we substitute in the equation $x = \frac{6}{y}$. When $y = 3.926$, for example, $x = 6 \div 3.926 = 1.528$. The four approximate solutions, $(1.5, 3.9)$, $(7.9, 0.8)$, $(-1.5, -3.9)$, $(-7.9, -0.8)$, are shown at the right.

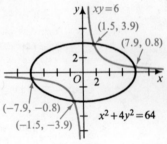

EXAMPLE 2. Solve the following equations simultaneously:

$$2x^2 + y^2 = 4 \quad \text{and} \quad x^2 - 2y^2 = 12.$$

SOLUTION: Use Method 2.

| Multiply by 2: |

$2x^2 + y^2 = 4 \quad\longrightarrow\quad 4x^2 + 2y^2 = 8$

$x^2 - 2y^2 = 12 \quad\longrightarrow\quad x^2 - 2y^2 = 12$

Add: $\quad 5x^2 \qquad = 20$

$\qquad\qquad x^2 \qquad\;\; = 4$

$\qquad\qquad\qquad x = \pm 2$

Now substitute $x = \pm 2$ in either equation to obtain $y = \pm 2i$. Since there is no solution (x, y) where both x and y are real numbers, the ellipse $2x^2 + y^2 = 4$ and the hyperbola $x^2 - 2y^2 = 12$ do not intersect. You can verify this result by graphing the given equations.

As noted in the introduction on page 360, the graph of a circle, ellipse, parabola, or hyperbola can degenerate into a single point, a line, a pair of lines, or even no points. By examining each of the following second-degree equations carefully, you should be able to spot these **degenerate conic sections.** Here are some examples:

1. $(x - 1)^2 + (y + 3)^2 = 0$ — A circle with center $(1, -3)$ and radius 0. In other words, the single point $(1, -3)$.

2. $4x^2 + y^2 = -4$ — No real solutions (x, y). Thus, no graph.

3. $x^2 - 3xy + 2y^2 = 0$ or $(x - 2y)(x - y) = 0$ — Two intersecting lines with equations $x - 2y = 0$ and $x - y = 0$.

EXAMPLE 3. Solve the system: $x^2 - xy - 2y^2 = 0$
$$y = x^2 - 4x + 2$$

SOLUTION: Factor the first equation.

$$x^2 - xy - 2y^2 = 0$$
$$(x - 2y)(x + y) = 0$$
$$x = 2y \quad \text{or} \quad x = -y$$

Substitute $2y$ for x in the second equation:

$$y = (2y)^2 - 4(2y) + 2$$
$$0 = 4y^2 - 9y + 2$$
$$0 = (4y - 1)(y - 2)$$

$y = \frac{1}{4}$ or $y = 2$
$x = 2(\frac{1}{4})$ | $x = 2(2)$
$= \frac{1}{2}$ | $= 4$

Substitute $-y$ for x in the second equation:

$$y = (-y)^2 - 4(-y) + 2$$
$$0 = y^2 + 3y + 2$$
$$0 = (y + 2)(y + 1)$$

$y = -2$ or $y = -1$
$x = -(-2)$ | $x = -(-1)$
$= 2$ | $= 1$

The four solutions, $(\frac{1}{2}, \frac{1}{4})$, $(4, 2)$, $(2, -2)$, and $(1, -1)$ are shown in the graph at the right. Notice that the graph of
$$x^2 - xy - 2y^2 = 0$$
is a pair of intersecting lines.

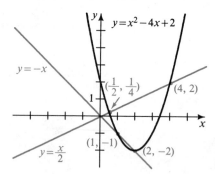

ORAL EXERCISES 11-4

1. Generally speaking, a system of two quadratic equations in two variables may have 4, 3, 2, 1, or 0 solutions. Illustrate each of these cases by sketching a circle and a parabola.

Describe the graph of each of the following degenerate conic sections. If there is no graph, say so.

2. $x^2 + 9y^2 = 0$

3. $x^2 - 9y^2 = 0$

4. $x^2 + 9y^2 + 1 = 0$

5. $4x^2 - y^2 = 0$

6. $4x^2 + y^2 = -1$

7. $4x^2 + y^2 = 0$

8. $(x - 3)(y - 2) = 0$

9. $xy = 0$

10. $y^2 - 3yx = 0$

11. $x^2 + xy = 0$ **12.** $x^2 - 4xy + 3y^2 = 0$ **13.** $x^2 - 4xy - 5y^2 = 0$

14. $x^2 + 4y^2 - 2x + 5 = 0$ **15.** $3x^2 + 4y^2 - 6x + 16y + 19 = 0$

WRITTEN EXERCISES 11-4

Sketch the graph of each system to determine the number of real solutions.
Then solve the system algebraically.

A **1.** $x^2 + 4y^2 = 16$ **2.** $4x^2 + y^2 = 16$ **3.** $x^2 + 9y^2 = 36$
      $x^2 + y^2 = 4$ $x^2 - y^2 = -4$ $x^2 - 2y = 4$

  4. $4x^2 + 4y^2 = 25$ **5.** $x^2 + y^2 = 25$ **6.** $y^2 - x^2 = 64$
      $2x + y^2 = 1$ $xy = -12$ $xy = 24$

  7. $x^2 - y^2 = 1$ **8.** $9x^2 + 4y^2 = 36$ **9.** $xy = 4$
      $y = -1 - x^2$ $x^2 + y^2 = 16$ $y = -4x^2$

  10. $9x^2 - 16y^2 = 144$ **11.** $(x + 3)^2 + y^2 = 1$ **12.** $x^2 + 6y^2 = 9$
       $x + y^2 = -4$ $x = -y^2$ $5x^2 + y^2 = 16$

B **13.** $x^2 + y^2 = 9$ **14.** $4x^2 + y^2 - 4y - 32 = 0$
       $x^2 + y^2 + 8x + 7 = 0$ $x^2 - y - 7 = 0$

  15. $2x^2 + 3xy - 2y^2 = 0$ **16.** $x^2 + y^2 = 40$
       $x^2 + 2y^2 = 6$ $x^2 + 2xy - 3y^2 = 0$

  17. $3x^2 - 4xy - 4y^2 = 0$ **18.** $x^2 + 2xy + y^2 = 4$
       $x - 2y^2 + 4 = 0$ $x^2 + y^2 - 2x + 4y - 20 = 0$

Graph the solution set of each inequality or system of inequalities.

  19. $x^2 - 9y^2 \leq 0$ **20.** $2xy - y^2 \geq 0$

  21. $x^2 + 4y^2 - 10x + 24y + 61 \leq 0$ **22.** $x^2 - y^2 + 2x - 2y \geq 0$

  23. $x^2 + 4y^2 \leq 16$ **24.** $9x^2 - y^2 \geq 0$
       $x^2 - 2xy \geq 0$ $x^2 + y^2 \leq 10$

Solve:

C **25.** $x^2 + xy + y^2 = 3$ **26.** $x^2 + 3y^2 = 3$
       $x^2 - y^2 = 3$ $3x^2 - xy = 6$

  27. A board 90 cm long just fits inside the bottom of a box 80 cm long and
  60 cm wide. How wide is the board?

CALCULATOR EXERCISE

Graph the system

$$x + y^2 = 11$$
$$x^2 + y = 7$$

to obtain approximate solutions. Then use your calculator to find the coordinates of each solution to the nearest hundredth.

11-5/COMMON PROPERTIES AND APPLICATIONS OF THE CONIC SECTIONS

As we saw on page 360, all of the conic sections can be obtained by slicing a double cone with a plane. Moreover, ellipses, hyperbolas, and parabolas can all be obtained from a single definition.

Common definition of ellipse, parabola, and hyperbola

Let F (the focus) be a fixed point not on a fixed line D (the directrix). Let P be a point in the plane of D and F, and let PN be the perpendicular distance from P to D. Now consider the set of points for which the ratio $PF:PN$ is the constant e. This set of points is:

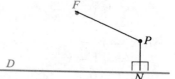

(1) an ellipse if $0 < e < 1$ (different values of e give ellipses of different shapes);
(2) a parabola if $e = 1$;
(3) a hyperbola if $e > 1$ (different values of e give hyperbolas of different shapes).

The number e is the eccentricity of the conic section. It is the same number e that we discussed earlier for ellipses and hyperbolas (pages 363 and 369). In discussions of conic sections e stands for eccentricity and should not be confused with the use of e as the base of the natural logarithm (page 179).

The diagram below suggests that once line D and point F are specified, the parabola with directrix D and focus F separates the set of ellipses and the set of hyperbolas having this same directrix and focus.

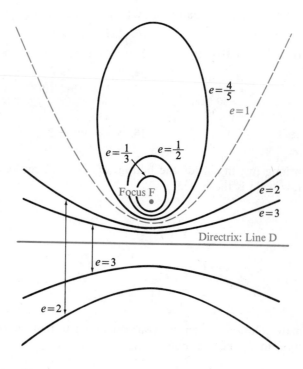

In Section 11-1 an ellipse was defined as a set of points P such that $PF_1 + PF_2 = $ a constant, and in our discussion in this section, an ellipse was defined as a set of points P such that $\dfrac{PF}{PN} = e$, a constant between 0 and 1. The next example shows that these different definitions yield the same equation for an ellipse.

EXAMPLE 1. Let the focus F be $(0, 3)$ and let the directrix D have equation $y = 12$. Find the equation of the set of points P for which $\dfrac{PF}{PN} = \dfrac{1}{2}$.

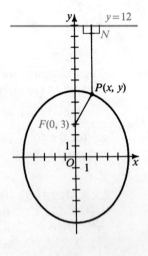

SOLUTION:

$$2(PF) = PN$$
$$2\sqrt{x^2 + (y-3)^2} = |12 - y|$$
$$4[x^2 + (y-3)^2] = (12 - y)^2$$
$$4x^2 + 4y^2 - 24y + 36 = 144 - 24y + y^2$$
$$4x^2 + 3y^2 = 108$$
$$\frac{x^2}{27} + \frac{y^2}{36} = 1$$

Thus, we have the ellipse shown, with vertices $(0, \pm6)$ and eccentricity $\frac{1}{2}$. Notice that this same ellipse could be obtained from a different initial condition, namely, that $PF_1 + PF_2 = 12$, where $F_1 = (0, 3)$ and $F_2 = (0, -3)$.

Analyzing the general equation

In our study of conic sections, we have considered primarily those with horizontal and vertical axes. The one exception has been the hyperbola of the form $xy = c$. If each axis of a conic is neither horizontal nor vertical, its equation will always contain an xy-term. It is possible to use trigonometry to analyze such equations, but we will not do so. Instead we will state without proof a theorem relating conic sections to second-degree equations.

THEOREM. Consider the second-degree equation

$$Ax^2 + Bxy + Cy^2 + Dx + Ey + F = 0.$$

If A, B, and C are not all 0 and if the graph is not degenerate, then the graph is:

(1) a circle or an ellipse if $B^2 - 4AC < 0$ (In a circle, $B = 0$ and $A = C$.)
(2) a parabola if $B^2 - 4AC = 0$
(3) a hyperbola if $B^2 - 4AC > 0$

(*Theorem continued on page 388*)

Conic sections   **387**

(4) Moreover, the following formula determines a **direction angle** α for the conic section, where α is an angle formed by the positive part of the x-axis and an axis of the conic section. If the conic section is a parabola, α may be an angle to the axis or a line perpendicular to the axis.

$$\alpha = \frac{\pi}{4} \qquad \text{if } A = C$$

$$\tan 2\alpha = \frac{B}{A - C} \quad \text{if } A \neq C, \, 0 < 2\alpha < \pi$$

In Example 2 we apply this formula to sketch some conic sections that are centered at the origin. We find the direction angle α and then sketch the lines $y = (\tan \alpha)x$ and $y = \dfrac{-1}{\tan \alpha}x$, which will be the axes (or axis and perpendicular) for the conic section. Then we use substitution and our knowledge of the particular conic to complete the graph.

EXAMPLE 2. Use the theorem just stated to identify the graph of the equation, find the angle α, and sketch the curve.
  a. $x^2 - 2xy + 3y^2 = 1$
  b. $x^2 - xy = -1$

SOLUTION: **a.** $B^2 - 4AC = (-2)^2 - 4(1)(3) = -8 < 0.$ $A \neq C$, so the graph is an ellipse.

$$\tan 2\alpha = \frac{B}{A - C} = \frac{-2}{1 - 3} = 1$$

Thus, $2\alpha = \dfrac{\pi}{4}$, and $\alpha = \dfrac{\pi}{8}$.

We sketch the axes

$$y = \left(\tan \frac{\pi}{8}\right)x \approx 0.41x$$

and $y = \dfrac{-1}{\tan \dfrac{\pi}{8}}x \approx -2.4x.$

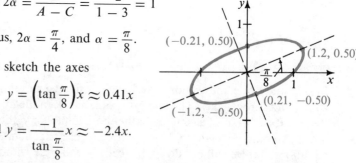

These axes intersect the graph at $(1.2, 0.50)$, $(-1.2, -0.50)$, $(0.21, -0.50)$, and $(-0.21, 0.50)$.

The graph intersects the x-axis and the y-axis at $(\pm 1, 0)$ and $\left(0, \pm\dfrac{\sqrt{3}}{3}\right)$, respectively. The graph is shown above.

b. $B^2 - 4AC = (-1)^2 - 4(1)(0) = 1 > 0$; the graph is a hyperbola.

$$\tan 2\alpha = \frac{B}{A - C} = \frac{-1}{1 - 0} = -1$$

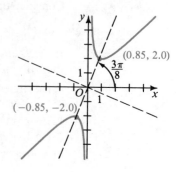

Thus $2\alpha = \dfrac{3\pi}{4}$ and $\alpha = \dfrac{3\pi}{8}$. We then sketch the axes

$$y = \left(\tan \frac{3\pi}{8}\right)x \approx 2.4x \quad \text{and} \quad y = \frac{-1}{\tan \dfrac{3\pi}{8}}x \approx -0.41x.$$

The graph intersects $y = 2.4x$ at $(0.85, 2.0)$ and $(-0.85, -2.0)$. A table of values gives the graph at the bottom of page 388.

Applications of conic sections

Many important scientific discoveries have been related to the conic sections. About 1590 Galileo discovered that a projectile fired horizontally from the top of a tower will fall to the earth along a parabolic path (if air resistance is ignored). And in 1609 Kepler hypothesized that all the planets move in elliptical orbits with the sun at one focus. Later that century, Newton used the elliptical-orbit hypothesis to formulate the Theory of Universal Gravitation, one of the greatest scientific discoveries of all time. Today it is known that satellites, comets, and atomic particles travel in paths that are very nearly elliptical, parabolic, or hyperbolic.

If a parabola is rotated about its axis, the surface it traces is called a paraboloid. A physical model of a paraboloid made of reflecting material is called a parabolic reflector. These reflectors are used in searchlights and automobile headlights. A light bulb at the focus of an automobile headlight reflects light in parallel "high beams" down the highway, as shown in the figure at the right. This figure shows a parabolic reflector being used to send light in parallel rays. Such reflectors can also be used to receive parallel light rays, waves, radio and sound waves. If the

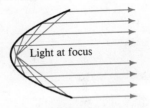

Light at focus

arrows were reversed in the figure, it would then illustrate the feeble light rays or radio waves coming from a distant star. In a parabolic receiver, the waves are all reflected to the focus, where their concentration can be recorded on film. Alternatively, the waves can be reflected by an antenna to some other point to be recorded. The same principle is used in the design of the solar furnace and the television equipment shown in the following photograph.

When watching a football game on TV, have you ever wondered how you can hear the quarterback's signals over the noise of the crowd? A parabolic reflector makes this possible. The reflector is pointed at the quarterback, whose voice is reflected to a microphone at the focus of the reflector.

Ellipses and hyperbolas have another reflection property. The two lines drawn from any point P on these curves to their foci will form equal angles with the tangent line at P. (See the figures that follow.) For the ellipse, this means that light or sound waves emitted from focus F_1 are reflected to focus F_2. This reflection property is the basis of "whispering chambers" in which the whispers of a person located at one focus can be heard clearly at the other focus even when the foci are far apart. There is such a chamber in the United States Capitol.

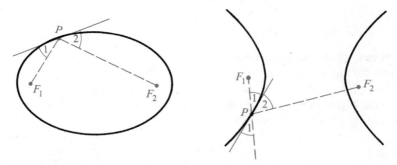

The reflecting property of hyperbolas is used in the Cassegrain telescope, which has two mirrors, one with a hyperbolic cross-section and the other with a parabolic cross-section. The following diagram shows a light ray from a star that is reflected toward F_1, the focus of the parabola and also one of the foci of the hyperbola. The ray is then reflected off the hyperbolic mirror to the eyepiece of the telescope, which is located at the other focus F_2 of the hyperbola.

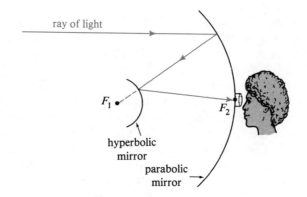

ORAL EXERCISES 11–5

Use the diagram on the next page for Exercises 1 and 2.

1. Find the distances of A, B, C, and P to:
   a. line D_1 with equation $x = 3$
   b. line D_2 with equation $y = -6$.

2. State an equation that expresses the fact that:
   a. $PO = 2 \cdot PA$ b. $PO = \frac{1}{2} \cdot PB$

c. the distance from P to C is equal to the distance from P to D_2.

d. the distance from P to B is twice the distance from P to D_2.

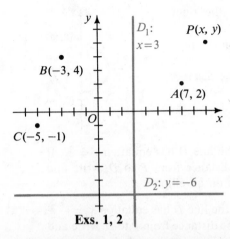

Exs. 1, 2

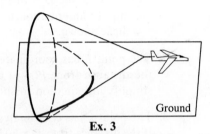

Ex. 3

3. The shock wave of a supersonic jet traveling parallel to the ground is in the shape of a cone trailing the jet. All points located on the intersection of this cone and the ground will receive a sonic boom at the same time. What kind of conic section will this intersection be?

4. The figure at the right represents an elliptical whispering chamber. If a person at focus F_1 whispers, all the resulting sound waves bounce off the ellipse and arrive at focus F_2. (For a proof of this property, see Written Exercises 15–18.) If the waves shown leave F_1 at the same time, explain why they will arrive at F_2 at the same time.

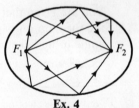

Ex. 4

5. Given that the following equations are not degenerate conics, tell whether the graph of each is a circle, an ellipse, a parabola, or a hyperbola.

a. $x^2 - 3xy + 2y^2 + 2x - y + 6 = 0$

b. $3x^2 - 4xy + 2y^2 - 3y = 0$

c. $x^2 - 6xy + 9y^2 + x - y - 1 = 0$

d. $144x^2 + 144y^2 - 216x + 96y - 47 = 0$

Experiments

1. Shine a flashlight against a wall so that the edge of the lighted area is **(a)** a circle, **(b)** an ellipse, **(c)** a parabola, and **(d)** a hyperbola. Relate this experiment to the diagrams of the conic sections on page 360.

2. Observe the light reflected from an ordinary pin-up lamp that reflects lighted areas on the ceiling and on the wall on which the lamp is hanging. Describe the boundaries of these lighted areas.

Conic sections   **391**

A **1. a.** By simply examining the ellipse at the right, write an equation for the ellipse.
   b. Find the coordinates of F, the focus on the positive x-axis.
   c. The distance from $P(x, y)$ to F is $\frac{1}{2}$ the distance from P to line D with equation $x = 4$. Write and simplify an equation to show that P lies on the ellipse shown.

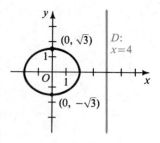

2. The point F has coordinates $(0, 8)$ and the line D has equation $y = \frac{25}{2}$. If the distance from $P(x, y)$ to F is $\frac{4}{5}$ the distance from P to D, write and simplify an equation to show that P lies on an ellipse.

3. The point F has coordinates $(-6, 0)$ and the line D has equation $x = -\frac{3}{2}$. If the distance from $P(x, y)$ to F is twice the distance from P to D, write and simplify an equation to show that $P(x, y)$ lies on a hyperbola.

4. The point F has coordinates $(0, -8)$ and the line D has equation $y = -2$. If the distance from $P(x, y)$ to F is twice the distance from P to D, show that P lies on a hyperbola.

You are given that the graphs of the following are not degenerate conics. Use the theorem on pages 387–388 to identify the graph of the equation, find the angle α, and sketch the curve.

5. $x^2 + xy + y^2 = 1$ **6.** $x^2 - xy + y^2 = 1$
7. $x^2 - 2xy - y^2 = 4$ **8.** $x^2 - xy + 2y^2 = 2$

B **9.** $x^2 - 2xy + y^2 - 4x\sqrt{2} - 4y\sqrt{2} = 0$
10. $x^2 - 2xy + y^2 - x\sqrt{2} - y\sqrt{2} = 0$

11. $y = x - \dfrac{1}{x}$ **12.** $x^2 - 3xy + y^2 = 5$

13. Consider the line l with equation $y = 2x - 4$ and the parabola with equation $4y = x^2$ and focus $(0, 1)$.
   a. By solving the equations simultaneously, verify that l is tangent to the parabola at $P(4, 4)$.
   b. The purpose of this exercise is to prove that $\angle 1 = \angle 2$. ($\angle 2$ is the angle formed by the tangent line l and the vertical line through P.)
   (1) Find the measures of $\angle 3$ and $\angle 4$. (*Hint:* $\tan \angle 3 = $ slope of $\overleftrightarrow{FP}$.)
   (2) Deduce the measures of $\angle 1$ and $\angle 2$.

14. As the focus B of an ellipse moves away from focus A, the eccentricity of the ellipse gets closer and closer to 1. For this reason, some people like to think of a parabola as "an ellipse with one focus at infinity." Using this point of view and the diagrams below, explain how the reflection property of the parabola (page 389) can be deduced from the reflection property of the ellipse.

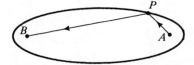

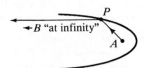

Exercises 15–18 explain why an ellipse has the reflection property shown on page 391. These problems are arranged sequentially, not odd/even.

15. One of the laws of physics states that light waves and radio waves reflect from a smooth surface in such a way that the angle of incidence is equal to the angle of reflection. Hence, in the diagram shown, $\angle 1 = \angle 2$. Suppose a person at A would like to shine a light in a mirror and have the light reflected to B. Show that if the person aims the light at the point B' that is on the opposite side of l from B and the same distance away from l, then $\angle 1 = \angle 2$ (and hence the light will reflect to B).

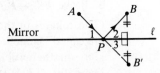

16. Use the figure for Exercise 15 to explain why:
    a. $AP + PB = AB'$
    b. the shortest distance from A to l to B is along the path of reflection shown. (*Hint:* Let Q be a point on l other than P. Show that $AQ + QB > AP + PB$.)
    Note: Since there is only one shortest path from A to l to B, part (b) shows that this shortest path is the path of reflection.

17. When a ray of light is reflected from a curved mirror, $\angle 1$ and $\angle 2$ are equal angles; $\angle 1$ and $\angle 2$ are the angles formed by the light rays and the tangent to the mirror at the point of reflection. Draw a large ellipse with foci at A and B using the method of Oral Exercise 5 on page 374. Then pick any point P on the ellipse and draw $\overline{AP}$, $\overline{BP}$, and the tangent to the ellipse at P. By measuring with a protractor, show that the angle between $\overline{AP}$ and the tangent is equal to the angle between $\overline{BP}$ and the tangent. Repeat the experiment for another point P.

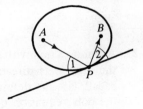

18. Line l is tangent to an ellipse at point P. This exercise shows that a ray of light traveling from focus A to point P will reflect to focus B.
    a. Explain why
    $$AQ + QB > AR + RB = AP + PB.$$
    b. Since $AQ + QB > AP + PB$, use Exercise 16(b) to explain why a ray of light traveling from A to P will reflect to B.

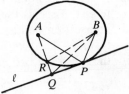

19. The diagram shows an ellipse formed by
a plane cutting through a cone. The diagram also shows two spheres which are
tangent to the plane of the ellipse at A
and B. These spheres are also tangent
to the cone, touching the cone in the parallel circles C_1 and C_2. Let P be any
point on the ellipse. (For visual simplicity, P is shown where it is.) The purpose
of this exercise is to prove that $PA + PB$
is a constant, thus proving that A and B
are the foci of the ellipse.

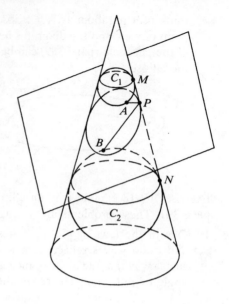

a. Explain why $PA = PM$ and $PB = PN$.
b. Prove $PA + PB$ is a constant.

(The spheres in the diagram at the right
are called **Dandelin spheres** after the Bel-
gian mathematician Germinal Pierre
Dandelin (1794–1847) who discovered
them in 1822.)

20. The diagram shows a hyperbola formed by a plane
cutting a double cone. Two spheres are tangent to
the plane of the hyperbola at A and B. These
spheres are also tangent to the cone, touching the
cone in the parallel circles C_1 and C_2. Let P be any
point on the hyperbola. (For visual simplicity, P is
shown where it is.) Use the diagram to show that
$PA - PB$ is a constant, thus proving that A and B
are the foci of the hyperbola.
(*Hint:* Show that $PA = PM$ and $PB = PN$.)

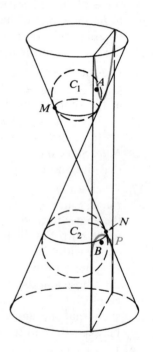

C **21.** A conic section with eccentricity e has its focus F at
the origin. Its directrix is the line $y = -p$, where
$p > 0$. Use the definition on page 386 to show that
the polar equation of the conic is

$$r = \frac{ep}{1 - e \sin \theta}$$

22. If you think about how two conic sections might intersect, it seems that there might be 0, 1, 2, 3, or 4
intersection points. Explain then why the 5 points
$(4, 0)$, $(-4, -4)$, $(0, 2)$, $(0, -2)$ and $(6, 1)$ all satisfy
the following two equations:

$$x^2 - 4xy + 4y^2 - 16 = 0$$
$$x^2 - 2xy - 4x = 0$$

23. The following theorem was known by the ancient Greeks: From a point O in the exterior region of a conic section, draw two tangents, $\overline{OR}$ and $\overline{OQ}$, to the conic. Let P be any other point and draw lines through P parallel to $\overline{OR}$ and $\overline{OQ}$, intersecting the conic in points A, B, C, and D, as shown. Then

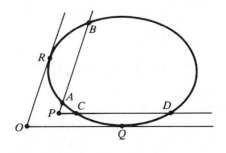

$$\frac{PA \cdot PB}{(RO)^2} = \frac{PC \cdot PD}{(QO)^2}.$$

a. If the ellipse shown above were a circle, then RO would equal __?__. What theorem about secants to a circle can then be deduced?

b. Illustrate the theorem when point P is in the interior of a circle. What theorem can you then deduce about intersecting chords within a circle?

c. Make a diagram illustrating the theorem for a parabola.

Chapter summary

1. Circles, ellipses, parabolas, and hyperbolas are called *conic sections* because it is possible to obtain these curves by slicing a cone (see figures on page 360). Some examples of these curves are given below and on page 396.

Equation	Curve
a. $y = ax^2 + bx + c$ $\quad a \neq 0$	Parabola with axis of symmetry $x = \dfrac{-b}{2a}$
$y = \pm\dfrac{1}{4p}x^2$	Parabola with vertex $(0, 0)$ and focus $(0, \pm p)$ (See diagams on pages 377 and 378.)
$x = \pm\dfrac{1}{4p}y^2$	Parabola with vertex $(0, 0)$ and focus $(\pm p, 0)$

Gertrude Rand Ferree received her Ph.D. in psychology in 1911. She later became interested in the study of human vision. She helped to develop many optical devices and instruments including the standard pictures used to identify color blindness. Rand was also involved in a study of night vision that was requested by the United States Navy.

	Equation	Curve

b. $(x - h)^2 + (y - k)^2 = r^2$ Circle with center (h, k) and radius r

c. $\dfrac{x^2}{a^2} + \dfrac{y^2}{b^2} = 1 \quad (a^2 \neq b^2)$ Ellipse with center $(0, 0)$, x-intercepts $\pm a$, y-intercepts $\pm b$

d. $\dfrac{x^2}{a^2} - \dfrac{y^2}{b^2} = 1$ Hyperbola with center $(0, 0)$, asymptotes $y = \pm \dfrac{b}{a}x$, x-intercepts $\pm a$

(See figure on page 369.)

$\dfrac{y^2}{a^2} - \dfrac{x^2}{b^2} = 1$ Hyperbola with center $(0, 0)$, asymptotes $y = \pm \dfrac{a}{b}x$, y-intercepts $\pm a$

$xy = c$ Hyperbola with center $(0, 0)$. The x- and y-axes are asymptotes.

e. $\dfrac{(x - h)^2}{a^2} \pm \dfrac{(y - k)^2}{b^2} = 1$ This graph is like the graph of $\dfrac{x^2}{a^2} \pm \dfrac{y^2}{b^2} = 1$ except its center is at (h, k).

2. The terminology and definitions for ellipses and hyperbolas are very similar and are summarized on page 368.

3. Given a line L (the directrix), a point F (the focus), and a positive constant e (the eccentricity), the set of points P such that $\dfrac{PF}{PN} = e$ is:

an ellipse if $0 < e < 1$.   (The bigger e is, the "skinnier" the ellipse.)

a parabola if $e = 1$.

a hyperbola if $e > 1$.   (The bigger e is, the more "open" the hyperbola.)

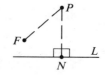

4. The graph of any equation of the form

$$Ax^2 + Bxy + Cy^2 + Dx + Ey + F = 0$$

is a conic section. The theorem on pages 387–388 provides a means of identifying what type of conic section the equation represents.

5. Two methods are shown on page 382 for solving two second-degree equations simultaneously.

Chapter test

1. Sketch each ellipse showing its vertices and foci. Also give the eccentricity of each.   11-1

   a. $4x^2 + 9y^2 = 36$ **b.** $x^2 + 9y^2 + 6x + 18y + 9 = 0$

2. Sketch each hyperbola showing vertices and foci. Give its eccentricity and the equations of its asymptotes.

11-2

   a. $\dfrac{x^2}{9} - \dfrac{y^2}{16} = 1$ **b.** $4y^2 - 25x^2 = 100$

3. Sketch each hyperbola and give the equations of its asymptotes.
   a. $xy = -12$ **b.** $(x - 4)(y + 3) = -12$

4. Sketch each parabola showing the vertex, focus, and directrix.

11-3

   a. $x = \frac{1}{8}y^2$ **b.** $y = 2x^2 - 4x + 7$

5. **a.** Solve simultaneously:

11-4

$$y^2 - x^2 = 6$$
$$xy = 4$$

   b. Sketch the graphs of the equations in part (a) to illustrate your solutions.

6. Given that the graph of the equation $-x^2 + 2xy - 3y^2 + 12 = 0$ is not a degenerate conic:

11-5

   a. Which type of conic does the equation represent?
   b. Find the angle α between the axis of the conic and the positive x-axis.
   c. Find the intercepts and use them to sketch a graph of the equation.

7. Given the point $F(0, 4)$ and the line $y = 2$. Consider the set of all points $P(x, y)$ such that $\dfrac{PF}{PN} = e$, where PN is the perpendicular distance from P to the line.

   a. Describe the set of points $P(x, y)$ if:
      (1) $0 < e < 1$ (2) $e > 1$
   b. Find the equation of the conic if $e = 1$.

CHAPTER TWELVE

Vectors and determinants

OBJECTIVES

1. To draw arrows representing sums, differences, and multiples of vectors given arrows representing the vectors.
2. To express a vector in polar or component form given its endpoints.
3. To perform vector operations in component form.
4. Given two points on a line, to find a direction vector, a vector equation, and a pair of parametric equations of the line.
5. To find the angle between two vectors or determine that they are parallel or perpendicular.
6. (Optional) To solve problems involving three-dimensional space using three-dimensional analogues of two-dimensional formulas.
7. To find the equation of a plane using vectors in three dimensions.
8. To use determinants to solve linear equations in two or more variables.

Properties and basic operations

12-1 / GEOMETRIC REPRESENTATION OF VECTORS

Vectors are quantities which are described by a *direction* and a *magnitude* (size). A force, for example, is a vector quantity because to describe it you must tell in which direction it is acting and with what strength. Another example of a vector quantity is velocity. The velocity of an airplane is described by giving the direction and speed of the airplane. Now let us consider a specific example.

Naum Gabo's Linear Construction No. 2 *is shown at the left. Gabo expressed the desire that the viewer conceive "the work's whole three-dimensional life in space."*

If an airplane is flying due east at 500 knots, its velocity can be represented by an arrow 500 units long pointing east.

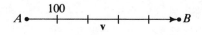

This velocity is called a **vector quantity,** or simply a **vector.** We shall denote this vector by the symbol **v.** (In handwritten work the vector could be denoted by $\vec{v}$ or $\underline{v}$.) If the arrow representing **v** extends from point A to point B, it is customary to write $\mathbf{v} = \overrightarrow{AB}$ and to refer to $\overrightarrow{AB}$ either as the velocity vector or as the arrow representing the velocity vector.*

There are, of course, many arrows which could represent the velocity **v.** For example, the arrows $\overrightarrow{CD}$ and $\overrightarrow{EF}$ shown below are both 500 units long and both point east. Hence, we may also write $\mathbf{v} = \overrightarrow{CD} = \overrightarrow{EF}$.

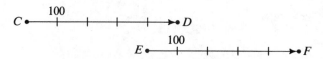

The statement $\overrightarrow{CD} = \overrightarrow{EF}$ does not mean that the two arrows are identical, but rather that they both represent the same velocity vector.

In general, any two arrows with the same length and the same direction represent the same vector. Hence, for parallelogram $ABCD$ shown at the right, we may write $\overrightarrow{AB} = \overrightarrow{DC}$ and $\overrightarrow{AD} = \overrightarrow{BC}$.

Addition of vectors

Let us consider again the airplane flying east at 500 knots. If it encounters a wind blowing toward the southeast at 100 knots, the resulting velocity of the plane is represented in the diagram by $\overrightarrow{AX}$, the diagonal arrow of parallelogram $ABXC$. Because the velocity $\overrightarrow{AX}$ results from combining the velocities $\overrightarrow{AB}$ and $\overrightarrow{AC}$, we shall write

$$\overrightarrow{AX} = \overrightarrow{AB} + \overrightarrow{AC},$$

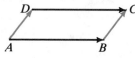

$$\begin{array}{ccc} \text{Plane's velocity} & \text{wind} & \text{Plane's resultant} \\ \text{with no wind} + & \text{velocity} = & \text{velocity} \end{array}$$

$$\overrightarrow{AB} \quad + \quad \overrightarrow{AC} \quad = \quad \overrightarrow{AX}$$

and call $\overrightarrow{AX}$ the **vector sum,** or **resultant,** of $\overrightarrow{AB}$ and $\overrightarrow{AC}$.

Forces, like velocities, also combine in a parallelogram pattern. In the diagram at the right below, $\overrightarrow{AB}$ and $\overrightarrow{AC}$ represent two forces acting on a body; $\overrightarrow{AB}$ represents a force of 20 newtons acting to the east and $\overrightarrow{AC}$ a force of 10 newtons acting to the northeast. (A newton (N) is the force required to bring a 1-kilogram object from rest to a speed of 1 meter per second in one second.) The resultant force is the force that has the same effect as these two forces combined. It is possible to verify experimentally that the resultant force is repre-

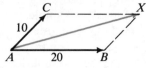

*The symbol $\overrightarrow{AB}$ is sometimes used to denote the ray that has vertex A and passes through point B. In this book, however, we shall reserve this type of notation for vectors.

sented by $\overrightarrow{AX}$, where X is the point that makes $ABXC$ a parallelogram. As before, $\overrightarrow{AX}$ is called the **vector sum** of $\overrightarrow{AB}$ and $\overrightarrow{AC}$, and we write $\overrightarrow{AX} = \overrightarrow{AB} + \overrightarrow{AC}$.

In the previous two diagrams, the arrows representing the vectors to be added both originated from the same point. It is often convenient to let these arrows be arranged consecutively, as shown at the left below. In this diagram, $\overrightarrow{PQ}$ represents a displacement of 10 kilometers east. This displacement is followed by a displacement of 5 kilometers northeast, represented by $\overrightarrow{QR}$. The net effect of these two consecutive displacements is the single displacement $\overrightarrow{PR}$, and we can write $\overrightarrow{PR} = \overrightarrow{PQ} + \overrightarrow{QR}$.

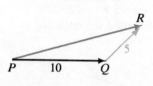

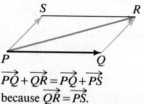

$$\overrightarrow{PQ} + \overrightarrow{QR} = \overrightarrow{PQ} + \overrightarrow{PS}$$
because $\overrightarrow{QR} = \overrightarrow{PS}$.

The diagram at the right above shows that this method of combining arrows is equivalent to the parallelogram method. Therefore, if you are given arrows representing **v** and **w**, then you can find an arrow representing **v** + **w** by either of the two methods shown below.

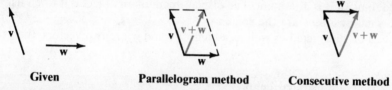

Given   **Parallelogram method**   **Consecutive method**

Note that in the parallelogram method the original arrows are placed "tail to tail," whereas in the consecutive method the original arrows are placed "head to tail," with the sum arrow extending from the "tail" of the first arrow to the "head" of the last arrow.

When three or more vectors are to be added, the consecutive method is the easier. (See the diagram below.)

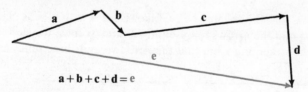

$$a + b + c + d = e$$

Absolute value of a vector

The **absolute value** (also called **magnitude** or **norm**) of a vector **v**, denoted by the symbol $|\mathbf{v}|$, is equal to the length of the arrow representing **v**; $|\mathbf{v}|$ can be interpreted in a variety of ways. For example:

1. If **v** is an airplane velocity vector, then $|\mathbf{v}|$ is the speed of the airplane.
2. If **v** is a force vector, then $|\mathbf{v}|$ is the strength of the force.
3. If $\mathbf{v} = \overrightarrow{AB}$ is a displacement vector, then $|\mathbf{v}|$ is the distance AB.

The diagram below shows that the absolute value of a vector on the number line with its tail at O is similar to the absolute value of a number.

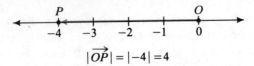

$$|\overrightarrow{OP}| = |-4| = 4$$

Negative of a vector; zero vector; vector subtraction

If an object moves from A to B, its displacement is the vector $\mathbf{v} = \overrightarrow{AB}$. The displacement $\overrightarrow{BA}$ that moves the object from B back to A is called the **negative** of vector $\mathbf{v}$, and we write

$$\overrightarrow{BA} = -\mathbf{v} = -\overrightarrow{AB}.$$

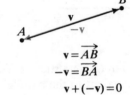

Since the displacement $\mathbf{v}$ followed by the displacement $-\mathbf{v}$ returns the object to its starting point, we write

$$\mathbf{v} + (-\mathbf{v}) = 0$$

and call $\mathbf{0}$ the **zero vector.** It is difficult to represent $\mathbf{0}$ geometrically. Perhaps the best way to think of it is as a point. The displacement of an object is $\mathbf{0}$ when its initial and final positions are the same.

Vectors can be subtracted as well as added. If $\mathbf{v}$ and $\mathbf{w}$ are two vectors, then

$$\mathbf{v} - \mathbf{w} \text{ means } \mathbf{v} + (-\mathbf{w}).$$

This subtraction is illustrated below.

Multiples of a vector

The vector sum $\mathbf{v} + \mathbf{v}$ is abbreviated as $2\mathbf{v}$. Likewise, $\mathbf{v} + \mathbf{v} + \mathbf{v} = 3\mathbf{v}$. The diagram below shows that the arrows representing $2\mathbf{v}$ and $3\mathbf{v}$ have the same direction as the arrow representing $\mathbf{v}$ but that they are two and three times as long, respectively. In general, if k is a positive real number, then $k\mathbf{v}$ is a vector with the same direction as $\mathbf{v}$ but with an absolute value k times as large. If $k < 0$, then $k\mathbf{v}$ has the same direction as $-\mathbf{v}$ and has an absolute value $|k|$ times as large. If $k \neq 0$, then $\dfrac{\mathbf{v}}{k}$ is defined to be equal to the vector $\dfrac{1}{k}\mathbf{v}$.

When working with vectors, it is customary to refer to real numbers as **scalars.** When this is done, the operation of multiplying a vector $\mathbf{v}$ by a scalar k is called **scalar multiplication.** This operation has the following properties, which you will be asked to prove as exercises in this section and the next. If $\mathbf{v}$ and $\mathbf{w}$ are vectors and k and m are scalars, then:

$$\left.\begin{array}{l} k(\mathbf{v} + \mathbf{w}) = k\mathbf{v} + k\mathbf{w} \\ (k + m)\mathbf{v} = k\mathbf{v} + m\mathbf{v} \end{array}\right\} \text{ Distributive Laws}$$

$$k(m\mathbf{v}) = (km)\mathbf{v} \qquad \text{Associative Law}$$

ORAL EXERCISES 12-1

1. **a.** If $ABCD$ is the parallelogram shown below, tell whether each of the following is true or false.

   (1) $\overrightarrow{BC} + \overrightarrow{BA} = \overrightarrow{BD}$ (2) $|\overrightarrow{BC}| + |\overrightarrow{BA}| = |\overrightarrow{BD}|$

   (3) $\overrightarrow{AO} = \overrightarrow{AC}$ (4) $\overrightarrow{AB} + \overrightarrow{CD} = \mathbf{0}$

   (5) $\overrightarrow{AO} = \overrightarrow{OC}$ (6) $\overrightarrow{AO} = \tfrac{1}{2}\overrightarrow{AC}$

   b. In statements (1) and (2) above, is the plus sign used in the same way? Explain.

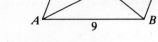

2. Complete the following statements if $ABCD$ is the parallelogram above.

   a. $\overrightarrow{AD} = ?$ **b.** $|\overrightarrow{AD}| = ?$ **c.** $\tfrac{1}{2}\overrightarrow{BD} = ?$

   d. $2\overrightarrow{AO} = ?$ **e.** $\overrightarrow{AB} + \overrightarrow{AD} = ?$ **f.** $\overrightarrow{AD} + \overrightarrow{DC} + \overrightarrow{CB} = ?$

   g. $\overrightarrow{AO} - \overrightarrow{DO} = \overrightarrow{AO} + ? = ?$ **h.** $\overrightarrow{BC} - \overrightarrow{BD} = \overrightarrow{BC} + ? = ?$

3. Explain the two different uses of the plus sign in the distributive law $(k + m)\mathbf{v} = k\mathbf{v} + m\mathbf{v}$.

4. In the associative law $k(m\mathbf{v}) = (km)\mathbf{v}$, two different kinds of multiplication are used on the right side of the equation. Explain.

5. Why is there no commutative law $k + \mathbf{v} = \mathbf{v} + k$?

WRITTEN EXERCISES 12-1

A   1. If $PQRS$ is a rhombus, which of the following statements are true?

   a. $\overrightarrow{PQ} = \overrightarrow{SR}$ **b.** $\overrightarrow{RQ} = \overrightarrow{QP}$

   c. $|\overrightarrow{RQ}| = |\overrightarrow{QP}|$ **d.** $\overrightarrow{PQ} + \overrightarrow{PS} = \overrightarrow{PR}$

2. If $ABCD$ is a parallelogram whose diagonals intersect at O, which of the following statements are true?

   a. $\overrightarrow{AB} = \overrightarrow{CD}$ **b.** $\overrightarrow{AB} = \overrightarrow{DC}$

   c. $\overrightarrow{AO} + \overrightarrow{OB} = \overrightarrow{AB}$ **d.** $\overrightarrow{BA} + \overrightarrow{BC} = \overrightarrow{BD}$

Complete each of the statements in Exercises 3 and 4 given that $PQRS$ and $PSTU$ are parallelograms.

3. **a.** $\overrightarrow{RS} + \overrightarrow{RQ} = ?$
   b. $\overrightarrow{SP} + \overrightarrow{ST} = ?$
   c. $\overrightarrow{TU} + \overrightarrow{UP} = ?$
   d. $\overrightarrow{TU} + \overrightarrow{UP} + \overrightarrow{PS} = ?$

4. **a.** $\overrightarrow{UT} + \overrightarrow{UP} = ?$
   b. $\overrightarrow{SR} + \overrightarrow{RQ} = ?$
   c. $\overrightarrow{PR} + \overrightarrow{RS} + \overrightarrow{ST} = ?$
   d. $\overrightarrow{RT} = ?$

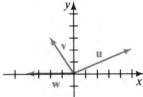

In Exercises 5–8 the vectors **u**, **v**, and **w** are represented by the arrows shown at the right. Use graph paper to draw the arrows corresponding to the specified vectors.

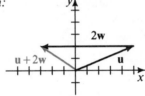

Example. **u** + 2**w**   *Solution:*

5. **a.** **u** + **v**   **b.** **v** + **u**   **c.** (**u** + **v**) + **w**   **d.** **u** + (**v** + **w**)
6. **a.** **u** − **v**   **b.** **v** − **u**   **c.** (**u** − **v**) − **w**   **d.** **u** − (**v** + **w**)
7. **a.** 3**v** + 3**w**   **b.** 3(**v** + **w**)   **c.** 2**u** − 2**v**   **d.** 2(**u** − **v**)
8. **a.** ½(**u** + **w**)   **b.** ½**u** + ½**w**   **c.** −**v** − **w**   **d.** −(**v** + **w**)

9. In the diagram, M and N are midpoints of $\overline{PQ}$ and $\overline{PR}$.
   a. Express $\overrightarrow{MN}$ in terms of **a** and **b**.
   b. Express $\overrightarrow{QR}$ in terms of **a** and **b**.
   c. Your answers to (a) and (b) show that $\overrightarrow{QR} = 2\,\overrightarrow{MN}$. What theorem about the segment joining the midpoints of two sides of a triangle does this equation suggest?

10. **a.** Complete the following equations given that $\overline{MN}$ is the median of trapezoid $ABCD$.
    (1) $\overrightarrow{AM} + \overrightarrow{MN} + \overrightarrow{ND} = ?$
    (2) $\overrightarrow{BM} + \overrightarrow{MN} + \overrightarrow{NC} = ?$
    (3) $\overrightarrow{AM} + \overrightarrow{BM} = ?$
    (4) $\overrightarrow{ND} + \overrightarrow{NC} = ?$

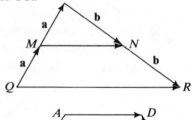

   b. Add equations (1) and (2), and by using (3) and (4), simplify the resulting equation to $2\overrightarrow{MN} = \overrightarrow{AD} + \overrightarrow{BC}$. What theorem about the median of a trapezoid does this equation suggest?

11. Make a diagram illustrating the velocity of an airplane heading east at 400 knots. Also illustrate a wind velocity of 50 knots blowing toward the northeast. If the airplane encounters this wind, illustrate its resultant velocity. Estimate the resultant speed and direction of the airplane.

12. Repeat Exercise 11 if the velocity of the airplane is 300 knots southwest and the velocity of the wind is 50 knots toward the east.

13. Make a diagram illustrating a force of 8 N (newtons) north and a force of 6 N west acting on a body. Also illustrate the resultant sum of these two forces and determine its strength. Then, using trigonometry, determine the approximate direction of this force (as a number of degrees west of north).

14. Repeat Exercise 13 if the forces are 4 N north and 6 N west.

15. Suppose the vectors **v** and **w** have the same direction. Express $|\mathbf{v} + \mathbf{w}|$ and $|\mathbf{v} - \mathbf{w}|$ in terms of $|\mathbf{v}|$ and $|\mathbf{w}|$.

16. Suppose the vectors **v** and **w** both have length 1 and have perpendicular directions. Let r and s be scalars. Express $|r\mathbf{v} + s\mathbf{w}|$ in terms of r and s.

17. If P is the midpoint of $\overline{AB}$ and $\mathbf{v} = \overrightarrow{AB}$, express the following vectors in terms of **v** (that is, in the form $t\mathbf{v}$ for some scalar t):

    a. $\overrightarrow{AP}$ b. $\overrightarrow{PA}$ c. $\overrightarrow{BA}$ d. $\overrightarrow{BP}$

18. Q divides $\overline{MN}$ in the ratio $2:1$; that is, $\dfrac{MQ}{QN} = 2$. If $\mathbf{v} = \overrightarrow{MN}$, express the following in terms of **v**:

    a. $\overrightarrow{MQ}$ b. $\overrightarrow{QN}$ c. $\overrightarrow{NQ}$

19. A divides $\overline{BC}$ in the ratio $2:3$; that is, $\dfrac{BA}{AC} = \dfrac{2}{3}$. If $\mathbf{v} = \overrightarrow{BA}$, express the following in terms of **v**:

    a. $\overrightarrow{CA}$ b. $\overrightarrow{BC}$ c. $\overrightarrow{AB}$

20. X divides $\overline{YZ}$ in the ratio $2:5$. If $\mathbf{v} = \overrightarrow{XY}$, express the following in terms of **v**:

    a. $\overrightarrow{ZX}$ b. $\overrightarrow{YZ}$ c. $\overrightarrow{ZY}$

21. In the given diagram, $\mathbf{x} = \overrightarrow{AB}$ and $\mathbf{y} = \overrightarrow{BD}$. E is the midpoint of $\overline{AD}$ and $\dfrac{BD}{DC} = \dfrac{1}{3}$. Express the following in the form $r\mathbf{x} + s\mathbf{y}$:

    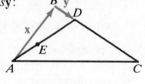

    Example. $\overrightarrow{AC}$

    Solution: $\overrightarrow{AC} = \overrightarrow{AB} + \overrightarrow{BC} = \mathbf{x} + 4\mathbf{y}$

    a. $\overrightarrow{AD}$ b. $\overrightarrow{AE}$ c. $\overrightarrow{BE}$ d. $\overrightarrow{EC}$

22. In the diagram, $AT:TC = 1:2$, $\overrightarrow{BA} = \mathbf{x}$, and $\overrightarrow{BC} = \mathbf{y}$. Express the following in the form $r\mathbf{x} + s\mathbf{y}$:

    a. $\overrightarrow{AB}$ b. $\overrightarrow{AC}$

    c. $\overrightarrow{AT}$ d. $\overrightarrow{BT}$

B 23. *ABCD* is a parallelogram with diagonals intersecting at *O*. If $\overrightarrow{AB} = \mathbf{x}$ and $\overrightarrow{AD} = \mathbf{y}$, express the following in terms of **x** and **y**:

a. $\overrightarrow{BC}$　b. $\overrightarrow{CD}$　c. $\overrightarrow{AC}$　d. $\overrightarrow{AO}$　e. $\overrightarrow{BO}$

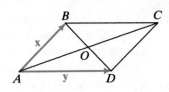

24. The diagonals of parallelogram *PQRS* intersect at *X*. If $\overrightarrow{RX} = \mathbf{a}$ and $\overrightarrow{RS} = \mathbf{b}$, express the following in terms of **a** and **b**:

a. $\overrightarrow{SX}$　　b. $\overrightarrow{SQ}$　　c. $\overrightarrow{PS}$

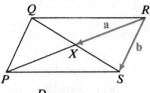

25. In the diagram at the right $DH:HF = 1:2$ and $EG:GH = 2:1$. If $\mathbf{v} = \overrightarrow{ED}$ and $\mathbf{w} = \overrightarrow{EF}$, express the following in terms of **v** and **w**:

a. $\overrightarrow{DF}$　　b. $\overrightarrow{DH}$　　c. $\overrightarrow{EH}$　　d. $\overrightarrow{DG}$

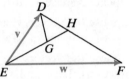

26. In $\triangle ABC$, $\dfrac{AP}{PB} = \dfrac{2}{1}$ and $\dfrac{PQ}{QC} = \dfrac{1}{3}$. If $\overrightarrow{PQ} = \mathbf{v}$ and $\overrightarrow{QB} = \mathbf{w}$, express the following terms of **v** and **w**:

a. $\overrightarrow{BP}$　　b. $\overrightarrow{PC}$　　c. $\overrightarrow{BC}$　　d. $\overrightarrow{AC}$

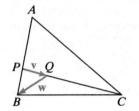

27. This exercise shows how to prove the associative law for vector addition. In the diagram at the right let $\mathbf{u} = \overrightarrow{AB}$, $\mathbf{v} = \overrightarrow{BC}$, and $\mathbf{w} = \overrightarrow{CD}$. Find arrows which represent the following vectors:

a. $\mathbf{u} + \mathbf{v}$
b. $(\mathbf{u} + \mathbf{v}) + \mathbf{w}$
c. $\mathbf{v} + \mathbf{w}$
d. $\mathbf{u} + (\mathbf{v} + \mathbf{w})$

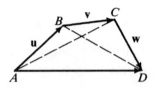

28. a. If $\overrightarrow{DA} = k\,\overrightarrow{BA}$ and $\overrightarrow{AE} = k\,\overrightarrow{AC}$, use geometry to explain why $\overrightarrow{DE} = k\,\overrightarrow{BC}$; that is, explain why $\overrightarrow{DE}$ is parallel to $\overrightarrow{BC}$ and k times as long.

b. Use part (a) and the diagram to prove that $k\mathbf{u} + k\mathbf{v} = k(\mathbf{u} + \mathbf{v})$.

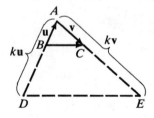

29. In the diagram, $\overrightarrow{AB}$ and $\overrightarrow{AC}$ represent forces of 5 and 3 N, respectively, which act at a 60° angle to each other. If *ABXC* is a parallelogram, find the measure of angle *B*, and use the Law of Cosines in $\triangle ABX$ to find the strength of the resultant force.

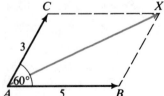

30. $\mathbf{F}_1$ is a force of 8 N pulling north and $\mathbf{F}_2$ is a force of 3 N pulling 60° east of north. Use the Law of Cosines to find $|\mathbf{F}_1 + \mathbf{F}_2|$.

31. The bearing of a vector **v** is the angle between $0°$ and $360°$, measured clockwise, that **v** makes with a vector pointing north. A plane flies 120 km due east and then 130 km due north. Find the distance and bearing of the plane from its starting point.

32. A plane flies 200 km due west and then 240 km due south. Find the distance and bearing of the plane from its starting point. (See Exercise 31.)

33. a. The diagram below shows the endpoints of several arrows from O. Copy this diagram and indicate the endpoints of the arrows from O representing $2\mathbf{a} + \mathbf{b}$, $2\mathbf{a} + 4\mathbf{b}$, and $-\mathbf{a} - \mathbf{b}$.

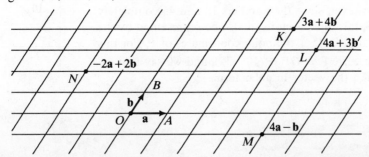

b. Because $\overrightarrow{OK} = 3\mathbf{a} + 4\mathbf{b}$, the point K may be labeled $(3, 4)$. The numbers 3 and 4 are called the *coordinates of K relative to* **a** *and* **b**. What are the coordinates of the points L, M, and N? of the points A and B?

c. If you are interested in the use of different coordinate systems in biology, see D'Arcy Thompson, *On Growth and Form*, Vol. II, Cambridge University Press, 1963, especially pages 1053–1083, from which the diagrams below came. These diagrams show two oceanic fish belonging to two different families. For the fish on the left, the coordinate system is rectangular. For the fish on the right, the coordinate system is skewed $70°$ to the horizontal. However, corresponding points of both fish have the same coordinates. For example, each eye has coordinates $(1, 2\frac{1}{2})$ and the top tip of each tail has the approximate coordinates $(6, 2\frac{3}{4})$.

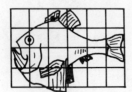
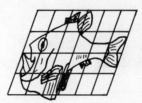

34. The diagram shows the graph of $y = x^2$ on a skewed-coordinate system. Make three copies of this coordinate system and show the graphs of:

a. $2x + 3y = 6$
b. $y = 4x - x^2$
c. $(x - 1)^2 + (y - 2)^2 = 4$

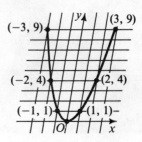

12-2 / ALGEBRAIC REPRESENTATION OF VECTORS

In the last section we represented vectors geometrically as arrows. In this section we shall represent vectors as *ordered pairs of numbers*. This can be done in the following two ways.

Polar form of a vector

If **F** is a force of 10 newtons acting at a 45° angle to the horizontal, then **F** can be represented by the arrow $\overrightarrow{OP}$ as shown. However, since the numbers 10 and 45 completely describe **F**, this force vector can also be represented by the ordered pair (10; 45°), the polar coordinates of *P*. For this reason, we write $\mathbf{F} = \overrightarrow{OP} = (10; 45°)$, and we say that we have written the vector in **polar form.**

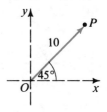

Niccolo Tartaglia (1500–1557) was raised in poverty by his mother and later used the name Tartaglia (meaning "stammerer") in his published works. A self-taught mathematician, he is best known for his solution of the general third-degree equation. It was over this solution that he came into conflict with Girolamo Cardano (see below). Tartaglia revealed his solution to Cardano in confidence and was outraged when Cardano published it six years later.

Girolamo Cardano (1501–1576) was a mathematician who studied the roots of equations extensively. He is credited with being the originator of the algebraic theory of equations, and his Ars Magna, *published in 1545, is the first great Latin work on the subject.*

Component form of a vector

The four arrows shown in the diagram at the left below all represent the same vector **u**. Viewed as a displacement, **u** consists of a change of 2 units in the *x*-direction and a change of 3 units in the *y*-direction. The numbers 2 and 3 are called the *x*- and *y-components* of **u**, and we write **u** = (2, 3).

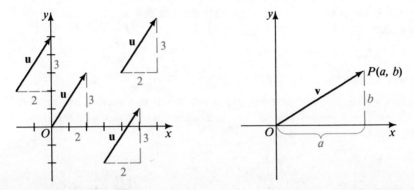

In general, we shall use the symbol (a, b) to represent a vector **v** consisting of steps of *a* units in the *x*-direction and *b* units in the *y*-direction. The numbers *a* and *b* are called the **x-** and **y-components,** or simply **components,** of **v**, and when we write **v** = (a, b) we are expressing **v** in its **component form.** Notice from the diagram above that the ordered pair (a, b) can represent the point *P* or the vector $\overrightarrow{OP}$. You can tell which meaning is intended from the context.

The diagram below shows that the *x*-component of $\overrightarrow{AB}$ is $8 - 3 = 5$ and the

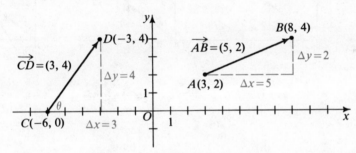

y-component is $4 - 2 = 2$. You can find the components of a vector $\overrightarrow{AB}$ from the coordinates of its endpoints by using the equation*

$$\overrightarrow{AB} = B - A.$$

In this case you have:

$$\overrightarrow{AB} = (8, 4) - (3, 2) = (8 - 3, 4 - 2) = (5, 2)$$

Similarly:

$$\overrightarrow{CD} = D - C = (-3, 4) - (-6, 0)$$
$$= (-3 - (-6), 4 - 0) = (3, 4)$$

*Actually, $\overrightarrow{AB} = \overrightarrow{OB} - \overrightarrow{OA}$. But since $\overrightarrow{OB}$ and $\overrightarrow{OA}$ are represented by the same ordered pairs as the points *B* and *A*, respectively, it is convenient to write $\overrightarrow{AB} = \overrightarrow{OB} - \overrightarrow{OA}$ as $\overrightarrow{AB} = B - A$.

Vectors and determinants   **409**

When a vector is expressed in component form, its absolute value can be found easily by using the Pythagorean Theorem. For the preceding example:

and
$$|\overrightarrow{AB}| = |(5, 2)| = \sqrt{5^2 + 2^2} = \sqrt{29}$$
$$|\overrightarrow{CD}| = |(3, 4)| = \sqrt{3^2 + 4^2} = 5$$

To determine the direction of a vector, you must often use trigonometry. For example, if θ is the measure of the angle that the vector $\overrightarrow{CD}$ in the preceding diagram makes with the horizontal, then $\tan \theta = \frac{4}{3} \approx 1.333$, and $\theta \approx 53.1°$.

Operations with vectors in component form

Operations with vectors are particularly simple when the vectors are represented in their component form. In the statements and figures which follow, $\mathbf{v} = (a, b)$, $\mathbf{u} = (c, d)$, and k is a scalar.

1. $|\mathbf{v}| = |(a, b)| = \sqrt{a^2 + b^2}$ See figure (a).
2. $k\mathbf{v} = k(a, b) = (ka, kb)$ See figure (b).
3. $|k\mathbf{v}| = |k|\,|\mathbf{v}|$ See figure (b).
4. $\mathbf{v} + \mathbf{u} = (a, b) + (c, d) = (a + c, b + d)$ See figure (c).
5. $\mathbf{v} - \mathbf{u} = \mathbf{v} + (-\mathbf{u}) = (a, b) - (c, d) = (a - c, b - d)$ See figure (d).

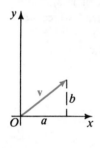

$$|\mathbf{v}| = \sqrt{a^2 + b^2}$$

(a)

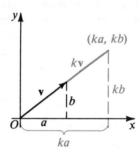

$$k\mathbf{v} = (ka, kb)$$
$$|k\mathbf{v}| = |k|\,|\mathbf{v}|$$

(b)

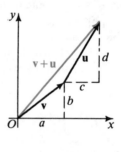

$$\mathbf{v} + \mathbf{u} = (a, b) + (c, d)$$
$$= (a + c, b + d)$$

(c)

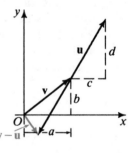

$$\mathbf{v} - \mathbf{u} = \mathbf{v} + (-\mathbf{u})$$
$$= (a, b) - (c, d)$$
$$= (a - c, b - d)$$

(d)

EXAMPLE 1. If $\mathbf{u} = (1, -3)$ and $\mathbf{v} = (2, 5)$, calculate the following:
   a. $\mathbf{u} + \mathbf{v}$ **b.** $2\mathbf{u} - 3\mathbf{v}$ **c.** the polar form of $\mathbf{u}$

SOLUTION: **a.** $\mathbf{u} + \mathbf{v} = (1, -3) + (2, 5) = (1 + 2, -3 + 5) = (3, 2)$

    b. $2\mathbf{u} - 3\mathbf{v} = 2(1, -3) - 3(2, 5) = (2, -6) - (6, 15)$
       $= (2 - 6, -6 - 15) = (-4, -21)$

    c. Note that
       $$|\mathbf{u}| = \sqrt{1^2 + (-3)^2} = \sqrt{10}.$$
    Refer to the diagram on the next page and note that θ is a fourth-quadrant angle.

Since

$$\tan \theta = \frac{-3}{1}, \text{ then,}$$

$\theta \approx -71.6°$, or
$\theta \approx 360° - 71.6° = 288.4°$.

Hence, the polar form of **u** is
$(\sqrt{10}; -71.6°)$, or $(\sqrt{10}; 288.4°)$.

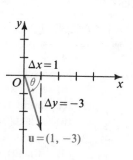

EXAMPLE 2. If $A = (0, 4)$ and $B = (6, 1)$, find the point P that is $\frac{2}{3}$ of the way from A to B.

SOLUTION: To get to P, start at A and travel $\frac{2}{3}\overrightarrow{AB}$.

$$
\begin{aligned}
P &= A + \tfrac{2}{3}\overrightarrow{AB} \\
&= (0, 4) + \tfrac{2}{3}(6, -3) \\
&= (0, 4) + (4, -2) \\
&= (4, 2)
\end{aligned}
$$

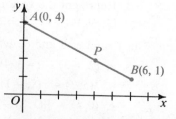

ORAL EXERCISES 12-2

1. a. What are the components of $\overrightarrow{AB}$? of $\overrightarrow{CD}$?
  b. Find $|\overrightarrow{AB}|$ and $|\overrightarrow{CD}|$.
  c. Find the point $\frac{1}{4}$ of the way from A to B.
  d. Find the point $\frac{3}{4}$ of the way from C to D.

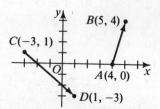

2. A force **F** of 2 N is represented in the diagram.

  a. Express **F** in polar form.
  b. Express **F** in component form.
  c. What is $|\mathbf{F}|$?

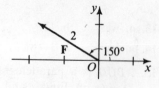

3. a. Study the diagram and give the component form of the vectors $\overrightarrow{AB}$, $\overrightarrow{BC}$, and $\overrightarrow{AC}$.
  b. Find $|\overrightarrow{AB}|$, $|\overrightarrow{BC}|$, and $|\overrightarrow{AC}|$.
  c. True or False?
    (1) $\overrightarrow{AB} + \overrightarrow{BC} = \overrightarrow{AC}$
    (2) $|\overrightarrow{AB}| + |\overrightarrow{BC}| = |\overrightarrow{AC}|$

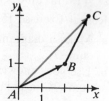

4. If $\mathbf{v} = (1, 2)$ and $\mathbf{u} = (3, 0)$, find:
  a. 3**v**   **b.** **v** + **u**   **c.** **v** − **u**   **d.** 2**v** + 3**u**

5. If $\mathbf{v} = (3; 40°)$, write 2**v** and −**v** in polar form.

6. a. If $\overrightarrow{AB} = (3, 2)$ and point $A = (4, 0)$, find point B.
  b. If $\overrightarrow{CD} = (4, -1)$ and point $D = (8, 8)$, find point C.

7. Justify the following statement: In general, if $\overrightarrow{AB} = (a, b)$, then $\overrightarrow{BA} = (-a, -b)$.

A
1. Represent the vector $(2, 5)$ by an arrow **(a)** with its tail at the origin, **(b)** with its tail at the point $(4, 1)$.

2. Represent the vector $(-3, 4)$ by an arrow **(a)** with its tail at the origin, **(b)** with its tail at the point $(-4, -5)$.

3. Represent the vector $(2, -1)$ by an arrow **(a)** with its tail at the point $(4, -6)$, **(b)** with its tip at the point $(3, 4)$.

4. Represent the vector $(-1, -3)$ by an arrow **(a)** with its tail at the point $(1, 4)$, **(b)** with its tip at the point $(1, 3)$.

In Exercises 5 and 6, evaluate the given expressions.

5. a. $(2, -3) - (-3, 4)$
  b. $\frac{1}{2}(5, 3) - \frac{3}{2}(1, -1)$
  c. $(0, 5) - \frac{1}{4}[(1, 7) - (-3, -1)]$

6. a. $(-5, 3) + (2, -3)$
  b. $\frac{1}{3}(3, 6) - \frac{1}{4}(0, -12)$
  c. $(-1, 2) + \frac{1}{3}[(2, 5) - (-1, 2)]$

In Exercises 7–12, **(a)** give the components of $\overrightarrow{AB}$, **(b)** find $|\overrightarrow{AB}|$, and **(c)** find the coordinates of the point P described.

7. $A = (0, 0)$, $B = (6, 3)$, and P is $\frac{1}{3}$ of the way from A to B.

8. $A = (1, 4)$, $B = (5, -4)$, and P is $\frac{1}{4}$ of the way from A to B.

9. $A = (7, -2)$, $B = (2, 8)$, and P is $\frac{4}{5}$ of the way from A to B.

10. $A = (-3, -4)$, $B = (-6, 1)$, and P is $\frac{1}{6}$ of the way from A to B.

11. $A = (-7, -4)$, $B = (-1, -1)$, and B is $\frac{2}{5}$ of the way from A to P.

12. $A = (3, -2)$, $B = (5, 1)$, and A is $\frac{2}{3}$ of the way from P to B.

In Exercises 13 and 14, let $\mathbf{v} = (2, 3)$ and $\mathbf{w} = (1, 2)$. Evaluate the given expression.

13. a. $|\mathbf{v}|$ **b.** $|2\mathbf{v}|$ **c.** $|\mathbf{v} + \mathbf{w}|$ **d.** $|5\mathbf{v} - 3\mathbf{w}|$

14. a. $|\mathbf{w}|$ **b.** $|-\mathbf{w}|$ **c.** $|\mathbf{v} - \mathbf{w}|$ **d.** $|-3\mathbf{v} + 4\mathbf{w}|$

15. $ABCD$ is a parallelogram with $A = (1, 2)$, $B = (3, 8)$, $C = (9, 10)$, and $D = (x, y)$. Find x and y.

16. Repeat Exercise 15 if $A = (-2, 1)$, $B = (-5, -3)$, and $C = (3, -2)$.

17. a. Make a diagram representing $\overrightarrow{OA} = (3; 0°)$, $\overrightarrow{OB} = (4; 90°)$, and their sum $\overrightarrow{OC}$.
  b. Use your knowledge of right triangles and trigonometry to express the sum $\overrightarrow{OC}$ in polar form.

18. Repeat Exercise 17 if $\overrightarrow{OA} = (5; 90°)$ and $\overrightarrow{OB} = (12; 180°)$.

19. a. The polar form of a vector is $(r; \theta)$ and its component form is (x, y). Consequently, $x = r \cos \theta$ and $y = \underline{\ ?\ }$.
  b. Find the x- and y-components of $(10; 152°)$ to the nearest tenth.

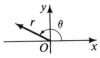

20. Given that $\mathbf{u} = (10; 30°)$ and $\mathbf{v} = (6; -30°)$, find $|\mathbf{u} + \mathbf{v}|$ to the nearest unit in two ways:

a. By converting **u** and **v** to component form and then finding **u** + **v** and |**u** + **v**|.

b. By using the Law of Cosines.

In Exercises 21 and 22, give angles to the nearest tenth of a degree and sides to 3 significant digits.

B **21.** A sailboat proceeds from A to B and then shifts course from B to C. If $\vec{AB} = (8; 30°)$ and $\vec{BC} = (4; 60°)$, find $\vec{AC}$ in polar form.

22. A sailboat takes a zigzag course from A to D as follows: $\vec{AB} = (10; 45°)$, $\vec{BC} = (8; -45°)$, and $\vec{CD} = (6; -90°)$. Find $|\vec{AD}|$.

23. a. Find r and s if $r(1, 2) + s(3, 0) = (9, 6)$. (*Hint:* Solve two simultaneous equations in r and s.)

b. How is your answer related to the diagram shown?

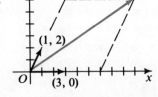

24. Find r and s if $r(1, 2) + s(3, 0) = (-1, 4)$. Illustrate your solution with a diagram like that in Exercise 23.

25. Find a scalar r such that $|r(3, 4)| = 1$.

26. Find a vector of length 1 in the same direction as $(-4, 3)$.

27. If $A = (4, 0)$, describe the set of points P such that $|\vec{AP}| = 2$.

28. If $A = (0, 0)$ and $B = (6, 8)$, describe the set of points P such that $\vec{AP} = t\vec{AB}$. Consider the following three cases:
(1) $0 \le t \le 1$ (2) $t \ge 1$ (3) $t \le 0$

29. The formula for the midpoint M of $\overline{AB}$ is sometimes written as $M = \frac{1}{2}(A + B)$. If $M = (\bar{x}, \bar{y})$, $A = (x_1, y_1)$, and $B = (x_2, y_2)$, this equation becomes

$$(\bar{x}, \bar{y}) = \frac{1}{2}[(x_1, y_1) + (x_2, y_2)].$$

Show how this last equation yields the midpoint formula given in Chapter 1 on page 4.

30. Suppose that the diagonals of quadrilateral $PQRS$ have the same midpoint. Then, by Exercise 29, $\frac{1}{2}(P + R) = \frac{1}{2}(S + Q)$. Use this equation to prove that $P - S = Q - R$ and $\vec{SP} = \vec{RQ}$. What geometry theorem does this prove?

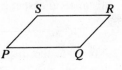

31. A theorem from geometry shows that the centroid G of a triangle is a point $\frac{2}{3}$ of the way from A to the midpoint M of $\overline{BC}$. Thus, $G = A + \frac{2}{3}(M - A)$ and $M = B + \frac{1}{2}(C - B)$.

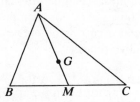

a. Use these two equations to prove that
$$G = \frac{1}{3}(A + B + C).$$

b. Find the centroid of $\triangle ABC$ given the points $A(-2, 5)$, $B(3, 7)$, and $C(5, -3)$.

A sailboat takes a zigzag course from A to E such that $\overrightarrow{AB} = \overrightarrow{CD} = (17; 35°)$, and $\overrightarrow{BC} = \overrightarrow{DE} = (11; -67°)$. Find $|\overrightarrow{AE}|$.

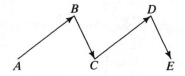

COMPUTER EXERCISES

Given two vectors $\mathbf{v}_1 = (r_1; \theta_1)$ and $\mathbf{v}_2 = (r_2; \theta_2)$, where θ_1 and θ_2 are between $0°$ and $90°$, write a program that prints out the following when you input the values of r_1, θ_1, r_2, and θ_2.

a. the component forms of $\mathbf{v}_1$, $\mathbf{v}_2$, and $\mathbf{v}_1 + \mathbf{v}_2$, and
b. the polar form of $\mathbf{v}_1 + \mathbf{v}_2$
when you input r_1, θ_1, r_2, and θ_2.

Applications of vectors

12-3/MOTION ON A LINE; VECTOR AND PARAMETRIC EQUATIONS OF A LINE

Motion on a line

If an object moves along a line with constant velocity $\mathbf{v}$, then its displacement $\mathbf{d}$ during time t is

$$\mathbf{d} = t\mathbf{v}.$$

In this case, $|\mathbf{d}|$ is the distance the object travels and $|\mathbf{v}|$ is its speed (rate). The distance and speed are related by the formula

$$|\mathbf{d}| = t|\mathbf{v}| \qquad \text{(distance equals rate times time).}$$

Note that the displacement $\mathbf{d}$ and the velocity $\mathbf{v}$ are vectors and that the distance $|\mathbf{d}|$ and the speed $|\mathbf{v}|$ are scalars.

In most scientific problems, time is measured from some moment assigned the time $t = 0$. For example, the time of liftoff of a rocket is called time $t = 0$. Events occurring before $t = 0$ occur at negative times and those after $t = 0$ at positive times.

EXAMPLE 1. At time $t = 0$, an object is located at $(3, 4)$. If it moves with constant velocity $\mathbf{v} = (2, 1)$, show on a graph its position at times $t = -3, -2, -1, 0, 1, 2,$ and 3.

SOLUTION:

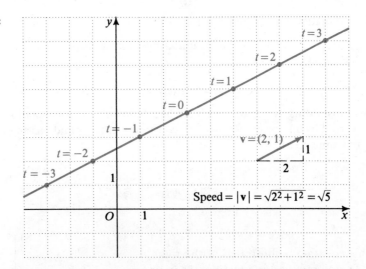

EXAMPLE 2. Suppose an object which is moving with constant velocity in the coordinate plane is located at $A = (5, 3)$ when the time $t = 0$ seconds and at $B = (-4, 15)$ when $t = 3$ seconds. Find its velocity and speed.

SOLUTION: The displacement of the object is:

$$\mathbf{d} = \overrightarrow{AB} = \text{Position } B - \text{Position } A$$
$$= (-4, 15) - (5, 3)$$
$$= (-4 - 5, 15 - 3)$$
$$= (-9, 12)$$

Since the elapsed time is 3 s, we substitute in the formula as follows:

$$\mathbf{d} = t\mathbf{v}$$
$$(-9, 12) = 3\mathbf{v}$$
$$\mathbf{v} = \tfrac{1}{3}(-9, 12) = (-3, 4)$$

The speed of the object is

$$|\mathbf{v}| = |(-3, 4)| = \sqrt{(-3)^2 + 4^2} = 5.$$

Vector and parametric equations of a line

Example 2 illustrates that the displacement $\mathbf{d}$ in moving from some initial point (x_0, y_0) to another point (x, y) can be found from the equation

displacement = new position − initial position.

$$\mathbf{d} \quad = \quad (x, y) \quad - \quad (x_0, y_0)$$

This equation can be rewritten as follows:

new position = initial position + displacement

$$(x, y) \quad = \quad (x_0, y_0) \quad + \quad \mathbf{d}$$

By substituting $t\mathbf{v}$ for $\mathbf{d}$, we obtain the following **vector equation** of the line traversed by an object moving with constant velocity $\mathbf{v}$ and passing through (x_0, y_0) at time $t = 0$.

$$\begin{pmatrix} \text{position at} \\ \text{time } t \end{pmatrix} = \begin{pmatrix} \text{position at} \\ \text{time } t = 0 \end{pmatrix} + (\text{time}) \times \begin{pmatrix} \text{constant} \\ \text{velocity} \end{pmatrix}$$

$$(x, y) \quad = \quad (x_0, y_0) \quad + \quad t\mathbf{v}$$

This equation is illustrated at the right.

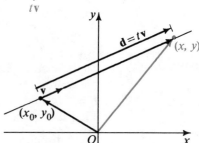

Therefore a vector equation of the line in Example 2 is

$$(x, y) = (5, 3) + t(-3, 4).$$

From this equation, we can obtain two equations which give the x- and y-coordinates of the moving object at time t.

These equations are called the **parametric equations** of the line and are derived as follows:

Vector equation of line: $(x, y) = (5, 3) + t(-3, 4)$

$$(x, y) = (5 - 3t, 3 + 4t)$$

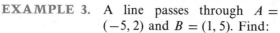

Parametric equations of line: $x = 5 - 3t$ and $y = 3 + 4t$

In this illustration, we have interpreted $\mathbf{v} = (-3, 4)$ as the velocity of an object moving along a line. The vector $\mathbf{v}$ can also be interpreted as a **direction vector** of the line. The slope of a nonvertical line in the coordinate plane can be calculated from a direction vector for the line:

$$\text{direction vector} = (\Delta x, \Delta y)$$

$$\text{slope} = \frac{\Delta y}{\Delta x}$$

EXAMPLE 3. A line passes through $A = (-5, 2)$ and $B = (1, 5)$. Find:
a. a direction vector of the line
b. a vector equation of the line
c. a pair of parametric equations of the line
d. a rectangular equation of the line

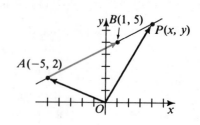

SOLUTION: **a.** The direction vector is
$$\vec{AB} = B - A = (1, 5) - (-5, 2) = (6, 3).$$

b. Think to yourself: "To get to P, first go to A, and then travel a multiple of $\vec{AB}$."

$$\vec{OP} = \vec{OA} + t\,\vec{AB}$$

$$(x, y) = (-5, 2) + t(6, 3) \leftarrow \text{vector equation}$$

c. From the vector equation, we get $(x, y) = (-5 + 6t, 2 + 3t)$. Hence:

$$\left.\begin{array}{l} x = -5 + 6t \\ y = 2 + 3t \end{array}\right\} \text{parametric equations}$$

d. To find a rectangular equation, we eliminate the parameter t from the parametric equations. Since $x = -5 + 6t$,

$$t = \frac{x + 5}{6}.$$

Therefore:

$$y = 2 + 3t$$
$$= 2 + 3\left(\frac{x + 5}{6}\right)$$
$$= 2 + \frac{3x}{6} + \frac{15}{6}$$
$$y = \frac{1}{2}x + \frac{9}{2} \leftarrow \text{rectangular equation}$$

(Of course, since we are given two points on the line, we could have found a rectangular equation using the methods of Chapter 1.)

Remark: A line has an infinite number of vector and parametric equations. For instance, we found the vector equation of the line in the last example to be $\vec{OP} = \vec{OA} + t\,\vec{AB}$. Instead, it could have been:

$$\vec{OP} = \vec{OB} + t\,\vec{BA}$$

$$(x, y) = (1, 5) + t(-6, -3)$$

The parametric equations would then have been:

$$x = 1 - 6t$$
$$y = 5 - 3t$$

You should check that these equations give the same rectangular equation as found in Example 3.

The parametric equations of a line are useful in finding the times and positions at which an object moving with constant velocity crosses a curve whose equation is known. This situation is illustrated in the next example.

EXAMPLE 4. An object moves along a line in such a way that its x- and y-coordinates at time t are

$$x = 1 - t \quad \text{and} \quad y = 1 + 2t.$$

When and where does the object cross the circle $x^2 + (y - 1)^2 = 25$?

SOLUTION: Substituting the parametric equations in the equation of the circle, we see that the time t of crossing must satisfy the following conditions:

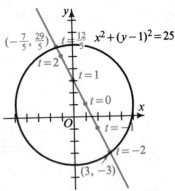

$$x^2 + (y - 1)^2 = 25$$
$$(1 - t)^2 + (1 + 2t - 1)^2 = 25$$
$$1 - 2t + t^2 + 4t^2 = 25$$
$$5t^2 - 2t - 24 = 0$$
$$(5t - 12)(t + 2) = 0$$

Thus the times of crossing are $t = \frac{12}{5}$ and $t = -2$.

To find the points of crossing, we substitute these times in the parametric equations:

When $t = \frac{12}{5}$,

$$x = 1 - \tfrac{12}{5} = -\tfrac{7}{5} \quad \text{and}$$
$$y = 1 + 2(\tfrac{12}{5}) = \tfrac{29}{5}.$$

When $t = -2$,

$$x = 1 - (-2) = 3 \quad \text{and}$$
$$y = 1 + 2(-2) = -3.$$

Thus, the points of crossing are $(-\tfrac{7}{5}, \tfrac{29}{5})$ and $(3, -3)$.

ORAL EXERCISES 12-3

1. An object moves from $A = (1, 0)$ to $B = (7, 8)$ in 2 seconds.
   a. What is its displacement **d** in that time?
   b. What is its velocity?
   c. What is its speed?

2. A line has vector equation $(x, y) = (2, -5) + t(1, 3)$.
   a. Name three points on the line.
   b. Find a direction vector of the line.
   c. Find the slope of the line.
   d. Give a pair of parametric equations of the line.

3. a. If $A = (0, 2)$ and $B = (3, 4)$ find the slope of line AB.
   b. Find a direction vector of line AB.
   c. Find a vector equation of this line.

A　**1.** An object moves with constant velocity. Find its velocity and speed given:

    a. when $t = 0$ it is at $(1, 3)$ and when $t = 3$ it is at $(10, 15)$.

    b. when $t = 1$ it is at $(-2, -3)$ and when $t = 4$ it is at $(4, 9)$.

2. An object moves with constant velocity. Find its velocity and speed given:

    a. when $t = 0$ it is at $(0, -2)$ and when $t = 4$ it is at $(5, 10)$.

    b. when $t = -1$ it is at $(1, 4)$ and when $t = 2$ it is at $(4, 1)$.

3. An object moves with velocity $(3, 1)$.

    a. If at time $t = 1$ it is at $(-1, 1)$, where is it at $t = 4$?

    b. If at time $t = -3$ it is at $(-6, -1)$, where is it at $t = -1$?

4. An object moves with velocity $(-2, 4)$.

    a. If at time $t = 2$ it is at $(3, 5)$, where is it at $t = 0$?

    b. If at time $t = -4$ it is at $(-6, 1)$, where is it at $t = 1$?

In Exercises 5–8 information is given about a particle moving along a line in the xy-plane. Use the information to find parametric equations for the x- and y-coordinates of the particle at an arbitrary time t.

5. Velocity $= (3, -1)$ and position at time $t = 0$ is $(2, 3)$.

6. Velocity $= (1, -1)$ and position at time $t = 0$ is $(1, -5)$.

7. Position at $t = 0$ is $(2, 0)$ and position at $t = 1$ is $(3, 4)$.

8. Position at $t = 0$ is $(1, 1)$ and position at $t = 2$ is $(5, 3)$.

9. A line has vector equation $(x, y) = (3, 2) + t(2, 4)$. Give a pair of parametric equations of the line and also a rectangular equation.

10. A line has parametric equations $x = 5 - t$ and $y = 4 + 2t$. Give a vector equation of the line and also a rectangular equation.

11. a. Describe the line having parametric equations $x = 2$, $y = t$.

    b. Give a direction vector of the line.

    c. What can you say about the slope of the line?

12. a. Describe the line having parametric equations $x = t$, $y = 3$.

    b. Give a direction vector of the line.

    c. What is the slope of the line?

13. The position of an object at time t is given by the parametric equations $x = 2 - 3t$ and $y = -1 + 2t$.

    a. What are the velocity and speed of the object?

    b. When and where does it cross the line $x + y = 2$?

14. The position of an object at time t is given by the parametric equations $x = 1 + 3t$ and $y = 2 - 4t$.

    a. What are its velocity and speed?

    b. When and where does it cross the x-axis?

15. An object moves with constant velocity so that its position at time t is $(x, y) = (1, 1) + t(-1, 1)$. When and where does it cross the circle $(x - 1)^2 + y^2 = 5$? Illustrate with a sketch.

Vectors and determinants　　**419**

16. An object moves with constant velocity so that its position at time t is $(x, y) = (2, 0) + t(1, -1)$. When and where does it cross the hyperbola $x^2 - 2y^2 = 4$? Illustrate with a sketch.

In Exercises 17–24, find a vector equation and the corresponding parametric equations for each specified line.

17. a. the line joining $(1, 0)$ and $(3, -4)$
   b. the line joining $(-2, 3)$ and $(5, 1)$

18. a. the line joining $(3, 1)$ and $(-4, -4)$
   b. the line joining $(0, 3)$ and $(6, 0)$

19. a. the line $y = 3x + 2$ **b.** the line $x + 3y = 3$

20. the line with x-intercept 4 and y-intercept -3

21. the line through $(7, 5)$ with inclination $45°$

22. the line through the origin with inclination $135°$

23. the horizontal line through (π, e)

24. the vertical line through $\left(\sqrt{2}, \sqrt{3}\right)$

25. a. A line has direction vector $(2, 3)$. What is its slope?
   b. A line has direction vector $(4, 6)$. What is its slope?
   c. Explain why the following lines are parallel:
$$(x, y) = (8, 1) + r(2, 3) \quad \text{and} \quad (x, y) = (2, 5) + s(4, 6)$$
   d. Find a vector equation of the line through $(7, 9)$ parallel to the line $(x, y) = (8, 1) + r(2, 3)$.

26. Find a vector equation of the line through $(2, 1)$ parallel to the line
$$(x, y) = (-2, 7) + t(3, 5).$$

27. A spider and a fly crawl so that their positions are as follows:
position of spider at time t: $(x, y) = (-2, 5) + t(1, -2)$
   position of fly at time t: $(x, y) = (1, 1) + t(-1, 1)$

   a. Make a sketch showing the spider and the fly at various times. Do they ever meet? Do their lines of travel meet?
   b. It is possible to determine algebraically whether the spider meets the fly. We just equate the x-coordinates of the positions of the two insects and also the y-coordinates. This gives the following equations:
$$x = -2 + t = 1 - t$$
$$y = 5 - 2t = 1 + t$$
   The spider will meet the fly if and only if there is a single value of t satisfying both equations. Is there?

28. Repeat Exercise 27 given the following information:
position of spider at time t: $(x, y) = (3, -2) + t(2, 1)$
   position of fly at time t: $(x, y) = (-1, 6) + t(4, -3)$

B   **29.** A particle travels along the line $3x + 4y = 12$ at the speed of 10 units per second. The particle intersects the y-axis when $t = 0$, and it moves down and to the right as shown. What is its position at time t?

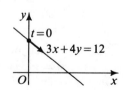

420   Chapter 12

30. A particle travels along the line $5x + 12y = 10$ at a speed of 13 units per second. It intersects the x-axis when $t = 0$, and it moves down and to the right. What is its position at time t?

31. An object travels for 1 s at velocity $(2, 1)$, for 3 s at velocity $(4, -1)$, and for 2 s at velocity $(-3, 6)$. At what velocity does it need to move to return to its starting point after 2 more seconds?

32. An object starts to move from the origin. It is to travel a certain amount of time with velocity $(1, 6)$ and then a certain amount of time with velocity $(3, 2)$. How much time is required at each velocity in order to reach the point $(15, 18)$?

Exercises 33–38 deal with parametric equations of a curve. In each exercise, sketch the curve with the given parametric equations. Also, give a rectangular equation of the curve. Use the method of the following example.

EXAMPLE. **a.** Sketch the curve with parametric equations
$$x = 2t \quad \text{and} \quad y = 1 - t^2.$$
b. Also, find a rectangular equation of the curve.

SOLUTION: **a.** First, make a table of values:

t	-3	-2	-1	0	1	2	3
$x = 2t$	-6	-4	-2	0	2	4	6
$y = 1 - t^2$	-8	-3	0	1	0	-3	-8

Then, use this table to plot points (x, y) on the curve.

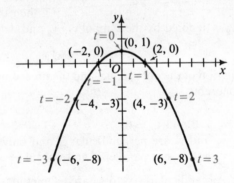

b. Since $x = 2t$, $t = \dfrac{x}{2}$. Thus:
$$y = 1 - t^2 = 1 - \left(\frac{x}{2}\right)^2 = 1 - \frac{x^2}{4}$$

33. $x = 2t,\ y = t^2 + 1$

34. $x = t^2 - 1,\ y = t - 1$

35. $x = 4 \cos t,\ y = 4 \sin t$

36. $x = \dfrac{t}{2},\ y = t(t^2 - 1)$

C **37.** $x = \cos t,\ y = \cos 2t$

38. $x = \cos 2t,\ y = \sin 2t$

12-4/PARALLEL AND PERPENDICULAR VECTORS; DOT PRODUCT

The diagram at the right shows that if $\overrightarrow{AB} = (x_1, y_1)$, then the slope of $\overrightarrow{AB}$ is $\dfrac{\Delta y}{\Delta x} = \dfrac{y_1}{x_1}$. The diagram also shows that if $\overrightarrow{CD} = k\overrightarrow{AB} = (kx_1, ky_1)$, then the

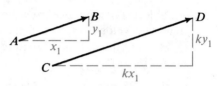

slope of $\overrightarrow{CD} = \dfrac{ky_1}{kx_1} = \dfrac{y_1}{x_1}$. Thus $\overrightarrow{AB}$ and $\overrightarrow{CD}$ are parallel, since they have the same slope. In such a case, it is customary to call the vectors, as well as the segments, **parallel.**

Similarly, it is customary to call the vectors $\overrightarrow{AB}$ and $\overrightarrow{CD}$ perpendicular if $\overrightarrow{AB}$ and $\overrightarrow{CD}$ are perpendicular. This occurs if $\overrightarrow{AB}$ and $\overrightarrow{CD}$ are horizontal and vertical respectively, or if the product of their slopes is -1. Thus, in the diagram below, v_1 and v_2 are perpendicular and

$$\left(\frac{y_1}{x_1}\right)\left(\frac{y_2}{x_2}\right) = -1,$$

$$y_1 y_2 = -x_1 x_2,$$

$$x_1 x_2 + y_1 y_2 = 0.$$

Notice that if the vectors are horizontal and vertical, respectively, so that $y_1 = 0$ and $x_2 = 0$, as illustrated, then we still have

$$x_1 x_2 + y_1 y_2 = 0.$$

The quantity $x_1 x_2 + y_1 y_2$ occurs often enough in vector work that it is given a special name: If $v_1 = (x_1, y_1)$ and $v_2 = (x_2, y_2)$, then the **dot product** of the vectors v_1 and v_2 is denoted by the symbol $v_1 \cdot v_2$ and defined by the equation

$$v_1 \cdot v_2 = x_1 x_2 + y_1 y_2.$$

From the definition of the dot product and the preceding discussion, we have the following theorem:

THEOREM.   If $v_1 \neq 0$ and $v_2 \neq 0$:
  a. v_1 and v_2 are perpendicular if and only if $v_1 \cdot v_2 = 0$.
  b. v_1 and v_2 are parallel if and only if $v_1 = kv_2$.

Remarks: 1. Perpendicular vectors are sometimes called **orthogonal vectors.**
  2. If $v_1 = kv_2$, it is customary to call v_1 *parallel* to v_2 even though the arrows representing v_1 and v_2 may be collinear.
  3. It is convenient to consider the zero vector to be both parallel and perpendicular to all vectors.

EXAMPLE 1.   If $u = (3, -6)$, $v = (4, 2)$, and $w = (-12, -6)$, find $u \cdot v$ and $v \cdot w$. Show that u and v are perpendicular and that v and w are parallel.

SOLUTION: $\mathbf{u} \cdot \mathbf{v} = (3, -6) \cdot (4, 2) = 3(4) + (-6)(2) = 0$. Hence, $\mathbf{u}$ and $\mathbf{v}$ are perpendicular.

$$\mathbf{v} \cdot \mathbf{w} = (4, 2) \cdot (-12, -6) = 4(-12) + 2(-6) = -60$$

The vectors $\mathbf{v}$ and $\mathbf{w}$ are parallel because

$$\mathbf{w} = (-12, -6) = -3(4, 2) = -3\mathbf{v}.$$

Properties of the dot product

The dot product has a number of properties that follow from its definition. Proofs of the first five properties listed below are left as exercises. The sixth property is proved below. This property provides an easy way to determine the angle between two nonzero vectors $\mathbf{u}$ and $\mathbf{v}$, that is, the angle between the arrows representing $\mathbf{u}$ and $\mathbf{v}$.

1. $\mathbf{u} \cdot \mathbf{v}$ is a scalar,
2. $\mathbf{u} \cdot \mathbf{v} = \mathbf{v} \cdot \mathbf{u}$
3. $\mathbf{u} \cdot \mathbf{u} = |\mathbf{u}|^2$
4. $k(\mathbf{u} \cdot \mathbf{v}) = (k\mathbf{u}) \cdot \mathbf{v}$
5. $\mathbf{u} \cdot (\mathbf{v} + \mathbf{w}) = \mathbf{u} \cdot \mathbf{v} + \mathbf{u} \cdot \mathbf{w}$
6. $\cos \theta = \dfrac{\mathbf{u} \cdot \mathbf{v}}{|\mathbf{u}| \, |\mathbf{v}|}$

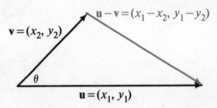

Proof of Property 6: The diagram at the right illustrates vectors $\mathbf{u}$ and $\mathbf{v}$ and the angle θ between these vectors. The sides of the triangle shown have the following lengths:

$|\mathbf{u}| = \sqrt{x_1^2 + y_1^2}$, $|\mathbf{v}| = \sqrt{x_2^2 + y_2^2}$, $|\mathbf{u} - \mathbf{v}| = \sqrt{(x_1 - x_2)^2 + (y_1 - y_2)^2}$

These equations, together with the Law of Cosines, yield the following:

$$\cos \theta = \frac{|\mathbf{u}|^2 + |\mathbf{v}|^2 - |\mathbf{u} - \mathbf{v}|^2}{2 \, |\mathbf{u}| \, |\mathbf{v}|}$$

$$= \frac{(x_1^2 + y_1^2) + (x_2^2 + y_2^2) - [(x_1 - x_2)^2 + (y_1 - y_2)^2]}{2 \, |\mathbf{u}| \, |\mathbf{v}|}$$

$$= \frac{x_1^2 + y_1^2 + x_2^2 + y_2^2 - [x_1^2 - 2x_1x_2 + x_2^2 + y_1^2 - 2y_1y_2 + y_2^2]}{2 \, |\mathbf{u}| \, |\mathbf{v}|}$$

$$= \frac{2x_1x_2 + 2y_1y_2}{2 \, |\mathbf{u}| \, |\mathbf{v}|}$$

$$= \frac{x_1x_2 + y_1y_2}{|\mathbf{u}| \, |\mathbf{v}|}$$

$$\cos \theta = \frac{\mathbf{u} \cdot \mathbf{v}}{|\mathbf{u}| \, |\mathbf{v}|}$$

EXAMPLE 2. Find the angle between the vectors $(1, 2)$ and $(-3, 1)$ to the nearest degree.

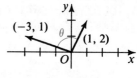

SOLUTION: $\cos \theta = \dfrac{(1, 2) \cdot (-3, 1)}{|(1, 2)| \, |(-3, 1)|}$

$= \dfrac{-3 + 2}{\sqrt{5} \sqrt{10}} = \dfrac{-1}{\sqrt{50}} = -\dfrac{1}{10} \sqrt{2} \approx -0.1414$

$\therefore \theta \approx 98°$

EXAMPLE 3. In $\triangle PQR$, $P = (2, 1)$, $Q = (4, 7)$, and $R = (-2, 4)$. Find the measure of $\angle P$ to the nearest ten minutes.

SOLUTION: First note that $\angle P$ is formed by $\overrightarrow{PQ}$ and $\overrightarrow{PR}$ and that

$\cos P = \dfrac{\overrightarrow{PQ} \cdot \overrightarrow{PR}}{|\overrightarrow{PQ}| \, |\overrightarrow{PR}|}.$

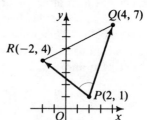

Since

$\overrightarrow{PQ} = (4, 7) - (2, 1) = (2, 6)$

and

$\overrightarrow{PR} = (-2, 4) - (2, 1) = (-4, 3),$

$\cos P = \dfrac{(2, 6) \cdot (-4, 3)}{|(2, 6)| \, |(-4, 3)|}$

$= \dfrac{10}{(\sqrt{40})(5)} = \dfrac{\sqrt{10}}{10} \approx 0.3162$

$\therefore \angle P \approx 71°30'$

ORAL EXERCISES 12-4

1. Which of the following three vectors are parallel and which are perpendicular?

$$\mathbf{u} = (4, -6) \qquad \mathbf{v} = (-2, 3) \qquad \mathbf{w} = (9, 6)$$

2. If $\mathbf{u} = (3, 4)$ and $\mathbf{v} = (-2, 2)$, calculate:

   a. $\mathbf{u} \cdot \mathbf{v}$ **b.** $2(\mathbf{u} \cdot \mathbf{v})$ **c.** $(2\mathbf{u}) \cdot \mathbf{v}$ **d.** $\mathbf{u} \cdot (2\mathbf{v})$

3. If $\mathbf{u} = (3, 4)$, evaluate: **a.** $|\mathbf{u}|$ **b.** $|\mathbf{u}|^2$ **c.** $\mathbf{u} \cdot \mathbf{u}$

4. If θ is the angle between $\mathbf{u} = (7, 1)$ and $\mathbf{v} = (5, 5)$, evaluate:

   a. $|\mathbf{u}|$ **b.** $|\mathbf{v}|$ **c.** $\mathbf{u} \cdot \mathbf{v}$ **d.** $\cos \theta$

5. For the figure shown, find $\cos \theta$ **(a)** by using the dot-product method, and **(b)** by using right-triangle trigonometry. Your results should agree.

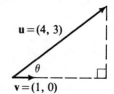

A   **1.** Evaluate: **a.** $(2, 3) \cdot (4, -5)$ **b.** $(3, -5) \cdot (7, 4)$

2. Evaluate: **a.** $(-3, 0) \cdot (5, 7)$ **b.** $\left(\frac{3}{5}, \frac{4}{5}\right) \cdot \left(\frac{1}{2}, -\frac{3}{2}\right)$

3. Find the value of k if the vectors $(4, 6)$ and $(k, 3)$ are **(a)** parallel, **(b)** perpendicular.

4. Find the value of k if the vectors $(6, -8)$ and $(4, k)$ are **(a)** parallel, **(b)** perpendicular.

5. If $\mathbf{u} = (-2, 3)$, evaluate: **a.** $\mathbf{u} \cdot \mathbf{u}$ **b.** $|\mathbf{u}|^2$

6. If $\mathbf{u} = (a, b)$, show that $\mathbf{u} \cdot \mathbf{u} = |u|^2$.

7. If $\mathbf{u} = (5, -3)$ and $\mathbf{v} = (3, 7)$, show that:
  a. $\mathbf{u} \cdot \mathbf{v} = \mathbf{v} \cdot \mathbf{u}$ **b.** $2(\mathbf{u} \cdot \mathbf{v}) = (2\mathbf{u}) \cdot \mathbf{v}$

8. If $\mathbf{u} = (1, 3)$ and $\mathbf{v} = (-4, 2)$, show that;
  a. $\mathbf{u} \cdot \mathbf{v} = \mathbf{v} \cdot \mathbf{u}$ **b.** $3(\mathbf{u} \cdot \mathbf{v}) = \mathbf{u} \cdot (3\mathbf{v})$

In Exercises 9 and 10 show that $\mathbf{u} \cdot (\mathbf{v} + \mathbf{w}) = \mathbf{u} \cdot \mathbf{v} + \mathbf{u} \cdot \mathbf{w}$ for the given vectors **u**, **v**, and **w**.

9. $\mathbf{u} = (-2, 5)$, $\mathbf{v} = (1, 3)$, and $\mathbf{w} = (-1, 2)$

10. $\mathbf{u} = (1, -4)$, $\mathbf{v} = (-2, -2)$, and $\mathbf{w} = (1, 5)$

11. Use Property 6 of the dot product to show that the angle between the vectors $\mathbf{u} = (1, 3)$ and $\mathbf{v} = (2, 1)$ is $45°$.

12. Show that the angle between the vectors $\mathbf{u} = (2, 3)$ and $\mathbf{v} = (1, -5)$ is $135°$.

13. Find the angle between the vectors $\mathbf{u} = (3, -4)$ and $\mathbf{v} = (3, 4)$.

14. Find the angle between the vectors $\mathbf{u} = (1, 3)$ and $\mathbf{v} = (-8, 6)$.

15. Given $A(1, 5)$, $B(4, 6)$, and $C(2, 8)$, find the measure of $\angle A$.

16. Given $A(-5, 1)$, $B(-3, 3)$, and $C(2, 2)$, find the measure of $\angle A$.

17. Given $P(0, 3)$, $Q(2, 4)$, and $R(3, 7)$, show that $\cos P = \dfrac{2\sqrt{5}}{5}$.

18. Given $R(0, -3)$, $S(2, 3)$, and $T(7, -2)$, show that $\cos R = \dfrac{\sqrt{5}}{5}$.

19. a. Given $A(1, -3)$, $B(-1, 3)$, and $C(6, 2)$, find $\cos C$ and $\sin C$.
  b. Use the formula Area $= \frac{1}{2} ab \sin C$ to find the area of $\triangle ABC$.

20. Repeat Exercise 19 for $A(5, 3)$, $B(3, 0)$, and $C(2, 2)$.

B   **21.** Let $\mathbf{u} = (x_1, y_1)$, $\mathbf{v} = (x_2, y_2)$, and $\mathbf{w} = (x_3, y_3)$. Show that
$$\mathbf{u} \cdot (\mathbf{v} + \mathbf{w}) = \mathbf{u} \cdot \mathbf{v} + \mathbf{u} \cdot \mathbf{w}$$
by expressing both sides of the equation in terms of the x- and the y-components.

22. Let $\mathbf{u} = (x_1, y_1)$, $\mathbf{v} = (x_2, y_2)$, and let k be a scalar. Show that
$$\mathbf{u} \cdot \mathbf{v} = \mathbf{v} \cdot \mathbf{u} \text{ and } (k\mathbf{u}) \cdot \mathbf{v} = k(\mathbf{u} \cdot \mathbf{v})$$
using the same technique as in Exercise 21.

23. The purpose of this exercise is to prove the Pythagorean Theorem using the dot product. In the right triangle shown, let $\mathbf{u} = \vec{AB}$ and $\mathbf{v} = \vec{BC}$.

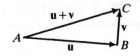

a. Note that $\vec{AC} = \mathbf{u} + \mathbf{v}$, so that
$$|\vec{AC}|^2 = (\mathbf{u} + \mathbf{v}) \cdot (\mathbf{u} + \mathbf{v}).$$
Show that this last equation simplifies to
$$|\vec{AC}|^2 = |\mathbf{u}|^2 + 2\mathbf{u} \cdot \mathbf{v} + |\mathbf{v}|^2.$$

b. Explain why $\mathbf{u} \cdot \mathbf{v} = 0$, and then prove that $|\vec{AC}|^2 = |\vec{AB}|^2 + |\vec{BC}|^2$.

24. Use dot products to prove that if $ABCD$ is a parallelogram, then
$$|\vec{AC}|^2 + |\vec{BD}|^2 = 2|\vec{AB}|^2 + 2|\vec{AD}|^2.$$
(*Hint:* Let $\mathbf{u} = \vec{AB}$ and $\mathbf{v} = \vec{AD}$. Then use dot products to express each term of the equation in terms of $\mathbf{u}$ and $\mathbf{v}$.)

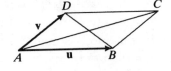

25. a. Show that $(\mathbf{u} + \mathbf{v}) \cdot (\mathbf{u} - \mathbf{v}) = 0$ if and only if $|\mathbf{u}| = |\mathbf{v}|$.

b. Use part (a) and the figure for Exercise 24 to prove that the diagonals of a parallelogram are perpendicular if and only if the parallelogram is a rhombus.

C 26. The purpose of this exercise is to use vectors to prove the theorem that the altitudes of a triangle are concurrent.

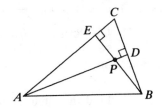

Given: $\vec{AD}$ and $\vec{BE}$ are altitudes of $\triangle ABC$.
(1) $\overleftrightarrow{AP} \perp \overline{BC}$ implies that $(P - A) \cdot (C - B) = 0$ which implies that
$$P \cdot C - A \cdot C - P \cdot B + A \cdot B = 0$$
(2) $\overleftrightarrow{BP} \perp \overline{CA}$ implies that $(P - B) \cdot (A - C) = 0$ which implies that
$$\underline{\quad ? \quad} = 0$$
(3) Add the final equations in (1) and (2), and factor the result.
(4) Show how your answer in (3) proves that $\overleftrightarrow{CP} \perp \overline{AB}$. Is the theorem proved?

27. Use the formula $\cos \theta = \dfrac{\mathbf{u} \cdot \mathbf{v}}{|\mathbf{u}| \, |\mathbf{v}|}$ to show that $-|\mathbf{u}| \, |\mathbf{v}| \leq \mathbf{u} \cdot \mathbf{v} \leq |\mathbf{u}| \, |\mathbf{v}|$. Then use this result and the result of Exercise 23(a) to prove that $|\mathbf{u} + \mathbf{v}| \leq |\mathbf{u}| + |\mathbf{v}|$.

CALCULATOR EXERCISE

Given line l_1 with inclination α and equation $(x, y) = (23, 37)t$ and line l_2 with inclination β and equation $(x, y) = (43, 19)t$. Find the angle between l_1 and l_2 by the following methods:
(1) using the fact that $\tan \alpha = $ slope of l_1 and $\tan \beta = $ slope of l_2
(2) using the dot product.

Three dimensions

12-5 / VECTORS IN THREE DIMENSIONS

One of the advantages of considering an ordered pair of numbers as a vector or as a point in a plane is that these ideas are easily generalized to three dimensions. The diagram shows a point $P = (2, 3, 4)$ in a three-dimensional coordinate system. The ordered triple of numbers $(2, 3, 4)$ also represents the vector $\overrightarrow{OP}$. To find $|\overrightarrow{OP}|$, note that in right $\triangle OAP$,

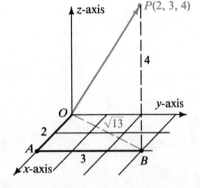

$$|\overrightarrow{OB}| = \sqrt{2^2 + 3^2} = \sqrt{13}$$

and that in right $\triangle OBP$,

$$|\overrightarrow{OP}| = \sqrt{(\sqrt{13})^2 + 4^2} = \sqrt{29}.$$

In general, if $A = (x_1, y_1, z_1)$ and $B = (x_2, y_2, z_2)$ are two points in space,

$$\overrightarrow{AB} = (x_2 - x_1, y_2 - y_1, z_2 - z_1).$$

The numbers $x_2 - x_1$, $y_2 - y_1$, and $z_2 - z_1$ are called the **x-, y-, and z-components** of $\overrightarrow{AB}$.

To find $|\overrightarrow{AB}|$, refer to the diagram at the right and note that in right $\triangle ADC$,

$$|\overrightarrow{AC}| = (x_2 - x_1)^2 + (y_2 - y_1)^2.$$

Consequently in right $\triangle ABC$,

$$|\overrightarrow{AB}|^2 = |\overrightarrow{AC}|^2 + |\overrightarrow{CB}|^2$$
$$|\overrightarrow{AB}|^2 = [(x_2 - x_1)^2 + (y_2 - y_1)^2] + (z_2 - z_1)^2$$
$$|\overrightarrow{AB}| = \sqrt{(x_2 - x_1)^2 + (y_2 - y_1)^2 + (z_2 - z_1)^2}$$

$$= \text{distance between } A \text{ and } B$$

The midpoint M of $\overline{AB}$ can be found as follows:

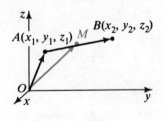

$$\overrightarrow{OM} = \overrightarrow{OA} + \tfrac{1}{2}\overrightarrow{AB}$$
$$\overrightarrow{OM} = (x_1, y_1, z_1) + \tfrac{1}{2}(x_2 - x_1, y_2 - y_1, z_2 - z_1)$$

$$M = \left(\frac{x_1 + x_2}{2}, \frac{y_1 + y_2}{2}, \frac{z_1 + z_2}{2}\right)$$

$$= \text{midpoint of } \overline{AB}$$

Notice that the distance and midpoint formulas for a three-dimensional coordinate system are generalizations of the two-dimensional formulas. These and other generalizations are summarized on the next page.

| Two dimensions. *If A, B, and P are points in a two-dimensional coordinate system with origin O.* | Three dimensions. *If A, B, and P are points in a three-dimensional coordinate system with origin O.* |

The points:

$A = (x_1, y_1), B = (x_2, y_2), P = (x, y)$ $A = (x_1, y_1, z_1), B = (x_2, y_2, z_2), P = (x, y, z)$

Distance between A and B:

$\sqrt{(x_2 - x_1)^2 + (y_2 - y_1)^2}$ $\sqrt{(x_2 - x_1)^2 + (y_2 - y_1)^2 + (z_2 - z_1)^2}$

Midpoint of $\overline{AB}$:

$\left(\dfrac{x_1 + x_2}{2}, \dfrac{y_1 + y_2}{2}\right)$ $\left(\dfrac{x_1 + x_2}{2}, \dfrac{y_1 + y_2}{2}, \dfrac{z_1 + z_2}{2}\right)$

Linear equation:

$ax + by = c$ $ax + by + cz = d$
  (a line)   (a plane)

Vector equation of line AB:

$\overrightarrow{OP} = \overrightarrow{OA} + t\,\overrightarrow{AB}$ $\overrightarrow{OP} = \overrightarrow{OA} + t\,\overrightarrow{AB}$

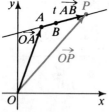

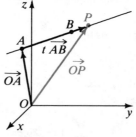

$|\overrightarrow{OP}| = r$: $|\overrightarrow{OP}| = r$:

Circle with center at O and radius r
$x^2 + y^2 = r^2$

Sphere with center at O and radius r
$x^2 + y^2 + z^2 = r^2$

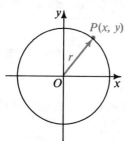

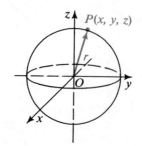

Two dimensions	Three dimensions

$|\overrightarrow{AP}| = r$:

Circle with center at
A and radius r
$(x - x_1)^2 + (y - y_1)^2 = r^2$

$|\overrightarrow{AP}| = r$:

Sphere with center at
A and radius r
$(x - x_1)^2 + (y - y_1)^2 + (z - z_1)^2 = r^2$

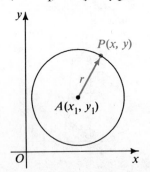

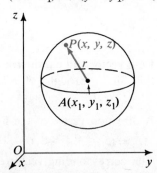

The rules for operations with three-dimensional vectors are analogous to those for two-dimensional vectors. If $\mathbf{u} = (x_1, y_1, z_1)$ and $\mathbf{v} = (x_2, y_2, z_2)$, then we have the following:

Vector Operations in Three Dimensions

1. Vector addition: $\mathbf{u} + \mathbf{v} = (x_1, y_1, z_1) + (x_2, y_2, z_2)$
$$= (x_1 + x_2, y_1 + y_2, z_1 + z_2)$$

2. Scalar multiplication: $k\mathbf{u} = k(x_1, y_1, z_1) = (kx_1, ky_1, kz_1)$

3. Dot product: $\mathbf{u} \cdot \mathbf{v} = (x_1, y_1, z_1) \cdot (x_2, y_2, z_2) = x_1x_2 + y_1y_2 + z_1z_2$

4. Absolute value of a vector: $|\mathbf{u}| = |(x_1, y_1, z_1)| = \sqrt{x_1^2 + y_1^2 + z_1^2}$

5. Cosine of angle between two vectors: $\cos\theta = \dfrac{\mathbf{u} \cdot \mathbf{v}}{|\mathbf{u}|\,|\mathbf{v}|}$

6. Perpendicular vectors: $\mathbf{u}$ and $\mathbf{v}$ are called perpendicular if and only if $\mathbf{u} \cdot \mathbf{v} = 0$.

7. Parallel vectors: If $\mathbf{u} \neq \mathbf{0}$ and $\mathbf{v} \neq \mathbf{0}$, $\mathbf{u}$ and $\mathbf{v}$ are called parallel if and only if $\mathbf{u} = k\mathbf{v}$.

Remarks: 1. Lines in space are perpendicular if they intersect and their direction vectors are perpendicular.
          2. Recall that we consider the zero vector to be both parallel and perpendicular to all vectors.

EXAMPLE 1. Write as a single vector: $2(3, 1, 4) + 3(1, -2, 0)$

   SOLUTION: $2(3, 1, 4) + 3(1, -2, 0) = (6, 2, 8) + (3, -6, 0)$
$$= (6 + 3, 2 + (-6), 8 + 0)$$
$$= (9, -4, 8)$$

Vectors and determinants   **429**

EXAMPLE 2. Find the angle θ between the vectors $(4, -5, 3)$ and $(7, 0, -1)$.

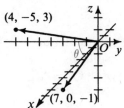

SOLUTION:
$$\cos \theta = \frac{(4, -5, 3) \cdot (7, 0, -1)}{|(4, -5, 3)| \, |(7, 0, -1)|}$$

$$\cos \theta = \frac{28 + 0 - 3}{\sqrt{50}\sqrt{50}} = \frac{25}{50} = \frac{1}{2}$$

$$\theta = 60°$$

EXAMPLE 3. Find vector and parametric equations of the line through $A = (2, 3, 1)$ and $B = (5, 4, 6)$.

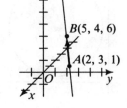

SOLUTION: A direction vector of the line is
$$\overrightarrow{AB} = (5 - 2, 4 - 3, 6 - 1) = (3, 1, 5).$$
The corresponding vector equation is:
$$\overrightarrow{OP} = \overrightarrow{OA} + t\,\overrightarrow{AB}$$
$$(x, y, z) = (2, 3, 1) + t(3, 1, 5)$$
Parametric equations of the line are;
$$x = 2 + 3t$$
$$y = 3 + t$$
$$z = 1 + 5t$$

EXAMPLE 4. Find a vector equation of the line through $(1, 5, -2)$ parallel to the line L with equation $(x, y, z) = (8, 0, 1) + t(4, 3, 2)$.

SOLUTION: From the equation of L, we see that $(4, 3, 2)$ is a direction vector of L. Thus $(4, 3, 2)$ must also be a direction vector of any line parallel to L. Hence, we want the equation of a line through $(1, 5, -2)$ with direction vector $(4, 3, 2)$. This equation is

$$(x, y, z) = (1, 5, -2) + t(4, 3, 2).$$

ORAL EXERCISES 12-5

1. Find the length and midpoint of $\overline{AB}$ if:
   a. $A = (0, 0, 0)$, $B = (1, -2, 3)$
   b. $A = (3, 0, -4)$, $B = (5, 4, 2)$

2. Imagine a rectangular box with one corner at the origin O of a three-dimensional coordinate system, as shown.
   a. If $G = (4, 5, 3)$, give the coordinates of points A, B, C, D, E, and F.
   b. Find $|\overrightarrow{OG}|$.

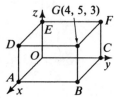

3. The axes of a two-dimensional coordinate system separate the plane into four regions called *quadrants*. Analogously, the axes of a three-dimensional coordinate system separate space into __?__ regions. Can you guess what these regions are called?

4. Find an equation of a sphere with radius 5 and center at the point **(a)** $(0, 0, 0)$, **(b)** $(1, 2, 3)$.

5. Simplify: **a.** $(3, 5, -2) + 2(1, 2, 3)$ **b.** $(3, 8, 1) \cdot (-4, 1, 4)$

6. Find k if the vectors $(2, 3, 4)$ and $(1, -2, k)$ are perpendicular.

7. Line L has vector equation $(x, y, z) = (2, 5, 1) + t(6, 7, 8)$.
   a. Name two points on L.   **b.** Find three parametric equations of L.
   c. Find a direction vector of L.
   d. Explain why L is parallel to the line $(x, y, z) = (8, -2, 2) + s(6, 7, 8)$.
   e. Explain why L is perpendicular to the line $(x, y, z) = (2, 5, 1) + s(4, 0, -3)$.

8. Explain how to find where the line $(x, y, z) = (-6, 0, 3) + t(4, 1, -1)$ intersects the sphere $x^2 + y^2 + z^2 = 9$.

WRITTEN EXERCISES 12-5

In Exercises 1–4, find the length and midpoint of $\overline{AB}$.

A   **1.** $A = (2, 5, -3)$, $B = (0, 3, 1)$ **2.** $A = (2, 8, -5)$, $B = (2, 0, -3)$
   3. $A = (3, -5, 0)$, $B = (-1, 1, 2)$ **4.** $A = (4, 1, 2)$, $B = (-4, -1, -2)$

5. Refer to Oral Exercise 2 and suppose that $G = (5, 6, 4)$. What are the coordinates of points A, B, C, D, E, and F? Find $|\overrightarrow{OG}|$.

6. Refer to Oral Exercise 2 and suppose that $G = (6, 7, 3)$. What are the coordinates of A, B, C, D, E, and F? Find $|\overrightarrow{OG}|$.

Simplify each expression.

7. **a.** $(3, 8, -2) + 2(4, -1, 2)$ **b.** $(1, -8, 6) \cdot (5, 2, 1)$ **c.** $|(3, 5, 1)|$
8. **a.** $(8, 7, 4) - (2, 0, 9) + (1, -2, 1)$ **b.** $(1, 4, 2) \cdot (3, 1, 2)$ **c.** $|(6, 2, 3)|$
9. If $\mathbf{u} = (1, 2, 1)$ and $\mathbf{v} = (2, -1, 1)$, find $\mathbf{u} + \mathbf{v}$ and $\mathbf{u} - \mathbf{v}$.
10. If $\mathbf{u} = (4, -1, 2)$ and $\mathbf{v} = (2, 3, -2)$, find $\mathbf{u} + \mathbf{v}$ and $\mathbf{u} - \mathbf{v}$.
11. Are the vectors $(3, -7, 1)$ and $(6, 3, 3)$ perpendicular?
12. Find k if $(2, k, -3)$ and $(4, 2, 6)$ are perpendicular.
13. Find an equation of a sphere with radius 2 and center at **(a)** the origin; **(b)** $(3, -1, 2)$.
14. Find an equation of the sphere with radius 7 and center $(1, 5, 3)$. Show that the point $(7, 7, 6)$ is on this sphere.
15. Find the center and radius of the sphere with equation
$$x^2 + y^2 + z^2 - 2x - 4y - 6z = 11$$
(*Hint:* By completing squares, rewrite the equation equivalently in the form $(x - x_1)^2 + (y - y_1)^2 + (z - z_1)^2 = r^2$.)

16. Find the center and radius of the sphere with equation
$$x^2 + y^2 + z^2 - 6x + 10y + 2z = 65.$$
(See the hint for Exercise 15.)

17. Find the angle between the vectors $(8, 6, 0)$ and $(2, -1, 2)$.

18. Find the angle between the vectors $(2, 2, 1)$ and $(3, 6, -2)$.

19. Let $A = (1, 3, 4)$, $B = (3, -1, 0)$, and $C = (3, 2, 6)$.
   a. Find the components of $\overrightarrow{AB}$ and $\overrightarrow{AC}$ and then show that these vectors are perpendicular.
   b. Find the area of right triangle ABC.

20. Repeat Exercise 19 if $A = (3, 7, -5)$, $B = (5, 9, -4)$, and $C = (7, 5, -9)$.

21. a. Given $A(0, 0, 0)$, $B(1, 1, -2)$, and $C(0, 1, -1)$, find $\cos A$ and $\sin A$.
   b. Find the area of $\triangle ABC$.

22. Repeat Exercise 21 given the points $A(1, 2, 3)$, $B(1, 5, 7)$, and $C(3, 3, 1)$.

23. Line L has vector equation $(x, y, z) = (-2, 0, 1) + t(4, -1, 1)$.
   a. Find three parametric equations of L.   b. Name two points on L.
   c. Which of the following points are on L: $(-10, 2, -1)$, $(18, -5, 6)$, $(14, -5, 4)$?
   d. Write a vector equation of the line through $(1, 2, 3)$ parallel to L.

24. Repeat Exercise 23 if the vector equation of L is
$$(x, y, z) = (6, -1, 0) + t(2, -1, 1).$$

25. Write vector and parametric equations of the line through $A = (4, 2, -1)$ and $B = (6, 3, 2)$.

26. Write vector and parametric equations of the line through $A = (2, 3, 1)$ and $B = (4, -1, 3)$.

27. Explain why the line $(x, y, z) = (4, -5, 2) + t(2, 1, 0)$ is parallel to the xy-plane.

28. Describe the line $(x, y, z) = (1, 2, 3) + t(0, 0, 1)$. Where does it intersect the xy-plane?

B   29. Given line L with equation $(x, y, z) = (5, 3, 0) + t(1, 4, -6)$.
   a. Find an equation of a line through $(1, 2, 3)$ parallel to L.
   b. Find an equation of a line through $(5, 3, 0)$ perpendicular to L.

30. Given line L with equation $(x, y, z) = (1, 3, -2) + t(2, 1, 1)$.
   a. Find an equation of a line through $(4, 5, 6)$ parallel to L.
   b. Find an equation of a line through $(1, 3, -2)$ perpendicular to L.

31. a. If $P = (x, y, z)$ is equidistant from $A = (0, 0, 0)$ and $B = (2, 3, 6)$, find and simplify an equation that must be satisfied by x, y, and z.
   b. Describe the set of points P in 3-dimensional space which are equidistant from A and B.

32. a. If $A = (3, 0, 4)$ and $B = (-1, 4, 6)$, find and simplify an equation that states $P = (x, y, z)$ is equidistant from A and B.

b. Verify that the coordinates of the midpoint of $\overline{AB}$ satisfy the equation derived in part **(a)**.

33. a. If A and B are given points in 3-dimensional space, describe the set of points P for which $\overrightarrow{PA} \cdot \overrightarrow{AB} = 0$.

   b. Give an equation of this set of points if $P = (x, y, z)$, $A = (3, 5, 2)$, and $B = (6, 6, 4)$.

34. a. If A and B are given points in 3-dimensional space, describe the set of points P for which $\overrightarrow{PA} \cdot \overrightarrow{PB} = 0$.

   b. Give a simplified equation of this set of points if $P = (x, y, z)$, $A = (2, 2, 1)$, and $B = (-2, -2, -1)$.

35. Where does the line $(x, y, z) = (3, 4, -2) + t(2, -1, 2)$ intersect the sphere $(x - 3)^2 + (y - 4)^2 + (z + 2)^2 = 36$?

36. All points in the xy-plane have z-coordinate 0. Consequently, the equation of this plane is $z = 0$. Where does the line

$$(x, y, z) = (6, -5, 4) + t(3, -2, 4)$$

intersect the xy-plane?

37. Reread Exercise 36 and then guess the equation of the xz-plane. Where does the line in Exercise 36 intersect the xz-plane?

38. Reread Exercise 36 and then guess the equation of the yz-plane. Where does the line in Exercise 36 intersect the yz-plane?

39. a. Show that the lines with vector equations

and
$$(x, y, z) = (3, 0, -4) + t(1, 2, 3)$$
$$(x, y, z) = (1, 5, 2) + s(2, 1, 4)$$

have no points in common.
(*Hint:* Set the x-coordinates equal, getting $3 + t = 1 + 2s$. Similarly, by equating the y-coordinates and equating the z-coordinates, you get two more equations in s and t. Show that there are no values of s and t that satisfy all three equations simultaneously.)

   b. Part (a) shows that the lines do not intersect. Are they parallel?

40. a. Show that the following lines intersect:

$$(x, y, z) = (-1, 5, 0) + t(1, 2, -1)$$
$$(x, y, z) = (0, 1, 3) + s(1, -1, 1)$$

   b. What is their intersection point?

41. a. Explain why the following lines are perpendicular:

$$(x, y, z) = (3, -1, 5) + t(2, 3, 6)$$
$$(x, y, z) = (3, -1, 5) + s(6, 6, -5)$$

   b. Explain why the line $(x, y, z) = (3, -1, 5) + t(2, 3, 6)$ is not perpendicular to

$$(x, y, z) = (2, -1, 5) + s(6, 6, -5).$$

42. Given $A(2, 0, 5)$, $B(6, 3, 2)$, $C(4, -1, -2)$, and $D(0, 6, -1)$ find the centroid of tetrahedron $ABCD$. (*Hint:* See Exercise 31 of Section 12-2.)

The equation of a plane

The graph of any equation of the form $ax + by + cz = d$, where a, b, and c are not all 0, is a **plane.** For example, the graph of $2x + 3y + 4z = 12$ is a plane. To picture this plane, we locate where it intersects the axes of a three-dimensional coordinate system.

$$2x + 3y + 4z = 12$$

Setting y and $z = 0$: $2x$ $= 12$ and $x = 6$

Setting x and $z = 0$: $3y$ $= 12$ and $y = 4$

Setting x and $y = 0$: $4z = 12$ and $z = 3$

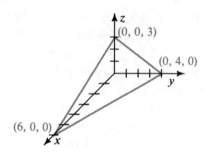

The diagram at the right shows that section of the plane contained in the first octant. We show the three points, since they are sufficient to determine the plane.

Parts of some other planes are shown below with their corresponding equations.

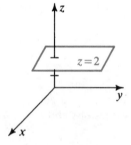 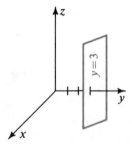 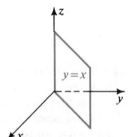

Vectors perpendicular to a plane

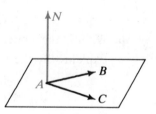

If a vector $\overrightarrow{AN}$ is perpendicular to a plane, as shown, it is perpendicular to $\overrightarrow{AB}$, $\overrightarrow{AC}$, and all other vectors in the plane. The next theorem gives a special relationship between a vector perpendicular to a plane and the equation of the plane.

THEOREM. If (a, b, c) is a vector perpendicular to a plane at the point (x_0, y_0, z_0), then an equation of the plane is $ax + by + cz = d$, where $d = ax_0 + by_0 + cz_0$.

The converse is also true.

Proof: Let $P = (x, y, z)$ be an arbitrary point in the plane and let $A = (x_0, y_0, z_0)$ be a known point in the plane. Then $\overrightarrow{AP} = (x - x_0, y - y_0, z - z_0)$.

If vector $\overrightarrow{AN} = (a, b, c)$ is perpendicular to the plane we have:

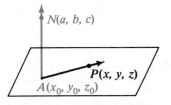

$$\overrightarrow{AN} \perp \overrightarrow{AP}$$
$$\overrightarrow{AN} \cdot \overrightarrow{AP} = 0$$
$$(a, b, c) \cdot (x - x_0, y - y_0, z - z_0) = 0$$
$$a(x - x_0) + b(y - y_0) + c(z - z_0) = 0$$
$$ax + by + cz = ax_0 + by_0 + cz_0$$
$$ax + by + cz = d, \text{ where } d = ax_0 + by_0 + cz_0$$

The converse can be proved by reversing the steps in the proof.

EXAMPLE. The vector $(3, 4, -2)$ is perpendicular to a plane at the point $A = (0, 1, 2)$. Find an equation of the plane.

SOLUTION: The vector perpendicular to the plane is $(3, 4, -2)$. Therefore, an equation of the plane is $3x + 4y - 2z = d$. To find d, substitute the coordinates of point $A = (0, 1, 2)$ in the equation $3x + 4y - 2z = d$. Then,

$$3 \cdot 0 + 4 \cdot 1 - 2 \cdot 2 = d \quad \text{and} \quad 0 = d$$

The equation of the plane is $3x + 4y - 2z = 0$.

(Notice that this plane passes through the origin because the coordinates of the origin satisfy the equation.)

ORAL EXERCISES 12-6

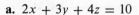

1. Find the x-, y-, and z- intercepts of the plane $3x + y + 2z = 6$.

2. A rectangular box is shown. Find an equation of the plane that contains:
   a. the top of the box
   b. the bottom of the box
   c. the front of the box
   d. the right end of the box

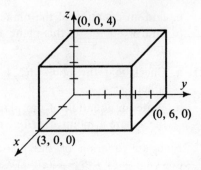

3. Name a vector perpendicular to each plane.
   a. $2x + 3y + 4z = 10$ b. $3x - 4z = 12$ c. $z = 5$

4. The vector $(1, 1, 1)$ is perpendicular to a plane at the point $(3, 4, 5)$. What is an equation of the plane?

5. The vector $(6, 7, 8)$ is perpendicular to a plane at the point $(0, 0, 0)$. What is an equation of the plane?

WRITTEN EXERCISES 12-6

Sketch each plane whose equation is given.

A
1. $2x + 3y + 6z = 12$ **2.** $3x + y + 2z = 6$
3. $5x - 2y + 2z = 10$ **4.** $2x + 4y - z = 8$
5. $z = 2$ **6.** $y = 4$

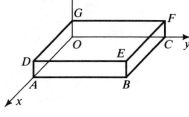

7. Let $E(4, 6, 1)$ be a vertex of the rectangular box shown. Find the equation of
    a. plane $DEFG$ **b.** plane $ABED$
    c. plane $BCFE$ **d.** plane $OADG$

8. Repeat Exercise 7 for $E(5, 7, 2)$.

Name a vector perpendicular to the plane whose equation is given.

  9. $3x + 4y + 6z = 12$ **10.** $3x - 5y + 4z = 0$
11. $x + y = 4$ **12.** $x - z = 0$
13. $z = 1$ **14.** $y = -3$

In Exercises 15–18, find an equation of the plane described.

15. The vector $(2, 3, 5)$ is perpendicular to the plane at point $A(3, 1, 7)$.
16. The vector $(1, -4, 2)$ is perpendicular to the plane at point $A(3, 0, 2)$.
17. The vector $(0, 0, 1)$ is perpendicular to the plane at point $A(1, 4, 5)$.
18. The vector $(2, 0, 3)$ is perpendicular to the plane at point $A(3, 8, -2)$.
19. Given the points $A(2, 2, 2)$ and $B(4, 6, 8)$:
    a. find an equation of the plane perpendicular to $\overline{AB}$ at its midpoint.
    b. show that the point $P(0, 1, 8)$ satisfies your equation in part (a).
    c. show that $PA = PB$.
20. Given the points $A(-2, 4, 1)$ and $B(0, 2, -1)$:
    a. find an equation of the plane perpendicular to $\overline{AB}$ at its midpoint.
    b. find two points in this plane and show that each is equidistant from A and B.

21. a. Check that the point $A(2, 1, 2)$ is on the sphere $x^2 + y^2 + z^2 = 9$.

    b. Write an equation of the plane tangent to the sphere at point A. (*Hint:* $\overrightarrow{OA} \perp$ plane.)

22. Find an equation of the plane tangent to $(x - 1)^2 + (y - 1)^2 + (z - 1)^2 = 49$ at the point $(7, -1, 4)$.

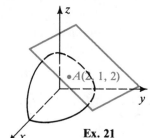

Ex. 21

B
23. The plane $z = 3$ intersects the sphere $x^2 + y^2 + z^2 = 25$ in a circle. Illustrate this with a sketch and find the area of the circle.

24. The plane $y = 5$ intersects the sphere $x^2 + y^2 + z^2 = 169$ in a circle. Illustrate this with a sketch and find the area of the circle.

25. Plane M_1 has equation $a_1x + b_1y + c_1z = d_1$. Plane M_2 has equation $a_2x + b_2y + c_2z = d_2$.

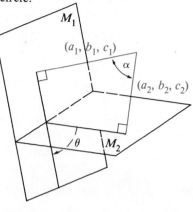

a. Explain why, in the diagram, $\theta = \alpha$.

b. The angle θ between the two planes is shown in the diagram. Explain why:

$$\cos \theta = \cos \alpha = \frac{(a_1, b_1, c_1) \cdot (a_2, b_2, c_2)}{|(a_1, b_1, c_1)| \, |(a_2, b_2, c_2)|}$$

26. Use the formula of Exercise 25(b) to find the measure of the angle between the planes
$$2x + y - z = 3 \quad \text{and} \quad x + 2y + z = 5.$$

27. Use the formula of Exercise 25(b) to find the measure of the angle between the planes
$$2x + 2y - z = 3 \quad \text{and} \quad 4x - 3y + 2z = 5.$$

28. Refer to Exercise 25.

a. What can you conclude about the two planes if
$$(a_1, b_1, c_1) \cdot (a_2, b_2, c_2) = 0?$$

b. Are the planes $3x + 4y + 2z = 5$ and $2x - y - z = 3$ perpendicular?

c. Are the planes $4x - 5y + 6z = 0$ and $3x - 2z = 7$ perpendicular?

29. a. Name a vector perpendicular to the plane $2x + 3y - z = 6$.

b. Is this same vector also perpendicular to the plane $2x + 3y - z = 12$?

c. Make a sketch showing both planes on the same axes. Are the two planes parallel?

d. Is the plane $4x + 6y - 2z = 4$ parallel to the first plane?

30. Which of the following planes are perpendicular and which are parallel?
(1) $3x + 2y - z = 6$ (2) $6x + 4y - 2z = 8$ (3) $4x - 2y + 8z = 7$

31. Given the plane M with equation $4x + y - 3z = 10$. Find an equation of a plane (a) parallel to M and (b) perpendicular to M.

32. Repeat Exercise 31 if plane M has equation $5x - 2y + z = 2$.

33. a. What is the direction vector of the line $(x, y, z) = (3, 1, 4) + t(4, -5, 2)$?

b. Explain why this line is parallel to the plane $2x + 2y + z = 7$.

34. The angle between a line and a plane is the angle α between the line and its projection in the plane. (See diagram.)

a. Suppose the line has direction vector (p, q, r), and that (a, b, c) is perpendicular to the plane. Show that
$$\sin \alpha = \frac{(a, b, c) \cdot (p, q, r)}{|(a, b, c)| \, |(p, q, r)|}.$$

b. Find the angle between the line $(x, y, z) = (2, 1, 1) + t(0, 1, -1)$ and the plane $x + 2y - 2z = 6$.

35. **a.** Explain why the line $(x, y, z) = (8, 9, 10) + t(3, 4, 5)$ is perpendicular to the plane $3x + 4y + 5z = 10$.

   b. Where does the line intersect the plane?

36. **a.** Explain why the line $(x, y, z) = (4, 2, 1) + t(2, 3, 4)$ is perpendicular to the plane $2x + 3y + 4z = 76$.

   b. Where does the line intersect the plane?

37. The purpose of this exercise is to find the distance from the point $(3, 1, 5)$ to the plane $2x + 2y + z = 4$.

   a. Find a vector equation of the line that contains $(3, 1, 5)$ and is perpendicular to the plane.

   b. Where does this line intersect the plane?

   c. Find the distance from the point to the plane.

C 38. Prove that the distance from (x_0, y_0, z_0) to the plane

$$ax + by + cz + k = 0 \quad \text{is} \quad \frac{|ax_0 + by_0 + cz_0 + k|}{\sqrt{a^2 + b^2 + c^2}}.$$

39. Study Exercise 38 and then guess a formula for the distance from (x_0, y_0) to the line $ax + by + k = 0$.

Determinants and applications

12-7 / DETERMINANTS

The expression $\begin{vmatrix} a_1 & b_1 \\ a_2 & b_2 \end{vmatrix}$ is called a 2 by 2 **determinant** (2 rows and 2 columns).
To find the value of this determinant:

first take the product of these numbers ⟶ $\begin{vmatrix} a_1 & & & b_1 \\ & & & \\ a_2 & & & b_2 \end{vmatrix}$

then subtract the product of these numbers ⟶

The value, then, of a 2 by 2 determinant is $a_1 b_2 - a_2 b_1$.

EXAMPLE 1. $\begin{vmatrix} 3 & 4 \\ 2 & 7 \end{vmatrix} = (3)(7) - (2)(4) = 13$

The general 3 by 3 determinant is shown at the right. Each element or number in this determinant is associated with a 2 by 2 determinant called its **minor**. The minor of an element is the determinant that remains when you delete the row and column containing that element.

$$\begin{vmatrix} a_1 & b_1 & c_1 \\ a_2 & b_2 & c_2 \\ a_3 & b_3 & c_3 \end{vmatrix}$$

$$\begin{vmatrix} a_1 & b_1 & c_1 \\ a_2 & b_2 & c_2 \\ a_3 & b_3 & c_3 \end{vmatrix} \rightarrow \text{ The minor of element } b_1 \text{ is } B_1 = \begin{vmatrix} a_2 & c_2 \\ a_3 & c_3 \end{vmatrix}.$$

The value of the 3 by 3 determinant shown above can be computed by expanding by the minors of one of its rows or columns. If you use the elements of the top row, for example, the value of the determinant is $a_1A_1 - b_1B_1 + c_1C_1$ where A_1 is the minor of a_1, B_1 is the minor of b_1, and C_1 is the minor of c_1.

EXAMPLE 2. $\begin{vmatrix} 5 & 2 & 8 \\ 3 & 4 & 1 \\ 7 & -1 & 6 \end{vmatrix} = 5\begin{vmatrix} 4 & 1 \\ -1 & 6 \end{vmatrix} - 2\begin{vmatrix} 3 & 1 \\ 7 & 6 \end{vmatrix} + 8\begin{vmatrix} 3 & 4 \\ 7 & -1 \end{vmatrix}$

$$= 5(25) - 2(11) + 8(-31) = -145$$

The determinant in Example 3 was expanded by the minors of its top row. You can expand the determinant by the minors of the elements in any row or any column and get the same final answer. Just consider the position of each element in the original determinant and find its corresponding position on the following "checkerboard" pattern. Add or subtract the corresponding term according to the $+$ or $-$ sign in that position.

$$\begin{vmatrix} + & - & + \\ - & + & - \\ + & - & + \end{vmatrix}$$

If we expand the determinant in Example 3 by the minors of its second column, we get

$$\begin{vmatrix} 5 & 2 & 8 \\ 3 & 4 & 1 \\ 7 & -1 & 6 \end{vmatrix} = -2\begin{vmatrix} 3 & 1 \\ 7 & 6 \end{vmatrix} + 4\begin{vmatrix} 5 & 8 \\ 7 & 6 \end{vmatrix} - (-1)\begin{vmatrix} 5 & 8 \\ 3 & 1 \end{vmatrix}$$

$$= -2(11) + 4(-26) + 1(-19) = -145$$

The evaluation of a determinant is simplified if you can expand by a row or column that contains several zeros, since terms in the expansion that have zero as a coefficient will drop out. The row and column operations described below can help you put a determinant in a form that has several zeros in a row or column.

1. If you factor a single row or column as shown, the value of the determinant is unchanged.

$$\begin{vmatrix} 4 & 8 \\ 3 & 5 \end{vmatrix} = 4\begin{vmatrix} 1 & 2 \\ 3 & 5 \end{vmatrix} \qquad \begin{vmatrix} 3 & 1 & 2 \\ 6 & 0 & 8 \\ -3 & 4 & 7 \end{vmatrix} = 3\begin{vmatrix} 1 & 1 & 2 \\ 2 & 0 & 8 \\ -1 & 4 & 7 \end{vmatrix}$$

2. If you multiply any row (or column) by a constant and add the resulting numbers to the elements of another row (or column) as shown on the following page, the value of the determinant is unchanged.

$$\begin{vmatrix} 5 & 3 & 7 \\ 1 & 1 & 2 \\ 4 & 4 & 9 \end{vmatrix} = \begin{vmatrix} 5 & 3 & 7 \\ 1 & 1 & 2 \\ (-4) \atop (-4)+4 & (-4) \atop (-4)+4 & 2 \atop (-8)+9 \end{vmatrix} = \begin{vmatrix} 5 & 3 & 7 \\ 1 & 1 & 2 \\ 0 & 0 & 1 \end{vmatrix}$$

—— Add $(-4 \times \text{Row 2})$ to Row 3. ——

The techniques used to evaluate 3 by 3 determinants can be extended to 4 by 4 and, in general, n by n determinants. For example, the minor of an element in a 4 by 4 determinant is the 3 by 3 determinant remaining when you delete the row and column containing that element. The "checkerboard" pattern used in expanding a 4 by 4 determinant by minors is shown at the right.

$$\begin{vmatrix} + & - & + & - \\ - & + & - & + \\ + & - & + & - \\ - & + & - & + \end{vmatrix}$$

EXAMPLE 3. Evaluate the determinant below after using row and column operations to simplify it.

$$\begin{vmatrix} 2 & 5 & 3 & 7 \\ 4 & 1 & 0 & 2 \\ 3 & 2 & 2 & 1 \\ -2 & 8 & 1 & 0 \end{vmatrix}$$

SOLUTION: Column 3 contains a 0 and also a 1, which can be used as a "pivot" to produce two more zeros in the column.

—— Add $(-3 \times \text{Row 4})$ to Row 1. ——

$$\begin{vmatrix} 2 & 5 & 3 & 7 \\ 4 & 1 & 0 & 2 \\ 3 & 2 & 2 & 1 \\ -2 & 8 & 1 & 0 \end{vmatrix} = \begin{vmatrix} (-3)(-2)+2 & (-3)8+5 & -3(1)+3 & 7 \\ 4 & 1 & 0 & 2 \\ (-2)(-2)+3 & (-2)8+2 & -2(1)+2 & 1 \\ -2 & 8 & 1 & 0 \end{vmatrix} = \begin{vmatrix} 8 & -19 & 0 & 7 \\ 4 & 1 & 0 & 2 \\ 7 & -14 & 0 & 1 \\ -2 & 8 & 1 & 0 \end{vmatrix}$$

—— Add $(-2 \times \text{Row 4})$ to Row 3. ——

Since Column 3 now contains 3 zeros, we expand by its minors, getting

$$0 \cdot \begin{vmatrix} 4 & 1 & 2 \\ 7 & -14 & 1 \\ -2 & 8 & 0 \end{vmatrix} - 0 \cdot \begin{vmatrix} 8 & -19 & 7 \\ 7 & -14 & 1 \\ -2 & 8 & 0 \end{vmatrix} + 0 \cdot \begin{vmatrix} 8 & -19 & 7 \\ 4 & 1 & 2 \\ -2 & 8 & 0 \end{vmatrix} - 1 \cdot \begin{vmatrix} 8 & -19 & 7 \\ 4 & 1 & 2 \\ 7 & -14 & 1 \end{vmatrix}$$

The last determinant can be evaluated directly or it can be simplified by more row and column operations. For example, you could use the 1 in Row 2 as the pivot to produce two zeros in the row by adding $-4 \times$ Column 2 to Column 1 and adding $-2 \times$ Column 2 to Column 3. You will complete the evaluation as Exercise 14 below.

440   **Chapter 12**

Evaluate the following determinants.

1. $\begin{vmatrix} 2 & 3 \\ 4 & 5 \end{vmatrix}$

2. $\begin{vmatrix} 8 & 5 \\ 4 & 4 \end{vmatrix}$

3. $\begin{vmatrix} 3 & -5 \\ 7 & 10 \end{vmatrix}$

4. $\begin{vmatrix} -5 & 4 \\ -4 & -3 \end{vmatrix}$

5. If you expand by minors of the top row, complete:

$$\begin{vmatrix} 3 & 2 & 4 \\ 5 & 7 & 1 \\ 8 & 6 & 9 \end{vmatrix} = \underline{?} \begin{vmatrix} 7 & 1 \\ 6 & 9 \end{vmatrix} - \underline{?} \begin{vmatrix} 5 & 1 \\ 8 & 9 \end{vmatrix} + \underline{?} \begin{vmatrix} 5 & 7 \\ 8 & 6 \end{vmatrix}$$

6. If you expand by minors of the third column, complete:

$$\begin{vmatrix} 3 & 2 & 4 \\ 5 & 7 & 1 \\ 8 & 6 & 9 \end{vmatrix} = 4 \begin{vmatrix} ? & ? \\ ? & ? \end{vmatrix} - 1 \begin{vmatrix} ? & ? \\ ? & ? \end{vmatrix} + 9 \begin{vmatrix} ? & ? \\ ? & ? \end{vmatrix}$$

7. Evaluate: $\begin{vmatrix} 1 & 2 & 3 \\ 4 & 5 & 6 \\ 0 & 0 & 1 \end{vmatrix}$

WRITTEN EXERCISES 12-7

Evaluate.

A 1. $\begin{vmatrix} 2 & 7 \\ 8 & 4 \end{vmatrix}$

2. $\begin{vmatrix} -3 & 5 \\ -7 & 2 \end{vmatrix}$

3. $\begin{vmatrix} 25 & 125 \\ 75 & 250 \end{vmatrix}$

4. Show that $\begin{vmatrix} ka & kb \\ c & d \end{vmatrix} = \begin{vmatrix} a & b \\ kc & kd \end{vmatrix} = k \cdot \begin{vmatrix} a & b \\ c & d \end{vmatrix}$.

5. Show that $\begin{vmatrix} a & b \\ c & d \end{vmatrix} = \begin{vmatrix} a-c & b-d \\ c & d \end{vmatrix}$, and use this result to evaluate $\begin{vmatrix} 387 & 411 \\ 385 & 410 \end{vmatrix}$.

6. True or false? $\begin{vmatrix} a & b \\ c & d \end{vmatrix} = \begin{vmatrix} a-kc & b-kd \\ c & d \end{vmatrix}$ for every real number k.

Evaluate.

7. $\begin{vmatrix} 4 & -7 & 3 \\ 2 & 0 & 0 \\ 5 & 1 & 6 \end{vmatrix}$

8. $\begin{vmatrix} -1 & 3 & 2 \\ 4 & 0 & 1 \\ 1 & 5 & 0 \end{vmatrix}$

9. $\begin{vmatrix} 2 & 4 & -4 \\ 3 & 1 & 6 \\ 0 & 2 & 5 \end{vmatrix}$

10. $\begin{vmatrix} 1 & -3 & 4 \\ 0 & 1 & 1 \\ 5 & -2 & 3 \end{vmatrix}$

11. $\begin{vmatrix} 5 & 5 & 10 \\ 1 & 4 & 3 \\ -1 & 2 & 6 \end{vmatrix}$

12. $\begin{vmatrix} 1 & 2 & 3 \\ 2 & 4 & 6 \\ 17 & 18 & 19 \end{vmatrix}$

13. In Exercise 12, one row is a multiple of another row and the value of the determinant is zero. Is it a coincidence that the value is zero?

14. Complete the evaluation of the determinant in Example 3 by using the suggestion in the last sentence of the example.

For Exercises 15 and 16, evaluate each determinant.

B **15.** $\begin{vmatrix} 2 & -4 & 7 & 3 \\ 0 & 5 & 1 & -2 \\ -2 & 1 & 0 & -3 \\ 0 & -6 & 4 & 2 \end{vmatrix}$ **16.** $\begin{vmatrix} -2 & 1 & 5 & 0 \\ 3 & 4 & -2 & 1 \\ 0 & 0 & 1 & 2 \\ 1 & 0 & 0 & -3 \end{vmatrix}$

C **17.** Consider the linear equations $a_1x + b_1y = c_1$, and $a_2x + b_2y = c_2$.

   a. Solve these equations for x and y, expressing your answers in terms of the a's, b's, and c's.

   b. Show that your answer can be written:

$$x = \frac{\begin{vmatrix} c_1 & b_1 \\ c_2 & b_2 \end{vmatrix}}{\begin{vmatrix} a_1 & b_1 \\ a_2 & b_2 \end{vmatrix}} \quad \text{and} \quad y = \frac{\begin{vmatrix} a_1 & c_1 \\ a_2 & c_2 \end{vmatrix}}{\begin{vmatrix} a_1 & b_1 \\ a_2 & b_2 \end{vmatrix}}$$

12-8/APPLICATIONS OF DETERMINANTS

Consider the following linear equations in two variables:

$$a_1x + b_1y = c_1$$
$$a_2x + b_2y = c_2$$

When these equations are solved algebraically for x and y, the answers are:

$$x = \frac{c_1b_2 - c_2b_1}{a_1b_2 - a_2b_1} \quad \text{and} \quad y = \frac{a_1c_2 - a_2c_1}{a_1b_2 - a_2b_1}$$

which can be written also as

$$x = \frac{\begin{vmatrix} c_1 & b_1 \\ c_2 & b_2 \end{vmatrix}}{\begin{vmatrix} a_1 & b_1 \\ a_2 & b_2 \end{vmatrix}} \quad \text{and} \quad y = \frac{\begin{vmatrix} a_1 & c_1 \\ a_2 & c_2 \end{vmatrix}}{\begin{vmatrix} a_1 & b_1 \\ a_2 & b_2 \end{vmatrix}}$$

There is an easy way to remember the determinants for x and y. Consider first the original equations and then their related array of numbers:

$$\begin{array}{ccc} a_1 & b_1 & c_1 \\ a_2 & b_2 & c_2 \end{array}$$

          x-coefficients y-coefficients   constant terms

The three determinants we need for x and y are easily remembered from the following array.

1. Determinant for denominators of both x and y $= \begin{vmatrix} a_1 & b_1 \\ a_2 & b_2 \end{vmatrix}$

2. Determinant for numerator of x $= \begin{vmatrix} c_1 & b_1 \\ c_2 & b_2 \end{vmatrix}$ In the first determinant, replace x-coefficients by the constants.

3. Determinant for numerator of y $= \begin{vmatrix} a_1 & c_1 \\ a_2 & c_2 \end{vmatrix}$ In the first determinant, replace y-coefficients by the constants.

EXAMPLE 1. Solve the following pair of equations: $\begin{aligned} 3x - 4y &= 8 \\ 11x + 9y &= 5 \end{aligned}$

SOLUTION: $\quad x = \dfrac{\begin{vmatrix} 8 & -4 \\ 5 & 9 \end{vmatrix}}{\begin{vmatrix} 3 & -4 \\ 11 & 9 \end{vmatrix}} = \dfrac{72 - (-20)}{27 - (-44)} = \dfrac{92}{71}$

$\quad y = \dfrac{\begin{vmatrix} 3 & 8 \\ 11 & 5 \end{vmatrix}}{\begin{vmatrix} 3 & -4 \\ 11 & 9 \end{vmatrix}} = \dfrac{15 - 88}{27 - (-44)} = \dfrac{-73}{71}$

Determinants can also be used to solve three equations in three variables. The method shown in Example 2 below is similar to the method used in Example 1.

EXAMPLE 2. Solve the system of equations: $\begin{aligned} 3x - y + 2z &= 4 \\ 2x + 3y - z &= 14 \\ 7x - 4y + 3z &= -4 \end{aligned}$

SOLUTION:

Let $D = \begin{vmatrix} 3 & -1 & 2 \\ 2 & 3 & -1 \\ 7 & -4 & 3 \end{vmatrix}$ $\quad$ Let $D_x = \begin{vmatrix} 4 & -1 & 2 \\ 14 & 3 & -1 \\ -4 & -4 & 3 \end{vmatrix}$

Let $D_y = \begin{vmatrix} 3 & 4 & 2 \\ 2 & 14 & -1 \\ 7 & -4 & 3 \end{vmatrix}$ $\quad$ Let $D_z = \begin{vmatrix} 3 & -1 & 4 \\ 2 & 3 & 14 \\ 7 & -4 & -4 \end{vmatrix}$

Notice that the numbers in D are the coefficients of x, y, and z in the three equations. To form D_x the first column of D is replaced by the constants on the right sides of the equations. Similarly, the second and third columns of D are replaced by the constants to get D_y and D_z, respectively. These determinants have the following values:

$D = -30 \qquad D_x = -30 \qquad D_y = -150 \qquad D_z = -90$

(Solution continued on page 444)

Vectors and determinants **443**

The values of x, y, and z are given by the following formulas:

$$x = \frac{D_x}{D} = \frac{-30}{-30} = 1 \qquad y = \frac{D_y}{D} = \frac{-150}{-30} = 5 \qquad z = \frac{D_z}{D} = \frac{-90}{-30} = 3$$

This method is known as **Cramer's Rule.** In general, Cramer's Rule uses n by n determinants to solve n equations in n variables. However, if n is larger than 3, a great deal of computation is involved. For $n > 4$, the solutions would probably have to be found by faster methods using a computer.

Determinants are useful in geometry as well as in algebra.

Area	Volume

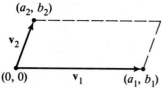

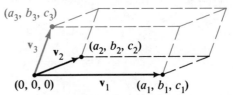

The area of a parallelogram in the xy-plane with sides determined by $\mathbf{v}_1 = (a_1, b_1)$ and $\mathbf{v}_2 = (a_2, b_2)$ is the absolute value of

$$\begin{vmatrix} a_1 & b_1 \\ a_2 & b_2 \end{vmatrix}.$$

The volume of a parallelepiped in xyz-space with sides determined by $\mathbf{v}_1 = (a_1, b_1, c_1)$, $\mathbf{v}_2 = (a_2, b_2, c_2)$, and $\mathbf{v}_3 = (a_3, b_3, c_3)$ is the absolute value of

$$\begin{vmatrix} a_1 & b_1 & c_1 \\ a_2 & b_2 & c_2 \\ a_3 & b_3 & c_3 \end{vmatrix}$$

EXAMPLE 3. Find the area of the triangle with vertices $P(1, 2)$, $Q(3, 6)$, and $R(6, 1)$.

SOLUTION: $\overrightarrow{PQ}$ and $\overrightarrow{PR}$ determine the sides of triangle PQR and parallelogram $PQSR$.

$$\overrightarrow{PQ} = Q - P = (3, 6) - (1, 2) = (2, 4)$$
$$\overrightarrow{PR} = R - P = (6, 1) - (1, 2) = (5, -1)$$

Area of $\triangle PQR = \frac{1}{2} \cdot$ area of $\square PQSR$

$$= \frac{1}{2} \cdot \text{absolute value of } \begin{vmatrix} 2 & 4 \\ 5 & -1 \end{vmatrix}$$

$$= \frac{1}{2} \cdot |-22|$$

$$= 11$$

The minus sign in the next-to-last step indicates that P, Q, and R have a negative (clockwise) orientation.

1. When determinants are used to solve the system of equations $\begin{aligned} 3x + 4y &= 7 \\ 5x + 6y &= 8 \end{aligned}$,

then $x = \dfrac{|?|}{|?|}$ and $y = \dfrac{|?|}{|?|}$.

2. When determinants are used to solve the system of equations

$\begin{aligned} 3x + y - 2z &= 4 \\ 2x - y + 4z &= 1 \\ x + 2y + 7z &= 3 \end{aligned}$ then $x = \dfrac{|?|}{|?|}, y = \dfrac{|?|}{|?|}, z = \dfrac{|?|}{|?|}$.

3. Why can't you use Cramer's Rule to find the solutions for the system of

equations $\begin{aligned} 2x + y &= 3 \\ 2x + y &= 5 \end{aligned}$?

4. What is the area of the parallelogram with sides determined by $v_1 = (1, 3)$ and $v_2 = (2, 4)$?

In Exercises 1–4, solve the given pair of equations by using determinants. Then, sketch the graphs of the two equations and label their intersection with the common solution of the equations.

A

1. $\begin{aligned} 5x - 4y &= 1 \\ 3x + 2y &= 5 \end{aligned}$

2. $\begin{aligned} 5x - 2y &= 11 \\ x + 3y &= 9 \end{aligned}$

3. $\begin{aligned} 3x + 2y &= -1 \\ 2x - y &= 4 \end{aligned}$

4. $\begin{aligned} 7x + y &= 7 \\ -x + 2y &= 14 \end{aligned}$

Solve the given pair of equations by using determinants.

5. $\begin{aligned} ax + by &= 1 \\ bx + ay &= 1 \end{aligned}$

6. $\begin{aligned} ax + by &= c \\ 3ax - 2by &= 4c \end{aligned}$

7. a. Draw the graphs of $9x - 6y = 3$ and $6x - 4y = 10$. Then solve the equations by using determinants.

   b. If $ax + by = c$ and $dx + ey = f$ have parallel line graphs, what determinant must be equal to zero?

8. a. Draw the graphs of $12x + 8y = -4$ and $6x + 4y = -2$. Then solve the equations by using determinants.

   b. If $ax + by = c$ and $dx + ey = f$ have the same graph, what determinants must be equal to zero?

For Exercises 9–14, find the area of each figure, given the points $P(4, 3)$, $Q(7, -1)$, $R(2, 3)$, $S(-3, 6)$, $T(-5, 4)$, and $V(-2, -5)$.

9. $\triangle PQR$ **10.** $\triangle PQS$ **11.** $\triangle RST$

12. $\triangle PSV$ **13.** $\square$ with sides $\overline{PR}, \overline{PS}$ **14.** $\square$ with sides $\overline{PS}, \overline{PT}$

Vectors and determinants   **445**

15. If the "area" of $\triangle LMN$ is zero, what can you say about points L, M, and N?

16. If a parallelepiped with sides $\overline{DE}$, $\overline{DF}$, and $\overline{DG}$ has "zero volume," what can you say about points D, E, F, and G?

For Exercises 17–18, solve each system of equations by using Cramer's Rule.

B 17. $x - 2y + 3z = 2$
        $2x - 3y + z = 1$
        $3x - y + 2z = 9$

18. $4x - 2y + 3z = 2$
    $5x - 6y + 2z = -1$
    $3x + 4y - 5z = 7$

19. $3x - 2y + z = 7$
    $2x + y - 3z = 1$
    $x + 2y + 2z = 4$

For Exercises 20–22, find the volume of each solid, given points $D(5, 1, 0)$, $E(3, 1, 4)$, $F(0, 2, -1)$, $G(5, 2, 0)$, and $H(3, 1, 3)$.

20. Parallelepiped with sides $\overline{DE}$, $\overline{DF}$, and $\overline{DG}$

21. Parallelepiped with sides $\overline{EF}$, $\overline{EG}$, and $\overline{EH}$

C 22. Pyramid with sides $\overline{DF}$, $\overline{DG}$, and $\overline{DH}$
        (*Hint:* Volume of pyramid $= \frac{1}{6}$ volume of parallelepiped)

12-9 / DETERMINANTS AND VECTORS IN THREE DIMENSIONS

The unit vectors in the positive x, y, and z directions are often denoted by the symbols $\mathbf{i}$, $\mathbf{j}$, and $\mathbf{k}$. This means that the vector $\mathbf{v} = (a, b, c)$ can be written also as

$$\mathbf{v} = a\mathbf{i} + b\mathbf{j} + c\mathbf{k}.$$

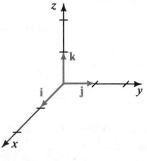

The dot product of two vectors is discussed in Section 12–4. Unlike the dot product, $\mathbf{v}_1 \cdot \mathbf{v}_2$, which is a scalar, the **cross product**, $\mathbf{v}_1 \times \mathbf{v}_2$ (read $\mathbf{v}_1$ cross $\mathbf{v}_2$), is another vector.

If $\mathbf{v}_1 = (a_1, b_1, c_1)$ and $\mathbf{v}_2 = (a_2, b_2, c_2)$, then the cross product of $\mathbf{v}_1$ and $\mathbf{v}_2$ is defined by the equation

$$\mathbf{v}_1 \times \mathbf{v}_2 = (b_1c_2 - b_2c_1)\mathbf{i} - (a_1c_2 - a_2c_1)\mathbf{j} + (a_1b_2 - a_2b_1)\mathbf{k},$$

which can be remembered by expanding the determinant $\begin{vmatrix} \mathbf{i} & \mathbf{j} & \mathbf{k} \\ a_1 & b_1 & c_1 \\ a_2 & b_2 & c_2 \end{vmatrix}$ by

minors along the top row.

PROPERTIES OF THE CROSS PRODUCT

1. $\mathbf{v} \times \mathbf{u} = -(\mathbf{u} \times \mathbf{v})$, so that $\mathbf{v} \times \mathbf{u}$ and $\mathbf{u} \times \mathbf{v}$ have opposite directions.

2. $\mathbf{u} \times (\mathbf{v} + \mathbf{w}) = (\mathbf{u} \times \mathbf{v}) + (\mathbf{u} \times \mathbf{w})$

3. $\mathbf{u} \times \mathbf{v}$ is perpendicular to $\mathbf{u}$ and to $\mathbf{v}$.

4. $|\mathbf{u} \times \mathbf{v}| = |\mathbf{u}|\,|\mathbf{v}| \sin\theta$, where θ is the angle between $\mathbf{u}$ and $\mathbf{v}$, so that the length of $\mathbf{u} \times \mathbf{v}$ is the same as the area of the parallelogram determined by $\mathbf{u}$ and $\mathbf{v}$.

5. If $\mathbf{u} = k\mathbf{v}$, so that $\mathbf{u}$ is parallel to $\mathbf{v}$, the angle θ between $\mathbf{u}$ and $\mathbf{v}$ is defined to be 0 and $\mathbf{u} \times \mathbf{v} = \mathbf{0}$.

The diagram accompanying Example 1 illustrates Properties 1, 3, and 4.

EXAMPLE. **a.** Find a vector perpendicular to the plane containing the points $P(1, 0, 3)$, $Q(2, 5, 0)$, and $R(3, 1, 4)$.
  b. Find the equation of the plane containing P, Q, and R.
  c. Find the area of the parallelogram with sides determined by $\overrightarrow{PQ}$ and $\overrightarrow{PR}$.

SOLUTION: **a.** Let $\mathbf{u} = \overrightarrow{PQ} = (1, 5, -3)$ and let $\mathbf{v} = \overrightarrow{PR} = (2, 1, 1)$. Then, by Property 3, $\mathbf{u} \times \mathbf{v}$ is the desired vector.

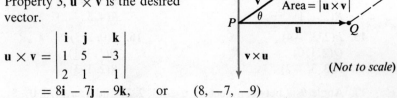

$$\mathbf{u} \times \mathbf{v} = \begin{vmatrix} \mathbf{i} & \mathbf{j} & \mathbf{k} \\ 1 & 5 & -3 \\ 2 & 1 & 1 \end{vmatrix}$$
$$= 8\mathbf{i} - 7\mathbf{j} - 9\mathbf{k}, \quad \text{or} \quad (8, -7, -9)$$

b. Since $(8, -7, -9)$ is perpendicular to the plane, the equation of the plane is $8x - 7y - 9z = d$ (by the theorem on page 434). To find d, substitute $P(1, 0, 3)$ (or Q or R) in the equation:

$$8 \cdot 1 - 7 \cdot 0 - 9 \cdot 3 = d$$
$$d = -19$$

Thus, $8x - 7y - 9z = -19$ is an equation of the plane.

c. By Property 4, the area of the parallelogram = $|\mathbf{u} \times \mathbf{v}| = \sqrt{8^2 + (-7)^2 + (-9)^2} = \sqrt{194}$.

If a parallelogram lies entirely in the xy-plane, then its area can be found by using 2 by 2 determinants, as in the preceding section. Otherwise, the method shown in Example 1(c) is used.

WRITTEN EXERCISES 12-9

For Exercises 1–6, let $\mathbf{u} = (4, 0, 1)$, $\mathbf{v} = (5, -1, 0)$, and $\mathbf{w} = (-3, 1, -2)$.

A **1.** Calculate $\mathbf{v} \times \mathbf{u}$ and $\mathbf{u} \times \mathbf{v}$. Do your results agree with Property 1?

2. Show that $\mathbf{u} \times (\mathbf{v} + \mathbf{w}) = (\mathbf{u} \times \mathbf{v}) + (\mathbf{u} \times \mathbf{w})$.

3. Show that $\mathbf{u} \times \mathbf{v}$ is perpendicular to both $\mathbf{u}$ and $\mathbf{v}$. (*Hint:* Use the dot product.)

4. Find a vector perpendicular to the plane of $\mathbf{u}$ and $\mathbf{v}$.

5. Find the area of a parallelogram with sides determined by **u** and **v**.

6. Find the area of a *triangle* with sides determined by **u** and **w**.

For Exercises 7–10, name a vector perpendicular to the plane.

7. $2x + 3y + 6z = 12$ **8.** $2x + 4y - z = 8$

9. $z = 3$ **10.** $x = 7$

For Exercises 11 and 12, find an equation of the plane described.

11. The plane perpendicular to $5\mathbf{i} - 2\mathbf{j} + 6\mathbf{k}$ at the point $(0, 1, 2)$.

12. The plane perpendicular to $7\mathbf{i} + 3\mathbf{j} - 4\mathbf{k}$ at the point $(2, -1, 3)$.

For Exercises 13–16, **(a)** find a vector perpendicular to the plane determined by P, Q, and R, **(b)** find the equation of the plane determined by P, Q, and R, and **(c)** find the area of $\triangle PQR$.

13. $P(1, 1, 0)$ **14.** $P(0, 0, 0)$
  $Q(-1, 0, 2)$ $Q(3, -1, 2)$
  $R(2, 1, 1)$ $R(4, 5, -2)$

15. $P(2, 3, -4)$ **16.** $P(0, 2, 3)$
  $Q(2, 1, 0)$ $Q(1, 1, 1)$
  $R(3, 3, -2)$ $R(2, 1, 3)$

B   **17.** Angle θ is between vectors $\mathbf{u} = (1, 2, 2)$ and $\mathbf{v} = (4, 3, 0)$.

  a. Find $\sin\theta$ using Property 4 of the Cross Product Properties.
  b. Find $\cos\theta$ using the dot product.
  c. Check that $\sin^2\theta + \cos^2\theta = 1$.

18. Repeat Exercise 17 for the vectors $\mathbf{u} = (1, 2, 1)$ and $\mathbf{v} = (1, 1, 2)$.

C   **19.** Let $\mathbf{u} = (a_1, b_1, c_1)$, $\mathbf{v} = (a_2, b_2, c_2)$, and $\mathbf{w} = (a_3, b_3, c_3)$.

  a. Use the definitions of cross product and dot product to prove that $|\mathbf{u} \times \mathbf{v}|^2 = |\mathbf{u}|^2 |\mathbf{v}|^2 - (\mathbf{u} \cdot \mathbf{v})^2$.
  b. Prove the five properties of the cross product listed on pages 446–447. (*Hint:* You will need part **(a)** to prove Property 4.)

COMPUTER EXERCISE

Write a computer program that will compute the coefficients and constant value for, and then print, the equation of the plane determined by three noncollinear points P, Q, and R when you input $P(x_1, y_1, z_1)$, $Q(x_2, y_2, z_2)$, and $R(x_3, y_3, z_3)$.

CALCULATOR EXERCISE

Suppose you are given the points $A(11, 17)$, $B(-8, 12)$, and $C(25, -13)$. You now can find the area of $\triangle ABC$ by using four different methods. Choose any two of the following methods and show that your answers agree.

Method 1. Use determinants.

Method 2. Use cross products.

Method 3. Use Hero's formula.

Method 4. Use the dot product or the Law of Cosines to find cos *C*. Then use $K = \frac{1}{2}ab \sin C$.

Chapter summary

1. A *vector* can be represented geometrically by an arrow and algebraically by an ordered pair of numbers. This ordered pair could give the *polar form* or the *component form* of the vector. The rules for vector operations in components are given on page 410.

2. **a.** If an object moves along a line in the plane with constant velocity $\mathbf{v} = (a, b)$, its displacement after time *t* is

$$\mathbf{d} = t\mathbf{v} = t(a, b).$$

   A *vector equation* of the line along which the object moves is:

   new position = initial position + displacement

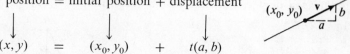

$$(x, y) \quad = \quad (x_0, y_0) \quad + \quad t(a, b)$$

   b. The vector (a, b) can be considered as a *direction vector* of the line as well as the velocity of the object moving along the line.

   c. From the vector equation of the line, we can obtain the two *parametric equations*

   $$x = x_0 + ta \quad \text{and} \quad y = y_0 + tb.$$

   These give the *x*- and *y*-coordinates of points on the line in terms of the parameter *t* (*t* is often considered to represent time).

3. **a.** If $\mathbf{u} = (x_1, y_1)$ and $\mathbf{v} = (x_2, y_2)$, then the *dot product* of $\mathbf{u}$ and $\mathbf{v}$ is

   $$\mathbf{u} \cdot \mathbf{v} = x_1 x_2 + y_1 y_2.$$

   Some properties of the dot product are listed on page 423. The angle between $\mathbf{u}$ and $\mathbf{v}$ is given by the formula

   $$\cos \theta = \frac{\mathbf{u} \cdot \mathbf{v}}{|\mathbf{u}| \, |\mathbf{v}|}.$$

   Vectors $\mathbf{u}$ and $\mathbf{v}$ are *perpendicular* if and only if $\mathbf{u} \cdot \mathbf{v} = 0$.

   b. Nonzero vectors $\mathbf{u}$ and $\mathbf{v}$ are *parallel* if and only if $\mathbf{u} = k\mathbf{v}$.

4. *Three dimensional vectors* are ordered triples of numbers. The rules for operations with these vectors are analogous to those for vectors in two dimensions as summarized on pages 428–429.

5. The graph of $ax + by + cz = d$ is a *plane* (provided *a*, *b*, and *c* are not all equal to 0). The vector (a, b, c) is perpendicular to this plane at the point (x_0, y_0, z_0), where $d = ax_0 + by_0 + cz_0$.

6. A *determinant* is a square array of numbers evaluated as shown on page 438. The value of an *n* by *n* determinant can be computed by expanding by the minors of one of its rows or columns.

7. *Cramer's Rule* involves using determinants to solve a system of *n* linear equations in *n* variables (pages 443–444). Determinants can also be used to find areas and volumes (page 444).

8. The *cross product* of the vectors $\mathbf{v}_1$ and $\mathbf{v}_2$ is defined by
$$\mathbf{v}_1 \times \mathbf{v}_2 = (b_1c_2 - b_2c_1)\mathbf{i} - (a_1c_2 - a_2c_1)\mathbf{j} + (a_1b_2 - a_2b_1)\mathbf{k}.$$

Properties of the cross product are given on pages 446–447. The cross product is useful in finding the equation of a plane. (See Example on page 447.)

Chapter test

1. A 3-unit force pulls north on an object at the same time that a 6-unit force pulls east. Find the strength of the resultant force and also its direction to the nearest degree. Illustrate with a vector diagram. **12–1**

2. In $\triangle ABC$,

$$AM = MB \quad \text{and} \quad \frac{AN}{NC} = \frac{1}{2}.$$

If $\mathbf{v} = \overrightarrow{BM}$ and $\mathbf{u} = \overrightarrow{AN}$, express each of the following in the form $r\mathbf{v} + s\mathbf{u}$:

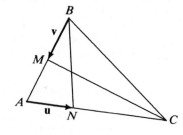

a. $\overrightarrow{BN}$ **b.** $\overrightarrow{BC}$ **c.** $\overrightarrow{CM}$

3. **a.** Illustrate on a vector diagram the vectors $\mathbf{u} = (1, 3)$, $\mathbf{v} = (4, -2)$, and $2\mathbf{u} - \mathbf{v}$. **12–2**

   b. Evaluate $|2\mathbf{u} - \mathbf{v}|$.

4. Find r and s if $r(3, 6) + s(3, -3) = (7, -4)$.

5. The position of an object at time t is given by the vector equation $(x, y) = (1, 5) + t(2, -1)$. **12–3**

   a. What is the velocity of the object and what is its speed?

   b. Find a pair of parametric equations of the line along which the object moves.

   c. When and where does the object cross the curve
   $$(x - 5)^2 + (y + 2)^2 = 25?$$

6. Find the value of k if the vectors $(8, -6)$ and $(k, 4)$ are: **12–4**

   a. perpendicular   **b.** parallel

7. In $\triangle ABC$, $A = (1, 2)$, $B = (3, 0)$, and $C = (8, 3)$. Find $\cos A$.

8. The equations of lines L_1 and L_2 are as follows: 12-5

L_1: $(x, y, z) = (2, 4, 1) + t(3, 5, -2)$
L_2: $(x, y, z) = (2, 4, 1) + s(8, -4, 2)$

a. Explain why L_1 is perpendicular to L_2.
b. Find a vector equation of the line through $(3, -1, 8)$ parallel to line L_1.

9. Find the center and radius of the sphere

$$x^2 + y^2 + z^2 - 2x + 6z = 6.$$

10. Find an equation of the plane through $(5, 0, -1)$ perpendicular to the vector $(1, 2, 3)$. 12-6

11. Evaluate the determinant: 12-7

$$\begin{vmatrix} 2 & 1 & 0 \\ 3 & 4 & 5 \\ -1 & 5 & 6 \end{vmatrix}$$

12. Use Cramer's Rule to solve: $ax - by = 1$ 12-8
$\qquad\qquad\qquad\qquad\quad 2ax + by = 0$

13. Let $\mathbf{u} = (1, 0, 3)$ and $\mathbf{v} = (2, 1, 4)$. 12-9

a. Find $\mathbf{u} \times \mathbf{v}$.
b. Find the area of the parallelogram determined by $\mathbf{u}$ and $\mathbf{v}$.

CHAPTER THIRTEEN

Sequences and series

OBJECTIVES

1. To identify an arithmetic or geometric sequence and find a formula for its nth term.
2. To find the sum of the first n terms of an arithmetic or geometric series.
3. To find or estimate the limit of an infinite sequence or to determine that the limit does not exist.
4. To find the sum of an infinite geometric series.
5. To represent series using sigma notation.
6. To identify certain power series and use these series in computation.
7. To use mathematical induction to prove a statement is true.

Finite sequences and series

13-1/ARITHMETIC AND GEOMETRIC SEQUENCES

We begin with two of the simplest types of sequences: *arithmetic* and *geometric*.

Arithmetic sequences

A sequence of numbers is called an **arithmetic sequence** if the *difference* of any two consecutive terms is constant.

EXAMPLES   1. 2, 6, 10, 14, 18, . . .   difference $d = 4$
               2. 17, 10, 3, -4, -11, -18, . . . $d = -7$
               3. $a, a + d, a + 2d, a + 3d, a + 4d, . . .$

Shown at the left is a magnetically levitated vehicle, called a maglev, *being operated on the Miyazaki Test Track in Kyushu, Japan. The maglev is propelled by a sequence of electrical impulses that activate magnets along the U-shaped guideway.*

Geometric sequences

A sequence of numbers is called a **geometric sequence** if the *ratio* of any two consecutive terms is constant.

EXAMPLES 4. 1, 3, 9, 27, 81, ...   ratio $r = 3$

5. 64, -32, 16, -8, 4, ... $r = -\frac{1}{2}$

6. a, ar, ar^2, ar^3, ar^4, ...

Whether a sequence is arithmetic or geometric or of some other type, the same notation can be used. The first term of a sequence is often denoted by t_1, the second and third terms by t_2 and t_3, and so on. The nth term of the sequence is then denoted by t_n. If you have a formula for t_n in terms of n, you can find the value of any term of the sequence. For example, suppose a sequence has the formula

$$t_n = n^2 + 1.$$

Then:
$$t_1 = 1^2 + 1 = 2$$
$$t_2 = 2^2 + 1 = 5$$
$$t_3 = 3^2 + 1 = 10$$
$$t_4 = 4^2 + 1 = 17, \text{ and so on}$$

Note that this sequence is neither arithmetic nor geometric.

Formulas for the nth terms of general arithmetic and geometric sequences are given below. Notice how similar these formulas are.

Arithmetic Sequence: $t_n \quad = \quad t_1 \quad + \quad (n-1)d$

To get the   start with   and add the
nth term,   the first   difference
                   term $n - 1$ times.

Geometric Sequence: $t_n \quad = \quad t_1 \quad \cdot \quad r^{n-1}$

To get the   start with   and multiply
nth term,   the first   by the ratio
                   term $n - 1$ times.

EXAMPLE 7.   Find formulas for the nth term of each sequence.
a. Arithmetic sequence: 8, 15, 22, 29, ...
b. Geometric sequence: 3, 6, 12, 24, ...

SOLUTION: **a.** In the arithmetic sequence, $t_1 = 8$ and the constant difference d is 7. Substituting in

$$t_n = t_1 + (n-1)d,$$

we find that the nth term is

$$t_n = 8 + (n-1)7 = 1 + 7n.$$

b. In the geometric sequence, $t_1 = 3$ and the constant ratio r is 2. Substituting in

$$t_n = t_1 \cdot r^{n-1},$$

we find that the nth term is

$$t_n = 3 \cdot 2^{n-1}.$$

EXAMPLE 8. A certain radioactive element has a half-life of 1 day. If 40 grams of the element are present now, how many grams will be present after n days?

SOLUTION 1: Using the half-life formula on page 174, we have immediately that the amount present after n days is $40(\frac{1}{2})^n$ grams.

SOLUTION 2: The amount present at the end of each day forms the following geometric sequence:

n = number of days	1	2	3	4
t_n = amount present (in grams)	20	10	5	$2\frac{1}{2}$

In this geometric sequence, the constant ratio $r = \frac{1}{2}$. Therefore, by substituting in the formula

$$t_n = t_1 \cdot r^{n-1},$$

we have $\qquad t_n = 20 \cdot (\frac{1}{2})^{n-1},$

or equivalently $\qquad t_n = 40(\frac{1}{2})^n.$

This answer agrees with the one given in Solution 1.

The purpose of showing both solutions to this example is to illustrate that the ideas of exponential growth and geometric sequences are closely related. Nevertheless, there is one contrast between the two solutions that you should note. In the half-life formula used in Solution 1, the number n of days is considered to be a continuous variable taking on all nonnegative real numbers. [See figure (a).] On the other hand, in Solution 2 the variable n represents the number of the term in a geometric sequence and, consequently, is a positive integer. [See figure (b).]

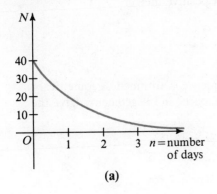

(a)

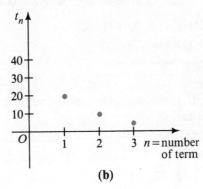

(b)

Sequences and series **455**

Explicit and recursive definitions

Sometimes a sequence is defined by giving the value of t_n in terms of the preceding term, t_{n-1}. For example, consider the sequence defined by the following formulas:

$$t_1 = 3 \quad \text{and} \quad t_n = 1 + 2 \cdot t_{n-1}$$

The second formula above states that the nth term is one more than twice the $(n-1)$st term. Knowing that the sequence begins with $t_1 = 3$, we can determine the first few terms of the sequence as follows:

$$t_n = 1 + 2 \cdot t_{n-1}$$
$$t_2 = 1 + 2 \cdot t_1 = 1 + 2 \cdot 3 = 7$$
$$t_3 = 1 + 2 \cdot t_2 = 1 + 2 \cdot 7 = 15$$
$$t_4 = 1 + 2 \cdot t_3 = 1 + 2 \cdot 15 = 31$$

The formulas $t_1 = 3$ and $t_n = 1 + 2 \cdot t_{n-1}$ give a *recursive definition* of the sequence. In a **recursive definition,** the first term is given, and then the general term t_n is given in terms of the preceding term(s), instead of explicitly in terms of n. The illustration below should help make clear the distinction between explicit and recursive definitions. We give a recursive definition and an arithmetic definition for the arithmetic sequence:

$$23, 20, 17, 14, \ldots.$$

Recursive definition: $\quad t_1 = 23 \quad \text{and} \quad t_n = t_{n-1} - 3$

Explicit definition: $\quad t_n = 26 - 3n$

What is a sequence?

By now you probably have a good intuitive idea of what a sequence is, but can you give a precise definition of *sequence*? In mathematics, a sequence is usually defined to be a function whose domain is the set of positive integers. For example, the sequence with nth term

$$t_n = 4n - 2$$

can be thought of as the function

$$t(n) = 4n - 2, \text{ where } n \text{ is a positive integer.}$$

ORAL EXERCISES 13-1

In Exercises 1–6, state whether the given sequence is arithmetic, geometric, or neither. If arithmetic, give the constant difference, and if geometric, give the constant ratio.

1. 3, 8, 13, 18

2. 4, 8, 16, 32

3. 2, 5, 10, 17

4. 23, 17, 11, 5

5. 27, −18, 12, −8

6. 1, −3, 5, −7

In Exercises 7–12, state the first four terms of the specified sequence. Then tell whether the sequence is arithmetic, geometric, or neither.

7. $t_n = 5n + 2$

8. $t_n = \dfrac{n+1}{n+2}$

9. $t_n = 3^n$

10. $t_1 = 7,\ t_n = t_{n-1} + 12$

11. $t_1 = 8,\ t_n = -2 \cdot t_{n-1}$

12. $t_1 = 1,\ t_2 = 1,\ t_n = t_{n-1} + t_{n-2}$
(Fibonacci sequence)

13. Find t_n for the following arithmetic sequences:
    a. 8, 10, 12, 14, ...   **b.** 30, 26, 22, 18, ...

14. Find t_n for the following geometric sequences:
    a. 4, 12, 36, 108, ...   **b.** 24, −12, 6, −3, ...

15. Does the sequence 2, 2, 2, ... satisfy the definition of:
    a. an arithmetic sequence?   **b.** a geometric sequence?

16. How does the graph of the sequence with nth term $t_n = 3n - 5$ differ from the graph of the straight line $y = 3x - 5$?

WRITTEN EXERCISES 13-1

Find the first four terms of the given sequence and state whether the sequence is arithmetic, geometric, or neither.

A   **1.** $t_n = 2n + 3$

2. $t_n = n^3 + 1$

3. $t_n = 3 \cdot 2^n$

4. $t_n = 3 - 7n$

5. $t_n = n + \dfrac{1}{n}$

6. $t_n = (-2)^n$

7. $t_1 = 6,\ t_n = 4 + t_{n-1}$

8. $t_1 = 9,\ t_n = \frac{1}{3} \cdot t_{n-1}$

9. $t_1 = 4,\ t_n = 5 \cdot t_{n-1} + 2$

10. $t_n = 16 \cdot 2^{2n}$

11. $t_1 = 1,\ t_n = (-\frac{1}{3})^n \cdot t_{n-1}$

12. $t_n = \cos n\pi$

State whether the given sequence is arithmetic, geometric, or neither. Try to find an explicit formula for t_n, the nth term of the sequence, in terms of n.

13. 17, 21, 25, 29, ...

14. 15, 7, −1, −9, ...

15. 8, 12, 18, 27, ...

16. 100, −50, 25, −12.5, ...

17. $\frac{1}{2}, \frac{2}{3}, \frac{3}{4}, \frac{4}{5}, \ldots$

18. 12, 15, 18, 21, ...

19. −8, 12, 32, 52, ...

20. 1, 4, 9, 16, ...

21. 11, 101, 1001, 10001, ...

22. $\dfrac{a}{9}, \dfrac{a^2}{18}, \dfrac{a^3}{36}, \dfrac{a^4}{72}, \ldots$

23. $2a - 2b,\ 3a - b,\ 4a,\ 5a + b, \ldots$

24. $\frac{2}{1}, \frac{3}{4}, \frac{4}{9}, \frac{5}{16}, \ldots$

25. $t_1 = 8,\ t_n = \frac{1}{2} \cdot t_{n-1}$

26. $t_1 = 6,\ t_n = t_{n-1} + 10$

27. $t_1 = 1,\ t_n = t_{n-1} + 2n - 1$

28. $t_1 = \dfrac{1}{2},\ t_n = \dfrac{n}{n+1}(t_{n-1} + 1)$

29. $2^{\frac{2}{3}}, 2^{\frac{5}{3}}, 2^{\frac{8}{3}}, \ldots$

30. $\sqrt{2}, \sqrt[3]{2}, \sqrt[4]{2}, \sqrt[5]{2}, \ldots$

Find the indicated term of each arithmetic sequence.

31. $t_1 = 15$, $t_2 = 21$, $t_{20} = ?$

32. $t_1 = 76$, $t_3 = 70$, $t_{101} = ?$

33. $t_3 = 8$, $t_5 = 14$, $t_{50} = ?$

34. $t_8 = 25$, $t_{20} = 61$, $t_1 = ?$

Find the indicated term of each geometric sequence.

35. $t_1 = 2^{-4}$, $t_2 = 2^{-3}$, $t_{12} = ?$

36. $t_1 = 2$, $t_2 = 2^{\frac{3}{2}}$, $t_{13} = ?$

37. $t_2 = 18$, $t_4 = 2$, $t_8 = ?$

38. $t_1 = 81$, $t_4 = 24$, $t_7 = ?$

39. How many terms are in the arithmetic sequence $18, 24, \ldots, 336$?

40. How many terms are in the arithmetic sequence $178, 170, \ldots, 2$?

B
41. Find the number of multiples of 7 between 30 and 300.

42. Find the number of multiples of 6 between 28 and 280.

43. How many 3-digit numbers are divisible by 4 and 6?

44. How many 4-digit numbers are *not* divisible by 11?

45. True or false? If the sequence a, b, c is arithmetic, then so is the sequence $\sin a, \sin b, \sin c$.

46. True or false? If the sequence a, b, c is arithmetic, then the sequence $2^a, 2^b, 2^c$ is geometric.

47. The numbers $3, 12, 48, 192$ form a geometric sequence. Find the common logarithms of these numbers and verify that they form an arithmetic sequence.

48. Explain why the logarithms of the terms of the geometric sequence $a, ar, ar^2, ar^3, \ldots$ form an arithmetic sequence.

49. Find x if the sequence $2, 8, 3x + 5$ is **(a)** arithmetic, **(b)** geometric.

50. Find x if the sequence $4, x, \frac{3}{2}x$ is **(a)** arithmetic, **(b)** geometric.

51. Find x and y if the sequence $y, 2x + y, 7y, 20, \ldots$ is arithmetic.

52. Find x and y if the sequence $2y, 2xy, 2, \dfrac{xy}{2}, \ldots$ is geometric.

In Exercises 53–58, give a recursive definition for the given sequence.

53. $9, 13, 17, 21, \ldots$

54. $81, 27, 9, 3, \ldots$

55. $1, 3, 7, 15, 31, \ldots$

56. $1, 2, 5, 26, 677, \ldots$

57. $1, 3, 6, 10, 15, 21, \ldots$

58. $1, 2, 6, 24, 120, 720, \ldots$

59. If the half-life of an element is 2 days, what fractional amount of the element remains after 2 days? 4 days? 10 days?

60. If P dollars is invested at 8% annual interest compounded quarterly, what is the value of the investment after 1 quarter? 2 quarters? q quarters? n years?

61. Find an arithmetic sequence none of whose terms is divisible by 2, 3, or 7.

62. Find an arithmetic sequence all of whose terms are multiples of 2 and 3 but not multiples of 4 or 5.

63. Consider the two arithmetic sequences:

$$\text{A. } 3, 14, 25, \ldots$$
$$\text{B. } 2, 9, 16, \ldots$$

Certain terms of sequence A are also terms of sequence B. Write the first five such terms.

64. The angle measures of $\triangle ABC$ are in arithmetic sequence, and the lengths of the sides are in geometric sequence. Use the Law of Cosines to prove that $\triangle ABC$ is equilateral.

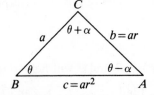

65. Find all right triangles having sides with integral lengths in an arithmetic sequence.

66. Find the nth term of a sequence whose first terms are:

$$t_1 = \frac{1+2}{2}, \quad t_2 = \frac{t_1+3}{2}, \quad t_3 = \frac{t_2+4}{2}, \quad t_4 = \frac{t_3+5}{2}, \ldots$$

In Exercises 67–73 use the following definitions as needed:

The **arithmetic mean** of the numbers a and b is $\dfrac{a+b}{2}$.

The **geometric mean** of the positive numbers a and b is $\sqrt{ab}$.

67. Find the arithmetic mean and the geometric mean of **(a)** 4 and 9, **(b)** 5 and 10.

68. Find x if the sequence a, x, b is **(a)** arithmetic, **(b)** geometric. Assume that a and b are positive.

C **69.** Prove that there are no right triangles having sides with integral lengths in a geometric sequence.

70. $\overline{CD}$ is the altitude to the hypotenuse of the right triangle ABC.
  a. Prove that $\triangle ADC \sim \triangle ACB$ and therefore that $\dfrac{x}{b} = \dfrac{b}{c}$; that is, b is the geometric mean of x and c.
  b. Likewise it can be shown that a is the geometric mean of _?_ and _?_.
  c. Use the results of parts (a) and (b) to prove that $a^2 + b^2 = c^2$.

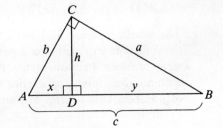

71. In the figure for Exercise 70, use similar triangles to prove that h is the geometric mean of x and y.

72. Prove that the arithmetic mean of two positive numbers is never less than their geometric mean.

73. Consider a circle with radius 6. Find A, the area of the inscribed equilateral triangle, and B, the area of the circumscribed equilateral triangle. Show that the geometric mean of A and B is the area of the inscribed regular hexagon.

The *Fibonacci sequence* is defined recursively by the equations

$$t_1 = 1, \quad t_2 = 1, \quad t_n = t_{n-1} + t_{n-2}.$$

Its first few terms are 1, 1, 2, 3, 5, 8, 13, 21, 34, 55.

The sequence has some surprising connections to the golden ratio,

$$R = \frac{\sqrt{5} + 1}{2} \approx 1.618033989.$$

Exercises 1 and 2 will help you discover two of these connections.

1. a. Find t_{11}, t_{12}, t_{13}, t_{14}, and t_{15}, the eleventh through fifteenth terms of the Fibonacci sequence.

b. Find $\dfrac{t_{11}}{t_{10}}$, $\dfrac{t_{12}}{t_{11}}$, $\dfrac{t_{13}}{t_{12}}$, $t\dfrac{14}{t_{13}}$, and $\dfrac{t_{15}}{t_{14}}$.

c. Describe how these ratios seem to be related to the golden ratio R.

2. A sequence is defined in terms of the golden ratio R as follows:

$$\text{If } n \text{ is odd:} \quad t_n = \frac{1}{\sqrt{5}}\left[R^n + \frac{1}{R^n}\right]$$

$$\text{If } n \text{ is even:} \quad t_n = \frac{1}{\sqrt{5}}\left[R^n - \frac{1}{R^n}\right]$$

Find the first five terms of this sequence and see if you can discover its relationship to the Fibonacci sequence.

13-2/ARITHMETIC AND GEOMETRIC SERIES AND THEIR SUMS

The words *sequence* and *series* are often used interchangeably in everyday conversation. For example, a person may refer to a *sequence of events* or to a *series of events*. In mathematics, however, the sequences and series studied deal with numbers instead of events, and a distinction is made between a sequence of numbers and a series of numbers. This distinction is best made by some examples.

Finite sequence: 2, 6, 10, 14

Related finite series: $2 + 6 + 10 + 14$

Infinite sequence: $\dfrac{1}{2}, \dfrac{1}{4}, \dfrac{1}{8}, \ldots, \dfrac{1}{2^n}, \ldots$

Related infinite series: $\dfrac{1}{2} + \dfrac{1}{4} + \dfrac{1}{8} + \cdots + \dfrac{1}{2^n} + \cdots$

In this section we shall consider finite arithmetic and geometric series and their sums, and compare the two.

THEOREM 1. The sum of the first n terms of an arithmetic series is

$$S_n = \frac{n(t_1 + t_n)}{2}.$$

Proof: We write the series for S_n twice, the second time with the order of the terms reversed. Then we add the two equations, term by term:

$$S_n = t_1 + (t_1 + d) + (t_1 + 2d) + \cdots + (t_n - d) + t_n$$
$$S_n = t_n + (t_n - d) + (t_n - 2d) + \cdots + (t_1 + d) + t_1$$
$$2S_n = (t_1 + t_n) + (t_1 + t_n) + (t_1 + t_n) + \cdots + (t_1 + t_n) + (t_1 + t_n)$$

Since there are n of these $(t_1 + t_n)$ terms,

$$2S_n = n(t_1 + t_n)$$
$$S_n = \frac{n(t_1 + t_n)}{2}$$

EXAMPLE 1. Find the sum of the first 25 terms of the arithmetic series

$$11 + 14 + 17 + 20 + \cdots$$

SOLUTION: *Step 1.* First find the 25th term.

$$t_{25} = t_1 + (n - 1)d = 11 + (25 - 1)3 = 83$$

Step 2. $S_{25} = \dfrac{n(t_1 + t_n)}{2}$

$$= \frac{25(11 + 83)}{2} = 1175$$

THEOREM 2. The sum of the first n terms of a geometric series is

$$S_n = \frac{t_1(1 - r^n)}{1 - r},$$

where r is the common ratio and $r \neq 1$.

Proof: Multiply the series for S_n by the constant ratio and then subtract the series from the original one, as shown below.

$$S_n = t_1 + t_1 r + t_1 r^2 + \cdots + t_1 r^{n-2} + t_1 r^{n-1}$$
$$rS_n = t_1 r + t_1 r^2 + \cdots + t_1 r^{n-2} + t_1 r^{n-1} + t_1 r^n$$
$$S_n - rS_n = t_1 + 0 + 0 + \cdots + 0 + 0 - t_1 r^n$$
$$S_n - rS_n = t_1 - t_1 r^n$$
$$S_n(1 - r) = t_1(1 - r^n)$$
$$S_n = \frac{t_1(1 - r^n)}{1 - r}, \quad r \neq 1$$

This formula is not defined for $r = 1$. If $r = 1$, however, the geometric series is simply a series of repeated numbers, like $a + a + \cdots + a$, whose sum is na.

EXAMPLE 2. Find the sum of the first 10 terms of the geometric series

$$2 - 6 + 18 - 54 + \cdots$$

SOLUTION: $S_{10} = \dfrac{t_1(1 - r^n)}{1 - r}$

$$= \dfrac{2(1 - (-3)^{10})}{1 - (-3)} = \dfrac{2(1 - 3^{10})}{4} = -29524$$

WRITTEN EXERCISES 13-2

For each of the arithmetic series in Exercises 1–8, find the specified sum.

A

1. S_{10}: $t_1 = 3$, $t_{10} = 39$
2. S_{200}: $t_1 = 18$, $t_{200} = 472$
3. S_{50}: $5 + 10 + 15 + \cdots$
4. S_{25}: $17 + 25 + 33 + \cdots$
5. S_{12}: $t_n = 5 + 3n$
6. S_{40}: $t_1 = 5$, $t_3 = 11$
7. $1 + 2 + 3 + \cdots + 1000$
8. $3 + 7 + 11 + \cdots + 99$

9. Find S_8 for the geometric series with $t_1 = 8$ and **(a)** $r = \frac{1}{2}$, **(b)** $r = -\frac{1}{2}$.

10. Show that the sum of the first 10 terms of the geometric series

$$1 + \tfrac{1}{3} + \tfrac{1}{9} + \tfrac{1}{27} + \cdots$$

is twice the sum of the first 10 terms of the series

$$1 - \tfrac{1}{3} + \tfrac{1}{9} - \tfrac{1}{27} + \cdots.$$

11. Show that $\sqrt{2} + 2 + 2\sqrt{2} + \cdots + 64 = \dfrac{63\sqrt{2}}{\sqrt{2} - 1}$.

12. Consider the series $1 + \sqrt[5]{3} + \sqrt[5]{3^2} + \sqrt[5]{3^3} + \cdots$. Show that

$$S_{15} = \dfrac{26}{3^{\frac{1}{5}} - 1}.$$

13. Find S_8 if the series $4.8 + 2.4 + \cdots$ is **(a)** arithmetic, **(b)** geometric.

14. Find S_{10} if the series $45 + 30 + \cdots$ is **(a)** arithmetic, **(b)** geometric.

15. Show that $1 + 2 + 4 + \cdots + 2^{n-1} = 2^n - 1$.

16. Show that $9 + 90 + 900 + \cdots + 9 \cdot 10^{n-1} = 10^n - 1$.

17. Find the sum of all multiples of 3 between 1 and 1000.

18. Find the sum of all positive 3-digit numbers divisible by 6.

19. Find the sum of all positive 3-digit numbers whose last digit is 3.

20. Find the sum of all positive odd numbers less than 400 which are divisible by 5.

21. Show that the sum of the first n positive integers is $\dfrac{n(n + 1)}{2}$.

22. A sequence is defined recursively by $t_1 = -2$, $t_n = (t_{n-1})^2 - 3$.
   a. Is the sequence arithmetic? geometric?
   b. Find S_{15}.

B

23. Find the sum of the series $1 - 3 + 5 - 7 + 9 - 11 + \cdots + 1001$.

24. Find the sum of the series $1 + 2 + 4 + 5 + 7 + 8 + 10 + 11 + \cdots + 299$, which is the sum of the integers except for multiples of 3.

25. To number the pages of a book consecutively 852 digits are required. How many pages are there?

26. If you save $10 one month, $12 the next month, $14 the next month, and so on, how long will you take to save a total of $400?

27. The originator of a chain letter writes 5 letters instructing each recipient to write 5 similar letters to additional people. Then these people each send 5 similar letters to other people. Determine the number of people who should receive letters if the chain continues unbroken for 12 steps. Explain why the process always fails. (There are laws forbidding chain letters which request money.)

28. Value Appliance Store has radios which can be purchased on a daily installment plan. You pay only 1 cent the first day, 2 cents the next day, 4 cents the next day, and so on, for 14 days. How much will the radio cost?

29. The diagram below shows the steps that might be taken in getting some oranges from a grower to you.

grower $\rightarrow$ regional market $\rightarrow$ trucker $\rightarrow$ wholesaler $\rightarrow$ trucker $\rightarrow$ retailer $\rightarrow$ you

   a. Assume that *every* person or organization in the chain above makes a 25% profit, that all oranges are sold, and that the grower sells oranges for $.20 per pound. How much will you have to pay per pound?

   b. If the grower receives $10,000 for oranges and sells to the regional market, what is the sum of all the profits made by the regional market, first trucker, wholesaler, second trucker, and retailer?

30. Repeat Exercise 29 assuming a 30% profit instead of a 25% profit. Compare your answers with those for Exercise 29 and note the cumulative effect of the 5% increase.

31. **a.** If you go back through ten generations, how many ancestors do you have? Count your parents as the first generation back, your four grandparents as your second generation, and so on. (Assume there are no duplications.)

   b. Through how many generations back must you go in order to have more than one million ancestors?

32. **a.** Suppose a doctor earns $40,000 during the first year of practice. Suppose also that each succeeding year the salary increases 10%. What is the total of the doctor's salaries over the first ten years?

   b. How many years must the doctor work if the salary total is to exceed a million dollars?

33. Let S_n be the sum of the first n positive odd integers.

   a. Evaluate S_1, S_2, S_3, and S_4.

   b. Suggest and prove a formula for S_n.

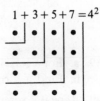

$1 + 3 + 5 + 7 = 4^2$

34. The number $T_n = 1 + 2 + 3 + \cdots + n = \dfrac{n(n+1)}{2}$ is

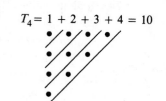

$T_4 = 1 + 2 + 3 + 4 = 10$

sometimes called a **triangular number** because it is possible to represent the number by a triangular array of dots, as shown.

a. Evaluate T_1, T_2, T_3, T_4, and T_5.

b. Add any two consecutive triangular numbers. Then make a conjecture and prove it.

35. Consider the series $1^3 + 2^3 + 3^3 + \cdots + n^3$. Evaluate S_1, S_2, S_3, S_4 and suggest a formula for S_n. (*Hint:* See Exercise 34.)

36. a. What generalization is suggested by the statements below?

$$1^2 - 0^2 = 1$$
$$3^2 - 1^2 = 8$$
$$6^2 - 3^2 = 27$$
$$10^2 - 6^2 = 64$$

b. If you extend the pattern above to n equations and then add all n equations together, what result do you get?

37. The sum of every row, column, and diagonal of a magic square equals a number S. For the 3-by-3 and 4-by-4 magic squares shown below, $S = 15$ and 34, respectively. Find S for an n-by-n magic square that contains consecutive integers starting with 1. (*Hint:* How many numbers are in an n-by-n square? What is their sum? Deduce the row sum S.)

$S = 15$

8	1	6
3	5	7
4	9	2

$S = 34$

1	15	14	4
12	6	7	9
8	10	11	5
13	3	2	16

38. A series of step diagrams is shown below. The values of n (the number of steps) and W (the number of segments in each diagram) are also given. Find the value of W in terms of the value of n.

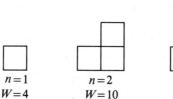

$n = 1$ $n = 2$ $n = 3$
$W = 4$ $W = 10$ $W = 18$

39. a. For a certain sequence, $S_n = n^3$. Is this sequence arithmetic? geometric?

b. Find a formula for t_n in this sequence. (*Hint:* $S_n = S_{n-1} + t_n$)

40. Repeat Exercise 39 if $S_n = 2n^2 + 8n$.

Use a calculator for Exercises 41 and 44.

41. If you invest $1000 per year every year for ten years and if your money is compounded annually at 12%, how much money will you have at the end of the tenth year?

C **42.** If you invest P dollars at an interest rate r compounded annually, after n years you will have $P(1 + r)^n$ dollars. Suppose you invest P dollars every year for n years. Show that at the time you make your nth investment, you will have

$$\frac{P[(1 + r)^n - 1]}{r} \text{ dollars.}$$

43. The purpose of this exercise is to develop a formula for the monthly payment P that is required to repay a loan of A dollars in n monthly installments, with interest on the unpaid balance equal to a monthly rate r. Let A_k = amount still owed after paying k installments. Then interest for the $(k + 1)$st month $= rA_k$ and the principal paid off in the $(k + 1)$st payment $= P - rA_k$. Thus,

$$A_{k+1} = A_k - (P - rA_k) = (1 + r)A_k - P$$

a. Use the equation above and the fact that $A_0 = A$ to find A_1, A_2, and A_3 in terms of A, r, and P.

b. Generalizing from A_1, A_2, and A_3, you get
$$A_n = (1 + r)^n A - [(1 + r)^{n-1} + (1 + r)^{n-2} + \cdots + (1 + r) + 1]P.$$
Since the bracketed quantity above is a geometric series, show that the equation can be rewritten

$$A_n = (1 + r)^n A - \left[\frac{(1 + r)^n - 1}{r}\right]P.$$

c. Since $A_n = 0$ (why?), use the last equation to show that $P = \dfrac{Ar(1 + r)^n}{(1 + r)^n - 1}$.

44. A *direct-reduction loan* is used primarily for mortgages on homes. It differs from other kinds of loans because you pay interest only on that portion of the loan which you have not repaid.

Suppose you take out a mortgage loan for $5000 at 10% interest to be repaid in 4 years. Your monthly payment will be $126.82.

a. Use the formula in Exercise 43(c) to show that $P = \$126.82$. Remember that $r =$ monthly rate $= \dfrac{10\%}{12}$.

The interest and principal reduction for the first two months are calculated as follows:

1st month: Interest = $5000 × 10% ÷ 12   = $41.67
          Principal reduction = $126.82 − $41.67 = $85.15
          New principal = $5000 − $85.15   = $4914.85

(*Exercise continued on page 466*)

44. (Continued)

2nd month: Interest = \$4914.85 × 10% ÷ 12   = \$40.96
Principal reduction = \$126.82 − \$40.96 = \$85.86
New principal = \$4914.85 − \$85.86   = \$4828.99

b. Calculate the interest and principal reduction for the third month of the loan.

c. Show that the principal reductions for the first three months form a geometric sequence. Find the common ratio of this sequence.

$$\left(\textit{Note: The common ratio should be } 1 + \frac{10\%}{12}. \textit{ Is it?} \right)$$

COMPUTER EXERCISE

Make a table showing the monthly repayment for a loan of \$1000 at various interest rates (8%, 10%, 12%, ... , 20%) for various periods of time (24, 36, 48, 60 months). Use the formula $P = \dfrac{Ar(1 + r)^n}{(1 + r)^n - 1}$, remembering that r is the monthly rate. Thus, if the annual rate is 8%, $r = \dfrac{0.08}{12}$.

Infinite sequences and series

13-3/LIMITS OF INFINITE SEQUENCES

Consider the infinite geometric sequence

$$\tfrac{1}{2}, \tfrac{1}{4}, \tfrac{1}{8}, \tfrac{1}{16}, \ldots, (\tfrac{1}{2})^n, \ldots.$$

The terms of this sequence are surely getting smaller, but how small do they get? With a calculator or logarithms, we can calculate that $t_{10} = (\tfrac{1}{2})^{10} \approx 0.001$ and $t_{100} = (\tfrac{1}{2})^{100} \approx 0.0000000000000000000000000000001$. By substituting larger and larger values of n, $t_n = (\tfrac{1}{2})^n$ becomes a smaller and smaller positive number. It never becomes zero; but we can make t_n come as close to zero as we like just by finding a large enough n.

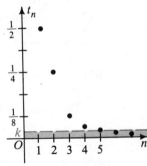

The graph at the right illustrates this idea. No matter how small a positive number k we choose, we can always make t_n be within k units of zero just by going far enough to the right on the graph.

The preceding discussion can be summarized by the following equation:

$$\lim_{n \to \infty} (\tfrac{1}{2})^n = 0$$

This is read "the limit of $(\tfrac{1}{2})^n$ as n approaches infinity equals zero."

As another illustration consider the sequence

$$1 - \frac{1}{1}, \; 1 + \frac{1}{2}, \; 1 - \frac{1}{3}, \; 1 + \frac{1}{4}, \ldots, \; 1 + \frac{(-1)^n}{n}, \ldots$$

The graph of this sequence illustrates that its limit (or target) is 1. You can make the terms of the sequence as close to 1 as you like just by going far enough to the right on the graph.

$$\lim_{n \to \infty} \left(1 + \frac{(-1)^n}{n} \right) = 1$$

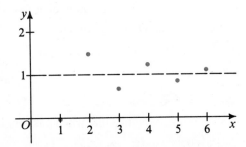

In everyday usage, the word "limit" sometimes suggests a barrier, but in mathematical usage it is better to think of a limit as a target. Thus the limit 1 is a target approached more and more closely by $1 + \dfrac{(-1)^n}{n}$ as n gets larger and larger.

You can usually estimate the limit of an infinite sequence by substituting large values of n in the formula for the nth term. This method is illustrated in the next example.

EXAMPLE 1. Find: **a.** $\displaystyle\lim_{n \to \infty} \sin \frac{1}{n}$ **b.** $\displaystyle\lim_{n \to \infty} (0.99)^n$

SOLUTION:

a. When $n = 100$, $\sin \dfrac{1}{n} = \sin \dfrac{1}{100} \approx 0.01$. As n becomes larger, $\dfrac{1}{n}$ gets nearer to 0 and so does $\sin \dfrac{1}{n}$. Hence, it appears that $\displaystyle\lim_{n \to \infty} \sin \dfrac{1}{n} = 0$.

b. We can evaluate $(0.99)^n$ for large n with logarithms or a calculator. For example,

$$(0.99)^{1000} \approx 4.3 \times 10^{-5}$$
$$(0.99)^{10000} \approx 2.2 \times 10^{-44}$$

Thus it appears that $\displaystyle\lim_{n \to \infty} (0.99)^n = 0$

We have just estimated $\displaystyle\lim_{n \to \infty} (0.99)^n$ as 0, and earlier we saw that $\displaystyle\lim_{n \to \infty} (\tfrac{1}{2})^n = 0$. These two examples are special cases of the following general theorem:

THEOREM 1. If $|r| < 1$, then $\displaystyle\lim_{n \to \infty} r^n = 0$.

In Example 1 we estimated limits by considering t_n for large values of n. In the next example, we show how to change the form of t_n to assist in finding a limit.

EXAMPLE 2. Find: **a.** $\lim\limits_{n \to \infty} \dfrac{n^2 + 1}{2n^2 - 3n}$ **b.** $\lim\limits_{n \to \infty} \dfrac{5n^2 + \sqrt{n}}{3n^3 + 7}$

SOLUTION: In both part (a) and part (b), we divide numerator and denominator by the highest power of n that occurs in the denominator.

a. Dividing numerator and denominator by n^2, we have:

$$\frac{n^2 + 1}{2n^2 - 3n} = \frac{1 + \dfrac{1}{n^2}}{2 - \dfrac{3}{n}}$$

Notice that when n is very large, $\dfrac{1}{n^2}$ and $\dfrac{3}{n}$ are very near 0. Therefore,

$$\frac{n^2 + 1}{2n^2 - 3n} \approx \frac{1}{2}$$

when n is very large. For this reason, we say:

$$\lim_{n \to \infty} \frac{n^2 + 1}{2n^2 - 3n} = \frac{1}{2}$$

b. Dividing numerator and denominator by n^3, we have:

$$\frac{5n^2 + \sqrt{n}}{3n^3 + 7} = \frac{\dfrac{5}{n} + \dfrac{1}{n^{\frac{5}{2}}}}{3 + \dfrac{7}{n^3}}$$

Notice that when n is very large, $\dfrac{5}{n}$, $\dfrac{1}{n^{\frac{5}{2}}}$, and $\dfrac{7}{n^3}$ are very near 0. Therefore,

$$\frac{5n^2 + \sqrt{n}}{3n^3 + 7} \approx \frac{0 + 0}{3 + 0} = 0$$

when n is very large. For this reason, we say:

$$\lim_{n \to \infty} \frac{5n^2 + \sqrt{n}}{3n^3 + 7} = 0$$

Situations in which a sequence has no limit

Not all sequences have limits. If the terms of a sequence do not "home in" on a single value, we say that the sequence has no limit or that the limit of the sequence does not exist. For example, the following sequence has no limit:

$$\frac{1}{2}, \ -\frac{2}{3}, \ \frac{3}{4}, \ -\frac{4}{5}, \ \ldots, \ \frac{(-1)^{n+1} \cdot n}{n + 1}, \ \ldots$$

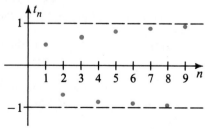

The diagram at the right shows the graph of this sequence. Notice that the odd-numbered terms form a sequence with limit 1. Similarly, the even-numbered terms form a sequence with limit -1.

Nevertheless, there is no single limiting number for all the terms of the sequence. Hence, we say that this sequence has no limit.

Infinite limits

Sometimes the terms of a sequence become arbitrarily large in absolute value. Here are two examples:

(a) $3, 7, 11, 15, \ldots, 4n - 1, \ldots$ (b) $-10, -100, -1000, \ldots, -10^n, \ldots$

The terms in sequence (a) become infinitely large. This sequence cannot approach a fixed number (or target) since no matter what fixed number is selected, the terms of the sequence will eventually exceed it. For this reason, we say that the limit of $4n - 1$ as n approaches infinity is infinity. We write this as

$$\lim_{n \to \infty} 4n - 1 = \infty.$$

Similarly, the terms in sequence (b) become infinitely negative; that is, they are farther and farther to the left of zero on the number line. We say that the limit of $-(10^n)$ as n approaches infinity is negative infinity and write

$$\lim_{n \to \infty} [-(10^n)] = -\infty.$$

It is important to realize that infinity is neither a number nor a place. When we say that "n approaches infinity," we mean that n is becoming arbitrarily large. When we say "the limit is infinity," we mean that the terms are becoming arbitrarily large, not that they are approaching some number.

EXAMPLE 3. Show that $\lim\limits_{n \to \infty} \dfrac{7n^3}{4n^2 - 5} = \infty$.

SOLUTION: Dividing numerator and denominator by n^2, the highest power of n in the denominator, we get:

$$\frac{7n^3}{4n^2 - 5} = \frac{7n}{4 - \dfrac{5}{n^2}}$$

When n is very large, $\dfrac{5}{n^2}$ is very near 0. Therefore, $\dfrac{7n^3}{4n^2 - 5} \approx \dfrac{7n}{4}$ when n is very large. Thus, since $\dfrac{7n}{4}$ becomes arbitrarily large as n does,

$$\lim_{n \to \infty} \frac{7n}{4} = \infty \quad \text{and} \quad \lim_{n \to \infty} \frac{7n^3}{4n^2 - 5} = \infty.$$

ORAL EXERCISES 13-3

Find the following limits:

1. $\lim\limits_{n \to \infty} \dfrac{n}{n + 1}$ **2.** $\lim\limits_{n \to \infty} \dfrac{n^2 - 1}{n^2}$

3. $\lim\limits_{n\to\infty} \dfrac{2n+1}{3n+1}$

4. $\lim\limits_{n\to\infty} \dfrac{8n^2-3n}{5n^2+7}$

5. $\lim\limits_{n\to\infty} \cos \dfrac{1}{n}$

6. $\lim\limits_{n\to\infty} \log\left(\cos \dfrac{1}{n}\right)$

7. $\lim\limits_{n\to\infty} (0.999)^n$

8. $\lim\limits_{n\to\infty} (1.001)^n$

9. $\lim\limits_{n\to\infty} \dfrac{n^4}{2n+1}$

10. $\lim\limits_{n\to\infty} \dfrac{n^2+9{,}999{,}999}{n^3}$

11. Do you think the sequence 1, 0, 1, 0, 1, 0, ... has a limit?

WRITTEN EXERCISES 13-3

In Exercises 1–12, find the given limit.

A   1. $\lim\limits_{n\to\infty} \dfrac{n+5}{n}$

2. $\lim\limits_{n\to\infty} \dfrac{n^2+1}{n^2}$

3. $\lim\limits_{n\to\infty} \left[1 + \dfrac{(-1)^n}{n}\right]$

4. $\lim\limits_{n\to\infty} \dfrac{4n-3}{2n+1}$

5. $\lim\limits_{n\to\infty} \dfrac{3n^2+5n}{8n^2}$

6. $\lim\limits_{n\to\infty} \dfrac{2n^4}{6n^5+7}$

7. $\lim\limits_{n\to\infty} \tan \dfrac{1}{n}$

8. $\lim\limits_{n\to\infty} \sec \dfrac{1}{n}$

9. $\lim\limits_{n\to\infty} \dfrac{\sqrt{n}}{n+1}$

10. $\lim\limits_{n\to\infty} \dfrac{5n^{\frac{2}{3}}-8n}{6n-1}$

11. $\lim\limits_{n\to\infty} \log_{10}\left(\dfrac{n+1}{n}\right)$

12. $\lim\limits_{n\to\infty} \log_{10} \sqrt[n]{10}$

In Exercises 13–18, find the limit of the specified sequence or state that the limit does not exist.

13. $\dfrac{1}{3}, -\dfrac{1}{9}, \dfrac{1}{27}, -\dfrac{1}{81}, \dfrac{1}{243}, \ldots$

14. $1, -4, 9, -16, 25, -36, \ldots$

15. $\dfrac{3}{2}, -\dfrac{4}{3}, \dfrac{5}{4}, -\dfrac{6}{5}, \dfrac{7}{6}, -\dfrac{8}{7}, \ldots$

16. $\dfrac{1}{10}, -\dfrac{2}{10^2}, \dfrac{3}{10^3}, -\dfrac{4}{10^4}, \dfrac{5}{10^5}, \ldots$

17. $t_n = \cos\left(\dfrac{n\pi}{2}\right)$

18. $t_n = \sin(n\pi)$

In Exercises 19–30, evaluate the given limit or state that the limit does not exist. If the sequence approaches ∞ or $-\infty$, so state.

19. $\lim\limits_{n\to\infty} \dfrac{5n^{\frac{5}{2}}-7n}{n^2+10n}$

20. $\lim\limits_{n\to\infty} \dfrac{5n}{n^{\frac{1}{2}}-3}$

21. $\lim\limits_{n\to\infty} \log_{10}\left(\dfrac{1}{n}\right)$

22. $\lim\limits_{n\to\infty} \dfrac{\sin n}{n}$

23. $\lim\limits_{n\to\infty} \dfrac{\cos(n\pi)}{n}$

24. $\lim\limits_{n\to\infty} e^{-n}$

25. $\lim\limits_{n\to\infty} \tan\left(\dfrac{\pi}{4} + n\pi\right)$

26. $\lim\limits_{n\to\infty} 2^n$

B   27. $\lim\limits_{n\to\infty} \log\left(\sec \dfrac{1}{n}\right)$

28. $\lim\limits_{n\to\infty} [(-1)^n - 1]$

29. $\lim\limits_{n\to\infty} \dfrac{\sqrt{n+1}}{\sqrt{n-1}}$

30. $\lim\limits_{n\to\infty} \dfrac{\sqrt[3]{8n^2-5n+1}}{\sqrt[3]{n^2+7n-3}}$

31. a. If the nth term of a geometric series is $t_n = \left(\dfrac{1}{3}\right)^{n-1}$, show that

$$S_n = \frac{3}{2}\left(1 - \left(\frac{1}{3}\right)^n\right).$$

    b. Find $\lim\limits_{n\to\infty} S_n$.

32. Consider the formula for the sum of the series $t_1 + t_1 r + \cdots + t_1 r^{n-1}$ and then suggest the formula for the sum of the infinite series.

$$t_1 + t_1 r + t_1 r^2 + \cdots, \quad \text{if } |r| < 1.$$

33. a. How is the number e defined? (Check page 179.)

    b. Evaluate $\lim\limits_{n\to\infty}\left(1 + \dfrac{2}{n}\right)^n$ by noting that

$$\left(1 + \frac{2}{n}\right)^n = \left(1 + \frac{1}{\frac{n}{2}}\right)^n = \left[\left(1 + \frac{1}{\frac{n}{2}}\right)^{\frac{n}{2}}\right]^2.$$

34. a. Show that $\lim\limits_{n\to\infty}\left(1 + \dfrac{3}{n}\right)^n = e^3$. [*Hint:* See Exercise 33(b).]

    b. Evaluate $\lim\limits_{n\to\infty}\left(1 + \dfrac{1}{2n}\right)^n$.

35. The area A under the curve $y = x^3$ between $x = 0$ and $x = 1$ is approximated by adding the areas of n rectangles as shown.

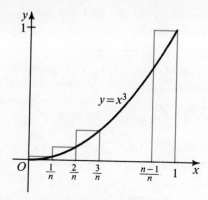

$$A \approx A_n = \left(\frac{1}{n}\right)^3 \cdot \frac{1}{n} + \left(\frac{2}{n}\right)^3 \cdot \frac{1}{n} + \left(\frac{3}{n}\right)^3 \cdot \frac{1}{n} + \cdots + \left(\frac{n}{n}\right)^3 \cdot \frac{1}{n}$$

$$= \frac{1}{n^4}[1^3 + 2^3 + 3^3 + \cdots + n^3]$$

(*Exercise continued on page 472*)

35. (Continued)

a. According to the results in Exercises 35 and 36, Section 13-2,

$$1^3 + 2^3 + \cdots + n^3 = \left[\frac{n(n + 1)}{2}\right]^2.$$ Use this formula to show that

$$A_n = \frac{n^2 + 2n + 1}{4n^2}.$$

b. As n becomes very large, what value is A_n approaching?

c. Find A by evaluating $\lim\limits_{n \to \infty} A_n$.

C **36.** Use the procedure in Exercise 35 to find the area under the curve $y = x^2$ between 0 and 1. You will need to know that

$$1^2 + 2^2 + 3^2 + \cdots + n^2 = \frac{n(n + 1)(2n + 1)}{6}.$$

CALCULATOR EXERCISES

1. a. Guess the value of $\lim\limits_{n \to \infty} \left(\sqrt{n + 1} - \sqrt{n}\right)$ by evaluating $\left(\sqrt{n + 1} - \sqrt{n}\right)$ for several large values of n.

b. Multiply $\left(\sqrt{n + 1} - \sqrt{n}\right)$ by $\dfrac{\sqrt{n + 1} + \sqrt{n}}{\sqrt{n + 1} + \sqrt{n}}$.

c. Determine what happens to the expression in (b) when n becomes very large.

2. A sequence is defined recursively by the equations

$$t_1 = 1, \quad t_n = \frac{t_{n-1}}{2} + \frac{1}{t_{n-1}}.$$

a. Find decimal approximations for the first five terms of the sequence.

b. Suggest a limit for this sequence.

3. a. Evaluate $\sqrt{1 + \sqrt{1 + \sqrt{1 + \sqrt{1 + \cdots}}}}$ by considering its value to be the limit of the following sequence as n approaches ∞:

$$\sqrt{1 + \sqrt{1}}, \quad \sqrt{1 + \sqrt{1 + \sqrt{1}}}, \quad \sqrt{1 + \sqrt{1 + \sqrt{1 + \sqrt{1}}}},$$
and so on.)

b. Compare your answers with the golden ratio $R = \dfrac{\sqrt{5} + 1}{2}$.

4. Evaluate $1 + \dfrac{1}{1 + \dfrac{1}{1 + \cdots}}$ by considering its value to be the limit of the

following sequence as n approaches ∞:

$$1 + \frac{1}{1}, \quad 1 + \frac{1}{1 + \dfrac{1}{1}}, \ldots$$

COMPUTER EXERCISE

A student leaves home to go to the movies. Halfway there, the student remembers some uncompleted homework and heads back home. Halfway back home, the student has a change of mind and heads back to the movies. You guessed it! Halfway back to the movies, the student, overcome by an attack of conscience, heads back to complete the homework. Suppose the student continues to vacillate in this fashion. Write a computer program that calculates how far from home the student is after each of the first twenty changes of mind. If you think of the student moving on a number line with home at 0 and the movies at 1, does the student appear to be approaching a limiting point on the number line?

13-4/SUMS OF INFINITE SERIES

The sum of an infinite series is very closely connected to the limit of an infinite sequence. To see this, consider the infinite geometric series

$$\tfrac{1}{2} + \tfrac{1}{4} + \tfrac{1}{8} + \tfrac{1}{16} + \cdots + (\tfrac{1}{2})^n + \cdots.$$

Associated with this series is a sequence called the *sequence of partial sums:*

$$S_1 = \tfrac{1}{2}$$
$$S_2 = \tfrac{1}{2} + \tfrac{1}{4} = \tfrac{3}{4}$$
$$S_3 = \tfrac{1}{2} + \tfrac{1}{4} + \tfrac{1}{8} = \tfrac{7}{8}$$
$$\vdots$$
$$S_n = \tfrac{1}{2} + \tfrac{1}{4} + \tfrac{1}{8} + \cdots + (\tfrac{1}{2})^n$$
$$= \frac{\tfrac{1}{2}(1 - (\tfrac{1}{2})^n)}{1 - \tfrac{1}{2}}, \text{ using the formula } S_n = \frac{t_1(1 - r^n)}{1 - r}$$
$$= 1 - (\tfrac{1}{2})^n$$

Since the sequence of partial sums $\tfrac{1}{2}, \tfrac{3}{4}, \tfrac{7}{8}, \ldots, 1 - (\tfrac{1}{2})^n$ has limit 1, we shall say that the infinite series has **limit** 1 or has the **sum** 1.

In general, for any infinite series $t_1 + t_2 + \cdots + t_n + \cdots,$

$$S_n = t_1 + t_2 + \cdots + t_n$$

is called the **nth partial sum.** If the **sequence of partial sums** $S_1, S_2, \ldots, S_n, \ldots$ has a finite limit S, then the infinite series is said to **converge** to the sum S. If the sequence of partial sums approaches infinity or has no finite limit, the infinite series is said to **diverge.**

Since we already have a formula for the nth partial sum of a geometric series, we can prove a theorem that tells when such series converge.

THEOREM. If $|r| < 1$, the infinite geometric series

$$t_1 + t_1 r + t_1 r^2 + \cdots + t_1 r^n + \cdots$$

converges to the sum

$$S = \frac{t_1}{1 - r}.$$

If $|r| \geq 1$ and $t_1 \neq 0$, then the series diverges.

Proof: The nth partial sum of the geometric series is

$$S_n = \frac{t_1(1 - r^n)}{1 - r}.$$

1. Therefore, if $|r| < 1$:

$$\lim_{n \to \infty} S_n = \lim_{n \to \infty} \frac{t_1(1 - r^n)}{1 - r}$$

$$= \frac{t_1(1 - 0)}{1 - r} \text{ since } \lim_{n \to \infty} r^n = 0 \text{ when } |r| < 1$$

$$= \frac{t_1}{1 - r}$$

2. However, if $|r| > 1$, r^n becomes infinite as n approaches infinity. Thus, S_n becomes infinite and the series diverges.

3. If $r = 1$, the series becomes the divergent series

$$t_1 + t_1 + t_1 + t_1 + \cdots.$$

4. If $r = -1$, the series becomes the divergent series

$$t_1 - t_1 + t_1 - t_1 + \cdots.$$

EXAMPLE 1. Find the sum of the infinite geometric series

$$9 - 6 + 4 - \cdots.$$

SOLUTION: Since $t_1 = 9$ and $r = -\frac{2}{3}$,

$$S = \frac{t_1}{1 - r} = \frac{9}{1 - (-\frac{2}{3})} = \frac{27}{5}.$$

EXAMPLE 2. For what values of x does the following infinite series converge?

$$1 + (x - 2) + (x - 2)^2 + (x - 2)^3 + \cdots$$

SOLUTION: This is an infinite geometric series with $r = x - 2$. By the theorem on page 473, the series converges when $|r| < 1$; that is, when $|x - 2| < 1$, or

$$1 < x < 3$$

This interval $1 < x < 3$ for which the series converges is called the *interval of convergence* for the series.

Our final example illustrates two important facts about repeating decimals. First, they can be written as infinite geometric series, and second, they represent rational numbers.

EXAMPLE 3. The infinite, repeating decimal $0.454545\ldots$ can be written as the infinite series

$$0.45 + 0.0045 + 0.000045 + \cdots.$$

What is the sum of this series?

SOLUTION: This is a geometric series with $t_1 = 0.45$ and $r = 0.01$. Therefore:

$$S = \frac{t_1}{1 - r} = \frac{0.45}{1 - 0.01} = \frac{0.45}{0.99} = \frac{5}{11}$$

ORAL EXERCISES 13-4

For each infinite geometric series, find S_1, S_2, S_3, and S_4. Also find the sum of the series if it converges.

1. $1 + \frac{1}{3} + \frac{1}{9} + \frac{1}{27} + \cdots$ **2.** $\frac{1}{2} - \frac{1}{4} + \frac{1}{8} - \frac{1}{16} + \cdots$

3. $1 + 3 + 5 + 7 + \cdots$ **4.** $1 + 0.1 + 0.01 + 0.001 + \cdots$

5. Find the interval of convergence for each series.

    a. $1 + x + x^2 + x^3 + \cdots$ **b.** $1 + 2x + 4x^2 + 8x^3 + \cdots$

6. Express $0.3333\ldots$ as an infinite geometric series. For this series, determine **(a)** t_1, **(b)** r, and **(c)** the sum.

7. Consider any infinite arithmetic series for which $a \neq 0$ and $d \neq 0$. Explain why this series diverges.

8. Do you think the series

$$\frac{1}{2} + \frac{2}{3} + \frac{3}{4} + \frac{4}{5} + \cdots + \frac{n}{n+1} + \cdots$$

converges or diverges? Tell why.

WRITTEN EXERCISES 13-4

In Exercises 1-8, find the sum of the given infinite geometric series.

A   **1.** $1 + \frac{1}{2} + \frac{1}{4} + \frac{1}{8} + \cdots$ **2.** $1 - \frac{1}{3} + \frac{1}{9} - \frac{1}{27} + \cdots$

    3. $24 - 12 + 6 - 3 + \cdots$ **4.** $\frac{1}{4} + \frac{1}{16} + \frac{1}{64} + \cdots$

    5. $5 + 5^{-1} + 5^{-3} + \cdots$ **6.** $\sqrt{27} + \sqrt{9} + \sqrt{3} + \cdots$

    7. $t_1 + t_2 + \cdots + t_n + \cdots$, where $t_n = 8(5)^{-n}$ **8.** $t_1 + t_2 + \cdots + t_n + \cdots$, where $t_n = (-2)^{1-n}$

9. Find the constant ratio of an infinite geometric series with sum 8 and first term 4.

10. Find the first three terms of an infinite geometric series with sum 81 and constant ratio $\frac{1}{3}$.

For each infinite geometric series, find **(a)** the interval of convergence and **(b)** the sum, expressed in terms of x.

11. $1 + x^2 + x^4 + x^6 + \cdots$ **12.** $1 + 3x + 9x^2 + \cdots$

13. $1 + (x - 3) + (x - 3)^2 + \cdots$ **14.** $1 - (x - 1) + (x - 1)^2 - \cdots$

15. $1 - \frac{2}{x} + \frac{4}{x^2} - \frac{8}{x^3} + \cdots$ **16.** $\frac{x^2}{3} - \frac{x^4}{6} + \frac{x^6}{12} - \cdots$

17. Show that the series $\sin^2 x + \sin^4 x + \sin^6 x + \cdots$ converges to $\tan^2 x$ if $x \neq \frac{\pi}{2} + n\pi.$

18. **a.** Show that the series $\tan^2 x - \tan^4 x + \tan^6 x - \cdots$ converges to $\sin^2 x$ if $-\frac{\pi}{4} < x < \frac{\pi}{4}.$

    b. Are there other values of x for which the series converges?

19. What is the value of x if the geometric series $1 + 2x + 4x^2 + \cdots$ converges to $\frac{3}{5}$?

20. What is the value of x if the geometric series $x^2 - x^3 + x^4 - \cdots$ converges to $\frac{x}{5}$?

21. Explain why there is no infinite geometric series with first term 10 and sum 4.

22. Explain why the sum of an infinite geometric series is positive if and only if the first term is positive.

In Exercises 23–28, express the given repeating decimal as a rational number. (See Example 3, pages 474–475.)

23. $0.777\ldots$ 24. $0.636363\ldots$ 25. $44.444\ldots$

26. $5.363636\ldots$ 27. $0.142857142857\ldots$ 28. $0.0123123123\ldots$

For the series in Exercises 29–32, find the first four partial sums, $S_1, S_2, S_3, S_4,$ and suggest a formula for S_n. Then find the sum of the infinite series by evaluating $\lim_{n\to\infty} S_n$. Note that since these series are not geometric, you cannot use the formula that was given earlier.

B   29. $\dfrac{1}{1\cdot 2} + \dfrac{1}{2\cdot 3} + \dfrac{1}{3\cdot 4} + \cdots + \dfrac{1}{n(n+1)} + \cdots$

     30. $\dfrac{1}{1\cdot 3} + \dfrac{1}{3\cdot 5} + \dfrac{1}{5\cdot 7} + \cdots + \dfrac{1}{(2n-1)(2n+1)} + \cdots$

     31. $\dfrac{1}{1\cdot 4} + \dfrac{1}{4\cdot 7} + \dfrac{1}{7\cdot 10} + \cdots + \dfrac{1}{(3n-2)(3n+1)} + \cdots$

     32. $\dfrac{3}{1\cdot 4} + \dfrac{5}{4\cdot 9} + \dfrac{7}{9\cdot 16} + \cdots + \dfrac{2n+1}{n^2(n+1)^2} + \cdots$

33. A ball is dropped from a height of 8 meters. Each time it hits the ground, it rebounds $\frac{3}{4}$ the distance it has fallen. In theory, how far will the ball travel before coming to rest?

34. Repeat Exercise 33 for a ball dropped from a height of 10 meters if it rebounds 95% of the distance it falls each time.

35. Each side of a square has length 12. The midpoints of the sides of the square are joined to form another square, and the midpoints of this square are joined to form still another square. If this process is continued indefinitely, what will be **(a)** the sum of the areas of all the squares and **(b)** the sum of the perimeters?

36. Each side of an equilateral triangle has length 12. The midpoints of the sides of the triangle are joined to form another equilateral triangle, and the midpoints of this triangle are joined to form still another triangle. If this process is continued, what will be **(a)** the sum of the areas of all the triangles and **(b)** the sum of the perimeters?

37. S_n is the nth partial sum and S is the limiting sum of the geometric series

$$1 + \tfrac{1}{2} + \tfrac{1}{4} + \tfrac{1}{8} + \cdots.$$

What is the smallest value of n for which $S - S_n < 0.0001$?

38. Repeat Exercise 37 for the geometric series $1 + \tfrac{2}{3} + \tfrac{4}{9} + \cdots$.

39. Here is an old paradox: Achilles races a turtle who has a 100-meter head start. If Achilles runs 10 meters per second and the turtle only 1 meter per second, when will Achilles overtake the turtle?

Erroneous Solution: When Achilles covers the 100-meter head start, the turtle has moved 10 meters ahead. And when Achilles covers this 10 meters, the turtle has moved 1 meter ahead. Every time Achilles runs to where the turtle was, the turtle has moved ahead. Thus, Achilles can *never* catch the turtle.

Correct Solution: Let $t_1 =$ the time for Achilles to cover the 100 meter head start; let $t_2 =$ the time to cover the next 10 meters; let $t_3 =$ the time for the next 1 meter, and so on. Find the first three terms and the sum of the infinite series $t_1 + t_2 + t_3 + \cdots$.

40. Comment on the following paradox: You can *never* leave the room in which you are sitting because in order to do so, you must first walk halfway to the door. Then you must walk half the remaining distance to the door, and then half the next remaining distance. Since you must continue to cover the halves of these remaining distances an infinite number of times, you can never leave the room.

Exercises 41–44 deal with sequences and series of complex numbers and should be done sequentially.

41. In an arithmetic sequence of complex numbers, $t_1 = 1 + 3i$ and $t_2 = 3 + 4i$. Find the next three terms. Also find t_{25} and S_{25}.

42. In a geometric sequence of complex numbers, $t_1 = i$ and the constant ratio is $r = 2i$. Find the next four terms of the sequence and represent all five terms graphically in an Argand diagram. Find the sum of these terms.

43. If $|r| < 1$, the formula

$$S = \frac{t_1}{1 - r}$$

holds for an infinite geometric series of complex numbers as well as real numbers. Use this formula to find the sum of the following series:

a. $1 + \dfrac{i}{2} + \dfrac{i^2}{4} + \dfrac{i^3}{8} + \dfrac{i^4}{16} + \cdots$

b. $27 - 9i + 3i^2 - i^3 + \cdots$

44. A bug leaves the origin and crawls 1 unit east, $\frac{1}{2}$ unit north, $\frac{1}{4}$ unit west, $\frac{1}{8}$ unit south, and so forth, as shown. Each segment of its journey can be considered as a complex number or vector. Hence, the bug's ultimate destination can be considered as the sum

$$S = t_1 + t_2 + t_3 + t_4 + \cdots$$

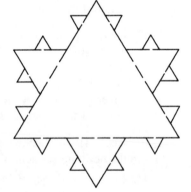

$t_3 = -\frac{1}{4} + 0i$

$t_4 = 0 - \frac{1}{8}i$

$t_2 = 0 + \frac{1}{2}i$

origin

$t_1 = 1 + 0i$

of an infinite geometric series.

a. Find r for this series.

b. Use the formula $S = \dfrac{t_1}{1 - r}$ to show that $S = \frac{4}{5} + \frac{2}{5}i$, so that the ultimate destination point is $(\frac{4}{5}, \frac{2}{5})$.

c. How far must the bug crawl to reach its ultimate destination?

C   **45.** Consider an infinite geometric series of positive terms that converges to S. What fractional part of S is the sum of the odd-numbered terms? the even-numbered terms? (Give your answers in terms of r, the constant ratio of the series.)

46. a. Prove that every rational number $\dfrac{a}{b}$ can be expressed as a repeating decimal. (For example, $\frac{2}{11} = 0.181818\ldots$ and $\frac{1}{4} = 0.25000\ldots$ or $0.24999\ldots$.)

     b. Prove that every repeating decimal represents a rational number.

Exercises 47 and 48 refer to the "snowflake curve" defined as follows: The sides of an equilateral triangle are trisected. A new equilateral triangle is placed on the middle third of each trisection. The sides common to the previous figure and the new triangles are then removed. This process continues indefinitely using the sides of the last figure obtained.

47. Find the area enclosed by the snowflake curve if the side of the initial equilateral triangle is one unit in length.

48. Show that the limit of the sequence of perimeters of the snowflake curve is infinite.

CALCULATOR EXERCISE

The infinite series $\frac{1}{5} + \frac{2}{25} + \frac{4}{125} + \frac{8}{625} + \cdots$ converges to $\frac{1}{3} = 0.3333\ldots$. Determine how many terms of the series must be added to make the sum:

a. correct in at least the first 2 decimal places, that is, 0.33.

b. correct in at least the first 3 decimal places, that is, 0.333.

c. correct in at least the first 4 decimal places, that is, 0.3333.

13-5/SIGMA NOTATION

The Greek letter $\sum$ (sigma) is often used in mathematics to express a series or its sum in abbreviated form. For example, $\displaystyle\sum_{k=1}^{100} k^2$ represents the series whose terms are obtained by evaluating k^2 first for $k = 1$, then for $k = 2$, then for $k = 3$, and so on until $k = 100$. That is:

$$\sum_{k=1}^{100} k^2 = 1^2 + 2^2 + 3^2 + \cdots + 100^2$$

The symbol on the left above may be read as "the sum of k^2 for values of k from 1 to 100." Similarly, the symbol $\displaystyle\sum_{k=5}^{10} 3k$ may be read as "the sum of $3k$ for values of k from 5 to 10." This symbol represents the series whose terms are obtained by evaluating $3k$ first for $k = 5$, then for $k = 6$, and so on, to $k = 10$. That is:

$$\sum_{k=5}^{10} 3k = 3 \cdot 5 + 3 \cdot 6 + 3 \cdot 7 + 3 \cdot 8 + 3 \cdot 9 + 3 \cdot 10 = 135$$

The symbol $3k$ is called the **summand,** the numbers 5 and 10 are called the **limits of summation,** and the symbol k is called the **index.** Any letter can be used for the index. For example:

$$\sum_{n=1}^{5} (-\tfrac{1}{2})^n = (-\tfrac{1}{2})^1 + (-\tfrac{1}{2})^2 + (-\tfrac{1}{2})^3 + (-\tfrac{1}{2})^4 + (-\tfrac{1}{2})^5$$

$$= -\tfrac{1}{2} + \tfrac{1}{4} - \tfrac{1}{8} + \tfrac{1}{16} - \tfrac{1}{32} = -\tfrac{11}{32}$$

Sigma notation can also be used to represent infinite series and their sums. For example:

$$\sum_{j=0}^{\infty} (\tfrac{1}{2})^j = (\tfrac{1}{2})^0 + (\tfrac{1}{2})^1 + (\tfrac{1}{2})^2 + (\tfrac{1}{2})^3 + \cdots$$

$$= 1 + \tfrac{1}{2} + \tfrac{1}{4} + \tfrac{1}{8} + \cdots = 2$$

The symbol $\displaystyle\sum_{j=0}^{\infty} (\tfrac{1}{2})^j$ is read, "the sum of $(\tfrac{1}{2})^j$ for values of j from 0 to infinity"; it represents both the infinite geometric series above and its sum, 2.

When you get used to using sigma notation, you will find it much easier than manipulating series that are written out in expanded form. For example, the following properties are consequences of the commutative, associative, and distributive properties of the real and imaginary numbers. These properties are proved in Exercises 29 and 30.

PROPERTIES OF FINITE SUMS

1. $\displaystyle\sum_{i=1}^{n} (a_i + b_i) = \sum_{i=1}^{n} a_i + \sum_{i=1}^{n} b_i$ **2.** $\displaystyle\sum_{i=1}^{n} ca_i = c \sum_{i=1}^{n} a_i$

The properties above can be used together with previously derived sums to derive the sums of many other series. Example 1 illustrates how these properties may be used with the following known sums.

$$\text{sum of integers} = \sum_{k=1}^{n} k = \frac{n(n+1)}{2} \qquad \text{(Proved in Section 13–2)}$$

$$\text{sum of squares} = \sum_{k=1}^{n} k^2 = \frac{n(n+1)(2n+1)}{6} \qquad \text{(To be proved in Section 13–7)}$$

$$\text{sum of cubes} = \sum_{k=1}^{n} k^3 = \left[\frac{n(n+1)}{2}\right]^2 \qquad \text{(Proved in Section 13–2)}$$

EXAMPLE 1.   Evaluate $1 \cdot 2 + 3 \cdot 4 + 5 \cdot 6 + \cdots + 199 \cdot 200$.

SOLUTION: $\displaystyle 1 \cdot 2 + 3 \cdot 4 + \cdots + 199 \cdot 200 = \sum_{k=1}^{100} (2k-1)(2k)$

$$= \sum_{k=1}^{100} (4k^2 - 2k)$$

$$= 4\sum_{k=1}^{100} k^2 - 2\sum_{k=1}^{100} k$$

$$= 4\left[\frac{100 \cdot 101 \cdot 201}{6}\right] - 2\left[\frac{100 \cdot 101}{2}\right]$$

$$= 1{,}353{,}400 - 10{,}100 = 1{,}343{,}300$$

ORAL EXERCISES 13–5

Give each series in expanded form.

1. $\displaystyle\sum_{k=1}^{4} 5k$ **2.** $\displaystyle\sum_{k=3}^{6} k^2$ **3.** $\displaystyle\sum_{j=2}^{8} (-1)^j$ **4.** $\displaystyle\sum_{n=1}^{\infty} \frac{1}{n}$

Express each of the following series using sigma notation.

5. $4 + 9 + 16 + 25 + 36$ **6.** $\frac{1}{2} + \frac{2}{3} + \frac{3}{4} + \frac{4}{5}$

7. the arithmetic series $3 + 6 + 9 + \cdots + 300$

8. the infinite geometric series $\frac{1}{3} + \frac{1}{9} + \frac{1}{27} + \frac{1}{81} + \cdots$

In Exercises 1–8, write the given series in expanded form and evaluate it.

A **1.** $\displaystyle\sum_{k=2}^{6} k$ **2.** $\displaystyle\sum_{k=1}^{10} 5k$ **3.** $\displaystyle\sum_{k=1}^{5} \frac{1}{k}$ **4.** $\displaystyle\sum_{n=3}^{7} (4n-7)$

5. $\displaystyle\sum_{n=0}^{\infty} 3^{1-n}$ **6.** $\displaystyle\sum_{j=2}^{\infty} j(-1)^{j}$ **7.** $\displaystyle\sum_{t=-2}^{2} 4^{t}$ **8.** $\displaystyle\sum_{s=0}^{6} |13-3s|$

In Exercises 9–16, express the given series using sigma notation.

9. $4 + 8 + 12 + 16 + 20$ **10.** $1 + 2 + 4 + 8 + 16 + 32$

11. $5 + 9 + 13 + \cdots + 101$ **12.** $2 + 4 + 6 + 8 + \cdots + 200$

13. $1 + \frac{1}{4} + \frac{1}{9} + \frac{1}{16} + \cdots$ **14.** $\frac{1}{2} + \frac{1}{4} + \frac{1}{6} + \frac{1}{8} + \cdots$

15. $\sin x + \sin 2x + \sin 3x + \cdots$ **16.** $48 + 24 + 12 + 6 + \cdots$

17. Show that $\displaystyle\sum_{t=1}^{4} \log t = \log 24$.   **18.** Show that $\displaystyle\sum_{k=1}^{4} k \log 2 = \log 2^{10}$.

19. Evaluate: $\displaystyle\sum_{k=1}^{100} \cos (k\pi)$ **20.** Evaluate: $\displaystyle\sum_{k=1}^{50} \sin \left(k \cdot \frac{\pi}{2}\right)$

B **21.** Evaluate $\displaystyle\sum_{k=1}^{100} \left[\sqrt{k}\right]$. (*Hint:* $[x]$ is the greatest integer in x. See page 292.)

22. Evaluate: **a.** $\displaystyle\sum_{n=0}^{\infty} (\frac{1}{2} + \frac{1}{2}i)^{n}$ **b.** $\displaystyle\sum_{n=0}^{\infty} |(\frac{1}{2} + \frac{1}{2}i)^{n}|$

23. Show that $\displaystyle\sum_{k=1}^{n} 2^{-k} = 1 - 2^{-n}$.

24. Show that $\displaystyle\sum_{k=1}^{8} (-1)^{k} \log k = \log \left(\frac{2 \cdot 4 \cdot 6 \cdot 8}{1 \cdot 3 \cdot 5 \cdot 7}\right)$.

In Exercises 25–28, express the given series using sigma notation.

25. a. $1 - \frac{1}{2} + \frac{1}{4} - \frac{1}{8} + \frac{1}{16} - \frac{1}{32}$ **b.** $-1 + \frac{1}{2} - \frac{1}{4} + \frac{1}{8} - \frac{1}{16} + \frac{1}{32}$

26. a. $27 - 9 + 3 - 1 + \frac{1}{3} - \frac{1}{9}$ **b.** $-27 + 9 - 3 + 1 - \frac{1}{3} + \frac{1}{9}$

27. $1 - 3 + 5 - 7 + \cdots - 99$ **28.** $-2 + 4 - 6 + \cdots + 100$

Show how the commutative, associative, and distributive properties are used to prove the following:

29. $\displaystyle\sum_{i=1}^{n} (a_i + b_i) = \sum_{i=1}^{n} a_i + \sum_{i=1}^{n} b_i$ **30.** $\displaystyle\sum_{i=1}^{n} c a_i = c \sum_{i=1}^{n} a_i$

Use the Properties of Finite Sums to prove the following:

31. $\displaystyle\sum_{i=1}^{n} (a_i + b_i)^2 = \sum_{i=1}^{n} a_i^2 + 2 \sum_{i=1}^{n} a_i b_i + \sum_{i=1}^{n} b_i^2$

32. $\displaystyle\sum_{x=1}^{n} (ax^2 + bx + c) = a \sum_{x=1}^{n} x^2 + b \sum_{x=1}^{n} x + cn$

Evaluate the following series using the technique in Example 1.

33. $1 \cdot 2 + 2 \cdot 3 + 3 \cdot 4 + \cdots + 100 \cdot 101$

34. $1 \cdot 3 + 3 \cdot 5 + 5 \cdot 7 + \cdots + 99 \cdot 101$

35. $1 \cdot 2 \cdot 3 + 2 \cdot 3 \cdot 4 + 3 \cdot 4 \cdot 5 + \cdots + 20 \cdot 21 \cdot 22$

36. $2 \cdot 4 + 6 \cdot 8 + 10 \cdot 12 + \cdots + 398 \cdot 400$

C

37. a. An 8-by-8 checkerboard has 64 little squares but also many other squares of various sizes. How many squares does it have in all?

b. Find the total number of squares in an n-by-n checkerboard.

38. a. The 3-by-3-by-3 cube shown has 27 little cubes but also many other cubes of various sizes. How many cubes does it have in all?

b. Find the total number of cubes in an n-by-n-by-n cube.

39. A stack of oranges is compactly arranged so the bottom layer consists of oranges in an equilateral triangle with n oranges on a side. The layer next to the bottom consists of an equilateral triangle of oranges with $n - 1$ oranges on a side. This pattern continues upward with one orange on top. How many oranges are there?

Fourth layer

Manuel Vallarta (1899–1977) was a mathematician and physicist. While working at the Massachusetts Institute of Technology, he received the prestigious Guggenheim Fellowship. This took him to Germany where he studied relativity with Albert Einstein and electromagnetic theory with Max Planck.

In 1946, Vallarta returned to his native Mexico where he founded the Instituto Nacional de Energiá Nuclear.

According to legend, the inventor of the game of chess requested the following reward: 1 grain of wheat on the first square of the chessboard, 2 grains on the second square, 4 grains on the third square, and so on until all 64 squares were accounted for. Although this appears to be a modest request, the exercises below will show otherwise.

1. How many grains will be on the 64th square? Give your answer in scientific notation.

2. What is the total number of grains of wheat?

3. If one bushel of wheat contains approximately 1×10^6 grains of wheat, how many bushels of wheat did the inventor request? Compare your answer with the world's current annual wheat production of approximately 5×10^9 bushels.

13-6/OTHER INFINITE SERIES (Optional)

Series of the form $\displaystyle\sum_{n=0}^{\infty} a_n x^n$ are called **power series.** When the coefficients are all the same (for all n, $a_n = a$), we get as a special case the infinite geometric series

$$\sum_{n=0}^{\infty} a x^n.$$

Some extremely useful power series are given below. The *interval of convergence* for each series is also given. Several of the series involve $n!$, read "n factorial," which is defined as follows:

$$n! = 1 \times 2 \times 3 \times \cdots \times n$$

For example, $4! = 1 \times 2 \times 3 \times 4 = 24$. We define $0!$ as 1.

(1) $\displaystyle e^x = 1 + \frac{x}{1!} + \frac{x^2}{2!} + \frac{x^3}{3!} + \cdots = \sum_{n=0}^{\infty} \frac{x^n}{n!}$, all real x

(2) $\displaystyle \cos x = 1 - \frac{x^2}{2!} + \frac{x^4}{4!} - \frac{x^6}{6!} + \cdots = \sum_{n=0}^{\infty} \frac{(-1)^n x^{2n}}{(2n)!}$, all real x

(3) $\displaystyle \sin x = x - \frac{x^3}{3!} + \frac{x^5}{5!} - \frac{x^7}{7!} + \cdots = \sum_{n=0}^{\infty} \frac{(-1)^n x^{2n+1}}{(2n + 1)!}$, all real x

(4) $\displaystyle \operatorname{Tan}^{-1} x = x - \frac{x^3}{3} + \frac{x^5}{5} - \frac{x^7}{7} + \cdots = \sum_{n=0}^{\infty} \frac{(-1)^n x^{2n+1}}{2n + 1}$, $-1 < x < 1$

(5) $\displaystyle \ln (1 + x) = x - \frac{x^2}{2} + \frac{x^3}{3} - \frac{x^4}{4} + \cdots = \sum_{n=0}^{\infty} (-1)^{n+1} \frac{x^n}{n}$, $-1 < x \le 1$

The preceding series are used by some calculators and computers to give approximations. For example, a computer has a subroutine that evaluates several terms of the series for sin x. If the computer encounters the number SIN (2) in a program, it will substitute $x = 2$ into the series. The computer is programmed to evaluate enough terms so the result will be accurate to a predetermined number of decimal places. The following diagram shows that sin 2 is quite closely approximated by just three terms of the sine series. To evaluate sin 4 or sin 5 accurately, more terms are needed.

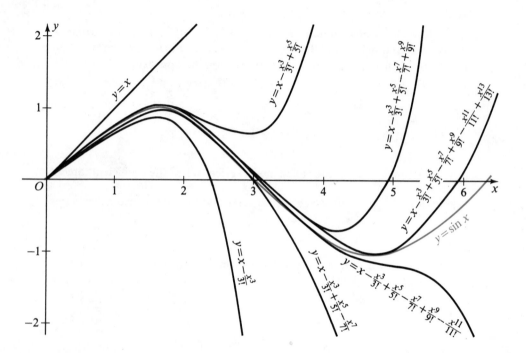

The following example shows how series (1) through (5) on page 483 can be used to find other series.

EXAMPLE 1.   Find the series for **(a)** sin $(-x)$ and **(b)** sin $2x$.

SOLUTION: **a.** In series (3) substitute $(-x)$ for x and obtain:

$$\sin(-x) = (-x) - \frac{(-x)^3}{3!} + \frac{(-x)^5}{5!} - \frac{(-x)^7}{7!} + \cdots$$

$$= -x + \frac{x^3}{3!} - \frac{x^5}{5!} + \frac{x^7}{7!} - \cdots$$

Notice that the series for sin $(-x)$ is the negative of the series for sin x. This agrees with the fact that sin $(-x) = -\sin x$.

b. In series (3) substitute $2x$ for x and obtain:

$$\sin 2x = 2x - \frac{(2x)^3}{3!} + \frac{(2x)^5}{5!} - \frac{(2x)^7}{7!} + \cdots$$

The next theorem, stated without proof, considers other series important in calculus:

THEOREM.   The series $\displaystyle\sum_{n=1}^{\infty} \frac{1}{n^k}$ converges if $k > 1$. It diverges if $k \le 1$.

The series corresponding to $k = 1$, $\displaystyle\sum_{n=1}^{\infty} \frac{1}{n}$, is called the **harmonic series.** Exercise 23 shows that it diverges.

EXAMPLE 2.   Show that $\displaystyle\sum_{n=1}^{\infty} \frac{n^2 + 3n}{n^3}$ diverges.

SOLUTION: When you are determining whether or not a series converges, you can ignore the beginning of the series. The sum of the first thousand, million, or billion terms is always finite. The real issue is whether the sum of all the rest of the terms is finite. In this case, when n is large,

$$\frac{n^2 + 3n}{n^3} = \frac{1 + \dfrac{3}{n}}{n} > \frac{1}{n}. \text{ Since } \sum_{n=1}^{\infty} \frac{1}{n} \text{ diverges, so does } \sum_{n=1}^{\infty} \frac{n^2 + 3n}{n^3}.$$

EXAMPLE 3.   Show that $\displaystyle\sum_{n=1}^{\infty} \frac{n}{n^3 + 5}$ converges.

SOLUTION: The argument is similar to that in Example 2, but in this case when n is large $\dfrac{n}{n^3 + 5} = \dfrac{1}{n^2 + \dfrac{5}{n}} < \dfrac{1}{n^2}$. Since $\displaystyle\sum_{n=1}^{\infty} \frac{1}{n^2}$ converges, so does

$$\sum_{n=1}^{\infty} \frac{n}{n^3 + 5}.$$

WRITTEN EXERCISES 13-6

In Exercises 1–18, refer to the series (1) through (5) listed on page 483.

A   **1.** Write infinite series in expanded form for $\cos \dfrac{\pi}{4}$ and $\sin\left(-\dfrac{\pi}{4}\right)$.

2. Write infinite series in expanded form for e and e^{-1}.

3. a. Substitute $x = 1$ in the series for $\ln(1 + x)$ and get an infinite series for $\ln 2$.
   b. Why can't you substitute $x = 2$ to get an infinite series for $\ln 3$?

4. Substitute $x = 1$ in the series for $\text{Tan}^{-1} x$ and get an infinite series for a number involving π.

In Exercises 5–10, obtain an infinite power series for each function. State the interval of convergence in each case. (Refer to series (1) through (5) as necessary.)

5. e^{-x} **6.** e^{-x^2} **7.** $\mathrm{Tan}^{-1} 2x$

8. $\ln(1 - x)$ **9.** $\sin x^2$ **10.** $\cos 2x$

B   **11.** Use the sine series to show that $\lim\limits_{x \to 0} \dfrac{\sin x}{x} = 1$.

      12. Use the cosine series to find $\lim\limits_{x \to 0} \dfrac{1 - \cos x}{x^2}$.

      13. Use series (5) to find $\lim\limits_{x \to 0} \dfrac{\ln(1 + x)}{x}$.

      14. Use series (1), (2), and (3) to prove that $e^{i\theta} = \cos\theta + i\sin\theta$.

      15. Use Exercise 14 to show that $e^{i\pi} = -1$ and $e^{2i\pi} = 1$. (The identity $e^{i\pi} = -1$ was discovered by the Swiss mathematician Leonhard Euler. Note the combination of arithmetic (the number -1), algebra (the number i), geometry (the number π), and analysis (the number e).)

      16. Use Exercise 14 to show that $\cos\theta = \dfrac{e^{i\theta} + e^{-i\theta}}{2}$.

      17. Use Exercise 14 to show that $\sin\theta = \dfrac{e^{i\theta} - e^{-i\theta}}{2i}$.

      18. Show that the expressions for $\sin\theta$ and $\cos\theta$ given in Exercises 16 and 17 satisfy the equation $(\sin\theta)^2 + (\cos\theta)^2 = 1$.

Use the technique of Examples 2 and 3 to see whether the following series converge or diverge.

19. $\displaystyle\sum_{n=1}^{\infty} \dfrac{4n - 3}{n^2 + 1}$ **20.** $\displaystyle\sum_{n=1}^{\infty} \dfrac{n}{3n^3 + 5}$

21. $\displaystyle\sum_{n=1}^{\infty} \dfrac{6n^{\frac{3}{2}}}{5n^{\frac{7}{2}} + 3n^{\frac{1}{2}}}$ **22.** $\displaystyle\sum_{n=1}^{\infty} \dfrac{n^5 + 3n^4 - 5n^3}{2n^6 + 1{,}000{,}000}$

C   **23.** Prove that the series

$$1 + \frac{1}{2} + \frac{1}{3} + \frac{1}{4} + \frac{1}{5} + \cdots + \frac{1}{n} + \cdots$$

diverges by rewriting the series as

$$1 + \tfrac{1}{2} + (\tfrac{1}{3} + \tfrac{1}{4}) + (\tfrac{1}{5} + \tfrac{1}{6} + \tfrac{1}{7} + \tfrac{1}{8}) + (\tfrac{1}{9} + \cdots + \tfrac{1}{16}) + \cdots$$

and proving that the terms within each pair of parentheses total more than $\tfrac{1}{2}$.

The series $\displaystyle\sum_{n=1}^{\infty} \frac{1}{n}$ diverges, but very slowly. Find how many terms are needed to make the partial sum $1 + \dfrac{1}{2} + \dfrac{1}{3} + \cdots + \dfrac{1}{n}$ greater than: **(a)** 3; **(b)** 4; **(c)** 10. (The answer is surprising!)

CALCULATOR EXERCISES

1. Use the first 8 terms of series (1) to approximate e.
2. Use the first 5 terms of series (2) to approximate cos 1.
3. Use the first 5 terms of series (3) to approximate sin 2.

13–7 / MATHEMATICAL INDUCTION

Observe the pattern in the following statements:

$$\frac{1}{1 \cdot 2} = \frac{1}{2}$$

$$\frac{1}{1 \cdot 2} + \frac{1}{2 \cdot 3} = \frac{2}{3}$$

$$\frac{1}{1 \cdot 2} + \frac{1}{2 \cdot 3} + \frac{1}{3 \cdot 4} = \frac{3}{4}$$

It *appears* that $\dfrac{1}{1 \cdot 2} + \dfrac{1}{2 \cdot 3} + \dfrac{1}{3 \cdot 4} + \cdots + \dfrac{1}{n(n+1)} = \dfrac{n}{n+1}$. Nevertheless, this appearance does not constitute a *proof* that the statement is true for all positive integers n. One way to prove the statement is to use a method called **mathematical induction.**

PROOF BY MATHEMATICAL INDUCTION

Let S be a statement in terms of a positive integer n.
1. Show that S is true for $n = 1$.
2. *Assume* that S is true for $n = k$, and then *prove* that S must be true for $n = k + 1$, where k is some positive integer.

If you can prove items (1) and (2) above, then you can conclude that S is true for *all* positive integers. The reason is that once you know S is true for $n = 1$, item (2) tells you it is true for $n = 1 + 1 = 2$. Applying item (2) again, S must be true for $n = 2 + 1 = 3$, and then for $n = 3 + 1 = 4$, and so on.

You might think of mathematical induction as something like setting up dominoes so they will all fall down. Set up the dominoes so that if any one domino falls, the next domino will fall. Then knock over the first domino, which will knock over the second, which will knock over the third, and so on. Thus, you can conclude that all the dominoes will fall down.

EXAMPLE 1. Prove that $\dfrac{1}{1\cdot 2}+\dfrac{1}{2\cdot 3}+\dfrac{1}{3\cdot 4}+\cdots+\dfrac{1}{n(n+1)}=\dfrac{n}{n+1}$ for all positive integers n.

SOLUTION: 1. Show that the statement is true for $n=1$.

$$\frac{1}{1(1+1)}=\frac{1}{1+1}. \quad \text{Yes, it's true.}$$

2. Assume that the statement is true for $n=k$, and then prove that it must be true for $n=k+1$.

Assume: $\dfrac{1}{1\cdot 2}+\dfrac{1}{2\cdot 3}+\cdots+\dfrac{1}{k(k+1)}=\dfrac{k}{k+1}$

Prove:

$$\frac{1}{1\cdot 2}+\frac{1}{2\cdot 3}+\cdots+\frac{1}{k(k+1)}+\frac{1}{(k+1)[(k+1)+1]}$$
$$=\frac{(k+1)}{(k+1)+1}, \quad \text{or}$$

$$\frac{1}{1\cdot 2}+\frac{1}{2\cdot 3}+\cdots+\frac{1}{k(k+1)}+\frac{1}{(k+1)(k+2)}=\frac{k+1}{k+2}$$

Basic strategy at this stage of proof: We will show that the left side of the "Prove" statement is equal to the right side. Take the left side and try to simplify it by using the commutative, associative, or distributive properties and the "Assume" statement.

Proof:

$$\frac{1}{1\cdot 2}+\frac{1}{2\cdot 3}+\cdots+\frac{1}{k(k+1)}+\frac{1}{(k+1)(k+2)}$$

$$=\underbrace{\frac{1}{1\cdot 2}+\frac{1}{2\cdot 3}+\cdots+\frac{1}{k(k+1)}}+\underbrace{\frac{1}{(k+1)(k+2)}}$$

To simplify, use the "Assume" statement.

$$=\qquad \frac{k}{k+1}\qquad + \qquad \frac{1}{(k+1)(k+2)}$$

$$=\qquad \frac{k(k+2)}{(k+1)(k+2)}\qquad + \qquad \frac{1}{(k+1)(k+2)}$$

$$=\qquad \frac{k^2+2k+1}{(k+1)(k+2)}$$

$$=\qquad \frac{(k+1)^2}{(k+1)(k+2)}$$

$$=\qquad \frac{k+1}{k+2}$$

EXAMPLE 2. Prove that $n^3 + 2n$ is a multiple of 3 for all positive integers n.

SOLUTION: 1. Show that the statement is true for $n = 1$.
$1^3 + 2 \cdot 1$ is a multiple of 3. Yes, it's true.

2. Assume that the statement is true for $n = k$, and then prove that it must be true for $n = k + 1$.

Assume: $k^3 + 2k$ is a multiple of 3.

Prove: $(k + 1)^3 + 2(k + 1)$ is a multiple of 3.
Use the same basic strategy as in Example 1.

Proof: $(k + 1)^3 + 2(k + 1) = k^3 + 3k^2 + 3k + 1 + 2k + 2$
$$= \underbrace{(k^3 + 2k)}_{\substack{\downarrow \\ \text{Use the} \\ \text{assumed} \\ \text{statement.} \\ \downarrow}} + \underbrace{3(k^2 + k + 1)}_{\downarrow}$$

$$= \overbrace{\text{a multiple of 3} + \text{a multiple of 3}}$$
$$= \text{a multiple of 3}$$

In the last line of the proof, we have used the fact that the sum of two multiples of 3, say $3i$ and $3j$, is another multiple of 3, namely $3(i + j)$.

WRITTEN EXERCISES 13-7

Use mathematical induction to prove the statements in these exercises. *Note that some of these statements can also be proved by other methods.*

For Exercises 1-12, prove that the statement is true for every positive integer n.

A
1. $1 + 2 + \cdots + n = \dfrac{n(n + 1)}{2}$

2. $1 + 3 + \cdots + (2n - 1) = n^2$

3. $\displaystyle\sum_{i=1}^{n} 2i = n^2 + n$

4. $\displaystyle\sum_{i=1}^{n} 2^{i-1} = 2^n - 1$

5. $\displaystyle\sum_{i=1}^{n} i^2 = \dfrac{n(n + 1)(2n + 1)}{6}$

6. $\displaystyle\sum_{i=1}^{n} i(i + 1) = \dfrac{n(n + 1)(n + 2)}{3}$

7. $\displaystyle\sum_{i=1}^{n} (2i - 1)^2 = \dfrac{n(2n - 1)(2n + 1)}{3}$

8. $(1 + x)^n \geq 1 + nx$ if $x > -1$.

9. $\displaystyle\sum_{i=1}^{n} (i \cdot 2^{i-1}) = 1 + (n - 1) \cdot 2^n$

10. $18^n - 1$ is a multiple of 17.

11. $11^n - 4^n$ is a multiple of 7.

12. $n(n^2 + 5)$ is a multiple of 6.

For Exercises 13 and 14, decide which values of n will make the inequality true, and then prove your answer correct.

B 13. $2^n > 2n$ 14. $n! > 2^n$

15. If, in a room with n people ($n \geq 2$), every person shakes hands once with every other person, prove that there will be $\dfrac{n^2 - n}{2}$ handshakes.

16. Prove that the sum of the cubes of any three consecutive integers is a multiple of 9.

17. Use the "triangle inequality" (Exercise 39, page 98) to prove that
$$|a_1 + a_2 + \cdots + a_n| \leq |a_1| + |a_2| + \cdots + |a_n|$$
for every positive integer n.

18. Prove De Moivre's Theorem (page 345) for every positive integer n.

19. Let
$$a_1 = \sqrt{2}, \; a_2 = \sqrt{2 + \sqrt{2}}, \; a_3 = \sqrt{2 + \sqrt{2 + \sqrt{2}}},$$
and so on. Prove that $a_n < 2$ for every positive integer n.

In Exercises 20–22, write out the first few terms of each sequence. Then suggest and prove a formula in terms of n for the nth term, a_n.

20. $a_1 = \dfrac{1}{2}, \; a_n = a_{n-1} - \dfrac{a_{n-1}}{n+1}$

21. $a_1 = 1, \; a_n = 2a_{n-1} + 1$ (*Hint:* Look at the sequence of powers of 2.)

22. $a_1 = \dfrac{1}{4}, \; a_n = a_{n-1} + \dfrac{1}{(3n-2)(3n+1)}$

In Exercises 23–25, write out the first few terms of each sequence. Then suggest and prove a formula in terms of n for the nth partial sum, S_n.

C 23. $S_n = \displaystyle\sum_{i=1}^{n} i^3$

24. $S_n = \displaystyle\sum_{i=1}^{n} (i \cdot i!)$ (*Hint:* Look at the sequence of factorials of the integers.)

25. $S_n = \displaystyle\sum_{i=1}^{n} i(i+1)(i+2)$ (*Hint:* See Exercises 1 and 6 for the sums $\displaystyle\sum_{i=1}^{n} i$ and $\displaystyle\sum_{i=1}^{n} i(i+1)$.)

26. Prove that a regular convex polygon with n sides ($n \geq 3$) has $\dfrac{n^2 - 3n}{2}$ diagonals.

27. Prove that it is possible to pay any debt of \$4, \$5, \$6, \$7, $\ldots$, \$$n$, and so on, by using only \$2 bills and \$5 bills. For example, a debt of \$11 can be paid with three \$2 bills and one \$5 bill, or $11 = 3 \cdot \$2 + 1 \cdot \$5$.

Chapter summary

1. A *sequence* is a function whose domain is the set of positive integers. A sequence can be defined explicitly by a formula such as $t_n = 3n + 5$ or recursively by a pair of formulas such as $t_1 = 8$ and $t_n = 3 + t_{n-1}$.

2. In an *arithmetic sequence* the difference d of any two consecutive terms is constant. In a *geometric sequence* the ratio r of any two consecutive terms is constant.

3. A *series* is an indicated sum of the terms of a sequence. For example, $8 + 11 + 14 + \cdots + (3n + 5)$ is an arithmetic series.

	$t_n = n$th term	$S_n =$ sum of n terms
Arithmetic	$t_n = t_1 + (n - 1)d$	$S_n = \dfrac{n}{2}(t_1 + t_n)$
Geometric	$t_n = t_1 \cdot r^{n-1}$	$S_n = \dfrac{t_1(1 - r^n)}{1 - r}$

4. If the nth term of a sequence becomes arbitrarily close to a number L as n gets larger and larger, we write $\lim\limits_{n \to \infty} t_n = L$ (read, "the limit as n approaches infinity of t_n equals L"). For example, $\lim\limits_{n \to \infty} \dfrac{n + 1}{n} = 1$. If there is no such limiting number L, we say that the sequence has no limit. A sequence whose terms increase (or decrease) without bound is said to have a limit of ∞ (or $-\infty$).

5. Associated with any infinite series $t_1 + t_2 + t_3 + \cdots$, there is a *sequence of partial sums:*

$$S_1 = t_1, \ S_2 = t_1 + t_2, \ S_3 = t_1 + t_2 + t_3, \text{ and so on.}$$

If this sequence has a limit L, we say that the series *converges* to the sum L. If the sequence of partial sums has no limit, the series *diverges*.

Leonhard Euler (1707–1783) was a Swiss mathematician who made major contributions to analysis, number theory, and mathematical physics. Euler created a tremendous amount of mathematics—much of it in the latter half of his life when he was blind. He introduced many of the symbols that we use today, including i, $f(x)$, Δy, and $\sum$. The symbol e for the base of natural logarithms was chosen in his honor.

Sequences and series   **491**

6. The infinite geometric series $t_1 + t_1 r + t_1 r^2 + \cdots$ converges to the sum $S = \dfrac{t_1}{1-r}$ if $|r| < 1$. The series diverges if $|r| \geq 1$ (and $t_1 \neq 0$).

7. Any infinite, repeating decimal can be written as an infinite series whose sum is a rational number. (See Example 3, page 474, and Exercise 46, page 478.)

8. The Greek letter $\sum$ (sigma) is often used to express a series or its sum. See the properties listed on page 480.)

9. (Optional) A list of five important power series is given on page 483. Also the series $\displaystyle\sum_{n=1}^{\infty} \dfrac{1}{n^k}$ diverges when $k \leq 1$ and converges when $k > 1$. These facts are useful in deciding whether or not other series converge.

10. A method called *mathematical induction* can be used to prove a statement is true for all positive integers. Proof by mathematical induction requires two steps: (1) Show that the statement is true for $n = 1$, and (2) prove that the statement is true for $n = k + 1$ whenever it is true for $n = k$.

Chapter test

1. State whether the given sequence is arithmetic, geometric, or neither. In each case give an explicit formula for t_n in terms of n.
   13–1
   a. 27, 18, 12, 8, ...   **b.** 4, 11, 18, 25, ...   **c.** 2, 5, 10, 17, ...

2. Find the first four terms of the sequence defined recursively by $t_1 = -4$ and $t_n = 2 \cdot t_{n-1} + 3$.

3. Find x if the sequence $10,\ x,\ \dfrac{x}{3}$ is **(a)** arithmetic; **(b)** geometric.

4. Find t_{25} and S_{25} for the arithmetic series $23 + 20 + 17 + \cdots$.
   13–2

5. Find (in terms of n) t_n and S_n for the geometric series $\frac{1}{4} + \frac{1}{8} + \frac{1}{16} + \frac{1}{32} + \cdots$.

6. Find the sum of the series $9 + 15 + 21 + 27 + \cdots + 603$.

7. Find the sum of the first 50 positive even integers.

8. For each of the following, state the value of the limit, if it exists. If the limit is infinite, so state.
   13–3

   a. $\displaystyle\lim_{n \to \infty} \dfrac{4n - 1}{7n}$ **b.** $\displaystyle\lim_{n \to \infty} \dfrac{2n^3 - 3n}{5n^2 + 1}$ **c.** $\displaystyle\lim_{n \to \infty} (0.75)^n$

9. Find the sum of the infinite geometric series $16 - 12 + 9 - \cdots$.
   13–4

10. For what values of x does the geometric series

$$1 + \frac{3x}{2} + \frac{9x^2}{4} + \cdots$$

converge?

11. Write in expanded form and then evaluate: 13-5

    a. $\displaystyle\sum_{j=0}^{3} (-1)^j(3j-1)$ **b.** $\displaystyle\sum_{k=1}^{\infty} \left(\frac{1}{5}\right)^k$

12. Express using sigma notation: $\frac{2}{3} + \frac{4}{5} + \frac{6}{7} + \cdots + \frac{50}{51}$

13. (Optional) Decide whether or not each series converges. 13-6

    a. $\displaystyle\sum_{n=1}^{\infty} \frac{n}{2n^2+5}$ **b.** $\displaystyle\sum_{n=2}^{\infty} \frac{1}{(n-1)(n+1)}$

14. (Optional) Obtain an infinite power series for $\text{Tan}^{-1} 2x$ and state the interval of convergence. (Refer to page 483.)

15. Prove by using mathematical induction that 13-7

$$2 + 7 + 12 + \cdots + (5n-3) = \frac{n(5n-1)}{2}.$$

494

CHAPTER FOURTEEN

Statistics

OBJECTIVES

1. To organize data into a frequency table and to display it graphically using a histogram or a frequency polygon.
2. To describe and analyze data using the mean, median, mode, variance, and standard deviation.
3. To use the standard normal distribution to study data known to have a normal distribution.
4. To use confidence intervals and sampling theory to draw conclusions from samples that are representative of the entire population.
5. To measure how closely two sets of data are related by finding their correlation coefficient. To use a regression line to predict trends.

14-1/TABLES, GRAPHS, AND AVERAGES

In this section, we will examine how to classify and organize data. Later, we will analyze the data by looking at measures of central tendency that are similar to a statistic you are already familiar with, the "average."

Shown at the left is a pine forest. A forester could measure the heights and circumferences of a relatively small number of trees and then use the methods of statistics to predict the average height and circumference of the trees in the forest.

The mathematics scores for the 170 seniors in a midwestern high school who took a standardized achievement test are as follows:

600	530	670	580	500	570	700	540	690	500
520	450	570	540	620	450	460	800	580	610
500	630	460	470	740	630	510	720	490	540
690	510	550	710	520	480	750	450	670	610
300	680	530	510	350	400	400	640	430	480
580	450	600	480	520	530	450	480	530	460
490	570	410	520	440	470	500	570	540	590
510	540	500	640	510	570	520	490	650	490
690	590	360	460	470	530	500	660	440	450
410	500	540	540	510	390	610	380	460	520
620	690	520	490	550	580	400	510	580	530
460	500	400	520	530	460	550	480	500	470
540	520	550	600	420	500	590	520	590	580
580	550	510	370	510	570	490	500	470	490
600	450	630	530	680	440	560	640	610	590
500	570	540	590	450	530	630	480	510	430
480	500	590	490	540	550	530	600	540	330

In order to make the data above easier to understand, we make a **frequency table,** as shown at the right. When the list of data is extensive, the construction of a frequency table is a job ideally suited to the computer. (See Computer Exercise 1 on page 502.) Whether you use a computer or not, you will need to do the following:

Class interval	Frequency
300–349	2
350–399	5
400–449	12
450–499	35
500–549	54
550–599	28
600–649	18
650–699	10
700–749	4
750–800	2

1. Decide how many class intervals your frequency table will have. In the standardized test example, the least score, 300, is subtracted from the highest score, 800, to obtain the *range* of scores, 500. Since the range is conveniently divisible by either 5, 10, or 20, we can choose to categorize the scores into 5, 10, or 20 classes. Arbitrarily, we choose 10 classes.
2. Determine the class width by dividing the range by the number of classes: class width = 500 ÷ 10 = 50.
The lower boundaries of the classes are determined by adding multiples of 50 to the lowest score consecutively. Thus, the lower boundaries of the classes are 300, 350, 400, 450, and so on.

A large set of data can be graphically displayed in a bar graph called a **histogram** or in the **frequency polygon** that is formed by connecting the midpoints of the tops of each bar of the histogram. A histogram and frequency polygon for the given frequency table are shown at the top of the next page.

Sometimes we use a single number, called *average,* to represent an entire set of data. In statistics we use three types of average, the *mean,* the *median,* and the *mode.* Each is a measure of the central tendency of the data.

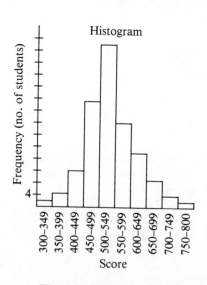

Histogram

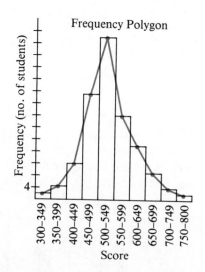

Frequency Polygon

We call the sum of a set of measurements divided by the number of measurements the **mean.** The mean is the familiar average from arithmetic. We use a bar to denote the mean, for example, $\bar{x}$ (read "x bar"). (Some books use μ (Greek mu)). We now state two formulas for the mean. If data is given in a list, $x_1, x_2, x_3, \ldots, x_n$:

$$\bar{x} = \frac{x_1 + x_2 + x_3 + \cdots + x_n}{n} = \frac{\displaystyle\sum_{i=1}^{n} x_i}{n}$$

If data is given in a frequency table,

measurement	x_1	x_2	$\cdots$	x_n
frequency	f_1	f_2	$\cdots$	f_n

$$\bar{x} = \frac{x_1 f_1 + x_2 f_2 + \cdots + x_n f_n}{f_1 + f_2 + \cdots + f_n} = \frac{\displaystyle\sum_{i=1}^{n} x_i f_i}{\displaystyle\sum_{i=1}^{n} f_i}$$

EXAMPLE 1. The 30 members of a statistics class recorded the number of hours they watched television on a particular Sunday. Use the given frequency table to calculate the mean number of hours spent watching television.

x_i = number of hours	0	1	2	3	4	5
f_i = frequency (number of students)	8	5	7	6	3	1

(*Solution follows on page 498*)

SOLUTION: $\bar{x} = \dfrac{\displaystyle\sum_{i=1}^{n} x_i f_i}{\displaystyle\sum_{i=1}^{n} f_i} = \dfrac{0 \cdot 8 + 1 \cdot 5 + 2 \cdot 7 + 3 \cdot 6 + 4 \cdot 3 + 5 \cdot 1}{8 + 5 + 7 + 6 + 3 + 1}$

$$= \frac{54}{30} = 1.8 \text{ hours}$$

If the data of Example 1 is displayed in a histogram, the mean, 1.8, is the *balancing point,* or *fulcrum,* of the histogram. Notice that the lower boundaries of the class intervals are -0.5, 0.5, 1.5, 2.5, and so on.

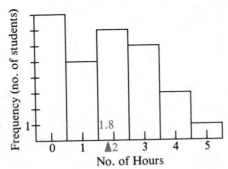

Two other averages in common use are the *median* and the *mode.* Sometimes one of these averages describes the central tendency better than the mean. The **median** of a set of measurements is found by arranging the data in increasing or decreasing order. If the number of measurements is odd, the median is the middle number in the ordered set. For example, the median of the data 1, 3, 3, 5, 6, 8, 9, is 5. If the number of measurements is even, the median is the mean of the two middle numbers. The median of the data 2, 3, 3, 5, 8, 9, is 4, which is the mean of 3 and 5.

The **mode** of a set of measurements is the one measurement that occurs most often. If there are two or more measurements that occur most often, the set of measurements is called *bimodal* (or *trimodal* or *multimodal*). If no measurement is repeated, we say that there is no mode. For example:

The set 2, 3, 3, 5, 8, 9 has mode 3, since 3 occurs twice
and the other numbers once.
The set 2, 3, 3, 5, 8, 8 has modes 3 and 8.
The set 1, 2, 3, 5, 6, 7 has no mode.

EXAMPLE 2. The frequency table below summarizes the numbers of brothers and sisters (siblings) for each of the 25 students in a statistics class. Find **(a)** the mean, **(b)** the median, and **(c)** the mode of the given data.

number of siblings	0	1	2	3	4	5
number of students	2	10	8	4	0	1

SOLUTION: **a.** mean $= \bar{x} = \dfrac{0 \cdot 2 + 1 \cdot 10 + 2 \cdot 8 + 3 \cdot 4 + 4 \cdot 0 + 5 \cdot 1}{25}$

$$= 1.72$$

b. Since the data consists of 25 numbers, the middle one is the thirteenth. Reading from the bottom row of the table, we see that twelve $(2 + 10)$ students have either 0 or 1 siblings. Thus, the thirteenth student has 2 siblings, which is the median number.

c. The mode of the given data is 1 because this number occurs most frequently (10 times).

ORAL EXERCISES 14-1

1. A set of 182 measurements ranges between 30 and 90. Give the first three class intervals if the measurements are to be divided into the following number of classes.
   a. 10   **b.** 12   **c.** 15

2. Give the mean, median, and mode of each set of data.
   a. 1, 2, 3, 5, 5   **b.** 3, 4, 4, 7, 8, 8, 10, 11

   c. 1, 2, 3

   d.

observation	0	1	2	3	4
frequency	3	2	5	8	2

3. Record the number of siblings each member of your class has. Tabulate the data and make a histogram. Then, calculate the mean, median, and mode.

4. **a.** Is it possible for the median of a set of integers not to be an integer?
   b. Is it possible for the mode of a set of integers not to be an integer?

WRITTEN EXERCISES 14-1

In Exercises 1–4, find the mean, median, and mode for each set of data.

A   1. The numbers of mice born in 10 different litters were:

$$5, 7, 6, 3, 8, 6, 4, 5, 6, 4$$

2. The test scores for a statistics student were:

$$85, 74, 92, 83, 87, 84, 78, 90$$

3. The numbers of items produced in one hour by 12 different machines were:

$$0, 1, 3, 5, 5, 5, 7, 9, 9, 11, 15, 99$$

4. The heights in centimeters of a college basketball team are:

$$185, 189, 191, 193, 193, 195, 196, 198, 198, 200$$

5. The number of trials required by 20 different puppies to learn a certain trick are as follows:

   9, 18, 12, 13, 10, 21, 7, 19, 11, 9, 4, 22, 16, 14, 28, 15, 16, 8, 13, 25

   a. Summarize this data in a histogram with 6 classes.
   b. What is the median number of trials required?

6. The minimum daily Celsius temperatures for each of the 30 days of September at the municipal airport were as follows:

   15°, 18°, 13°, 11°, 14°, 15°, 13°, 11°, 10°, 9°, 9°, 12°, 11°, 10°, 11°, 8°, 9°, 12°, 13°, 12°, 11°, 8°, 7°, 6°, 8°, 10°, 8°, 7°, 4°, 3°

   a. Summarize this data in a histogram with 5 classes.
   b. What are the median and mode of the data?

7. The table below gives the results of a driver-education experiment. The experiment measures the time between the appearance of a stimulus on a screen and the student's reaction of depressing a brake pedal.

time to the nearest tenth of a second	0.1	0.2	0.3	0.4	0.5	0.6	0.7
frequency	1	4	5	4	3	1	2

   a. Draw a frequency polygon for the data.
   b. Find the mean, median, and mode.

8. The following table gives the number of questions answered correctly by 30 students on the written portion of a driver education test. (Perfect score = 20.)

number of correct answers	12	13	14	15	16	17	18	19	20
number of students	1	0	3	3	3	5	6	5	4

   a. Draw a frequency polygon for the data.
   b. Find the mean, median, and mode.

9. A consumer organization collected last week's prices for a chef's salad at 100 restaurants. The prices were as follows:

price in cents	number of restaurants	price in cents	number of restaurants
51–100	8	301–350	10
101–150	10	351–400	6
151–200	30	401–450	2
201–250	20	451–500	1
251–300	12	500–551	1

   a. Draw a frequency polygon for the data.
   b. Calculate the mean, median, and mode using the middle price in each category as the representative price for the category.

10. Find a frequency table in a newspaper or magazine. Calculate the mean, median, and any modes.

11. Tell whether the "average" referred to is the mean, median, or mode.
a. The "average" family in Missouri has 2.1 children.
b. The "average" student comes from a family of 4 people.

B   **12.** For a group of 10 teenagers, the mean age is 17.1, the median is 16.5, and the mode is 16. If a 21-year-old joins the group, give the mean, median, and mode for the ages of the 11 persons.

13. A pollster reports that for 100 families interviewed, the mean number of children per family was 2.038, the median was 1.9, and the mode was 1.82. Explain why each of these figures must be wrong.

14. The distribution of salaries in a company are shown below.

job	5 executives	15 supervisors	80 production workers
salary range	$40,200–$87,500	$15,800–$25,000	$9,200–$18,700

Given that the median salary is $14,500 and the mean salary is $17,800, answer the following.
a. If the 15 supervisors are each given a $1000 raise, what are the new mean and median?
b. If the 80 production workers each get a $1000 raise, and no one else gets a raise, what are the new mean and median?

If $x_1, x_2, \ldots, x_n$ are assigned weights $w_1, w_2, \ldots, w_n$, then the *weighted*

average is $\dfrac{\sum\limits_{i=1}^{n} x_i w_i}{\sum\limits_{i=1}^{n} w_i}$. Use this formula in Exercises 15–18.

15. A grade in a certain course is determined by a weighted average of the scores of the tests taken. The mid-semester test counts twice as much as a regular test and the semester test counts three times as much as a regular test. What is the weighted average of the following scores?
regular tests: 85, 70, 76, 78, 86
mid-semester test: 82
semester test: 87

16. Five liters of water at 70° are mixed with 3 L of water at 80° and 2 L of water at 90°. What is the temperature of the mixture?

In Exercises 17 and 18, find the center of gravity of the system described.

17. A 2 kg mass at $A = (3, 4)$ and a 3 kg mass at $B = (-2, -6)$.
(*Hint:* Find the weighted averages for each coordinate separately.)

18. A 1 kg mass at $A = (2, -3)$, a 2 kg mass at $B = (5, 3)$ and a 3 kg mass at $C = (4, 5)$.

CALCULATOR EXERCISE

Find to the nearest whole number the mean math score in the frequency table on page 496. Since such scores are reported in multiples of 10, the first class labeled 300–349 will contain only the scores 300, 310, 320, 330, and 340. Assume that all of the scores are 320, the score in the middle of the class. Likewise assume that all scores in the next class labeled 350–399 are 370, and so on.

COMPUTER EXERCISES

1. Write a computer program that makes a frequency table for a set of data.
2. Write a computer program that will compute the mean of a set of data given in a frequency table. In the next section, you will be asked to modify this program to compute more information.

14-2 / VARIABILITY

The mean, median, and mode are **statistics** used to describe the center of a set of data. In this section we shall discuss statistics that are used to describe the dispersion of data about the mean. The need for such statistics can be seen by comparing the following two histograms. Each presents algebra test scores for a class of 25 students. The mean for both classes is between 76 and 80, but the scores for the second class are spread more widely from the mean.

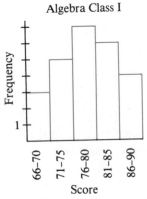

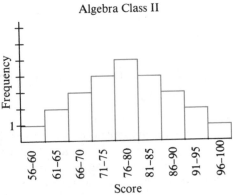

Recall from the last section that the range of a set of data is the difference between the least and greatest numbers in the set.

EXAMPLE 1. Find the range of ages for the two groups below.

Group A: 8, 13, 13, 14, 14, 14, 15, 15, 15, 20
Group B: 7, 7, 8, 9, 11, 13, 15, 15, 17, 18

SOLUTION: For Group A, the range is $20 - 8 = 12$. For Group B, the range is $18 - 7 = 11$.

Example 1 illustrates that while the range is easy to compute, by itself it can be a misleading measure of dispersion. Although Group A has a larger range, the group is really more homogeneous than Group B. The dispersion of the data away from the mean is much greater in Group B.

The *variance* and the *standard deviation* are measures that better describe the spread of data away from the mean. The **variance,** denoted s^2 or σ^2 (sigma squared), is the mean of the squares of the deviations from $\bar{x}$.

$$s^2 = \frac{(x_1 - \bar{x})^2 + (x_2 - \bar{x})^2 + \cdots + (x_n - \bar{x})^2}{n} = \frac{\sum\limits_{i=1}^{n} (x_i - \bar{x})^2}{n} \qquad (1)$$

The **standard deviation,** denoted s or σ, is the positive square root of the variance.

$$s = \sqrt{\text{variance}} = \sqrt{\frac{\sum\limits_{i=1}^{n} (x_i - \bar{x})^2}{n}} \qquad (2)$$

These formulas can be expressed in an equivalent form that is often easier to use.*

$$\text{variance: } s^2 = \frac{\sum\limits_{i=1}^{n} x_i^2 - n(\bar{x})^2}{n} \qquad (1a)$$

$$\text{standard deviation: } s = \sqrt{\frac{\sum\limits_{i=1}^{n} x_i^2 - n(\bar{x})^2}{n}} \qquad (2a)$$

EXAMPLE 2. Calculate the variance and standard deviation for the data:

3, 5, 6, 7, 9

SOLUTION: First, find the mean $\bar{x} = \dfrac{3 + 5 + 6 + 7 + 9}{5} = 6$. Since $\bar{x}$ is an integer, formulas (1) and (1a) are equally easy to use. We shall illustrate both.
formula (1):

x_i	$x_i - \bar{x}$	$(x_i - \bar{x})^2$
3	-3	9
5	-1	1
6	0	0
7	1	1
9	3	9

20 = sum

$$\text{variance} = s^2 = \frac{\sum\limits_{i=1}^{n} (x_i - \bar{x})^2}{n}$$

$$= \frac{20}{5} = 4$$

$$\text{standard deviation} = s = \sqrt{4} = 2$$

(*Solution continued on page 504*)

*These formulas are often written with a denominator of $n - 1$ instead of n. The reasons for this are best left to a full course in statistics. If you have a calculator that gives means and standard deviations, check your instruction booklet to see which is used.

formula (1a):

x_i	x_i^2
3	9
5	25
6	36
7	49
9	81

$200 = \text{sum}$

$$\text{variance} = s^2 = \frac{\sum_{i=1}^{n} x_i^2 - n(\bar{x})^2}{n}$$

$$= \frac{200 - 5 \cdot 6^2}{5} = 4$$

$$\text{standard deviation} = s = \sqrt{4} = 2$$

If the data appear in a frequency table instead of a list, then formula (1a) can be modified. When the n pieces of data are listed as r distinct points $x_1, \ldots, x_r$, and each x_i occurs with frequency f_i, then the formula becomes:

$$\text{variance: } s^2 = \frac{\sum_{i=1}^{r} x_i^2 \cdot f_i - n(\bar{x})^2}{\sum_{i=1}^{r} f_i} \qquad (1b)$$

We can use the mean and standard deviation of a set of data to determine a *standard score* for each data entry. The standard score shows how many standard deviations the data entry is from the mean. The **standard score** z corresponding to a data entry x from a set of data with mean $\bar{x}$ and standard deviation s is given by the following formula:

$$\text{standard score} = \frac{\text{data entry} - \text{mean}}{\text{standard deviation}}$$

$$z = \frac{x - \bar{x}}{s}$$

EXAMPLE 3. Give the standard score of each of the measurements in Example 2.

SOLUTION: In Example 2, $\bar{x} = 6$ and $s = 2$. Thus, the standard scores are as follows:

$$3 \rightarrow \frac{3 - 6}{2} = -1.5$$

$$5 \rightarrow \frac{5 - 6}{2} = -0.5$$

$$6 \rightarrow \frac{6 - 6}{2} = 0$$

$$7 \rightarrow \frac{7 - 6}{2} = 0.5$$

$$9 \rightarrow \frac{9 - 6}{2} = 1.5$$

1. **a.** What does the following information tell you about two different basketball teams?

   Team A: mean height = 192 cm; standard deviation = 4 cm
   Team B: mean height = 192 cm; standard deviation = 8 cm

   b. Which team do you think has a better chance of getting rebounds?

2. Complete this table in order to find the standard deviation of 1, 2, 5, 7, 10:

x_i	x_i^2
1	?
2	?
5	?
7	?
10	?
?	? = sum

   mean = $\bar{x}$ = ?

   $$\text{variance} = s^2 = \frac{\sum_{i=1}^{n} x_i^2 - n(\bar{x})^2}{n} = ?$$

   standard deviation = $s = \sqrt{?} \approx ?$

3. The mean and standard deviation of a set of data are $\bar{x} = 10$ and $s = 4$. Give the standard scores of the following numbers in the set of data.

   a. 14   **b.** 18   **c.** 20   **d.** 8   **e.** 10

4. Is it possible to have a variance of zero? [*Hint:* Look at formula (1).]

In Exercises 1 and 2, calculate the mean and standard deviation for each set of data.

A

1. The number of grammatical errors on 5 daily French quizzes: 10, 8, 7, 5, 5

2. The number of fish caught on each day of vacation: 3, 8, 7, 0, 4, 7, 13

3. **a.** Each number in row B is 10 more than the corresponding number in row A. Compare the mean and standard deviation of each row of numbers.

row A	1	2	3	4	5
row B	11	12	13	14	15

   b. Estimate the mean and standard deviation of the numbers: 21, 22, 23, 24, 25.

4. **a.** Each number in row B is four times the corresponding daily number in row A. Compare the mean and standard deviation of each row of numbers.

row A	5	5	7	7	8	10
row B	20	20	28	28	32	40

   b. Estimate the mean and standard deviation of the numbers: 50, 50, 70, 70, 80, 100.

5. A set of test scores has a mean of 78 and a standard deviation of 6. Give the standard score of each of the following test scores:
  a. 90   **b.** 75   **c.** 78   **d.** 70

6. For all of the apartments in a large building, the number of kilowatt hours of electricity used last month had a mean of 760 and a standard deviation of 80. Give the standard score for each of the following usages:
  a. 720   **b.** 860   **c.** 600   **d.** 776

7. The math achievement scores of 45 physics students are compared with the math achievement scores of 45 students selected at random. Which group of students do you think will have the greater mean score? The greater standard deviation in its scores?

8. The typing rates (in words per minute) of 25 professional typists are compared with the typing rates of 25 randomly selected people who own typewriters. Which group do you think will have the greater standard deviation of typing rates?

9. Four different people think of an integer from 1 to 9.
  a. What is the smallest possible standard deviation of their four numbers?
  b. What is the largest possible standard deviation?

10. Make two lists of five integers from 1 to 9 with repetitions allowed so that one list has **(a)** the largest possible mean and the smallest possible standard deviation and the other list has **(b)** a single mode equal to 1 and the largest possible standard deviation.

B   **11.** Thirty samples of milk purchased at various stores were tested for bacteria. The bacteria count per milliliter in these 30 samples is given below. Find the mean and standard deviation of this data. (Use formula (1b).)

x_i = bacteria count	3	4	5	6	7	8
f_i = frequency	1	3	6	9	7	4

12. Twenty puppies were taught to sit and stay on command. Find the mean and standard deviation of the number of trials required before they learned to do this.

Number of trials	7	8	9	10	11	12
Number of puppies	1	2	5	4	4	4

13. A survey was conducted to determine the number of radios owned in 25 households. Find the mean and standard deviation of the following data collected.

Number of radios	0	1	2	3	5	6	7
Number of households	1	3	4	8	6	2	1

14. Prove that if you add a constant C to each piece of data, its standard deviation is unchanged.

15. Derive formula (1a) from formula (1).

CALCULATOR EXERCISE

Some calculators have special routines for calculating the mean, variance, and standard deviation. Usually, there is a key that will give you the variance or standard deviation using a formula with denominator n. (See the footnote on page 503.) Use the appropriate keys and methods for entering data for your calculator to find the mean and standard deviation of the data in Exercise 11.

COMPUTER EXERCISE

Modify your program for the computer exercise of Section 14-1 so that it will compute the mean and standard deviation of data given in a frequency table. Run your program for the data in Exercise 12.

14-3/THE NORMAL DISTRIBUTION

As we have seen in previous sections, histograms have various shapes according to the distribution of data. The distribution of the last digits of 50 different telephone numbers shown in figure (a) is an example of a *uniform distribution*. Each digit occurs about 10% of the time.

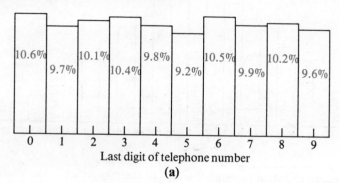

Last digit of telephone number

(a)

Figure (b) shows the distribution of precipitation in a midwestern city. This distribution is an example of a *skewed distribution* with a long tail to the right.

In this section, we will consider a third type of distribution, called the *normal distribution*. The normal distribution is extremely important in that it has applications to heights and weights of people, dimensions of manufactured goods, test scores, blood cholesterol levels, and times for marathon races, to name just a few.

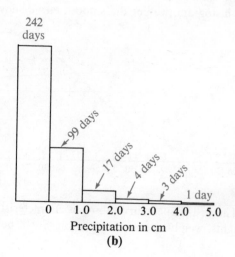

Precipitation in cm

(b)

Consider the following experiment: A psychologist asks 100 people to solve a certain puzzle and records their solution times. The results are recorded in the histogram shown as figure (c). The number inside each rectangle shows the percentage of people that completed the puzzle in the indicated time interval.

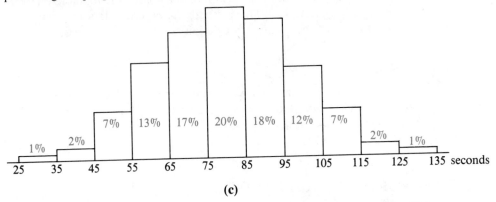

(c)

Each of the class intervals in the preceding histogram is 10 units wide. If the psychologist had observed a much larger population, the data would be more specific, and it would make sense to use smaller class intervals, as shown in figure (d) following.

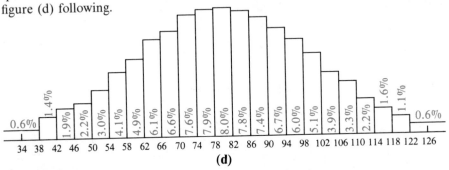

(d)

If a smooth curve is drawn through the tops of the rectangles in the preceding histogram, we get the smooth curve shown as figure (e).

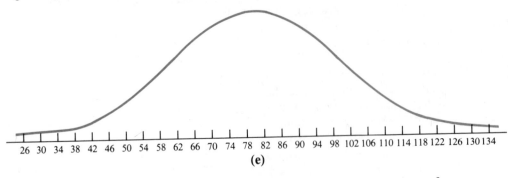

(e)

The bell-shaped curve above is characteristic of many, many distributions of heights, weights, scores, and so on. Such, bell-shaped distributions are called

normal distributions. The maximum point of the curve of a normal distribution is at the *mean*. Moreover, all normal distributions have these important properties:

About 68% of the distribution is within 1 standard deviation of the mean.
About 95% of the distribution is within 2 standard deviations of the mean.
About 99% of the distribution is within 3 standard deviations of the mean.

EXAMPLE 1. The results of the puzzle-solving experiment described on the previous page are normally distributed with mean $\bar{x} = 80$ seconds and a standard deviation $s = 20$ seconds. Make a sketch showing the percentages of times within 1, 2, and 3 standard deviations of the mean.

SOLUTION:

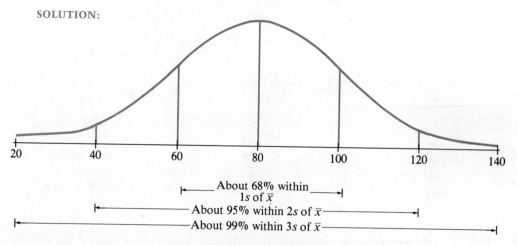

EXAMPLE 2. Over the last 10 years, the mean weight $\bar{x}$ of newborn babies in a large metropolitan hospital has been 3.4 kg and the standard deviation s has been 0.4 kg. Show this information on a normal curve. Also, show the standard scores of the weights.

SOLUTION:

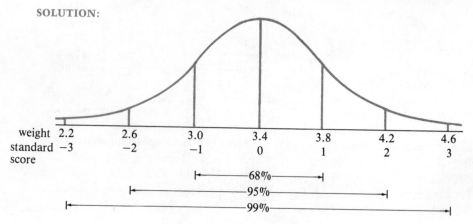

The **standard normal distribution** is the normal distribution with mean 0 and standard deviation 1. The standard normal distribution is particularly important because any normal distribution can be related to it through the use of standard scores. The graph of the *standard normal curve* is shown below.

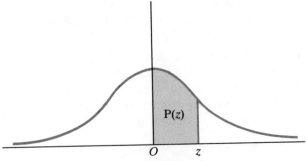

Shaded Area = P(z) = Percentage of
distribution between 0 and z

The equation of the standard normal curve is

$$y = \frac{1}{\sqrt{2\pi}} e^{\frac{-x^2}{2}}$$

The standard normal curve has the important property that the total area under the graph of the curve is 1. Also, the area under the curve between 0 and z is the percentage of the population having standard scores between 0 and z. These percentages are given in Table 14–1.

z	P(z)	z	P(z)	z	P(z)
0.1	4.0%	1.1	36.4%	2.1	48.2%
0.2	7.9%	1.2	38.5%	2.2	48.6%
0.3	11.8%	1.3	40.3%	2.3	48.9%
0.4	15.5%	1.4	41.9%	2.4	49.2%
0.5	19.1%	1.5	43.3%	2.5	49.4%
0.6	22.6%	1.6	44.5%	2.6	49.5%
0.7	25.8%	1.7	45.5%	2.7	49.6%
0.8	28.8%	1.8	46.4%	2.8	49.7%
0.9	31.6%	1.9	47.1%	2.9	49.8%
1.0	34.1%	2.0	47.7%	3.0	49.9%

Table 14–1

Notice that when $z = 1$, the table lists $P(z) = 34.1\%$. Because the distribution is symmetric about the mean 0, doubling this gives approximately 68% of the distribution between $z = -1$ and $z = 1$. Likewise, the table lists $P(2) = 47.7\%$, and doubling this gives approximately 95% of the distribution between $z = -2$ and $z = 2$. Example 3 shows how the table can be used to find other percentages.

EXAMPLE 3. Use the data of Example 2 and predict what percentage of newborn babies have weights greater than 4.0 kg.

SOLUTION:

Step 1. Convert 4.0 to a standard score:

$$z = \frac{4.0 - \bar{x}}{s} = \frac{4.0 - 3.4}{0.4} = 1.5$$

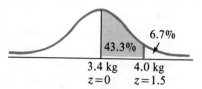

Step 2. Using Table 14–1, we find that the value of $P(1.5)$ is 43.3%. Thus, 43.3% of the newborns have weights corresponding to standard scores between $z = 0$ and $z = 1.5$. Since 50% have weights corresponding to standard scores greater than 0, we know that $50 - 43.3 = 6.7$ percent have weights corresponding to standard scores greater than 1.5. Thus, about 6.7% of the newborns have weights greater than 4 kg.

EXAMPLE 4. Use the data of Example 2 to predict what percentage of newborn babies have weights between 3.6 kg and 4.4 kg.

SOLUTION:

Step 1. Convert 3.6 and 4.4 to standard scores:

$$z = \frac{3.6 - \bar{x}}{s} = \frac{3.6 - 3.4}{0.4} = 0.5$$

$$z = \frac{4.4 - \bar{x}}{s} = \frac{4.4 - 3.4}{0.4} = 2.5$$

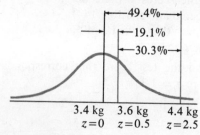

Step 2. Using Table 14–1, we find that $P(0.5) = 19.1\%$ and $P(2.5) = 49.4\%$. Thus, 19.1% of the newborns have weights corresponding to standard scores between $z = 0$ and $z = 0.5$; and 49.4% of the newborns have weights corresponding to standard scores between $z = 0$ and $z = 2.5$. Thus, the percentage of the newborns with weights corresponding to standard scores between $z = 0.5$ and $z = 2.5$, is $49.4 - 19.1 = 30.3$ percent. Thus, 30.3% of the newborns have weights between 3.6 kg and 4.4 kg.

Percentiles

Sometimes a set of data is arranged in ascending order and divided into 100 equal parts. The 99 points that divide the data are called percentiles. Percentiles are often used in reporting scores on a standardized test. If you score in the 70th percentile, for instance, you know that 70 percent of all people taking the test had a score less than or equal to yours. Similarly the 25th percentile separates the bottom 25% of the test scores. The 25th percentile, 50th percentile (the median), and 75th percentile are also called the first quartile, the second quartile, and the third quartile. As the diagram below shows, the quartiles divide the population into quarters.

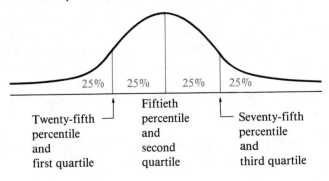

25% | 25% | 25% | 25%

Twenty-fifth percentile and first quartile

Fiftieth percentile and second quartile

Seventy-fifth percentile and third quartile

EXAMPLE 5. A standardized test has a mean score of 50 and a standard deviation of 10. Find **(a)** the ninetieth percentile **(b)** the first quartile.

SOLUTION:

a. 1. Let z represent the ninetieth percentile. Then as the diagram suggests 40% of the scores will be between 0 and z. Table 14–1 shows that $P(z) = 40\%$ when $z \approx 1.3$.

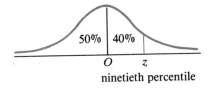

50% | 40%

O z

ninetieth percentile

2. Find the test score that corresponds to a standard score $z = 1.3$.

$$z = \frac{x - \overline{x}}{s}$$

$$1.3 = \frac{x - 50}{10} : x = 63$$

b. 1. Table 14–1 shows that $P(z) = 25\%$ when $z \approx 0.7$. This corresponds to the third quartile, and by symmetry $z \approx -0.7$ corresponds to the first quartile.

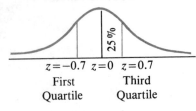

25%

$z = -0.7$ $z = 0$ $z = 0.7$

First Quartile

Third Quartile

2. Convert the standard score $z = -0.7$ to a test score.

$$z = \frac{x - \bar{x}}{s}$$

$$-0.7 = \frac{x - 50}{10} : x = 43$$

ORAL EXERCISES 14-3

Use Table 14-1 to find the percentage of the normal distribution satisfying the given inequality.

1. $0 < z < 0.5$ **2.** $0 < z < 1.8$ **3.** $-1.2 < z < 0$

4. $-1.2 < z < 1.2$ **5.** $z > 2$ **6.** $z < 1$

7. $z < 1.4$ **8.** $z > -1.4$ **9.** $z < 0.7$

10. In a certain large school district, the set of all standardized math scores is normally distributed with mean $\bar{x} = 500$ and standard deviation $s = 100$.
   a. Find approximately the percentage of scores that are:
      (1) between 400 and 600   (2) between 300 and 700
      (3) greater than 700   (4) greater than 650
   b. Find the score that is approximately the 90th percentile.
   c. Find the score that is approximately the third quartile.

11. a. The reaction times of all people in a psychology experiment were normally distributed with a mean $\bar{x}$ of 2 seconds and a standard deviation s of 0.5 seconds. What percentage of the people had reaction times
      (1) between 2 and 3 seconds?   (2) between 1.5 and 2.5 seconds?
      (3) more than 2.5 seconds?   (4) less than 0.5 seconds?
   b. Find the reaction time at the 20th percentile.

WRITTEN EXERCISES 14-3

When necessary to use Table 14-1, round data to the nearest tenth.

A   **1.** On a standardized aptitude test, scores are normally distributed with mean $\bar{x} = 100$ and standard deviation $s = 10$. What percentage of scores are:
   a. between 90 and 110?   **b.** between 80 and 120?
   c. between 70 and 130?   **d.** greater than 115?
   e. less than 92?   **f.** between 105 and 125?

2. The standardized verbal scores of students entering a large university are normally distributed with mean $\bar{x} = 600$ and standard deviation $s = 80$. What percentage of these scores are:
   a. between 600 and 700?   **b.** between 760 and 800?
   c. greater than 650?   **d.** less than 500?

3. In a very large factory, employees must report to work at 8 A.M. The arrival times are normally distributed with mean 7:52 A.M. and standard deviation 4 minutes.
   a. On a typical day, what percent of the employees will be late?
   b. What percent of the employees arrive between 7:59 and 8:00 A.M.?
   c. What percent of the employees arrive before 7:45 A.M.?

4. A light bulb manufacturing company claims that a 100 watt bulb has a mean life of 750 hours with a standard deviation of 70 hours. What percentage of their light bulbs will have a life
  a. less than 610 hours? **b.** more than 940 hours?
  c. between 700 and 800 hours? **d.** between 730 and 770 hours?

5. A manufacturer makes ball bearings with diameters between 17.0 mm and 18.0 mm. Quality control procedures indicate that the bearings produced have a mean diameter of 17.6 mm, and a standard deviation of 0.3 mm. What percentage of these ball bearings fail to meet the specifications?

6. On a tree farm, the heights of the trees are normally distributed with a mean of 200 cm and a standard deviation of 20 cm. What percentage of these trees will be taller than **(a)** 150 cm? **(b)** 224 cm?

7. The daily number of letters handled by a certain post office is normally distributed with mean 20,000 and standard deviation 100. On what percentage of the days will the post office handle **(a)** more than 20,200 letters? **(b)** fewer than 19,900 letters?

8. An airline has room for 262 passengers on its New York to Los Angeles flight. Because some people who book the flight do not show up, the airline books 290 people on the flight. If the mean number of "no-shows" is 45 people with a standard deviation of 10, what percentage of these flights will be unable to seat all the passengers who do show up?

9. Suppose that in a large school district, the standardized math test scores have a mean $\bar{x}$ of 500 and a standard deviation s of 100. What score is **(a)** the 80th percentile and **(b)** the first quartile?

10. At a certain university, the combined math plus verbal standardized test scores of incoming freshmen had a mean of 1250 and a standard deviation of 70. Find approximately **(a)** the 90th percentile and **(b)** the third quartile.

B **11.** The professor teaching a large freshman course decides to give A grades to the top 10% of the final exam scores and B grades to the next 30%. If the mean score was 72 and the standard deviation was 9, what were the minimum scores needed for an A and a B?

12. Refer to Exercise 11 and suppose the professor decided that 5% will fail and 10% will get D grades. What is **(a)** the minimum passing grade and **(b)** the minimum grade needed to get a C?

13. A 2-liter beverage bottle is filled from a machine that fills the bottle with a mean average of 2.01 liters and a standard deviation of 0.002 liters.
  a. What percentage of the bottles will actually contain more than 2 liters of beverage?
  b. Ten percent of the bottles will contain less than _?_ liters.
  c. Eighty percent of the bottles will contain more than _?_ liters.

14. The weights of all students in a certain grade in a school district are normally distributed with $\bar{x} = 56$ kg and $s = 5$ kg.
  a. The "middle half" have weights between _?_ kg and _?_ kg.
  b. The lightest 10% have weights less than _?_ kg.
  c. The heaviest 5% have weights greater than _?_ kg.

15. Reaction times for a particular task are normally distributed with mean 0.92 seconds and standard deviation 0.24 seconds. A reaction time in excess of 1.2 seconds is considered too slow. What percent of the people tested are too slow to perform the task?

C 16. The state marathon race last year attracted such a large number of runners, that entrants this year must first qualify by running in a regional marathon. The times of the 3750 regional runners are normally distributed with mean 198 minutes 36 seconds and standard deviation 23 minutes 14 seconds. If there are to be only 600 runners in the state marathon, what is the slowest time that will qualify a regional runner for the state race?

CALCULATOR EXERCISES

1. Show that the maximum y value on the standard normal curve $y = \dfrac{1}{\sqrt{2\pi}} e^{-\frac{x^2}{2}}$ is approximately 0.4.

   $$y = \frac{1}{\sqrt{2\pi}} e^{-\frac{x^2}{2}}$$

2. If z is small, the area under the curve from $x = 0$ to $x = z$ is approximately the same as the area of the rectangle shown in the diagram. Find the rectangular areas for $z = 0.1, 0.2, 0.3,$ and 0.4 and compare your answers with $P(0.1)$, $P(0.2)$, $P(0.3)$, and $P(0.4)$ given in Table 14–1. (Remember that $P(z)$ is a percentage.)

COMPUTER EXERCISE

Write a computer program to find the area under the standard normal curve $y = \dfrac{1}{\sqrt{2\pi}} e^{-\frac{x^2}{2}}$ from $x = 0$ to $x = z$. Run your program for $z = 1$, $z = 2$, and $z = 3$, and compare your answers with $P(1)$, $P(2)$, and $P(3)$ in Table 14–1.

14-4 / CONFIDENCE INTERVALS

In statistics, the word *population* does not necessarily refer to people. It may, for example, refer to a set of stereos or a set of suitcases carried aboard an airplane. A **population** is the entire set of individuals or objects in which we are interested, whether they be stereos whose prices are recorded or suitcases that are weighed, and so on. A subset of the population is called a **sample.**

When you know all the relevant information about a population, you can use probability theory to make deductions about a sample. For example, knowing the contents of a standard 52-card deck, you can make probability predictions about a 5-card sample that you are about to deal. On the other hand, there are many situations in which you know all about a sample, but not about the population from which it comes. In this case, you can use statistics to make inferences about the population based on the sample.

In everyday situations, samples often take the form of a survey or poll. For example, suppose that in a recent state poll of 100 randomly selected voters, 56 indicated that they were going to vote in favor of Proposition 1. Now, 100 voters are not very many compared with an entire state's voting population. If 560 out of 1000 had favored Proposition 1 instead of just 56 out of 100, we could be more confident of the poll's result. Suppose also that a different poll showed that as many as 60 out of 100 voters favor Proposition 1, and another showed that only 49 out of 100 favor it. The question is, how close is the sample proportion favoring Proposition 1 to the population proportion favoring it? At this point, we need to introduce some notation:

Let p = proportion of an entire population having a given characteristic.
Let $\bar{p}$ = proportion of those sampled having a given characteristic.

If 60 out of 100 voters sampled favor Proposition 1, for example, then $\bar{p} = \dfrac{60}{100} = 0.60$. If 573 out of 1000 voters sampled favor Proposition 1, then $\bar{p} = \dfrac{573}{1000} = 0.573$.

You can see that $\bar{p}$ will vary from sample to sample and will depend partly on the sample size n. Moreover, an advanced theorem in statistics states three things about the values of $\bar{p}$:

(1) The values of $\bar{p}$ are normally distributed.
(2) The mean of this distribution is p.
(3) The standard deviation of this distribution is approximately $\sqrt{\dfrac{\bar{p}(1-\bar{p})}{n}}$ where n represents the sample size.

Since $\bar{p}$ is normally distributed, we can apply the 68%-95%-99% normal rule given on page 509. For instance, 95% of the time, $\bar{p}$ is within 2 standard deviations of its mean, p. Since we use $\bar{p}$ to estimate p the following equivalent statement is more useful: 95% of the time, p is within 2 standard deviations of $\bar{p}$. This interval in which p lies 95% of the time is called a *95% confidence interval*.

DEFINITION. **A 95% confidence interval** for p is the interval:

$$\bar{p} - 2 \times (\text{standard deviation}) < p < \bar{p} + 2 \times (\text{standard deviation})$$

$$\bar{p} - 2\sqrt{\frac{\bar{p}(1-\bar{p})}{n}} < p < \bar{p} + 2\sqrt{\frac{\bar{p}(1-\bar{p})}{n}}$$

EXAMPLE 1. Suppose that 56 out of 100 voters polled favor Proposition 1. What is the 95% confidence interval for p, the unknown proportion of the population favoring Proposition 1?

SOLUTION: Step 1. $\bar{p} = \dfrac{56}{100} = 0.56$

Step 2. The standard deviation of $\bar{p}$ is

$$\sqrt{\frac{\overline{p}(1 - \overline{p})}{n}} = \sqrt{\frac{(0.56)(0.44)}{100}} \approx 0.05.$$

Step 3. The 95% confidence interval for p is

$$0.56 - 2(0.05) < p < 0.56 + 2(0.05)$$
$$0.46 < p < 0.66$$

In Example 1, other samples of 100 voters will yield other 95% confidence intervals. Most of these confidence intervals (about 95% of them) will contain p, but a few of them (about 5%) will not. For this reason, we can be 95% confident that the unknown value of p is between 0.46 and 0.66. The 95% confidence region from 0.46 to 0.66, however, is rather wide and does not pinpoint p to any great extent. Another difficulty is that we want to know whether or not $p > 0.5$ since this is an election. Example 2 shows that we can obtain a narrower confidence interval by taking a larger sample.

EXAMPLE 2. Suppose 560 out of 1000 voters polled favor Proposition 1. Find a 95% confidence interval for p.

SOLUTION: Step 1. $\overline{p} = \dfrac{560}{1000} = 0.56$

Step 2. The standard deviation of $\overline{p}$ is

$$\sqrt{\frac{\overline{p}(1 - \overline{p})}{n}} = \sqrt{\frac{(0.56)(0.44)}{1000}} \approx 0.016$$

Step 3. The 95% confidence interval for p is

$$0.56 - 2(0.016) < p < 0.56 + 2(0.016)$$
$$0.528 < p < 0.592$$

The results of Examples 1 and 2 are summarized below. Notice that the sample of size 1000 gives a much narrower confidence interval than the sample of size 100. However, the larger sample is more costly and time consuming.

Sample size	95% confidence interval
100	$0.46 < p < 0.66$
1000	$0.528 < p < 0.592$

Sometimes, a 95% confidence interval is not enough. In testing new medical drugs or procedures, a *99% confidence interval* may be required before the new drug or procedure is approved for general use. A 99% confidence interval is based on 3 standard deviations from the normal distribution of $\overline{p}$.

DEFINITION. **A 99% confidence interval** for p is the interval:

$$\overline{p} - 3 \times (\text{standard deviation}) < p < \overline{p} + 3 \times (\text{standard deviation})$$

$$\overline{p} - 3\sqrt{\frac{\overline{p}(1 - \overline{p})}{n}} < p < \overline{p} + 3\sqrt{\frac{\overline{p}(1 - \overline{p})}{n}}$$

EXAMPLE 3. A new, medically approved wheelchair was judged to be an improvement over a standard wheelchair by 320 out of 400 patients in a survey.

   a. Find a 99% confidence interval for p, the proportion of the whole population already using wheelchairs for whom the new wheelchair would be an improvement.

   b. If the standard wheelchair has been acceptable to 72% of the population already using wheelchairs, should the new wheelchair be manufactured exclusively?

SOLUTION: **a. Step 1.** $\bar{p} = \dfrac{320}{400} = 0.80$

   Step 2. The standard deviation of $\bar{p}$ is

   $$\sqrt{\frac{\bar{p}(1-\bar{p})}{n}} = \sqrt{\frac{(0.80)(0.20)}{400}} = 0.02$$

   Step 3. The 99% confidence interval for p is

   $$0.80 - 3(0.02) < p < 0.80 + 3(0.02)$$
   $$0.74 \quad < p < \quad 0.86$$

   b. Since 0.72 lies outside the 99% confidence interval, the new wheelchair is probably a little better than the old. However, costs, availability, and other medically related questions about both wheelchairs must be considered in deciding whether to replace the old with the new.

ORAL EXERCISES 14-4

1. **a.** If $\bar{p} = 0.4$ and $n = 24$, find the standard deviation of $\bar{p}$. Also, find a 95% confidence interval for p.
   b. Find a 99% confidence interval for p in part (a).

2. **a.** If $\bar{p} = 0.5$ and $n = 25$, find the standard deviation of $\bar{p}$. Also, find a 95% confidence interval for p.
   b. Find a 99% confidence interval for p in part (a).

3. **a.** If 50 out of 100 people surveyed favor brand X, find $\bar{p}$ and the standard deviation of $\bar{p}$.
   b. Find a 95% confidence interval for p in part (a), the proportion of the population favoring brand X.

4. Does the formula for the standard deviation of $\bar{p}$ depend on the sample size or the population size?

5. Consider these two polls:
   In Poll A, 23 out of 100 randomly chosen students in Central High School favor having an additional gym class in place of a study period. In Poll B, 23 out of 100 randomly chosen high school students from around the state favor having an additional gym class in place of a study period. Is there any difference in the 95% confidence intervals of these two polls?

In Exercises 1–6, find $\bar{p}$ and the standard deviation of $\bar{p}$. Also, give a 95% confidence interval for p, the proportion of the entire population having the indicated characteristic.

A

1. Seven out of 10 dentists surveyed recommend Plain-Old toothpaste.

2. Six out of 10 doctors polled recommend eating apples.

3. Thirty-six out of 100 voters surveyed favor Proposition 4.

4. Sixteen out of 25 students questioned prefer summer to winter.

5. Twenty-five out of 400 flashbulbs tested were defective.

6. Eighty out of 100 students have after-school jobs.

Find a 99% confidence interval for p in Exercises 7–12.

7. Of 1000 adults randomly selected for reading tests, 30 were illiterate.

8. Of 1600 eligible voters who were polled, 1280 favored the proposal.

9. Out of 100 workers surveyed, 9 have been promoted.

10. In a television poll, 10 out of 100 TV sets were tuned to the concert.

11. Of 500 students interviewed, 200 prefer tennis to racketball.

12. In a survey, 150 out of 200 people preferred brand A to other brands.

13. If you wish to halve the standard deviation of $\bar{p}$, you must multiply the sample size by   ? .

14. If you wish to divide the standard deviation of $\bar{p}$ by 10, you must multiply the sample size by   ? .

B

15. A poll by a national magazine reported in an article that 36% of the 900 voters surveyed approved the immigration quotas. The article also reported a "margin of error of plus or minus 3 percentage points." Is this 3% margin of error associated with a 95% or a 99% confidence interval?

16. Suppose, in Exercise 15, the article had reported a "margin of error of plus or minus 2 percentage points." Is this 2% margin of error associated with a 68% or a 95% confidence interval?

17. In a poll of 240 college freshmen, 144 exercised regularly.
    a. Find a 95% confidence interval for the population p of freshmen who exercised regularly.
    b. Find a 99% confidence interval for p.

18. An audit of 840 randomly selected tax returns shows that 30% require payment of additional taxes. Find a 95% and a 99% confidence interval for the proportion of tax returns requiring additional payments.

19. a. Make a graph of the function $f(\bar{p}) = \bar{p}(1 - \bar{p})$.
    b. What is the maximum value of this function?
    c. Explain why the standard deviation of $\bar{p}$ is always less than or equal to
$$\sqrt{\frac{1}{4n}}.$$

20. Use Exercise 19(c) to prove that the widest possible 95% confidence interval is

$$\bar{p} - \sqrt{\frac{1}{n}} < p < \bar{p} + \sqrt{\frac{1}{n}}.$$

EXAMPLE. A researcher wants to ensure that 95% of the time $\bar{p}$ is 0.1 of p. How large a sample is needed?

SOLUTION: In a 95% confidence interval, the maximum possible difference between p and $\bar{p}$ is $\sqrt{\frac{1}{n}}$. (See Exercise 20.)

$$\sqrt{\frac{1}{n}} \leq 0.1$$

$$\frac{1}{n} \leq 0.01$$

$$n \geq 100$$

Thus a sample size of at least 100 is needed.

21. A pollster wants a confidence interval in which $\bar{p}$ is within 0.01 of p. How large a sample is needed for a 95% confidence interval?

22. A pollster seeking to find the population of voters who favor an increased budget wants a 95% confidence interval of the form

$$\bar{p} - 0.03 < p < \bar{p} + 0.03.$$

How large a sample is needed?

23. A quality control inspector wishes to estimate the percentage p of computer chips that are defective. Suppose that $|p - \bar{p}| < 0.05$ is a requirement. How large a sample is needed for a 95% confidence interval?

C   **24.** Find an 80% confidence interval for the proportion of voters favoring Article VII, if only 18 out of 50 favor it.

14-5/SAMPLING

In Section 14-4 we saw that in order to deduce facts about a large population, we can take a poll or survey to sample a small part of the population. The process of selecting a sample that is representative of the total population is called **sampling. Sampling theory** is the branch of statistics that deals with the questions and decisions that arise when a sample is taken. How should the sample be selected? How large should the sample size be? How reliable are the conclusions drawn from the sample? The answers are often dependent upon the circumstances under which the survey is conducted, the resources available to conduct the survey, and the results desired from the survey.

We can distinguish between *probability sampling* and *nonprobability sampling*. **Nonprobability** sampling involves methods in which selection is not done by chance procedures. Examples of nonprobability sampling are *convenience sampling, judgment sampling,* and *sampling by questionnaire.* Because the selection process is subjective, there is a possibility that a sample will be biased; that is, the sample and the total population will differ in important characteristics. Consider, for example, a quality control inspector who inspects the top few apples in several crates to determine the percentage of the fruit that is bruised. This is called **convenience sampling** because it is more convenient to sample the top apples than the bottom ones. The resulting sample is not representative of all the apples because apples near the bottom are more likely to be bruised.

In **judgment sampling** an expert or group of experts select a representative sample according to their subjective judgment. It is most likely that no two experts will agree on what is truly "representative."

Questionnaires also may lead to problems since conclusions are based on voluntary responses. This problem occurs with other voluntary responses, such as phoned responses to broadcast questions. Usually only those who feel strongly about an issue will respond. Another problem is that the language of a questionnaire may be biased. Consider whether you agree with statements (a) and (b) below.

> (a) It is best to buy the least expensive brand.
> (b) It is best to buy the cheapest brand.

Some people might agree with (a) but disagree with (b) because "cheapest" also may mean "poorest quality."

A famous example of a biased poll occurred in 1936. A popular magazine telephoned some of its subscribers asking whether they favored Roosevelt or Landon for president. The poll showed that Landon would be victorious by a landslide, only to have Roosevelt win the actual election. The problem, of course, was that the magazine's subscribers and telephone owners were not representative of the population. The sample was biased toward voters who preferred Landon. Can you see why this is an example of convenience sampling?

In **probability sampling** the selection of a sample is objective. Each sample unit is selected with a known probability through some random process. The most basic type of probability sampling, called simple random sampling, is a procedure in which every possible sample of n units from the whole population has an equal chance of being chosen. This can be done by any chance method, such as flipping a coin or drawing lots, but random numbers generated by a computer or taken from a table are most frequently used.

The table on the following page shows 1000 random digits. It can be used to select a random sample from a set of data.

EXAMPLE 1. Choose a simple random sample of 6 test scores from the 170 scores given on page 496.

SOLUTION: First we number the data from 0 to 169, counting from left to right beginning with the first row. (The first row is numbered 0 to 9, the second row is numbered 10 to 19, and so on.) Next we choose a row at random from the table of

(*Solution continued on page 522*)

random digits, say row 4, and mark off the digits in periods of three. We then subtract 200, 400, 600, or 800 from any numbers greater than 199 to produce numbers from 0 to 199 (inclusive). We eliminate duplications, and in this case, we eliminate any remaining numbers greater than 169 because the scores are numbered from 0 to 169.

Digits from table: 052, 198, 161, 910, 651, 670, 799, 251, 159
After subtraction: 52, 198, 161, 110, 51, 70, 199, 51, 159

After eliminating duplications and numbers greater than 169, the list is 52, 161, 110, 51, 70, and 159. The scores numbered 51, 52, 70, 110, 159, and 161 in the list on page 496 are 450, 600, 510, 460, 430, and 500.

Table 14-2   One Thousand Random Digits*

00	54463	22662	65905	70639	79365	67382	29085	69831	47058	08186
01	15389	85205	18850	39226	42249	90669	96325	23248	60933	26927
02	85941	40756	82414	02015	13858	78030	16269	65978	01385	15345
03	61149	69440	11286	88218	58925	03638	52862	62733	33451	77455
04	05219	81619	10651	67079	92511	59888	84502	72095	83463	75577
05	41417	98326	87719	92294	46614	50948	64886	20002	97365	30976
06	28357	94070	20652	35774	16249	75019	21145	05217	47286	76305
07	17783	00015	10806	83091	91530	36466	39981	62481	49177	75779
08	40950	84820	29881	85966	62800	70326	84740	62660	77379	90279
09	82995	64157	66164	41180	10089	41757	78258	96488	88629	37231
10	96754	17676	55659	44105	47361	34833	86679	23930	53249	27083
11	34357	88040	53364	71726	45690	66334	60332	22554	90600	71113
12	06318	37403	49927	57715	50423	67372	63116	48888	21505	80182
13	62111	52820	07243	79931	89292	84767	85693	73947	22278	11551
14	47534	09243	67879	00544	23410	12740	02540	54440	32949	13491
15	98614	75993	84460	62846	59844	14922	48730	73443	48167	34770
16	24856	03648	44898	09351	98795	18644	39765	71058	90368	44104
17	96887	12479	80621	66223	86085	78285	02432	53342	42846	94771
18	90801	21472	42815	77408	37390	76766	52615	32141	30268	18106
19	55165	77312	83666	36028	28420	70219	81369	41943	47366	41067

*Reprinted by permission from *STATISTICAL METHODS* by George W. Snedecor and William G. Cochran (c) 1980 by The Iowa State University Press, Ames, Iowa 50010.

After a random sample is found, it can be used to estimate characteristics of the total population.

EXAMPLE 2. Estimate the mean of the 170 scores given on page 496.

SOLUTION: We find the sample mean to estimate the mean of the entire population.

$$\bar{x} = \frac{\sum_{i=1}^{6} x_i}{6} = \frac{2950}{6} \approx 492$$

The mean of the scores is about 492.

A simple random sample is one type of probability sampling. Another type is the **stratified random sample.** In this sample, the entire population is broken into groups, called **strata,** chosen because the individuals in each **stratum** have some common property. For example, you might have two strata (male and female) or four strata (first-year students, sophomores, juniors, and seniors). The only requirement for choosing strata is that you know what percent of the entire population each stratum represents. You then take a simple random sample of each stratum and combine the results as shown in the next example.

EXAMPLE 3. Southbrook High School has 960 students and 40 teachers. The school newspaper interviews 60 students and 12 teachers to see whether they favor candidate A or B in a coming local election. The results are given below.

Stratum	Population size	Sample size	Number favoring candidate A
Students	960	60	32
Teachers	40	12	4
Total	1000	72	36

Estimate the percentage of the school population favoring candidate A.

SOLUTION: First note that 50% of the sample (36 out of 72) favored candidate A. Nevertheless, it would be a mistake to estimate that 50% of the school population favors A because the teachers are disproportionately represented in the sample. Instead, we use these steps.

Step 1. Note the proportions of each stratum in the sample who favor A:

$\bar{p}_1$ = sample proportion of students favoring A = $\frac{32}{60} = \frac{8}{15}$

$\bar{p}_2$ = sample proportion of teachers favoring A = $\frac{4}{12} = \frac{1}{3}$

Step 2. To estimate p, the proportion of the whole population favoring A, take a weighted average of $\bar{p}_1$ and $\bar{p}_2$ assigning weights according to the numbers in each stratum:

$\bar{p}$ = estimate of $p = \frac{960}{1000}\bar{p}_1 + \frac{40}{1000}\bar{p}_2$

$= \frac{96}{100} \cdot \frac{8}{15} + \frac{4}{100} \cdot \frac{1}{3} = \frac{788}{1500} \approx 52.5\%$

Thus, about 52.5% of the students and teachers favor candidate A.

In Exercises 1-6, state whether the sampling procedure described is an example of **(a)** convenience sampling, **(b)** judgment sampling, **(c)** sampling by questionnaire, **(d)** simple random sampling, or **(e)** stratified random sampling. Mention any problems that the sampling procedure might cause.

1. A television rating organization mails questionnaires to subscribers of a television magazine. On the basis of responses to this questionnaire, it names the most popular shows.

2. A popular financial magazine would like to survey the techniques of successful managers. A managerial consultant is asked to select a group of fifty successful managers to participate in the survey.

3. A manufacturing company would like to determine the approximate market share of a certain product. A representative of the company is asked to stand in front of a certain grocery store and ask the first one hundred people who go into the store whether they use the product.

4. A television cable company with 20,000 subscribers wishes to determine its most popular services. To do so, the company electronically monitors 1000 of its subscribers. The subscribers were chosen as follows: Ping pong balls numbered 1 to 20,000 were mixed in a bin and 1000 were drawn at random. The subscribers whose numbers corresponded to the number drawn participated in the survey.

5. In order to determine the public response to a proposed change in regulations concerning small businesses, a questionnaire is being developed. Two alternative questions differing by just one word are suggested for the survey.
   (a) Do you regret the disappearance of the friendly corner grocery store?
   (b) Do you regret the disappearance of the outmoded corner grocery store?

6. An automobile manufacturer sends out questionnaires to all known owners of its vehicles and finds that 90% of the vehicles belonging to people who responded are still in use. The manufacturer cites these results as evidence of vehicle durability. (If most of those vehicles were manufactured within the last three years, are your conclusions different?)

7. Senior citizens are 20% of Littleton's voting population. In a poll of 100 citizens, half of whom were senior citizens, 30 senior citizens voted yes and 20 nonsenior citizens voted yes. Estimate the percentage of the whole voting population who would vote yes.

WRITTEN EXERCISES 14-5

In Exercises 1-4, state any errors you think might occur in the sampling situation discussed.

A 1. A senator explains a vote in favor of a new bill by stating that 60% of the mail received favored the bill.

2. A city newspaper invites readers to mail in a form from the newspaper and

choose one of three ways in which improvements to the zoo could be financed. On the basis of these responses, the newspaper reports that financing the improvements by raising the admission charge is the first choice of the majority of the citizens.

3. A radio talk show host invites listeners to telephone the station and talk about their feelings on a proposed highway to be built in their county.

4. A newspaper reporter randomly stops people going in a grocery store and asks, "Does your family use newspaper coupons?"

5. A random sample of 5 pieces of data are to be drawn from a list of 80 pieces of data, which are numbered from 0 to 79. Use the table of random digits to find out which pieces of data to select. Start with row 10 of the table. (*Hint:* Read the digits in pairs ignoring the numbers 80 to 99 and any duplications.)

6. Select numbers for a random sample of 8 pieces of data from the 80 pieces of data described in Exercise 5; use row 7 of the table of random digits.

B 7. Using the table of random digits, choose a sample of 10 pieces of data from the list of scores on page 496. Compute the sample mean.

8. Using a different row from the one used in Exercise 7, choose another sample of 10 pieces of data from the list of scores on page 496. Compute the sample mean.

9. In Farmington High School, there are 360 ninth- or tenth-grade students and 320 eleventh- or twelfth-grade students. In a poll of 40 ninth- or tenth-grade students, 12 indicate that they intend to vote for Lahey as student council president. The same poll shows that 24 out of 40 eleventh- or twelfth-grade students intend to vote for Lahey. Estimate the percentage of the student body favoring Lahey.

10. Twenty-five percent of a city's employees live in the city and 75% live in the surrounding suburbs. The mayor conducts a small survey asking 10 employees who live in the city and 10 who live in the suburbs if they would prefer increases in pay or increases in fringe benefits. Eight of the 10 city dwellers and four of the 10 suburbanites preferred an increase in pay. Estimate the percentage of city employees preferring an increase in pay to an increase in fringe benefits.

11. Estimate the total number of years of service of 200 factory workers based on a sample of size $n = 5$ where the years of service reported are 15, 8, 20, 5, and 12.

12. Suppose a factory packaged 150 boxes of light bulbs on a certain day. Estimate the total number of defective light bulbs packaged that day based on a sample of 8 boxes that contained the following number of defective bulbs: 5, 3, 5, 1, 2, 9, 4, 3.

13. The State Conservation Department wants to estimate the number of deer in a mountainous area. It captures and tags 80 deer and then releases them. Later it captures 156 deer and finds that 12 are tagged. What is your estimate of the number of deer in the area?

14. The number of fish in a pond is unknown; call it x. Initially, c fish are caught, marked, and released. Later on, d fish are caught, and it is observed that m of them are marked. Estimate the value of x in terms of s, d, and m.

The table below gives the distribution of students in a fictitious small college and responses of those sampled on two questions, one about meals and one about the required English course.

Stratum	Stratum size	Sample size	Pleased with meals	Pleased with English course
Freshmen	300	25	15	20
Sophomores	250	25	12	14
Juniors	250	25	10	18
Seniors	200	25	9	16

15. Estimate the percentage of the student body pleased with the meals at the college.

16. Estimate the percentage of the student body pleased with their required English course.

14-6/CORRELATION COEFFICIENT

Statisticians often look for a linear relationship between two variables. For example, is income at age 25 related to years of education? To answer this question, suppose you took a sample of ten people aged 25 years and obtained the following data:

x = years of education	8	10	12	12	13	14	16	16	18	19
y = income in thousands	8	11	10	14	18	24	28	34	30	32

You could then plot the points (x, y) and get the following *scatter plot*.

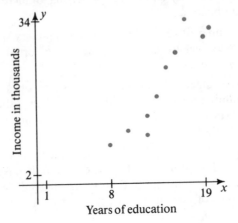

Although the points of the scatter plot do not lie on a straight line, they are clustered about a line. In order to measure how closely points tend to cluster about a line, statisticians use a *correlation coefficient, r*. If the data fits perfectly on a line with positive slope, the correlation coefficient is $r = +1$. If the data fits perfectly on a line with negative slope, the correlation coefficient is $r = -1$. If the data does not fit perfectly on a line, but is clustered around it, then the correlation coefficient r will be between 1 and -1. If the line has positive (or negative) slope, the data is said to have positive (or negative) correlation. The closer the data fits the line, the closer the value of r is to $+1$ or to -1. The correlation coefficient for the scatter plot of the figure just given is $r \approx 0.93$. Other scatter plots and their correlation coefficients are shown below.

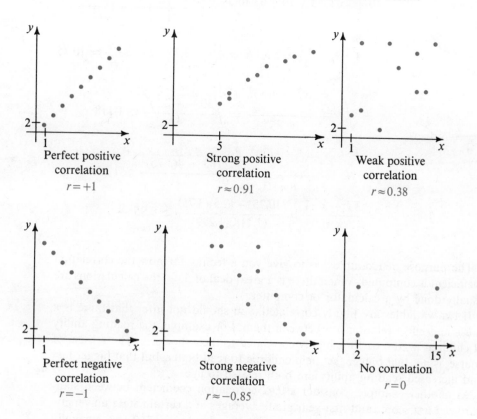

Perfect positive
correlation
$r = +1$

Strong positive
correlation
$r \approx 0.91$

Weak positive
correlation
$r \approx 0.38$

Perfect negative
correlation
$r = -1$

Strong negative
correlation
$r \approx -0.85$

No correlation
$r = 0$

The correlation coefficient for a set of data $(x_1, y_1), (x_2, y_2), \ldots, (x_n, y_n)$ is

$$r = \frac{\overline{xy} - \overline{x} \cdot \overline{y}}{s_x s_y}.$$

In words,

$$r = \frac{\text{mean of the products} - \text{product of the means}}{\text{product of the standard deviations}}$$

When we figure the correlation coefficient for a set of data, it is often helpful to organize the data in a table as shown in Example 1 on the following page.

EXAMPLE 1. Make a scatter plot and find the correlation coefficient for the data $(x, y) = (1, 2), (2, 3), (3, 5),$ and $(4, 5)$.

SOLUTION: We show the data in a table.

x	y	xy	x^2	y^2
1	2	2	1	4
2	3	6	4	9
3	5	15	9	25
4	5	20	16	25
10	15	43	30	63 totals

$$n = 4 \quad \overline{x} = \frac{10}{4} = 2.5 \quad \overline{y} = \frac{15}{4} = 3.75 \quad \overline{xy} = \frac{43}{4} = 10.75$$

$$s_x = \sqrt{\frac{\sum_{i=1}^{n} x_i^2 - n(\overline{x})^2}{n}} = \sqrt{\frac{30 - 4(2.5)^2}{4}} \approx 1.118$$

$$s_y = \sqrt{\frac{\sum_{i=1}^{n} y_i^2 - n(\overline{y})^2}{n}} = \sqrt{\frac{63 - 4(3.75)^2}{4}} \approx 1.299$$

$$r = \frac{\overline{xy} - \overline{x} \cdot \overline{y}}{s_x s_y} = \frac{10.75 - (2.5)(3.75)}{(1.118)(1.299)} \approx 0.95$$

The purpose of Example 1 is to give you a feeling for how the correlation coefficient is computed. When there is a great deal of data, the calculations are usually done by a calculator or computer.

If two variables are highly correlated, you should not infer that there is a cause-and-effect relationship between them. For example, the reading ability of children aged 4 to 10 is highly correlated to their shoe size. The reason, of course, is not that bigger feet help children to read better, but that bigger feet and increased reading ability are both related to age.

As another example, consider a 0.62 correlation coefficient between standardized test scores and first-year grade averages at a certain state university. The positive correlation indicates that higher standardized scores generally yield higher first-year averages, but good test scores do not themselves cause good averages. Rather, the scores and the averages are both influenced by other variables such as knowledge, ability, and desire. In short, the 0.62 correlation between standardized test scores and first-year averages does not indicate a cause-and-effect relationship. Nevertheless, it does indicate a relationship and is therefore useful, along with other data, in predicting academic success in college.

To show how such a prediction might be done, consider the scatter plot below. A relatively strong correlation, such as 0.62, indicates that the points tend to be around a line called a *regression line*. If we know the equation of the line, we can predict y (first-year grade average) from x (test score). The regression line below has an equation $y = 0.0024x - 0.2632$. If a candidate for admission to the university has a total test score of 1200, then, the predicted first-year average is 2.6. This predicted value is approximate and should be regarded as the center of a range of values. Advanced statistics books present formulas for finding the range of values. For our purposes it is sufficient to be aware that this prediction and those called for in Exercises 13–16 are approximate.

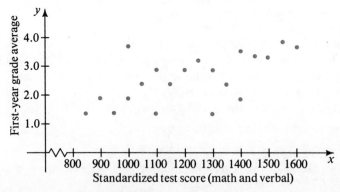

The equation of the regression line can be found by using two of its properties:

1. The line passes through $(\bar{x}, \bar{y})$.

2. The line has slope $r\left(\dfrac{s_y}{s_x}\right)$.

EXAMPLE 2. Find the equation of the regression line for the data in Example 1.

SOLUTION: Step 1. The line passes through $(\bar{x}, \bar{y}) = (2.5, 3.75)$.

Step 2. The slope is $r\left(\dfrac{s_y}{s_x}\right) = 0.95\left(\dfrac{1.299}{1.118}\right) \approx 1.104$.

Step 3. The equation is $y - 3.75 = 1.104(x - 2.5)$ or
$$y = 1.102x + 0.994$$

ORAL EXERCISES 14-6

1. Six scatter plots are shown below. Match each to one of the estimated values of r listed below.
$r = 0.8, 0.6, 0.2, 0, -0.2, -0.6, -0.8$

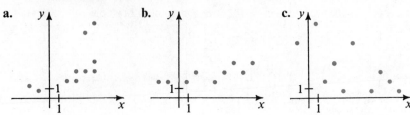

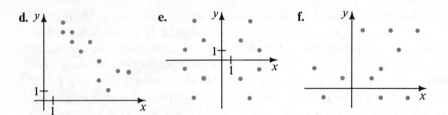

d. **e.** **f.**

2. For each pair of variables, tell whether you think the correlation will be positive, negative, or approximately zero.
   a. The height and weight of a person
   b. The height and intelligence test score of a person
   c. The height of a person and the average height of his or her parents
   d. The temperature and the atmospheric pressure
   e. The value of an automobile and its age
   f. The incidence of flu and the outside temperature

3. **a.** Find r given the following information:
$$\bar{x} = 4, \bar{y} = 3, \overline{xy} = 16, s_x = 3, s_y = 2.$$
   b. Give the equation of the regression line for part (a).

WRITTEN EXERCISES 14-6

For each pair of variables, tell whether you think the correlation will be positive, negative, or approximately zero. Briefly give your reasons.

A 1. The number of hours spent studying statistics and the grade on a statistics test

2. The total number of absences in a chemistry course and the average of all the grades in that course

3. The weight of a car and its fuel economy

4. The blood pressure of a person and his or her weight

5. The blood pressure of a person and his or her age

6. Standardized test mathematics scores and standardized test verbal scores

7. The supply of an item and the demand for that item

8. The number of hours of sleep and the ability to memorize

Display each set of data in a scatter plot and find the correlation coefficient. Also, find the equation of the regression line and draw it on your scatter plot.

9. x	y	10. x	y	11. x	y	12. x	y
1	1	1	4	1	6	-2	-1
1	2	2	3	2	6	-1	-1
3	3	2	1	3	5	0	0
3	4	3	0	4	3	2	3
				5	0	6	4

B **13.** In a ten-week course, the final grades A, B, C, D, and F were given the scores $y = 4, 3, 2, 1$, and 0, respectively. The regression line for predicting scores from x, the number of absences, was $y = 3.5 - 0.5x$. What average letter grade can you predict for a student with **(a)** 1 absence? **(b)** 3 absences? **(c)** 8 absences?

14. When the sales volume in hundreds of units is plotted against x, the money spent on advertising in thousands of dollars, researchers obtained a regression line with the equation $y = 14x + 0.7$. What average sales volumes (in hundreds of units) can you predict for the following amounts (in thousands of dollars) spent on advertising? **a.** 10 **b.** 5 **c.** 7

Find the equation of the regression line and use it to predict the missing entry.

15.

x = amount of fertilizer (kilogram per hectare)	0	20	40	60	80	100
y = crop yield (metric ton per hectare)	4.6	5.8	6.6	7.2	7.9	?

16.

x = standardized test math score	400	500	500	600	700	800	550
y = college first-year math average	50	65	70	80	85	95	?

C **17.** The diagram below shows a scatter plot and an arbitrary line through $(\bar{x}, \bar{y})$. For each point (x_i, y_i), let d_i represent the vertical distance between the point and the line: $d_i = y_i - y = y_i - (\bar{y} + m(x_i - \bar{x})) = (y_i - \bar{y}) - m(x_i - \bar{x})$. Finally let $D = \sum_{i=1}^{n} d_i^2$.

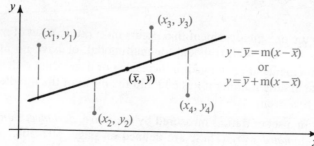

$y - \bar{y} = m(x - \bar{x})$

or

$y = \bar{y} + m(x - \bar{x})$

a. Show that $D = \sum_{i=1}^{n} (y_i - \bar{y})^2 - 2m \sum_{i=1}^{n} (y_i - \bar{y})(x_i - \bar{x}) + m^2 \sum_{i=1}^{n} (x_i - \bar{x})^2$.

b. Note that $\bar{x}$, $\bar{y}$, and each x_i, y_i are known so that D is a quadratic function of m, the slope of the line. Show that D is minimized when

$$m = \dfrac{\sum\limits_{i=1}^{n} (y_i - \bar{y})(x_i - \bar{x})}{\sum\limits_{i=1}^{n} (x_i - \bar{x})^2}.$$

(*Exercise 17 continued*)

c. In part (b), divide the numerator and denominator by n and show that the value of m that minimizes D is $r\left(\dfrac{s_y}{s_x}\right)$.

Hint: Remember that $\bar{x}$ and $\bar{y}$ are constants. Thus,

$$\frac{\sum\limits_{i=1}^{n} \bar{x}y_i}{n} = \frac{\bar{x}\sum\limits_{i=1}^{n} y_i}{n} = \bar{x}\cdot\bar{y}.$$

COMPUTER EXERCISE

Write a computer program that will print the correlation coefficient for a set of data. Run your program for the data concerning education and income in the table on page 526.

Optional. Also have your program calculate the slope of the regression line and predict a value of y when you input a value of x.

Chapter summary

1. A table of data can be visually displayed in a *histogram* or *frequency polygon*.

2. For a set of measurements $x_1, x_2, x_3, \ldots, x_n$:

   a. The *mean* $\bar{x} = \sum\limits_{i=1}^{n} \dfrac{x_i}{n}$.

   b. The *mode* is the one measurement that occurs most often. A set of measurements can be bimodal, trimodal, or multimodal, or have no mode.

   c. If the measurements are arranged in increasing or decreasing order, the *median* is the middle number if n is odd and the mean of the middle two numbers if n is even.

3. The variability in a set of data is measured by the *range, variance, standard deviation,* and *standard score.* These are defined on pages 502–504.

4. A *normal distribution* of data has a bell-shaped histogram in which: about 68% of the distribution is within 1 standard deviation of the mean; about 95% of the distribution is within 2 standard deviations of the mean; about 99% of the distribution is within 3 standard deviations of the mean. A set of measurements arranged in ascending order can be divided into 100 equal parts by *percentiles.*

5. To determine how close a sample proportion with a given characteristic is to the population proportion with the same characteristic, we can use the concept of *confidence intervals.* See pages 515–518.

6. A sample of a population must be representative of the whole population. We can distinguish between *nonprobability sampling* such as *convenience sampling, judgment sampling,* and *sampling by questionnaire,* and *probability sampling* such as *simple random sampling* and *stratified random sampling.* In a simple random sample every possible sample of n units from the whole population has the same chance of being chosen. A table of random digits offers one way to choose such a sample. (See Example 1 on page 521.)

7. A scatter plot offers a visual picture of any linear relationship between two variables. (See page 527.) The *correlation coefficient* is a number between -1 and 1 that measures how closely the points in the scatter plot tend to cluster about a line. (See page 527.) The line that best fits a scatter plot is called a *regression line.* (See page 528.)

Chapter test

1. Display the following set of data in a histogram and find the mean, median, and mode:

   0, 1, 2, 2, 2, 3, 3, 4, 6, 7.

   14–1

2. Calculate the **(a)** range, **(b)** variance, and **(c)** standard deviation for the data in Exercise 1.

   14–2

3. Give the standard score of each piece of data in Exercise 1.

4. At a certain university, the mathematics plus verbal standardized test scores of incoming students are normally distributed with mean score = 1200 and standard deviation = 100. Use the table on page 508 to find the percentage of incoming students whose total scores are **(a)** more than 1300. **(b)** less than 1000. Find **(c)** the 90th percentile and **(d)** the third quartile of test scores.

   14–3

5. In a survey, 60 out of 100 florists recommended growing a particular variety of plants from cuttings rather than from seeds. From a 95% confidence interval, find the proportion of all florists who favor using cuttings.

   14–4

6. Choose a random sample of 4 test scores from a list of 50 scores numbered from 0 to 49. Use the table of random numbers on page 522 and start with row 9.

   14–5

7. An actuary wants to estimate the total number of people in 500 households. Estimate the total number if a random sample of 10 households is selected and they are found to have the following number of inhabitants:

   5, 1, 3, 2, 3, 7, 4, 6, 1, 4

8. A statistics class correlated heights and weights of the entire class. Letting x stand for heights and y for weights, they found the following statistics: $\bar{x} = 160$ cm, $\bar{y} = 60$ kg, $\overline{xy} = 9635$, $s_x = 10$, $s_y = 5$.
   a. Calculate the correlation coefficient.
   b. Give the equation of the regression line.

   14–6

534

CHAPTER FIFTEEN

Probability

OBJECTIVES

1. To use the Multiplication and Addition Principles to find the number of possible outcomes of two or more actions.
2. To solve problems using conditional probability.
3. To solve problems involving permutations, combinations, and probabilities.
4. To expand binomials and use the Binomial Theorem to solve probability problems.

Permutations and combinations

15-1/THE MULTIPLICATION AND ADDITION PRINCIPLES

Have you ever guessed at several answers on a multiple-choice test and then wondered what your chances were of passing the test? Have you ever wondered how many different 13-card hands can be dealt from a 52-card deck? Or what your chances are of getting all 4 aces in your hand? In this chapter, we shall see how to answer questions like these. We begin our discussion by showing how the following multiplication principle can be used to determine the number of possible outcomes of two or more consecutive actions.

THE MULTIPLICATION PRINCIPLE. If an action can be performed in n_1 ways and then, for each of these ways, another action can be performed in n_2 ways, the two actions can be performed consecutively in $n_1 n_2$ ways.

The solar electricity generating plant shown produces one third of the electricity required by the town of Bigelow, California. Such plants function best in locations where the probability of sunny weather is high.

EXAMPLE 1. If you have 4 sweaters and 2 pairs of jeans, how many different sweater-and-jeans combinations can you make?

SOLUTION: There are two actions to perform: (1) choosing a sweater and (2) choosing a pair of jeans. Since there are 4 sweaters from which to choose and 2 pairs of jeans from which to choose, there are $4 \times 2 = 8$ ways to choose a sweater-and-jeans combination. If we call the sweaters A, B, C, and D, and the pairs of jeans 1 and 2, the so-called tree diagram at the right shows the 8 possible combinations.

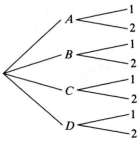

The Multiplication Principle can be extended to three or more consecutive actions, as illustrated by the following examples.

EXAMPLE 2. How many license plates can be made using 2 letters followed by 3 digits?

SOLUTION: The following diagram represents the five spaces of the license plate.

☐ ☐ ☐ ☐ ☐

The first space can be filled in 26 ways with one of the letters of the alphabet. After the first space is filled, there are 26 ways of filling the second space. Thus, there are 26×26 ways of filling the first 2 spaces.

| 26 | 26 | ☐ | ☐ | ☐ |

Now the third space can be filled with any of the 10 digits. The same is true for the fourth and fifth spaces.

| 26 | 26 | 10 | 10 | 10 |

Thus, by the Multiplication Principle, the number of license plates is
$$26 \cdot 26 \cdot 10 \cdot 10 \cdot 10 = 676,000.$$

(Note that sometimes the digit zero is not permitted as the first of the three digits. If this is the case, then the number of possible license plates would be $26 \cdot 26 \cdot 9 \cdot 10 \cdot 10 = 608,400$.)

EXAMPLE 3. In how many ways can 8 people line up in a cafeteria line?

SOLUTION: The first place in line can be filled by any of the 8 people. The second place in line, then, can be filled by any of the remaining 7 people, the third place by any of the remaining 6 people, and so on. The diagram below illustrates this reasoning.

| 8 | 7 | 6 | 5 | 4 | 3 | 2 | 1 |

By the Multiplication Principle, the answer is
$$8 \cdot 7 \cdot 6 \cdot 5 \cdot 4 \cdot 3 \cdot 2 \cdot 1 = 40,320.$$

The product $8 \cdot 7 \cdot 6 \cdot 5 \cdot 4 \cdot 3 \cdot 2 \cdot 1$ appearing in Example 3 can be abbreviated as 8! (read "8 factorial"). Recall the general definition of $n!$:

$$n! = n \cdot (n - 1) \cdots 3 \cdot 2 \cdot 1$$
$$0! = 1$$

You have seen that the Multiplication Principle applies when you perform consecutive actions, such as filling a license plate with 3 letters *followed* by 4 digits. Suppose, however, that you want to know the number of ways to make a license plate using *either* 3 letters *or* 4 digits. These two actions are **mutually exclusive;** that is, they cannot both be performed at the same time. In situations involving mutually exclusive actions, the following Addition Principle applies.

THE ADDITION PRINCIPLE. If two actions are mutually exclusive, and the first can be done in n_1 ways and the second in n_2 ways, then one action *or* the other can be done in $n_1 + n_2$ ways.

EXAMPLE 4.   How many 2- or 3-digit numbers can be formed from the digits 1, 2, 3, 4, 5 if there are to be no repeated digits in a number?

SOLUTION: We can form $5 \cdot 4 = 20$ 2-digit numbers,

or we can form $5 \cdot 4 \cdot 3 = 60$ 3-digit numbers.

In all, we can form

$$20 + 60 = 80$$

numbers having 2 or 3 digits.

Like the Multiplication Principle, the Addition Principle can also be extended to more than two actions.

ORAL EXERCISES 15-1

1. Evaluate: **(a)** 2!   **(b)** 3!   **(c)** 4!.
2. Suppose you are told that 10! = 3,628,800. What is 9!?
3. **a.** If a girl has 6 different skirts and 10 different blouses, how many different skirt-and-blouse outfits are possible?
   b. If she also has 3 different sweaters, how many skirt-blouse-and-sweater outfits are there?
4. A boy has 2 sports coats and 4 sweaters.
   a. How many coat-and-sweater combinations can he wear?
   b. Suppose he decides to wear either a sports coat or a sweater, but not both. How many choices does he have?
5. If 10 runners compete in a race, in how many different ways can prizes be awarded for first, second, and third places?

A

1. Evaluate: **(a)** 5! **(b)** 6! **(c)** 7! **(d)** 0!

2. Evaluate: **(a)** $\dfrac{10!}{9!}$ **(b)** $\dfrac{20!}{18!}$

3. In how many different orders can you arrange 5 books on a shelf?

4. In how many different orders can 9 people stand in a line?

5. Many radio stations have 4-letter call signs beginning with K. How many such call signs are possible **(a)** if letters can be repeated? **(b)** if letters cannot be repeated?

6. How many different 3-digit numbers can be formed using the digits 4, 5, 6, 7, 8 **(a)** if digits can be repeated? **(b)** if digits cannot be repeated?

7. In how many ways can 4 people be seated in 12 chairs?

8. In how many ways can 4 different prizes be given to any 4 of 10 people if no person receives more than 1 prize?

9. **a.** A high-school coach must decide on the batting order for a baseball team of 9 players. Show that there are 362,880 different orders from which the coach can choose.

   b. Show that there are 40,320 different batting orders with the pitcher batting last.

10. A track coach must choose a 4-person 400 meter relay team and a 4-person 800 meter relay team from a squad of 7 sprinters, any of whom can run on either team. If the fastest sprinter of the 7 runs last in both races, in how many ways can the coach form the two teams if each of the 6 remaining sprinters runs only once and each different order is counted as a different team?

11. How many numbers consisting of 1, 2, or 3 digits (without repetitions) can you form using the digits 1, 2, 3, 4, 5, 6?

12. If you have 5 signal flags and can send messages by hoisting one or more flags on a flagpole, how many messages can you send?

13. In some states license plates consist of 3 letters followed by 2 or 3 digits (for example, ABC-055 or RRK-54). How many such possibilities are there for those plates with 2 digits? for those with 3 digits? In all, how many license plates are possible?

14. How many possibilities are there for a license plate with 2 letters and 3 or 4 nonzero digits?

15. How many 6-digit numbers begin with an odd digit and end with an even digit?

16. How many integers between 1000 and 9999, inclusive, contain no 7's?

17. In how many different ways can you answer 10 true-false questions?

18. In how many different ways can you answer 10 multiple-choice questions if each question has 5 choices?

19. a. How many 6-letter "words" can be formed using the letters of the word SENIOR? (*Note:* We allow any arrangement of letters, such as RENISO, to count as a "word." We also assume that each letter in the original word is used once.)

   b. How many 5-letter "words" can be formed?

20. a. How many 5-letter "words" can be formed using the letters of SPINACH?

   b. How many of these will have no vowels?

   c. How many will have at least one vowel?

B **21.** Show that you can form 120 5-letter "words" from GREAT but only 60 5-letter "words" from GREET. (In a GREET "word," there are two E's, one G, one R, and one T.)

22. Explain why the number of 7-letter "words" you can form from ALXBYMZ is 7!, but the number of 7-letter "words" you can form from ALABAMA is $\dfrac{7!}{4!}$.

23. Explain why the number of 5-letter "words" you can form from ABCDE is 5!, but the number of 5-letter "words" you can form from ABABA is $\dfrac{5!}{2!3!}$.

24. How many "words" can be formed using all the letters of STATISTICS? (*Hint:* See Exercises 22 and 23.)

25. a. Show that $10 \cdot 9 \cdot 8 \cdot 7 = \dfrac{10!}{6!}$.

   b. Show that $n \cdot (n - 1) \cdot (n - 2) \cdots (n - r + 1) = \dfrac{n!}{(n - r)!}$.

C **26.** In how many zeros does the number 100! end?

CALCULATOR EXERCISE

Evaluate $\log_{10} 9!$. Then, without using your calculator, find the value of $\log_{10} 10!$. Use your calculator to check your answer.

15-2/PERMUTATIONS AND COMBINATIONS

In some situations involving choices, the order in which the choices are made is important, whereas in others it is not. Suppose, for example, that a club of 12 members wishes to choose a president, a vice president, and a treasurer. In this case, the order of the choices is important since the order A as president, B as vice president, and C as treasurer, for instance, is different from the order B, A, and C for these 3 offices, respectively. From Section 15-1, we know that the number of ways of filling the 3 offices is

$$12 \cdot 11 \cdot 10.$$

Suppose, on the other hand, that the club of 12 members merely wants to choose a committee of 3. In this case, the order of selection is *not* important, since a selection of A, B, and C is the same as a selection of B, A, and C. Now for each committee of 3 that can be chosen (for example, A, B, C), there are 3! different slates of officers (ABC, ACB, BAC, BCA, CAB, CBA). Thus:

$$\text{number of committees} \times 3! = \text{number of slates of officers}$$

$$\text{number of committees} = \frac{\text{number of slates of officers}}{3!}$$

$$= \frac{12 \cdot 11 \cdot 10}{1 \cdot 2 \cdot 3} = 220$$

In the first situation, in which 3 out of 12 people are to be chosen as officers, the order of selection is important. The selection is called a **permutation** of 3 people from a set of 12; the number of such selections is denoted as $P(12, 3)$.

In the second situation, in which 3 out of 12 people are to be chosen as a committee, the order of selection is not important. The selection is called a **combination** of 3 people from a set of 12; the number of such combinations is denoted as $C(12, 3)$. In general, the symbols $P(n, r)$ and $C(n, r)$ denote the number of permutations and the number of combinations, respectively, of r things chosen from n things.

Permutation (order important)

$$P(12, 3) = 12 \cdot 11 \cdot 10$$

$$P(n, r) = n \cdot (n - 1)(n - 2) \cdots$$
$$(n - r + 1)$$

$$= \frac{n!}{(n - r)!}$$

(This formula was derived in Exercise 25 on page 539.)

Combination (order not important)

$$C(12, 3) = \frac{12 \cdot 11 \cdot 10}{1 \cdot 2 \cdot 3}$$

$$C(n, r) = \frac{n(n - 1)(n - 2) \cdots (n - r + 1)}{1 \cdot 2 \cdot 3 \cdots r}$$

$$= \frac{n!}{(n - r)! \, r!}$$

(This formula is to be derived in Exercise 19 on page 543.)

EXAMPLE 1. A company advertises two job openings, one for a copywriter and one for an artist. If 10 people who are qualified for either position apply, in how many ways can the openings be filled?

SOLUTION: Since the jobs are different, the order of selecting people matters: X as copywriter and Y as artist is different from Y as copywriter and X as artist. The solution is found by counting permutations:

$$P(10, 2) = 10 \cdot 9 = 90$$

EXAMPLE 2. A company advertises two job openings for computer programmers, both with the same salary and job description. In how many ways can the openings be filled if 10 people apply?

SOLUTION: Since the two jobs are identical, the order of selecting people is not important. Choosing X and Y for the positions is the same as choosing Y and X. Thus, the answer is a number of combinations that is half as large as the number of permutations in Example 1.

$$C(10, 2) = \frac{10 \cdot 9}{1 \cdot 2} = 45$$

If you have a calculator with a factorial key, you can use it to help solve permutation and combination problems that involve greater numbers, such as the next example. The example refers to a standard bridge deck of 52 cards. The deck consists of four suits: spades and clubs, which are black, and hearts and diamonds, which are red. Each suit consists of 13 cards: an ace, 2, 3, 4, 5, 6, 7, 8, 9, 10, jack, queen, and king. Jacks, queens, and kings are often referred to as *face cards*.

EXAMPLE 3. How many different ways are there to deal 13 cards from a standard deck of 52 cards (a) if the order in which the cards are dealt is important? (b) if the order in which the cards are dealt is not important?

SOLUTION: **(a)** $P(52, 13) = 52 \cdot 51 \cdot 50 \cdot 49 \cdots 40 = \dfrac{52!}{39!} \approx 3.95 \times 10^{21}$

(b) $C(52, 13) = \dfrac{52 \cdot 51 \cdot 50 \cdot 49 \cdots 40}{1 \cdot 2 \cdot 3 \cdot 4 \cdots 13} = \dfrac{52!}{39!13!} \approx 6.35 \times 10^{11}$

ORAL EXERCISES 15-2

Evaluate.

1. a. $P(5, 2)$ **b.** $C(5, 2)$ **2. a.** $P(6, 3)$ **b.** $C(6, 3)$

3. a. $P(10, 3)$ **b.** $C(10, 3)$ **4. a.** $P(4, 4)$ **b.** $C(4, 4)$

5. In how many ways can a club of 10 members choose a president, vice president, and treasurer?

6. In how many ways can a club of 10 members choose a committee of 3?

7. A lock on a safe has a dial with 50 numbers on it. To open it, you must turn the dial left, then right, then left to 3 different numbers. Explain why such a lock should be called a permutation lock instead of a combination lock.

8. a. Four people (A, B, C, and D) apply for three jobs (clerk, secretary, and receptionist). If each person is qualified for each job, make a list of the number of ways the jobs can be filled. For example, ABC means that A is clerk, B is secretary, and C is receptionist. This is different from BAC and CBA.

b. Your list should contain $P(4, 3)$ entries. How many of the entries involve persons A, B, and C? How many of the entries involve persons A, C, and D?

c. If A, B, C, and D apply for three identical job openings as a clerk, make a list of the number of ways the openings can be filled.

A   **1. a.** In how many ways can a club of 20 members choose a president and a vice president?

   b. In how many ways can they choose a 2-person committee?

2. a. In how many ways can a club of 13 members choose 4 different officers?

   b. In how many ways can they choose a 4-person committee?

3. a. In how many ways can a host couple choose 5 couples to invite for dinner from a group of 10 couples?

   b. A basketball coach must choose a center, a left guard, a right guard, a left forward, and a right forward from a team of 10 players. In how many ways can this be done?

4. A teacher has a collection of 20 true-false questions and wishes to choose 5 of them for a quiz. How many quizzes can be made if the order of the questions is considered **(a)** important? **(b)** unimportant?

5. Each of the 200 students attending a school dance has a ticket number for a door prize. If 3 different numbers are selected, how many ways are there for awarding the prizes **(a)** assuming that the 3 prizes are identical? **(b)** assuming that the 3 prizes are different?

6. Suppose you bought 4 books to give as gifts to four friends. In how many ways can the books be given if they are **(a)** all different? **(b)** all identical?

7. Eight people apply for 3 job positions. In how many ways can the 3 positions be filled if the positions are **(a)** all different? **(b)** all the same?

8. How many different ways are there to deal a hand of 5 cards from a standard deck of 52 cards if the order in which the cards are dealt is **(a)** important? **(b)** not important?

9. a. A hiker wishes to invite 7 friends to go on a trip but has room for only 4 of them. In how many ways can they be chosen?

   b. If there were room for only 3 friends, in how many ways could they be chosen? How is your answer related to the answer for part (a)? Why?

10. a. In how many ways can you choose 3 letters from the word LOGARITHM if you are not concerned with the order of the letters?

   b. In how many ways can you choose 6 letters from LOGARITHM if the order of letters is unimportant? Compare with part (a).

11. Show that $C(100, 2) = C(100, 98)$ by **(a)** using the formula for $C(n, r)$; **(b)** explaining how choosing 2 out of 100 is related to choosing 98 out of 100.

12. a. Show that $C(11, 3) = C(11, 8)$.

   b. Study part (a) and also Exercise 11. Then make a generalization and prove it.

13. a. Evaluate $C(5, 0)$ to find how many ways you can select no objects from a group of 5.

   b. Evaluate $C(5, 5)$ to find how many ways you can select 5 objects from a group of 5.

14. How does the formula for $C(5, 5)$ suggest the definition: $0! = 1$?

15. In how many ways can 6 hockey players be chosen from a group of 12 if the playing positions **(a)** are considered? **(b)** are not considered?

16. Of the 12 players on a school's basketball team, the coach must choose 5 players to be in the starting lineup. In how many ways can this be done if the playing positions **(a)** are considered? **(b)** are not considered?

B **17.** Solve for n: $C(n, 2) = 45$

18. Solve for n: $C(n, 2) = P(n - 1, 2)$

19. Show that $\dfrac{n(n - 1)(n - 2) \cdots (n - r + 1)}{1 \cdot 2 \cdot 3 \cdots r} = \dfrac{n!}{(n - r)!\,r!}$.

20. Give the values of the combinations shown in the following triangular array. What patterns can you discover in this array?

$$
\begin{array}{ccccccc}
& & & C(1, 0) & C(1, 1) & & \\
& & C(2, 0) & C(2, 1) & C(2, 2) & & \\
& C(3, 0) & C(3, 1) & C(3, 2) & C(3, 3) & & \\
C(4, 0) & C(4, 1) & C(4, 2) & C(4, 3) & C(4, 4) & & \\
C(5, 0) & C(5, 1) & C(5, 2) & C(5, 3) & C(5, 4) & C(5, 5) &
\end{array}
$$

C **21.** In the World Series, two teams, A and B, play each other until one team has won 4 games. For example, the symbol ABBAAA represents a 6-game series in which A wins games 1, 4, 5, and 6.

a. Explain why there are $C(5, 3)$ different 6-game series won by team A.

b. Without actually listing the various series between A and B, show that there are 70 different sequences of games possible.

CALCULATOR EXERCISE

If the 52 cards of a standard deck are dealt to four people, 13 cards at a time, the first person can receive $C(52, 13)$ possible hands. Then the second person can receive 13 of the remaining 39 cards in $C(39, 13)$ possible ways. The third person can receive 13 of the remaining 26 cards in $C(26, 13)$ ways, and the fourth person can receive 13 of the remaining 13 cards in $C(13, 13)$ ways. Thus, the total number of ways of distributing the 52 cards into four 13-card hands is

$$C(52, 13) \cdot C(39, 13) \cdot C(26, 13) \cdot C(13, 13).$$

Without using a calculator, show that the above product simplifies to $\dfrac{52!}{(13!)^4}$. Then use a calculator to evaluate this number.

ASSORTED EXERCISES 15-1 AND 15-2

The following exercises are counting problems. They test your ability to distinguish between permutation problems, combination problems, and problems that involve the Multiplication Principle or the Addition Principle.

1. Three identical door prizes are to be given to three lucky people in a crowd of 100. In how many ways can this be done?

2. The license plates in a certain state consist of 3 letters followed by 3 nonzero digits. How many such licenses are possible?

3. How many numbers between 1000 and 9999, inclusive, **(a)** contain no zeros? **(b)** contain no ones? **(c)** begin with an even digit and end with an odd digit?

4. A lock has a dial with 50 numbers on it. To open it, you must turn left to a number, right to a number, then left to a number. How many possibilities are there if **(a)** the 3 numbers must be different? **(b)** the 3 numbers are not necessarily different?

5. A student must take four final exams that are to be scheduled by computer during the morning and afternoon testing periods on Monday through Friday. How many ways are there to schedule the four exams?

6. **a.** A railway has 30 stations. On each ticket, the departure station and the destination station are printed. How many different kinds of tickets are there?

   b. If a ticket could be used in either direction between two stations, how many different kinds of tickets would be needed?

7. How many 6-letter "words" can be formed by using all of the letters of **(a)** the word RADISH? **(b)** the word SQUASH?

8. There are 3 roads from town A to town B, 5 roads from town B to town C, and 4 roads from town C to town D. How many ways are there to go from A to D via B and C? How many different round trips are possible?

9. In the Morse code, letters, digits, and various punctuation marks are represented by a sequence of dots and dashes. Sequences can be from 1 unit to 5 units in length. How many such sequences are possible?

10. A teacher must pick 3 high school students from a class of 30 to prepare and serve food at the junior high school picnic. How many selections are possible?

Ernest Everett Just (1883–1941) was an outstanding scientist who did most of his research at the Marine Biological Laboratories in Woods Hole, Massachusetts. He received international recognition for his work in marine biology.

11. A town council consists of 8 members including the mayor.
    a. How many different committees of 4 can be chosen from this council?
    b. How many of these committees will include the mayor?
    c. How many will not include the mayor?
    d. Verify that the answer to part (a) is the sum of the answers to part (b) and part (c).

12. Repeat Exercise 11 if the council has 9 members including the mayor.

13. If you have a $1 bill, a $5 bill, a $10 bill, and a $20 bill, how many different sums of money can you make using one or more of these bills?

14. The Pizza Place offers pepperoni, mushrooms, sausages, onions, anchovies, and peppers as toppings for their regular plain pizza. How many different pizzas can be made?

15. **a.** How many 4-letter "words" can be formed by using the 8 letters of TRIANGLE?
    b. How many of the "words" formed in part (a) have no vowels?
    c. How many of the "words" formed in part (a) have at least one vowel?

16. How many 5-digit numbers contain at least one 3? (*Hint:* How many contain no 3's?)

17. Five boys and 5 girls stand in a line. How many arrangements are possible if **(a)** all of the boys stand in succession? **(b)** the boys and girls stand alternately?

18. Answer Exercise 17 if the 5 boys and 5 girls stand in a circle instead of a line.

Probability; the binomial theorem

15-3/PROBABILITY

If a single die is rolled, each of the six faces has the same chance of turning up. Hence, we say that the *probability* of rolling a "3" is $\frac{1}{6}$. Likewise, when a coin is flipped, a "head" is as likely to occur as a "tail," so we say that the probability of "heads" is $\frac{1}{2}$.

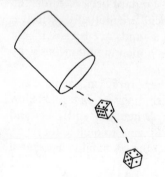

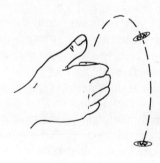

The tables below show the 4 possible outcomes when a coin is tossed twice and the 36 possible outcomes when a die is rolled twice. From the first table, we see that 2 of the 4 possibilities correspond to getting 1 head and 1 tail. Hence, we can say the probability of getting 1 head and 1 tail in 2 tosses of a coin is $\frac{2}{4} = \frac{1}{2}$. From the second table, we can see that 3 of the 36 possibilities will give a sum of 11 or 12. Consequently, we say that the probability of an 11 or a 12 is $\frac{3}{36} = \frac{1}{12}$.

Table 1

second toss

	H	T
first toss H	H, H	H, T
T	T, H	T, T

Table 2

second roll

	1	2	3	4	5	6
1	1, 1	1, 2	1, 3	1, 4	1, 5	1, 6
2	2, 1	2, 2	2, 3	2, 4	2, 5	2, 6
first roll 3	3, 1	3, 2	3, 3	3, 4	3, 5	3, 6
4	4, 1	4, 2	4, 3	4, 4	4, 5	4, 6
5	5, 1	5, 2	5, 3	5, 4	5, 5	5, 6
6	6, 1	6, 2	6, 3	6, 4	6, 5	6, 6

sum = 11 sum = 12

In general, if an *experiment* (such as tossing a coin or rolling a die) can lead to n equally likely outcomes, then the set of all these outcomes is called the **sample space** for the experiment, and any subset of this sample space is called an **event**. If m of the n outcomes of an experiment correspond to some event A, then the **probability** of event A is

$$P(A) = \frac{m}{n}.$$

If all n outcomes correspond to event A, then A is certain to occur and

$$P(A) = \frac{n}{n} = 1.$$

If event A is impossible, $P(A) = 0$. The event "not A" occurs when event A does not. Since their probabilities total 1,

$$P(\text{not } A) = 1 - P(A).$$

Two events are said to be *mutually exclusive* if they cannot both occur simultaneously. For example, if a card is chosen from a standard deck of cards, the events "black card is chosen" and "diamond is chosen" are mutually exclusive. On the other hand, the events "black card is chosen" and "ace is chosen" are not mutually exclusive because both events can happen simultaneously by choosing a black ace.

The circular regions in the diagrams on the next page represent events A and B. The rule for finding $P(A \text{ or } B)$ is given below each diagram. The rule below the diagram at the right needs some explanation. When you add $P(A)$ and $P(B)$, the region inside both circles labeled "A and B" gets accounted for twice. To compensate for this double counting, we subtract the probability associated with this region, "A and B."

Mutually Exclusive Events	Intersecting Events
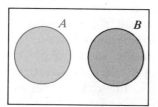	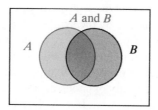
$P(A \text{ or } B) = P(A) + P(B)$	$P(A \text{ or } B) = P(A) + P(B) - P(A \text{ and } B)$

ADDITION RULE. $P(A \text{ or } B) = P(A) + P(B) - P(A \text{ and } B)$

SPECIAL CASE. If A and B are mutually exclusive, then

$$P(A \text{ or } B) = P(A) + P(B)$$

EXAMPLE 1. Each student in a class of 30 studies one foreign language, French or Spanish, and one science, chemistry or biology. Their choices are shown in the table below.

	Chemistry	Biology	Total
French	4	3	7
Spanish	13	10	23
Total	17	13	30

If a student is selected at random from the class, what is the probability that the student studies chemistry or French?

SOLUTION: Let C stand for the event "student studies chemistry."
Let F stand for the event "student studies French."
Since 17 of the 30 students study chemistry,

$$P(C) = \frac{17}{30}.$$

Since 7 of the 30 students study French,

$$P(F) = \frac{7}{30}.$$

$$P(C \text{ or } F) = P(C) + P(F) - P(C \text{ and } F)$$

$$= \frac{17}{30} + \frac{7}{30} - \frac{4}{30}$$

$$= \frac{20}{30} = \frac{2}{3}$$

The result of Example 1 can be verified by observing from the table that $13 + 4 + 3$, or 20, of the 30 students study chemistry or French.

Conditional probability

If an event A is known to have happened, then the probability of event B happening is denoted by the symbol $P(B|A)$, which is read "the probability of B, given A." This probability is called a **conditional probability** because the probability of B is determined under the *condition* that A has occurred.

Referring to the events in Example 1, the symbol $P(F|C)$ stands for the conditional probability of choosing a student who studies French given that the student studies chemistry. To calculate this probability, we need only consider the 17 students who study chemistry, not all 30 students. Of these 17 students, 4 study French. Therefore, $P(F|C) = \frac{4}{17}$. On the other hand, $P(C|F)$ stands for the conditional probability of choosing a student who studies chemistry given that the student studies French. In this case, we need to consider only the 7 students who study French to see that $P(C|F) = \frac{4}{7}$.

Independent and dependent events

Events A and B are said to be **independent** if $P(B|A) = P(B)$. In other words, A and B are independent if the occurrence of A does not affect the probability of B's occurrence. For example, when a die is rolled twice, the events "first roll is a 3" and "second roll is a 2" are independent because the first toss doesn't affect the number rolled on the second toss.

On the other hand, the events "first roll is a 3" and "sum of both rolls is a 4" are not independent. The occurrence of one event definitely affects the occurrence of the other as the following calculations show:

$P(\text{sum is } 4) = \frac{3}{36} = \frac{1}{12}$ (From Table 2, page 546)

$P(\text{sum is } 4 | \text{first roll is } 3) = \frac{1}{6}$ (If first roll is 3, second roll must be 1.)

Two events like these that are not independent are called **dependent events.** We are now ready to state the multiplication rule for finding $P(A \text{ and } B)$.

MULTIPLICATION RULE. $P(A \text{ and } B) = P(A) \cdot P(B|A)$

SPECIAL CASE.   If A and B are independent, then

$$P(A \text{ and } B) = P(A) \cdot P(B)$$

EXAMPLE 2.   A junior and a senior each simultaneously pick an integer at random from 1 to 10, inclusive. Find the probability that "the junior's number is even and the senior's number is divisible by 3."

SOLUTION: Let J be the event "the junior's number is even."
Let S be the event "the senior's number is divisible by 3."
Since the two events are independent:

$$P(J \text{ and } S) = P(J) \cdot P(S)$$

$$= \frac{5}{10} \cdot \frac{3}{10} = \frac{3}{20}$$

EXAMPLE 3. Two cards are dealt from a standard deck. Find the probability of the following events.

a. Both cards are red.   **b.** Neither card is an ace.

SOLUTION:

a. $P\left(\begin{array}{c}\text{first is red and}\\\text{second is red}\end{array}\right) = P\left(\begin{array}{c}\text{first}\\\text{is red}\end{array}\right) \cdot P\left(\begin{array}{c}\text{second}\\\text{is red}\end{array}\Big|\begin{array}{c}\text{first}\\\text{is red}\end{array}\right)$

$$= \frac{26}{52} \cdot \frac{25}{51}$$

$$= \frac{25}{102}$$

b. $P\left(\begin{array}{c}\text{first is not an ace and}\\\text{second is not an ace}\end{array}\right) = P\left(\begin{array}{c}\text{first is}\\\text{not an ace}\end{array}\right) \cdot P\left(\begin{array}{c}\text{second is}\\\text{not an ace}\end{array}\Big|\begin{array}{c}\text{first is}\\\text{not an ace}\end{array}\right)$

$$= \frac{48}{52} \cdot \frac{47}{51}$$

$$= \frac{188}{221}$$

Notice in Example 3 that we assumed that each card in the deck is equally likely to be drawn. We assumed earlier that when a die is rolled, each of the six faces is equally likely to show. You may make similar assumptions in the text and exercises throughout the rest of this book.

Examples 4 and 5 show how the multiplication rules can be applied to more than two events.

EXAMPLE 4. Three cards are chosen at random from a deck. Find the probability that all three cards are clubs.

SOLUTION:

$$P\left(\begin{array}{c}\text{all 3}\\\text{are clubs}\end{array}\right) = P\left(\begin{array}{c}\text{first is}\\\text{a club}\end{array}\right) \cdot P\left(\begin{array}{c}\text{second}\\\text{is a club}\end{array}\Big|\begin{array}{c}\text{first is}\\\text{a club}\end{array}\right) \cdot P\left(\begin{array}{c}\text{third is}\\\text{a club}\end{array}\Big|\begin{array}{c}\text{first two}\\\text{are clubs}\end{array}\right)$$

$$= \frac{13}{52} \cdot \frac{12}{51} \cdot \frac{11}{50} = \frac{11}{850}$$

EXAMPLE 5. Suppose that in Example 4 each card is replaced in the deck before the next card is chosen at random. Find the probability that all three cards are clubs.

SOLUTION: In Example 4 the events "first card is a club," "second card is a club," and "third card is a club" were dependent. In this example, these events are independent because the deck is returned to its original condition after the first and second drawings.

$$P\left(\begin{array}{c}\text{all three}\\\text{are clubs}\end{array}\right) = P\left(\begin{array}{c}\text{first is}\\\text{a club}\end{array}\right) \cdot P\left(\begin{array}{c}\text{second is}\\\text{a club}\end{array}\right) \cdot P\left(\begin{array}{c}\text{third is}\\\text{a club}\end{array}\right)$$

$$= \frac{13}{52} \cdot \frac{13}{52} \cdot \frac{13}{52} = \frac{1}{64}$$

In Example 5 the drawing is said to be **with replacement,** whereas in Example 4 the drawing was **without replacement.** In the text and exercises throughout the rest of the chapter you may assume, as we have, that drawing is without replacement, unless otherwise stated.

ORAL EXERCISES 15-3

1. If a card is drawn from a standard deck of 52 cards, what is the probability that it is:
   a. the queen of hearts?   **b.** a heart?   **c.** a queen?
   d. Are the events "card is a queen" and "card is a heart" dependent or independent?

2. Study Table 1 on page 546 and give the probability of tossing a coin twice and getting:
   a. no heads   **b.** at least one head

3. Study Table 2 on page 546 and give the probability of rolling a die twice and getting:
   a. a sum of 3   **b.** a sum of 4   **c.** a sum of 3 or 4

4. If two dice are rolled, what is the probability that both will show the same number?

5. If the probability of rain tomorrow is 40%, what is the probability of no rain tomorrow?

6. If the probability of no accidents during one month in a manufacturing plant is 0.82, what is the probability of at least one accident during one month?

7. A card is drawn from a standard deck.
   Event A is "card is black."
   Event B is "card is an 8."
   a. Find $P(A)$ and $P(A|B)$.
   b. Find $P(B)$ and $P(B|A)$.
   c. Are events A and B independent?

8. Two cards are drawn from a standard deck.
   Event A is "first card is a 7."
   Event B is "second card is a 7."
   a. Find $P(B)$, $P(B|A)$, and $P(B|\text{not } A)$.
   b. Are events A and B independent?

9. Suppose that in Exercise 8 the first card is replaced in the deck before the second card is randomly chosen. Are the events A and B independent?

10. Give an example of two mutually exclusive events.

11. Explain why the formula
$$P(A \text{ or } B) = P(A) + P(B) - P(A \text{ and } B)$$
can be simplified if A and B are mutually exclusive.

A   **1.** If a card is drawn from a standard deck of 52 cards, what is the probability that it is:
     a. a spade?   **b.** a black card?   **c.** a jack, queen, or king?

     2. One of the first 10 positive integers is picked at random. What is the probability that it is:
     a. even?   **b.** divisible by 3?   **c.** a prime?
                                            (*Note:* 1 is *not* a prime.)

     3. Two dice are rolled. What is the probability that their sum is:
     a. 6?   **b.** 7?   **c.** 8?

     4. Two dice are rolled. What is the probability that their sum is even?

Exercises 5–16 refer to the following experiment and events: Two cards are drawn from a standard deck.

A: first card is a club B: first card is a 7

C: second card is black D: second card is a heart

Find the following probabilities.

     5. $P(A|B)$ **6.** $P(B|A)$ **7.** $P(C|A)$ **8.** $P(\text{not } C|A)$

     9. $P(C|D)$ **10.** $P(D|A)$ **11.** $P(A \text{ or } B)$ **12.** $P(C \text{ or } D)$

     13. $P(B \text{ or not } A)$ **14.** $P(A \text{ and } D)$ **15.** $P(A \text{ and } C)$ **16.** $P(A \text{ and not } C)$

     17. Suppose someone rolls two dice and announces that their sum is 6. What is the probability that both dice show "3"?

     18. What is the probability of getting a sum of 10 when a die is rolled twice? What is the probability of *not getting* a sum of 10 when a die is rolled twice?

     19. Suppose dice were made as regular dodecahedrons with the faces numbered from 1 to 12. If 2 such dice are rolled what is the probability that the sum is 2? 3? 4? What sum is most likely to appear?

     20. Two different numbers are picked from the numbers 1, 2, 3, and 4. What is the probability that their sum is greater than 4?

     21. A coin is flipped twice. What is the probability of getting at least one "tail"? (*Hint:* First consider the probability of getting no "tails.")

     22. A die is rolled twice. What is the probability of getting:
     a. no sixes?   **b.** at least one six?

     23. According to insurance tables, the probability that A dies before age 80 is 0.4, and the probability that B dies before age 80 is 0.3. Find the probabilities of the following events. (Assume that the ages at death for A and B are independent.)
     a. Both A and B die before age 80.
     b. Neither dies before age 80.
     c. A dies before age 80, but B does not.
     d. At least one of A and B dies before age 80.

24. Three people are chosen at random. Find the probabilities of the following events.

  a. All were born on Mondays.

  b. None was born on a Monday.

  c. They were born on three different days of the week.

In Exercises 25–30, use the table below, which gives the fields of concentration of the 400 students in a small college.

Field of Concentration	Class				
	Freshman	Sophomore	Junior	Senior	Totals
Natural sciences	50	35	33	29	147
Social sciences	20	25	28	24	97
Humanities	40	40	39	37	156
	110	100	100	90	400

25. If a student is selected at random, what is the probability that:

  a. the student's field of concentration is the natural sciences?

  b. the student is a freshman in the social sciences?

26. If a sophomore is selected at random, what is the probability that his or her field is the humanities?

27. a. What is the probability that a senior's field of concentration is the natural sciences?

  b. What is the probability that a student in the natural sciences is a senior?

28. a. What is the probability that a junior is concentrating in the humanities?

  b. What is the probability that a student in the humanities is a junior?

29. A student is selected at random. Are the events "student is a junior" and "student is in humanities" independent?

30. A student is selected at random. Are the events "student is a junior" and "student is in natural sciences" independent?

B **31.** Three cards are chosen from a standard deck. What are the probabilities of the following events?

  a. All hearts   **b.** All red cards   **c.** No aces

  d–f. What are the probabilities of the events in parts (a)–(c) if each card is replaced before the next is drawn?

32. Two cards are drawn from a standard deck. What are the probabilities of the following events?

  a. Two aces   **b.** Two red aces   **c.** No red aces

  d–f. What are the probabilities of the events in parts (a)–(c) if the first card is replaced before the second is drawn?

33. Suppose you shuffle together the 12 face cards from a deck. If 3 of these 12 cards are drawn, what is the probability that each of them is a:

  a. jack?   **b.** spade?   **c.** black card?

34. Repeat Exercise 33 if you pick just 2 of the 12 cards.

35. Three letters are chosen at random from the word DRAWING. What is the probability of choosing:
a. no vowels? **b.** at least one vowel?

36. A 3-letter word is randomly formed from 3 of the 6 letters of WHALES. What is the probability that the 3-letter word begins and ends with a consonant?

37. A number between 1000 and 9999 inclusive is chosen at random. What is the probability that it:
a. has no zeros? **b.** begins and ends with an odd digit?

38. A die is rolled 3 times. What is the probability that the 3 numbers are all different?

CALCULATOR OR COMPUTER EXERCISE

The Birthday Problem: What is the probability that in a group of 10 randomly chosen people, at least two will have the same birthday? To answer this question, it is easiest to find first, the probability of *no* birthdays in common. If you go from person to person writing down their birthdays then:

$$P(\text{no match}) = P\left(\begin{array}{l}\text{second person}\\\text{does not match}\\\text{first}\end{array}\right) \times P\left(\begin{array}{l}\text{third person}\\\text{does not match}\\\text{either of first two}\end{array}\right) \times \cdots \times P\left(\begin{array}{l}\text{tenth person}\\\text{does not match}\\\text{any of first nine}\end{array}\right)$$

$$= \frac{364}{365} \times \frac{363}{365} \times \frac{362}{365} \times \cdots \times \frac{356}{365} = \frac{P(364, 9)}{365^9}$$

Thus $P(\text{at least 1 match}) = 1 - \dfrac{P(364, 9)}{365^9}$

a. Evaluate this probability of a match.
b. Repeat the problem if there are 20 people instead of 10.
c. How many people are needed before there is a 50% chance that there is at least one match of birthdays?

Maria Gaetana Agnesi (1718–1799) was a remarkable Italian mathematician, a child prodigy who, by the age of 20, had mastered the mathematics of Newton, Fermat, Descartes, Euler, and Bernoulli. Her most important work was Analytical Institutions, *a classic and comprehensive text on calculus and analysis.*

15-4/PROBLEMS SOLVED WITH COMBINATIONS

Example 1 following considers the experiment of drawing five cards from a standard deck. We show two methods of finding the probability that all five cards are hearts. Method 1 uses the approach of conditional probability, which was presented in Section 15–3. Method 2, which uses combinations, is presented because it can be used to solve many problems that are not readily solved using conditional probability.

EXAMPLE 1. Five cards are chosen at random from a standard deck. Find the probability that all are hearts.

SOLUTION:

Method 1. $P(\text{all hearts}) = \dfrac{13}{52} \cdot \dfrac{12}{51} \cdot \dfrac{11}{50} \cdot \dfrac{10}{49} \cdot \dfrac{9}{48} = \dfrac{33}{66,640} \approx 4.95 \times 10^{-4}$

Method 2. The experiment of choosing 5 cards from 52 has $C(52, 5)$ equally likely outcomes. We are interested in the combinations consisting of picking 5 (of the 13) hearts. There are $C(13, 5)$ such combinations. Therefore:

$$P(\text{all hearts}) = \frac{C(13, 5)}{C(52, 5)} = \frac{1287}{2,598,960} \approx 4.95 \times 10^{-4}$$

EXAMPLE 2. Five cards are chosen at random from a standard deck. Find the probability that exactly 2 cards are hearts.

SOLUTION: In this example it is impractical to use a card-by-card approach based on conditional probability, as in Method 1 of Example 1. The difficulty is that we do not know whether the hearts will appear as the first and second cards, as the first and third cards, or as some other two cards. Instead, we determine the number of 5-card combinations that contain exactly 2 hearts. There are $C(13, 2)$ choices for the 2 hearts, and there are $C(39, 3)$ choices for the other 3 cards. According to the Multiplication Principle (page 535), there are $C(13, 2) \cdot C(39, 3)$ combinations of 5 cards containing exactly 2 hearts. Since there are $C(52, 5)$ combinations of 5 cards, we have:

$$P\left(\begin{array}{c}2 \text{ hearts and } 3 \\ \text{cards not hearts}\end{array}\right) = \frac{C(13, 2) \cdot C(39, 3)}{C(52, 5)}$$

$$= \frac{78 \cdot 9139}{2,598,960} \approx 2.74 \times 10^{-1}$$

EXAMPLE 3. Thirteen cards are dealt from a standard deck. What is the probability that the 13 cards contain exactly 4 aces and exactly 3 kings?

SOLUTION: 1. There are $C(4, 4)$ choices for the aces.

2. There are $C(4, 3)$ choices for the kings.

3. The remaining 6 cards must be chosen from the 44 cards that are neither aces nor kings. Thus, there are $C(44, 6)$ choices for the remaining cards.

4. There are $C(52, 13)$ choices for the 13 cards.

5. Using these results and the Multiplication Principle, we find:

$$P\left(\begin{array}{l}\text{13 cards with exactly}\\\text{4 aces and exactly 3 kings}\end{array}\right) = \frac{C(4, 4) \cdot C(4, 3) \cdot C(44, 6)}{C(52, 13)}$$

$$= \frac{1 \cdot 4 \cdot 7{,}059{,}052}{C(52,13)} \approx 4.45 \times 10^{-5}$$

ORAL EXERCISES 15-4

1. Three marbles are picked at random from a bag containing 4 red marbles and 5 white marbles. Match each event with its probability.

   Events:

   a. All 3 marbles are red.
   b. Exactly 2 marbles are red.
   c. Exactly 1 marble is red.
   d. No marble is red.

   Probabilities:

   $$\frac{C(4, 1) \cdot C(5, 2)}{C(9, 3)}$$

   $$\frac{C(4, 2) \cdot C(5, 1)}{C(9, 3)}$$

   $$\frac{C(4, 3) \cdot C(5, 0)}{C(9, 3)}$$

   $$\frac{C(4, 0) \cdot C(5, 3)}{C(9, 3)}$$

2. Five cards are drawn from a standard deck. Match each event with its probability.

   Events:

   a. All 4 aces are chosen.
   b. No aces are chosen.
   c. Exactly 4 diamonds are chosen.
   d. Four aces and 1 jack are chosen.

   Probabilities:

   $$\frac{C(4, 0) \cdot C(48, 5)}{C(52, 5)}$$

   $$\frac{C(4, 4) \cdot C(48, 1)}{C(52, 5)}$$

   $$\frac{C(4, 4) \cdot C(4, 1)}{C(52, 5)}$$

   $$\frac{C(13, 4) \cdot C(39, 1)}{C(52, 5)}$$

3. Five cards are chosen from a standard deck.
   a. You can find the probability of at least 1 ace by calculating the following sum:

   $$P\left(\begin{array}{l}\text{exactly}\\\text{1 ace}\end{array}\right) + P\left(\begin{array}{l}\text{exactly}\\\text{2 aces}\end{array}\right) + P\left(\begin{array}{l}\text{exactly}\\\text{3 aces}\end{array}\right) + P\left(4 \text{ aces}\right)$$

   It is far easier, however, to find $1 - P(?)$.
   b. Find the probability of at least 1 ace using the method suggested in part (a).

In Exercises 1 and 2, leave your answers in terms of factorials unless directed otherwise by your instructor.

A   **1.** Five cards are dealt from a standard deck. What is the probability of getting:
   a. all four aces?   **b.** no aces?   **c.** at least one ace?

2. Thirteen cards are dealt from a standard deck. What is the probability of getting:
   a. all clubs?   **b.** no clubs?   **c.** at least 1 club?

3. There are 5 red marbles and 3 white marbles in a bag. If 2 marbles are drawn at random, what is the probability that the number of red marbles is 0? 1? 2? Check to see that your answers total 1.

4. Repeat Exercise 3 if the bag contains 6 red marbles and 2 white marbles; that is, find the probabilities for all possible numbers of red marbles.

5. Free concert tickets are distributed to 4 students chosen at random from 8 juniors and 12 seniors in the school band. Find the probability that free tickets were received by:
   a. 4 seniors   **b.** exactly 3 seniors   **c.** exactly 2 seniors
   d. exactly 1 senior   **e.** no seniors

6. A town council is composed of 8 Democrats, 7 Republicans, and 5 Independents. A committee of 3 is chosen by selecting names from a hat. What is the probability that the committee has:
   a. 2 Democrats and 1 Republican?
   b. 3 Independents?
   c. no Independents?
   d. 1 Democrat, 1 Republican, and 1 Independent?

B   **7.** Thirteen cards are dealt from a standard deck. What is the probability of getting:
   a. all red cards?   **b.** 7 diamonds and 6 hearts?
   c. At least 1 face card (jack, queen, or king)?

8. Thirteen cards are dealt from a standard deck. What is the probability of getting:
   a. all of the 12 face cards?
   b. 7 spades, 3 hearts, and 3 clubs?
   c. at least one diamond?

9. A quality control inspector randomly inspects 4 transistors in every box of 100. If 1 or more transistors are defective, the whole box is not accepted for shipment. Suppose a box contains 10 defective transistors and 90 good ones. What is the probability it will not be accepted?

10. A lot of 20 T.V. sets consists of 6 defective sets and 14 good ones. If a sample of 3 sets is chosen, what is the probability that the sample will contain:
   a. all defective sets?   **b.** at least one defective set?

11. A committee of 4 is chosen at random from 5 married couples. What is the probability that the committee will contain no two people who are married to each other?

12. The letters of the word ABRACADABRA are written on cards and placed in a hat. If 5 cards are chosen at random, what is the probability of getting:
   a. all A's?   **b.** at least one B?   **c.** one of each letter?

15-5/THE BINOMIAL THEOREM

Our goal in this section is to derive a formula for expanding $(a + b)^n$ for positive integers n. For small values of n we get the following series:

$$(a + b)^1 = 1a + 1b$$
$$(a + b)^2 = 1a^2 + 2ab + 1b^2$$
$$(a + b)^3 = 1a^3 + 3a^2b + 3ab^2 + 1b^3$$
$$(a + b)^4 = 1a^4 + 4a^3b + 6a^2b^2 + 4ab^3 + 1b^4$$
$$(a + b)^5 = 1a^5 + 5a^4b + 10a^3b^2 + 10a^2b^3 + 5ab^4 + 1b^5$$

From these examples we can see that the first term of the series for $(a + b)^n$ is always $1a^nb^0$. In successive terms the exponents of a decrease by 1 and the exponents of b increase by 1, so that the sum of the two exponents in a term is always n. The coefficients of the terms also have a pattern which can be seen by studying the array of numbers below. The first five rows of this array are like the array above except that the a's, b's, and plus signs have been omitted.

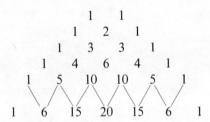

```
        1   1
      1   2   1
    1   3   3   1
  1   4   6   4   1
1   5   10   10   5   1
1   6   15   20   15   6   1
```

This array is called **Pascal's Triangle,** named for the French mathematician Blaise Pascal. (See the biography on page 553.) Notice that (except for the 1's) each number is the sum of the two numbers just above it. Hence, from the fifth row of the triangle, we can quickly form the sixth row, as shown. You should now be able to write the series expansion for $(a + b)^6$.

So far, it might seem that to get the numbers in the sixth row of Pascal's Triangle you must first know the numbers in the fifth row, but this is not necessary. Each number in the sixth row can be calculated directly by formula.

	$C(6, 0)$	$C(6, 1)$	$C(6, 2)$	$C(6, 3)$	$C(6, 4)$	$C(6, 5)$	$C(6, 6)$
6th row	1	$\dfrac{6}{1}$	$\dfrac{6 \cdot 5}{1 \cdot 2}$	$\dfrac{6 \cdot 5 \cdot 4}{1 \cdot 2 \cdot 3}$	$\dfrac{6 \cdot 5 \cdot 4 \cdot 3}{1 \cdot 2 \cdot 3 \cdot 4}$	$\dfrac{6 \cdot 5 \cdot 4 \cdot 3 \cdot 2}{1 \cdot 2 \cdot 3 \cdot 4 \cdot 5}$	$\dfrac{6!}{6!}$
	1	6	15	20	15	6	1

You may be wondering why combinations have anything to do with the sixth row of Pascal's Triangle and $(a + b)^6$. To see why, we write $(a + b)^6$ in factored form:

$$(a + b)^6 = (a + b)(a + b)(a + b)(a + b)(a + b)(a + b)$$

The term a^6 in the series expansion for $(a + b)^6$ is obtained by multiplying the a's in the 6 factors. The term containing a^5b is obtained by multiplying an a from 5 of the factors and a b from 1 of them. Since there are $C(6, 1) = 6$ ways to choose the factor that contributes the b, there will be 6 of these a^5b terms. Likewise, the term containing a^4b^2 is obtained by multiplying an a from 4 of the factors and a b from 2 of the factors. Since there are $C(6, 2) = 15$ ways to choose the 2 factors that contribute the b's, there will be 15 of these a^4b^2 terms. A similar argument can be used to prove the following theorem:

THE BINOMIAL THEOREM.   If n is a positive integer, then:

$$(a + b)^n = C(n, 0)a^nb^0 + C(n, 1)a^{n-1}b^1 + C(n, 2)a^{n-2}b^2$$
$$+ C(n, 3)a^{n-3}b^3 + \cdots + C(n, n)a^0b^n.$$

Equivalently: $(a + b)^n = 1a^nb^0 + \dfrac{n}{1}a^{n-1}b^1 + \dfrac{n(n - 1)}{1 \cdot 2}a^{n-2}b^2$

$$+ \dfrac{n(n - 1)(n - 2)}{1 \cdot 2 \cdot 3}a^{n-3}b^3 + \cdots + 1a^0b^n$$

EXAMPLE.   Give the first four terms in the series expansion for $(x - 2y)^{10}$ in simplified form.

SOLUTION: We first use the Binomial Theorem to expand $(a + b)^{10}$:

$$(a + b)^{10} = a^{10} + 10a^9b + \dfrac{10 \cdot 9}{1 \cdot 2}a^8b^2 + \dfrac{10 \cdot 9 \cdot 8}{1 \cdot 2 \cdot 3}a^7b^3 + \cdots + b^{10}$$

$$(a + b)^{10} = a^{10} + 10a^9b + 45a^8b^2 + 120a^7b^3 + \cdots + b^{10}$$

Then, substituting x for a and $-2y$ for b in the equation above, we obtain:

$$(x - 2y)^{10} = x^{10} + 10x^9(-2y) + 45x^8(-2y)^2 + 120x^7(-2y)^3 + \cdots + (-2y)^{10}$$

$$(x - 2y)^{10} = x^{10} - 20x^9y + 180x^8y^2 - 960x^7y^3 + \cdots + 1024y^{10}$$

ORAL EXERCISES 15-5

1. Study Pascal's Triangle (page 557). Complete the next row of the triangle.

2. **a.** Find the first 4 numbers in the eighth row of Pascal's Triangle.
   b. State the first 4 terms of the series expansion $(x + y)^8$.

Use Pascal's Triangle to give the series expansion of the following.

3. $(a - b)^3$ 4. $(a + b)^4$ 5. $(a - b)^4$

6. Find **(a)** the third term in the expansion of $(x + y)^6$, and **(b)** the fourth term in the expansion of $(x + y)^9$.

In Exercises 1–10, give the series expansion of the given expression. Simplify your answers.

A

1. **a.** $(a + b)^5$ **b.** $(a - b)^5$

2. **a.** $(p + q)^6$ **b.** $(p - q)^6$

3. **a.** $(x + y)^7$ **b.** $(x - y)^7$ **c.** $(2x - y)^7$

4. **a.** $(a + b)^3$ **b.** $(a - 3b)^3$ **c.** $(a^2 - 3b)^3$

5. $\left(1 + \dfrac{x}{2}\right)^3$ **6.** $(2x^2 - 1)^4$ **7.** $(x^2 + 1)^5$

8. $(x^2 - y^2)^3$ **9.** $(30 + 1)^4$ **10.** $(30 - 1)^4$

In Exercises 11–14, find the first four terms of the series expansion of the given expression. Do *not* simplify your answers.

11. $(a^2 - b)^{100}$ **12.** $(3p + 2q)^{20}$

13. $(\sin x + \sin y)^{10}$ **14.** $(\sin x - \cos y)^{30}$

15. Find the value of $(1.01)^5$ to the nearest hundredth by considering the series expansion of $(1 + 0.01)^5$.

16. Find the value of $(0.99)^5$ to the nearest hundredth by considering the series expansion of $(1 - 0.01)^5$.

B

17. In the expansion of $(a + b)^{12}$, what is the coefficient of a^8b^4? of a^4b^8?

18. In the expansion of $(a + b)^{20}$, what is the coefficient of $a^{17}b^3$? of a^3b^{17}?

19. In the expansion of $\left(x^2 + \dfrac{2}{x}\right)^{12}$, find the term that contains no x.

20. In the expansion of $(a^3 - 2)^{10}$, find the term containing a^{18}.

21. A municipal council consists of a mayor and n councilors.
    a. How many committees of k members can be chosen from this council?
    b. How many committees can be chosen if the mayor is on the committee? if the mayor is not on the committee?

22. Use your answers to Exercise 21 to complete the following equation:
$$C(n + 1, k) = C(?, k - 1) + C(?, k)$$

23. **a.** Show that the series expansion for $(\cos \theta + i \sin \theta)^3$ simplifies to
$$(\cos^3 \theta - 3 \cos \theta \sin^2 \theta) + i(3 \sin \theta \cos^2 \theta - \sin^3 \theta).$$
    b. What does De Moivre's Theorem say about $(\cos \theta + i \sin \theta)^3$?
    c. Use parts (a) and (b) to show that:
$$\cos 3\theta = \cos^3 \theta - 3 \cos \theta \sin^2 \theta$$
$$\sin 3\theta = 3 \sin \theta \cos^2 \theta - \sin^3 \theta$$

24. Use the series expansion for $(\cos \theta + i \sin \theta)^4$ to show that:
$$\cos 4\theta = \cos^4 \theta - 6 \cos^2 \theta \sin^2 \theta + \sin^4 \theta$$
$$\sin 4\theta = 4 \cos^3 \theta \sin \theta - 4 \cos \theta \sin^3 \theta$$

25. Let S_n be the sum of the numbers in the nth row of Pascal's Triangle. Guess a formula for S_n. Can you prove that your formula is correct?

26. Rewrite the Binomial Theorem using sigma (Σ) notation.

In the Binomial Theorem, if you let $a = 1$ and $b = x$, you get:

$$(1 + x)^n = 1 + nx + \frac{n(n-1)}{1 \cdot 2}x^2 + \frac{n(n-1)(n-2)}{1 \cdot 2 \cdot 3}x^3 + \cdots$$

If n is not a positive integer, the right side is an infinite series which has the sum $(1 + x)^n$ when $|x| < 1$. (A proof of this requires calculus.) Use this fact to do Exercises 27–30.

C   **27.** Show that $(1 + x)^{-1} = 1 - x + x^2 - x^3 + \cdots$. (As a check on your work, notice that the right side of the equation is an infinite geometric series with first term 1 and ratio $-x$. Thus (recall page 473) the series has sum $\dfrac{1}{1 - (-x)} = (1 + x)^{-1}$ when $|x| < 1$.)

28. Show that $(1 + x)^{-2} = 1 - 2x + 3x^2 - 4x^3 + \cdots$.

29. Show that $(1 + x)^{\frac{1}{2}}$ is approximately $1 + \frac{1}{2}x$ when $|x|$ is small. Use your result to approximate $\sqrt{1.04}$ and $\sqrt{0.98}$.

30. Show that $(1 + x)^{\frac{1}{3}}$ is approximately $1 + \frac{1}{3}x$ when $|x|$ is small. Use your result to approximate $\sqrt[3]{1.12}$ and also $\sqrt[3]{67} = \sqrt[3]{64(1 + \frac{3}{64})} = 4\sqrt[3]{1 + \frac{3}{64}}$.

15–6/APPLYING THE BINOMIAL SERIES TO PROBABILITY

If a die is rolled 3 times, each roll is *independent* of the others; that is, no outcome of any roll affects the outcome of any other roll. When the successive outcomes of an experiment are independent in this way, we can multiply the probabilities of each outcome to obtain the probability of the entire sequence of outcomes. For example, the probability of rolling a die 3 times and getting 2 consecutive sixes (S) followed by a non-six (N) is

$$P(SSN) = \tfrac{1}{6} \cdot \tfrac{1}{6} \cdot \tfrac{5}{6} = \tfrac{5}{216}.$$

A complete list of the various ways in which sixes can appear with 3 rolls of a die is given in the table below.

	3 sixes	2 sixes			1 six			0 sixes
Outcome	SSS	SSN	SNS	NSS	SNN	NSN	NNS	NNN
Probability	$\tfrac{1}{6} \cdot \tfrac{1}{6} \cdot \tfrac{1}{6}$	$\tfrac{1}{6} \cdot \tfrac{1}{6} \cdot \tfrac{5}{6}$	$\tfrac{1}{6} \cdot \tfrac{5}{6} \cdot \tfrac{1}{6}$	$\tfrac{5}{6} \cdot \tfrac{1}{6} \cdot \tfrac{1}{6}$	$\tfrac{1}{6} \cdot \tfrac{5}{6} \cdot \tfrac{5}{6}$	$\tfrac{5}{6} \cdot \tfrac{1}{6} \cdot \tfrac{5}{6}$	$\tfrac{5}{6} \cdot \tfrac{5}{6} \cdot \tfrac{1}{6}$	$\tfrac{5}{6} \cdot \tfrac{5}{6} \cdot \tfrac{5}{6}$
	$(\tfrac{1}{6})^3$	$3(\tfrac{1}{6})^2(\tfrac{5}{6})$			$3(\tfrac{1}{6})(\tfrac{5}{6})^2$			$(\tfrac{5}{6})^3$

Now compare:

$$(S + N)^3 \;=\; S^3 \;+\; 3S^2N \;+\; 3SN^2 \;+\; N^3$$

If the die were rolled 4 times instead of 3, we would have the following binomial distribution of probabilities:

Outcome	4 sixes	3 sixes	2 sixes	1 six	0 sixes
Probability	$(\frac{1}{6})^4$	$4(\frac{1}{6})^3(\frac{5}{6})$	$6(\frac{1}{6})^2(\frac{5}{6})^2$	$4(\frac{1}{6})(\frac{5}{6})^3$	$(\frac{5}{6})^4$

$$(S + N)^4 = S^4 + 4S^3N + 6S^2N^2 + 4SN^3 + N^4$$

The two previous illustrations can be generalized to give the following theorem:

THE BINOMIAL PROBABILITY THEOREM. Suppose an experiment consists of a sequence of n repeated independent trials, each trial having two possible outcomes, either A or B. If on each trial, $P(A) = a$ and $P(B) = b$, then the binomial expansion of $(a + b)^n$ gives the following probabilities for the occurrences of A:

Outcome	n A's	$(n - 1)$ A's	$(n - 2)$ A's	$\cdots$	k A's	$\cdots$
Probability	a^n	$C(n, n - 1)a^{n-1}b$	$C(n, n - 2)a^{n-2}b^2$	$\cdots$	$C(n, k)a^k b^{n-k}$	$\cdots$

$$(a + b)^n = a^n + C(n, n - 1)a^{n-1}b + C(n, n - 2)a^{n-2}b^2 + \cdots + C(n, k)a^k b^{n-k} + \cdots$$

EXAMPLE. A coin is tossed 10 times. What is the probability of exactly 4 "heads"?

SOLUTION: We use the Binomial Probability Theorem with $n = 10$, $k = 4$, and $a = b = \frac{1}{2}$.

$$P(4 \text{ out of 10 result in "heads"}) = C(10, 4)(\tfrac{1}{2})^4(\tfrac{1}{2})^6$$

$$= 210(\tfrac{1}{2})^{10} = \frac{210}{1024} \approx 0.205$$

ORAL EXERCISES 15-6

1. If $H = T = \frac{1}{2}$, then:
$$(H + T)^4 = (\tfrac{1}{2})^4 + 4(\tfrac{1}{2})^3(\tfrac{1}{2}) + 6(\tfrac{1}{2})^2(\tfrac{1}{2})^2 + 4(\tfrac{1}{2})(\tfrac{1}{2})^3 + (\tfrac{1}{2})^4$$
The first term of this series is the probability of getting 4 heads in 4 tosses of a coin. What probabilities do the other terms of the series represent?

2. In your own words, tell how to go about finding the probability of having 3 out of 4 rolled dice come up "5."

3. Suppose a die is rolled twice and comes up "3" on the first roll. What is the probability that the second roll also shows "3"?

4. A coin is tossed 9 times and comes up "heads" each time. What is the probability that it will come up "heads" on the next toss?

A   **1.** Make a table, like the table on page 560, showing the 8 different ways in which "heads" (*H*) and "tails" (*T*) can occur if a coin is tossed 3 times. Find the probability that the number of "heads" is:
   a. 3   **b.** 2   **c.** 1   **d.** 0

2. Repeat Exercise 1, assuming that the coin is bent and that the probability of its coming up "heads" on any toss is $\frac{2}{5}$.

3. Make a table, like the table on page 560, showing the 16 different ways in which sixes (*S*) and non-sixes (*N*) can occur when a die is rolled 4 times. Find the probability that the number of sixes is:
   a. 4   **b.** 3   **c.** 2   **d.** 1   **e.** 0

4. Consider the set of all families with exactly 4 children. If one of these families is picked at random, find the probability that it has:
   a. 4 boys   **b.** 3 boys   **c.** 2 boys   **d.** 1 boy   **e.** 0 boys
   Assume that $P(\text{boy}) = \frac{1}{2}$.

5. What is the probability of getting 2 "fives" in 4 rolls of a die?

6. What is the probability of getting 1 "three" in 7 rolls of a die?

7. a. If a card is drawn from a well-shuffled deck, what is the probability of drawing a spade?
   b. If a card is drawn from each of two decks, what is the probability of drawing 0 spades?   1 spade?   2 spades?

8. If a card is drawn from each of 3 decks, what is the probability of drawing 3 spades?   2 spades?   1 spade?   0 spades?

9. A quiz has 7 multiple-choice questions, each with 4 choices. If you guess at every question, what is the probability of getting them all right?   of getting 6 out of 7 right?

10. A mathematics quiz consists of 7 multiple-choice questions, with 3 choices for the answer to each. A student guesses at the answers throughout the test. What is the probability of getting all 7 answers correct? of getting 6 answers correct out of 7?

11. The integers 1, 2, 3, 4, and 5 are written on cards that are placed in a hat. Three cards are then drawn at random. Copy and complete the table below concerning this experiment.

	Drawing with replacement	Drawing without replacement
$P(\text{no odd integers})$	?	?
$P(\text{exactly 1 odd integer})$	?	?
$P(\text{exactly 2 odd integers})$	?	?
$P(\text{exactly 3 odd integers})$	?	?

12. Suppose the integers 1, 2, 3, . . . , 10 are written on cards that are placed in a hat. Four cards are then drawn at random. Make a table for this experiment like the table in Exercise 11. Show the probabilities of choosing exactly 0, 1, 2, 3, and 4 odd integers.

B 13. If 8 out of every 10 nutritionists recommend Brand X, what is the probability that nutritionists A, B, and C all recommend Brand X? What is the probability that none of them do? That at least one of them does?

14. In a certain high school, one third of the senior boys are at least 6 feet tall. What is the probability that in a randomly selected group of 7 senior boys all are less than 6 feet tall? That all but one are less than 6 feet tall?

15. A bag contains 4 oranges and 3 apples. Suppose you pull out a piece of fruit and then replace it and mix up the contents of the bag. If you do this twice more, what is the probability that the number of oranges chosen is 0? 1? 2? 3?

16. **a.** A baseball player's batting average is .270. What is the probability that the player gets exactly 1 hit in the next 5 times at bat?
    b. When you use the Binomial Probability Theorem to do part (a), you are assuming that the probability of a hit is always .270. Explain why this assumption may not be valid.

17. In a given package of tomato seeds, 9 seeds out of 10 will sprout on the average. What is the probability that out of the first 10 seeds planted, 1 will not sprout?

Chapter summary

1. *The Multiplication Principle:* If an action can be performed in n_1 ways and then, for each of these ways, another action can be performed in n_2 ways, the two actions can be performed consecutively in $n_1 n_2$ ways.

2. *The Addition Principle:* If two actions are mutually exclusive, and the first can be done in n_1 ways and the second in n_2 ways, then one action or the other can be done in $n_1 + n_2$ ways.

3. The choice of r objects out of n objects is called a *permutation* $P(n, r)$ if the order of choosing is important and a *combination* $C(n, r)$ if the order is unimportant.

$$P(n, r) = \frac{n!}{n - r!} \qquad C(n, r) = \frac{n!}{(n - r)! r!}$$

4. If an experiment has n *equally likely outcomes* and m of these correspond to an event A happening, then the *probability* of A is $P(A) = \frac{m}{n}$. The equation $P(A) = 1 - P(\text{not } A)$ is often useful. For example, the probability of at least one ace is $1 - P(\text{no aces})$.

5. Two events A and B are *mutually exclusive if* they both cannot occur simultaneously; that is $P(A$ and $B) = 0$.

6. Two events A and B are *independent if* the occurrence of one does not affect the probability of the other; that is, $P(A|B) = P(A)$.

7. *Addition Rule:* $P(A$ or $B) = P(A) + P(B) - P(A$ and $B)$
   Special Case: $P(A$ or $B) = P(A) + P(B)$ if A and B are mutually exclusive

8. *Multiplication Rule:* $P(A$ and $B) = P(A) \cdot P(B|A)$
   Special Case: $P(A$ and $B) = P(A) \cdot P(B)$ if A and B are independent

9. *Pascal's Triangle* and the *Binomial Theorem* are useful in expanding $(a + b)^n$.

10. The *Binomial Probability Theorem on* page 561 should be reviewed.

Chapter test

1. How many 4-digit numbers can be formed using the digits 1, 2, 3, 4, 5, 6:   15–1
   a. if the number is to contain no repeated digits?
   b. if repeated digits are allowed?

2. In how many ways can you answer 5 true-false questions?

3. In how many ways can a president, a vice-president, and a treasurer be elected from a club of 12 members?

4. In how many ways can a committee of 3 be elected from a club of 12   15–2
   members?

5. Two dice are rolled. What is the probability that their sum is 7 or 11?   15–3

6. Two cards are dealt from a 52-card deck. What is the probability that both cards are **(a)** aces? **(b)** hearts?

7. A bag contains 4 red balls and 2 white balls. If 2 different balls are picked at random, what is the probability of getting:
   a. both red?   **b.** both white?   **c.** one of each color?

8. The table below shows some data for a small high school.

Average	Freshman	Sophomore	Junior	Senior	Total
80 or better	22	25	24	29	100
Below 80	34	27	24	15	100
	56	52	48	44	200

Four events using data given in the table are defined at the top of the following page. Using the table and the events defined, answer questions (a)–(g), which concern a student picked at random.

F: a freshman is chosen
J: a junior is chosen
A: a student with an average of 80 or better is chosen
C: a student with an average below 80 is chosen

Find the following probabilities.
a. $P(F)$ **b.** $P(F|A)$ **c.** $P(A|F)$ **d.** $P(A|\text{not } F)$ **e.** $P(A \text{ or } F)$
f. Are the events F and A independent?
g. Are the events J and C independent?

9. Four cards are drawn from a standard deck. Find the probability that the four cards chosen are face cards from four different suits.

   15-4

10. **a.** Write the first 4 rows of Pascal's Triangle.
   15-5
   b. Give the series expansion for $(2x - y)^4$.

11. Write the first 4 terms in the series expansion for $(a + b)^{25}$.

12. A wooden cube is painted red on 2 of its 6 sides. If the cube is rolled 3 times, find the probability that a red face comes up:

   15-6

   a. 3 times   **b.** 2 times   **c.** once   **d.** not at all

13. What is the probability of getting at least 7 heads when a coin is flipped 8 times?

566

CHAPTER SIXTEEN

Introduction to calculus

OBJECTIVES

1. To find limits for functions or quotients of functions.
2. To graph functions that are quotients of polynomials, showing asymptotes for the graphs.
3. To find derivatives of functions.
4. To graph functions using derivatives.
5. To solve extreme-value problems using derivatives.
6. To find instantaneous velocities and accelerations.

16-1/LIMITS OF FUNCTIONS

The limit as x approaches infinity or negative infinity

In Chapter 13 we discussed limits of infinite sequences and saw, for example, that the following sequence has limit 1.

$$\frac{2}{1}, \frac{3}{2}, \frac{4}{3}, \frac{5}{4}, \ldots, \frac{n+1}{n}, \ldots$$

The Space Shuttle, shown at left, uses a great deal of fuel to attain its initial acceleration. Section 16-6 shows how an object's position, velocity, and acceleration are related.

We expressed this fact by writing:

(1)
$$\lim_{n \to \infty} \frac{n+1}{n} = 1$$

Likewise, for the function $f(x) = \frac{x+1}{x}$, we write

(2)
$$\lim_{x \to \infty} \frac{x+1}{x} = 1$$

and say that "the limit of $\frac{x+1}{x}$ as x approaches infinity is 1." The only difference between statements (1) and (2) is that the sequence $t_n = \frac{n+1}{n}$ is defined only for positive integers n, whereas the function $f(x) = \frac{x+1}{x}$ is defined for all real numbers $x \neq 0$.

Because $f(x)$ is defined for negative as well as positive numbers, we can also talk about "the limit of $f(x)$ as x approaches negative infinity." For the case in which $f(x) = \frac{x+1}{x}$, this limit is again 1, and we write

$$\lim_{x \to -\infty} \frac{x+1}{x} = 1.$$

These symbols mean that $\frac{x+1}{x}$ can be made as close to 1 as we like just by considering negative values of x with large enough absolute value. For example, when $x = -1,000,000$, $\frac{x+1}{x} = \frac{-999,999}{-1,000,000} \approx 1.$

If the absolute values of a function become arbitrarily large as x approaches infinity or negative infinity, then the function has no finite limit but will approach $-\infty$ or ∞. A formal definition is given on page 572, but Example 1 should make these ideas clear.

EXAMPLE 1. Evaluate: **a.** $\lim_{x \to \infty} x^{\frac{1}{3}}$ **b.** $\lim_{x \to -\infty} x^{\frac{1}{3}}$

SOLUTION: **a.** Since values of $x^{\frac{1}{3}}$ become arbitrarily large as x becomes arbitrarily large, $\lim_{x \to \infty} x^{\frac{1}{3}} = \infty$.

b. As x approaches negative infinity, values of $x^{\frac{1}{3}}$ are negative and $|x^{\frac{1}{3}}|$ becomes arbitrarily large. Thus, $\lim_{x \to -\infty} x^{\frac{1}{3}} = -\infty$.

EXAMPLE 2. Explain why $\lim_{x \to \infty} x \sin x \neq \infty$.

SOLUTION: As x approaches infinity, $\sin x$ oscillates between 1 and -1. This means that $x \sin x$ equals zero infinitely often as x approaches infinity. Therefore there is no x beyond which the values of $x \sin x$ continually increase without bound. This means that $\lim_{x \to \infty} x \sin x \neq \infty$.

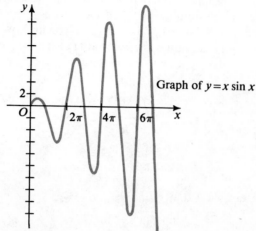

Graph of $y = x \sin x$

The limit as x approaches c

To determine the behavior of a function as x approaches a finite value c, we consider the following two limits:

(1) $\lim_{x \to c^+} f(x)$, read "the limit of $f(x)$ as x approaches c from the right."

(2) $\lim_{x \to c^-} f(x)$, read "the limit of $f(x)$ as x approaches c from the left."

The meaning of limits of this type is discussed in the following two examples.

EXAMPLE 3. The graph of $f(x)$ is shown at the right. Find $\lim_{x \to 2^+} f(x)$ and $\lim_{x \to 2^-} f(x)$.

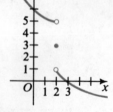

SOLUTION: The fact that $f(2) = 3$ has nothing to do with the solution. When evaluating $\lim_{x \to 2^+} f(x)$, we are concerned with the values of $f(x)$ for values of x near, but greater than, 2. Since these values are getting nearer and nearer to 1 as x approaches 2 from the right (that is, from values greater than 2), we have:

$$\lim_{x \to 2^+} = 1$$

Likewise, the graph shows that the values of $f(x)$ are getting nearer and nearer to 5 as x approaches 2 from the left. Hence we have:

$$\lim_{x \to 2^-} = 5$$

EXAMPLE 4. If $f(x) = \dfrac{x^2 - 4}{x - 2}$, describe the behavior of $f(x)$ near $x = 2$.

(*Solution follows on page 570*)

Introduction to calculus   **569**

SOLUTION: The fact that $f(2)$ is undefined has nothing to do with the solution. The problem is to determine whether the values of $f(x)$ are getting close to any number as x gets close to 2.

By substituting values of x near 2, we get the values of $f(x)$ below. These values suggest that $f(x)$ is approaching 4 as x approaches 2 from the right or from the left.

$$f(2.1) \quad = 4.1 \qquad\qquad f(1.9) \quad = 3.9$$
$$f(2.01) \quad = 4.01 \qquad\qquad f(1.99) \quad = 3.99$$
$$f(2.001) = 4.001 \qquad\qquad f(1.999) = 3.999$$
$$\downarrow \qquad\qquad\qquad\qquad\qquad \downarrow$$
$$\lim_{x \to 2^+} f(x) = 4 \qquad\qquad \lim_{x \to 2^-} f(x) = 4$$

Notice in Example 4 that:

$$\lim_{x \to 2^+} f(x) = \lim_{x \to 2^-} f(x) = 4$$

In this case, we can speak of "*the* limit of $f(x)$ as x approaches 2" and write

$$\lim_{x \to 2} f(x) = 4.$$

In Example 3, however,

$$\lim_{x \to 2^+} f(x) = 1 \quad \text{and} \quad \lim_{x \to 2^-} f(x) = 5.$$

Since the right-hand and left-hand limits are different, we say that $\lim_{x \to 2} f(x)$ does not exist. A formal definition of a limit is given on page 572.

Often the easiest limits to evaluate are those involving a *continuous function*. Roughly speaking, a function is continuous if you can draw its graph without lifting your pencil from the paper. A formal definition of a **continuous function** is as follows: A function $f(x)$ is continuous at c if $\lim_{x \to c} f(x) = f(c)$.

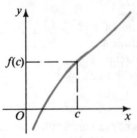

$f(x)$ is continuous at c.
$\lim_{x \to c} f(x) = f(c)$

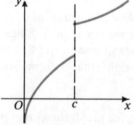

$f(x)$ is discontinuous at c.
Either $f(c)$ is undefined
or $\lim_{x \to c} f(x)$ does not exist

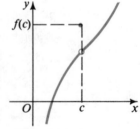

$f(x)$ is discontinuous at c.
$\lim_{x \to c} f(x) \neq f(c)$

Limit problems often involve a quotient of two functions. For many of these problems, the following theorem is useful. (A proof of the theorem is difficult and will not be given.)

THEOREM 1. If $\lim f(x)$ and $\lim g(x)$ both exist, and $\lim g(x) \neq 0$ then,

$$\lim \frac{f(x)}{g(x)} = \frac{\lim f(x)}{\lim g(x)}$$

This theorem applies whether we consider the limit as x approaches ∞, $-\infty$, or a real number c. For this reason, the theorem refers to "lim," instead of $\lim\limits_{x \to \infty}$, $\lim\limits_{x \to -\infty}$ or $\lim\limits_{x \to c}$.

EXAMPLE 5. $\lim\limits_{x \to 2} \dfrac{\sin x}{\sqrt{x^2 + 1}} = \dfrac{\lim\limits_{x \to 2} \sin x}{\lim\limits_{x \to 2} \sqrt{x^2 + 1}} = \dfrac{\sin 2}{\sqrt{5}}.$

Techniques for evaluating $\lim \dfrac{f(x)}{g(x)}$

1. Use the quotient theorem given above, if possible.
2. If $\lim f(x) = 0$ and $\lim g(x) = 0$, try the following techniques.

   a. Factor $g(x)$ and $f(x)$ and reduce $\dfrac{f(x)}{g(x)}$ to lowest terms.

   b. If $f(x)$ or $g(x)$ involves a square root, then multiplying both $f(x)$ and $g(x)$ by the conjugate of the square root expression is sometimes helpful.

EXAMPLE 6. Applying 2(a), $\lim\limits_{x \to 2} \dfrac{x^2 - 4}{x - 2} = \lim\limits_{x \to 2} \dfrac{(x - 2)(x + 2)}{(x - 2)} = \lim\limits_{x \to 2} (x + 2) = 4.$
This agrees with what we found in Example 4.

EXAMPLE 7. Applying 2(b), $\lim\limits_{x \to 0} \dfrac{1 - \sqrt{1 + x}}{x} = \lim\limits_{x \to 0} \dfrac{1 - \sqrt{1 + x}}{x} \left(\dfrac{1 + \sqrt{1 + x}}{1 + \sqrt{1 + x}} \right)$

$$= \lim\limits_{x \to 0} \frac{-x}{x\left(1 + \sqrt{1 + x}\right)}$$

$$= \lim\limits_{x \to 0} \frac{-1}{1 + \sqrt{1 + x}} = -\frac{1}{2}$$

3. If $\lim f(x) \neq 0$ and $\lim g(x) = 0$, then either statement (a) or (b) below is true.

   a. $\lim \dfrac{f(x)}{g(x)}$ does not exist.

   b. $\lim \dfrac{f(x)}{g(x)} = +\infty$ or $-\infty$.

Examples 8 and 9 are limits of quotients as just described.

EXAMPLE 8. $\lim\limits_{x \to 1} \dfrac{x}{x - 1}$ does not exist because

$$\lim\limits_{x \to 1^+} \frac{x}{x - 1} = \infty \text{ and } \lim\limits_{x \to 1^-} \frac{x}{x - 1} = -\infty.$$

EXAMPLE 9. $\displaystyle\lim_{x\to 1}\frac{2}{(x-1)^2}=\infty$

4. If x is approaching infinity or negative infinity, divide numerator and denominator by the highest power of x in any term of the denominator.

EXAMPLE 10. $\displaystyle\lim_{x\to\infty}\frac{x^3-4x}{2x^4+5}=\lim_{x\to\infty}\frac{\dfrac{1}{x}-\dfrac{4}{x^3}}{2+\dfrac{5}{x^4}}=\frac{0-0}{2+0}=0$

5. If all else fails, you can guess $\displaystyle\lim_{x\to\infty}\frac{f(x)}{g(x)}$ by evaluating $\dfrac{f(x)}{g(x)}$ for very large values of x. Also, you can guess $\displaystyle\lim_{x\to c}\frac{f(x)}{g(x)}$ by evaluating $\dfrac{f(x)}{g(x)}$ for values of x very near $x=c$. This was done in Example 4.

Although the ideas of a limit and of a continuous function are fairly easy to understand on an intuitive level, they are quite difficult to express formally. If you do not understand the precise definitions given below, do not become discouraged. Most people practice with them before they understand them.

DEFINITIONS

1. $\displaystyle\lim_{x\to c}f(x)=L$ means that for any small positive number ε (Greek epsilon), there is a positive number δ (Greek delta) such that
$$|f(x)-L|<\varepsilon$$
whenever $0<|x-c|<\delta$.

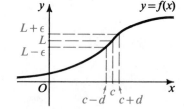

2. $\displaystyle\lim_{x\to\infty}f(x)=L$ means that for any small positive number ε there is a value of x, call it x_1, such that
$$|f(x)-L|<\varepsilon$$
whenever $x>x_1$.

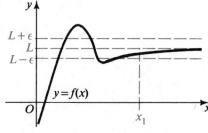

3. $\displaystyle\lim_{x\to\infty}f(x)=\infty$ means that for any large positive number M, there is a value of x, call it x_1, such that
$$f(x)>M$$
whenever $x>x_1$.

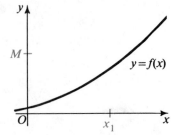

ORAL EXERCISES 16-1

1. How do you read each of the following equations?

   a. $\lim\limits_{x\to\infty} \dfrac{2x^2}{x^2+1} = 2$

   b. $\lim\limits_{x\to-\infty} \dfrac{3x+1}{2x-5} = \dfrac{3}{2}$

   c. $\lim\limits_{x\to 2^+} f(x) = 3$

   d. $\lim\limits_{x\to 2^-} g(x) = \infty$

2. The graph of $y = f(x)$ is shown at the right. What is:

   a. $\lim\limits_{x\to\infty} f(x)$?   **b.** $\lim\limits_{x\to-\infty} f(x)$?

   c. $\lim\limits_{x\to 2} f(x)$?   **d.** $f(2)$

   e. Explain why the function is not continuous at $x = 2$.

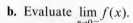

3. **a.** Refer to the graph for Exercise 2 and evaluate $\lim\limits_{x\to 0^+} f(x)$.

   b. Evaluate $\lim\limits_{x\to 0^-} f(x)$.   **c.** Explain why $\lim\limits_{x\to 0} f(x)$ does not exist.

4. Evaluate the following limits:

   a. $\lim\limits_{x\to\infty} \dfrac{5x^2-3x+1}{7x^2+9}$

   b. $\lim\limits_{x\to-\infty} \dfrac{3x^3+\sin x}{4x^3+\cos x}$

   c. $\lim\limits_{x\to 1} \dfrac{x(x-1)}{2(x-1)}$

   d. $\lim\limits_{x\to 1^+} \dfrac{1}{x-1}$

   e. $\lim\limits_{x\to 1^-} \dfrac{1}{x-1}$

   f. $\lim\limits_{x\to 4} \dfrac{x^2-16}{x-4}$

WRITTEN EXERCISES 16-1

In Exercises 1–26, evaluate the given limit or state that it does not exist.

A   **1.** $\lim\limits_{x\to\infty} \dfrac{3x-5}{4x+9}$

2. $\lim\limits_{x\to\infty} \dfrac{2x^2-7x}{3x^2+5}$

3. $\lim\limits_{x\to-\infty} \dfrac{8x^2-7x+5}{4x^2+9}$

4. $\lim\limits_{x\to-\infty} \dfrac{5x^3}{7x^3+8x^2}$

5. $\lim\limits_{x\to\infty} \dfrac{(x^2+1)(x^2-1)}{2x^4}$

6. $\lim\limits_{x\to\infty} \dfrac{x^2\cos\frac{1}{x}}{2x^2-1}$

7. $\lim\limits_{x\to 1} \dfrac{x^2-1}{x-1}$

8. $\lim\limits_{x\to 3} \dfrac{2x^2-6x}{x-3}$

9. $\lim\limits_{t\to 2} \dfrac{t^2+t-6}{2t-4}$

10. $\lim\limits_{x\to 4} \dfrac{x-4}{x^2-x-12}$

11. $\lim\limits_{x\to 0^+} \dfrac{x+1}{x}$

12. $\lim\limits_{t\to-2^+} \dfrac{t}{t+2}$

13. $\lim\limits_{x\to 3^+} \dfrac{x-4}{x-3}$

14. $\lim\limits_{x\to 3^-} \dfrac{2x-6}{x^2-3x}$

15. a. $\lim\limits_{x \to 0^+} \dfrac{|x|}{x}$ **b.** $\lim\limits_{x \to 0^-} \dfrac{|x|}{x}$ **c.** $\lim\limits_{x \to 0} \dfrac{|x|}{x}$

16. a. $\lim\limits_{x \to 1^+} \dfrac{x-3}{x^2-1}$ **b.** $\lim\limits_{x \to 1^-} \dfrac{x-3}{x^2-1}$ **c.** $\lim\limits_{x \to 1} \dfrac{x-3}{x^2-1}$

17. a. $\lim\limits_{x \to 2^+} \dfrac{x-2}{\sqrt{x^2-4}}$ **b.** $\lim\limits_{x \to 2^-} \dfrac{x-2}{\sqrt{x^2-4}}$ **c.** $\lim\limits_{x \to 2} \dfrac{x-2}{\sqrt{x^2-4}}$

18. a. $\lim\limits_{x \to 3^+} [x]$ **b.** $\lim\limits_{x \to 3^-} [x]$ **c.** $\lim\limits_{x \to 3} [x]$

(*Note:* $[x]$ is the greatest integer that is contained in x. See page 292.)

19. $\lim\limits_{x \to 0} \dfrac{x}{1-\sqrt{1-x}}$ **20.** $\lim\limits_{x \to 0} \dfrac{2-\sqrt{4-x}}{x}$

B   **21.** $\lim\limits_{x \to 0} \dfrac{1-\sqrt{x^2+1}}{x^2}$ **22.** $\lim\limits_{x \to -\infty} \left(\sqrt{4x^2-4x} - 2x \right)$

23. $\lim\limits_{x \to \infty} \left(\sqrt{x+1} - \sqrt{x} \right)$ **24.** $\lim\limits_{x \to \infty} \left(\sqrt{x^2+2x} - x \right)$

25. a. $\lim\limits_{h \to 0} \dfrac{(1+h)^2-1}{h}$ **b.** $\lim\limits_{h \to 0} \dfrac{(2+h)^2-4}{h}$ **c.** $\lim\limits_{h \to 0} \dfrac{(x+h)^2-x^2}{h}$

26. a. $\lim\limits_{h \to 0} \dfrac{(1+h)^3-1}{h}$ **b.** $\lim\limits_{h \to 0} \dfrac{(2+h)^3-8}{h}$ **c.** $\lim\limits_{h \to 0} \dfrac{(x+h)^3-x^3}{h}$

Use the graph of the function to evaluate the following limits.

a. $\lim\limits_{x \to \infty} f(x)$ **b.** $\lim\limits_{x \to -\infty} f(x)$ **c.** $\lim\limits_{x \to 2^+} f(x)$ **d.** $\lim\limits_{x \to 2^-} f(x)$

27.

28.

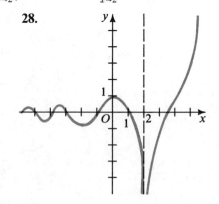

Evaluate each limit or state that it does not exist.

29. $\lim\limits_{x \to \infty} (43{,}987)^{\frac{1}{x}}$ **30.** $\lim\limits_{x \to 0} 2^{\frac{1}{x}}$

31. $\lim\limits_{x \to 0} \dfrac{3^x - 3^{-x}}{3^x + 3^{-x}}$ **32.** $\lim\limits_{x \to -\infty} \dfrac{3^x - 3^{-x}}{3^x + 3^{-x}}$

33. $\lim\limits_{x \to 0^+} \sin \dfrac{1}{x}$ (x in radians)   **34.** $\lim\limits_{x \to \infty} \sin \dfrac{1}{x}$

35. a. In the diagram at the right, explain why arc PQ has length θ, if θ is in radians.

b. Use the diagram to explain why

$$\lim_{\theta \to 0} \frac{\sin \theta}{\theta} = 1.$$

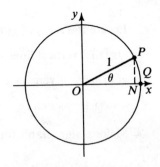

Evaluate the limit. In some cases the result of Exercise 35(b) is helpful.

36. $\displaystyle\lim_{\theta \to 0} \frac{\tan \theta}{\theta}$

$\left(Hint: \tan \theta = \dfrac{\sin \theta}{\cos \theta}.\right)$

37. $\displaystyle\lim_{\theta \to 0} \frac{\sin 2\theta}{\theta}$

($Hint: \sin 2\theta = 2 \sin \theta \cos \theta.$)

38. $\displaystyle\lim_{\theta \to 0} \cot \theta \cdot \sin \theta$

39. $\displaystyle\lim_{x \to \infty} x \sin \frac{1}{x}$

$\left(Hint: \text{Let } t = \dfrac{1}{x}. \text{ As } x \to \infty, t \to 0.\right)$

40. $\displaystyle\lim_{t \to 0} \frac{1 - \cos t}{t^2}$ $\left(Hint: \text{Multiply by } \dfrac{1 + \cos t}{1 + \cos t}.\right)$

C **41.** Recall that each rational number can be expressed as a fraction in lowest terms $\dfrac{p}{q}$, where p is an integer and q is a nonzero integer.

Let $f(x) = \begin{cases} 0 & \text{if } x \text{ is irrational;} \\ \dfrac{1}{q} & \text{if } x \text{ is a rational number.} \end{cases}$

For example, $0.\overline{6} = \frac{2}{3}$, so $f(0.\overline{6}) = \frac{1}{3}$.

Find **a.** $\displaystyle\lim_{x \to 0} f(x)$ **b.** $\displaystyle\lim_{x \to 1} f(x)$ **c.** $\displaystyle\lim_{x \to \sqrt{2}} f(x)$

16-2/GRAPHS OF RATIONAL FUNCTIONS

In this section we shall study the techniques for drawing graphs of rational functions. **Rational functions** are functions of the form $y = \dfrac{f(x)}{g(x)}$ where $f(x)$ and $g(x)$ are polynomials. The special case $y = \dfrac{1}{f(x)}$ was studied in Section 8–4. (You might wish to review this section before going farther.)

PROCEDURE FOR GRAPHING THE RATIONAL FUNCTION $y = \dfrac{f(x)}{g(x)}$

1. Locate the zeros of $f(x)$. A zero of $f(x)$ will be an x-intercept of the graph of $y = \dfrac{f(x)}{g(x)}$ unless it is also a zero of $g(x)$.

2. Locate the zeros of $g(x)$. These values are not in the domain of $y = \dfrac{f(x)}{g(x)}$. Usually, vertical lines through these values will be asymptotes. If $f(x)$ and $g(x)$ have a zero in common, however, then $\dfrac{f(x)}{g(x)}$ should be reduced to lowest terms before we can determine vertical asymptotes.

3. Make a sign graph for $\dfrac{f(x)}{g(x)}$, which should be in lowest terms, using the techniques of Section 3–2. (Essentially, the sign of $\dfrac{f(x)}{g(x)}$ changes at each zero zero of $f(x)$ or $g(x)$, unless the zero is a double zero, in which case the sign remains the same.)

4. If $x = c$ is a vertical asymptote, then the graph will approach infinity as it nears the asymptote in a region where the values of the function are positive. The graph will approach negative infinity as it nears the asymptote in a region where the values of the function are negative.

5. Nonvertical asymptotes can be found by considering $\displaystyle\lim_{x\to\infty} \dfrac{f(x)}{g(x)}$ and $\displaystyle\lim_{x\to-\infty} \dfrac{f(x)}{g(x)}$. If $\displaystyle\lim_{x\to\pm\infty} \dfrac{f(x)}{g(x)} = c$, then $y = c$ is a horizontal asymptote.

EXAMPLE 1. Sketch the graph of $y = \dfrac{x - 3}{x^2 - 1}$.

SOLUTION: We use the procedure just described, as follows.

Step 1. The numerator $x - 3$ has a zero at $x = 3$, which is not a zero of the denominator. Hence 3 is an x-intercept of the graph.

Step 2. The denominator $x^2 - 1$ has zeros at $x = -1$ and $x = 1$. Hence the graph has vertical asymptotes $x = -1$ and $x = 1$.

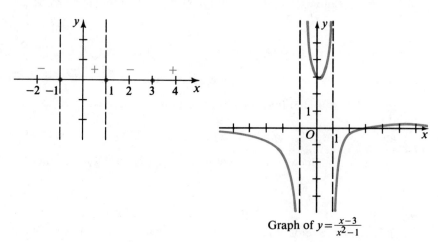

Graph of $y = \dfrac{x-3}{x^2-1}$

Step 3. The sign of $\dfrac{x - 3}{x^2 - 1}$ changes at -1, 1, and 3.

Step 4. Since the values of the function are negative in the region immediately to the right of the asymptote $x = 1$, the graph approaches negative infinity as it nears $x = 1$ from the right. Since the values of the function are positive in the region to the left of the asymptote, the graph approaches infinity as it nears $x = 1$ from the left.

Similar reasoning shows that the graph approaches infinity as it nears the asymptote $x = -1$ from the right, and that the graph approaches negative infinity as it nears $x = -1$ from the left.

Step 5. $\lim\limits_{x \to \infty} \dfrac{x - 3}{x^2 - 1} = 0$ and $\lim\limits_{x \to -\infty} \dfrac{x - 3}{x^2 - 1} = 0$. Hence the graph has the x-axis as a horizontal asymptote at the left and at the right. The graph is shown on the previous page at the right.

EXAMPLE 2. Sketch the graph of $y = \dfrac{x^2 - 2x - 3}{x^2 - 4}$.

SOLUTION: A short form of the solution is shown.

Steps 1–3.
$$\frac{x^2 - 2x - 3}{x^2 - 4} = \frac{(x - 3)(x + 1)}{(x - 2)(x + 2)}$$

Steps 4–5.
$$\lim_{x \to -\infty} \frac{x^2 - 2x - 3}{x^2 - 4} = 1$$
$$\lim_{x \to \infty} \frac{x^2 - 2x - 3}{x^2 - 4} = 1$$

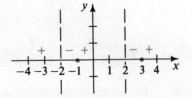

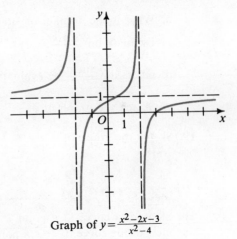

Graph of $y = \dfrac{x^2 - 2x - 3}{x^2 - 4}$

ORAL EXERCISES 16–2

For each function give the x- and y-intercepts. Also, find the equations of any vertical or horizontal asymptotes.

1. $y = \dfrac{x - 2}{(x - 1)(x - 3)}$

2. $y = \dfrac{x}{x^2 - 4}$

3. $y = \dfrac{x^2 - 9}{x - 3}$

4. $y = \dfrac{x^2 - 9}{x^2 + 9}$

5. $y = \dfrac{2x^2}{x^2 - 6x}$

6. $y = \dfrac{x^3}{x^2 - 4}$

Sketch the graph of each function. Show vertical and horizontal asymptotes and all intercepts.

A

1. $y = \dfrac{x - 1}{x + 1}$

2. $y = \dfrac{x}{x - 2}$

3. $y = \dfrac{x}{x^2 - 1}$

4. $y = \dfrac{2x^2}{x^2 - 9}$

5. $y = \dfrac{x^2 + 4}{x^2 - 4}$

6. $y = \dfrac{x^2 - 4}{x^2 + 4}$

7. $y = \dfrac{x - 4}{(x - 1)(x + 2)}$

8. $y = \dfrac{x + 3}{(x + 1)(x - 3)}$

9. $y = \dfrac{12}{x^2 + 2}$

10. $y = \dfrac{12x}{x^2 + 2}$

11. $y = \dfrac{12x^2}{x^2 + 2}$

12. $y = \dfrac{3x^2}{x^2 - 3x}$

B

13. $y = \dfrac{x^2 - 2x}{x^2 - 4}$

14. $y = \dfrac{x^2 - 16}{x^2 - 6x + 8}$

15. $y = \dfrac{x^2 - x - 2}{x^2 - 2x + 1}$

16. Explain why the line $y = x$ is an asymptote for the graph of $y = \dfrac{x^3}{x^2 + 1}$. Sketch the graph of this function.

17. Explain why the line $y = -x$ is an asymptote for the graph of $y = \dfrac{1 - x^2}{x}$. Sketch the graph of this function.

C

18. Consider the function $f(x) = \dfrac{\sin x}{x}$, where x is in radians.

   a. What is the domain of $f(x)$?   **b.** What are the zeros of $f(x)$?

   c. Evaluate $\lim\limits_{x \to \infty} \dfrac{\sin x}{x}$.

   d. Guess $\lim\limits_{x \to 0} \dfrac{\sin x}{x}$ by evaluating $\dfrac{\sin x}{x}$ for several small values of x.

   e. Sketch the graph of $y = \dfrac{\sin x}{x}$.

19. Sketch the graph of $y = e^{-x^2}$. (Recall that the number e is approximately 2.718.)

16-3 / THE SLOPE OF A CURVE

The slope of a curve at a point P is a measure of the steepness of the curve. If Q is a point on the curve near P, then the slope of the curve at P is approximately the slope of line segment $\overline{PQ}$. The approximation becomes better as Q moves along the curve toward P. The **slope of the curve** at P is defined to be the limit of the slope of $\overline{PQ}$ as Q approaches P along the curve. In symbols:

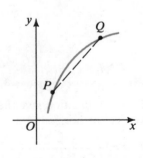

$$\text{slope of curve at } P = \lim_{Q \to P} (\text{slope of } \overline{PQ})$$

Suppose we want to find the slope of the curve $y = x^2$ at $P = (1, 1)$. We can let Q be a point on the curve near P by choosing the x-coordinate of Q to be

$1 + h$, which differs from the x-coordinate of P by h units. The y-coordinate of Q is $(1 + h)^2$ according to the equation of the curve, as shown in the diagram below. We now calculate the slope of $\overline{PQ}$.

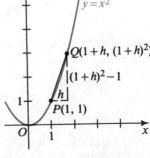

$$\text{slope of } \overline{PQ} = \frac{(1 + h)^2 - 1}{h}$$

$$= \frac{(1 + 2h + h^2) - 1}{h}$$

$$= \frac{h(2 + h)}{h} = 2 + h$$

As Q approaches P, h approaches 0. Thus:

$$\text{slope of curve at } (1, 1) = \lim_{Q \to P} (\text{slope of } \overline{PQ})$$

$$= \lim_{h \to 0} (2 + h) = 2$$

Therefore, the slope of the curve $y = x^2$ at $P(1, 1)$ is 2.

To find the slope of the curve $y = x^2$ at $P(2, 4)$, we proceed similarly:

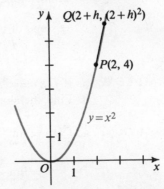

$$\text{slope of } \overline{PQ} = \frac{(2 + h)^2 - 4}{h}$$

$$= \frac{4 + 4h + h^2 - 4}{h}$$

$$= \frac{4h + h^2}{h}$$

$$= 4 + h$$

$$\lim_{Q \to P} (\text{slope of } \overline{PQ}) = \lim_{h \to 0} (4 + h) = 4.$$

Therefore, the slope of the curve at $(2, 4)$ is 4.

We have seen that the slope of the curve $y = x^2$ is 2 when $x = 1$ and is 4 when $x = 2$. In general, the slope of the curve $y = x^2$ at the point (x, x^2) is a function of x. In the following example, we shall discover what this function is.

EXAMPLE 1. Find the slope of the curve $y = x^2$ at the point $P = (x, x^2)$.

SOLUTION: Let Q be any other point on the graph, as shown in the diagram. Then:

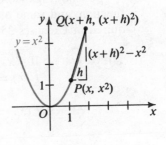

$$\text{slope of } \overline{PQ} = \frac{(x + h)^2 - x^2}{h}$$

$$= \frac{x^2 + 2xh + h^2 - x^2}{h}$$

$$= \frac{2xh + h^2}{h} = 2x + h$$

(*Solution continued on page 580*)

Introduction to calculus   **579**

$$\lim_{Q \to P} (\text{slope of } \overline{PQ}) = \lim_{h \to 0} (2x + h) = 2x$$

Therefore, the slope of the curve $y = x^2$ at (x, x^2) is $2x$.

Using this expression, $2x$, for the slope, we can see that when $x = 1$, the slope is $2 \cdot 1 = 2$, and when $x = 2$, the slope is $2 \cdot 2 = 4$. Note that these results agree with our earlier work.

We have seen that the slope of the graph of $y = f(x)$ varies from point to point. In fact, the slope of $y = f(x)$ is itself a function of x. This slope function, called the **derivative of $f(x)$** and denoted $f'(x)$, is defined as follows:

$$f'(x) = \lim_{h \to 0} \frac{f(x + h) - f(x)}{h}$$

For a given value of x, the value of $f'(x)$ is the slope of the curve at $(x, f(x))$ for the following reason:

$$\lim_{h \to 0} \frac{f(x + h) - f(x)}{h} = \lim_{Q \to P} (\text{slope of } \overline{PQ})$$

$$= \text{slope of curve at } P(x, f(x))$$

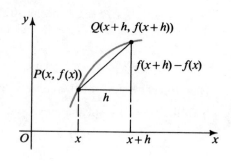

EXAMPLE 2. **a.** If $f(x) = x^3$, find $f'(x)$.
b. Find the slope of the curve $y = f(x)$ at the point $(2, 8)$.

SOLUTION: **a.** $f'(x) = \lim\limits_{h \to 0} \dfrac{f(x + h) - f(x)}{h} = \lim\limits_{h \to 0} \dfrac{(x + h)^3 - x^3}{h}$

$$= \lim_{h \to 0} \frac{x^3 + 3x^2h + 3xh^2 + h^3 - x^3}{h}$$

$$= \lim_{h \to 0} \frac{3x^2h + 3xh^2 + h^3}{h} = \lim_{h \to 0} (3x^2 + 3xh + h^2)$$

$$= 3x^2$$

b. $f'(2) = 3(2)^2 = 12$
The slope at $(2, 8)$ is 12.

The following theorems enable us to find the derivative of every polynomial function, and some other functions as well. Proofs of Theorems 3–5 are required in Exercises 35–37.

THEOREM 2. If $f(x)$ is a constant function (that is, $f(x) = c$ for all x), then $f'(x) = 0$ for all x.

Proof: $f'(x) = \lim\limits_{h \to 0} \dfrac{f(x + h) - f(x)}{h}$

$$= \lim_{h \to 0} \frac{c - c}{h} = \lim_{h \to 0} \frac{0}{h} = \lim_{h \to 0} 0 = 0$$

580   Chapter 16

THEOREM 3. Let n be any nonzero real number. If $f(x) = x^n$, then
$$f'(x) = nx^{n-1}.$$

The results of Examples 1 and 2, where we saw that the derivative of $f(x) = x^2$ is $f'(x) = 2x$ and the derivative of $f(x) = x^3$ is $f'(x) = 3x^2$, agree with this theorem.

THEOREM 4. Let n be any nonzero real number. If $f(x) = cx^n$, then
$$f'(x) = cnx^{n-1}.$$

THEOREM 5. If $f(x) = g(x) + h(x)$, then $f'(x) = g'(x) + h'(x)$.

EXAMPLE 3. Find the derivative of each function.

    a. $f(x) = x^5$ **b.** $f(x) = 3x^5$

    c. $f(x) = 12x^2 - 4x + 7$ **d.** $f(x) = 8\sqrt{x} - \dfrac{7}{x^2}$

SOLUTION: **a.** To find the derivative of $f(x) = x^5$, we apply Theorem 3 with $n = 5$:
$$f'(x) = 5x^4$$

    b. To find the derivative of $f(x) = 3x^5$, we apply Theorem 4 with $c = 3$ and $n = 5$:
$$f'(x) = 3 \cdot 5x^4 = 15x^4$$

    c. We regard $f(x) = 12x^2 - 4x + 7$ as the sum of functions defined by its terms and use Theorem 4.
$$f'(x) = 12 \cdot 2x - 4 \cdot 1x^0 + 0$$
$$= 24x - 4$$

    d. We rewrite $f(x) = 8\sqrt{x} - \dfrac{7}{x^2}$ as $f(x) = 8x^{\frac{1}{2}} - 7x^{-2}$ and proceed as in part (c) of this example.
$$f'(x) = 8 \cdot \frac{1}{2}x^{\frac{1}{2}-1} - 7(-2)x^{-2-1}$$
$$= 4x^{-\frac{1}{2}} - 7x^{-3}$$

The line **tangent** to a curve at a point on the curve is defined to be the line that passes through the point and that has a slope that is the same as the slope of the curve at that point.

EXAMPLE 4. Write an equation of the line that is tangent to the curve $y = \sqrt{x}$ at the point $(9, 3)$.

SOLUTION: $f(x) = \sqrt{x} = x^{\frac{1}{2}}$
$$f'(x) = \frac{1}{2}x^{-\frac{1}{2}} = \frac{1}{2\sqrt{x}}$$

(*Solution continued on page 582*)

$$\text{slope of tangent at } (9, 3) = \text{slope of curve at } (9, 3)$$
$$= f'(9)$$
$$= \frac{1}{2\sqrt{9}} = \frac{1}{6}$$

The tangent line has slope $\frac{1}{6}$ and contains the point $(9, 3)$. An equation of the line is

$$\frac{y - 3}{x - 9} = \frac{1}{6}$$

or $y = \frac{1}{6}x + \frac{3}{2}$.

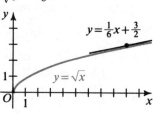

ORAL EXERCISES 16-3

Find the derivative of each function. Express the derivative without using fractional or negative exponents.

1. $f(x) = x^8$ **2.** $f(x) = x^{12}$

3. $f(x) = x^{-4}$ **4.** $g(x) = 3x^7$

5. $h(x) = x^{\frac{1}{2}}$ **6.** $f(x) = 4x^{\frac{3}{2}}$

7. $g(x) = \sqrt[3]{x}$ **8.** $h(x) = \dfrac{1}{\sqrt[3]{x}}$

9. $f(x) = 8x^3 - 7x^2 + 3x - 2$ **10.** $g(x) = 4x^{-5} - 2x^{-3} + 9x^{-1} + 5$

11. a. If $f(x) = 4x^2$, then $f'(x) = \underline{\ ?\ }$ and $f'(-1) = \underline{\ ?\ }$.
     b. What is the slope of $y = 4x^2$ at the point $(-1, 4)$?

12. What is the slope of $y = x^4$ at $(2, 16)$?

WRITTEN EXERCISES 16-3

A   **1.** $P(1, 1)$ is a point on the graph of $y = x^3$. Find the slope of $\overline{PQ}$ if Q is:
     a. $(2, 8)$ **b.** $(1.5, 3.375)$ **c.** $(1.1, 1.331)$
     d. If $f(x) = x^3$, find $f'(x)$ and $f'(1)$.

   2. $P(2, \frac{1}{2})$ is a point on the graph of $y = \dfrac{1}{x}$. Find the slope of $\overline{PQ}$ if Q is:

     a. $(3, f(3))$ **b.** $(2.5, f(2.5))$ **c.** $(2.1, f(2.1))$

     d. If $f(x) = \dfrac{1}{x}$, find $f'(x)$ and $f'(2)$.

Find the derivative of the given function. Express the derivative without using fractional or negative exponents.

3. $f(x) = 2x^5$ **4.** $f(x) = 3x^6$

5. $f(x) = -2x^3$ **6.** $f(x) = 6x^{\frac{2}{3}}$

7. $f(x) = \dfrac{3}{x^2}$

8. $g(x) = \dfrac{4}{\sqrt{x}}$

9. $f(x) = \sqrt{5x}$

（*Hint:* $\sqrt{5x} = \sqrt{5} \cdot \sqrt{x}$）

10. $f(x) = \dfrac{1}{4x^4}$

11. $g(x) = \dfrac{7}{2}x^2 - 5x + 3 - \dfrac{1}{x}$

12. $f(x) = \dfrac{4}{3}x^3 - \dfrac{2}{x} - \dfrac{5}{x^2} + \pi$

Find the slope of each curve at the given point P.

13. $y = x^3$; $P(-1, -1)$

14. $y = \dfrac{x^4}{2}$; $P\left(-1, \dfrac{1}{2}\right)$

15. $y = 3x^2 - 2x + 1$; $P(0, 1)$

16. $y = 2\sqrt{x}$; $P(9, 6)$

17. $y = 8x^4 - 7x^2 + 5x + 6$; $P(-1, 2)$

18. $y = x^3 - 5x^2 + 4x + 2$; $P(2, -2)$

B **19.** $y = \dfrac{2}{x} - \dfrac{4}{x^2}$; $P(2, 0)$

20. $y = \sqrt[3]{x}$; $P(0, 0)$

21. If $f(x) = mx + k$, find $f'(1)$, $f'(2)$, and $f'(3)$.

22. If $f(x) = ax^2 + bx + c$, $a \neq 0$, find $f'\left(-\dfrac{b}{2a}\right)$.

Find an equation of the line tangent to the given curve at the given point P.

23. $y = x^3$; $P(-2, -8)$

24. $y = 3x^2 - 5x$; $P(2, 2)$

25. $y = \dfrac{1}{x^2}$; $P(-1, 1)$

26. $y = \dfrac{1}{\sqrt[3]{x}}$; $P\left(8, \dfrac{1}{2}\right)$

27. a. Sketch the graph of $f(x) = 6x^2 - x^3$.
   b. For what values of x does $f'(x) = 0$?
   c. On your graph of $f(x)$, indicate the two points where $f'(x) = 0$.

28. a. Sketch the graph of $f(x) = x^3 - 9x$.
   b. For what values of x does $f'(x) = 0$?
   c. On your graph of $f(x)$, indicate the two points where $f'(x) = 0$.

Find a function that has the given derivative.

29. $f'(x) = 4x^3$

30. $g'(x) = 5x^4$

31. $h'(x) = 3x^5$

32. $f'(x) = x^7$

33. $f'(x) = 3\sqrt{x}$

34. $g'(x) = \sqrt[3]{x}$

C **35. a.** Use the Binomial Theorem (page 558) to write the first three terms and the last term in the expansion of $(x + h)^n$.
   b. Use part (a) and the definition of $f'(x)$ to prove Theorem 3 on page 581 for positive integral values of n.
   (*Note:* The Binomial Theorem can be generalized to apply to any nonzero real number n. This form of the Binomial Theorem can be used to give a general proof of Theorem 3.)

36. Prove Theorem 4 on page 581. [*Hint:* Use the definition of $f'(x)$.]

37. Prove Theorem 5 on page 581. [*Hint:* Use the definition of $f'(x)$.]

Consider the function $f(x) = 6x - x^2$, whose derivative is $f'(x) = 6 - 2x$. Values of x, $f(x)$, and $f'(x)$ are given in the following table.

x	0	1	2	3	4	5	6
$f(x) = 6x - x^2$	0	5	8	9	8	5	0
$f'(x) = 6 - 2x$	6	4	2	0	-2	-4	-6

Much can be learned about the graph of $f(x)$ by plotting each point $(x, f(x))$ and by drawing the tangent to the curve at each point plotted. The tangent in each case is the line through $(x, f(x))$ that has slope $f'(x)$. Figure (a) below shows these points and tangent lines. Figure (b) shows a smooth curve drawn through the points and having the corresponding tangent lines. The smooth curve is the graph of $f(x)$.

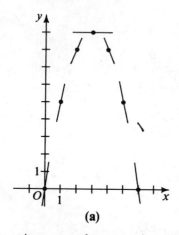

(a)

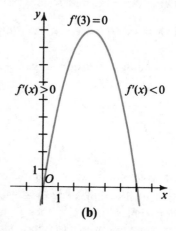

(b)

As x increases, the curve rises whenever $f'(x) > 0$, and the curve falls whenever $f'(x) < 0$. When $f'(x) = 0$ the curve has a horizontal tangent and $f(x)$ attains its maximum value.

A similar analysis can be made for the graph of

$$f(x) = x^3 - 12x^2 + 36x + 10.$$

We have:

$$f'(x) = 3x^2 - 24x + 36$$
$$= 3(x^2 - 8x + 12)$$
$$= 3(x - 2)(x - 6)$$

The sign of $f'(x)$ is given in the following table.

x	$x < 2$	$x = 2$	$2 < x < 6$	$x = 6$	$x > 6$
$f'(x)$	$+$	0	$-$	0	$+$

We see that $f(x)$ is rising whenever $f'(x) > 0$, $f(x)$ is falling whenever $f'(x) < 0$, and $f(x)$ has a horizontal tangent when $f'(x) = 0$. A point such as $(2, 42)$, where $f'(x) = 0$ and the curve changes from rising to falling as x increases, is called a *local maximum point*. There is no nearby point for which $f(x) > 42$. Similarly, a point such as $(6, 10)$, where $f'(x) = 0$ and the curve changes from falling to rising as x increases, is called a *local minimum point*. There is no nearby point for which $f(x) < 10$.

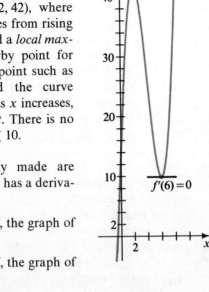

The observations previously made are valid for any function $f(x)$ that has a derivative on an interval I.

1. If $f'(x) > 0$ on an interval I, the graph of $f(x)$ rises as x increases.

2. If $f'(x) < 0$ on an interval I, the graph of $f(x)$ falls as x increases.

3. If $f'(c) = 0$, the graph of $f(x)$ has a horizontal tangent at $x = c$, that is, at the point $(c, f(c))$. The function may have a local maximum, a local minimum, or neither.

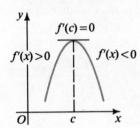

Local maximum at $x = c$

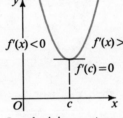

Local minimum at $x = c$

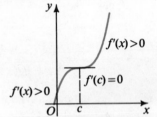

Neither maximum nor minimum; $f'(x)$ is positive or negative on both sides of $x = c$.

EXAMPLE 1. Graph $f(x) = x^4 - 4x^3 + 14$. Identify any local maximum and minimum points.

SOLUTION: To determine where local maximum and minimum points may occur, we find $f'(x)$ and set it equal to 0.

$$f(x) = x^4 - 4x^3 + 14$$
$$f'(x) = 4x^3 - 12x^2$$
$$= 4x^2(x - 3)$$
$$f'(x) = 0 \text{ when } x = 0 \text{ or } x = 3.$$

(*Solution continued on page 586*)

x	$x < 0$	$x = 0$	$0 < x < 3$	$x = 3$	$x > 3$
$f'(x)$	$-$	0	$-$	0	$+$

$x = 0$ does not give a maximum or minimum since $f'(x)$ is negative on both sides of $x = 0$. $x = 3$ gives a local minimum since $f'(3) = 0$ and the curve changes from falling to rising at $x = 3$. There is a local minimum point at $(3, f(3)) = (3, -13)$.

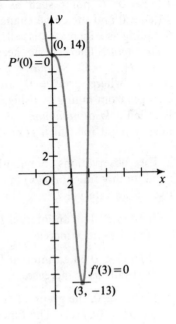

The procedure that has been outlined for finding local maximums and minimums can be used with all polynomial functions. It can also be used with any other function whose derivative exists everywhere the function is defined.

EXAMPLE 2. Graph $f(x) = x + x^{-1}$. Identify any local maximum and minimum points.

SOLUTION: $f(x) = x + x^{-1}$

$$f'(x) = 1 - x^{-2} = 1 - \frac{1}{x^2}$$

$f'(x) = 0$ when $x = 1$ or $x = -1$.

x	$x < -1$	$x = -1$	$-1 < x < 1, x \neq 0$	$x = 1$	$x > 1$
$f'(x)$	$+$	0	$-$	0	$+$

Neither $f(x)$ nor $f'(x)$ is defined at $x = 0$. A local maximum occurs at $(-1, -2)$. A local minimum occurs at $(1, 2)$. Observe that for large $|x|$, $y = x + x^{-1}$ is approximately the same as $y = x$. This indicates that $y = x$ is an asymptote. Also, since $\lim_{x \to 0^+} f(x) = \infty$

and $\lim_{x \to 0^-} f(x) = -\infty$, the y-axis is an asymptote.

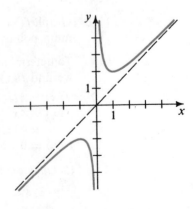

1. Is it possible for a function to have a local maximum value that is less than a local minimum value? Explain.

2. A second-degree polynomial function $f(x)$ has $f'(1) = 4$, $f'(2) = 0$, and $f'(3) = -4$. Does $x = 2$ give a local maximum, a local minimum, or neither?

3. A polynomial function $f(x)$ is such that $f'(-3) = -2, f'(1) = 0, f'(3) = 3$, and $f'(x) \neq 0$ except at 1. Must $x = 1$ be a local minimum of $f(x)$? Explain.

4. The derivative of a polynomial function $f(x)$ has values as given in the table below. Give the values of x that correspond to local maximums and minimums.

x	$x < -1$	$x = -1$	$-1 < x < 0$	$x = 0$	$0 < x < 2$	$x = 2$	$x > 2$
$f'(x)$	$+$	0	$-$	0	$-$	0	$+$

5. Give the x-coordinate of the local maximum and minimum of the function $f(x) = \frac{1}{3}x^3 - 4x$.

6. Let $f(x) = |x|$.
   a. Find the minimum of $f(x)$.
   b. What is the slope of $f(x)$ at any point where $x > 0$?
   c. What is the slope of $f(x)$ at any point where $x < 0$?
   d. Explain why $f'(0)$ does not exist.

WRITTEN EXERCISES 16-4

Graph each function. Identify any local maximum and minimum points.

A
1. $f(x) = x^3 - 12x$
2. $f(x) = x^3 - 3x^2$
3. $f(x) = 9x - x^3$
4. $f(x) = x^3 + 4x^2 - 3x - 12$
5. $g(x) = 2x^2 - x^4$
6. $h(x) = 40 + 8x - 5x^2 - x^3$
7. $g(x) = x^4 - 6x^2 + 8$
8. $f(x) = x^4 - 8x^2 - 9$
9. $f(t) = 3t^4 - 4t^3 - 12t^2$
10. $g(t) = 8t^3 - 3t^4$
11. $h(x) = 5x^4 - 4x^5$
12. $F(x) = x^3 - 3x^2 - 9x + 27$

Use the derivative to find the vertex of each parabola.

13. $y = x^2 - 6x + 4$
14. $y = 5 + 8x - 4x^2$

B
15. $y = (x - a)^2$
16. $y = ax^2 + bx + c$

17. Show that the function $f(x) = x^3 - 3x^2 + 3x - 5$ has neither a local maximum nor a local minimum.

18. The derivative of $f(x) = \sin x$ is $f'(x) = \cos x$. Use this fact to show that the function $g(x) = x + \sin x$ has an infinite number of local maximums and local minimums.

19. The symmetry point of the graph of the equation $y = ax^3 + bx^2 + cx + d$ occurs at $x = -\dfrac{b}{3a}$.

    a. Find the symmetry point S of the graph of $y = 2x^3 - 6x^2 - 9x + 8$.

    b. Find the coordinates of the maximum point P and the minimum point Q of the graph of $y = 2x^3 - 6x^2 - 9x + 8$.

    c. Show that P, S, and Q are collinear.

    d. Show that S is the midpoint of $\overline{PQ}$.

Graph each function. Identify local maximum and minimum points and asymptotes.

    20. $y = x + 2x^{-1}$ **21.** $y = x^2 + x^{-2}$

C   **22.** $y = x^2 + 3x + \dfrac{1}{x}$ **23.** $y = x^2 + \dfrac{1}{x}$

16-5/EXTREME VALUE PROBLEMS

In many applications of mathematics, great interest is focused on finding the *maximum* or *minimum value* of a function. In many such problems the domain of the variable, say x, is given by an inequality in the form $a \le x \le b$, where a and b are called *endpoints*. In such problems a local maximum or minimum can occur at an endpoint so we must check these values, as shown in the next example.

EXAMPLE 1.   Congruent squares are cut from the corners of a 1 m square piece of tin, and the edges are then turned up to make an open box. How large should the squares cut from the corners be in order to maximize the volume of the box?

SOLUTION: Let $x =$ length of a side of the square cut from a corner; then $0 \le x \le \frac{1}{2}$.

    The volume V of the box can be expressed as a function of x.

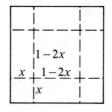

$$V = f(x) = (1 - 2x)(1 - 2x)x$$
$$= x - 4x^2 + 4x^3$$

If $f(x)$ has a maximum or minimum, it will occur at an endpoint or where $f'(x) = 0$. Since $f(0) = f(\frac{1}{2}) = 0$, we eliminate these values. Next we set $f'(x) = 0$.

$$f'(x) = 1 - 8x + 12x^2 = 0$$
$$(1 - 6x)(1 - 2x) = 0$$

                $x = \frac{1}{6}$, $x = \frac{1}{2}$ eliminated previously

If $x < \frac{1}{6}$, $f'(x) > 0$; if $\frac{1}{6} < x < \frac{1}{2}$, $f'(x) < 0$.

Thus a maximum of $f(x)$ occurs when $x = \frac{1}{6}$. The squares cut from each corner should be $\frac{1}{6}$ m by $\frac{1}{6}$ m for the box to have its greatest volume. Since $f(\frac{1}{6}) = \frac{2}{27}$, the maximum volume is $\frac{2}{27}$ m³.

EXAMPLE 2. A manufacturer produces cardboard boxes that have a square base. The top of each box is a double flap that opens as shown. The bottom of the box has a double layer of cardboard for strength. If each box must have a volume of 12 cubic meters, what dimensions will minimize the amount of cardboard used?

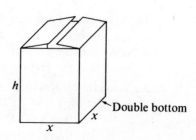

Double bottom

SOLUTION:

1. Let the dimensions of the box be x, x, and h as shown. Express the amount of cardboard C as a function of x and h.

$$C(x, h) = \text{double bottom} + \text{top} + 4 \text{ sides}$$
$$= 2x^2 + x^2 + 4xh$$
$$= 3x^2 + 4xh$$

2. Express C as a function of one variable, x, by using the fact that the volume is fixed at 12 cubic meters.

$$\text{Volume} = x^2 h = 12$$

$$h = \frac{12}{x^2}$$

$$C(x) = 3x^2 + 4x\left(\frac{12}{x^2}\right) = 3x^2 + \frac{48}{x}$$

3. Take the derivative and set it equal to zero. We can ignore the endpoint at 0 since no corresponding value of h can be found.

$$C'(x) = 6x - 48x^{-2} = 0$$

$$6x = \frac{48}{x^2}$$

$$6x^3 = 48$$

$$x = 2$$

When $x = 2$, $h = \dfrac{12}{x^2} = 3$.

4. The dimensions of the box should be 2 m by 2 m by 3 m. For further assurance that $x = 2$ gives a minimum of $C(x)$ rather than a maximum, observe that

$$C'(x) = 6x - 48x^{-2}$$

$$= \frac{6x^3 - 48}{x^2}$$

$$= \frac{6(x^3 - 8)}{x^2}$$

For $0 < x < 2$, $C'(x) < 0$, while for $x > 2$, $C'(x) > 0$. Thus, since $C'(2) = 0$, we have a minimum at $x = 2$.

A   **1.** The lower side of a rectangle is on the x-axis and the upper two vertices are on the parabola $y = 3 - x^2$.

     a. Show that the area of the rectangle is $A(x) = 6x - 2x^3,\ x > 0$.

     b. Show that the rectangle has the greatest area when $x = 1$.

   2. An isosceles trapezoid is inscribed in the parabola $y = 9 - x^2$ with the base of the trapezoid on the x-axis.

     a. Show that the area of the trapezoid is $A(x) = \frac{1}{2}(9 - x^2)(6 + 2x),\ x > 0$.

     b. Show that the trapezoid has its greatest area when $x = 1$.

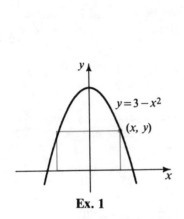

Ex. 1

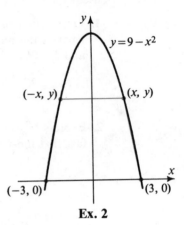

Ex. 2

   3. A rectangle has one vertex at the origin, another on the positive x-axis, another on the positive y-axis, and the fourth vertex on the line $y = 8 - 2x$. What is the greatest area the rectangle can have?

   4. A rectangle has area A, where A is a constant. The length of the rectangle is x.

     a. Express the width of the rectangle in terms of A and x.

     b. Express the perimeter in terms of A and x.

     c. Find the length and width that will minimize the perimeter.

     d. Of all rectangles with a given area, which one has the least perimeter?

B   **5.** Congruent squares are cut from the corners of a rectangular sheet of metal 8 cm wide and 15 cm long. The edges are then turned up to make an open box. Let x be the length of a side of the square cut from each corner.

     a. Write an equation that expresses the volume of the box, $V(x)$, as a function of x.

     b. What value of x will maximize the volume?

   6. A 1 m by 2 m sheet of metal is cut and folded, as indicated, to make a box with a top.

     a. Express the volume of the box, $V(x)$, as a function of x.

     b. What value of x maximizes the volume?

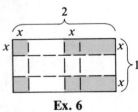

Ex. 6

7. The graphs of $y = x$ and $y = x^3$ intersect three times.
   a. Find the coordinates of the points of intersection.
   b. For what positive value of x between the intersection points is the vertical distance between the graphs the greatest?

8. A manufacturer produces metal boxes that have a square base, no top, and a volume of 4000 cm³.

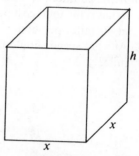

   a. If the base edges are x cm long, express the height h in terms of x.
   b. Show that the area of the metal used in the box is given by the function
$$A(x) = x^2 + 16000x^{-1}.$$

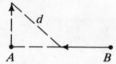

   c. What values of x and h will minimize the amount of metal used to make the box?

Ex. 8

9. A toy company can make x Handy-Kid tool sets a day at a total cost of $300 + 12x + 0.2x^{\frac{3}{2}}$ dollars. It can sell each set for $18. How many sets should be made each day in order to maximize the profit?

10. B is 100 km east of A. At noon a truck leaves A, traveling north at 60 km/h, and a car leaves B traveling west at 80 km/h.

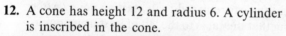

   a. Write an expression for the distance d between the car and the truck after t hours of travel.
   b. When will the distance between the two vehicles be a minimum? (*Hint:* For $a > 0$, $\sqrt{a}$ will be a minimum when a is a minimum.)
   c. What is the minimum distance between the vehicles?

C 11. A box company will produce a box with a square base and no top that has a volume of 8 m³. Material for the bottom costs $6/m² and material for the sides costs $3/m². Find the dimensions of the box that will minimize the cost.

12. A cone has height 12 and radius 6. A cylinder is inscribed in the cone.

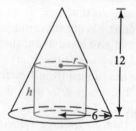

   a. Use similar triangles to show that $h = 12 - 2r$, $0 \le r < 6$.
   b. Find the maximum volume of the inscribed cylinder.

13. A cone is inscribed in a sphere with radius 6.

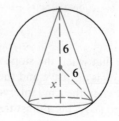

   a. Show that the volume of the cone is
$$V(x) = \tfrac{1}{3}\pi(36 - x^2)(6 + x) \text{ for } 0 \le x < 6.$$
   b. Find the maximum volume of the cone.

14. Find the maximum volume of a cylinder inscribed in a sphere with radius 10.

15. It costs $(6 + 0.004x)$ dollars per mile to operate a truck at x miles per hour. In addition, it costs $14.40 per hour to pay the driver.

   a. What is the total cost per mile if the truck is driven at 30 miles per hour? at 40 miles per hour?

   b. At what speed should the truck be driven to minimize the total cost per mile?

16-6/VELOCITY AND ACCELERATION

Imagine a toy car that can move forward or backward along a long straight track. The directed distance of the car from its starting point is a function of time. This *position function* is illustrated in the graph below.

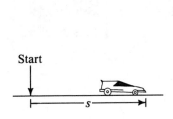

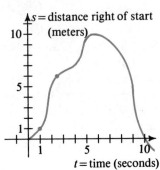

We can tell a great deal about the motion of the car from the graph. For example, the car apparently slowed and nearly stopped at time $t = 4$ seconds, and at time $t = 6$ seconds the car stopped moving forward and began to move backward. At time $t = 10$ seconds, the car returned to its starting point. After 10 seconds, s is negative, indicating that the car is to the left of its starting point.

   We can also learn about the velocity of the car from the graph. For example, if we consider the time interval from $t = 1$ second to $t = 3$ seconds, we see that the car has moved from 1 meter to 6 meters, a distance of 5 meters in 2 seconds. Thus, its *average velocity* in that 2-second interval is $\frac{5}{2} = 2.5$ meters per second. In general, if the positions of the car at times t_1 and t_2 are s_1 and s_2, respectively, then its **average velocity** over this time interval is:

$$\text{average velocity} = \frac{\text{change in distance}}{\text{change in time}} = \frac{\Delta s}{\Delta t} = \frac{s_2 - s_1}{t_2 - t_1}$$
$$= \text{slope of line joining } (t_1, s_1) \text{ and } (t_2, s_2)$$

   Notice that when the slope of the line joining (t_1, s_1) and (t_2, s_2) is negative, the velocity is negative and the car is moving backward. (See figure (a).) Notice also that when the time interval between t_1 and t_2 is very small, the line joining (t_1, s_1) and (t_2, s_2) is almost the same as the line tangent to the graph at (t_1, s_1), as shown in figure (b). The slope of this tangent line is called the *instantaneous velocity* of the car at the time t_1 to distinguish it from the average velocity of the car over the time interval containing t_1.

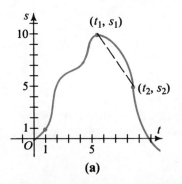

(a)

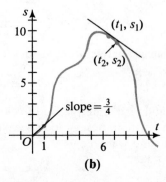

(b)

If the position of an object is given as a function $s(t)$ of time, then the **instantaneous velocity** at time t is defined as follows:

$$\text{instantaneous velocity} = \lim_{\Delta t \to 0} \frac{s(t + \Delta t) - s(t)}{\Delta t} = s'(t)$$

The instantaneous velocity is the value of the derivative of the position function.

If a ball is thrown vertically upward with an initial speed of 30 m/s, its height in meters t seconds later is approximately:

$$h(t) = 30t - 5t^2$$

The average velocity of the ball between times $t = 1$ and $t = 2$ is:

$$\text{average velocity} = \frac{h(2) - h(1)}{2 - 1}$$

$$= \frac{40 - 25}{2 - 1} = 15 \text{ m/s}$$

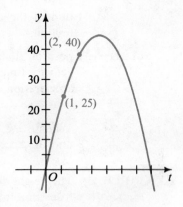

Notice that the average velocity is the slope of the chord joining the points $(1, 25)$ and $(2, 40)$.

For contrast, we can find the instantaneous velocity of the ball at $t = 1$ and $t = 2$.

$$\text{instantaneous velocity} = h'(t)$$
$$h(t) = 30t - 5t^2$$
$$h'(t) = 30 - 10t$$
$$h'(1) = 20 \qquad h'(2) = 10$$

The instantaneous velocity of the ball is 20 m/s at $t = 1$ and 10 m/s at $t = 2$. These values are, of course, the slopes of the tangents to the graph of the function at the points where $t = 1$ and $t = 2$. The velocity of the ball is 0 when $t = 3$. At this time the ball has reached its maximum height of $h(3) = 45$ m. The ball "stops" for an instant, changing its direction of motion from upward $(h'(t) > 0)$ to downward $(h'(t) < 0)$.

Acceleration is a measure of the rate of change of the velocity of an object. If the position function of the object is $s(t)$, then the velocity function is $v(t) = s'(t)$. The **average acceleration** from time t_1 to t_2 is:

$$\text{average acceleration} = \frac{v(t_2) - v(t_1)}{t_2 - t_1} = \frac{s'(t_2) - s'(t_1)}{t_2 - t_1}$$

Instantaneous acceleration at time t is defined as:

$$\text{instantaneous acceleration} = \lim_{\Delta t \to 0} \frac{v(t + \Delta t) - v(t)}{\Delta t} = v'(t)$$

Acceleration is the derivative of velocity, and velocity is the derivative of the position function. Acceleration is called the *"second derivative"* of the position function, and we write

$$\text{acceleration} = s''(t).$$

Since acceleration is the rate of change of velocity over time, it is given in units such as meters per second per second, (m/s)/s, or feet per second per second, (ft/s)/s. These are often abbreviated as m/s² and ft/s² respectively.

EXAMPLE 1. An object moves along a straight line so that after t seconds the distance in meters of the object to the right of its starting point is $s(t) = t^3 - 3t^2 + 5t$. Find the velocity and acceleration of the object at the end of the first two seconds of motion.

SOLUTION: position function, $s(t) = t^3 - 3t^2 + 5t$
velocity function, $s'(t) = 3t^2 - 6t + 5$
acceleration function, $s''(t) = 6t - 6$

$$s'(2) = 5, \quad s''(2) = 6$$

When $t = 2$, the velocity of the object is 5 m/s to the right, and the acceleration is 6 m/s² to the right.

EXAMPLE 2. The height above the ground of an object propelled upward from a rooftop 100 m high with an initial velocity of 80 m/s is given by the function

$$h(t) = 100 + 80t - 5t^2.$$

a. What is the greatest height reached by the object?
b. Show that the acceleration of the object is always the same, that is, that the acceleration is constant.

SOLUTION: **a.** $h(t) = 100 + 80t - 5t^2$
$h'(t) = 80 - 10t$
Maximum height is reached when $h'(t) = 0$.

$$80 - 10t = 0$$
$$t = 8$$

Since $h(8) = 420$, the object reaches a maximum height of 420 m.

b. $h(t) = 100 + 80t - 5t^2$

$h'(t) = 80 - 10t$

acceleration $= h''(t) = -10$

Acceleration is always -10 m/s^2. This means that the velocity decreases by 10 m/s each second.

ORAL EXERCISES 16-6

1. A ball is thrown upward so that its height in feet at time t seconds is:

$$h(t) = 48t - 16t^2$$

   a. What is the velocity at time t? at time $t = 1$?

   b. What is the acceleration at time t?

2. A distance-time graph for a toy car is shown below.

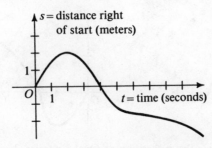

   a. How far does the toy car travel in the first two seconds?

   b. What is its average velocity during the first two seconds?

   c. When does the car move to the right? to the left?

   d. When does the car return to its starting point?

   e. What is the instantaneous velocity of the car at $t = 1$? $t = 2$? $t = 3$?

   f. At what time does the car move fastest to the right? to the left?

   g. What is the farthest the car moves to the right during the first 6 seconds?

3. A particle moves along a line so that its distance s in meters to the right of the origin after t seconds is $s = t^2 - 5t$.

   a. Where is the particle after 1 second?

   b. Does the particle begin moving left or right from the origin?

   c. How far does the particle move in the first two seconds?

   d. What is its average velocity over the first two seconds?

   e. When does the particle return to the origin?

WRITTEN EXERCISES 16-6

A   1. A ball is thrown upward from the top of an 80 ft building so that its height in feet above the ground t seconds later is $h(t) = 80 + 64t - 16t^2$.

   a. Find the average velocity over the interval from $t = 0$ to $t = 2$.

   b. What is the instantaneous velocity at time t? at time $t = 1$?

(Exercises continued on page 596)

1. (*Continued*)

    c. When is the velocity zero?

    d. What is the ball's maximum height above the ground?

    e. For what values of t is the ball falling?

    f. When does the ball hit the ground?

    g. What is the acceleration at time t?

    h. Graph the function $h(t)$.

2. A helicopter climbs vertically from the top of a 98 m building so that its height above the ground, in meters, after t seconds is $h(t) = 98 + 49t - 4.9t^2$. Answer the questions of Exercise 1 for the helicopter.

In Exercises 3 and 4 answer the questions of Oral Exercise 2 for the graphs shown below. In each graph, s is the distance (in meters, m) to the right of the starting point and t is the time (in seconds, s).

3.
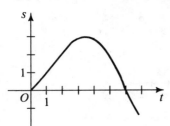

4.
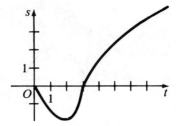

In Exercises 5 and 6 answer the questions of Oral Exercise 3 assuming that s is related to t by the given equation.

    5. $s = 8t - t^3$ **6.** $s = t^3 - 3t$

In Exercises 7 and 8 read the given description of the motion of a particle along a line. Then make a rough graph (like that in Exercises 3 and 4) showing the distance of the particle to the right of the origin as a function of time.

B   **7.** A particle leaves the origin, traveling right with velocity 1 m/s. Then it gradually slows until it reverses its direction at time $t = 3$. It returns to the origin at $t = 5$, traveling at a velocity of 2 m/s and then gradually slows until it stops to the left of the origin at time $t = 10$.

    8. A particle leaves the origin, traveling right very slowly, and gradually increasing its velocity until it reaches 2 m/s at time $t = 4$. Then the velocity begins to decrease and finally becomes zero at $t = 8$. At this time, the particle begins to move slowly left, and at $t = 20$ it finally returns to its starting point and stops.

    9. If a ball is thrown upward from the top of a 10 meter building with a velocity of 24.5 m/s, its height in meters above the ground t seconds later is $h(t) = 10 + 24.5t - 4.9t^2$. -

    a. How long does it take the ball to reach its maximum height?

    b. What is the maximum height of the ball above the ground?

    c. How long does it take for the ball to fall from its maximum height to the ground?

10. A particle (small object) moves along a number line so that its distance to the right of the origin at time t is $2t^3 - 6t^2 + 8$.
   a. At what times is the particle at the origin?
   b. At what times is the particle not moving?
   c. At what time is the velocity of the particle neither increasing nor decreasing; that is, at what time is the particle neither accelerating nor decelerating?

11. Suppose a car is traveling so that its distance west of Durham, in miles, at a time t hours after noon is $s(t) = 10t^{\frac{3}{2}} - 15t + 20$.
   a. How fast is the car going, and in which direction, at 12:15 P.M.?
   b. Where is the car, and what is the time, when its velocity is zero?

12. A ball is thrown horizontally from the top of a 100 ft building. The initial speed of the ball is 85 ft/s. The height h in feet of the ball above the ground at time t seconds is $h = 100 + 85t - 16t^2$.
   a. How long does it take the ball to hit the ground?
   b. How fast is the ball moving when it is halfway to the ground?
   c. How fast is the ball moving when it hits the ground?
   d. Suppose the horizontal component of the ball's velocity remains constant until the ball reaches the ground. How far is the ball from the base of the building when it lands?

Chapter summary

1. *Limits* are used to describe the behavior of a function $f(x)$ as x approaches a finite point c or as x approaches infinity or negative infinity. These are typical examples.

   a. $\displaystyle\lim_{x\to\infty} \frac{4x-3}{2x+1} = 2$ means the values of $\dfrac{4x-3}{2x+1}$ can be made as close to 2 as we like just by considering larger and larger values of x.

   b. $\displaystyle\lim_{x\to-\infty} \frac{3}{x-5} = 0$ means the values of $\dfrac{3}{x-5}$ can be made as close to 0 as we like just by considering negative values of x which have larger and larger absolute values.

   c. $\displaystyle\lim_{x\to2^+} \frac{1}{x-2} = \infty$ means the values of $\dfrac{1}{x-2}$ can be made arbitrarily large by approaching 2 nearer and nearer from the right.

   d. $\displaystyle\lim_{x\to2^-} \frac{3x^2-6x}{x-2} = 6$ means the values of $\dfrac{3x^2-6x}{x-2}$ can be made as close to 6 as we like just by approaching 2 nearer and nearer from the left.

2. If $\lim\limits_{x \to c^+} f(x) = L$ and also $\lim\limits_{x \to c^-} f(x) = L$, then we can speak of the limit as x approaches c and write $\lim\limits_{x \to c} f(x) = L$. If $\lim\limits_{x \to c^+} f(x) \neq \lim\limits_{x \to c^-} f(x)$, then $\lim\limits_{x \to c} f(x)$ does not exist.

3. A summary of techniques for finding limits is given on pages 571–572.

4. A summary of the steps used in graphing rational functions is given on pages 575–576.

5. The *derivative* of the function $f(x)$ is denoted $f'(x)$ and it gives the slope of the graph of $y = f(x)$ at the point (x, y). By definition, $f'(x) = \lim\limits_{h \to 0} \dfrac{f(x + h) - f(x)}{h}$. Theorems 1, 2, and 3 offer short cuts for calculating derivatives.

6. A point $(c, f(c))$ is a local *maximum* or *minimum* point of a polynomial function $f(x)$ if $f'(c) = 0$ and $f'(x)$ changes sign at $x = c$.

7. If $s(t)$ is the position of an object at time t, then the *instantaneous velocity* at time t is

$$\lim_{\Delta t \to 0} \frac{s(t + \Delta t) - s(t)}{\Delta t} = s'(t).$$

8. If $s(t)$ and $v(t)$ are, respectively, the position function and instantaneous velocity function of an object at time t, then the *instantaneous acceleration* at time t is

$$\lim_{\Delta t \to 0} \frac{v(t + \Delta t) + v(t)}{\Delta t} = v'(t) = s''(t).$$

Chapter test

1. Evaluate each limit.　　　　　　　　　　　　　　　　　　　　　　　16–1

  a. $\lim\limits_{x \to \infty} \dfrac{x^2 - x}{x^2 - 1}$　　b. $\lim\limits_{x \to 1} \dfrac{x^2 - x}{x^2 - 1}$　　c. $\lim\limits_{x \to -\infty} \cos\left(\dfrac{1}{x}\right)$　　d. $\lim\limits_{x \to 9} \dfrac{\sqrt{x} - 3}{x - 9}$

2. Explain why $\lim\limits_{x \to 0} \dfrac{|x|}{x}$ does not exist.

3. Sketch the graph of each equation showing all horizontal and vertical asymptotes.　　16–2

  a. $y = \dfrac{x}{x - 4}$　　　　　　　b. $y = \dfrac{x^2 + 1}{x^2 - 1}$

4. Find the derivative of the following functions.　　16–3

  a. $x^3 - 5x^2 + 7x - 2$　　b. $\dfrac{1}{x^2} - \dfrac{3}{x} + 8$

  c. $\sqrt[3]{x^2} + \sqrt[3]{\dfrac{1}{x}}$　　　　d. $\dfrac{1}{x^8} + x^8$

5. **a.** Define $f'(x)$ as a limit.

   b. Use this *definition* (and no other theorems) to prove that if $f(x) = x^2$, then $f'(x) = 2x$.

6. Sketch the graph of $y = 3x^2 - x^3$. Label with both coordinates all local maximum and minimum points. 16-4

7. Suppose you are told that $x = 2$ is the only positive root of the equation $f'(x) = 0$. If $f'(1) > 0$, and $f'(3) < 0$, what conclusions can you make about the function at $x = 2$? 16-5

8. **a.** A rectangular solid has a square base. Express its surface area in terms of x and h.

   b. If the solid has volume 1000 express h in terms of x.

   c. Use (a) and (b) to express the surface area in terms of x alone.

   d. Show that the surface area is minimized when $x = h$; that is, when the solid is a cube.

9. The position function of an object moving a distance in meters to the right along a straight line is $s(t) = 2t^2 - t + 1$. Find the velocity and acceleration of the object when $t = 3$ seconds. 16-6

Cumulative review   (Chapters 11–16)

1. For each of the following equations, state whether its graph is: an ellipse, a hyperbola, a parabola, a circle, or none of these.

   a. $x^2 + 4y^2 - 9y = 0$

   b. $x^2 - 4y - 8x = 0$

   c. $(x - 4)(y - 9) = 0$

   d. $4x^2 + 4y^2 + 9x + 9y = 0$

   e. $4y^2 - 4x^2 = 9$

2. Write an equation of the ellipse with center at the origin, eccentricity 0.75, and a vertex at $(0, -8)$.

3. Sketch the graph of $x^2 + 4y^2 - 2x + 24y + 33 = 0$, labeling the coordinates of its center and vertices.

4. Sketch the graphs of $xy = 6$ and $x^2 - y^2 = 9$ on the same axes. Then determine algebraically where the two graphs intersect.

5. Write an equation of the parabola with horizontal axis of symmetry and containing the points $(4, 0)$, $(3, -1)$, and $(0, 2)$. Also find the coordinates of the focus and the equation of the directrix for this parabola.

6. An airplane is heading west at a velocity of 450 knots and encounters a southwesterly wind with a velocity of 40 knots. Use vectors to show the resultant velocity of the airplane when it encounters this wind, and use trigonometry to find the resulting speed of the plane.

7. If $\mathbf{u} = (2, -1)$ and $\mathbf{v} = (-3, 4)$:
   a. Calculate $|\mathbf{v} - \mathbf{u}|$.
   b. Write the polar form of $2\mathbf{u} + \mathbf{v}$.

8. Write both the vector equation and the parametric equations for the line $2x + 3y = 15$.

9. Find the angle between the vectors $\mathbf{u} = (2, 4)$ and $\mathbf{v} = (-1, 2)$.

10. Given line l with equation $(x, y, z) = (-2, 1, 3) + t(2, -1, -4)$, find an equation of a line through $(3, 2, 1)$ parallel to l.

11. Find an equation of the plane perpendicular to the vector $(4, 7, 1)$ at the point $(0, 2, 3)$.

12. Evaluate the determinant: $\begin{vmatrix} 5 & 2 & 3 \\ 2 & 1 & 0 \\ 1 & -2 & 2 \end{vmatrix}$

13. Solve the given pair of equations by using determinants.
$$2x - y = 5$$
$$3x - 2y = 4$$

14. If $\mathbf{u} = (2, 3, -1)$ and $\mathbf{v} = (1, 4, 3)$ calculate $\mathbf{u} \times \mathbf{v}$.

15. Find the first four terms of the given sequence and tell whether the sequence is arithmetic, geometric, or neither.
   a. $t_n = (-1)^n + 5$ b. $t_n = 3n + 1$ c. $t_n = 4 - n$

16. Find S_5 if the series $8 + 12 + \cdots$ is (a) arithmetic, (b) geometric.

17. Evaluate the given limit or state that the limit does not exist.
   a. $\lim\limits_{n \to \infty} \dfrac{3n}{n^2 - 4}$ b. $\lim\limits_{n \to \infty} \cos n$
   c. $\lim\limits_{n \to \infty} \log \dfrac{10n^2 + 3}{n^2}$ d. $\lim\limits_{n \to \infty} \dfrac{\sin n}{n}$

18. For the geometric series $\dfrac{x}{4} + \dfrac{x^3}{8} + \dfrac{x^5}{16} + \cdots$, find (a) the interval of convergence and (b) the sum, expressed in terms of x.

19. Show that $\sum\limits_{k=1}^{3} \log_6 k = 1$.

20. Use mathematical induction to prove $\sum\limits_{i=1}^{n} 2i = n(n + 1)$.

21. The following table gives the distance in kilometers that the employees of a company must commute each day to work.

distance	5	15	18	22	31	45
number of people	3	3	2	4	6	2

   a. Calculate the mean, median, and mode for these data.
   b. Find the standard deviation.

22. A cafeteria must supply enough lunches to feed all of a school enrollment of 400 students. On any particular day, the mean number of students absent is 35 with a standard deviation of 5. How many lunches should the staff prepare if they intend to have enough food for all the students 99.5% of the time?

23. Out of 25 voters, 18 favored Proposition 2. Find $\bar{p}$ and give a 95% confidence interval for p.

24. The weather bureau for a certain city has correlated the annual temperature in Fahrenheit and annual precipitation in inches for the last 40 years. Letting x stand for temperatures and y for precipitation, they obtained the following facts: $\bar{x} = 55$, $\bar{y} = 40$, $\overline{xy} = 2240$, $s_x = 7$, and $s_y = 8$.
    a. Calculate the correlation coefficient.
    b. Give the equation of the regression line.

25. In how many ways can 8 people finish in the first 3 positions of a race?

26. In how many ways can you arrange the letters in the word MATH?

27. A committee of 5 people is to be selected from a club that has 25 members. How many different committees can there be?

28. Suppose someone rolls two dice and announces that their sum is 8. What is the probability that one of the dice shows a "5"?

29. Five cards are dealt from a standard deck. What is the probability that all of the cards are in the same suit?

30. Give the series expansion of the expression $(a - b)^5$.

31. A certain coin is bent so that the probability of heads on any toss is $\frac{1}{3}$. If the coin is tossed 7 times, what is the probability of getting exactly 3 heads?

32. Evaluate the given limit or state that it does not exist.
    a. $\lim\limits_{x \to 3} \dfrac{x^2 - 9}{x - 3}$ b. $\lim\limits_{x \to \infty} \dfrac{3x^2 + 2x}{x^3 - 4}$
    c. $\lim\limits_{x \to 2^-} \dfrac{x + 4}{x - 2}$ d. $\lim\limits_{x \to 1} \dfrac{3x^2 + 1}{x - 1}$

33. Sketch the graph of the following function. Show vertical and horizontal asymptotes and all intercepts.
$$y = \frac{x^2 - 4x + 3}{x^2 - x - 2}$$

34. Find the slope of the curve $y = \dfrac{-\sqrt{x}}{3}$ at the point $P(4, -\frac{2}{3})$.

35. Graph the function $f(x) = 4x^3 - 3x + 5$. Identify any local maximum and minimum points.

36. A rectangular plot is to be enclosed by using part of an existing fence as one side and 120 m of fencing for the other three sides. What is the greatest area the rectangular plot can have?

37. The position of an object moving a distance in meters to the left of the origin is $s(t) = 2t^3 + 4t^2$. Find the velocity and acceleration when $t = 2$.

Cumulative test (Chapters 1–7)

No tables are needed for this test. Leave answers involving irrational numbers in terms of π and simple radicals.

In Exercises 1–12, simplify the given expressions.

1. $\cos 210°$

2. $\tan \pi$

3. $\csc \dfrac{5\pi}{6}$

4. $\cot 315°$

5. $\operatorname{Sin}^{-1}\left(-\dfrac{\sqrt{3}}{2}\right)$

6. $\sec\left(\operatorname{Tan}^{-1}\left(-\dfrac{3}{4}\right)\right)$

7. $e^{\ln 1}$

8. $8^{-\frac{5}{3}}$

9. $\log_{10} \sqrt[5]{100}$

10. $\dfrac{1}{1-i}$

11. $(3^{-1}-2^{-2})^{-2}$

12. $\sqrt{-27}-\dfrac{1}{\sqrt{-3}}$

Sketch the graph of the equation, inequality, or system.

13. $y = \tan x$

14. $x^2 + 6x + y^2 = -5$

15. $y \geq -2x^2 + 4x - 5$

16. $y = -\log_3 x$

17. $y = x^3 + 3x^2 - x - 3$

18. $3x - 2y \leq 2$
$5x + 4y > -15$

19. If $A = (-4, 5)$, $B = (1, 2)$ and $C = (-2, -3)$, **(a)** show that $\triangle ABC$ is an isosceles right triangle; **(b)** find an equation for the perpendicular bisector of the hypotenuse.

20. Solve: **a.** $5x + 4y = -1$
$4x - y = 2$
b. $x^2 + y^2 = 1$
$2x + y = 2$

21. Find the zeros of $f(x) = 2x^4 + 4x^2 - 32$. Also find $f(i)$.

22. Use the Rational Roots Theorem to find all real and imaginary roots of $5x^3 + 21x^2 + 21x + 5 = 0$.

23. Solve: $\log_5 50 = \frac{1}{2} \log_5 4 = 4^x$.

24. If $g(x) = \dfrac{1}{\sqrt{x + 1}}$, find $g^{-1}(x)$, its domain, and its range.

25. If $f(x) = \left| \dfrac{1}{2}x \right|$ and $g(x) = \sin x$, find $f\left(g\left(-\dfrac{\pi}{3}\right)\right)$ and $g\left(f\left(-\dfrac{\pi}{3}\right)\right)$.

26. If you invest some money at 9.6% interest compounded monthly, find an expression for t, the number of years needed for the investment to quadruple.

27. Express as a single trigonometric function of x: $\cot x(\sec x - \cos x)$.

28. Solve for $0 \leq x < 2\pi$: $2 \sin x \cot x = 3 \tan x$.

29. If $\angle A$ is obtuse and $\csc A = \frac{29}{20}$, find $\sec A$.

30. For what values of t is $t^3 + 2t^2 - 4t < 0$?

31. In $\triangle RST$, $r = 10\sqrt{2}$, $\angle S = 30°$, and $s = 10$. Find the possible values of $\angle R$ and $\angle T$.

32. In $\triangle ABC$, $\angle A = 23°$, $b = 8$, and $c = 10$. Approximate the value of a and the triangle's area to the nearest tenth. ($\sin 23° \approx 0.39$; $\cos 23° \approx 0.92$)

Cumulative test on trigonometry (Chapters 6–10)

In Exercises 1–18, do not use trigonometric tables or a calculator. Leave answers involving irrational numbers in terms of π and simplified radicals.

Evaluate.

1. $\csc 120°$

2. $\cos(-\pi)$

3. $\cot \dfrac{4\pi}{3}$

4. $\text{Tan}^{-1}(-1)$

5. $\sec(\text{Sin}^{-1} \frac{5}{13})$

6. $\cos 22.5°$

7. $\tan 345°$

8. $\text{Cos}^{-1}\left(\cos\left(-\dfrac{\pi}{6}\right)\right)$

9. $1 - 2\sin^2(-15°)$

Simplify.

10. $\tan^2 x(1 - \csc^2 x)$

11. $\dfrac{\csc \theta}{\csc(90° - \theta)}$

12. $\dfrac{\sin(-\theta)}{1 - \cos^2 \theta}$

13. A sector has area 24 and a central angle of 3.75 radians. Find its arclength.

14. If α is a third-quadrant angle and $\cos \alpha = -\frac{2}{3}$, find the other five trigonometric functions of α.

15. Solve for x, $0 < x < 2\pi$:   **a.** $2\sin x - 2\csc x = 3$ **b.** $\sec x \sin 2x = \tan \dfrac{x}{2}$

16. In $\triangle PQR$, $p = 3$, $q = 8$, and $\angle R = 60°$. Find the value of r.

17. Draw the graph of $y = -3\sin\left(x + \dfrac{\pi}{2}\right)$.

18. If $\tan \theta = 2$, evaluate $\tan(\theta - 60°)$ and $\tan 2\theta$.

Use tables or a calculator in Exercises 19–24. Give angle measures to the nearest tenth of a degree.

19. Find the inclination of the line joining $(-2, 3)$ and $(6, 1)$.

20. A rhombus has an angle of 52° and sides of length 10 cm. Find the lengths of the diagonals to the nearest tenth.

21. Find the area of $\triangle ABC$ if $a = 7$, $b = 8$, and $\angle C = 125°$.

22. In $\triangle ABC$, $\angle A = 30°$, $a = 5$, and $b = 9$. Find the value of c and the measures of $\angle B$ and $\angle C$.

23. A triangle has sides of lengths 4, 5, and 2. Find the measure of the largest angle.

24. Find the measure of an acute angle formed by the lines with equations $3x + y = 6$ and $x - 2y = 4$.

25. Sketch the polar graph of $r = 4\sin \theta$. Find a rectangular equation for the graph.

26. If $z_1 = -2$ and $z_2 = 1 - i\sqrt{3}$, express z_1, z_2, and $z_1 z_2$ in polar form. Show the three complex numbers in an Argand diagram.

27. If $z = \dfrac{1}{2} - \dfrac{\sqrt{3}}{2}i$, find z^{10} and the three cube roots of z. Express each answer in the form $x + yi$.

Cumulative test (Chapters 8–16)

Test A: Exercises 1–16, which test Chapters 8–13.

Test B: Exercises 1–21, which test Chapters 8–16.

1. Sketch the graph of $y = -2 \cos \frac{1}{2}(x - \pi)$.

2. Determine the line and point symmetry of the graph of $x^3 + y^3 = 0$.

3. If $\tan \alpha = -\frac{21}{20}$ and $0 \le \alpha < \pi$, find $\cos 2\alpha$, $\tan \frac{\alpha}{2}$, and $\sin\left(\frac{2\pi}{3} + \alpha\right)$.

4. Use formulas to find the exact values of $\cos 255°$ and $\sin \frac{13\pi}{12} \cos \frac{13\pi}{12}$.

5. Solve for x, $0 \le x < 2\pi$: $\sin^2 \frac{x}{2} = -\cos x$.

6. If $z_1 = -2\sqrt{3} - 2i$ and $z_2 = 3i$, express z_1, z_2, $z_1 z_2$, and $(z_1)^5$ in polar form.

7. Find each cube root of $27i$ in the form $x + yi$.

8. Sketch the graph of each system. Then solve it algebraically.
   a. $4x^2 - 9y^2 = 144$
      $9x^2 - 9y^2 = 189$
   b. $x = -2y^2 + 8$
      $xy = -6$

9. Find an equation of the parabola with focus $(0, -1)$ and directrix $y = 3$.

10. The position of an object at time t is given by the parametric equations $x = -1 - t$ and $y = 3 + 7t$. Find the velocity and speed of the object.

11. If $\mathbf{u} = (3, -1)$ and $\mathbf{v} = (6, 2)$, find the angle between $\mathbf{u}$ and $\mathbf{v}$.

12. If $R = (0, -3, 7)$, $S = (3, -2, 9)$, and $T = (-5, -3, 6)$, use the cross product of vectors to find the area of $\triangle RST$.

13. Find S_6 if the series $2 + 6 + \cdots$ is (a) arithmetic; (b) geometric.

14. Find the sum of the infinite geometric series $48 - 12 + 3 - \cdots$.

15. Use mathematical induction to show that $\displaystyle\sum_{k=1}^{n} \frac{1}{2^k} = 1 - \frac{1}{2^n}$.

16. Given the data 2, 3, 3, 3, 5, 6, 6, 7, 7, 8, find the mean, median, mode, variance, and standard deviation.

17. Out of 100 appliances tested, 64 were found to be energy-efficient. Find a 95% confidence interval for the total number of appliances in the population that are energy-efficient.

18. In how many ways can 4 students be chosen from a group of 300 for 4 identical scholarships? for 4 different scholarships?

19. A die is rolled 10 times. What is the probability of getting a "4" at least 8 times?

20. Evaluate: a. $\displaystyle\lim_{x \to 2} \frac{x^3 - 8}{x - 2}$ b. $\displaystyle\lim_{x \to 0} \frac{\cos 2x}{|x|}$

21. If $f(x) = x^4 - 2x^3 + x^2$, find (a) the slope of $f(x)$ at $(-1, 4)$; (b) each maximum and minimum point.

Geometry review

1. Conditions for congruent triangles

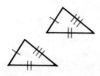

SSS: 3 pairs of corre-
sponding sides
congruent

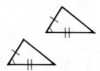

SAS: 2 pairs of corre-
sponding sides
and included an-
gles congruent

ASA: 2 pairs of corre-
sponding angles
and included
sides congruent

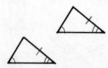

AAS: 2 pairs of corresponding
angles and a pair of cor-
responding nonincluded
sides congruent

HL: in a right triangle, one pair
of corresponding legs and
hypotenuse of each con-
gruent.

2. Conditions for similar triangles

SSS: 3 pairs of corre-
sponding sides
proportional

SAS: 2 pairs of corre-
sponding sides
proportional, and
included angles
congruent

AA: 2 pairs of corre-
sponding angles
congruent

3. Special triangle relationships

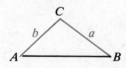

$a = b$ if and only if $\angle A = \angle B$.

$\angle A + \angle B + \angle C = 180°$.

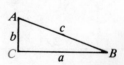

$c^2 = a^2 + b^2$ if and only if $\angle C = 90°$;
otherwise: $c^2 = a^2 + b^2 - 2ab \cos C$.

605

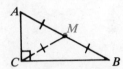

The midpoint of the hypotenuse of a right triangle is equidistant from the 3 vertices.

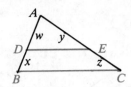

$DE \parallel BC$ if and only if

(1) $\dfrac{w}{x} = \dfrac{y}{z}$, or

(2) $\dfrac{w}{w + x} = \dfrac{y}{y + z} = \dfrac{DE}{BC}$.

If $w = x$ and $y = z$, then $DE = \frac{1}{2}BC$.

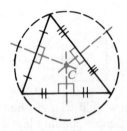

Perpendicular bisectors meet at circumcenter C.

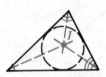

Angle bisectors meet at incenter I.

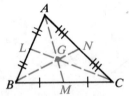

Medians meet at centroid G;
$AG = \frac{2}{3}AM$, $BG = \frac{2}{3}BN$,
$CG = \frac{2}{3}CL$.

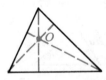

Altitudes meet at orthocenter O.

4. Parallel lines

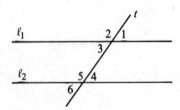

$l_1 \parallel l_2$ if and only if
(1) $\angle 1 \cong \angle 6$, or
(2) $\angle 3 \cong \angle 4$, or
(3) $\angle 3$ and $\angle 5$ supplements.

5. Special quadrilaterals

Parallelogram Properties

  Opposite sides ∥ and ≅
  Opposite angles ≅
  Diagonals bisect each other.
  If a rhombus: Diagonals ⊥
  If a rectangle: Diagonals ≅

Trapezoid Properties

  Length of median half the
    sum of bases
  Median ∥ bases
  If isosceles: Base angles ≅

6. Area formulas

Triangle: $\quad A = \frac{1}{2}bh = \frac{1}{2}ab \sin C$ Circle: $\quad A = \pi r^2$

Parallelogram: $A = bh = ab \sin c$ Sphere: $\quad A = 4\pi r^2$

Trapezoid: $\quad A = \frac{1}{2}h(b_1 + b_2)$ Cylinder: $A = 2\pi r(r + h)$

Cone: $\qquad A = \pi r(r + s)$ (s = slant height)

Pyramid: $\quad A = B + \frac{1}{2}nas$ (B = area of base, n = number of base sides,
  a = length of one base side, s = slant height)

7. Volume formulas

Prism: $\quad V = Bh$ (B = area of base)   Sphere: $\quad V = \frac{4}{3}\pi r^3$

Cone: $\quad V = \frac{1}{3}\pi r^2 h$ Cylinder: $V = \pi r^2 h$

Pyramid: $V = \frac{1}{3}Bh$ (B = area of base)

8. Circles and angle relationships

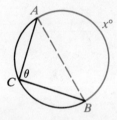

(1) $\theta = \frac{1}{2}x$
(2) $\theta = 90°$ if $\overline{AB}$ is a diameter.

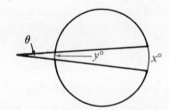

$\theta = \frac{1}{2}(x - y)$

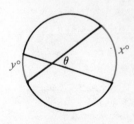

$\theta = \frac{1}{2}(x + y)$

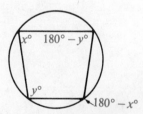

Opposite angles of inscribed
quadrilateral are supplementary.

9. Circles and chord, secant, and tangent relationships

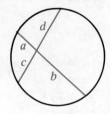

$$a \cdot b = c \cdot d$$

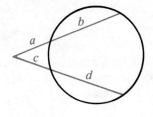

$$a \cdot (a + b) = c \cdot (c + d)$$

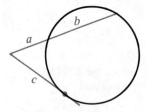

$$a \cdot (a + b) = c^2$$

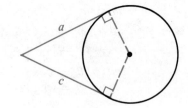

(1) $a = c$
(2) radius $\perp$ tangent

Formulas and laws from trigonometry

DEFINITIONS

$$\tan \theta = \frac{\sin \theta}{\cos \theta}$$

$$\cot \theta = \frac{1}{\tan \theta} = \frac{\cos \theta}{\sin \theta}$$

$$\sec \theta = \frac{1}{\cos \theta}$$

$$\csc \theta = \frac{1}{\sin \theta}$$

PYTHAGOREAN FORMULAS

$$\sin^2 \theta + \cos^2 \theta = 1$$
$$1 + \tan^2 \theta = \sec^2 \theta$$
$$1 + \cot^2 \theta = \csc^2 \theta$$

ADDITION FORMULAS

$$\sin (\alpha + \beta) = \sin \alpha \cos \beta + \cos \alpha \sin \beta$$
$$\sin (\alpha - \beta) = \sin \alpha \cos \beta - \cos \alpha \sin \beta$$
$$\cos (\alpha + \beta) = \cos \alpha \cos \beta - \sin \alpha \sin \beta$$
$$\cos (\alpha - \beta) = \cos \alpha \cos \beta + \sin \alpha \sin \beta$$

$$\tan (\alpha + \beta) = \frac{\tan \alpha + \tan \beta}{1 - \tan \alpha \tan \beta}$$

$$\tan (\alpha - \beta) = \frac{\tan \alpha - \tan \beta}{1 + \tan \alpha \tan \beta}$$

DOUBLE ANGLE FORMULAS

$$\sin 2\alpha = 2 \sin \alpha \cos \alpha$$
$$\cos 2\alpha = \cos^2 \alpha - \sin^2 \alpha$$
$$\qquad = 2 \cos^2 \alpha - 1$$
$$\qquad = 1 - 2 \sin^2 \alpha$$
$$\tan 2\alpha = \frac{2 \tan \alpha}{1 - \tan^2 \alpha}$$

HALF-ANGLE FORMULAS

$$\sin \frac{\alpha}{2} = \pm \sqrt{\frac{1 - \cos \alpha}{2}}$$

$$\cos \frac{\alpha}{2} = \pm \sqrt{\frac{1 + \cos \alpha}{2}}$$

$$\tan \frac{\alpha}{2} = \pm \sqrt{\frac{1 - \cos \alpha}{1 + \cos \alpha}} = \frac{\sin \alpha}{1 + \cos \alpha}$$

$$\qquad = \frac{1 - \cos \alpha}{\sin \alpha}$$

AREA OF TRIANGLE $\quad K = \dfrac{1}{2} ab \sin C$

LAW OF SINES $\quad \dfrac{\sin A}{a} = \dfrac{\sin B}{b} = \dfrac{\sin C}{c}$

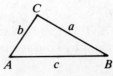

LAW OF COSINES $\quad c^2 = a^2 + b^2 - 2ab \cos C, \quad \cos C = \dfrac{a^2 + b^2 - c^2}{2ab}$

$$\cos C = \frac{\vec{CA} \cdot \vec{CB}}{|\vec{CA}||\vec{CB}|} \quad (\textit{Vector form of law})$$

SECTORS	θ in radians	θ in degrees
Arclength	$s = r\theta$	$s = \dfrac{\theta}{360} \cdot 2\pi r$
Area	$K = \frac{1}{2}r^2\theta = \frac{1}{2}rs$	$K = \dfrac{\theta}{360} \cdot \pi r^2$

College entrance examinations

The Scholastic Aptitude Test (SAT)

There are two 30-minute mathematical sections in the SAT. The questions (a total of 60) in these sections are designed to measure fundamental quantitative abilities closely related to college-level work.

In some questions you will be asked to apply graphic, spatial, numerical, symbolic, and logical techniques to problems that may be similar to exercises in your textbooks. Other questions may require you to do some original thinking.

The mathematical content is restricted to what is typically taught in the ninth grade or earlier. The arithmetic includes the four basic operations of addition, subtraction, multiplication, and division; percent; average; properties of odd and even numbers; prime numbers; and divisibility of numbers. The algebra includes negative numbers; simplifying algebraic expressions; linear equations; inequalities; simple quadratic equations; positive integer exponents; and roots. The geometry includes concepts of area, perimeter, volume, and circumference; special properties of isosceles, equilateral, and right triangles; the relationship between the lengths of the sides in 30°-60°-90° triangles and 45°-45°-90° triangles; properties of parallel and perpendicular lines; and locating points on a coordinate grid.

Two kinds of multiple-choice questions are in the mathematical sections:

(1) Standard multiple-choice questions (approximately $\frac{2}{3}$ of the test)
(2) Quantitative comparison questions (approximately $\frac{1}{3}$ of the test)

Some sample questions of both types are given on the following pages. Answers to these questions appear at the end of the book.

Note: The directions given for both types of question, as well as the examples given for the quantitative comparison questions, are printed as they would appear in an actual test booklet. Please disregard references made to using extra space for scratchwork and filling in answers on answer sheets.

The questions and accompanying text are reprinted, with permission, from copyrighted publications of the College Entrance Examination Board and Educational Testing Service, the copyright owner of the sample questions. Permission to reprint SAT Test and Achievement Test material does not constitute review or endorsement by Educational Testing Service or the College Board of this publication.

STANDARD MULTIPLE-CHOICE QUESTIONS

Directions: In this section solve each problem, using any available space on the page for scratchwork. Then decide which is the best of the choices given and blacken the corresponding space on the answer sheet.

The following information is for your reference in solving some of the problems:

Circle of radius r: Area $= \pi r^2$; Circumference $= 2\pi r$

The number of degrees of arc in a circle is 360.

The measure in degrees of a straight angle is 180.

Triangle: The sum of the measures in degrees of the angles of a triangle is 180.

If $\angle CDA$ is a right angle, then

(1) area of $\triangle ABC = \dfrac{AB \times CD}{2}$

(2) $AC^2 = AD^2 + DC^2$

Definitions of symbols:

$=$ is equal to	$\neq$ is unequal to
$<$ is less than	$\leq$ is less than or equal to
$>$ is greater than	$\geq$ is greater than or equal to
$\perp$ is perpendicular to	$\parallel$ is parallel to

Note: Figures which accompany problems in this test are intended to provide information useful in solving the problems. They are drawn as accurately as possible EXCEPT when it is stated in a specific problem that its figure is not drawn to scale. All figures lie in a plane unless otherwise indicated. In this test, all numbers used are real numbers.

1. If $2a + b = 5$, then $4a + 2b =$

(A) $\frac{5}{4}$ (B) $\frac{5}{2}$ (C) 10   (D) 20   (E) 25

2. If $16 \cdot 16 \cdot 16 = 8 \cdot 8 \cdot P$, then $P =$

(A) 4   (B) 8   (C) 32   (D) 48   (E) 64

3. The town of Mason is located on Eagle Lake. The town of Canton is west of Mason. Sinclair is east of Canton, but west of Mason. Dexter is east of Richmond, but west of Sinclair and Canton. Assuming all these towns are in the United States, which town is farthest west?

(A) Mason   (B) Dexter   (C) Canton   (D) Sinclair   (E) Richmond

4. If the average of seven x's is 7, what is the average of fourteen x's?

(A) $\frac{1}{7}$ (B) $\frac{1}{2}$ (C) 1   (D) 7   (E) 14

5. If an asterisk (*) between two expressions indicates that the expression on the right exceeds the expression on the left by 1, which of the following is (are) true for all real numbers x?

I. $x(x + 2) * (x + 1)^2$ III. $\dfrac{x}{y} * \dfrac{x + 1}{y + 1}$

II. $x^2 * (x + 1)^2$

(A) None   (B) I only   (C) II only   (D) III only   (E) I and III

6. If a car travels x kilometers of a trip in h hours, in how many hours can it travel the next y kilometers at this rate?

(A) $\dfrac{xy}{h}$ (B) $\dfrac{hy}{x}$ (C) $\dfrac{hx}{y}$ (D) $\dfrac{h + y}{x}$ (E) $\dfrac{x + y}{h}$

7. If 90 percent of p is 30 percent of q, then q is what percent of p?

(A) 3%   (B) 27%   (C) 30%   (D) 270%   (E) 300%

8. The figure at the right shows a piece of paper in the shape of a parallelogram with measurements as indicated. If the paper is tacked at its center to a flat surface and then rotated about its center, the points covered by the paper will be a circular region of diameter

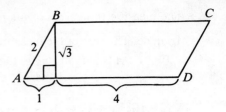

(A) $\sqrt{3}$ (B) 2   (C) 5   (D) $\sqrt{28}$ (E) $\sqrt{39}$

9. An accurate 12-hour clock shows 3 o'clock at a certain instant. Exactly 11,999,999,995 hours later, what time does the clock show?

(A) 5 o'clock   (B) 7 o'clock   (C) 8 o'clock
(D) 9 o'clock   (E) 10 o'clock

10. In the triangles at the right, if AB, CD, and EF are line segments, what is the sum of the measures of the six marked angles?

(A) 180°   (B) 360°
(C) 540°   (D) 720°
(E) It cannot be determined from the information given.

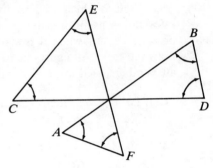

11. If the area of a square is 16 and the coordinates of one corner are $(3, 2)$, which of the following could be the coordinates of the diagonally opposite corner?

(A) $(7, 2)$ (B) $(6, 7)$ (C) $(3, 6)$ (D) $(-1, 2)$ (E) $(-1, 6)$

12. A box with a square base is filled with grass seed. How many cubic feet of seed will it contain when full, if the height of the box is 2 feet and one side of the base measures 18 *inches*?

(A) 3 ft³   (B) $4\frac{1}{2}$ ft³   (C) 6 ft³   (D) 36 ft³   (E) 648 ft³

13. The houses on the east side of a street are numbered with the consecutive even integers from 256 to 834 inclusive. How many houses are there on the east side of the street?

(A) 287   (B) 288   (C) 289   (D) 290   (E) 291

14. If the 9-inch by 12-inch piece of paper shown in the figure at right were folded flat along the line PQ, then R would be how many inches closer to S?

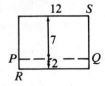

(A) 4   (B) 2   (C) $\sqrt{193} - \sqrt{153}$
(D) $15 - \sqrt{193}$ (E) $\sqrt{193} - 13$

15. Which of the following fractions is greater than $\frac{2}{3}$ and less than $\frac{3}{4}$?

(A) $\frac{5}{8}$ (B) $\frac{4}{5}$ (C) $\frac{7}{12}$ (D) $\frac{9}{16}$ (E) $\frac{7}{10}$

QUANTITATIVE COMPARISON QUESTIONS

Quantitative comparison questions emphasize the concepts of equalities, inequalities, and estimation. They generally involve less reading, take less time to answer, and require less computation than regular multiple-choice questions.

Directions: Each of the following questions consists of two quantities, one in Column A and one in Column B. You are to compare the two quantities and on the answer sheet blacken space

A   if the quantity in *Column A* is the greater;

B   if the quantity in *Column B* is the greater;

C   if the two quantities are equal;

D   if the relationship cannot be determined from the information given.

(*Reminder:* Since Questions 16–22 which follow are merely samples, no answer spaces are provided in this text.)

Notes: 1. In certain questions, information concerning one or both of the quantities to be compared is centered above the two columns.

2. A symbol that appears in both columns represents the same thing in *Column A* as it does in *Column B*.

3. Letters such as x, n, and k stand for real numbers.

EXAMPLES

	Column A	Column B	Answers
E1.	2×6	$2 + 6$	Ⓐ Ⓑ Ⓒ Ⓓ

E2.	$x + y$	$180 - z$	Ⓐ Ⓑ **Ⓒ** Ⓓ
E3.	$p - q$	$q - p$	Ⓐ Ⓑ Ⓒ **Ⓓ**

	Column A		Column B
		$a < 0$	
		$b < 0$	
16.	$a + b$		$a - b$
17.	The average of $(30 - 4)$ and $(30 + 4)$		The average of 28, 29, 30, 31, and 32

Column A	Column B

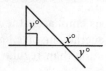

| 18. | $x - y$ | 90 |

$$0 < m < 6$$
$$0 < n < 8$$

| 19. | m | n |

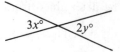

20.	x	y
21.	$x + 1$	$2x + 1$
22.	Area of a triangle with altitude 4	Area of a triangle with base 5

How the Tests Are Scored

The SAT Test, and also the Achievement Tests described next, are scored as follows: You get one point for each question that you answer correctly, and you lose a fraction of a point for each question you answer incorrectly. You neither gain nor lose points for questions you omit. If you mark more than one answer for a question, it is scored like an omitted question.

For questions with 4 answer-choices,

$$\text{your raw score} = \text{number right} - \tfrac{1}{3} (\text{number wrong}).$$

For questions with 5 answer-choices,

$$\text{your raw score} = \text{number right} - \tfrac{1}{4} (\text{number wrong}).$$

Your total raw score is the sum of these raw scores. The College Board converts your raw score for the entire SAT test or achievement test to a scaled score between 200 and 800.

Because of the way the test is scored, haphazard or random guessing on questions you know nothing about is not recommended. When you know that one or more choices can be eliminated, however, guessing from the remaining choices should be to your advantage.

The Achievement Tests

The College Board offers two Achievement Tests in mathematics, Mathematics Level I and Mathematics Level II. Mathematics Level I is the College Board's principal mathematics Achievement Test for admission to college. Candidates who have had three years of college preparatory mathematics should generally take Level I. Those candidates who have had more than three years should strongly consider taking Level II; however, the Level II Test is suitable for able students who have had three years of a strong mathematics curriculum.

The Level I examination is a broad-range, cumulative test. Questions are based on topics usually covered in a college preparatory mathematics sequence. About half the questions deal with topics from algebra and plane Euclidean geometry; the rest cover some aspects of coordinate geometry, right triangle trigonometry, functions and functional notation for composition and inverse, space perception of simple solids (including area and volume formulas), mathematical reasoning, and the nature of proof.

This wide range of topics provides an opportunity for students with different backgrounds to demonstrate their understanding and achievement in the areas of mathematics that they have studied. It is not expected that all students will be familiar with all of the topics included.

The Level II test is narrower in scope than the Level I test and stresses those aspects of mathematics that are prerequisites for a course in calculus similar to the Calculus BC course of the College Board's Advanced Placement Program. The content of the Level II test overlaps that of Level I; however, the questions in the Level II test concentrate on more advanced work and call for greater depth of understanding and sophistication. The Level II test is composed of approximately equal parts of algebra, geometry (including both coordinate and synthetic in two and three dimensions and vectors), trigonometry, functions, and a miscellaneous category consisting of such topics as sequences and limits, logic and proof, probability and statistics, and number theory. The trigonometry questions in Level II stress properties and graphs as well as the inverse of trigonometric functions.

The diagrams below show approximately how the questions in each test are distributed among the major curriculum areas. Comparison of the two diagrams should help you to decide which test you are better prepared to take.

Level I

Algebra	Geometry			Trigonometry	Functions	Miscellaneous
	Plane Euclidean	Solid	Coordinate			

Level II

Algebra	Geometry		Trigonometry	Functions	Miscellaneous
	Solid	Coordinate			

A student electing to take Level II need not be familiar with every topic tested. The authors of this text recommend Level II for students who have studied most of the chapters of this text and for honor students who have studied through Chapter 8. For other students the Level I test will generally be the better choice. In addition to these general guidelines, the advice of your mathematics teacher or guidance counselor should assist you in your choice of level.

Both Level I and Level II tests consist of approximately 50 multiple-choice questions with a one hour time limit.

Some sample questions for the Achievement Tests follow. Answers to these questions appear at the end of the book. In studying these sample questions you should be aware that there is considerable overlap in the content of the two tests. For example, a number of the sample Level II questions would be appropriate for use on the Level I test. However, not all such questions would appear on the same Level I test because of the different emphases in terms of topics covered and mathematical sophistication required. Also a large number of the Level I sample questions would be appropriate for use on a Level II test but for the same reasons as before not all such questions would appear on the same Level II test.

As a general rule, do not try to work a problem by testing each of the answer choices. Such a method is time-consuming and, in many instances, will not work. However, you should look at the answer choices while studying the question. The form of the answer choices may help to clarify the problem.

An attempt is made to construct questions that can be understood by all students regardless of the mathematics curriculums they have studied. Symbolism has been kept simple; for example, the symbol PQ may be used to denote a line, a ray, a segment, or the measure (length) of a segment; the particular interpretation of the symbol is indicated by its context in the problem. Throughout these tests, if the rule of a function f (considered to be a real-valued function unless otherwise stated) is specified but the domain is not, then the domain of f is understood to be the set of all real numbers x for which $f(x)$ is a real number.

Note: Figures that accompany problems in the College Board achievement tests are intended to provide information useful in solving the problems. They are drawn as accurately as possible EXCEPT when it is stated in a specific problem that its figure is not drawn to scale. All figures lie in a plane unless otherwise indicated.

The following sample questions are reprinted, with permission, from copyrighted publications of the College Board and Educational Testing Service, the copyright owner of the questions.

1. If $\dfrac{3^x}{3^2} = 3^8$, then $x =$

   (A) 4   (B) 6   (C) 10   (D) 16   (E) 64

2. The figure consists of two equilateral triangles, LMP and RST.

   If $RS = \frac{1}{3}LM$, then $\dfrac{\text{area } \triangle RST}{\text{area } \triangle LMP} =$

   (A) $\dfrac{1}{9}$ (B) $\dfrac{1}{6}$ (C) $\dfrac{1}{4}$ (D) $\dfrac{1}{3}$ (E) $\dfrac{1}{\sqrt{3}}$

3. Let $f(x) = x - 1$ and $g(x) = 1 - x$. If $f(x) = g(x) + 1$, then x is

   (A) $-\frac{3}{2}$ (B) 0   (C) 1   (D) $\frac{3}{2}$ (E) any number

4. In the figure, if $\dfrac{PQ}{PR} = \dfrac{2}{3}$, then $\sin \angle P =$

   (A) $\dfrac{2}{5}$ (B) $\dfrac{2}{3}$ (C) $\dfrac{\sqrt{5}}{3}$ (D) $\dfrac{\sqrt{5}}{2}$ (E) $\dfrac{3}{2}$

5. If $i^2 = -1$, which of the following is an expression for $\sqrt{-49} - \sqrt{-25}$ in the form $a + bi$, where a and b are real?

   (A) $0 + (\sqrt{24})i$ (B) $0 + (\sqrt{74})i$ (C) $0 + 24i$
   (D) $0 + 2i$ (E) $2 + 0i$

6. If $x + 2 = y$, what is the value of $|x - y| + |y - x|$?

   (A) -4 (B) 0   (C) 2   (D) 4
   (E) It cannot be determined from the information given.

7. Two rectangular solids have the dimensions 4, 6, h, and 8, 2, $(2h - 1)$, respectively. Their volumes are equal when $h =$

   (A) $\frac{1}{8}$ (B) $\frac{4}{5}$ (C) 1   (D) 2   (E) 4

8. If $f(x) = 5x + 6$, then $f(x) < 16$ if and only if

   (A) $x < 2$ (B) $x > 2$ (C) $x = 2$
   (D) $-2 < x < 2$ (E) x is any real number

9. If x is the measure of an acute angle such that $\tan x = \dfrac{k}{3}$, then $\sin x =$

   (A) $\dfrac{k}{3 + k}$ (B) $\dfrac{3}{\sqrt{9 - k^2}}$ (C) $\dfrac{k}{\sqrt{9 - k^2}}$

   (D) $\dfrac{3}{\sqrt{9 + k^2}}$ (E) $\dfrac{k}{\sqrt{9 + k^2}}$

10. The circle in the figure has center at O. If PQ and QR are secants and if $x = 40$, what is y?

    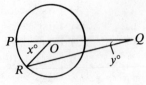

    (A) 10   (B) 20   (C) 30   (D) 40
    (E) It cannot be determined from the information given.

11. On the curve shown in the figure, determine the y-coordinate of each point at which $y = 2x$.

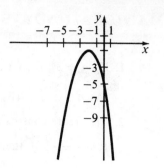

(A) There is no such point.
(B) -1
(C) -2 only
(D) -5
(E) -2 and -10

12. In how many points do the graphs of $x^2 + y^2 = 9$ and $x^2 = 8y$ intersect?

(A) One   (B) Two   (C) Three   (D) Four   (E) More than four

13. If $f(x) = 2x + 1$ and $g(x) = 3x - 1$, then $f(g(x)) =$

(A) $6x - 1$ (B) $6x + 2$ (C) $x - 2$ (D) $5x$ (E) $6x^2 + x - 1$

14. If $h, k, m,$ and n are positive numbers, k is greater than m, and n is greater than h, which of the following are true?

I. $k + h$ is greater than $m + n$.
II. $k + n$ is greater than $m + h$.
III. $k + m$ is greater than $n + h$.

(A) None
(B) I only
(C) II only
(D) I and II only
(E) I, II, and III

15. The following instructions to a computer are carried out in the order specified.

1. LET S = 0
2. LET X = 5
3. LET THE NEW VALUE OF S EQUAL THE OLD VALUE OF S PLUS THE VALUE OF X
4. INCREASE THE VALUE OF X BY 2
5. IF X < 8, GO BACK TO INSTRUCTION 3. OTHERWISE GO ON TO INSTRUCTION 6
6. WRITE THE FINAL VALUE OF S

What value of S should be written in instruction 6?

(A) 0   (B) 5   (C) 7   (D) 9   (E) 12

16. $\log_2 25$ is between what pair of consecutive integers?

(A) 1 and 2   (B) 2 and 3   (C) 4 and 5   (D) 5 and 6   (E) 12 and 13

17. Which quadrants of the plane contain points of the graph $2x - y > 4$?

(A) First, second, and third only
(B) First, second, and fourth only
(C) First, third, and fourth only
(D) Second, third, and fourth only
(E) First, second, third, and fourth

18. If a and b are positive integers and if 9 is a factor of their product, which of the following must be true?

(A) 9 is a factor of both a and b.
(B) 9 is a factor of one number but not of the other.
(C) 9 is a factor of at least one of the numbers a, b.
(D) 3 is a factor of both a and b.
(E) 3 is a factor of at least one of the numbers a, b.

19. If $x = bc$ and $y = bd$, then $y - x =$

(A) $bc(1 - d)$ (B) $bd(1 - c)$ (C) $d(b - c)$ (D) $b(d - c)$ (E) $c(b - d)$

20. The graph of a certain linear function has a negative slope. If its x-intercept is negative, then its y-intercept must be

(A) positive   (B) negative   (C) non-negative
(D) less than its x-intercept   (E) greater than its x-intercept

MATHEMATICS LEVEL II TEST—SAMPLE QUESTIONS

1. If $f(x) = x^2 - x^3$, then $f(-1) =$

(A) 2   (B) 1   (C) 0   (D) -1 (E) -2

2. The base of an equilateral triangle lies on the x-axis. What is the sum of the slopes of the three sides?

(A) -1 (B) 0   (C) 1   (D) $2\sqrt{3}$ (E) $1 + 2\sqrt{3}$

3. The set of all real x such that $\sqrt{x^2} = -x$ consists of

(A) zero only   (B) nonpositive real numbers only
(C) positive real numbers only   (D) all real numbers
(E) no real numbers

4. Which of the following is the set of all real numbers x such that

$$|x| - 5 \leq |x - 5|?$$

(A) The set of all real numbers   (B) $\{5\}$ (C) $\{-5\}$
(D) $\{x: x \leq 0\}$ (E) $\{x: x \geq 0\}$

5. If $\sin x = 2\cos x$, what is $\tan x$?

(A) $\dfrac{1}{2}$ (B) $\dfrac{\sqrt{2}}{2}$ (C) $\sqrt{2}$ (D) 2

(E) It cannot be determined from the information given.

6. For what real numbers x is $y = 2^{-x}$ a negative number?

(A) All real x (B) $x > 0$ only   (C) $x \geq 0$ only
(D) $x < 0$ only   (E) No real x

7. If f and g are functions such that $f(x) = 2x - 3$ and $f(g(x)) = x$, then $g(x) =$

(A) $2x + 3$ (B) $3x + 2$ (C) $3x - 2$ (D) $\dfrac{x + 3}{2}$ (E) $\dfrac{3x - 1}{2}$

8. If f and g are inverse functions and if $(5, 0)$ is a point on the graph of $y = g(x)$, which of the following could define f?

I. $f(x) = x - 5$
II. $f(x) = 3x + 5$
III. $f(x) = x^3 + 5$

(A) I only   (B) II only   (C) III only
(D) I and II only   (E) II and III only

9. For $x > 1$, the graph of $y = \log_x x$ intersects the line $x = \pi$ at $y =$

(A) $-\pi$ (B) 0 (C) $\dfrac{1}{\pi}$ (D) 1 (E) π

10. For all x and y such that $xy \neq 0$, let $f(x, y) = \dfrac{xy}{x^2 + y^2}$. Then $f(x, -x) =$

(A) $-x^2$ (B) $-\dfrac{1}{x^2}$ (C) $-\dfrac{1}{2}$ (D) 0 (E) $\dfrac{1}{2}$

11. $\lim\limits_{n \to \infty} \dfrac{3n + 12}{2n - 12} =$

(A) -6 (B) $-\frac{3}{2}$ (C) -1 (D) $\frac{2}{3}$ (E) $\frac{3}{2}$

12. $\dfrac{\cos 50°}{\sin 40°} =$

(A) -1 (B) 0 (C) 1 (D) $\tan 10°$ (E) $\cot 10°$

13. If $0 < y < x < \dfrac{\pi}{2}$, which of the following are true?

I. $\sin y < \sin x$
II. $\cos y < \cos x$
III. $\tan y < \tan x$

(A) None   (B) I and II only   (C) I and III only
(D) II and III only (E) I, II, and III

14. Lines l_1 and l_2 both intersect line l_3 at right angles, but do not necessarily lie in the same plane. Which of the following statements about l_1 and l_2 is correct?

(A) They are necessarily parallel to each other.
(B) They cannot be parallel to each other.
(C) They are necessarily either parallel or perpendicular to each other.
(D) They are necessarily perpendicular to each other.
(E) They are not necessarily parallel or perpendicular to each other.

15. If a complex number, $a + bi$, is plotted in a coordinate plane as the point (a, b), then the set of points such that $|a + bi| \leq 1$ is a region whose boundary is

(A) two parallel lines (B) an equilateral triangle (C) a square
(D) an ellipse (E) a circle

16. If m and n are real numbers, under which of the following conditions will $\dfrac{m + ni}{1 + i}$ be a real number different from zero?

(A) $m = 0, n \neq 0$ (B) $m \neq 0, n = 0$ (C) $m = -n \neq 0$
(D) $m = n \neq 0$ (E) $m = 2n \neq 0$

17. Which of the following lines are asymptotes of the graph of $y = \dfrac{x - 2}{x - 3}$?

(A) $x = 3$ only (B) $y = 3$ only (C) $x = 1$ and $y = 3$
(D) $x = 3$ and $y = 3$ (E) $x = 3$ and $y = 1$

18. A polynomial equation, $P(x) = 0$, with real coefficients has three roots. If two of the roots are 0 and i, then $P(x)$ could be

(A) $x^2 + 1$ (B) $x^3 - 1$ (C) $x^3 + 1$ (D) $x^3 - x$ (E) $x^3 + x$

19. Which of the following equations has the graph shown in the figure?

(A) $y = \sin \dfrac{x}{2} + 1$

(B) $y = \sin 2x$

(C) $y = 2 \sin \dfrac{x}{2}$

(D) $y = 2 \sin x$

(E) $y = 2 \sin 2x$

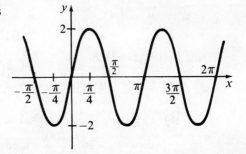

20. The graph of $\begin{cases} x = 4t - 2 \\ y = 4t^2 \end{cases}$ in the xy-plane is

(A) a circle (B) a parabola (C) an ellipse that is not a circle
(D) a hyperbola (E) a straight line

21. Use the graphs of $y = x^2$ and $y = 2^x$ to determine the number of real solutions of the equation $x^2 = 2^x$.

(A) None (B) One (C) Three (D) Four (E) Infinitely many

22. If n is a positive integer, then $n!$ is divisible by 9 if and only if

(A) $n \geq 3$ (B) $n \geq 6$ (C) $n \geq 9$
(D) n is a multiple of 3 (E) n is a multiple of 9

23. The probability of obtaining more heads than tails in five tosses of a fair coin is

(A) $\frac{5}{16}$ (B) $\frac{2}{5}$ (C) $\frac{1}{2}$ (D) $\frac{3}{5}$ (E) $\frac{5}{8}$

24. In the figure, S is the set of all points in the shaded region. Which of the following shows the set T consisting of all points $(x, y - x)$ where (x, y) is a point in S?

(A)

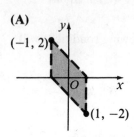

(B)

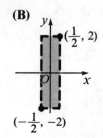

(C)

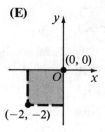

(D)

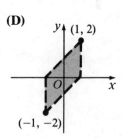

(E)

25. $\sin (\text{Arctan } 1) =$

(A) 0 (B) $\dfrac{1}{2}$ (C) $\dfrac{\sqrt{2}}{2}$ (D) $\dfrac{\sqrt{3}}{2}$ (E) 1

26. If $f(x) = 10^x$, where x is real, and if the inverse function of f is denoted by f^{-1}, then what is $\dfrac{f^{-1}(a)}{f^{-1}(b)}$ where $a > 1$ and $b > 1$?

(A) $\log_{10} a - \log_{10} b$ (B) $\log_{10} (a - b)$ (C) $\dfrac{\log_{10} a}{\log_{10} b}$ (D) $\dfrac{10^b}{10^a}$

(E) None of the above

27. In the figure, R and T are the midpoints of two adjacent edges of the cube. If the length of each edge of the cube is h, what is the volume of the pyramid $PRST$?

(A) $\dfrac{h^3}{24}$ (B) $\dfrac{h^3}{12}$ (C) $\dfrac{h^3}{8}$

(D) $\dfrac{h^3}{6}$ (E) $\dfrac{h^3}{4}$

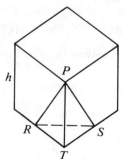

28. Which of the following represents the rectangular-coordinate graph of the given equations?

$$x = 2 \cos \theta$$
$$y = 4 \sin \theta$$

(A)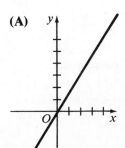

(B)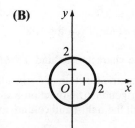

(C)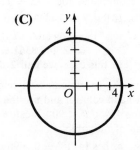

(D)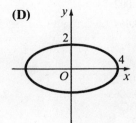

(E)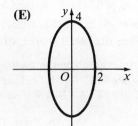

29. Which of the following represents the graph of $|3x - 5| \leq y$?

(A)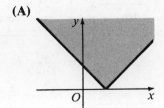

(B)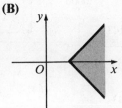

(C)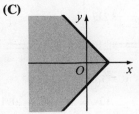

(D)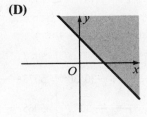

(E)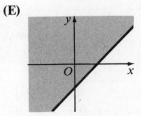

30. $\dfrac{(n + 1)!}{n!} - n =$

(A) 0   (B) 1   (C) n (D) $n + 1$ (E) $n!$

Using the tables

Table 1. Example: $\sqrt{830} = 10\sqrt{8.3} \approx 28.81$

Table 2. Example **a.** $\log 4.72 = 0.6739$

Example **b.** If $\log N = 0.4843$, then $N = 3.05$

Example **c.** $\log 718 = \log(100 \cdot 7.18) = \log 10^2 + \log 7.18$
$$= 2 + 0.8561 = 2.8561$$
In this case, we call 2 the characteristic and 0.8561 the mantissa.

Tables 3 and 4. To evaluate a trigonometric function of θ when $0° \le \theta \le 45°$, locate θ in the left-hand column and use the trigonometric-function headings at the top of the table. When $45° < \theta \le 90°$, locate θ in the right-hand column and use the headings at the bottom of the table.

Examples: **a.** $\cos 14°10' = 0.7696$ **b.** $\tan 51.9° = 1.275$

Table 5. Examples: **a.** $\sin 0.95 = 0.8134$ **b.** $\csc 1.51 = 1.002$

Table 6. Example **a.** $e^{2.6} = e^{2.5} \cdot e^{0.1} \approx (12.182)(1.1052) \approx 13.464$

Example **b.** $e^{-3.21} = e^{-3} \cdot e^{-0.21} \approx (0.0498)(0.8106) \approx 0.0404$

Interpolation. Examples: Use interpolation to find:
a. $\log 8.217$ **b.** x if $\log x = 0.5725$
c. $\tan 68°34'$ **d.** θ if $\sin \theta = 0.5065$

a. From Table 2:

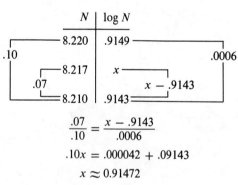

$$\frac{.07}{.10} = \frac{x - .9143}{.0006}$$
$$.10x = .000042 + .09143$$
$$x \approx 0.91472$$

b. From Table 2:

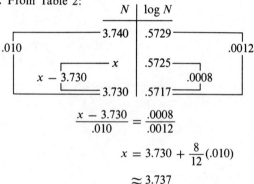

$$\frac{x - 3.730}{.010} = \frac{.0008}{.0012}$$
$$x = 3.730 + \frac{8}{12}(.010)$$
$$\approx 3.737$$

c. From Table 3:

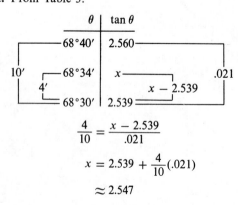

$$\frac{4}{10} = \frac{x - 2.539}{.021}$$
$$x = 2.539 + \frac{4}{10}(.021)$$
$$\approx 2.547$$

d. From Table 3:

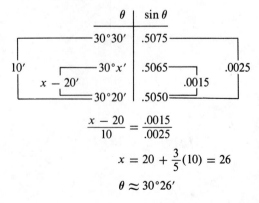

$$\frac{x - 20}{10} = \frac{.0015}{.0025}$$
$$x = 20 + \frac{3}{5}(10) = 26$$
$$\theta \approx 30°26'$$

Table 1   Squares and square roots

N	N^2	$\sqrt{N}$	$\sqrt{10N}$	N	N^2	$\sqrt{N}$	$\sqrt{10N}$
1.0	1.00	1.000	3.162	5.5	30.25	2.345	7.416
1.1	1.21	1.049	3.317	5.6	31.36	2.366	7.483
1.2	1.44	1.095	3.464	5.7	32.49	2.387	7.550
1.3	1.69	1.140	3.606	5.8	33.64	2.408	7.616
1.4	1.96	1.183	3.742	5.9	34.81	2.429	7.681
1.5	2.25	1.225	3.873	6.0	36.00	2.449	7.746
1.6	2.56	1.265	4.000	6.1	37.21	2.470	7.810
1.7	2.89	1.304	4.123	6.2	38.44	2.490	7.874
1.8	3.24	1.342	4.243	6.3	39.69	2.510	7.937
1.9	3.61	1.378	4.359	6.4	40.96	2.530	8.000
2.0	4.00	1.414	4.472	6.5	42.25	2.550	8.062
2.1	4.41	1.449	4.583	6.6	43.56	2.569	8.124
2.2	4.84	1.483	4.690	6.7	44.89	2.588	8.185
2.3	5.29	1.517	4.796	6.8	46.24	2.608	8.246
2.4	5.76	1.549	4.899	6.9	47.61	2.627	8.307
2.5	6.25	1.581	5.000	7.0	49.00	2.646	8.367
2.6	6.76	1.612	5.099	7.1	50.41	2.665	8.426
2.7	7.29	1.643	5.196	7.2	51.84	2.683	8.485
2.8	7.84	1.673	5.292	7.3	53.29	2.702	8.544
2.9	8.41	1.703	5.385	7.4	54.76	2.720	8.602
3.0	9.00	1.732	5.477	7.5	56.25	2.739	8.660
3.1	9.61	1.761	5.568	7.6	57.76	2.757	8.718
3.2	10.24	1.789	5.657	7.7	59.29	2.775	8.775
3.3	10.89	1.817	5.745	7.8	60.84	2.793	8.832
3.4	11.56	1.844	5.831	7.9	62.41	2.811	8.888
3.5	12.25	1.871	5.916	8.0	64.00	2.828	8.944
3.6	12.96	1.897	6.000	8.1	65.61	2.846	9.000
3.7	13.69	1.924	6.083	8.2	67.24	2.864	9.055
3.8	14.44	1.949	6.164	8.3	68.89	2.881	9.110
3.9	15.21	1.975	6.245	8.4	70.56	2.898	9.165
4.0	16.00	2.000	6.325	8.5	72.25	2.915	9.220
4.1	16.81	2.025	6.403	8.6	73.96	2.933	9.274
4.2	17.64	2.049	6.481	8.7	75.69	2.950	9.327
4.3	18.49	2.074	6.557	8.8	77.44	2.966	9.381
4.4	19.36	2.098	6.633	8.9	79.21	2.983	9.434
4.5	20.25	2.121	6.708	9.0	81.00	3.000	9.487
4.6	21.16	2.145	6.782	9.1	82.81	3.017	9.539
4.7	22.09	2.168	6.856	9.2	84.64	3.033	9.592
4.8	23.04	2.191	6.928	9.3	86.49	3.050	9.644
4.9	24.01	2.214	7.000	9.4	88.36	3.066	9.695
5.0	25.00	2.236	7.071	9.5	90.25	3.082	9.747
5.1	26.01	2.258	7.141	9.6	92.16	3.098	9.798
5.2	27.04	2.280	7.211	9.7	94.09	3.114	9.849
5.3	28.09	2.302	7.280	9.8	96.04	3.130	9.899
5.4	29.16	2.324	7.348	9.9	98.01	3.146	9.950
5.5	30.25	2.345	7.416	10	100.00	3.162	10.000

Table 2 Common logarithms of numbers*

N	0	1	2	3	4	5	6	7	8	9
1.0	0000	0043	0086	0128	0170	0212	0253	0294	0334	0374
1.1	0414	0453	0492	0531	0569	0607	0645	0682	0719	0755
1.2	0792	0828	0864	0899	0934	0969	1004	1038	1072	1106
1.3	1139	1173	1206	1239	1271	1303	1335	1367	1399	1430
1.4	1461	1492	1523	1553	1584	1614	1644	1673	1703	1732
1.5	1761	1790	1818	1847	1875	1903	1931	1959	1987	2014
1.6	2041	2068	2095	2122	2148	2175	2201	2227	2253	2279
1.7	2304	2330	2355	2380	2405	2430	2455	2480	2504	2529
1.8	2553	2577	2601	2625	2648	2672	2695	2718	2742	2765
1.9	2788	2810	2833	2856	2878	2900	2923	2945	2967	2989
2.0	3010	3032	3054	3075	3096	3118	3139	3160	3181	3201
2.1	3222	3243	3263	3284	3304	3324	3345	3365	3385	3404
2.2	3424	3444	3464	3483	3502	3522	3541	3560	3579	3598
2.3	3617	3636	3655	3674	3692	3711	3729	3747	3766	3784
2.4	3802	3820	3838	3856	3874	3892	3909	3927	3945	3962
2.5	3979	3997	4014	4031	4048	4065	4082	4099	4116	4133
2.6	4150	4166	4183	4200	4216	4232	4249	4265	4281	4298
2.7	4314	4330	4346	4362	4378	4393	4409	4425	4440	4456
2.8	4472	4487	4502	4518	4533	4548	4564	4579	4594	4609
2.9	4624	4639	4654	4669	4683	4698	4713	4728	4742	4757
3.0	4771	4786	4800	4814	4829	4843	4857	4871	4886	4900
3.1	4914	4928	4942	4955	4969	4983	4997	5011	5024	5038
3.2	5051	5065	5079	5092	5105	5119	5132	5145	5159	5172
3.3	5185	5198	5211	5224	5237	5250	5263	5276	5289	5302
3.4	5315	5328	5340	5353	5366	5378	5391	5403	5416	5428
3.5	5441	5453	5465	5478	5490	5502	5514	5527	5539	5551
3.6	5563	5575	5587	5599	5611	5623	5635	5647	5658	5670
3.7	5682	5694	5705	5717	5729	5740	5752	5763	5775	5786
3.8	5798	5809	5821	5832	5843	5855	5866	5877	5888	5899
3.9	5911	5922	5933	5944	5955	5966	5977	5988	5999	6010
4.0	6021	6031	6042	6053	6064	6075	6085	6096	6107	6117
4.1	6128	6138	6149	6160	6170	6180	6191	6201	6212	6222
4.2	6232	6243	6253	6263	6274	6284	6294	6304	6314	6325
4.3	6335	6345	6355	6365	6375	6385	6395	6405	6415	6425
4.4	6435	6444	6454	6464	6474	6484	6493	6503	6513	6522
4.5	6532	6542	6551	6561	6571	6580	6590	6599	6609	6618
4.6	6628	6637	6646	6656	6665	6675	6684	6693	6702	6712
4.7	6721	6730	6739	6749	6758	6767	6776	6785	6794	6803
4.8	6812	6821	6830	6839	6848	6857	6866	6875	6884	6893
4.9	6902	6911	6920	6928	6937	6946	6955	6964	6972	6981
5.0	6990	6998	7007	7016	7024	7033	7042	7050	7059	7067
5.1	7076	7084	7093	7101	7110	7118	7126	7135	7143	7152
5.2	7160	7168	7177	7185	7193	7202	7210	7218	7226	7235
5.3	7243	7251	7259	7267	7275	7284	7292	7300	7308	7316
5.4	7324	7332	7340	7348	7356	7364	7372	7380	7388	7396

*Decimal points omitted.

Table 2 Common logarithms of numbers

N	0	1	2	3	4	5	6	7	8	9
5.5	7404	7412	7419	7427	7435	7443	7451	7459	7466	7474
5.6	7482	7490	7497	7505	7513	7520	7528	7536	7543	7551
5.7	7559	7566	7574	7582	7589	7597	7604	7612	7619	7627
5.8	7634	7642	7649	7657	7664	7672	7679	7686	7694	7701
5.9	7709	7716	7723	7731	7738	7745	7752	7760	7767	7774
6.0	7782	7789	7796	7803	7810	7818	7825	7832	7839	7846
6.1	7853	7860	7868	7875	7882	7889	7896	7903	7910	7917
6.2	7924	7931	7938	7945	7952	7959	7966	7973	7980	7987
6.3	7993	8000	8007	8014	8021	8028	8035	8041	8048	8055
6.4	8062	8069	8075	8082	8089	8096	8102	8109	8116	8122
6.5	8129	8136	8142	8149	8156	8162	8169	8176	8182	8189
6.6	8195	8202	8209	8215	8222	8228	8235	8241	8248	8254
6.7	8261	8267	8274	8280	8287	8293	8299	8306	8312	8319
6.8	8325	8331	8338	8344	8351	8357	8363	8370	8376	8382
6.9	8388	8395	8401	8407	8414	8420	8426	8432	8439	8445
7.0	8451	8457	8463	8470	8476	8482	8488	8494	8500	8506
7.1	8513	8519	8525	8531	8537	8543	8549	8555	8561	8567
7.2	8573	8579	8585	8591	8597	8603	8609	8615	8621	8627
7.3	8633	8639	8645	8651	8657	8663	8669	8675	8681	8686
7.4	8692	8698	8704	8710	8716	8722	8727	8733	8739	8745
7.5	8751	8756	8762	8768	8774	8779	8785	8791	8797	8802
7.6	8808	8814	8820	8825	8831	8837	8842	8848	8854	8859
7.7	8865	8871	8876	8882	8887	8893	8899	8904	8910	8915
7.8	8921	8927	8932	8938	8943	8949	8954	8960	8965	8971
7.9	8976	8982	8987	8993	8998	9004	9009	9015	9020	9025
8.0	9031	9036	9042	9047	9053	9058	9063	9069	9074	9079
8.1	9085	9090	9096	9101	9106	9112	9117	9122	9128	9133
8.2	9138	9143	9149	9154	9159	9165	9170	9175	9180	9186
8.3	9191	9196	9201	9206	9212	9217	9222	9227	9232	9238
8.4	9243	9248	9253	9258	9263	9269	9274	9279	9284	9289
8.5	9294	9299	9304	9309	9315	9320	9325	9330	9335	9340
8.6	9345	9350	9355	9360	9365	9370	9375	9380	9385	9390
8.7	9395	9400	9405	9410	9415	9420	9425	9430	9435	9440
8.8	9445	9450	9455	9460	9465	9469	9474	9479	9484	9489
8.9	9494	9499	9504	9509	9513	9518	9523	9528	9533	9538
9.0	9542	9547	9552	9557	9562	9566	9571	9576	9581	9586
9.1	9590	9595	9600	9605	9609	9614	9619	9624	9628	9633
9.2	9638	9643	9647	9652	9657	9661	9666	9671	9675	9680
9.3	9685	9689	9694	9699	9703	9708	9713	9717	9722	9727
9.4	9731	9736	9741	9745	9750	9754	9759	9763	9768	9773
9.5	9777	9782	9786	9791	9795	9800	9805	9809	9814	9818
9.6	9823	9827	9832	9836	9841	9845	9850	9854	9859	9863
9.7	9868	9872	9877	9881	9886	9890	9894	9899	9903	9908
9.8	9912	9917	9921	9926	9930	9934	9939	9943	9948	9952
9.9	9956	9961	9965	9969	9974	9978	9983	9987	9991	9996

Table 3　Trigonometric functions of θ
(θ in degrees and minutes)

θ Degrees	θ Radians	sin θ	cos θ	tan θ	cot θ	sec θ	csc θ		
0° 00′	.0000	.0000	1.0000	.0000	Undefined	1.000	Undefined	1.5708	90° 00′
10′	.0029	.0029	1.0000	.0029	343.8	1.000	343.8	1.5679	50′
20′	.0058	.0058	1.0000	.0058	171.9	1.000	171.9	1.5650	40′
30′	.0087	.0087	1.0000	.0087	114.6	1.000	114.6	1.5621	30′
40′	.0116	.0116	.9999	.0116	85.94	1.000	85.95	1.5592	20′
50′	.0145	.0145	.9999	.0145	68.75	1.000	68.76	1.5563	10′
1° 00′	.0175	.0175	.9998	.0175	57.29	1.000	57.30	1.5533	89° 00′
10′	.0204	.0204	.9998	.0204	49.10	1.000	49.11	1.5504	50′
20′	.0233	.0233	.9997	.0233	42.96	1.000	42.98	1.5475	40′
30′	.0262	.0262	.9997	.0262	38.19	1.000	38.20	1.5446	30′
40′	.0291	.0291	.9996	.0291	34.37	1.000	34.38	1.5417	20′
50′	.0320	.0320	.9995	.0320	31.24	1.001	31.26	1.5388	10′
2° 00′	.0349	.0349	.9994	.0349	28.64	1.001	28.65	1.5359	88° 00′
10′	.0378	.0378	.9993	.0378	26.43	1.001	26.45	1.5330	50′
20′	.0407	.0407	.9992	.0407	24.54	1.001	24.56	1.5301	40′
30′	.0436	.0436	.9990	.0437	22.90	1.001	22.93	1.5272	30′
40′	.0465	.0465	.9989	.0466	21.47	1.001	21.49	1.5243	20′
50′	.0495	.0494	.9988	.0495	20.21	1.001	20.23	1.5213	10′
3° 00′	.0524	.0523	.9986	.0524	19.08	1.001	19.11	1.5184	87° 00′
10′	.0553	.0552	.9985	.0553	18.07	1.002	18.10	1.5155	50′
20′	.0582	.0581	.9983	.0582	17.17	1.002	17.20	1.5126	40′
30′	.0611	.0610	.9981	.0612	16.35	1.002	16.38	1.5097	30′
40′	.0640	.0640	.9980	.0641	15.60	1.002	15.64	1.5068	20′
50′	.0669	.0669	.9978	.0670	14.92	1.002	14.96	1.5039	10′
4° 00′	.0698	.0698	.9976	.0699	14.30	1.002	14.34	1.5010	86° 00′
10′	.0727	.0727	.9974	.0729	13.73	1.003	13.76	1.4981	50′
20′	.0756	.0756	.9971	.0758	13.20	1.003	13.23	1.4952	40′
30′	.0785	.0785	.9969	.0787	12.71	1.003	12.75	1.4923	30′
40′	.0814	.0814	.9967	.0816	12.25	1.003	12.29	1.4893	20′
50′	.0844	.0843	.9964	.0846	11.83	1.004	11.87	1.4864	10′
5° 00′	.0873	.0872	.9962	.0875	11.43	1.004	11.47	1.4835	85° 00′
10′	.0902	.0901	.9959	.0904	11.06	1.004	11.10	1.4806	50′
20′	.0931	.0929	.9957	.0934	10.71	1.004	10.76	1.4777	40′
30′	.0960	.0958	.9954	.0963	10.39	1.005	10.43	1.4748	30′
40′	.0989	.0987	.9951	.0992	10.08	1.005	10.13	1.4719	20′
50′	.1018	.1016	.9948	.1022	9.788	1.005	9.839	1.4690	10′
6° 00′	.1047	.1045	.9945	.1051	9.514	1.006	9.567	1.4661	84° 00′
10′	.1076	.1074	.9942	.1080	9.255	1.006	9.309	1.4632	50′
20′	.1105	.1103	.9939	.1110	9.010	1.006	9.065	1.4603	40′
30′	.1134	.1132	.9936	.1139	8.777	1.006	8.834	1.4573	30′
40′	.1164	.1161	.9932	.1169	8.556	1.007	8.614	1.4544	20′
50′	.1193	.1190	.9929	.1198	8.345	1.007	8.405	1.4515	10′
7° 00′	.1222	.1219	.9925	.1228	8.144	1.008	8.206	1.4486	83° 00′
10′	.1251	.1248	.9922	.1257	7.953	1.008	8.016	1.4457	50′
20′	.1280	.1276	.9918	.1287	7.770	1.008	7.834	1.4428	40′
30′	.1309	.1305	.9914	.1317	7.596	1.009	7.661	1.4399	30′
40′	.1338	.1334	.9911	.1346	7.429	1.009	7.496	1.4370	20′
50′	.1367	.1363	.9907	.1376	7.269	1.009	7.337	1.4341	10′
8° 00′	.1396	.1392	.9903	.1405	7.115	1.010	7.185	1.4312	82° 00′
10′	.1425	.1421	.9899	.1435	6.968	1.010	7.040	1.4283	50′
20′	.1454	.1449	.9894	.1465	6.827	1.011	6.900	1.4254	40′
30′	.1484	.1478	.9890	.1495	6.691	1.011	6.765	1.4224	30′
40′	.1513	.1507	.9886	.1524	6.561	1.012	6.636	1.4195	20′
50′	.1542	.1536	.9881	.1554	6.435	1.012	6.512	1.4166	10′
9° 00′	.1571	.1564	.9877	.1584	6.314	1.012	6.392	1.4137	81° 00′
		cos θ	sin θ	cot θ	tan θ	csc θ	sec θ	θ Radians	θ Degrees

Table 3 Trigonometric functions of θ
(θ in degrees and minutes)

θ Degrees	θ Radians	$\sin \theta$	$\cos \theta$	$\tan \theta$	$\cot \theta$	$\sec \theta$	$\csc \theta$		
9° 00′	.1571	.1564	.9877	.1584	6.314	1.012	6.392	1.4137	81° 00′
10′	.1600	.1593	.9872	.1614	6.197	1.013	6.277	1.4108	50′
20′	.1629	.1622	.9868	.1644	6.084	1.013	6.166	1.4079	40′
30′	.1658	.1650	.9863	.1673	5.976	1.014	6.059	1.4050	30′
40′	.1687	.1679	.9858	.1703	5.871	1.014	5.955	1.4021	20′
50′	.1716	.1708	.9853	.1733	5.769	1.015	5.855	1.3992	10′
10° 00′	.1745	.1736	.9848	.1763	5.671	1.015	5.759	1.3963	80° 00′
10′	.1774	.1765	.9843	.1793	5.576	1.016	5.665	1.3934	50′
20′	.1804	.1794	.9838	.1823	5.485	1.016	5.575	1.3904	40′
30′	.1833	.1822	.9833	.1853	5.396	1.017	5.487	1.3875	30′
40′	.1862	.1851	.9827	.1883	5.309	1.018	5.403	1.3846	20′
50′	.1891	.1880	.9822	.1914	5.226	1.018	5.320	1.3817	10′
11° 00′	.1920	.1908	.9816	.1944	5.145	1.019	5.241	1.3788	79° 00′
10′	.1949	.1937	.9811	.1974	5.066	1.019	5.164	1.3759	50′
20′	.1978	.1965	.9805	.2004	4.989	1.020	5.089	1.3730	40′
30′	.2007	.1994	.9799	.2035	4.915	1.020	5.016	1.3701	30′
40′	.2036	.2022	.9793	.2065	4.843	1.021	4.945	1.3672	20′
50′	.2065	.2051	.9787	.2095	4.773	1.022	4.876	1.3643	10′
12° 00′	.2094	.2079	.9781	.2126	4.705	1.022	4.810	1.3614	78° 00′
10′	.2123	.2108	.9775	.2156	4.638	1.023	4.745	1.3584	50′
20′	.2153	.2136	.9769	.2186	4.574	1.024	4.682	1.3555	40′
30′	.2182	.2164	.9763	.2217	4.511	1.024	4.620	1.3526	30′
40′	.2211	.2193	.9757	.2247	4.449	1.025	4.560	1.3497	20′
50′	.2240	.2221	.9750	.2278	4.390	1.026	4.502	1.3468	10′
13° 00′	.2269	.2250	.9744	.2309	4.331	1.026	4.445	1.3439	77° 00′
10′	.2298	.2278	.9737	.2339	4.275	1.027	4.390	1.3410	50′
20′	.2327	.2306	.9730	.2370	4.219	1.028	4.336	1.3381	40′
30′	.2356	.2334	.9724	.2401	4.165	1.028	4.284	1.3352	30′
40′	.2385	.2363	.9717	.2432	4.113	1.029	4.232	1.3323	20′
50′	.2414	.2391	.9710	.2462	4.061	1.030	4.182	1.3294	10′
14° 00′	.2443	.2419	.9703	.2493	4.011	1.031	4.134	1.3265	76° 00′
10′	.2473	.2447	.9696	.2524	3.962	1.031	4.086	1.3235	50′
20′	.2502	.2476	.9689	.2555	3.914	1.032	4.039	1.3206	40′
30′	.2531	.2504	.9681	.2586	3.867	1.033	3.994	1.3177	30′
40′	.2560	.2532	.9674	.2617	3.821	1.034	3.950	1.3148	20′
50′	.2589	.2560	.9667	.2648	3.776	1.034	3.906	1.3119	10′
15° 00′	.2618	.2588	.9659	.2679	3.732	1.035	3.864	1.3090	75° 00′
10′	.2647	.2616	.9652	.2711	3.689	1.036	3.822	1.3061	50′
20′	.2676	.2644	.9644	.2742	3.647	1.037	3.782	1.3032	40′
30′	.2705	.2672	.9636	.2773	3.606	1.038	3.742	1.3003	30′
40′	.2734	.2700	.9628	.2805	3.566	1.039	3.703	1.2974	20′
50′	.2763	.2728	.9621	.2836	3.526	1.039	3.665	1.2945	10′
16° 00′	.2793	.2756	.9613	.2867	3.487	1.040	3.628	1.2915	74° 00′
10′	.2822	.2784	.9605	.2899	3.450	1.041	3.592	1.2886	50′
20′	.2851	.2812	.9596	.2931	3.412	1.042	3.556	1.2857	40′
30′	.2880	.2840	.9588	.2962	3.376	1.043	3.521	1.2828	30′
40′	.2909	.2868	.9580	.2994	3.340	1.044	3.487	1.2799	20′
50′	.2938	.2896	.9572	.3026	3.305	1.045	3.453	1.2770	10′
17° 00′	.2967	.2924	.9563	.3057	3.271	1.046	3.420	1.2741	73° 00′
10′	.2996	.2952	.9555	.3089	3.237	1.047	3.388	1.2712	50′
20′	.3025	.2979	.9546	.3121	3.204	1.048	3.357	1.2683	40′
30′	.3054	.3007	.9537	.3153	3.172	1.049	3.326	1.2654	30′
40′	.3083	.3035	.9528	.3185	3.140	1.049	3.295	1.2625	20′
50′	.3113	.3062	.9520	.3217	3.108	1.050	3.265	1.2595	10′
18° 00′	.3142	.3090	.9511	.3249	3.078	1.051	3.236	1.2566	72° 00′
		$\cos \theta$	$\sin \theta$	$\cot \theta$	$\tan \theta$	$\csc \theta$	$\sec \theta$	θ Radians	θ Degrees

Table 3 Trigonometric functions of θ
(θ in degrees and minutes)

θ Degrees	θ Radians	sin θ	cos θ	tan θ	cot θ	sec θ	csc θ		
18° 00′	.3142	.3090	.9511	.3249	3.078	1.051	3.236	1.2566	72° 00′
10′	.3171	.3118	.9502	.3281	3.047	1.052	3.207	1.2537	50′
20′	.3200	.3145	.9492	.3314	3.018	1.053	3.179	1.2508	40′
30′	.3229	.3173	.9483	.3346	2.989	1.054	3.152	1.2479	30′
40′	.3258	.3201	.9474	.3378	2.960	1.056	3.124	1.2450	20′
50′	.3287	.3228	.9465	.3411	2.932	1.057	3.098	1.2421	10′
19° 00′	.3316	.3256	.9455	.3443	2.904	1.058	3.072	1.2392	71° 00′
10′	.3345	.3283	.9446	.3476	2.877	1.059	3.046	1.2363	50′
20′	.3374	.3311	.9436	.3508	2.850	1.060	3.021	1.2334	40′
30′	.3403	.3338	.9426	.3541	2.824	1.061	2.996	1.2305	30′
40′	.3432	.3365	.9417	.3574	2.798	1.062	2.971	1.2275	20′
50′	.3462	.3393	.9407	.3607	2.773	1.063	2.947	1.2246	10′
20° 00′	.3491	.3420	.9397	.3640	2.747	1.064	2.924	1.2217	70° 00′
10′	.3520	.3448	.9387	.3673	2.723	1.065	2.901	1.2188	50′
20′	.3549	.3475	.9377	.3706	2.699	1.066	2.878	1.2159	40′
30′	.3578	.3502	.9367	.3739	2.675	1.068	2.855	1.2130	30′
40′	.3607	.3529	.9356	.3772	2.651	1.069	2.833	1.2101	20′
50′	.3636	.3557	.9346	.3805	2.628	1.070	2.812	1.2072	10′
21° 00′	.3665	.3584	.9336	.3839	2.605	1.071	2.790	1.2043	69° 00′
10′	.3694	.3611	.9325	.3872	2.583	1.072	2.769	1.2014	50′
20′	.3723	.3638	.9315	.3906	2.560	1.074	2.749	1.1985	40′
30′	.3752	.3665	.9304	.3939	2.539	1.075	2.729	1.1956	30′
40′	.3782	.3692	.9293	.3973	2.517	1.076	2.709	1.1926	20′
50′	.3811	.3719	.9283	.4006	2.496	1.077	2.689	1.1897	10′
22° 00′	.3840	.3746	.9272	.4040	2.475	1.079	2.669	1.1868	68° 00′
10′	.3869	.3773	.9261	.4074	2.455	1.080	2.650	1.1839	50′
20′	.3898	.3800	.9250	.4108	2.434	1.081	2.632	1.1810	40′
30′	.3927	.3827	.9239	.4142	2.414	1.082	2.613	1.1781	30′
40′	.3956	.3854	.9228	.4176	2.394	1.084	2.595	1.1752	20′
50′	.3985	.3881	.9216	.4210	2.375	1.085	2.577	1.1723	10′
23° 00′	.4014	.3907	.9205	.4245	2.356	1.086	2.559	1.1694	67° 00′
10′	.4043	.3934	.9194	.4279	2.337	1.088	2.542	1.1665	50′
20′	.4072	.3961	.9182	.4314	2.318	1.089	2.525	1.1636	40′
30′	.4102	.3987	.9171	.4348	2.300	1.090	2.508	1.1606	30′
40′	.4131	.4014	.9159	.4383	2.282	1.092	2.491	1.1577	20′
50′	.4160	.4041	.9147	.4417	2.264	1.093	2.475	1.1548	10′
24° 00′	.4189	.4067	.9135	.4452	2.246	1.095	2.459	1.1519	66° 00′
10′	.4218	.4094	.9124	.4487	2.229	1.096	2.443	1.1490	50′
20′	.4247	.4120	.9112	.4522	2.211	1.097	2.427	1.1461	40′
30′	.4276	.4147	.9100	.4557	2.194	1.099	2.411	1.1432	30′
40′	.4305	.4173	.9088	.4592	2.177	1.100	2.396	1.1403	20′
50′	.4334	.4200	.9075	.4628	2.161	1.102	2.381	1.1374	10′
25° 00′	.4363	.4226	.9063	.4663	2.145	1.103	2.366	1.1345	65° 00′
10′	.4392	.4253	.9051	.4699	2.128	1.105	2.352	1.1316	50′
20′	.4422	.4279	.9038	.4734	2.112	1.106	2.337	1.1286	40′
30′	.4451	.4305	.9026	.4770	2.097	1.108	2.323	1.1257	30′
40′	.4480	.4331	.9013	.4806	2.081	1.109	2.309	1.1228	20′
50′	.4509	.4358	.9001	.4841	2.066	1.111	2.295	1.1199	10′
26° 00′	.4538	.4384	.8988	.4877	2.050	1.113	2.281	1.1170	64° 00′
10′	.4567	.4410	.8975	.4913	2.035	1.114	2.268	1.1141	50′
20′	.4596	.4436	.8962	.4950	2.020	1.116	2.254	1.1112	40′
30′	.4625	.4462	.8949	.4986	2.006	1.117	2.241	1.1083	30′
40′	.4654	.4488	.8936	.5022	1.991	1.119	2.228	1.1054	20′
50′	.4683	.4514	.8923	.5059	1.977	1.121	2.215	1.1025	10′
27° 00′	.4712	.4540	.8910	.5095	1.963	1.122	2.203	1.0996	63° 00′
		cos θ	sin θ	cot θ	tan θ	csc θ	sec θ	θ Radians	θ Degrees

Table 3 Trigonometric functions of θ
(θ in degrees and minutes)

θ Degrees	θ Radians	sin θ	cos θ	tan θ	cot θ	sec θ	csc θ		
27° 00′	.4712	.4540	.8910	.5095	1.963	1.122	2.203	1.0996	63° 00′
10′	.4741	.4566	.8897	.5132	1.949	1.124	2.190	1.0966	50′
20′	.4771	.4592	.8884	.5169	1.935	1.126	2.178	1.0937	40′
30′	.4800	.4617	.8870	.5206	1.921	1.127	2.166	1.0908	30′
40′	.4829	.4643	.8857	.5243	1.907	1.129	2.154	1.0879	20′
50′	.4858	.4669	.8843	.5280	1.894	1.131	2.142	1.0850	10′
28° 00′	.4887	.4695	.8829	.5317	1.881	1.133	2.130	1.0821	62° 00′
10′	.4916	.4720	.8816	.5354	1.868	1.134	2.118	1.0792	50′
20′	.4945	.4746	.8802	.5392	1.855	1.136	2.107	1.0763	40′
30′	.4974	.4772	.8788	.5430	1.842	1.138	2.096	1.0734	30′
40′	.5003	.4797	.8774	.5467	1.829	1.140	2.085	1.0705	20′
50′	.5032	.4823	.8760	.5505	1.816	1.142	2.074	1.0676	10′
29° 00′	.5061	.4848	.8746	.5543	1.804	1.143	2.063	1.0647	61° 00′
10′	.5091	.4874	.8732	.5581	1.792	1.145	2.052	1.0617	50′
20′	.5120	.4899	.8718	.5619	1.780	1.147	2.041	1.0588	40′
30′	.5149	.4924	.8704	.5658	1.767	1.149	2.031	1.0559	30′
40′	.5178	.4950	.8689	.5696	1.756	1.151	2.020	1.0530	20′
50′	.5207	.4975	.8675	.5735	1.744	1.153	2.010	1.0501	10′
30° 00′	.5236	.5000	.8660	.5774	1.732	1.155	2.000	1.0472	60° 00′
10′	.5265	.5025	.8646	.5812	1.720	1.157	1.990	1.0443	50′
20′	.5294	.5050	.8631	.5851	1.709	1.159	1.980	1.0414	40′
30′	.5323	.5075	.8616	.5890	1.698	1.161	1.970	1.0385	30′
40′	.5352	.5100	.8601	.5930	1.686	1.163	1.961	1.0356	20′
50′	.5381	.5125	.8587	.5969	1.675	1.165	1.951	1.0327	10′
31° 00′	.5411	.5150	.8572	.6009	1.664	1.167	1.942	1.0297	59° 00′
10′	.5440	.5175	.8557	.6048	1.653	1.169	1.932	1.0268	50′
20′	.5469	.5200	.8542	.6088	1.643	1.171	1.923	1.0239	40′
30′	.5498	.5225	.8526	.6128	1.632	1.173	1.914	1.0210	30′
40′	.5527	.5250	.8511	.6168	1.621	1.175	1.905	1.0181	20′
50′	.5556	.5275	.8496	.6208	1.611	1.177	1.896	1.0152	10′
32° 00′	.5585	.5299	.8480	.6249	1.600	1.179	1.887	1.0123	58° 00′
10′	.5614	.5324	.8465	.6289	1.590	1.181	1.878	1.0094	50′
20′	.5643	.5348	.8450	.6330	1.580	1.184	1.870	1.0065	40′
30′	.5672	.5373	.8434	.6371	1.570	1.186	1.861	1.0036	30′
40′	.5701	.5398	.8418	.6412	1.560	1.188	1.853	1.0007	20′
50′	.5730	.5422	.8403	.6453	1.550	1.190	1.844	.9977	10′
33° 00′	.5760	.5446	.8387	.6494	1.540	1.192	1.836	.9948	57° 00′
10′	.5789	.5471	.8371	.6536	1.530	1.195	1.828	.9919	50′
20′	.5818	.5495	.8355	.6577	1.520	1.197	1.820	.9890	40′
30′	.5847	.5519	.8339	.6619	1.511	1.199	1.812	.9861	30′
40′	.5876	.5544	.8323	.6661	1.501	1.202	1.804	.9832	20′
50′	.5905	.5568	.8307	.6703	1.492	1.204	1.796	.9803	10′
34° 00′	.5934	.5592	.8290	.6745	1.483	1.206	1.788	.9774	56° 00′
10′	.5963	.5616	.8274	.6787	1.473	1.209	1.781	.9745	50′
20′	.5992	.5640	.8258	.6830	1.464	1.211	1.773	.9716	40′
30′	.6021	.5664	.8241	.6873	1.455	1.213	1.766	.9687	30′
40′	.6050	.5688	.8225	.6916	1.446	1.216	1.758	.9657	20′
50′	.6080	.5712	.8208	.6959	1.437	1.218	1.751	.9628	10′
35° 00′	.6109	.5736	.8192	.7002	1.428	1.221	1.743	.9599	55° 00′
10′	.6138	.5760	.8175	.7046	1.419	1.223	1.736	.9570	50′
20′	.6167	.5783	.8158	.7089	1.411	1.226	1.729	.9541	40′
30′	.6196	.5807	.8141	.7133	1.402	1.228	1.722	.9512	30′
40′	.6225	.5831	.8124	.7177	1.393	1.231	1.715	.9483	20′
50′	.6254	.5854	.8107	.7221	1.385	1.233	1.708	.9454	10′
36° 00′	.6283	.5878	.8090	.7265	1.376	1.236	1.701	.9425	54° 00′
		cos θ	sin θ	cot θ	tan θ	csc θ	sec θ	θ Radians	θ Degrees

Table 3 Trigonometric functions of θ
(θ in degrees and minutes)

θ Degrees	θ Radians	sin θ	cos θ	tan θ	cot θ	sec θ	csc θ		
36° 00′	.6283	.5878	.8090	.7265	1.376	1.236	1.701	.9425	54° 00′
10′	.6312	.5901	.8073	.7310	1.368	1.239	1.695	.9396	50′
20′	.6341	.5925	.8056	.7355	1.360	1.241	1.688	.9367	40′
30′	.6370	.5948	.8039	.7400	1.351	1.244	1.681	.9338	30′
40′	.6400	.5972	.8021	.7445	1.343	1.247	1.675	.9308	20′
50′	.6429	.5995	.8004	.7490	1.335	1.249	1.668	.9279	10′
37° 00′	.6458	.6018	.7986	.7536	1.327	1.252	1.662	.9250	53° 00′
10′	.6487	.6041	.7969	.7581	1.319	1.255	1.655	.9221	50′
20′	.6516	.6065	.7951	.7627	1.311	1.258	1.649	.9192	40′
30′	.6545	.6088	.7934	.7673	1.303	1.260	1.643	.9163	30′
40′	.6574	.6111	.7916	.7720	1.295	1.263	1.636	.9134	20′
50′	.6603	.6134	.7898	.7766	1.288	1.266	1.630	.9105	10′
38° 00′	.6632	.6157	.7880	.7813	1.280	1.269	1.624	.9076	52° 00′
10′	.6661	.6180	.7862	.7860	1.272	1.272	1.618	.9047	50′
20′	.6690	.6202	.7844	.7907	1.265	1.275	1.612	.9018	40′
30′	.6720	.6225	.7826	.7954	1.257	1.278	1.606	.8988	30′
40′	.6749	.6248	.7808	.8002	1.250	1.281	1.601	.8959	20′
50′	.6778	.6271	.7790	.8050	1.242	1.284	1.595	.8930	10′
39° 00′	.6807	.6293	.7771	.8098	1.235	1.287	1.589	.8901	51° 00′
10′	.6836	.6316	.7753	.8146	1.228	1.290	1.583	.8872	50′
20′	.6865	.6338	.7735	.8195	1.220	1.293	1.578	.8843	40′
30′	.6894	.6361	.7716	.8243	1.213	1.296	1.572	.8814	30′
40′	.6923	.6383	.7698	.8292	1.206	1.299	1.567	.8785	20′
50′	.6952	.6406	.7679	.8342	1.199	1.302	1.561	.8756	10′
40° 00′	.6981	.6428	.7660	.8391	1.192	1.305	1.556	.8727	50° 00′
10′	.7010	.6450	.7642	.8441	1.185	1.309	1.550	.8698	50′
20′	.7039	.6472	.7623	.8491	1.178	1.312	1.545	.8668	40′
30′	.7069	.6494	.7604	.8541	1.171	1.315	1.540	.8639	30′
40′	.7098	.6517	.7585	.8591	1.164	1.318	1.535	.8610	20′
50′	.7127	.6539	.7566	.8642	1.157	1.322	1.529	.8581	10′
41° 00′	.7156	.6561	.7547	.8693	1.150	1.325	1.524	.8552	49° 00′
10′	.7185	.6583	.7528	.8744	1.144	1.328	1.519	.8523	50′
20′	.7214	.6604	.7509	.8796	1.137	1.332	1.514	.8494	40′
30′	.7243	.6626	.7490	.8847	1.130	1.335	1.509	.8465	30′
40′	.7272	.6648	.7470	.8899	1.124	1.339	1.504	.8436	20′
50′	.7301	.6670	.7451	.8952	1.117	1.342	1.499	.8407	10′
42° 00′	.7330	.6691	.7431	.9004	1.111	1.346	1.494	.8378	48° 00′
10′	.7359	.6713	.7412	.9057	1.104	1.349	1.490	.8348	50′
20′	.7389	.6734	.7392	.9110	1.098	1.353	1.485	.8319	40′
30′	.7418	.6756	.7373	.9163	1.091	1.356	1.480	.8290	30′
40′	.7447	.6777	.7353	.9217	1.085	1.360	1.476	.8261	20′
50′	.7476	.6799	.7333	.9271	1.079	1.364	1.471	.8232	10′
43° 00′	.7505	.6820	.7314	.9325	1.072	1.367	1.466	.8203	47° 00′
10′	.7534	.6841	.7294	.9380	1.066	1.371	1.462	.8174	50′
20′	.7563	.6862	.7274	.9435	1.060	1.375	1.457	.8145	40′
30′	.7592	.6884	.7254	.9490	1.054	1.379	1.453	.8116	30′
40′	.7621	.6905	.7234	.9545	1.048	1.382	1.448	.8087	20′
50′	.7650	.6926	.7214	.9601	1.042	1.386	1.444	.8058	10′
44° 00′	.7679	.6947	.7193	.9657	1.036	1.390	1.440	.8029	46° 00′
10′	.7709	.6967	.7173	.9713	1.030	1.394	1.435	.7999	50′
20′	.7738	.6988	.7153	.9770	1.024	1.398	1.431	.7970	40′
30′	.7767	.7009	.7133	.9827	1.018	1.402	1.427	.7941	30′
40′	.7796	.7030	.7112	.9884	1.012	1.406	1.423	.7912	20′
50′	.7825	.7050	.7092	.9942	1.006	1.410	1.418	.7883	10′
45° 00′	.7854	.7071	.7071	1.000	1.000	1.414	1.414	.7854	45° 00′
		cos θ	sin θ	cot θ	tan θ	csc θ	sec θ	θ Radians	θ Degrees

Table 4 Trigonometric functions of θ (θ in decimal degrees)

θ Degrees	θ Radians	$\sin \theta$	$\cos \theta$	$\tan \theta$	$\cot \theta$	$\sec \theta$	$\csc \theta$		
0.0	.0000	.0000	1.0000	.0000	undefined	1.000	undefined	1.5708	**90.0**
0.1	.0017	.0017	1.0000	.0017	573.0	1.000	573.0	1.5691	89.9
0.2	.0035	.0035	1.0000	.0035	286.5	1.000	286.5	1.5673	89.8
0.3	.0052	.0052	1.0000	.0052	191.0	1.000	191.0	1.5656	89.7
0.4	.0070	.0070	1.0000	.0070	143.2	1.000	143.2	1.5638	89.6
0.5	.0087	.0087	1.0000	.0087	114.6	1.000	114.6	1.5621	89.5
0.6	.0105	.0105	.9999	.0105	95.49	1.000	95.49	1.5603	89.4
0.7	.0122	.0122	.9999	.0122	81.85	1.000	81.85	1.5586	89.3
0.8	.0140	.0140	.9999	.0140	71.62	1.000	71.62	1.5568	89.2
0.9	.0157	.0157	.9999	.0157	63.66	1.000	63.66	1.5551	89.1
1.0	.0175	.0175	.9998	.0175	57.29	1.000	57.30	1.5533	**89.0**
1.1	.0192	.0192	.9998	.0192	52.08	1.000	52.09	1.5516	88.9
1.2	.0209	.0209	.9998	.0209	47.74	1.000	47.75	1.5499	88.8
1.3	.0227	.0227	.9997	.0227	44.07	1.000	44.08	1.5481	88.7
1.4	.0244	.0244	.9997	.0244	40.92	1.000	40.93	1.5464	88.6
1.5	.0262	.0262	.9997	.0262	38.19	1.000	38.20	1.5446	88.5
1.6	.0279	.0279	.9996	.0279	35.80	1.000	35.81	1.5429	88.4
1.7	.0297	.0297	.9996	.0297	33.69	1.000	33.71	1.5411	88.3
1.8	.0314	.0314	.9995	.0314	31.82	1.000	31.84	1.5394	88.2
1.9	.0332	.0332	.9995	.0332	30.14	1.001	30.16	1.5376	88.1
2.0	.0349	.0349	.9994	.0349	28.64	1.001	28.65	1.5359	**88.0**
2.1	.0367	.0366	.9993	.0367	27.27	1.001	27.29	1.5341	87.9
2.2	.0384	.0384	.9993	.0384	26.03	1.001	26.05	1.5324	87.8
2.3	.0401	.0401	.9992	.0402	24.90	1.001	24.92	1.5307	87.7
2.4	.0419	.0419	.9991	.0419	23.86	1.001	23.88	1.5289	87.6
2.5	.0436	.0436	.9990	.0437	22.90	1.001	22.93	1.5272	87.5
2.6	.0454	.0454	.9990	.0454	22.02	1.001	22.04	1.5254	87.4
2.7	.0471	.0471	.9989	.0472	21.20	1.001	21.23	1.5237	87.3
2.8	.0489	.0488	.9988	.0489	20.45	1.001	20.47	1.5219	87.2
2.9	.0506	.0506	.9987	.0507	19.74	1.001	19.77	1.5202	87.1
3.0	.0524	.0523	.9986	.0524	19.08	1.001	19.11	1.5184	**87.0**
3.1	.0541	.0541	.9985	.0542	18.46	1.001	18.49	1.5167	86.9
3.2	.0559	.0558	.9984	.0559	17.89	1.002	17.91	1.5149	86.8
3.3	.0576	.0576	.9983	.0577	17.34	1.002	17.37	1.5132	86.7
3.4	.0593	.0593	.9982	.0594	16.83	1.002	16.86	1.5115	86.6
3.5	.0611	.0610	.9981	.0612	16.35	1.002	16.38	1.5097	86.5
3.6	.0628	.0628	.9980	.0629	15.89	1.002	15.93	1.5080	86.4
3.7	.0646	.0645	.9979	.0647	15.46	1.002	15.50	1.5062	86.3
3.8	.0663	.0663	.9978	.0664	15.06	1.002	15.09	1.5045	86.2
3.9	.0681	.0680	.9977	.0682	14.67	1.002	14.70	1.5027	86.1
4.0	.0698	.0698	.9976	.0699	14.30	1.002	14.34	1.5010	**86.0**
4.1	.0716	.0715	.9974	.0717	13.95	1.003	13.99	1.4992	85.9
4.2	.0733	.0732	.9973	.0734	13.62	1.003	13.65	1.4975	85.8
4.3	.0750	.0750	.9972	.0752	13.30	1.003	13.34	1.4957	85.7
4.4	.0768	.0767	.9971	.0769	13.00	1.003	13.03	1.4940	85.6
4.5	.0785	.0785	.9969	.0787	12.71	1.003	12.75	1.4923	85.5
4.6	.0803	.0802	.9968	.0805	12.43	1.003	12.47	1.4905	85.4
4.7	.0820	.0819	.9966	.0822	12.16	1.003	12.20	1.4888	85.3
4.8	.0838	.0837	.9965	.0840	11.91	1.004	11.95	1.4870	85.2
4.9	.0855	.0854	.9963	.0857	11.66	1.004	11.71	1.4853	85.1
5.0	.0873	.0872	.9962	.0875	11.43	1.004	11.47	1.4835	**85.0**
5.1	.0890	.0889	.9960	.0892	11.20	1.004	11.25	1.4818	84.9
5.2	.0908	.0906	.9959	.0910	10.99	1.004	11.03	1.4800	84.8
5.3	.0925	.0924	.9957	.0928	10.78	1.004	10.83	1.4783	84.7
5.4	.0942	.0941	.9956	.0945	10.58	1.004	10.63	1.4765	84.6
5.5	.0960	.0958	.9954	.0963	10.39	1.005	10.43	1.4748	84.5
5.6	.0977	.0976	.9952	.0981	10.20	1.005	10.25	1.4731	84.4
5.7	.0995	.0993	.9951	.0998	10.02	1.005	10.07	1.4713	84.3
5.8	.1012	.1011	.9949	.1016	9.845	1.005	9.895	1.4696	84.2
5.9	.1030	.1028	.9947	.1033	9.677	1.005	9.728	1.4678	84.1
6.0	.1047	.1045	.9945	.1051	9.514	1.006	9.567	1.4661	**84.0**
		$\cos \theta$	$\sin \theta$	$\cot \theta$	$\tan \theta$	$\csc \theta$	$\sec \theta$	θ Radians	θ Degrees

Table 4 Trigonometric functions of θ (θ in decimal degrees)

θ Degrees	θ Radians	sin θ	cos θ	tan θ	cot θ	sec θ	csc θ		
6.0	.1047	.1045	.9945	.1051	9.514	1.006	9.567	1.4661	**84.0**
6.1	.1065	.1063	.9943	.1069	9.357	1.006	9.411	1.4643	83.9
6.2	.1082	.1080	.9942	.1086	9.205	1.006	9.259	1.4626	83.8
6.3	.1100	.1097	.9940	.1104	9.058	1.006	9.113	1.4608	83.7
6.4	.1117	.1115	.9938	.1122	8.915	1.006	8.971	1.4591	83.6
6.5	.1134	.1132	.9936	.1139	8.777	1.006	8.834	1.4574	83.5
6.6	.1152	.1149	.9934	.1157	8.643	1.007	8.700	1.4556	83.4
6.7	.1169	.1167	.9932	.1175	8.513	1.007	8.571	1.4539	83.3
6.8	.1187	.1184	.9930	.1192	8.386	1.007	8.446	1.4521	83.2
6.9	.1204	.1201	.9928	.1210	8.264	1.007	8.324	1.4504	83.1
7.0	.1222	.1219	.9925	.1228	8.144	1.008	8.206	1.4486	**83.0**
7.1	.1239	.1236	.9923	.1246	8.028	1.008	8.091	1.4469	82.9
7.2	.1257	.1253	.9921	.1263	7.916	1.008	7.979	1.4451	82.8
7.3	.1274	.1271	.9919	.1281	7.806	1.008	7.870	1.4434	82.7
7.4	.1292	.1288	.9917	.1299	7.700	1.008	7.764	1.4416	82.6
7.5	.1309	.1305	.9914	.1317	7.596	1.009	7.661	1.4399	82.5
7.6	.1326	.1323	.9912	.1334	7.495	1.009	7.561	1.4382	82.4
7.7	.1344	.1340	.9910	.1352	7.396	1.009	7.463	1.4364	82.3
7.8	.1361	.1357	.9907	.1370	7.300	1.009	7.368	1.4347	82.2
7.9	.1379	.1374	.9905	.1388	7.207	1.010	7.276	1.4329	82.1
8.0	.1396	.1392	.9903	.1405	7.115	1.010	7.185	1.4312	**82.0**
8.1	.1414	.1409	.9900	.1423	7.026	1.010	7.097	1.4294	81.9
8.2	.1431	.1426	.9898	.1441	6.940	1.010	7.011	1.4277	81.8
8.3	.1449	.1444	.9895	.1459	6.855	1.011	6.927	1.4259	81.7
8.4	.1466	.1461	.9893	.1477	6.772	1.011	6.845	1.4242	81.6
8.5	.1484	.1478	.9890	.1495	6.691	1.011	6.765	1.4224	81.5
8.6	.1501	.1495	.9888	.1512	6.612	1.011	6.687	1.4207	81.4
8.7	.1518	.1513	.9885	.1530	6.535	1.012	6.611	1.4190	81.3
8.8	.1536	.1530	.9882	.1548	6.460	1.012	6.537	1.4172	81.2
8.9	.1553	.1547	.9880	.1566	6.386	1.012	6.464	1.4155	81.1
9.0	.1571	.1564	.9877	.1584	6.314	1.012	6.392	1.4137	**81.0**
9.1	.1588	.1582	.9874	.1602	6.243	1.013	6.323	1.4120	80.9
9.2	.1606	.1599	.9871	.1620	6.174	1.013	6.255	1.4102	80.8
9.3	.1623	.1616	.9869	.1638	6.107	1.013	6.188	1.4085	80.7
9.4	.1641	.1633	.9866	.1655	6.041	1.014	6.123	1.4067	80.6
9.5	.1658	.1650	.9863	.1673	5.976	1.014	6.059	1.4050	80.5
9.6	.1676	.1668	.9860	.1691	5.912	1.014	5.996	1.4032	80.4
9.7	.1693	.1685	.9857	.1709	5.850	1.015	5.935	1.4015	80.3
9.8	.1710	.1702	.9854	.1727	5.789	1.015	5.875	1.3998	80.2
9.9	.1728	.1719	.9851	.1745	5.730	1.015	5.816	1.3980	80.1
10.0	.1745	.1736	.9848	.1763	5.671	1.015	5.759	1.3963	**80.0**
10.1	.1763	.1754	.9845	.1781	5.614	1.016	5.702	1.3945	79.9
10.2	.1780	.1771	.9842	.1799	5.558	1.016	5.647	1.3928	79.8
10.3	.1798	.1788	.9839	.1817	5.503	1.016	5.593	1.3910	79.7
10.4	.1815	.1805	.9836	.1835	5.449	1.017	5.540	1.3893	79.6
10.5	.1833	.1822	.9833	.1853	5.396	1.017	5.487	1.3875	79.5
10.6	.1850	.1840	.9829	.1871	5.343	1.017	5.436	1.3858	79.4
10.7	.1868	.1857	.9826	.1890	5.292	1.018	5.386	1.3840	79.3
10.8	.1885	.1874	.9823	.1908	5.242	1.018	5.337	1.3823	79.2
10.9	.1902	.1891	.9820	.1926	5.193	1.018	5.288	1.3806	79.1
11.0	.1920	.1908	.9816	.1944	5.145	1.019	5.241	1.3788	**79.0**
11.1	.1937	.1925	.9813	.1962	5.097	1.019	5.194	1.3771	78.9
11.2	.1955	.1942	.9810	.1980	5.050	1.019	5.148	1.3753	78.8
11.3	.1972	.1959	.9806	.1998	5.005	1.020	5.103	1.3736	78.7
11.4	.1990	.1977	.9803	.2016	4.959	1.020	5.059	1.3718	78.6
11.5	.2007	.1994	.9799	.2035	4.915	1.020	5.016	1.3701	78.5
11.6	.2025	.2011	.9796	.2053	4.872	1.021	4.973	1.3683	78.4
11.7	.2042	.2028	.9792	.2071	4.829	1.021	4.931	1.3666	78.3
11.8	.2059	.2045	.9789	.2089	4.787	1.022	4.890	1.3648	78.2
11.9	.2077	.2062	.9785	.2107	4.745	1.022	4.850	1.3631	78.1
12.0	.2094	.2079	.9781	.2126	4.705	1.022	4.810	1.3614	**78.0**
		cos θ	sin θ	cot θ	tan θ	csc θ	sec θ	θ Radians	θ Degrees

Table 4 Trigonometric functions of θ (θ in decimal degrees)

θ Degrees	θ Radians	sin θ	cos θ	tan θ	cot θ	sec θ	csc θ		
12.0	.2094	.2079	.9781	.2126	4.705	1.022	4.810	1.3614	**78.0**
12.1	.2112	.2096	.9778	.2144	4.665	1.023	4.771	1.3596	77.9
12.2	.2129	.2113	.9774	.2162	4.625	1.023	4.732	1.3579	77.8
12.3	.2147	.2130	.9770	.2180	4.586	1.023	4.694	1.3561	77.7
12.4	.2164	.2147	.9767	.2199	4.548	1.024	4.657	1.3544	77.6
12.5	.2182	.2164	.9763	.2217	4.511	1.024	4.620	1.3526	77.5
12.6	.2199	.2181	.9759	.2235	4.474	1.025	4.584	1.3509	77.4
12.7	.2217	.2198	.9755	.2254	4.437	1.025	4.549	1.3491	77.3
12.8	.2234	.2215	.9751	.2272	4.402	1.025	4.514	1.3474	77.2
12.9	.2251	.2233	.9748	.2290	4.366	1.026	4.479	1.3456	77.1
13.0	.2269	.2250	.9744	.2309	4.331	1.026	4.445	1.3439	**77.0**
13.1	.2286	.2267	.9740	.2327	4.297	1.027	4.412	1.3422	76.9
13.2	.2304	.2284	.9736	.2345	4.264	1.027	4.379	1.3404	76.8
13.3	.2321	.2300	.9732	.2364	4.230	1.028	4.347	1.3387	76.7
13.4	.2339	.2317	.9728	.2382	4.198	1.028	4.315	1.3369	76.6
13.5	.2356	.2334	.9724	.2401	4.165	1.028	4.284	1.3352	76.5
13.6	.2374	.2351	.9720	.2419	4.134	1.029	4.253	1.3334	76.4
13.7	.2391	.2368	.9715	.2438	4.102	1.029	4.222	1.3317	76.3
13.8	.2409	.2385	.9711	.2456	4.071	1.030	4.192	1.3299	76.2
13.9	.2426	.2402	.9707	.2475	4.041	1.030	4.163	1.3282	76.1
14.0	.2443	.2419	.9703	.2493	4.011	1.031	4.134	1.3265	**76.0**
14.1	.2461	.2436	.9699	.2512	3.981	1.031	4.105	1.3247	75.9
14.2	.2478	.2453	.9694	.2530	3.952	1.032	4.077	1.3230	75.8
14.3	.2496	.2470	.9690	.2549	3.923	1.032	4.049	1.3212	75.7
14.4	.2513	.2487	.9686	.2568	3.895	1.032	4.021	1.3195	75.6
14.5	.2531	.2504	.9681	.2586	3.867	1.033	3.994	1.3177	75.5
14.6	.2548	.2521	.9677	.2605	3.839	1.033	3.967	1.3160	75.4
14.7	.2566	.2538	.9673	.2623	3.812	1.034	3.941	1.3142	75.3
14.8	.2583	.2554	.9668	.2642	3.785	1.034	3.915	1.3125	75.2
14.9	.2601	.2571	.9664	.2661	3.758	1.035	3.889	1.3107	75.1
15.0	.2618	.2588	.9659	.2679	3.732	1.035	3.864	1.3090	**75.0**
15.1	.2635	.2605	.9655	.2698	3.706	1.036	3.839	1.3073	74.9
15.2	.2653	.2622	.9650	.2717	3.681	1.036	3.814	1.3055	74.8
15.3	.2670	.2639	.9646	.2736	3.655	1.037	3.790	1.3038	74.7
15.4	.2688	.2656	.9641	.2754	3.630	1.037	3.766	1.3020	74.6
15.5	.2705	.2672	.9636	.2773	3.606	1.038	3.742	1.3003	74.5
15.6	.2723	.2689	.9632	.2792	3.582	1.038	3.719	1.2985	74.4
15.7	.2740	.2706	.9627	.2811	3.558	1.039	3.695	1.2968	74.3
15.8	.2758	.2723	.9622	.2830	3.534	1.039	3.673	1.2950	74.2
15.9	.2775	.2740	.9617	.2849	3.511	1.040	3.650	1.2933	74.1
16.0	.2793	.2756	.9613	.2867	3.487	1.040	3.628	1.2915	**74.0**
16.1	.2810	.2773	.9608	.2886	3.465	1.041	3.606	1.2898	73.9
16.2	.2827	.2790	.9603	.2905	3.442	1.041	3.584	1.2881	73.8
16.3	.2845	.2807	.9598	.2924	3.420	1.042	3.563	1.2863	73.7
16.4	.2862	.2823	.9593	.2943	3.398	1.042	3.542	1.2846	73.6
16.5	.2880	.2840	.9588	.2962	3.376	1.043	3.521	1.2828	73.5
16.6	.2897	.2857	.9583	.2981	3.354	1.043	3.500	1.2811	73.4
16.7	.2915	.2874	.9578	.3000	3.333	1.044	3.480	1.2793	73.3
16.8	.2932	.2890	.9573	.3019	3.312	1.045	3.460	1.2776	73.2
16.9	.2950	.2907	.9568	.3038	3.291	1.045	3.440	1.2758	73.1
17.0	.2967	.2924	.9563	.3057	3.271	1.046	3.420	1.2741	**73.0**
17.1	.2985	.2940	.9558	.3076	3.251	1.046	3.401	1.2723	72.9
17.2	.3002	.2957	.9553	.3096	3.230	1.047	3.382	1.2706	72.8
17.3	.3019	.2974	.9548	.3115	3.211	1.047	3.363	1.2689	72.7
17.4	.3037	.2990	.9542	.3134	3.191	1.048	3.344	1.2671	72.6
17.5	.3054	.3007	.9537	.3153	3.172	1.049	3.326	1.2654	72.5
17.6	.3072	.3024	.9532	.3172	3.152	1.049	3.307	1.2636	72.4
17.7	.3089	.3040	.9527	.3191	3.133	1.050	3.289	1.2619	72.3
17.8	.3107	.3057	.9521	.3211	3.115	1.050	3.271	1.2601	72.2
17.9	.3124	.3074	.9516	.3230	3.096	1.051	3.254	1.2584	72.1
18.0	.3142	.3090	.9511	.3249	3.078	1.051	3.236	1.2566	**72.0**
		cos θ	sin θ	cot θ	tan θ	csc θ	sec θ	θ Radians	θ Degrees

Table 4 Trigonometric functions of θ (θ in decimal degrees)

θ Degrees	θ Radians	sin θ	cos θ	tan θ	cot θ	sec θ	csc θ		
18.0	.3142	.3090	.9511	.3249	3.078	1.051	3.236	1.2566	**72.0**
18.1	.3159	.3107	.9505	.3269	3.060	1.052	3.219	1.2549	71.9
18.2	.3177	.3123	.9500	.3288	3.042	1.053	3.202	1.2531	71.8
18.3	.3194	.3140	.9494	.3307	3.024	1.053	3.185	1.2514	71.7
18.4	.3211	.3156	.9489	.3327	3.006	1.054	3.168	1.2497	71.6
18.5	.3229	.3173	.9483	.3346	2.989	1.054	3.152	1.2479	71.5
18.6	.3246	.3190	.9478	.3365	2.971	1.055	3.135	1.2462	71.4
18.7	.3264	.3206	.9472	.3385	2.954	1.056	3.119	1.2444	71.3
18.8	.3281	.3223	.9466	.3404	2.937	1.056	3.103	1.2427	71.2
18.9	.3299	.3239	.9461	.3424	2.921	1.057	3.087	1.2409	71.1
19.0	.3316	.3256	.9455	.3443	2.904	1.058	3.072	1.2392	**71.0**
19.1	.3334	.3272	.9449	.3463	2.888	1.058	3.056	1.2374	70.9
19.2	.3351	.3289	.9444	.3482	2.872	1.059	3.041	1.2357	70.8
19.3	.3368	.3305	.9438	.3502	2.856	1.060	3.026	1.2339	70.7
19.4	.3386	.3322	.9432	.3522	2.840	1.060	3.011	1.2322	70.6
19.5	.3403	.3338	.9426	.3541	2.824	1.061	2.996	1.2305	70.5
19.6	.3421	.3355	.9421	.3561	2.808	1.062	2.981	1.2287	70.4
19.7	.3438	.3371	.9415	.3581	2.793	1.062	2.967	1.2270	70.3
19.8	.3456	.3387	.9409	.3600	2.778	1.063	2.952	1.2252	70.2
19.9	.3473	.3404	.9403	.3620	2.762	1.064	2.938	1.2235	70.1
20.0	.3491	.3420	.9397	.3640	2.747	1.064	2.924	1.2217	**70.0**
20.1	.3508	.3437	.9391	.3659	2.733	1.065	2.910	1.2200	69.9
20.2	.3526	.3453	.9385	.3679	2.718	1.066	2.896	1.2182	69.8
20.3	.3543	.3469	.9379	.3699	2.703	1.066	2.882	1.2165	69.7
20.4	.3560	.3486	.9373	.3719	2.689	1.067	2.869	1.2147	69.6
20.5	.3578	.3502	.9367	.3739	2.675	1.068	2.855	1.2130	69.5
20.6	.3595	.3518	.9361	.3759	2.660	1.068	2.842	1.2113	69.4
20.7	.3613	.3535	.9354	.3779	2.646	1.069	2.829	1.2095	69.3
20.8	.3630	.3551	.9348	.3799	2.633	1.070	2.816	1.2078	69.2
20.9	.3648	.3567	.9342	.3819	2.619	1.070	2.803	1.2060	69.1
21.0	.3665	.3584	.9336	.3839	2.605	1.071	2.790	1.2043	**69.0**
21.1	.3683	.3600	.9330	.3859	2.592	1.072	2.778	1.2025	68.9
21.2	.3700	.3616	.9323	.3879	2.578	1.073	2.765	1.2008	68.8
21.3	.3718	.3633	.9317	.3899	2.565	1.073	2.753	1.1991	68.7
21.4	.3735	.3649	.9311	.3919	2.552	1.074	2.741	1.1973	68.6
21.5	.3752	.3665	.9304	.3939	2.539	1.075	2.729	1.1956	68.5
21.6	.3770	.3681	.9298	.3959	2.526	1.076	2.716	1.1938	68.4
21.7	.3787	.3697	.9291	.3979	2.513	1.076	2.705	1.1921	68.3
21.8	.3805	.3714	.9285	.4000	2.500	1.077	2.693	1.1903	68.2
21.9	.3822	.3730	.9278	.4020	2.488	1.078	2.681	1.1886	68.1
22.0	.3840	.3746	.9272	.4040	2.475	1.079	2.669	1.1868	**68.0**
22.1	.3857	.3762	.9265	.4061	2.463	1.079	2.658	1.1851	67.9
22.2	.3875	.3778	.9259	.4081	2.450	1.080	2.647	1.1833	67.8
22.3	.3892	.3795	.9252	.4101	2.438	1.081	2.635	1.1816	67.7
22.4	.3910	.3811	.9245	.4122	2.426	1.082	2.624	1.1798	67.6
22.5	.3927	.3827	.9239	.4142	2.414	1.082	2.613	1.1781	67.5
22.6	.3944	.3843	.9232	.4163	2.402	1.083	2.602	1.1764	67.4
22.7	.3962	.3859	.9225	.4183	2.391	1.084	2.591	1.1746	67.3
22.8	.3979	.3875	.9219	.4204	2.379	1.085	2.581	1.1729	67.2
22.9	.3997	.3891	.9212	.4224	2.367	1.086	2.570	1.1711	67.1
23.0	.4014	.3907	.9205	.4245	2.356	1.086	2.559	1.1694	**67.0**
23.1	.4032	.3923	.9198	.4265	2.344	1.087	2.549	1.1676	66.9
23.2	.4049	.3939	.9191	.4286	2.333	1.088	2.538	1.1659	66.8
23.3	.4067	.3955	.9184	.4307	2.322	1.089	2.528	1.1641	66.7
23.4	.4084	.3971	.9178	.4327	2.311	1.090	2.518	1.1624	66.6
23.5	.4102	.3987	.9171	.4348	2.300	1.090	2.508	1.1606	66.5
23.6	.4119	.4003	.9164	.4369	2.289	1.091	2.498	1.1589	66.4
23.7	.4136	.4019	.9157	.4390	2.278	1.092	2.488	1.1572	66.3
23.8	.4154	.4035	.9150	.4411	2.267	1.093	2.478	1.1554	66.2
23.9	.4171	.4051	.9143	.4431	2.257	1.094	2.468	1.1537	66.1
24.0	.4189	.4067	.9135	.4452	2.246	1.095	2.459	1.1519	**66.0**
		cos θ	sin θ	cot θ	tan θ	csc θ	sec θ	θ Radians	θ Degrees

Table 4   Trigonometric functions of θ (θ in decimal degrees)

θ Degrees	θ Radians	sin θ	cos θ	tan θ	cot θ	sec θ	csc θ		
24.0	.4189	.4067	.9135	.4452	2.246	1.095	2.459	1.1519	**66.0**
24.1	.4206	.4083	.9128	.4473	2.236	1.095	2.449	1.1502	65.9
24.2	.4224	.4099	.9121	.4494	2.225	1.096	2.439	1.1484	65.8
24.3	.4241	.4115	.9114	.4515	2.215	1.097	2.430	1.1467	65.7
24.4	.4259	.4131	.9107	.4536	2.204	1.098	2.421	1.1449	65.6
24.5	.4276	.4147	.9100	.4557	2.194	1.099	2.411	1.1432	65.5
24.6	.4294	.4163	.9092	.4578	2.184	1.100	2.402	1.1414	65.4
24.7	.4311	.4179	.9085	.4599	2.174	1.101	2.393	1.1397	65.3
24.8	.4328	.4195	.9078	.4621	2.164	1.102	2.384	1.1380	65.2
24.9	.4346	.4210	.9070	.4642	2.154	1.102	2.375	1.1362	65.1
25.0	.4363	.4226	.9063	.4663	2.145	1.103	2.366	1.1345	**65.0**
25.1	.4381	.4242	.9056	.4684	2.135	1.104	2.357	1.1327	64.9
25.2	.4398	.4258	.9048	.4706	2.125	1.105	2.349	1.1310	64.8
25.3	.4416	.4274	.9041	.4727	2.116	1.106	2.340	1.1292	64.7
25.4	.4433	.4289	.9033	.4748	2.106	1.107	2.331	1.1275	64.6
25.5	.4451	.4305	.9026	.4770	2.097	1.108	2.323	1.1257	64.5
25.6	.4468	.4321	.9018	.4791	2.087	1.109	2.314	1.1240	64.4
25.7	.4485	.4337	.9011	.4813	2.078	1.110	2.306	1.1222	64.3
25.8	.4503	.4352	.9003	.4834	2.069	1.111	2.298	1.1205	64.2
25.9	.4520	.4368	.8996	.4856	2.059	1.112	2.289	1.1188	64.1
26.0	.4538	.4384	.8988	.4877	2.050	1.113	2.281	1.1170	**64.0**
26.1	.4555	.4399	.8980	.4899	2.041	1.114	2.273	1.1153	63.9
26.2	.4573	.4415	.8973	.4921	2.032	1.115	2.265	1.1135	63.8
26.3	.4590	.4431	.8965	.4942	2.023	1.115	2.257	1.1118	63.7
26.4	.4608	.4446	.8957	.4964	2.014	1.116	2.249	1.1100	63.6
26.5	.4625	.4462	.8949	.4986	2.006	1.117	2.241	1.1083	63.5
26.6	.4643	.4478	.8942	.5008	1.997	1.118	2.233	1.1065	63.4
26.7	.4660	.4493	.8934	.5029	1.988	1.119	2.226	1.1048	63.3
26.8	.4677	.4509	.8926	.5051	1.980	1.120	2.218	1.1030	63.2
26.9	.4695	.4524	.8918	.5073	1.971	1.121	2.210	1.1013	63.1
27.0	.4712	.4540	.8910	.5095	1.963	1.122	2.203	1.0996	**63.0**
27.1	.4730	.4555	.8902	.5117	1.954	1.123	2.195	1.0978	62.9
27.2	.4747	.4571	.8894	.5139	1.946	1.124	2.188	1.0961	62.8
27.3	.4765	.4586	.8886	.5161	1.937	1.125	2.180	1.0943	62.7
27.4	.4782	.4602	.8878	.5184	1.929	1.126	2.173	1.0926	62.6
27.5	.4800	.4617	.8870	.5206	1.921	1.127	2.166	1.0908	62.5
27.6	.4817	.4633	.8862	.5228	1.913	1.128	2.158	1.0891	62.4
27.7	.4835	.4648	.8854	.5250	1.905	1.129	2.151	1.0873	62.3
27.8	.4852	.4664	.8846	.5272	1.897	1.130	2.144	1.0856	62.2
27.9	.4869	.4679	.8838	.5295	1.889	1.132	2.137	1.0838	62.1
28.0	.4887	.4695	.8829	.5317	1.881	1.133	2.130	1.0821	**62.0**
28.1	.4904	.4710	.8821	.5340	1.873	1.134	2.123	1.0804	61.9
28.2	.4922	.4726	.8813	.5362	1.865	1.135	2.116	1.0786	61.8
28.3	.4939	.4741	.8805	.5384	1.857	1.136	2.109	1.0769	61.7
28.4	.4957	.4756	.8796	.5407	1.849	1.137	2.103	1.0751	61.6
28.5	.4974	.4772	.8788	.5430	1.842	1.138	2.096	1.0734	61.5
28.6	.4992	.4787	.8780	.5452	1.834	1.139	2.089	1.0716	61.4
28.7	.5009	.4802	.8771	.5475	1.827	1.140	2.082	1.0699	61.3
28.8	.5027	.4818	.8763	.5498	1.819	1.141	2.076	1.0681	61.2
28.9	.5044	.4833	.8755	.5520	1.811	1.142	2.069	1.0664	61.1
29.0	.5061	.4848	.8746	.5543	1.804	1.143	2.063	1.0647	**61.0**
29.1	.5079	.4863	.8738	.5566	1.797	1.144	2.056	1.0629	60.9
29.2	.5096	.4879	.8729	.5589	1.789	1.146	2.050	1.0612	60.8
29.3	.5114	.4894	.8721	.5612	1.782	1.147	2.043	1.0594	60.7
29.4	.5131	.4909	.8712	.5635	1.775	1.148	2.037	1.0577	60.6
29.5	.5149	.4924	.8704	.5658	1.767	1.149	2.031	1.0559	60.5
29.6	.5166	.4939	.8695	.5681	1.760	1.150	2.025	1.0542	60.4
29.7	.5184	.4955	.8686	.5704	1.753	1.151	2.018	1.0524	60.3
29.8	.5201	.4970	.8678	.5727	1.746	1.152	2.012	1.0507	60.2
29.9	.5219	.4985	.8669	.5750	1.739	1.154	2.006	1.0489	60.1
30.0	.5236	.5000	.8660	.5774	1.732	1.155	2.000	1.0472	**60.0**
		cos θ	sin θ	cot θ	tan θ	csc θ	sec θ	θ Radians	θ Degrees

Table 4 Trigonometric functions of θ (θ in decimal degrees)

θ Degrees	θ Radians	sin θ	cos θ	tan θ	cot θ	sec θ	csc θ		
30.0	.5236	.5000	.8660	.5774	1.732	1.155	2.000	1.0472	**60.0**
30.1	.5253	.5015	.8652	.5797	1.725	1.156	1.994	1.0455	59.9
30.2	.5271	.5030	.8643	.5820	1.718	1.157	1.988	1.0437	59.8
30.3	.5288	.5045	.8634	.5844	1.711	1.158	1.982	1.0420	59.7
30.4	.5306	.5060	.8625	.5867	1.704	1.159	1.976	1.0402	59.6
30.5	.5323	.5075	.8616	.5890	1.698	1.161	1.970	1.0385	59.5
30.6	.5341	.5090	.8607	.5914	1.691	1.162	1.964	1.0367	59.4
30.7	.5358	.5105	.8599	.5938	1.684	1.163	1.959	1.0350	59.3
30.8	.5376	.5120	.8590	.5961	1.678	1.164	1.953	1.0332	59.2
30.9	.5393	.5135	.8581	.5985	1.671	1.165	1.947	1.0315	59.1
31.0	.5411	.5150	.8572	.6009	1.664	1.167	1.942	1.0297	**59.0**
31.1	.5428	.5165	.8563	.6032	1.658	1.168	1.936	1.0280	58.9
31.2	.5445	.5180	.8554	.6056	1.651	1.169	1.930	1.0263	58.8
31.3	.5463	.5195	.8545	.6080	1.645	1.170	1.925	1.0245	58.7
31.4	.5480	.5210	.8535	.6104	1.638	1.172	1.919	1.0228	58.6
31.5	.5498	.5225	.8526	.6128	1.632	1.173	1.914	1.0210	58.5
31.6	.5515	.5240	.8517	.6152	1.625	1.174	1.908	1.0193	58.4
31.7	.5533	.5255	.8508	.6176	1.619	1.175	1.903	1.0175	58.3
31.8	.5550	.5270	.8499	.6200	1.613	1.177	1.898	1.0158	58.2
31.9	.5568	.5284	.8490	.6224	1.607	1.178	1.892	1.0140	58.1
32.0	.5585	.5299	.8480	.6249	1.600	1.179	1.887	1.0123	**58.0**
32.1	.5603	.5314	.8471	.6273	1.594	1.180	1.882	1.0105	57.9
32.2	.5620	.5329	.8462	.6297	1.588	1.182	1.877	1.0088	57.8
32.3	.5637	.5344	.8453	.6322	1.582	1.183	1.871	1.0071	57.7
32.4	.5655	.5358	.8443	.6346	1.576	1.184	1.866	1.0053	57.6
32.5	.5672	.5373	.8434	.6371	1.570	1.186	1.861	1.0036	57.5
32.6	.5690	.5388	.8425	.6395	1.564	1.187	1.856	1.0018	57.4
32.7	.5707	.5402	.8415	.6420	1.558	1.188	1.851	1.0001	57.3
32.8	.5725	.5417	.8406	.6445	1.552	1.190	1.846	.9983	57.2
32.9	.5742	.5432	.8396	.6469	1.546	1.191	1.841	.9966	57.1
33.0	.5760	.5446	.8387	.6494	1.540	1.192	1.836	.9948	**57.0**
33.1	.5777	.5461	.8377	.6519	1.534	1.194	1.831	.9931	56.9
33.2	.5794	.5476	.8368	.6544	1.528	1.195	1.826	.9913	56.8
33.3	.5812	.5490	.8358	.6569	1.522	1.196	1.821	.9896	56.7
33.4	.5829	.5505	.8348	.6594	1.517	1.198	1.817	.9879	56.6
33.5	.5847	.5519	.8339	.6619	1.511	1.199	1.812	.9861	56.5
33.6	.5864	.5534	.8329	.6644	1.505	1.201	1.807	.9844	56.4
33.7	.5882	.5548	.8320	.6669	1.499	1.202	1.802	.9826	56.3
33.8	.5899	.5563	.8310	.6694	1.494	1.203	1.798	.9809	56.2
33.9	.5917	.5577	.8300	.6720	1.488	1.205	1.793	.9791	56.1
34.0	.5934	.5592	.8290	.6745	1.483	1.206	1.788	.9774	**56.0**
34.1	.5952	.5606	.8281	.6771	1.477	1.208	1.784	.9756	55.9
34.2	.5969	.5621	.8271	.6796	1.471	1.209	1.779	.9739	55.8
34.3	.5986	.5635	.8261	.6822	1.466	1.211	1.775	.9721	55.7
34.4	.6004	.5650	.8251	.6847	1.460	1.212	1.770	.9704	55.6
34.5	.6021	.5664	.8241	.6873	1.455	1.213	1.766	.9687	55.5
34.6	.6039	.5678	.8231	.6899	1.450	1.215	1.761	.9669	55.4
34.7	.6056	.5693	.8221	.6924	1.444	1.216	1.757	.9652	55.3
34.8	.6074	.5707	.8211	.6950	1.439	1.218	1.752	.9634	55.2
34.9	.6091	.5721	.8202	.6976	1.433	1.219	1.748	.9617	55.1
35.0	.6109	.5736	.8192	.7002	1.428	1.221	1.743	.9599	**55.0**
35.1	.6126	.5750	.8181	.7028	1.423	1.222	1.739	.9582	54.9
35.2	.6144	.5764	.8171	.7054	1.418	1.224	1.735	.9564	54.8
35.3	.6161	.5779	.8161	.7080	1.412	1.225	1.731	.9547	54.7
35.4	.6178	.5793	.8151	.7107	1.407	1.227	1.726	.9529	54.6
35.5	.6196	.5807	.8141	.7133	1.402	1.228	1.722	.9512	54.5
35.6	.6213	.5821	.8131	.7159	1.397	1.230	1.718	.9495	54.4
35.7	.6231	.5835	.8121	.7186	1.392	1.231	1.714	.9477	54.3
35.8	.6248	.5850	.8111	.7212	1.387	1.233	1.710	.9460	54.2
35.9	.6266	.5864	.8100	.7239	1.381	1.235	1.705	.9442	54.1
36.0	.6283	.5878	.8090	.7265	1.376	1.236	1.701	.9425	**54.0**
		cos θ	sin θ	cot θ	tan θ	csc θ	sec θ	θ Radians	θ Degrees

Table 4 Trigonometric functions of θ (θ in decimal degrees)

θ Degrees	θ Radians	sin θ	cos θ	tan θ	cot θ	sec θ	csc θ		
36.0	.6283	.5878	.8090	.7265	1.376	1.236	1.701	.9425	**54.0**
36.1	.6301	.5892	.8080	.7292	1.371	1.238	1.697	.9407	53.9
36.2	.6318	.5906	.8070	.7319	1.366	1.239	1.693	.9390	53.8
36.3	.6336	.5920	.8059	.7346	1.361	1.241	1.689	.9372	53.7
36.4	.6353	.5934	.8049	.7373	1.356	1.242	1.685	.9355	53.6
36.5	.6370	.5948	.8039	.7400	1.351	1.244	1.681	.9338	53.5
36.6	.6388	.5962	.8028	.7427	1.347	1.246	1.677	.9320	53.4
36.7	.6405	.5976	.8018	.7454	1.342	1.247	1.673	.9303	53.3
36.8	.6423	.5990	.8007	.7481	1.337	1.249	1.669	.9285	53.2
36.9	.6440	.6004	.7997	.7508	1.332	1.250	1.666	.9268	53.1
37.0	.6458	.6018	.7986	.7536	1.327	1.252	1.662	.9250	**53.0**
37.1	.6475	.6032	.7976	.7563	1.322	1.254	1.658	.9233	52.9
37.2	.6493	.6046	.7965	.7590	1.317	1.255	1.654	.9215	52.8
37.3	.6510	.6060	.7955	.7618	1.313	1.257	1.650	.9198	52.7
37.4	.6528	.6074	.7944	.7646	1.308	1.259	1.646	.9180	52.6
37.5	.6545	.6088	.7934	.7673	1.303	1.260	1.643	.9163	52.5
37.6	.6562	.6101	.7923	.7701	1.299	1.262	1.639	.9146	52.4
37.7	.6580	.6115	.7912	.7729	1.294	1.264	1.635	.9128	52.3
37.8	.6597	.6129	.7902	.7757	1.289	1.266	1.632	.9111	52.2
37.9	.6615	.6143	.7891	.7785	1.285	1.267	1.628	.9093	52.1
38.0	.6632	.6157	.7880	.7813	1.280	1.269	1.624	.9076	**52.0**
38.1	.6650	.6170	.7869	.7841	1.275	1.271	1.621	.9058	51.9
38.2	.6667	.6184	.7859	.7869	1.271	1.272	1.617	.9041	51.8
38.3	.6685	.6198	.7848	.7898	1.266	1.274	1.613	.9023	51.7
38.4	.6702	.6211	.7837	.7926	1.262	1.276	1.610	.9006	51.6
38.5	.6720	.6225	.7826	.7954	1.257	1.278	1.606	.8988	51.5
38.6	.6737	.6239	.7815	.7983	1.253	1.280	1.603	.8971	51.4
38.7	.6754	.6252	.7804	.8012	1.248	1.281	1.599	.8954	51.3
38.8	.6772	.6266	.7793	.8040	1.244	1.283	1.596	.8936	51.2
38.9	.6789	.6280	.7782	.8069	1.239	1.285	1.592	.8919	51.1
39.0	.6807	.6293	.7771	.8098	1.235	1.287	1.589	.8901	**51.0**
39.1	.6824	.6307	.7760	.8127	1.230	1.289	1.586	.8884	50.9
39.2	.6842	.6320	.7749	.8156	1.226	1.290	1.582	.8866	50.8
39.3	.6859	.6334	.7738	.8185	1.222	1.292	1.579	.8849	50.7
39.4	.6877	.6347	.7727	.8214	1.217	1.294	1.575	.8831	50.6
39.5	.6894	.6361	.7716	.8243	1.213	1.296	1.572	.8814	50.5
39.6	.6912	.6374	.7705	.8273	1.209	1.298	1.569	.8796	50.4
39.7	.6929	.6388	.7694	.8302	1.205	1.300	1.566	.8779	50.3
39.8	.6946	.6401	.7683	.8332	1.200	1.302	1.562	.8762	50.2
39.9	.6964	.6414	.7672	.8361	1.196	1.304	1.559	.8744	50.1
40.0	.6981	.6428	.7660	.8391	1.192	1.305	1.556	.8727	**50.0**
40.1	.6999	.6441	.7649	.8421	1.188	1.307	1.552	.8709	49.9
40.2	.7016	.6455	.7638	.8451	1.183	1.309	1.549	.8692	49.8
40.3	.7034	.6468	.7627	.8481	1.179	1.311	1.546	.8674	49.7
40.4	.7051	.6481	.7615	.8511	1.175	1.313	1.543	.8657	49.6
40.5	.7069	.6494	.7604	.8541	1.171	1.315	1.540	.8639	49.5
40.6	.7086	.6508	.7593	.8571	1.167	1.317	1.537	.8622	49.4
40.7	.7103	.6521	.7581	.8601	1.163	1.319	1.534	.8604	49.3
40.8	.7121	.6534	.7570	.8632	1.159	1.321	1.530	.8587	49.2
40.9	.7138	.6547	.7559	.8662	1.154	1.323	1.527	.8570	49.1
41.0	.7156	.6561	.7547	.8693	1.150	1.325	1.524	.8552	**49.0**
41.1	.7173	.6574	.7536	.8724	1.146	1.327	1.521	.8535	48.9
41.2	.7191	.6587	.7524	.8754	1.142	1.329	1.518	.8517	48.8
41.3	.7208	.6600	.7513	.8785	1.138	1.331	1.515	.8500	48.7
41.4	.7226	.6613	.7501	.8816	1.134	1.333	1.512	.8482	48.6
41.5	.7243	.6626	.7490	.8847	1.130	1.335	1.509	.8465	48.5
41.6	.7261	.6639	.7478	.8878	1.126	1.337	1.506	.8447	48.4
41.7	.7278	.6652	.7466	.8910	1.122	1.339	1.503	.8430	48.3
41.8	.7295	.6665	.7455	.8941	1.118	1.341	1.500	.8412	48.2
41.9	.7313	.6678	.7443	.8972	1.115	1.344	1.497	.8395	48.1
42.0	.7330	.6691	.7431	.9004	1.111	1.346	1.494	.8378	**48.0**
		cos θ	sin θ	cot θ	tan θ	csc θ	sec θ	θ Radians	θ Degrees

Table 4 Trigonometric functions of θ (θ in decimal degrees)

θ Degrees	θ Radians	$\sin\theta$	$\cos\theta$	$\tan\theta$	$\cot\theta$	$\sec\theta$	$\csc\theta$		
42.0	.7330	.6691	.7431	.9004	1.111	1.346	1.494	.8378	**48.0**
42.1	.7348	.6704	.7420	.9036	1.107	1.348	1.492	.8360	47.9
42.2	.7365	.6717	.7408	.9067	1.103	1.350	1.489	.8343	47.8
42.3	.7383	.6730	.7396	.9099	1.099	1.352	1.486	.8325	47.7
42.4	.7400	.6743	.7385	.9131	1.095	1.354	1.483	.8308	47.6
42.5	.7418	.6756	.7373	.9163	1.091	1.356	1.480	.8290	47.5
42.6	.7435	.6769	.7361	.9195	1.087	1.359	1.477	.8273	47.4
42.7	.7453	.6782	.7349	.9228	1.084	1.361	1.475	.8255	47.3
42.8	.7470	.6794	.7337	.9260	1.080	1.363	1.472	.8238	47.2
42.9	.7487	.6807	.7325	.9293	1.076	1.365	1.469	.8221	47.1
43.0	.7505	.6820	.7314	.9325	1.072	1.367	1.466	.8203	**47.0**
43.1	.7522	.6833	.7302	.9358	1.069	1.370	1.464	.8186	46.9
43.2	.7540	.6845	.7290	.9391	1.065	1.372	1.461	.8168	46.8
43.3	.7557	.6858	.7278	.9424	1.061	1.374	1.458	.8151	46.7
43.4	.7575	.6871	.7266	.9457	1.057	1.376	1.455	.8133	46.6
43.5	.7592	.6884	.7254	.9490	1.054	1.379	1.453	.8116	46.5
43.6	.7610	.6896	.7242	.9523	1.050	1.381	1.450	.8098	46.4
43.7	.7627	.6909	.7230	.9556	1.046	1.383	1.447	.8081	46.3
43.8	.7645	.6921	.7218	.9590	1.043	1.386	1.445	.8063	46.2
43.9	.7662	.6934	.7206	.9623	1.039	1.388	1.442	.8046	46.1
44.0	.7679	.6947	.7193	.9657	1.036	1.390	1.440	.8029	**46.0**
44.1	.7697	.6959	.7181	.9691	1.032	1.393	1.437	.8011	45.9
44.2	.7714	.6972	.7169	.9725	1.028	1.395	1.434	.7994	45.8
44.3	.7732	.6984	.7157	.9759	1.025	1.397	1.432	.7976	45.7
44.4	.7749	.6997	.7145	.9793	1.021	1.400	1.429	.7959	45.6
44.5	.7767	.7009	.7133	.9827	1.018	1.402	1.427	.7941	45.5
44.6	.7784	.7022	.7120	.9861	1.014	1.404	1.424	.7924	45.4
44.7	.7802	.7034	.7108	.9896	1.011	1.407	1.422	.7906	45.3
44.8	.7819	.7046	.7096	.9930	1.007	1.409	1.419	.7889	45.2
44.9	.7837	.7059	.7083	.9965	1.003	1.412	1.417	.7871	45.1
45.0	.7854	.7071	.7071	1.0000	1.000	1.414	1.414	.7854	**45.0**
		$\cos\theta$	$\sin\theta$	$\cot\theta$	$\tan\theta$	$\csc\theta$	$\sec\theta$	θ Radians	θ Degrees

Table 5 Trigonometric functions of θ (θ in radians)

θ Radians	θ Degrees	sin θ	cos θ	tan θ	cot θ	sec θ	csc θ
0.00	0° 00′	0.0000	1.000	0.0000	Undefined	1.000	Undefined
.01	0° 34′	.0100	1.000	.0100	100.0	1.000	100.0
.02	1° 09′	.0200	0.9998	.0200	49.99	1.000	50.00
.03	1° 43′	.0300	0.9996	.0300	33.32	1.000	33.34
.04	2° 18′	.0400	0.9992	.0400	24.99	1.001	25.01
0.05	2° 52′	0.0500	0.9988	0.0500	19.98	1.001	20.01
.06	3° 26′	.0600	.9982	.0601	16.65	1.002	16.68
.07	4° 01′	.0699	.9976	.0701	14.26	1.002	14.30
.08	4° 35′	.0799	.9968	.0802	12.47	1.003	12.51
.09	5° 09′	.0899	.9960	.0902	11.08	1.004	11.13
0.10	5° 44′	0.0998	0.9950	0.1003	9.967	1.005	10.02
.11	6° 18′	.1098	.9940	.1104	9.054	1.006	9.109
.12	6° 53′	.1197	.9928	.1206	8.293	1.007	8.353
.13	7° 27′	.1296	.9916	.1307	7.649	1.009	7.714
.14	8° 01′	.1395	.9902	.1409	7.096	1.010	7.166
0.15	8° 36′	0.1494	0.9888	0.1511	6.617	1.011	6.692
.16	9° 10′	.1593	.9872	.1614	6.197	1.013	6.277
.17	9° 44′	.1692	.9856	.1717	5.826	1.015	5.911
.18	10° 19′	.1790	.9838	.1820	5.495	1.016	5.586
.19	10° 53′	.1889	.9820	.1923	5.200	1.018	5.295
0.20	11° 28′	0.1987	0.9801	0.2027	4.933	1.020	5.033
.21	12° 02′	.2085	.9780	.2131	4.692	1.022	4.797
.22	12° 36′	.2182	.9759	.2236	4.472	1.025	4.582
.23	13° 11′	.2280	.9737	.2341	4.271	1.027	4.386
.24	13° 45′	.2377	.9713	.2447	4.086	1.030	4.207
0.25	14° 19′	0.2474	0.9689	0.2553	3.916	1.032	4.042
.26	14° 54′	.2571	.9664	.2660	3.759	1.035	3.890
.27	15° 28′	.2667	.9638	.2768	3.613	1.038	3.749
.28	16° 03′	.2764	.9611	.2876	3.478	1.041	3.619
.29	16° 37′	.2860	.9582	.2984	3.351	1.044	3.497
0.30	17° 11′	0.2955	0.9553	0.3093	3.233	1.047	3.384
.31	17° 46′	.3051	.9523	.3203	3.122	1.050	3.278
.32	18° 20′	.3146	.9492	.3314	3.018	1.053	3.179
.33	18° 55′	.3240	.9460	.3425	2.920	1.057	3.086
.34	19° 29′	.3335	.9428	.3537	2.827	1.061	2.999
0.35	20° 03′	0.3429	0.9394	0.3650	2.740	1.065	2.916
.36	20° 38′	.3523	.9359	.3764	2.657	1.068	2.839
.37	21° 12′	.3616	.9323	.3879	2.578	1.073	2.765
.38	21° 46′	.3709	.9287	.3994	2.504	1.077	2.696
.39	22° 21′	.3802	.9249	.4111	2.433	1.081	2.630
0.40	22° 55′	0.3894	0.9211	0.4228	2.365	1.086	2.568
.41	23° 30′	.3986	.9171	.4346	2.301	1.090	2.509
.42	24° 04′	.4078	.9131	.4466	2.239	1.095	2.452
.43	24° 38′	.4169	.9090	.4586	2.180	1.100	2.399
.44	25° 13′	.4259	.9048	.4708	2.124	1.105	2.348
0.45	25° 47′	0.4350	0.9004	0.4831	2.070	1.111	2.299
.46	26° 21′	.4439	.8961	.4954	2.018	1.116	2.253
.47	26° 56′	.4529	.8916	.5080	1.969	1.122	2.208
.48	27° 30′	.4618	.8870	.5206	1.921	1.127	2.166
.49	28° 05′	.4706	.8823	.5334	1.875	1.133	2.125

Table 5   Trigonometric functions of θ (θ in radians)

θ Radians	θ Degrees	sin θ	cos θ	tan θ	cot θ	sec θ	csc θ
0.50	28° 39′	0.4794	0.8776	0.5463	1.830	1.139	2.086
.51	29° 13′	.4882	.8727	.5594	1.788	1.146	2.048
.52	29° 48′	.4969	.8678	.5726	1.747	1.152	2.013
.53	30° 22′	.5055	.8628	.5859	1.707	1.159	1.978
.54	30° 56′	.5141	.8577	.5994	1.668	1.166	1.945
0.55	31° 31′	0.5227	0.8525	0.6131	1.631	1.173	1.913
.56	32° 05′	.5312	.8473	.6269	1.595	1.180	1.883
.57	32° 40′	.5396	.8419	.6410	1.560	1.188	1.853
.58	33° 14′	.5480	.8365	.6552	1.526	1.196	1.825
.59	33° 48′	.5564	.8309	.6696	1.494	1.203	1.797
0.60	34° 23′	0.5646	0.8253	0.6841	1.462	1.212	1.771
.61	34° 57′	.5729	.8196	.6989	1.431	1.220	1.746
.62	35° 31′	.5810	.8139	.7139	1.401	1.229	1.721
.63	36° 06′	.5891	.8080	.7291	1.372	1.238	1.697
.64	36° 40′	.5972	.8021	.7445	1.343	1.247	1.674
0.65	37° 15′	0.6052	0.7961	0.7602	1.315	1.256	1.652
.66	37° 49′	.6131	.7900	.7761	1.288	1.266	1.631
.67	38° 23′	.6210	.7838	.7923	1.262	1.276	1.610
.68	38° 58′	.6288	.7776	.8087	1.237	1.286	1.590
.69	39° 32′	.6365	.7712	.8253	1.212	1.297	1.571
0.70	40° 06′	0.6442	0.7648	0.8423	1.187	1.307	1.552
.71	40° 41′	.6518	.7584	.8595	1.163	1.319	1.534
.72	41° 15′	.6594	.7518	.8771	1.140	1.330	1.517
.73	41° 50′	.6669	.7452	.8949	1.117	1.342	1.500
.74	42° 24′	.6743	.7385	.9131	1.095	1.354	1.483
0.75	42° 58′	0.6816	0.7317	0.9316	1.073	1.367	1.467
.76	43° 33′	.6889	.7248	.9505	1.052	1.380	1.452
.77	44° 07′	.6961	.7179	.9697	1.031	1.393	1.437
.78	44° 41′	.7033	.7109	.9893	1.011	1.407	1.422
.79	45° 16′	.7104	.7038	1.009	.9908	1.421	1.408
0.80	45° 50′	0.7174	0.6967	1.030	0.9712	1.435	1.394
.81	46° 25′	.7243	.6895	1.050	.9520	1.450	1.381
.82	46° 59′	.7311	.6822	1.072	.9331	1.466	1.368
.83	47° 33′	.7379	.6749	1.093	.9146	1.482	1.355
.84	48° 08′	.7446	.6675	1.116	.8964	1.498	1.343
0.85	48° 42′	0.7513	0.6600	1.138	0.8785	1.515	1.331
.86	49° 17′	.7578	.6524	1.162	.8609	1.533	1.320
.87	49° 51′	.7643	.6448	1.185	.8437	1.551	1.308
.88	50° 25′	.7707	.6372	1.210	.8267	1.569	1.297
.89	51° 00′	.7771	.6294	1.235	.8100	1.589	1.287
0.90	51° 34′	0.7833	0.6216	1.260	0.7936	1.609	1.277
.91	52° 08′	.7895	.6137	1.286	.7774	1.629	1.267
.92	52° 43′	.7956	.6058	1.313	.7615	1.651	1.257
.93	53° 17′	.8016	.5978	1.341	.7458	1.673	1.247
.94	53° 52′	.8076	.5898	1.369	.7303	1.696	1.238
0.95	54° 26′	0.8134	0.5817	1.398	0.7151	1.719	1.229
.96	55° 00′	.8192	.5735	1.428	.7001	1.744	1.221
.97	55° 35′	.8249	.5653	1.459	.6853	1.769	1.212
.98	56° 09′	.8305	.5570	1.491	.6707	1.795	1.204
.99	56° 43′	.8360	.5487	1.524	.6563	1.823	1.196

$\dfrac{\pi}{6}$ → (pointing to row .53)

$\dfrac{\pi}{4}$ → (pointing to row .79)

Table 5   Trigonometric functions of θ (θ in radians)

θ Radians	θ Degrees	sin θ	cos θ	tan θ	cot θ	sec θ	csc θ
1.00	57° 18'	0.8415	0.5403	1.557	0.6421	1.851	1.188
1.01	57° 52'	.8468	.5319	1.592	.6281	1.880	1.181
1.02	58° 27'	.8521	.5234	1.628	.6142	1.911	1.174
1.03	59° 01'	.8573	.5148	1.665	.6005	1.942	1.166
1.04	59° 35'	.8624	.5062	1.704	.5870	1.975	1.160
1.05	60° 10'	0.8674	0.4976	1.743	0.5736	2.010	1.153
1.06	60° 44'	.8724	.4889	1.784	.5604	2.046	1.146
1.07	61° 18'	.8772	.4801	1.827	.5473	2.083	1.140
1.08	61° 53'	.8820	.4713	1.871	.5344	2.122	1.134
1.09	62° 27'	.8866	.4625	1.917	.5216	2.162	1.128
1.10	63° 02'	0.8912	0.4536	1.965	0.5090	2.205	1.122
1.11	63° 36'	.8957	.4447	2.014	.4964	2.249	1.116
1.12	64° 10'	.9001	.4357	2.066	.4840	2.295	1.111
1.13	64° 45'	.9044	.4267	2.120	.4718	2.344	1.106
1.14	65° 19'	.9086	.4176	2.176	.4596	2.395	1.101
1.15	65° 53'	0.9128	0.4085	2.234	0.4475	2.448	1.096
1.16	66° 28'	.9168	.3993	2.296	.4356	2.504	1.091
1.17	67° 02'	.9208	.3902	2.360	.4237	2.563	1.086
1.18	67° 37'	.9246	.3809	2.428	.4120	2.625	1.082
1.19	68° 11'	.9284	.3717	2.498	.4003	2.691	1.077
1.20	68° 45'	0.9320	0.3624	2.572	0.3888	2.760	1.073
1.21	69° 20'	.9356	.3530	2.650	.3773	2.833	1.069
1.22	69° 54'	.9391	.3436	2.733	.3659	2.910	1.065
1.23	70° 28'	.9425	.3342	2.820	.3546	2.992	1.061
1.24	71° 03'	.9458	.3248	2.912	.3434	3.079	1.057
1.25	71° 37'	0.9490	0.3153	3.010	0.3323	3.171	1.054
1.26	72° 12'	.9521	.3058	3.113	.3212	3.270	1.050
1.27	72° 46'	.9551	.2963	3.224	.3102	3.375	1.047
1.28	73° 20'	.9580	.2867	3.341	.2993	3.488	1.044
1.29	73° 55'	.9608	.2771	3.467	.2884	3.609	1.041
1.30	74° 29'	0.9636	0.2675	3.602	0.2776	3.738	1.038
1.31	75° 03'	.9662	.2579	3.747	.2669	3.878	1.035
1.32	75° 38'	.9687	.2482	3.903	.2562	4.029	1.032
1.33	76° 12'	.9711	.2385	4.072	.2456	4.193	1.030
1.34	76° 47'	.9735	.2288	4.256	.2350	4.372	1.027
1.35	77° 21'	0.9757	0.2190	4.455	0.2245	4.566	1.025
1.36	77° 55'	.9779	.2092	4.673	.2140	4.779	1.023
1.37	78° 30'	.9799	.1994	4.913	.2035	5.014	1.021
1.38	79° 04'	.9819	.1896	5.177	.1931	5.273	1.018
1.39	79° 39'	.9837	.1798	5.471	.1828	5.561	1.017
1.40	80° 13'	0.9854	0.1700	5.798	0.1725	5.883	1.015
1.41	80° 47'	.9871	.1601	6.165	.1622	6.246	1.013
1.42	81° 22'	.9887	.1502	6.581	.1519	6.657	1.011
1.43	81° 56'	.9901	.1403	7.055	.1417	7.126	1.010
1.44	82° 30'	.9915	.1304	7.602	.1315	7.667	1.009
1.45	83° 05'	0.9927	0.1205	8.238	0.1214	8.299	1.007
1.46	83° 39'	.9939	.1106	8.989	.1113	9.044	1.006
1.47	84° 14'	.9949	.1006	9.887	.1011	9.938	1.005
1.48	84° 48'	.9959	.0907	10.98	.0911	11.03	1.004
1.49	85° 22'	.9967	.0807	12.35	.0810	12.39	1.003

$\frac{\pi}{3}$ → (pointing to row 1.04)

Table 5 Trigonometric functions of θ (θ in radians)

θ Radians	θ Degrees	$\sin \theta$	$\cos \theta$	$\tan \theta$	$\cot \theta$	$\sec \theta$	$\csc \theta$
1.50	85° 57′	0.9975	0.0707	14.10	0.0709	14.14	1.003
1.51	86° 31′	.9982	.0608	16.43	.0609	16.46	1.002
1.52	87° 05′	.9987	.0508	19.67	.0508	19.70	1.001
1.53	87° 40′	.9992	.0408	24.50	.0408	24.52	1.001
1.54	88° 14′	.9995	.0308	32.46	.0308	32.48	1.000
1.55	88° 49′	0.9998	0.0208	48.08	0.0208	48.09	1.000
1.56	89° 23′	.9999	.0108	92.62	.0108	92.63	1.000
1.57	89° 57′	1.000	.0008	1256	.0008	1256	1.000

$\dfrac{\pi}{2}$ →

Table 6 e^x and e^{-x}

x	e^x	e^{-x}	x	e^x	e^{-x}
0.00	1.0000	1.0000	0.36	1.4333	0.6977
0.01	1.0101	0.9900	0.37	1.4477	0.6907
0.02	1.0202	0.9802	0.38	1.4623	0.6839
0.03	1.0305	0.9704	0.39	1.4770	0.6771
0.04	1.0408	0.9608	0.40	1.4918	0.6703
0.05	1.0513	0.9512	0.41	1.5068	0.6637
0.06	1.0618	0.9418	0.42	1.5220	0.6570
0.07	1.0725	0.9324	0.43	1.5373	0.6505
0.08	1.0833	0.9231	0.44	1.5527	0.6440
0.09	1.0942	0.9139	0.45	1.5683	0.6376
0.10	1.1052	0.9048	0.46	1.5841	0.6313
0.11	1.1163	0.8958	0.47	1.6000	0.6250
0.12	1.1275	0.8869	0.48	1.6161	0.6188
0.13	1.1388	0.8781	0.49	1.6323	0.6126
0.14	1.1503	0.8694	0.50	1.6487	0.6065
0.15	1.1618	0.8607	1.0	2.7183	0.3679
0.16	1.1735	0.8521	1.5	4.4817	0.2231
0.17	1.1853	0.8437	2.0	7.3891	0.1353
0.18	1.1972	0.8353	2.5	12.182	0.0821
0.19	1.2092	0.8270	3.0	20.086	0.0498
0.20	1.2214	0.8187	3.5	33.115	0.0302
0.21	1.2337	0.8106	4.0	54.598	0.0183
0.22	1.2461	0.8025	4.5	90.017	0.0111
0.23	1.2586	0.7945	5.0	148.41	0.0067
0.24	1.2712	0.7866	5.5	244.69	0.0041
0.25	1.2840	0.7788	6.0	403.43	0.0025
0.26	1.2969	0.7711	6.5	665.14	0.0015
0.27	1.3100	0.7634	7.0	1096.6	0.0009
0.28	1.3231	0.7558	7.5	1808.0	0.0006
0.29	1.3364	0.7483	8.0	2981.0	0.0003
0.30	1.3499	0.7408	8.5	4914.8	0.0002
0.31	1.3634	0.7334	9.0	8103.1	0.0001
0.32	1.3771	0.7261	9.5	13359	0.0001
0.33	1.3910	0.7189	10.0	22026	.00005
0.34	1.4049	0.7118			
0.35	1.4191	0.7047			

Appendix 1

The Field Axioms for the Real Number System

Axioms, or **postulates,** for the real numbers are statements that are accepted as true without proof. The axioms for the real numbers are the basis for computation in arithmetic and algebra. A set of real numbers, together with the operations of addition and multiplication, that satisfies all the axioms of equality, addition, and multiplication, as well as the distributive axiom, is an example of what mathematicians call a **field.** In the field of real numbers, the following axioms of equality are assumed to be true.

AXIOMS OF EQUALITY

For all real numbers a, b, and c:

Reflexive Property	$a = a$
Symmetric Property	If $a = b$, then $b = a$.
Transitive Property	If $a = b$ and $b = c$, then $a = c$.

The following *Substitution Property* is also useful.

Substitution Property For all real numbers a and b, if $a = b$, then a can be replaced by b in any expression without changing the value of the expression.

Throughout our discussion we will use the symbol $\Re$ to denote the set of real numbers. The symbol $\in$ means "is a member of" or "belongs to." Thus, $a \in \Re$ means "a belongs to the set of real numbers $\Re$." The following axioms are true for all a, b, and $c \in \Re$ unless noted otherwise.

REAL NUMBER AXIOMS

Axiom Name	Addition	Multiplication
Closure	$a + b$ is a unique real number.	ab is a unique real number.
Commutative	$a + b = b + a$	$ab = ba$
Associative	$(a + b) + c = a + (b + c)$	$(ab)c = a(bc)$
Identity	There is a unique real number 0 such that $a + 0 = 0 + a = a.$	There is a unique real number 1 such that $a \cdot 1 = 1 \cdot a = a.$
Inverse	For every $a \in \Re$ there is a unique $-a \in \Re$ such that $a + (-a) = (-a) + a = 0.$	For every nonzero $a \in \Re$ there is a unique $\frac{1}{a} \in \Re$ such that $a\left(\frac{1}{a}\right) = \left(\frac{1}{a}\right)a = 1.$
Distributive	$a(b + c) = ab + ac$	

The operations subtraction and division can be defined in terms of addition and multiplication, respectively.

DEFINITION OF SUBTRACTION: For all $a, b \in \mathcal{R}$,

$$a - b = a + (-b).$$

DEFINITION OF DIVISION: For all $a, b \in \mathcal{R}$, $b \neq 0$,

$$a \div b = a \cdot \left(\frac{1}{b}\right).$$

These definitions and field axioms can be used to prove theorems that state additional properties of real numbers. Although you are probably already familiar with many of the properties these theorems state, it is important to see that they can be proved step-by-step from the Axioms of Equality and the Real Number Axioms on page 645. Two theorems are given below with their respective proofs. Notice that each step is justified by a reason. You are asked to prove other theorems in the exercises. Note that, for the sake of brevity, proofs often do not include the Closure Axiom in the *Reasons* column.

THEOREM 1. For all a, b, and $c \in \mathcal{R}$, if $a + c = b + c$, then $a = b$.

Proof:

	Statements	Reasons
1.	$a + c = b + c$	Given
2.	$(a + c) + (-c) = (a + c) + (-c)$	Reflexive Property of Equality
3.	$(a + c) + (-c) = (b + c) + (-c)$	Substitution Property
4.	$a + [c + (-c)] = b + [c + (-c)]$	Associative Axiom for Addition
5.	$a + 0 = b + 0$	Inverse Axiom for Addition
6.	$a = b$	Identity Axiom for Addition

Once a theorem has been proved it can be used in a subsequent proof. Notice that Theorem 1 is used in the proof of Theorem 2.

THEOREM 2. For all $a \in \mathcal{R}$, $a \cdot 0 = 0$.

Proof:

	Statements	Reasons
1.	$a \cdot a = 0 + a \cdot a$	Identity Axiom for Addition
2.	$a = 0 + a$	Identity Axiom for Addition
3.	$a \cdot (0 + a) = 0 + a \cdot a$	Substitution Property
4.	$a \cdot 0 + a \cdot a = 0 + a \cdot a$	Distributive Axiom
5.	$a \cdot 0 = 0$	Theorem 1

EXERCISES

Replace each _?_ with a variable or numeral so that a true statement results. Name the axiom that makes the statement true.

1. $a + 2 = 2 + \underline{?}$
2. $(4c)d = \underline{?}(cd)$
3. $\underline{?}(\frac{1}{3}) = 1$
4. $3 + \underline{?} = 0$
5. $2(a + 3) = 2a + \underline{?}$
6. $(2 + 1) + c = \underline{?} + (1 + c)$

Give the reason for each step in the proof. Where necessary, complete the proof. Any previously proved theorem can be used as a reason.

7. Theorem 3.   For all $c \in \Re$, $(-1)c = -c$.

　　Proof:

Statements	Reasons
1.　$(-1)c + c = c(-1) + c$	_?_
2.　　　　　　　$= c(-1) + c(1)$	_?_
3.　　　　　　　$= c[(-1) + (1)]$	_?_
4.　　　　　　　$= c(0)$	_?_
5.　　　　　　　$= 0$	_?_
6.　$(-1)c + c = 0$	Transitive Property of Equality
7.　$(-c) + c = 0$	_?_
8.　$(-1)c + c = (-c) + c$	Substitution Property
9.　　　$(-1)c = -c$	_?_

8. Theorem 4.   For all $a, c \in \Re$, $a(-c) = -(ac) = (-a)c$.

　　Proof:

Statements	Reasons
1. $a(-c) = (-c)a$	_?_
2.　　　　$= [(-1)c]a$	Theorem _?_
3.　　　　$= (-1)(ca)$	_?_
4.　　　　$= -(ca)$	Theorem _?_
5.　　　　$= -(ac)$	_?_
6. $a(-c) = -(ac)$	_?_

(You are asked to prove that $-(ac) = (-a)c$ in Exercise 9.)

Prove each of the following statements. Assume that each variable represents a real number.

9. $-(ac) = (-a)c$
10. If $ac = bc$ and $c \neq 0$, then $a = b$.
11. $-(a + b) = (-a) + (-b)$
12. $(x + y)z = xz + yz$
13. If $a \neq 0$ and $b \neq 0$, then $\dfrac{1}{ab} = \dfrac{1}{a} \cdot \dfrac{1}{b}$. $\left(\text{Hint: Show that } \dfrac{1}{ab} \cdot ab = 1 \text{ and }\right.$

that $\left. \left(\dfrac{1}{a} \cdot \dfrac{1}{b}\right)ab = 1.\right)$

14. $(ab)^2 = a^2b^2$ (*Hint:* Use the definition $x^2 = x \cdot x$.)

Appendix 2

Descartes' Rule of Signs and Bounds for Real Roots

The problem of determining the roots of polynomial equations occurs frequently in mathematics. Information that narrows down the set of *possible* roots of a polynomial equation greatly reduces the work involved in finding the roots. **Descartes' Rule of Signs** gives us information about the number of real roots of polynomial equations with real coefficients.

First consider a simplified polynomial $P(x)$ written in order of decreasing powers of x. Whenever the coefficients of two adjacent terms have opposite signs, we say that $P(x)$ has a *variation in sign*. Thus, the polynomial

$$x^6 - 3x^4 - 2x^2 + 4x - 5$$

$$\underbrace{\qquad}_{1} \quad \underbrace{\qquad}_{2} \; \underbrace{\qquad}_{3}$$

has three variations in sign. (Notice that you ignore terms having coefficient 0.)

DESCARTES' RULE OF SIGNS

Let $P(x)$ be a polynomial with real coefficients; then

1. the number of positive roots of $P(x) = 0$ is either equal to the number of variations in sign of $P(x)$ or is less than this number by a positive even integer;
2. the number of negative roots of $P(x) = 0$ is either equal to the number of variations in sign of $P(-x)$ or is less than this number by a positive even integer.

Sometimes Descartes' Rule gives complete information about the numbers of different types of roots. For example, $P(x) = x^4 + 11x - 6$ has *one* variation in sign and thus $P(x) = 0$ has *one positive root*. Since $P(-x) = (-x)^4 + 11(-x) - 6 = x^4 - 11x - 6$ has *one* variation in sign, $P(x) = 0$ has *one negative root*. Since $P(x)$ is of degree four, $P(x) = 0$ has exactly four roots. (See the Fundamental Theorem of Algebra, p. 84.) Since the equation has one positive real root and one negative real root, the remaining two roots must be imaginary.

More frequently, Descartes' Rule leaves us with several possibilities.

EXAMPLE 1. List the possibilities for the nature of the roots (positive, negative, and imaginary) of $P(x) = 0$ for the polynomial

$$P(x) = -2x^4 + x^3 - 7x^2 + 4x + 1.$$

SOLUTION: The given polynomial has three variations in sign, so the number of positive roots of $P(x) = 0$ is 3 or 1.

Examine $P(-x)$ to find the possible numbers of negative roots.

$$P(-x) = -2x^4 - x^3 - 7x^2 - 4x + 1$$

Since $P(-x)$ has one variation in sign, there is one negative root.

Since there are exactly four roots, we can make a chart to show the possible combinations of roots.

Number of positive real roots	Number of negative real roots	Number of imaginary roots
3	1	0
1	1	2

Descartes' Rule of Signs is used to determine the *number* of positive and negative roots of a polynomial equation $P(x) = 0$. The following theorem can be used to narrow the region of possible real roots down to an interval on the number line, that is, to find *bounds* for the real roots of $P(x) = 0$.

THEOREM   Let $P(x)$ be a polynomial with real coefficients and positive leading coefficient.

    1. Let M be a nonnegative real number. If the coefficients of the quotient and remainder obtained by dividing $P(x)$ by $x - M$ are all nonnegative, then $P(x) = 0$ has no roots greater than M.

    2. Let L be a nonpositive real number. If the coefficients of the quotient and remainder obtained by dividing $P(x)$ by $x - L$ are alternately nonnegative and nonpositive, then $P(x) = 0$ has no roots less than L.

The numbers M and L are called **upper** and **lower bounds,** respectively, for the real roots of $P(x) = 0$. To apply the theorem, use synthetic division for different values of x until the last line of numbers satisfies the conditions for M or L as stated in the theorem.

EXAMPLE 2.   Find the least nonnegative integral upper bound and the greatest non-positive integral lower bound for the roots of $3x^3 - 4x^2 - 4x + 4 = 0$. Graph the set of possible real roots on a number line.

SOLUTION: Use synthetic substitution with $x = 1$, 2, 3, ... and $x = -1, -2, -3, \ldots$ until numbers that satisfy the respective conditions for upper and lower bounds are found. The results are shown in the table.

x	coefficients of quotient and remainder			
1|	3	−1	−5	−1
2|	3	2	0	4	← all ≥ 0
−1|	3	−7	3	1	
−2|	3	−10	16	−28	← alternately ≥ 0 and ≤ 0.

Thus, $M = 2$ is the upper bound and $L = -2$ is the lower bound. The real roots of $3x^3 - 4x^2 - 4x + 4 = 0$ must therefore lie on the interval graphed below.

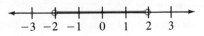

EXERCISES

Use Descartes' Rule to make a chart summarizing the possible combinations of positive, negative, and imaginary roots of each polynomial equation.

1. $y^2 - y - 4 = 0$ **2.** $-y^2 + y - 3 = 0$

3. $y^3 - y^2 + 2y + 3 = 0$ **4.** $x^3 + 6x^2 - x - 12 = 0$

5. $2x^4 - 3x^3 - x^2 + x + 1 = 0$ **6.** $3y^4 - y^3 + 4y^2 + 8 = 0$

7. $5y^4 - y^3 + y^2 + 2y - 4 = 0$ **8.** $-4x^4 - 3x^3 + x^2 - x + 5 = 0$

For each polynomial equation (a) find the least nonnegative integral upper bound and the greatest nonpositive integral lower bound of the real roots, and (b) write an interval in which the real roots lie.

9. $x^3 - 2x^2 + 3x + 5 = 0$ **10.** $x^3 - 4x^2 + 8x - 5 = 0$

11. $2x^3 - 3x^2 - 3x + 3 = 0$ **12.** $2x^3 - 2x^2 + x + 8 = 0$

13. $y^4 + y^3 + 16 = 0$ **14.** $y^4 - 2y - 3 = 0$

15. $y^4 - 2y^3 - 3y^2 + 4y + 8 = 0$ **16.** $y^4 + 2y^3 - y^2 + 7y - 5 = 0$

Appendix 3

More Operations and Properties of Functions

In Chapter 4 we discussed the *composition of functions,* an operation that combines two functions to produce a third one. You can also combine two functions to produce another by adding, subtracting, multiplying, and dividing the functions.

The **sum $s(x)$ of two functions $f(x)$ and $g(x)$** is defined as

$$s(x) = f(x) + g(x),$$

for all x in the domains of both $f(x)$ and $g(x)$. The graph of the sum function can be obtained from the graphs of $f(x)$ and $g(x)$ as shown below.

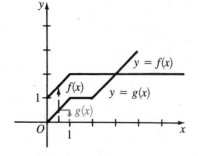
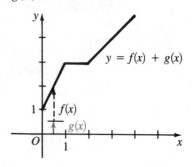

Similarly, we can define the product and difference of two functions $f(x)$ and

$g(x)$. For all x in the domains of both $f(x)$ and $g(x)$, the **product $p(x)$ of $f(x)$ and $g(x)$** is defined as

$$p(x) = f(x) \cdot g(x),$$

and the **difference $d(x)$ of $f(x)$ and $g(x)$** as

$$d(x) = f(x) - g(x).$$

Defining the quotient of two functions requires a little extra care, since the divisor must not be zero. The **quotient $q(x)$ of $f(x)$ divided by $g(x)$** is defined as

$$q(x) = \frac{f(x)}{g(x)},$$

for all x in the domains of both $f(x)$ and $g(x)$ for which $g(x) \neq 0$.

EXAMPLE 1. If $f(x) = 2x$ and $g(x) = x^2$, find each of the following and state any necessary restrictions.

    a. $s(x) = f(x) + g(x)$ **b.** $d(x) = f(x) - g(x)$

    c. $p(x) = f(x) \cdot g(x)$ **d.** $q(x) = \dfrac{f(x)}{g(x)}$

SOLUTION: **a.** $s(x) = f(x) + g(x) = 2x + x^2$
           b. $d(x) = f(x) - g(x) = 2x - x^2$
           c. $p(x) = f(x) \cdot g(x) = 2x \cdot x^2 = 2x^3$
           d. $q(x) = \dfrac{f(x)}{g(x)} = \dfrac{2x}{x^2} = \dfrac{2}{x}, \; x \neq 0$

Certain types of functions have been given special names. A function $f(x)$ is said to be **increasing on an interval I** if whenever $x_2 > x_1$ in that interval, $f(x_2) > f(x_1)$. That is, as the x-values increase in that interval, so do the function values. The graph at the left below illustrates an increasing function on the interval I. Notice that the graph of an increasing function *rises* as the x-values increase.

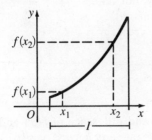 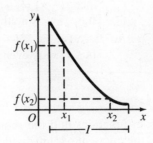

A function $f(x)$ is said to be **decreasing on an interval I** if whenever $x_2 > x_1$ in that interval, $f(x_2) < f(x_1)$. As the x-values increase in that interval, the function values decrease. The graph at the right above illustrates a decreasing function on the interval I. The graph of a decreasing function *falls* as the x-values increase.

A function $f(x)$ is called an **even function** if $f(-x) = f(x)$ for all x in the domain of $f(x)$. (It is assumed that $-x$ is in the domain whenever x is.) The graph in diagram (a) below is that of an even function. A function $f(x)$ is called an **odd function** if $f(-x) = -f(x)$ for all x in the domain of $f(x)$. The graph in diagram (b) below is that of an odd function.

(a)

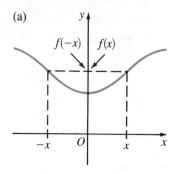

(b)

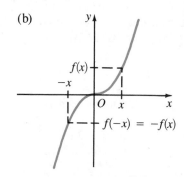

As the graph in diagram (a) illustrates, the graph of an *even* function is symmetric about the y-axis. The graph in diagram (b) illustrates that the graph of an *odd* function is symmetric about the *origin*.

EXAMPLE 2. Determine whether each function is even, odd, or neither.

    a. $f(x) = 2x^3 - 4x$
    b. $f(x) = x^2 - 3x + 1$
    c. $f(x) = 7 - 5x^2$

SOLUTION: For each function compare $f(-x)$ to $f(x)$ and to $-f(x)$.

    a. $f(-x) = 2(-x)^3 - 4(-x) = -2x^3 + 4x = -f(x)$
        Thus, $f(x)$ is odd.
    b. $f(-x) = (-x)^2 - 3(-x) + 1 = x^2 + 3x + 1$
        Since $f(-x) \neq f(x)$, $f(x)$ is not even.
        Since $f(-x) \neq -f(x)$, $f(x)$ is not odd.
        Thus, $f(x)$ is neither even nor odd.
    c. $f(-x) = 7 - 5(-x)^2 = 7 - 5x^2 = f(x)$
        Thus, $f(x)$ is even.

EXERCISES

For each pair of functions find the sum, product, and difference of $f(x)$ and $g(x)$, and find the quotient of $f(x)$ divided by $g(x)$. State any necessary restrictions.

1. $f(x) = x^2 - 1$; $g(x) = 2x + 3$

2. $f(x) = 2x + 5$; $g(x) = x^2$

3. $f(x) = \dfrac{x-1}{x}$; $g(x) = \dfrac{3}{x-3}$

4. $f(x) = \dfrac{x}{x+4}$; $g(x) = \dfrac{1}{x}$

5. $f(x) = 2x^3 - x + 5$; $g(x) = x^2 + x$

6. $f(x) = \sqrt{x+4}$; $g(x) = \sqrt{x+4}$

7. If $f(x) = x + 1$ and $g(x) = x^2$, sketch the graphs of $f(x)$, $g(x)$, and $f(x) + g(x)$.

State whether each graph is that of an increasing function, decreasing function, or neither over the interval I shown.

8.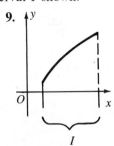

9.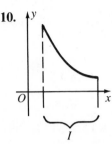

10.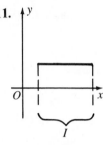

11.

In each exercise, the given points lie on the graph of a function. Tell whether the function is increasing, decreasing, or neither over the interval $-10 \le x \le 10$.

12. $(-4, 1)$, $(10, 9)$, $(-10, -3)$, and $(5, 6)$

13. $(3, 6)$, $(10, 16)$, $(-10, 4)$, and $(8, 16)$

14. $(2, 7)$, $(-10, -1)$, $(-4, 1)$, and $(10, 9)$

15. $(-8, 4)$, $(0, 0)$, $(10, -3)$, and $(-10, 5)$

State whether each function is even, odd, or neither.

16. $f(x) = x^2$ **17.** $f(x) = 3x - x^2$ **18.** $f(x) = 2x + x^3$

19. $f(x) = x^4 + 1$ **20.** $f(x) = -|x|$ **21.** $f(x) = 5x^3 + 3$

Appendix 4

The Wrapping Function

Suppose a vertical number line is tangent to the unit circle $x^2 + y^2 = 1$ at the point $R(1, 0)$ and has the same scale, or unit length, as the x- and y-axes. Let the origin of the number line be at R as shown. Imagine that the number line can be wrapped around and around the circle so that every real number t corresponds to a point $P(x, y)$ on the circle. For positive real numbers, the number line is wrapped in the counterclockwise direction; for negative real numbers, the number line is wrapped in the clockwise direction. The function that pairs with each real number a point on the circle is called the **wrapping function**. We will denote the wrapping function by W. Thus, in the diagram, $W(t) = P(x, y)$.

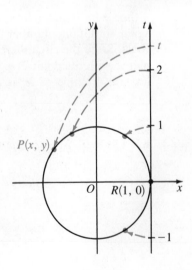

If θ is the measure of $\angle ROP$ (see diagram below), then according to the definitions in Chapter 6, the cosine of θ is the x-coordinate of P and the sine of θ is the y-coordinate of P. That is,

$$P(x, y) = (\cos \theta, \sin \theta).$$

The cosine and sine of the real number t such that $W(t) = P(x, y)$ are defined similarly:

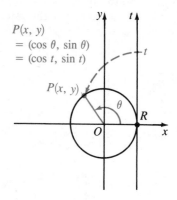

$$\cos t = x$$
$$\sin t = y$$

That is,

$$P(x, y) = (\cos t, \sin t).$$

Thus, we can find sines and cosines of real numbers as well as of angles.

EXAMPLE 1. Use the wrapping function to find $\cos \pi$ and $\sin \pi$.

SOLUTION: Since π is half the circumference of the unit circle, the wrapping function must wrap the segment of the number line from R to π exactly halfway around the unit circle. Thus,

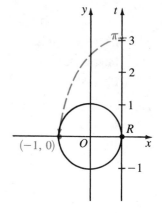

$$W(\pi) = (-1, 0).$$

Therefore, since $W(t) = P(x, y) = (\cos t, \sin t)$ for all real numbers t,

$$W(\pi) = (-1, 0) = (\cos \pi, \sin \pi).$$

Hence

$$\cos \pi = -1 \quad \text{and} \quad \sin \pi = 0.$$

Since the number line can be wrapped around the unit circle again and again, an infinite number of real numbers are paired with each point on the circle. Since the circumference of the unit circle is 2π, if the real number t is paired with the point $P(x, y)$, then any real number that differs from t by an integral multiple of 2π is also paired with $P(x, y)$. Thus, for any integer k,

$$W(t) = W(t + 2k\pi).$$

EXAMPLE 2. Use the wrapping function to find $\cos \dfrac{7\pi}{2}$ and $\sin \dfrac{7\pi}{2}$.

SOLUTION: Since $\dfrac{7\pi}{2} = \dfrac{3\pi}{2} + 2\pi$,

$$W\!\left(\frac{7\pi}{2}\right) = W\!\left(\frac{3\pi}{2} + 2\pi\right) = W\!\left(\frac{3\pi}{2}\right).$$

To determine $W\left(\dfrac{3\pi}{2}\right)$, notice that $\dfrac{3\pi}{2} = \dfrac{3}{4}(2\pi)$. Thus $\dfrac{3\pi}{2}$ is wrapped onto the point three quarters of the way around the unit circle. This point is $(0, -1)$. Therefore,

$$\cos \frac{7\pi}{2} = \cos \frac{3\pi}{2} = 0$$

and

$$\sin \frac{7\pi}{2} = \sin \frac{3\pi}{2} = -1.$$

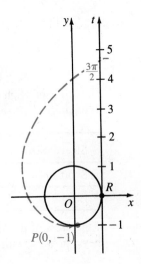

EXERCISES

1. a. What is the domain of the wrapping function W?
   b. What is its range?   **c.** What is its period?

Find each of the following.

2. $W(-\pi)$ **3.** $W\left(\dfrac{\pi}{2}\right)$ **4.** $W\left(-\dfrac{\pi}{2}\right)$ **5.** $W\left(-\dfrac{3\pi}{2}\right)$

6. $W(2\pi)$ **7.** $W(3\pi)$ **8.** $W(98\pi)$ **9.** $W(101\pi)$

Given each $W(t) = P(x, y)$, find $\cos t$ and $\sin t$.

10. $\left(\dfrac{\sqrt{2}}{2}, -\dfrac{\sqrt{2}}{2}\right)$ **11.** $\left(-\dfrac{3}{5}, \dfrac{4}{5}\right)$ **12.** $(0, -1)$ **13.** $\left(-\dfrac{\sqrt{3}}{2}, -\dfrac{1}{2}\right)$

Consider another wrapping function, denoted by $\mathcal{W}$, that wraps a vertical number line around a unit square as shown. Given a real number t, $\mathcal{W}(t)$ is a point $S(x, y)$ on the square. Define two functions, cas and san, as follows:

$$\text{cas } t = x\text{-coordinate of } S \text{ and}$$
$$\text{san } t = y\text{-coordinate of } S.$$

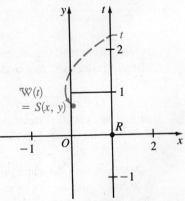

Find each of the following.

14. $\mathcal{W}(2)$ **15.** $\mathcal{W}(3)$ **16.** $\mathcal{W}(4)$

17. Find cas t and san t for $t = 0$, $\frac{1}{2}$, 1, and $\frac{3}{2}$.

18. What is the period of $\mathcal{W}(t)$?

Appendix 5

Matrices

The Jefferson High School Pep Club sold blue T-shirts and used the following table to organize their sales data.

Sleeve\Size	Small	Medium	Large
Short	1	5	3
Long	4	6	2

If the users of this data understand what each number in the table represents, then they can work with the array:

$$\begin{bmatrix} 1 & 5 & 3 \\ 4 & 6 & 2 \end{bmatrix}$$

This array, or *matrix*, has two rows representing sleeve length and three columns representing size. (We say this is a 2 × 3 (read "2 by 3") matrix.)

Definition of matrix (plural: matrices). A **matrix** is a rectangular array of numbers. Each entry in the array is called an **element** of the matrix. (We usually enclose the elements within brackets, double lines, or parentheses.) The **dimensions** of the matrix are determined by the number of rows and columns: rows × columns

The T-shirt matrix has dimensions 2 × 3. The dimensions of the following matrices are 2 × 1, 2 × 2, and 1 × 2, respectively.

$$\begin{bmatrix} 3 \\ 2 \end{bmatrix} \qquad \begin{bmatrix} 1 & -4 \\ 2 & 3 \end{bmatrix} \qquad [2 \quad 4]$$

Notation: We often use an upper-case letter (with or without subscripts) such as B or $B_{2\times3}$ to represent a matrix and lower-case letters with subscripts to represent elements of a matrix. In the general 2 × 3 matrix,

$$\begin{bmatrix} a_{11} & a_{12} & a_{13} \\ a_{21} & a_{22} & a_{23} \\ a_{31} & a_{32} & a_{33} \end{bmatrix},$$

a_{ij} represents the element in the ith row and jth column. For example, a_{13} is the element in the first row and third column.

Definition. If two matrices have the same dimensions, elements in the same row and column are called corresponding elements. Two matrices are **equal** if and only if they have the same dimensions and their corresponding elements are equal.

Matrices with the same dimensions can be added by finding the sums of corresponding elements. For example, let B represent the matrix of blue T-shirts and let the following matrix W represent the sale of white T-shirts in various sleeve lengths and sizes.

$$W = \begin{bmatrix} 3 & 4 & 2 \\ 1 & 2 & 4 \end{bmatrix}$$

Then $B + W = \begin{bmatrix} 1 & 5 & 3 \\ 4 & 6 & 2 \end{bmatrix} + \begin{bmatrix} 3 & 4 & 2 \\ 1 & 2 & 4 \end{bmatrix} = \begin{bmatrix} 4 & 9 & 5 \\ 5 & 8 & 6 \end{bmatrix}$.

In B and W, corresponding entries represent the same sleeve length and size. Therefore, each entry in $B + W$ represents the total number of blue and white T-shirts in that length and size sold.

Definition. If A and B are $m \times n$ matrices with elements a_{ij} and b_{ij} respectively, then $A + B$ is the $m \times n$ matrix with elements $a_{ij} + b_{ij}$ and $A - B$ is the $m \times n$ matrix with elements $a_{ij} - b_{ij}$.

EXAMPLE 1.
$$\begin{bmatrix} 2 & -3 \\ -1 & 4 \end{bmatrix} + \begin{bmatrix} -4 & 2 \\ -3 & 5 \end{bmatrix} = \begin{bmatrix} 2 + (-4) & -3 + 2 \\ -1 + (-3) & 4 + 5 \end{bmatrix}$$
$$= \begin{bmatrix} -2 & -1 \\ -4 & 9 \end{bmatrix}$$

EXAMPLE 2.
$$[8 \quad -2] - [-3 \quad 7] = [8 - (-3) \quad -2 - 7]$$
$$= [11 \quad -9]$$

Scalar multiplication for matrices is the multiplication of a matrix by a real number (the scalar). This product is found by multiplying each element of the matrix by the scalar. Using the pep club's matrix for blue T-shirts, if each person who bought a blue T-shirt doubled the order, the new matrix for sales would be the product of the scalar 2 and the matrix B.

$$2B = 2 \begin{bmatrix} 1 & 5 & 3 \\ 4 & 6 & 2 \end{bmatrix} = \begin{bmatrix} 2 & 10 & 6 \\ 8 & 12 & 4 \end{bmatrix}$$

Definition. For any scalar r, a real number, and $m \times n$ matrix A with elements a_{ij}, the scalar product of r and A, written rA, is the $m \times n$ matrix with elements ra_{ij}.

EXAMPLE 3.
$$-4 \begin{bmatrix} 5 \\ -2 \end{bmatrix} = \begin{bmatrix} (-4)(5) \\ (-4)(-2) \end{bmatrix} = \begin{bmatrix} -20 \\ 8 \end{bmatrix}$$

The product of two matrices requires a procedure different from those used to find sums and products by scalars. Referring once again to the matrix for blue T-shirts, suppose that short-sleeved shirts cost $6 each and that long-sleeved shirts cost $8 each. The following multiplication yields a matrix that represents the total costs for the three sizes of blue T-shirts.

$$[6 \quad 8] \begin{bmatrix} 1 & 5 & 3 \\ 4 & 6 & 2 \end{bmatrix} = [6 \cdot 1 + 8 \cdot 4 \quad 6 \cdot 5 + 8 \cdot 6 \quad 6 \cdot 3 + 8 \cdot 2] = [38 \quad 78 \quad 34]$$

The product matrix represents the following sales information: $38 for small T-shirts (1 short-sleeved and 4 long-sleeved), $78 for medium T-shirts, and $34 for large T-shirts.

Definition. The product of an $m \times n$ matrix A and an $n \times p$ matrix B is the $m \times p$ matrix C whose entry in row i and column j is given by

$$c_{ij} = a_{i1}b_{1j} + a_{i2}b_{2j} + \cdots + a_{in}b_{nj}.$$

The following examples illustrate how to find the product of two matrices.

EXAMPLE 4. Let $A = \begin{bmatrix} 2 & 3 \\ 5 & 4 \end{bmatrix}$ and $B = \begin{bmatrix} 7 & 1 \\ 8 & 6 \end{bmatrix}$. Find AB and BA.

SOLUTION: The product C of A and B is also a 2×2 matrix. To find c_{11}, multiply each element in the first row of A by each element in the first column of B.

$$\begin{bmatrix} 2 & 3 \\ 5 & 4 \end{bmatrix}\begin{bmatrix} 7 & 1 \\ 8 & 6 \end{bmatrix} = \begin{bmatrix} 2\cdot 7 + 3\cdot 8 & \\ & \end{bmatrix} = \begin{bmatrix} 38 & \\ & \end{bmatrix}$$

To find c_{12}, use the first row of A and the second column of B.

$$\begin{bmatrix} 2 & 3 \\ 5 & 4 \end{bmatrix}\begin{bmatrix} 7 & 1 \\ 8 & 6 \end{bmatrix} = \begin{bmatrix} & 2\cdot 1 + 3\cdot 6 \\ & \end{bmatrix} = \begin{bmatrix} & 20 \\ & \end{bmatrix}$$

The completed multiplication is shown below.

$$AB = \begin{bmatrix} 2\cdot 7 + 3\cdot 8 & 2\cdot 1 + 3\cdot 6 \\ 5\cdot 7 + 4\cdot 8 & 5\cdot 1 + 4\cdot 6 \end{bmatrix} = \begin{bmatrix} 38 & 20 \\ 67 & 29 \end{bmatrix}$$

Proceed the same way to find BA which is a 2×2 matrix.

$$BA = \begin{bmatrix} 7\cdot 2 + 1\cdot 5 & 7\cdot 3 + 1\cdot 4 \\ 8\cdot 2 + 6\cdot 5 & 8\cdot 3 + 6\cdot 4 \end{bmatrix} = \begin{bmatrix} 19 & 25 \\ 46 & 48 \end{bmatrix}$$

Notice that $AB \neq BA$. In general, matrix multiplication is NOT commutative.

EXAMPLE 5. Find each product, if possible.

a. $[1 \quad -3 \quad 4]\begin{bmatrix} 2 & 4 \\ 0 & 6 \\ -1 & 5 \end{bmatrix}$ **b.** $\begin{bmatrix} 2 & 4 \\ 0 & 2 \\ -1 & 5 \end{bmatrix}[1 \quad -3 \quad 4]$

SOLUTION: **a.** The product is defined and has dimensions 1×2.

$$[1 \quad -3 \quad 4]\begin{bmatrix} 2 & 4 \\ 0 & 6 \\ -1 & 5 \end{bmatrix} = [-2 \quad 6]$$

b. The product of a 3×2 matrix and a 1×3 matrix is not defined because the number of columns in the left matrix does not equal the number of rows in the right matrix.

EXERCISES

Exercises 1–20: $A = \begin{bmatrix} 3 & -4 \\ 2 & 6 \end{bmatrix}$, $B = \begin{bmatrix} 5 & 0 \\ -1 & -2 \end{bmatrix}$, $C = \begin{bmatrix} 1 & -2 & 4 \\ 2 & 0 & -3 \end{bmatrix}$,

$D = \begin{bmatrix} -2 & 3 \\ 0 & 1 \\ 4 & 2 \end{bmatrix}$, $E = [2 \quad 3]$, and $F = \begin{bmatrix} 4 \\ -1 \end{bmatrix}$.

1. **a.** State the dimensions of D.
   b. Which element in D would be represented by d_{12}?

2. If $[x + 3 \quad 2y - 5] = E$, find x and y.

In Exercises 3–17, find each matrix, if possible.

3. $A + B$	4. $5C$	5. $3A - 2B$	6. $C + D$	7. $-E$
8. AB	9. BA	10. CD	11. BD	12. DF
13. EF	14. FE	15. AA	16. $BC + 2C$	17. $3A - F$

In Exercises 18–20, find each matrix X.

18. $A + X = B$	19. $2X = D$	20. $3X - B = A$

The Inverse of a Matrix

The real number 1 is the multiplicative identity for the set of real numbers because multiplying any real number a by 1 does not change its value: $a \times 1 = a$. Similarly, the matrix $I = \begin{bmatrix} 1 & 0 \\ 0 & 1 \end{bmatrix}$ is the **multiplicative identity** for the set of 2×2 matrices:

$$\begin{bmatrix} a & b \\ c & d \end{bmatrix} \begin{bmatrix} 1 & 0 \\ 0 & 1 \end{bmatrix} = \begin{bmatrix} a & b \\ c & d \end{bmatrix}.$$

Two real numbers, such as 3 and $\frac{1}{3}$ are multiplicative inverses of one another because $3 \cdot \frac{1}{3} = 1$. Similarly, two matrices A and B such that $AB = BA = I$ are called multiplicative inverses, or simply **inverses,** of one another. For example, $\begin{bmatrix} 8 & 5 \\ 3 & 2 \end{bmatrix}$ and $\begin{bmatrix} 2 & -5 \\ -3 & 8 \end{bmatrix}$ are inverses since

$$\begin{bmatrix} 8 & 5 \\ 3 & 2 \end{bmatrix} \begin{bmatrix} 2 & -5 \\ -3 & 8 \end{bmatrix} = \begin{bmatrix} 2 & -5 \\ -3 & 8 \end{bmatrix} \begin{bmatrix} 8 & 5 \\ 3 & 2 \end{bmatrix} = \begin{bmatrix} 1 & 0 \\ 0 & 1 \end{bmatrix} = I.$$

In Exercise 13, you are asked to show that: If $A = \begin{bmatrix} a & b \\ c & d \end{bmatrix}$ and if $ad - bc \neq 0$, then the inverse of A, denoted A^{-1}, exists and is given by

$$A^{-1} = \frac{1}{D} \begin{bmatrix} d & -b \\ -c & a \end{bmatrix} \text{ where } D = ad - bc \neq 0.$$

The quantity $D = ad - bc$ is called the **determinant of A** and is denoted det A.

EXAMPLE 1. Find the inverse of $A = \begin{bmatrix} -2 & -5 \\ 3 & 6 \end{bmatrix}$, if it exists.

SOLUTION: Since det $A = (-2)(6) - (-5)(3) \neq 0$, A^{-1} exists.

$$A^{-1} = \frac{1}{3} \begin{bmatrix} 6 & 5 \\ -3 & -2 \end{bmatrix} = \begin{bmatrix} 2 & \frac{5}{3} \\ -1 & -\frac{2}{3} \end{bmatrix}$$

Check: $\begin{bmatrix} 2 & \frac{5}{3} \\ -1 & -\frac{2}{3} \end{bmatrix} \begin{bmatrix} -2 & -5 \\ 3 & 6 \end{bmatrix} = \begin{bmatrix} -2 & -5 \\ 3 & 6 \end{bmatrix} \begin{bmatrix} 2 & \frac{5}{3} \\ -1 & -\frac{2}{3} \end{bmatrix} = \begin{bmatrix} 1 & 0 \\ 0 & 1 \end{bmatrix}$

EXAMPLE 2. Solve $\begin{bmatrix} -2 & -5 \\ 3 & 6 \end{bmatrix} X = \begin{bmatrix} 4 & -1 \\ -9 & 3 \end{bmatrix}$ for the 2 × 2 matrix X.

SOLUTION: Method: Think of the given matrix equation as $AX = B$. To solve for X, multiply both sides of the equation (on the left) by A^{-1}. Then, we get $A^{-1}AX = A^{-1}B$ and $X = A^{-1}B$. The matrix A in this example is the matrix given in Example 1. Hence,

$$X = \begin{bmatrix} 2 & \frac{5}{3} \\ -1 & -\frac{2}{3} \end{bmatrix} \begin{bmatrix} 4 & -1 \\ -9 & 3 \end{bmatrix} = \begin{bmatrix} -7 & 3 \\ 2 & -1 \end{bmatrix}$$

Note: Since matrix multiplication is, in general, not commutative, the product BA^{-1} would not give X.

The inverse of an $n \times n$ matrix A where $n > 2$ is more difficult to calculate than the inverse of a 2 × 2 matrix. However, A^{-1} exists whenever det $A \ne 0$. For a discussion of determinants see page 438.

In 1750, Gabriel Cramer published his method for solving linear systems using determinants. Arthur Cayley is usually given credit for the invention of matrices in 1857. He used them for linear transformations such as rotations, dilatations, and reflections about a coordinate axis. Today, matrices and determinants are used in many fields, including computer science, engineering, genetics, statistics, and various social sciences.

EXERCISES

Find the inverse of each matrix A, if it exists.

1. $\begin{bmatrix} 9 & -7 \\ 5 & -4 \end{bmatrix}$ **2.** $\begin{bmatrix} 2 & -4 \\ -5 & 11 \end{bmatrix}$ **3.** $\begin{bmatrix} -8 & 2 \\ 16 & -4 \end{bmatrix}$ **4.** $\begin{bmatrix} -2 & 1 \\ 1 & 1 \end{bmatrix}$

Solve each matrix equation for X.

5. $\begin{bmatrix} 4 & 9 \\ 3 & 7 \end{bmatrix} X = \begin{bmatrix} 3 & -1 \\ 2 & 0 \end{bmatrix}$ **6.** $\begin{bmatrix} -4 & 7 \\ 2 & -3 \end{bmatrix} X = \begin{bmatrix} -2 & 5 \\ 1 & 3 \end{bmatrix}$

7. $\begin{bmatrix} 2 & -3 \\ 1 & 1 \end{bmatrix} X - \begin{bmatrix} 4 & -1 \\ -3 & 7 \end{bmatrix} = \begin{bmatrix} -2 & 2 \\ 6 & -5 \end{bmatrix}$

8. $\begin{bmatrix} 2 & 3 \\ 4 & 4 \end{bmatrix} X + \begin{bmatrix} -1 & 5 \\ 3 & -2 \end{bmatrix} = \begin{bmatrix} 2 & 1 \\ -2 & 4 \end{bmatrix}$

In Exercises 9–11, solve each matrix equation for X. Assume that each matrix is $n \times n$ and that no determinant has value 0.

9. $BX = C$ **10.** $AX = BA$ **11.** $3X - B = X$

12. Verify that $\begin{bmatrix} 1 & 1 & 3 \\ -2 & 5 & 4 \\ 3 & -6 & -4 \end{bmatrix}$ and $\begin{bmatrix} -4 & 14 & 11 \\ -4 & 13 & 10 \\ 3 & -9 & -7 \end{bmatrix}$ are inverses.

13. Follow these steps to derive the formula for A^{-1}.

a. If $A = \begin{bmatrix} a & b \\ c & d \end{bmatrix}$ and $A^{-1} = \begin{bmatrix} w & x \\ y & z \end{bmatrix}$,

then $I = \begin{bmatrix} a & b \\ c & d \end{bmatrix}\begin{bmatrix} w & x \\ y & z \end{bmatrix} = \begin{bmatrix} ? & ? \\ ? & ? \end{bmatrix}$.

b. Using your answer to part (a), you can write four equations in the unknowns w, x, y, and z. Two of these equations are

$$aw + by = 1 \quad \text{and} \quad cw + dy = 0.$$

What are the other two equations?

c. Solve the equations given in part (b) to show that $w = \dfrac{d}{D}$ and that $y = -\dfrac{c}{D}$

where $D = ad - bc$.

d. Using the two equations you supplied in part (b), find expressions for x and z.

14. A point (x, y) can be rotated about the origin through an angle θ to its image (u, v) by using the matrix multiplication:

$$\begin{bmatrix} \cos\theta & -\sin\theta \\ \sin\theta & \cos\theta \end{bmatrix}\begin{bmatrix} x \\ y \end{bmatrix} = \begin{bmatrix} u \\ v \end{bmatrix}$$

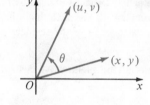

a. If $\theta = 90°$, the rotation matrix is $\begin{bmatrix} 0 & -1 \\ 1 & 0 \end{bmatrix}$.

Find the image of $(2, 5)$ under a $90°$ rotation.

b. Find the rotation matrix for a rotation of $-90°$.

What is the relationship between the rotation matrices for $90°$ and $-90°$?

c. Find the image of $(-4, 6)$ under a rotation of $30°$. Give an exact answer and an approximation to the nearest hundredth.

15. According to Exercise 14, a rotation about the origin through an angle α can be represented by the matrix $A = \begin{bmatrix} \cos\alpha & -\sin\alpha \\ \sin\alpha & \cos\alpha \end{bmatrix}$.

a. A rotation about the origin through an angle β can be represented by the matrix $B = ?$.

b. Find and simplify the four entries of AB.

c. What rotation does the matrix product represent?

The transpose of a matrix A, denoted A^T, is found by making each element a_{ij} of A the element a_{ji} of A^T. Find the transpose of each matrix in Exercises 16–18.

16. $\begin{bmatrix} 1 & 2 \\ 3 & 4 \end{bmatrix}$ **17.** $\begin{bmatrix} 5 & 4 & -9 \\ -2 & 3 & 6 \end{bmatrix}$ **18.** $\begin{bmatrix} 3 & -2 \\ -1 & 5 \\ 2 & 4 \end{bmatrix}$

Glossary

Absolute value $|x|$ The absolute value of a number x, interpreted as its distance from zero on the number line. (p. 95)

Absolute value of a vector (also called magnitude or norm)   Equals length of arrow representing $\mathbf{v}$. (p. 401)

Acceleration   Rate of change of velocity over time. (p. 594)

Amplitude of a periodic function   One half the difference between the maximum and minimum values of the function. (p. 274)

Angle of depression   Angle by which observer's line of sight must be depressed from horizontal to point observed. (p. 242)

Angle of elevation   Angle by which observer's line of sight must be elevated from horizontal to point observed. (p. 241)

Antilogarithm of x The number N such that $\log N = x$. (p. 166)

Apparent size   Measure of angle subtended at one's eye by object being observed. (p. 196)

Argand diagrams   Diagrams that represent complex numbers geometrically. (p. 340)

Asymptote   Name given to a line that a curve approaches more and more closely until the distance between the curve and line almost vanishes. (p. 294)

Average   A single number used to represent an entire set of data. (p. 496)

Average velocity   Distance divided by time. (p. 592)

Axis of symmetry   A line such that it is possible to pair the points of a graph so that the line is the perpendicular bisector of the segment joining each pair. (p. 281)

Bracket-and-halving method   A method of approximating roots of polynomial equations. (p. 76)

Cofunctions   Sine, cosine; tangent, cotangent; secant, cosecant. (p. 219)

Combination   An arrangement of a set of objects in which order is not important. (p. 538)

Common logarithms   Logarithms to base 10. (p. 161)

Complex number   Any number of the form $a + bi$ where a and b are real numbers. (p. 21)

Complex number plane   The plane in which points represent complex numbers. (p. 340)

Complex number system   Consists of the real numbers and the imaginary numbers. (p. 21)

Composite of two functions   A new function produced by combining the given functions. (p. 127)

Composition   The operation that combines two functions. (p. 128)

Conditional probability   The situation in which the probability of one event is determined under the condition that another event has occurred. (p. 546)

Conic sections   Cross sections resulting from slicing of a double cone by a plane. *Degenerate* conic section is the name used when the result is a single point, a line, or a pair of lines. (p. 360)

Cosecant $\csc \theta = \dfrac{1}{\sin \theta}$, $\sin \theta \neq 0$. (p. 212)

Cotangent $\cot \theta = \dfrac{1}{\tan \theta}$, $\tan \theta \neq 0$. (p. 212)

Constant   A polynomial of degree 0. (p. 48)

Continuous function $f(x)$ is continuous at c if $\lim\limits_{x \to c} f(x) = f(c)$. (p. 570)

Coordinates   An ordered pair of numbers associated with a point in the plane. (p. 1)

Correlation coefficient   A number between -1 and 1 that measures how closely points plotted tend to cluster about a line displaying a relationship between two variables. (p. 526)

Cosine $\cos\theta = \dfrac{x}{r}$ given $P(x, y)$ a point on circle $x^2 + y^2 = r^2$ and θ an angle in standard position with terminal ray OP. (p. 200)

Coterminal angles   Two angles in standard position having the same terminal ray. (p. 192)

Cross product   The operation $\mathbf{v}_1 \times \mathbf{v}_2$ (read $\mathbf{v}_1$ cross $\mathbf{v}_2$). (p. 446)

Cubic   A polynomial of degree 3. (p. 48)

Degree   Unit for measuring angles. (p. 190)

Degree of a polynomial in x   The highest power of x that has a nonzero coefficient. (p. 47)

Dependent events   Two events that are not independent. (p. 546)

Derivative of $f(x)$ Slope function $f'(x)$. (p. 580)

Determinant   A square array of numbers useful in solving systems of equations. (p. 438)

Directrix of parabola   A fixed line such that all points in a parabola are equidistant from that line and a fixed point (focus). (p. 377)

Discriminant   The quantity $b^2 - 4ac$ that appears beneath the radical sign in the quadratic formula. (p. 26)

Displacement   If an object moves from A to B, its displacement is $\mathbf{v} = \overrightarrow{AB}$. (p. 401)

Domain of a function   The set of real numbers for which the function is defined. (p. 118)

Dot product $\mathbf{v}_1 \cdot \mathbf{v}_2 = x_1x_2 + y_1y_2$. (p. 422)

Doubling time   Time required for a quantity growing exponentially to double. (p. 173)

e   The irrational number which is the limit, as n approaches infinity, of $\left(1+\dfrac{1}{n}\right)^n$. (p. 179). It is approximately $2.718\ldots$ (p. 179)

Eccentricity e   For ellipse and hyperbola, $e = \dfrac{\text{distance from center to focus}}{\text{distance from center to vertex}}$. (pp. 363, 369)

   For parabola, $e = 1$. (p. 386)

   For circle, $e = 0$. (p. 363)

Ellipse   (1) The set of all points P in the plane such that $PF_1 + PF_2 = 2a$, where F_1 and F_2 are fixed points and a is a positive constant.

   (2) $\dfrac{x^2}{a^2} + \dfrac{y^2}{b^2} = 1$ where center is at origin and the major and minor axes are horizontal and vertical. (p. 361)

Event   Any subset of a sample space. (p. 546)

Exponential equation   An equation in which the variable occurs as an exponent. (p. 169)

Exponential function with base b   Any function of the form $f(x) = ab^x$, $a > 0$, $b > 0$, $b \neq 1$. (p. 157)

Extreme values   Maximum and minimum values of a function. (p. 132)

Factor Theorem $x - a$ is a factor of the polynomial $P(x)$ if and only if $P(a) = 0$. (p. 169)

Frequency polygon   Formed by connecting midpoints of top of each bar in a histogram. (p. 496)

Frequency table   Table that organizes data. (p. 496)

Function   (1) A correspondence or rule that assigns to each value of x a unique number. (p. 48)
   (2) Consists of a set of real numbers, called the domain of the function, and a rule that assigns to each element in the domain exactly one real number. (p. 118)

Geometric sequence   A sequence of numbers having a constant ratio for any two consecutive terms. (p. 454)

Graph of an equation   The set of all points in the plane corresponding to ordered-pair solutions. (p. 2)

Half-life   Period during which half of a quantity decaying exponentially decays. (p. 174)

Histogram   Bar graph used to display a set of data. (p. 496)

Hyperbola   The set of all points P for which $|PF_1 - PF_2| = 2a$, where F_1 and F_2 are the foci and $0 < a$. (p. 369)

Identity function   Its rule assigns each number to itself. (p. 139)

Imaginary number   The square root of a negative number (p. 21). A complex number for which $b \neq 0$. (p. 21)

Imaginary unit i   The number $\sqrt{-1}$ with property that $i^2 = -1$ (p. 21)

Inclination of a line   Measure of the angle from the positive direction of the x-axis to the line. (p. 215)

Independent events   The occurrence of an event A does not affect the probability of an event B's occurrence. (p. 546)

Infinite series   The sum of an infinite sequence. (p. 460)

Instantaneous velocity   Derivative of position function with respect to time. (p. 593)

Inverse functions $f(x)$ and $g(x)$ provided $g(f(x)) = x$ for all x in the domain of $f(x)$, $f(g(x)) = x$ for all x in the domain of $g(x)$. (p. 139)

Law of Cosines   In a $\triangle ABC$, $c^2 = a^2 + b^2 - 2ab \cos C$. (p. 253)

Law of Sines   The sines of the angles of a triangle are proportional to the lengths of the opposite sides. (p. 248)

Leading coefficient   The coefficient of the term with the highest power of x in a polynomial in x. (p. 47)

Limit   See p. 572 for formal definition.

Line of symmetry   For any two points, the line of symmetry is the perpendicular bisector of the segment joining the points. (p. 140)

Linear equation   Any equation of the form $ax + by = c$ where a and b are not both zero. (p. 2)

Linear function   A function that can be defined by a linear polynomial. (p. 48)

Linear polynomial   A polynomial of degree 1. (p. 48)

Linear–interpolation method   A method of approximating roots of polynomial equations by replacing part of the graph of the polynomial with a line segment. (p. 77)

Linear programming   The branch of mathematics that can be used to determine maximum profits, for example. (p. 105)

Logarithm to base b of a positive number N   The exponent k such that $b^k = N$, given $b > 0$, $b \neq 1$. (p. 160)

Logarithmic function with base b   Inverse of exponential function with base b. (p. 160)

Mathematical induction   A method of proof used to show that a statement is true for all positive integers. (pp. 487, 492)

Mean ($\bar{x}$)   A statistical average found by dividing the sum of a set of measurements by the number of measurements. (p. 497)

Median   A statistical average found by arranging the data in increasing or decreasing order. It will be either the middle number or the mean of the two middle ones. (p. 498)

Mode   A statistical average that is the one measurement occurring most often. (p. 498)

Natural logarithm of x $\text{Log}_e x$, or $\ln x$. (p. 180)

Normal distributions   Distributions of data along a bell-shaped curve. (p. 508)

One-to-one functions   Functions that have inverses. (p. 141)

Ordered pair   A pair of numbers, as $(4, -2)$, in which the first number, 4, represents the x-coordinate of a point, and the second number, -2, represents the y-coordinate. (p. 2)

Origin   Point of intersection of x- and y-axes. (p. 2)

Parabola The graph of a quadratic function. (p. 56)

Pascal's Triangle An array of exponents in the expansion of $(a + b)^n$. (p. 555)

Percentiles The points that divide a set of data into 100 equal parts, the data being arranged in ascending order. (p. 512)

Periodic function $f(x)$ if it has domain D and there is a number $p > 0$ such that $f(x + p) = f(x)$ for all x in D. If p is the smallest such positive number, it is called the period of the function. (p. 274)

Permutation An arrangement of a set of objects in which order is important. (p. 538)

Point of symmetry A point of a graph for which it is possible to pair the points of the graph in such a way that the point is the midpoint of the segment joining each pair. (p. 282)

Polar axis Name given to reference ray in polar coordinate system; usually coincides with x-axis. (p. 333)

Polar coordinates Used to describe a point $P(r; \theta)$ where r is the distance from the origin (the *polar distance*) and θ is the angle measured from the polar axis (the *polar angle*) to ray OP. (p. 334)

Polar equation Equation of a curve given in terms of r and θ (p. 334)

Pole Name given to origin in polar coordinates. (p. 333)

Polynomial in x An expression of the form $a_n x^n + a_{n-1} x^{n-1} + \cdots + a_2 x^2 + a_1 x + a_0$ where n is a nonnegative integer. (p. 47) The numbers $a_n, a_{n-1}, \ldots a_1, a_0$ are called the coefficients of the polynomial. (p. 47)

Polynomial equation An equation that can be written in the form $P(x) = 0$, where $P(x)$ is a polynomial. (p. 51)

Polynomial inequalities $P(x) < 0$ and $P(x) > 0$ where $P(x)$ is a polynomial. (p. 98)

Population In statistics, the entire set of objects being studied. (p. 515)

Power series Series of the form $\Sigma a_n x^n$. (p. 483)

Probability of an event $P(A) = \dfrac{m}{n}$, where m of n outcomes of an experiment correspond to some event A. (p. 546)

Pure imaginary numbers Imaginary numbers for which $a = 0$, such as $3i$, $-i$, and $i \sqrt{7}$. (p. 22)

Quadrants The coordinate plane is divided into four quadrants by the x- and y- axes. (p. 2)

Quadrantal angle The angle whose terminal ray lies on either axis when the angle is in standard position. (p. 192)

Quadratic A polynomial of degree 2. (p. 48)

Quadratic equation Any equation that can be written in the form $ax^2 + bx + c = 0$, where $a \neq 0$. (p. 24)

Quartic A polynomial of degree 4. (p. 48)

Quintic A polynomial of degree 5. (p. 48)

Radian Unit of angle measurement. (p. 190)

Range of a function The set of real numbers assigned by the function rule to the elements in the domain. (p. 118)

Rational numbers Numbers that are the ratio of two integers. (p. 20)

Real number system Consists of zero, all positive and negative integers, rational numbers, and irrational numbers. (p. 21)

Reference angle The acute positive angle that the terminal ray of an angle makes with the positive or negative part of the x-axis. (p. 206)

Remainder Theorem When a polynomial $P(x)$ is divided by $x - a$, the remainder $R = P(a)$. (p. 68)

Resultant Vector sum of two vectors. (p. 400)

Revolution Unit for measuring angles. (p. 189)

Sample In statistics, a subset of the population. (p. 515)

Sample space The set of all equally likely outcomes from an experiment. (p. 544)

Scalars Term used to refer to real numbers when working with vectors. (p. 403)

Secant $\sec\theta = \dfrac{1}{\cos\theta}$, $\cos\theta \neq 0$. (p. 212)

Sector of a circle Formed by two radii and their included arc. (p. 195)

Sequence A function whose domain is the set of positive integers. (p. 456)

Sign graph A sketch that shows when the value of a polynomial function is positive, negative, or zero. (p. 99)

Sine $\sin\theta = \dfrac{y}{r}$ given $P(x, y)$ a point on circle $x^2 + y^2 = r^2$ and θ an angle in standard position with terminal ray OP. (p. 200)

Slope function, $f'(x)$ Derivative of $f(x)$. (p. 580)

Slope of curve (at a point) The value of the derivative at the point. (p. 578)

Slope of a line Tangent of inclination of the line. (p. 216)

Slope of a nonvertical line A number measuring the steepness of the line relative to the x-axis. (p. 8)

Standard deviation The positive square root of the variance. (p. 503)

Standard position Position of angle whose vertex is at the origin and whose initial ray is the positive x-axis. (p. 192)

Synthetic division A short-cut method of dividing a polynomial by a divisor of the form $x - a$. (p. 67)

Tangent $\tan\theta = \dfrac{\sin\theta}{\cos\theta}$, $\cos\theta \neq 0$. (p. 212)

Tangent to a curve at a point on the curve The line that passes through the point and that has a slope the same as the slope of the curve at that point. (p. 581)

Trigonometric equations Equations that are true for some but not all values of an angle. (p. 226)

Trigonometric identities Equations that are true for all values of an angle, except those values for which either side of the equation is undefined. (p. 220)

Unit circle Circle $x^2 + y^2 = 1$. (p. 202)

Variance The mean of the squares of the deviations of data away from the mean. It describes the dispersion of data away from the mean. (p. 503)

Vector (1) A quantity described by both direction and magnitude (p. 399) (2) A quantity named by an ordered pair of numbers. (p. 408)

x-axis and y-axis Perpendicular lines, usually horizontal and vertical respectively, used in setting up a coordinate system. (p. 2)

x-intercept The x-coordinate of a point where a graph intersects the x-axis. (p. 2)

y-intercept The y-coordinate of a point where a graph intersects the y-axis. (p. 2)

Zero of a function A value of the variable that causes the value of a function to be zero. (p. 49)

Zero vector Displacement of an object is 0 when initial and final positions are the same. (p. 402)

Index

674   Index

ACKNOWLEDGMENTS

Mechanical art by ANCO of Boston
Portrait art by Rob Bolster

Photographs:

Page xii	Jack and Betty Cheetham, MAGNUM PHOTOS
Page 46	© Sea World of Florida
Page 92	courtesy of The New Chrysler Corporation
Page 116	Fred Ward, BLACK STAR
Page 146	"Departure of the Warrior," Greek vase from the Attic Period, circa 500 B.C. Gift of T. J. Appleton, Museum of Fine Arts, Boston
Page 188	© Peter Menzel, STOCK BOSTON
Page 198	PHOTO RESEARCHERS, INC.
Page 236	Klaus D. Francke, PETER ARNOLD, INC.
Page 272	© Gordon S. Smith, PHOTO RESEARCHERS, INC.
Page 308	Millgard Photographers, courtesy of the Selmer Company, Elkhart, Indiana
Page 332	© Douglas Faulkner
Page 358	NASA
Page 389	Herb Weitman, NFL Properties, Inc.
Page 398	"Linear Construction Number 2," 1949, by Naum Gabo, Hirshhorn Museum and Sculpture Garden, Smithsonian Institution
Page 452	courtesy of the Japanese National Railways
Page 494	Tony Castelvecchi, BLACK STAR
Page 534	© Peter Menzel, STOCK BOSTON
Page 566	NASA

Answers to selected exercises

Oral Exercises 1-1, page 5 1. a. 10 **b.** (4, 3)
2. a. $2\sqrt{5}$ **b.** (5, 4) **3. a.** $6\sqrt{2}$ **b.** (0, 1) **4. a.** 8
b. (7, −5) **5.** (a), (b), (d) **6.** $(7\frac{1}{2}, 0)$, (0, 5) **7.** (2, 3)
Answers may vary for Exercises 8 and 9.
8. a. Examples: (0, 4), (2, 4), (4, 4) **b.** $y = 4$
9. a. Examples: (8, 0), (8, 2), (8, 4) **b.** $x = 8$

Written Exercises 1-1, pages 6-8 1. a. 10 **b.** (4, 4)
3. a. 17 **b.** $(-\frac{1}{2}, 1)$ **5. a.** 6.5 **b.** $(-\frac{3}{4}, \frac{3}{2})$ **7. a.** 5
b. (4.8, −0.3) **9. a.** a, c **11.** Area of $\triangle OPQ = 3$
13. (5, 3); $y = 3$, $x = 5$ **15.** (3, 0) **17.** $(-\frac{5}{4}, -\frac{7}{4})$
19. parallelogram **21.** parallelogram **23. b.** 2:1
25. a. $PA = PB = 5$ **b.** −2 **27.** (−3, 0), (9, 0)
29. (0, 2), (0, 10) **31. a.** $\angle C = 90°$ **b.** (0, 5) **33.** No
35. $A(2, -1)$, $B(12, 7)$, $C(8, 11)$

Calculator Exercise, page 8 316.5

Oral Exercises 1-2, page 12 1. $\frac{7}{5}$ **2.** 6 **3.** −1 **4.** $\frac{1}{2}$
5. 0 **6.** No slope **7.** 3; 4 **8.** $\frac{3}{5}$; −3 **9.** $-\frac{4}{3}$; 3
10. 0; −2 **11. a.** $\frac{4}{5}$ **b.** $-\frac{5}{4}$ **12. a.** $-\frac{3}{2}$ **b.** $\frac{2}{3}$
13. a. Slope of $\overline{AB}$ = slope of $\overline{CD}$ = $\frac{4}{3}$; slope of
$\overline{AD}$ = slope of $\overline{BC}$ = $-\frac{3}{4}$. By Thm. 2b, opposite sides
are parallel. **b.** Using the values from part (a) and
Thm. 3b, adjacent sides are perpendicular. **14.** The
change in x-values is zero, so definition of slope would
involve division by zero.

Written Exercises 1-2, pages 12-14 1. $\frac{3}{5}$ **3.** $-\frac{2}{3}$ **5.** 0
7. −2 **9.** $\frac{a - b}{b - a} = -1$ **11. a.** $\frac{2}{7}$ **b.** $-\frac{7}{2}$ **13.** 3; 5
15. $\frac{3}{2}$; $-\frac{1}{2}$ **17.** $-\frac{1}{2}$; $\frac{7}{6}$ **19.** 0; −3 **21. a.** $\frac{3}{4}$; $-\frac{7}{3}$
23. Given lines have slopes 1 and −1 and thus are
perpendicular. **25. a.** 0 **b.** $8\frac{1}{2}$ **27.** (a) and (b) are
parallel, (c) is perpendicular to both. **29. a.** Slope of
$\overline{AB}$ and $\overline{CD} = \frac{5}{3}$; slope of $\overline{AD}$ and $\overline{BC} = \frac{1}{3}$ **b.** Both
have midpoint (2, 0). **31.** (−4, −5) and (5, 1); (−3, 0)
and (0, 2) **35. b.** x is undefined; the lines coincide or
do not intersect. **41.** $C(-2, 14)$, $D(-8, 6)$; $C(14, 2)$,
$D(8, -6)$

Calculator Exercise, page 14 a. Area = 6.37

Oral Exercises 1-3, page 17 1. $y = 3x + 7$
2. $y = \frac{5}{3}x - 2$ **3.** $\frac{y - 6}{x + 4} = 3$ **4. a.** $\frac{y - 2}{x - 1} = 4$
b. $\frac{y - 2}{x - 1} = -\frac{1}{4}$ **5.** Find the slope of $\overline{AB}$ and the
coordinates of M. Use the point-slope form.

Written Exercises 1-3, pages 17-20 1. $y = -2x + 8$
3. $y = 2x + 4$ **5.** $\frac{y - 7}{x - 1} = 4$ **7.** $\frac{y - 8}{x - 2} = \frac{1}{2}$

9. $\frac{y - 4}{x + 1} = \frac{2}{3}$ or $\frac{y - 8}{x - 5} = \frac{2}{3}$ **11.** $y = -7$ **13.** $x = 0$
15. $x = 2$ **17.** $\frac{y - 3}{x - 4} = -\frac{5}{4}$ **19.** $\frac{y + 2}{x - 8} = \frac{1}{2}$
21. $\frac{y - 4}{x + 2} = 2$ **23.** $y = x - 6$ **27. a, b.** $\frac{y - 9}{x - 9} = \frac{3}{4}$
c. Yes; yes **29. a.** $y = x + 1$, $y = -2x + 7$, $y = 3$
b. $G(2, 3)$ **31. a.** $y = x$, $x = 6$, $y = -\frac{1}{2}x + 9$
b. $O(6, 6)$ **33. a.** $G(5, \frac{8}{3})$ **b.** $C(7, 0)$ **c.** $O(1, 8)$
d. Slope $\overrightarrow{OG} = -\frac{4}{3}$, slope $\overline{GC} = -\frac{4}{3}$; $\overline{GC}$ is on $\overrightarrow{OG}$.
35. −3 **37. a.** $5\sqrt{10}$; $y = 3x + 13$ **b.** $y = -\frac{1}{3}x - \frac{1}{3}$
c. (−4, 1) **d.** $2\sqrt{10}$ **e.** 50 **39. a.** $\frac{y - y_1}{x - x_1} = \frac{b}{a}$ **d.** 5
41. b. $x + y = -2$; $7x - 7y = -10$

Oral Exercises 1-4, page 22 1. $7 + 7i$ **2.** $-1 - 5i$
3. $2 + 4i$ **4.** $28 - 6i$ **5.** 17 **6.** 34 **7.** 3 **8.** $a^2 + b^2$
9. $x = 2$, $y = -5$

Written Exercises 1-4, pages 23-24 1. $7i$ **3.** −3
5. 2 **7.** $-2 + 5i$ **9.** $8 + 32i$ **11.** 37 **13.** 30
15. $31 - 34i$ **17.** $-9 - 40i$ **19.** $\frac{2}{29} - \frac{5}{29}i$
21. $\frac{12}{13} + \frac{5}{13}i$ **23.** $\frac{19}{51} + \frac{10i\sqrt{2}}{51}$ **25.** $-5i$ **27.** i **29.** i
31. $x = 2$, $y = -3$ **33. a.** $79^2 = 6241$
b. $(3 - i)^2 = 8 - 6i$ **35.** $\left[\frac{\sqrt{2}}{2}(1 + i)\right]^2 = \frac{1}{2}(2i) = i$
37. $(a + bi) + (a - bi) = 2a + 0i = 2a$ **39.** Real

Calculator Exercise, page 24 $42660913 + 30082416i$

Oral Exercises 1-5, pages 26-27 1. 16 **2.** 25 **3.** $\frac{49}{4}$
4. a^2 **5.** Completing the square **6.** Factoring
7. Factoring **8.** Quadratic formula **9.** Quadratic
formula **10.** Completing the square **11.** $\sqrt{b^2 - 4ac}$ is
imaginary if $b^2 - 4ac < 0$ **12.** $\sqrt{b^2 - 4ac} = 0$ if
$b^2 - 4ac = 0$, so (double) root is $-\frac{b}{2a}$.
13. $\sqrt{b^2 - 4ac}$ is a whole number if $b^2 - 4ac$ is the
square of an integer. **14.** No; b isn't an integer.
15. Yes **16.** Yes

Written Exercises 1-5, pages 27-29 1. $x = \frac{7}{3}$ or
$x = -1$ **3.** $x = -\frac{9}{2}$ or $x = 2$ **5.** $x = 21$ or $x = -19$
7. $x = -35$ or $x = 45$ **9.** $z = 34$ or $z = -26$
11. $x = -3 \pm i$ **13.** $x = 3 \pm 2\sqrt{3}$
15. $x = \frac{-1 \pm \sqrt{6}}{5}$ **17.** $t = 2 \pm i$ **19.** $v = 8$ or $v = 2$
21. −47; $x = \frac{5 \pm i\sqrt{47}}{12}$ **23.** 109; $y = \frac{-5 \pm \sqrt{109}}{6}$
25. $x = \frac{13}{3}$ or $x = -3$ **27.** $z = \pm3\sqrt{2}$ **29.** $n = 2$ or
$n = -6$ **31.** $x = 6$ **33.** $x = \frac{-7 + \sqrt{1169}}{2}$

1

35. a. $64 - 16k$ **b.** $k = 4$ **c.** $k < 4$ **d.** $k > 4$
e. Examples: $k = 4$, $k = \frac{7}{4}$, $k = 3$ **39.** $-\frac{5}{3}$; $\frac{1}{3}$ **41.** $\frac{7}{5}$;
$-\frac{9}{5}$ **43. b.** $x^2 - 7x + 12 = 0$ **45.** $x^2 - 6x - 16 = 0$
47. $x^2 + 8x + 6 = 0$ **49.** $x^2 - 10x + 29 = 0$

51. $x = \dfrac{\sqrt{5} \pm 3}{4}$ **53.** $x = 0$ or $x = 1$ **55.** $x = -2i$ or

$x = -i$ **57. a.** 1, 2, 5, 7, 12 **b.** Examples: 117, 126

Calculator Exercise, page 29 **1.** $x = 1.025$ or
$x = -0.469$

Oral Exercises 1-6, page 31 **1.** $(0, 0)$; 4 **2.** $(2, 7)$; 6
3. $(4, -7)$; $\sqrt{7}$ **4.** $(0, -6)$; 6 **5.** $(1, 3)$; $\sqrt{19}$
6. $(0, 0)$; 3 **7.** $(x - 7)^2 + (y - 3)^2 = 36$
8. $(x + 5)^2 + (y - 4)^2 = 2$ **9.** $x^2 + y^2 = 169$

Written Exercises 1-6, pages 31-33
1. $(x - 4)^2 + (y - 3)^2 = 4$
3. $(x + 4)^2 + (y + 9)^2 = 9$ **5.** $(x - 6)^2 + y^2 = 15$
7. $(x - 2)^2 + (y - 3)^2 = 18$
9. $(x - 4)^2 + (y - 3)^2 = 25$
11. $(x - 5)^2 + (y + 4)^2 = 16$
13. $(x - 6)^2 + (y - 2)^2 = 1$
15. $(x - 4)^2 + (y + 5)^2 = 25$
19. $(x - 1)^2 + (y - 4)^2 = 1$; $C(1, 4)$, $r = 1$
21. $x^2 + (y - 6)^2 = 11$; $C(0, 6)$, $r = \sqrt{11}$
23. $(x - \frac{5}{2})^2 + (y - \frac{9}{2})^2 = 27$; $C(\frac{5}{2}, \frac{9}{2})$, $r = 3\sqrt{3}$
25. $(x - 1)^2 + (y + \frac{1}{2})^2 = \frac{31}{12}$; $C\left(1, -\dfrac{1}{2}\right)$, $r = \dfrac{\sqrt{93}}{6}$
27. One point **29.** No points **31.** Two points
39. The required graph is the portion of the graph of
$x^2 + y^2 = 9$ in Quadrants **a.** I, II; **b.** III, IV; **c.** I,
IV; **d.** II, III **41.** $x^2 + y^2 = 100$
43. $(x - 1)^2 + (y - 1)^2 = 2$
45. $(x - 4)^2 + (y - 5)^2 = 25$ **47.** $(-1, -1)$ **49.** The
points on the circles $x^2 + y^2 = 1$ and $x^2 + y^2 = 4$

Written Exercises 1-7, pages 35-36 **1.** $(8, 15)$, $(15, 8)$
3. No intersection **5.** $(-5, 12)$, $(\frac{135}{29}, -\frac{352}{29})$ **7.** $(0, 0)$,
$(2\sqrt{3}, 6)$ **9.** $(2, \pm 2\sqrt{3})$ **11.** $(\frac{4}{5}, -\frac{22}{5})$, $(4, 2)$
13. Tangent at $(2, -1)$ **17. a.** $x^2 + y^2 = 13$
b. $2x + 3y = 13$ **19.** 10 **21.** The graph of
$x^2 + y^2 = 9$ is the circle with center $(0, 0)$ and radius 3.
The graph of $(x + y)^2 = 9$ is two lines, $y = -x + 3$
and $y = -x - 3$. **23.** $P\left(3 + \dfrac{3\sqrt{2}}{2}, \dfrac{3\sqrt{2}}{2}\right)$;
$y = (-1 + \sqrt{2})x$

Calculator Exercise, page 36 131,370

Oral Exercises 1-8, page 41 **1.** (a), (c)

Chapter Test, pages 44-45 **1. a.** $2\sqrt{17}$ **b.** $(-3, -2)$
3. $(-\frac{1}{2}, 1)$ **5.** (b) and (c) are parallel, (a) is
perpendicular to both. **7.** $4x + 3y = 35$
9. $3x - 4y = -4$ **11.** $-5 + 12i$ **13.** $\dfrac{2 - 3i}{13}$ **15.** i

17. -20; $\dfrac{1 \pm i\sqrt{5}}{3}$ **19.** $(x + 3)^2 + (y - 4)^2 = 16$

Oral Exercises 2-1, page 49 **1.** quadratic; 5
2. quartic; 5 **3.** cubic; -4 **4.** quintic; -2 **5.** 23
6. 7 **7.** 11 **8.** 3 **9.** $2n^2 + 4n + 7$ **10.** $18x^2 + 5$
11. $\dfrac{2}{x^2} + 5$ **12.** $2a^2 + 4ab + b^2 + 5$ **13.** -5 and 3
14. ± 4 and $\pm 5i$ **15.** 3 **16.** $\dfrac{-b \pm \sqrt{b^2 - 4ac}}{2a}$ if
$a \neq 0$

Written Exercises 2-1, pages 49-51 **1.** Yes; 4 and 2
3. Yes; $\frac{17}{3}$ **5.** No; 4 and -1 **7.** Yes; none **9.** Yes;
0 and -1 **11.** No; none **13. a.** 51 **b.** -41
c. $8x + 11$ **d.** $\dfrac{4}{a} - 5$ **15. a.** 13 **b.** $-2 - 10i$
c. $-i + 1$ **d.** $18a^2 - 15a + 6$ **17. a.** $\dfrac{79\sqrt{2}}{27}$
b. $-12i\sqrt{3}$ **c.** $\dfrac{x^3}{27} - 3x$ **d.** $x^3 - 9x^2 + 18x$
19. $\pm\sqrt{3}$ **21.** $-\frac{95}{2}$ **23.** 2; -8 **25.** $f(x) = 2x + 5$
27. $h(x) = \frac{4}{3}x - \frac{25}{3}$ **29. a.** 7 **b.** 7 **31.** $f(x)$ increases
by 28. **35. a.** Yes **b-d.** No **37. a.** The differences
increase by 2. **b.** The differences increase by 4.
c. Yes

Calculator Exercise, page 51 213

Oral Exercises 2-2, pages 54-55 **1.** (3) **2.** (2) **3.** (1)
4. (2) **5.** (1) **6.** 3 **7.** A root (-4) was lost.
8. $x = \pm 6$ **9.** $x = 2$ or $x = \frac{1}{4}$ **10.** $x^3 + x^2 - 2x$
11. $x^3 - 2x$ **12.** $x^3 - 2x^2 + x - 2$

Written Exercises 2-2, page 55 **1.** $x = \pm\sqrt{6}$ or
$x = \pm i\sqrt{2}$ **3.** $x = 0$, $x = \frac{5}{3}$, or $x = -4$ **5.** $x = 0$,
$x = 6$, or $x = 1$ **7.** $x = \pm\sqrt{7}$ or $x = \pm 2i$
9. $x = \pm 2i$ or $x = \frac{3}{2}$ **11.** $s = 0$ or $s = 3$ **13.** $x = 0$
or $x = \dfrac{-3 \pm \sqrt{17}}{4}$ **15.** $t = \dfrac{\pm\sqrt{-3 \pm \sqrt{17}}}{2}$;
$t \approx \pm 0.53$ or $t \approx \pm 1.33i$ **17.** $x = 1$ or $x = -\frac{5}{4}$
19. $w = 0$, $w = \frac{1}{4}$, or $w = -\frac{1}{2}$ **21.** $x = 0$, $x = \frac{2}{3}b$, or
$x = b$ **23.** $x = 0$ **25.** $t = 3$ **27.** $v = 8$ or $v = 5$
29. $x = \frac{2}{9}$ **31.** $x = \pm\sqrt{5}$ or $\pm\dfrac{\sqrt{2}}{2}$

Calculator Exercise, page 55 $x \approx \pm 0.904$ or ± 1.257

Oral Exercises 2-3, pages 59-60 **1.** $(1, 6)$
2. $(-1, -3)$ **3. a.** upward **b.** one **4. a.** upward
b. no **5. a.** downward **b.** two **6. a.** downward
b. two **7. a.** upward **b.** no **8. a.** upward **b.** two
9. $(5, 4)$ **10.** $(7, 8)$ **11.** $(-2, -3)$ **12.** (h, k)
13. $(0, 5)$ **14.** $(0, -10)$ **15.** 6 and 4 **16.** 2
17. $-1 \pm \sqrt{3}$ **18.** None

Written Exercises 2-3, pages 60-62

1. a. Axis of symmetry:
$x = 3$

(0, 0) (6, 0) x
(3, -9)

b. Axis of symmetry:
$x = 3$

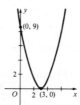

(0, 9)
O 2 (3, 0) x

c. Axis of symmetry:
$x = 3$

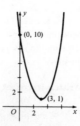

(0, 10)
(3, 1)
O 2 x

3. Axis of symmetry:
$x = -4$

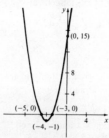

(0, 15)
8
4
(-5, 0) (-3, 0)
(-4, -1)

5. Axis of symmetry:
$x = 0$

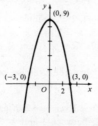

(0, 9)
(-3, 0) (3, 0)
O 2 x

7. Axis of symmetry
$x = 1$

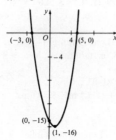

(-3, 0) O 4 (5, 0) x
-4
(0, -15)
(1, -16)

9. Axis of symmetry:
$x = -\frac{1}{2}$

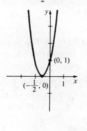

(0, 1)
$(-\frac{1}{2}, 0)$ 1 x

11. Axis of symmetry:
$x = -2$

x-intercepts: $-2 \pm \dfrac{\sqrt{22}}{2}$

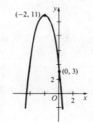

(-2, 11)
(0, 3)
O 2 x

13. Axis of symmetry:
$x = 4$

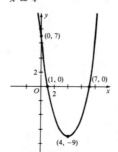

(0, 7)
2
(1, 0) (7, 0)
O 2 x
(4, -9)

15. Axis of symmetry:
$x = -1$

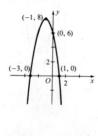

(-1, 8) (0, 6)
(-3, 0) 2 (1, 0)
O 2 x

17. Axis of symmetry:
$x = 1$
x-intercepts: $1 \pm 2\sqrt{2}$

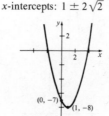

2
2 x
(0, -7) (1, -8)

19. Axis of symmetry:
$x = -3$
x-intercepts: $-3 \pm \sqrt{8}$

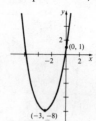

2 (0, 1)
-2 2 x
(-3, -8)

21. Axis of symmetry: $x = 1$; vertex: $(1, -2)$;
x-intercepts: $1 \pm \dfrac{\sqrt{2}}{2}$; y-intercept: 2   **23.** Axis of
symmetry: $x = -4$; vertex: $(-4, 0)$; x-intercept: -4;
y-intercept: 8   **25.** Axis of symmetry: $x = -1$; vertex:
$(-1, 5)$; x-intercepts: $-1 \pm \dfrac{\sqrt{10}}{2}$; y-intercept: 3

27. Axis of symmetry: $x = 2$; vertex: $(2, -1)$;
x-intercepts: $2 \pm \frac{1}{2}$; y-intercept: 15   **29.** Axis of
symmetry: $x = -1$; vertex: $(-1, 16)$; no x-intercepts;
y-intercept: 17   **31.** Axis of symmetry: $x = -3$; vertex:
$(-3, 6)$; no x-intercepts; y-intercept: 24   **33.** Axis of
symmetry: $x = 0$; vertex: $(0, 6)$; x-intercepts: $\pm 2\sqrt{6}$
35. $(2, 0)$ **37.** $(-6, 0), (-1, -5)$
39. $y = -3(x - 2)(x + 1)$ **41.** $y = -\frac{1}{2}(x - 4)^2 + 8$
43. $y = \frac{3}{8}(x - 4)^2$ **45.** $y = \frac{7}{4}(x - 3)^2 - 5$

Oral Exercises 2-4, pages 64-65   **1.** (9) 3; 3 is a double
root; (10) 4; 3 is a double root; (11) 5; -2 and 3 are
double roots; (12) 3; 2 is a triple root; (13) 3; 2 is a
triple root; (14) 4; 3 is a triple root
3. a. $y = (x - 1)(x - 4)^2$ **b.** $y = -x(x + 1)(x - 3)$
c. $y = -(x - 1)^2$ **d.** $y = x(x - 2)^3$

Answers to selected exercises   **3**

Written Exercises 2–4, pages 65–67

1.

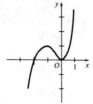

3.

5.

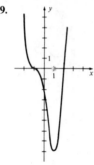

7.

9.

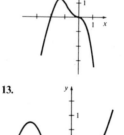

11.

13.

17. $y = -(x + 3)(x + 1)(x - 1)$
19. $y = (x + 3)^2(x + 1)(x - 1)$ **23.** $(0, 0)$, $(1, -3)$, $(-1, 3)$ **25.** $(\sqrt{2}, 4)$, $(-\sqrt{2}, 4)$
27. $y = x(x - 1)(x - 2)$ **29.** $y = -\frac{3}{8}(x + 4)(x - 2)$
31. $y = -2x^2(x - 3)$ **33.** $y = -\frac{1}{2}(x + 1)^2(x - 2)^2$
35. $P(x) = -(x + 3)(x + 1)(x - 2)$

Oral Exercises 2–5, pages 69–70 1. Yes; yes;
$P(2) = P(-2) = 0$ **2.** No; yes; $P(2) \neq 0$, $P(-2) = 0$
3. 6; 4 **4. a.** $x^4 - 8x^2 + 5x - 1$; $x + 3$
b. $x^3 - 3x^2 + x + 2$; -7 **c.** -7
5. a. $2x^3 - 3x^2 - 2x + 1$; $x + \frac{1}{2}$ **b.** $2x^2 - 4x$; 1
c. 1 **6. a.** $x^2 + 2x - 1$ **b.** Yes; find the zeros of $Q(x)$
using the quadratic formula.

Written Exercises 2–5, pages 70–71 1. a. Yes **b.** No
3. a. -4 **b.** -2 **5.** $x^2 - x + 4$; 5
7. $t^3 - 3t^2 + 3t + 2$; 0

9. $y^4 + 3y^3 + 10y^2 + 30y + 91$; 273
11. $3x^2 - 8x + 21$; $-41x + 1$ **13.** $P(\frac{1}{3}) = 3$;
$P(-\frac{2}{3}) = -\frac{7}{3}$ **15.** 8 **19.** $2x^3 - x^2 + 7x + 7$
21. $x = -1$, $x = \frac{1}{2}$ **23.** $x = 4$, $x = \frac{1}{2}$ **25.** $x = \frac{1}{2}$,
$x = -\frac{1}{2}$ **29.** $a = -8$, $b = 9$ **31.** $a = \pm 1$ or $a = 2$

Calculator Exercise, page 71 $f(7) = 3138$; Method 2

Oral Exercises 2–6, page 74 1. ± 1 **2.** ± 1, ± 2
3. ± 1, ± 3, ± 9 **4.** ± 1, ± 2, ± 4, $\pm \frac{1}{2}$ **5.** ± 1, ± 2,
$\pm \frac{1}{3}$, $\pm \frac{2}{3}$ **6.** ± 1, ± 2, $\pm \frac{1}{6}$, $\pm \frac{1}{3}$, $\pm \frac{2}{3}$, $\pm \frac{1}{2}$ **7.** ± 1, ± 3,
± 5, ± 15 **8.** ± 1, ± 2, ± 3, ± 4, ± 6, ± 12, $\pm \frac{1}{2}$, $\pm \frac{3}{2}$
9. ± 1, ± 2, ± 4, ± 8, $\pm \frac{1}{3}$, $\pm \frac{2}{3}$, $\pm \frac{4}{3}$, $\pm \frac{8}{3}$ **10.** ± 1, ± 2,
$\pm \frac{1}{8}$, $\pm \frac{1}{4}$, $\pm \frac{1}{2}$ **11.** No; yes **12. a.** $P(x)$ changes sign
between -2 and -1. **b.** Irrational **c.** If $x > 0$, every
term of $P(x)$ is positive and the sum, therefore, is
positive.

Written Exercises 2–6, page 75 1. $x = \pm 1$
3. $x = \pm 1$ or $x = \pm 3$ **5.** $x = -1$, $x = 2$, or $x = \frac{1}{3}$
7. $x = -1$, $x = 3$, or $x = -5$ **9.** $x = -1$, $x = 2$, or
$x = \frac{4}{3}$ **11.** $x = -\frac{1}{2}$ or $x = -1 \pm \sqrt{2}$ **13.** $x = 4$ or
$x = \pm 2i$ **15.** $x = -1$ or $x = \pm \sqrt{3}$ **17.** $x = -2$ or
$x = 1 \pm \dfrac{\sqrt{2}}{2}$ **21.** intersect at $(2, 2)$, tangent at
$(-1, 2)$ **23.** intersect at $(0, 0)$, $(-2, -6)$, $(2, 6)$
25. $h = 2$ or $h = -1 + \sqrt{6}$
29. $7 \text{ cm} \times 10 \text{ cm} \times 16 \text{ cm}$

Calculator Exercise, page 76 a. greater than **b.** less
than **c.** The root is between 0.68 and 0.69.

Oral Exercises 2–7, page 78 1. between 1 and 2
2. between 1 and 2 **3.** between -1 and 0, 2 and 3
4. between 0 and 1, -1 and 0, -5 and -4 **5.** -2.2,
0.4, 2.8

Written Exercises 2–7, page 79 1. 1.5 **3.** 2.6 **5.** 1.7
7. 0.5 **9.** 0.5 **11. a.** $\frac{1}{3}x^2(x + 2)$ **b.** 6.1

Oral Exercises 2–8, page 81 1. a. minimum **b.** $x = 4$
2. a. maximum **b.** $x = 2$ **3. a.** minimum **b.** $x = \frac{3}{2}$

Written Exercises 2–8, pages 81–83
1. $A(x) = x(120 - 2x)$; 30 **3.** 433.5 m² **9.** 45 m
11. $2.40 **13.** 6×9

Calculator Exercises, page 83 1. piece for circle:
$\dfrac{40\pi}{4 + \pi} \approx 17.6$ cm; piece for square: $\dfrac{160}{4 + \pi} \approx 22.4$ cm

Oral Exercises 2–9, page 87 1. $\sqrt{3} - 2i$
2. $-3 - \sqrt{2}$, $2i$ **3. a.** $-\frac{5}{2}$; $-\frac{3}{2}$ **b.** $\frac{1}{2}$; $\frac{3}{2}$ **c.** -2; 0
d. 0; -32

Written Exercises 2–9, pages 87–89 1. F **3.** F **5.** F
7. T **9.** $\frac{3}{4}$; $\frac{3}{2}$ **11.** $-\frac{5}{3}$; $\frac{2}{3}$ **13.** $x^2 - 2x + 2 = 0$
15. $x^2 - 6x + 7 = 0$ **17.** $2 - i\sqrt{5}$, -4

4   **Answers to selected exercises**

19. $x^3 - 10x^2 + 33x - 34 = 0$
21. $4x^3 - 12x^2 + 3x + 19 = 0$
23. $x^4 - 10x^3 + 29x^2 - 10x + 28 = 0$
25. $c = -6$; $d = -4$ **27.** $c = -27$

Chapter Test, pages 90–91 1. a. $-16i$
b. $x^3 - 6x^2 + 8x$ **c.** $x = 0$ or $x = \pm 2$
3. a.

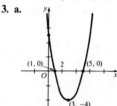

b.

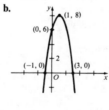

5. $(1, -3)$, $(-1, -5)$, $(4, 0)$ **7.** -4 **9.** $x = \frac{1}{2}$ or
$x = \pm i$ **11.** -1.1

Oral Exercises 3-1, pages 96–97 1. $x > -4$
2. $x < 4$ **3.** $x > -5$ **4.** $-2 < x < 2$ **5.** $3 < x < 7$
6. $x < -4$ or $x > -2$ **7.** f **8.** a **9.** e **10.** c **11.** b
12. d

Written Exercises 3-1, pages 97–98 1. $x < 3$
3. $x < 0$ **5.** $x \le 11$ **7.** $x > 6$ **9.** $x < 4$
11. $-3 < x < 3$ **13.** $1 < x < 7$ **15.** $x \le -10$ or
$x \ge -4$ **17.** $x = 12$ or $x = 4$ **19.** $-\frac{1}{2} \le x \le \frac{9}{2}$
21. $-\frac{17}{4} \le x \le \frac{1}{4}$ **23.** $-4 \le x \le -2$ or $2 \le x \le 4$
25. $1 < x < 3$ or $5 < x < 7$ **27.** $5 < x < 7$ or
$7 < x < 9$ **29.** $x = \pm 4$ or $x = \pm 2$ **31.** $x = 0$ or
$x = \pm 3$ **33.** $x < 0$ or $x > \frac{1}{2}$ **35.** $0 \le x \le 2$
39. a. $a = 1, b = 2$; $a = -1, b = -2$; $a = -1$,
$b = 1$ **b.** Yes

Oral Exercises 3-2, page 100 1. $-2 < x < 2$ or
$x > 4$ **2.** $x \le 1$ **3.** $1 < x < 2$ or $x > 4$ **4.** $x < 2$ or
$4 < x < 7$ **5.** $x \le 5$ **6.** No such x

Written Exercises 3-2, pages 100–101 1. $x < -4$ or
$x > 3$ **3.** $1 < x < 2$ or $x > 4$ **5.** $x < 4$; $x \ne 2$,
$x \ne 3$ **7.** $-3 < x < 5$ **9.** $x \le -1$ or $x \ge 1.5$
11. $-\frac{7}{2} \le x \le 1$ **13.** $x < -\sqrt{5}$ or $x > \sqrt{5}$
15. $a > 2$ **17.** $n < -3$ or $1 < n < 2$ **19.** $x < -2$ or
$\frac{1}{2} < x < 1$ **21.** $-2 < x < 2$ **23.** $x < -1$ or $x > 1$
25. $x < 3$ or $4 < x < 5$ **27.** $-4 < x \le \frac{5}{2}$ or $x > 7$
29. All real x **33.** $x < -\frac{5}{2}$ or $x > \frac{4}{3}$ **35.** No such real x

Oral Exercises 3-3, pages 103–104 1. above
2. below **3.** on **4.** on **5.** above **6.** below
7. a. $x > 0$ and $y > 0$ **b.** $x < 0$ and $y > 0$; $x < 0$ and
$y < 0$; $x > 0$ and $y < 0$ **8.** $x \ge 0, y \ge 0$, and
$y \le -x + 2$ **9.** $x \ge 0, 0 \le y \le 3$, and $y \le 4 - 2x$
10. $y > x^2$ and $y < x + 2$

Written Exercises 3-3, pages 104–105 1. Points on or
above the line $y = x$ **3.** Points below the line

$3x + 4y = 12$ **5.** Points on or below the parabola
$y = x^2$ **7.** Points below the parabola
$y = 2x^2 - 4x + 1$ **9.** Points on or between the lines
$x = 0$ and $x = 2$ **11.** Points below the line $y = -1$ or
above the line $y = 1$ **13.** Points between the lines
$x = 1$ and $x = 5$ **15.** Points above the curve
$y = x^3 - 9x$ **17.** Points on or inside the circle with
center at the origin and radius 2 **19.** Points on or
outside the circle with center $(0, 1)$ and radius 1

21.

23.

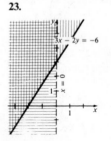

25.

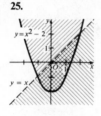

27.

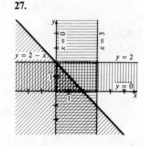

29.

31.

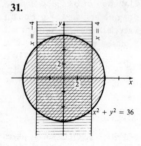

33.

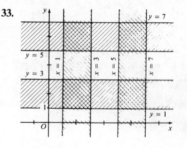

35. 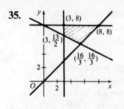 **37.**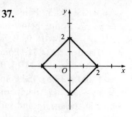

Written Exercises 3–4, pages 108–110 **1. a.** 5 consoles, 4 portables **3.** 5 consoles, 3 portables **5. b.** $x = 100$, $y = 700$

Oral Exercises 3–5, page 112 **1.** $a = k + 1$, $b = 3k$, $c = k - 2$ **2.** $a = 4$, $b = 3 + k$, $c = -5$ **3.** $a = p^2$, $b = -p$, $c = -2$ **4.** $a = 4 - r$, $b = 1 - s$, $c = 1 - t$ **5.** h and k

Written Exercises 3–5, pages 112–113 **1.** $k = 5$ **3.** $k = 3 \pm 2\sqrt{2}$ **5.** $y = \dfrac{\sqrt{3}}{3}x$ or $y = -\dfrac{\sqrt{3}}{3}x$ **7.** $-10 \leq k \leq 10$ **9. b.** $-1 < m < 1$ **11. a.** $y = 4x + k$ **b.** $k = 1$ **13.** $r \geq \sqrt{5}$ **17.** $x^2 + (y - 4)^2 = \frac{15}{4}$

Chapter Test, pages 114–115 **1. a.** $x > \frac{1}{2}$ **b.** $-2 < x < 8$ **c.** $-5 \leq x \leq 3$ **3. a.** $-2 \leq x \leq \frac{3}{4}$ **b.** $x < -\frac{1}{2}$ or $0 < x < 1$ **5.** $1 \leq x \leq 3$, $y \geq 2$, $x + 2y \leq 9$ **7. a.** $y = 2x + k$ **b.** $-5 \leq k \leq 5$

Oral Exercises 4–1, pages 120–121 **1. a.** speed; time **b.** the car shifted **c.** 1 s **2. a.** $\{r \mid 200 \leq r \leq 800\}$ **b.** 90% **c.** 500 **3. a.** Area is a function of the radius. **b.** Volume is a function of the radius. **c.** Area is a function of base and height. **d.** Distance is a function of rate and time. **e.** Force is a function of mass and acceleration. **f.** Amount is a function of principal, interest rate, and time. **4. a.** 8 **b.** -1 **c.** -4 **d.** -7 **5. a.** -5 **b.** -1 **c.** 0 **d.** 20 **6. a.** $\{x \mid x \neq -2\}$ **b.** $\{s \mid s \neq 1\}$ **c.** $\{y \mid y \neq \pm 2\}$ **7. a.** $\{t \mid t \geq 0\}$ **b.** $\{t \mid t \geq 4\}$ **c.** $\{t \mid |t| \geq 2\}$ **8. a.** Calgary, Atlanta **b.** Any cities with pressure ≤ 75.90 cm

Written Exercises 4–1, pages 121–124 **1.** length of a side **3.** length of base edge; height **5. a.** $\{x \mid x \neq 9\}$ **b.** $\{x \mid x \neq 0\}$ **c.** $\{x \mid x \neq -2, x \neq -3\}$ **d.** all real numbers **e.** $\{x \mid x \leq 5\}$ **f.** $\{x \mid |x| \geq 5\}$ **7.** $y = 2x + 3$ **9. a.** $S = 180(n - 2)$ **b.** $\{n \mid n$ is an integer $\geq 3\}$ **11. a.** 50 km/h **b.** 30 km/h **13. a.** $n = 8C - 28$ **c.** 18°C **15. a.** 8π; 64π **19. a.** $V = s^2h$ **b.** Examples: $V = 4$ through $(\frac{1}{2}, 16)$, $(1, 4)$, and $(2, 1)$; $V = 16$ through $(1, 16)$, $(2, 4)$, and $(4, 1)$

Calculator Exercise, page 124 No

Oral Exercises 4–2, pages 125–126 **1. a.** all real numbers; all real numbers; 0 **b.** all real numbers; $\{y \mid y \geq 0\}$; 0 **c.** all real numbers; $\{y \mid y \geq -4\}$; ± 2 **2.** $\{x \mid x \geq -2\}$; $\{y \mid -2 \leq y \leq 1\}$; -2, 1 **3.** all real numbers; $\{y \mid y \geq 5\}$; none **4. a.** No; fails vertical-line test **b.** Yes; passes vertical-line test **c.** No; fails vertical-line test **5.** all real numbers **6.** No; the vertical line $x = 4$ intersects the graph in more than one point.

Written Exercises 4–2, pages 126–127 **1.** a, c **3. a.** all real numbers; $\{y \mid y \geq 0\}$; 0 **b.** all real numbers; $\{y \mid y \geq -2\}$; ± 2 **5. a.** all real numbers; $3 \pm \sqrt{2}$ **b.** parabola through $(3 - \sqrt{2}, 0)$ and $(3 + \sqrt{2}, 0)$ with vertex $(3, 2)$ **c.** $\{y \mid y \leq 2\}$ **7. a.** all real numbers; 2 **b.** 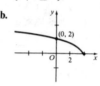 **c.** all real numbers

9. a. $\{x \mid x \leq 4\}$; 4 **b.** 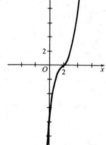 **c.** $\{y \mid y \geq 0\}$

11. a. all real numbers; $\{x \mid x \geq 3\}$ **c.** $\{y \mid y \leq 0\}$ **b.**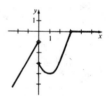

13. a. -30 **b.** $f(x)$ decreases by 30 **15.** domain: $\{x \mid 0 \leq x \leq 70\}$; range: $\{y \mid 31{,}200 \leq y \leq 34{,}225\}$; maximum: $f(x) = 34{,}225$ **17.** d

Oral Exercises 4–3, page 129 **1.** 8 **2.** 32 **3.** 0 **4.** 6 **5.** -1 **6.** -2 **7.** $f(g(x)) = \dfrac{x^3}{8}$ **8.** $g(f(x)) = \dfrac{x^3}{2}$ **9.** $f(h(x)) = (x - 2)^3$ **10.** $h(f(x)) = x^3 - 2$ **11.** $g(h(x)) = \dfrac{x - 2}{2}$ **12.** $h(g(x)) = \dfrac{x - 4}{2}$

Written Exercises 4–3, pages 129–131 **1.** $\frac{16}{3}$ **3.** $\frac{8}{3}$ **5.** 4 **7.** $f(g(x)) = \dfrac{4}{x^3}$ **9.** $f(h(x)) = \dfrac{4}{4x - 1}$ **11.** $g(h(x)) = (4x - 1)^3$ **13.** 3 **15.** 0 **17.** $f(g(h(x))) = \sqrt{2x - 3}$ **19.** $h(f(g(x))) = \dfrac{\sqrt{6x - 3}}{3}$ **21.** $x = \pm\dfrac{1}{2}$ **23.** 0 **25.** $k(f(h(x)))$ **27.** $f(k(h(x)))$

29. a. $r = \dfrac{C}{2\pi}$ **b.** $A = \dfrac{C^2}{4\pi}$ **31.** $s = 331 + \tfrac{1}{3}(F - 32)$

33. a. $y = 18z - 11$ **b.** $z = \dfrac{y + 11}{18}$ **35.** 1.5 m/s;

10 L/min **37.** $f(g(x)) = 2\sqrt{16 - x^2}$, domain:

$\{x \mid |x| \le 4\}$; $g(f(x)) = 2\sqrt{4 - x^2}$, domain:

$\{x \mid |x| \le 2\}$ **39.** $f(g(x)) = 1 - x$, domain:

$\{x \mid x \le 1\}$; $g(f(x)) = \sqrt{1 - x^2}$, domain: $\{x \mid |x| \le 1\}$

41. $g(g(1)) = \tfrac{5}{2}$, $g(g(g(1))) = \tfrac{11}{4}$, $g(g(g(g(1)))) = \tfrac{23}{8}$

43. a. $E = \dfrac{130}{(1 + t)^2}$ **b.** 9 s

Calculator Exercise, page 131 4.32 seconds will be lost.

Oral Exercises 4-4, pages 134-135 **1. a.** $V(e) = e^3$

b. $e(V) = \sqrt[3]{V}$ **2. a.** $d(s) = s\sqrt{2}$ **b.** $s(d) = \dfrac{d\sqrt{2}}{2}$

c. $A(d) = \dfrac{d^2}{2}$ **3. a.** $AP + PB = \sqrt{x^2 + 16} + 8 - x$

b. $\{x \mid 0 \le x \le 8\}$ **4. a.** $d(xy) = \sqrt{(x - 2)^2 + y^2}$

b. $d(x) = \sqrt{(x - 2)^2 + x^4}$ **5.** $10t$

6. $\sqrt{100t^2 + 720t + 3600}$ **7.** $V(w) = \dfrac{w - w^3}{2}$

8. a. \$101; $k + 1$ dollars **b.** $k - 1$ dollars

Written Exercises 4-4, pages 135-139 **1.** $A(h) = \dfrac{h^2\sqrt{3}}{8}$

3. $t(n) = \dfrac{11n}{36}$ **5.** $p(n) = 6n$ **7.** S. A.$(h) = \dfrac{5\pi h^2}{8}$

9. $s(d) = \dfrac{3d}{2}$ **11.** $V(w) = \dfrac{3w - 2w^3}{4}$ **13. a.** $C(t) = \pi t$

b. $A(t) = \dfrac{\pi t^2}{4}$ **15.** $C(x) = \dfrac{8x^3 + 192}{x}$

17. $C(r) = \dfrac{4\pi r^3 + 800\pi}{r}$

19. a. $d(t) = \sqrt{13{,}000t^2 - 11{,}700t + 4225}$

21. a. $d(y) = \sqrt{y^2 - y + 1}$ **b.** $\dfrac{\sqrt{3}}{2}$

23. a. $d(x) = 2\sqrt{25 - 4x}$ **b.** domain: $\{x \mid |x| \le 6\}$;

range: $\{y \mid 2 \le y \le 14\}$ **25. a.** $C(x) = 4x + \dfrac{180}{x}$

b. $24\sqrt{5} \approx \$53.67$ **27.** $d(t) = 9\sqrt{9t^2 + 6t + 101}$;

domain: $\{t \mid 0 \le t \le 3\}$; range: $\{y \mid 9\sqrt{101} \le y \le 90\sqrt{2}\}$

29. a. $A(w) = (-\tfrac{1}{8}\pi - \tfrac{1}{2})w^2 + 150w$ **b.** $w = \dfrac{600}{\pi + 4}$

31. $A(x) = \sqrt{16 - x^2}\,(x + 4)$

33. $V(r) = \dfrac{2\pi r^4}{3(r^2 - 1)}$

Oral Exercises 4-5, pages 141-142 **1. a.** 2 **b.** 3 **c.** 2

2. a. $\dfrac{x}{4}$ **b.** $\dfrac{x - 2}{3}$ **c.** $\dfrac{x + 1}{2}$ **3. a.** (8, 2), (1, 1),

$(-1, -1)$, (0, 0), $(-8, -2)$ **b.** $f^{-1}(x) = \sqrt[3]{x}$ **4.** No

5. $g(x)$; $g(x)$ **6.** b, d

Written Exercises 4-5, pages 142-143 **1. a.** 2 **b.** 3
c. 7

3. a.

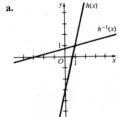

3. b. $h^{-1}(x) = \dfrac{x + 3}{4}$

5. Yes

7. No

9. Yes; $f^{-1}(x) = \dfrac{x + 5}{3}$

11. No

13. Yes; $f^{-1}(x) = \dfrac{1}{x}$

15. Yes; $f^{-1}(x) = 5 - x^2$, $x \ge 0$

17. No

19.

$g^{-1}(x) = \sqrt{x - 2}$, $x \ge 2$

21.

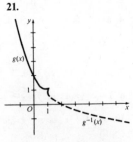

$g^{-1}(x) = 1 - \sqrt{x - 1}$, $x \ge 1$

23.

25.

27. a. (i) $f(g(x)) = \dfrac{x - 5}{3}$, inverse of

$f(g(x)) = 3x + 5$, $f^{-1}(x) = 3x$, $g^{-1}(x) = x + 5$,

$g^{-1}(f^{-1}(x)) = 3x + 5$; (ii) $f(g(x)) = \dfrac{1}{2x + 1}$, inverse

of $f(g(x)) = \dfrac{1 - x}{2x}$, $f^{-1}(x) = \dfrac{1}{x}$, $g^{-1}(x) = \dfrac{x - 1}{2}$,

$g^{-1}(f^{-1}(x)) = \dfrac{1 - x}{2x}$; (iii) $f(g(x)) = 27x^3$, inverse of

$f(g(x)) = \dfrac{\sqrt[3]{x}}{3}$, $f^{-1}(x) = \sqrt[3]{x}$, $g^{-1}(x) = \dfrac{x}{3}$,

$g^{-1}(f^{-1}(x)) = \dfrac{\sqrt[3]{3}}{3}$ **29.** $f^{-1}(x + 1) - f^{-1}(x) > 0$

Calculator Exercises, page 143 **1.** $f^{-1}(x) = g(x)$,

$g^{-1}(x) = f(x)$ **2.** Yes **3.** $\{x \mid x > 0\}$

Chapter Test, page 145 **1. a.** Cost is a function of

distance. **b.** \$36 **3.** $f(x) = -\dfrac{1}{2}x + 2$

5. a.

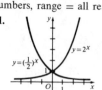

b. domain: $\{x \mid -3 \le x \le 3\}$; range: $\{y \mid 0 \le y \le 3\}$; zeros: ± 3

7. a. $g(h(f(x)))$ **b.** $f(k(h(x)))$ **c.** $k(g(h(x)))$
9. a. $d(x) = \sqrt{-6x + 10}$ **b.** domain: $\{x \mid -1 \le x \le 1\}$; range: $\{y \mid 2 \le y \le 4\}$ **11. a.** $g(x)$
b. $f(x)$ is not one-to-one.

Oral Exercises 5-1, page 150 **1. a.** $\frac{1}{8}$ **b.** $\frac{1}{64}$ **c.** $\frac{3}{2}$ **d.** $\frac{9}{4}$
2. a. $\frac{1}{9}$ **b.** $\frac{1}{8}$ **c.** $\frac{5}{2}$ **d.** $\frac{125}{8}$ **3. a.** $\frac{4}{9}$ **b.** $\frac{1}{144}$ **c.** 2 **d.** 1
4. a. x^8 **b.** $\frac{1}{x^{15}}$ **c.** x^2 **d.** $\frac{1}{a^4}$ **5. a.** $\frac{r^2}{3}$ **b.** $\frac{25}{a^4}$
c. $8b^3$ **d.** $16x^4$ **6. a.** $\frac{1}{x^3}$ **b.** x^{11} **c.** $\frac{x^4}{2}$ **d.** $\frac{1}{r^5}$
7. a. 8 **b.** 6^n **c.** 2 **d.** 1 **8. a.** $\frac{4}{3}$ **b.** 8 **c.** $\frac{9}{2}$ **d.** 27
9. a. $x^5 + x^4$ **b.** $2x + 2$ **c.** $x + 2$ **d.** $3 - \frac{9}{x^2}$
10. a. $n^4 + 1$ **b.** n^2 **c.** $3a^4 - a^7$ **d.** $3a^{10}$

Written Exercises 5-1, pages 150–151 **1. a.** $\frac{1}{16}$
b. $-\frac{1}{16}$ **c.** 81 **3. a.** $\frac{1}{32}$ **b.** $\frac{1}{256}$ **c.** $\frac{1}{4}$ **5. a.** $\frac{1}{3}$ **b.** $\frac{1}{16}$
c. $\frac{1}{16}$ **7. a.** $\frac{1}{m^3 n^6}$ **b.** $-\frac{a^{15}}{64}$ **c.** $\frac{8}{x^2}$ **9. a.** $\frac{81}{a}$ **b.** $48x^4$
c. $729n^{12}$ **11. a.** $\frac{16}{a^4}$ **b.** $a^5 b^7$ **c.** $-\frac{1}{n^2}$
13. a. $2x^2 - 4$ **b.** $x^5 + 2x^4$ **c.** $4n^6 - 2$ **15. a.** $\frac{8}{3}$
b. 32 **c.** $\frac{ab}{b - a}$ **17. a.** 3 **b.** 5 **c.** $2^{\frac{5n + 7}{2}}$ **19. a.** b^n
b. $\frac{1}{b^2}$ **c.** $\frac{1}{b^{2n-1}}$ **21. a.** $\frac{4}{3}$ **b.** $\frac{1}{80}$ **23. a.** $\frac{1}{x} + \frac{1}{y}$
b. $\frac{1}{y + 1}$ **31.** $x = 0$ or $x = 1$ **33.** $x = 1$
35. $x = \pm 1$ **37.** $x = \pm 1$

Oral Exercises 5-2, pages 153–154 **1. a.** 2 **b.** $\frac{1}{2}$ **c.** 8
d. $\frac{1}{8}$ **2. a.** -3 **b.** $-\frac{1}{3}$ **c.** 4 **d.** $\frac{1}{4}$ **3. a.** 1000 **b.** $\frac{1}{4}$
c. $\frac{1}{3}$ **d.** $\frac{1}{9}$ **4. a.** $\frac{5}{3}$ **b.** $\frac{8}{27}$ **c.** 6 **d.** 4 **5. a.** $4x^8$ **b.** $\frac{3}{n^2}$
c. $\frac{8}{x}$ **d.** $\frac{x^2}{y^6}$ **6. a.** $\frac{5}{x^2}$ **b.** a^2 **c.** $\frac{1}{n^4}$ **d.** $b^{\frac{1}{3}}$ **7. a.** 64
b. 9 **c.** 1 **d.** $8x$ **8. a.** 8 **b.** $\frac{1}{2}$ **c.** $\frac{1}{2}x$ **d.** $\frac{1}{8}$
9. a. $2x - 1$ **b.** $1 + a$ **c.** $x + 2x^2$ **10. a.** $x^{2n} + 1$
b. 25 **c.** 8 **11. a.** $x = 6$ **b.** $x = \frac{5}{2}$ **c.** $x = \pm\sqrt{5}$
12. a. $x = 27$ **b.** $x = \frac{1}{16}$ **c.** $x = \frac{1}{32}$

Written Exercises 5-2, pages 154–155 **1. a.** $x^{\frac{5}{2}}$ **b.** $y^{\frac{2}{3}}$
c. $x^{\frac{5}{8}}$ **d.** $(2a)^{\frac{3}{2}}$ or $8^{\frac{1}{2}}a^{\frac{3}{2}}$ **3. a.** $\frac{5b^{\frac{1}{3}}}{a^{\frac{2}{3}}}$ **b.** $6c^{\frac{1}{2}}d^{\frac{2}{3}}$ **c.** $\frac{2^{\frac{1}{4}}a^{\frac{4}{3}}}{b^{\frac{2}{3}}}$

d. $9n^{\frac{1}{2}}$ **5. a.** $\frac{\sqrt{5x}}{x}$ **b.** $\sqrt[3]{6x^2}$ **c.** $\sqrt[5]{x^2 y^3}$ **d.** $\sqrt[7]{\frac{a^4}{b^4}}$
7. a. $\frac{5}{3}$ **b.** $\frac{125}{27}$ **c.** $\frac{b}{2}$ **d.** $\frac{9}{a^4}$ **9. a.** $\frac{1}{8}$ **b.** 5 **c.** a^2
d. $2x^2$ **11. a.** 36 **b.** 729 **c.** $\frac{12}{5}$ **d.** 12 **13. a.** $\frac{729}{64}$
b. 8 **c.** 3 **15. a.** x **b.** $\frac{1}{x^4}$ **c.** $x + 2x^3 + x^5$
17. a. $a^2 - 2a$ **b.** $2n + 2$ **c.** $x^2 - 2x + 1$
19. a. $x - 2$ **b.** $y - 3y^2$ **c.** $4a^2 - 2a^2 b$
21. a. $x = 2$ **b.** $x = \frac{1}{2}$ **c.** $x = 2$ **23. a.** $x = \frac{4}{9}$
b. $x = \frac{2}{3}$ **c.** $x = 6$ **25. a.** $x = \frac{1}{32}$ **b.** $x = \frac{1}{2}$
c. $x = -7\frac{3}{4}$ **27. a.** $x = 16$ **b.** $x = \frac{1999}{2000}$ **c.** $x = \frac{1999}{400}$
29. a. $a^{\frac{1}{2}}b^{\frac{1}{2}}(a - b)$ **b.** $a^{\frac{1}{2}}b^{-\frac{1}{2}}(1 - ab)$ **31. a.** $(x^2 + 1)^{\frac{1}{2}}$
b. $2(x^2 + 2)^{-\frac{1}{2}}$ **33. a.** $-(2x - 1)^{-\frac{2}{3}}$
b. $-(x + 1)^{-\frac{1}{2}}(x + 2)^{-\frac{1}{2}}$ **35.** $x = 3$ or $x = -1$
37. $x = -12$ **39.** $x = 125$ or $x = -1$ **41.** $x = 9$ or
$x = 4$ **43.** $x = \frac{1}{2}$

Calculator Exercises, page 156 **1.** $D \approx 293.7$,
$D^{\#} \approx 311.1$, $E \approx 329.6$, $F \approx 349.2$, $F^{\#} \approx 370.0$,
$G \approx 392.0$, $G^{\#} \approx 415.3$, $A \approx 440$

Oral Exercises 5-3, pages 157–158 **1.** $f(-3) = \frac{1}{8}$,
$f^{-1}(16) = 4$ **2.** $f(2) = 9, f^{-1}(\frac{1}{3}) = -1$ **3.** $b = 10$
4. $a = 5, b = 3$ **5.** $a = 7$ **6.** $0 < b < 1$
7. a. domain = all real numbers, range = positive real
numbers **b.** domain = positive real numbers,
range = all real numbers **8.** 200

Written Exercises 5-3, pages 158–159 **1.** $f(x) = 3(\frac{5}{3})^x$
3. $f(x) = \frac{1}{2} \cdot 6^x$ **5.** $f(x) = 64(\frac{1}{4})^x$ **7.** $f(x) = 4 \cdot 2^x$
9. $f(x) = 3 \cdot 4^x$ **11.** $f(x) = 48 \cdot 4^x$ **13.** $f(x) = 2 \cdot 3^x$
15. $f^{-1}(12) = 1, f^{-1}(\frac{3}{4}) = -1, f^{-1}(\frac{3}{64}) = -3$ **17. a.** 3
b. 81

19. a. $f(-2) = \frac{1}{4}, f(-1) = \frac{1}{2}$,
$f(0) = 1, f(1) = 2, f(2) = 4$
b, c. See graph at right.
d. $f(x)$: domain = all real
numbers, range = positive
real numbers; $f(x)^{-1}$:
domain = positive real
numbers, range = all real numbers.

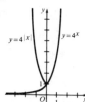

21.

23. a.

b. For $x < 0$,
$4^{|x|} = \frac{1}{4^x}$

29. a.

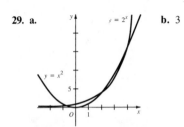

b. 3

Calculator Exercises, page 159 **1. a.** 8.5742 **b.** 8.8152
c. 8.8214 **d.** 8.8244 **e.** 8.8250 **3.** 2.18 **5. a.** $a = 5.3$,
$b \approx 1.027$ **b.** 1092 million, 4139 million

Oral Exercises 5-4, page 162 **1.** $5^2 = 25$ **2.** $5^3 = 125$
3. $5^{-1} = \frac{1}{5}$ **4.** $5^{\frac{1}{2}} = \sqrt{5}$ **5.** $4^{\frac{3}{2}} = 8$ **6.** $8^{\frac{2}{3}} = 4$ **7.** 2
8. 4 **9.** -2 **10.** -1 **11.** $-\frac{1}{2}$ **12.** $\frac{2}{3}$ **13.** $x = 625$
14. $x = 1$ **15.** $x = \frac{1}{25}$ **16.** $x = 11$ **17.** $x = 4$
18. $x = 16$ **19.** 32 **20.** 112 **21.** 115

Written Exercises 5-4, pages 162-164 **1.** $8^{-\frac{1}{3}} = \frac{1}{2}$
3. $6^0 = 1$ **5.** $(\sqrt{3})^4 = 9$ **7. a.** 2 **b.** 5 **c.** 6 **d.** 10
9. a. -1 **b.** -3 **c.** $\frac{1}{3}$ **d.** 0 **11. a.** 2 **b.** $\frac{1}{2}$ **c.** $\frac{3}{2}$
d. $-\frac{1}{3}$ **13. a.** $\frac{5}{2}$ **b.** -3 **c.** -2 **d.** 4 **15.** $x = 1000$
b. $x = \pm 1000$ **17. a.** $x = 8$ **b.** $x = 2$ **19. a.** $x = 4$
b. $x = 9$ **21. a.** $x = \pm 4$ **b.** $x = 4$ or $x = -1$
23. a. $x = 3$ **b.** $x = 16$

25. a. See graph at right.
b. (1) $f(x) = \log_3 x$; domain:
positive real numbers; range:
all real numbers; zeros: $x = 1$
(2) $g(x) = 3^x$; domain: all real
numbers; range: positive real
numbers; zeros: none

27. a. 68 **b.** 30 **29. a.** 65 decibels **b.** 67 decibels
31. Human gastric juice: $pH = 2$, acidic; acid rain:
$pH = 4.5$, acidic; pure water: $pH = 7$, neutral; good
soil for potatoes: $pH = 7.3$, alkaline; sea water:
$pH = 9$, alkaline **33. a.** $\log_4 16 = 2$, $\log_{16} 4 = \frac{1}{2}$
b. $\log_9 27 = \frac{3}{2}$, $\log_{27} 9 = \frac{2}{3}$

Calculator Exercises, page 164 **1. a.** $\log_2 25$ is between
4 and 5 **b.** 6.03

Oral Exercises 5-5, pages 166-167
1. $2 \log_b M + \log_b N$ **2.** $2 \log_b M - \log_b N$
3. $\frac{1}{2} \log_b M - 2 \log_b N$ **4.** $\frac{1}{3} \log_b M + \frac{1}{3} \log_b N$
5. $\log_b M + \frac{1}{2} \log_b N$ **6.** $2 \log_b M - 3 \log_b N$ **7.** $\log_5 6$
8. $\log_3 20$ **9.** 1 **10.** $\log_7 9$ **11.** 2 **12.** $\frac{1}{2}$ **13.** $\log MN^2$
14. $\log \dfrac{P^2}{Q}$ **15.** $\log MNP$ **16.** $\log \dfrac{MN}{P^3}$ **17.** $\log \sqrt{\dfrac{M}{N}}$
18. $R\sqrt[3]{S}$ **19.** no **20.** no

Written Exercises 5-5, pages 167-169
1. $2 \log_b M + 2 \log_b N$ **3.** $\frac{1}{3} \log_b M - \frac{1}{3} \log_b N$
5. $2 \log_b M + \frac{1}{2} \log_b N$ **7.** $\log_5 24$ **9.** $\log_6 15$ **11.** 2

13. $\log \dfrac{M}{N^3}$ **15.** $\log_b \dfrac{AB^2}{C^3}$ **17.** $\log \sqrt[3]{\dfrac{M^2}{NP}}$ **19.** $\log_2 \pi r^2$
21. 2 **23.** 2.4393 **25.** 0.2386 **27.** $N = 552,000$
29. 0.0541 **31. a.** $y = x^2$ **b.** $y = 5x^3$ **c.** $y = 49x$
d. $y = \dfrac{4}{x^2}$ **33. a.** $r + 2s$ **b.** $2r + 2s$ **c.** $r - 2s$
d. $r - 2$ **35.** $x = \frac{6}{5}$ **37.** $x = 2$ **39.** $x = 5$
41. a. $f(1) = 0$ **f.** $f(x) = \log_b x$, $b > 0$, $b \neq 1$
g. $f(100) = 2$, $f(1000) = 3$ **43.** 1230 km

Oral Exercises 5-6, page 170 **1.** $x = \sqrt[3]{81} = 3\sqrt[3]{3}$
2. $x = 4$ **3.** $x = 1.5$ **4.** $x = \pm\sqrt[4]{8}$ **5.** $x = 0.4771$
6. $x = 0.9085$ **7.** $x = 2.4082$ **8.** $x = 1.2400$

Written Exercises 5-6, page 171 **1.** $x = 2.26$
3. $x = 18.9$ **5.** $x = -4.17$ **7.** $x = 7.12$
9. a. $x = 2.25$ **b.** $x = 2.16$ **11. a.** $x = 0$
b. $x = 0.22$ **13.** 1.37 **15.** -0.315 **19.** 1 **21.** $\frac{1}{2}$
23. 2 **25. a.** $\dfrac{1}{x}$ **b.** $\dfrac{2}{x}$ **c.** $\dfrac{1}{x + y}$
27. $x = \log_2 3 = 1.58$ **29.** $x = 0$ **31.** $x = 0.5$ or
$x = 1.5$ **35. a.** 49.5 cm **b.** 279.4 cm

Calculator Exercises, page 172 **1.** $x = 1.85$

Oral Exercises 5-7, page 175 **1. a.** 10.5% **b.** 1.03,
1.15, 1.046, 2.2 **2. a.** 0.32% **b.** 0.88, 0.925, 0.2, 0
3. a. $200(1.05)^t$ **b.** $20(1.08)^t$ **c.** $4.5(1.105)^t$ **d.** 6%
4. a. $9800(0.80)^t$ **b.** $2200(0.85)^t$ **c.** 25% **5. a.** $23\frac{1}{3}$
years **b.** 14 years **c.** $8\frac{3}{4}$ years

Written Exercises 5-7, pages 176-178 **1.** \$1214; \$4910
3. \$1.97; \$3.87 **5.** \$14,766; \$6229 **7.** \$0.83; \$0.69
9. a. 1600 g **b.** 800 g **c.** 100 g **d.** $3200(\frac{1}{2})^{\frac{1}{4}}$
11. \$8000; \$6400; \$5100;
\$4100; \$3300 See graph
at right. **13.** \$112.68
15. About \$55 billion
17. $23\frac{1}{3}$ years **19.** $11\frac{2}{3}$
years **21.** 7% **23.** 6%
25. $A_0(\frac{1}{2})^{10}$ **27.** 308; 154
29. no **31.** \$5657 **33.** 35
years **35.** 1.0%
37. a. $a = 10$, $b \approx 0.9963$ **b.** 1:42 P.M.

Calculator Exercises, pages 178-179 **1.** 169 million km
3. 6 billion years

Oral Exercises 5-8, page 180 **1.** $e^{1.8} = 6$ **2.** $e^{1.1} = 3$
3. $e^{4.6} = 100$ **4.** 1 **5.** 2 **6.** 1 **7.** 2 **8.** 5 **9.** -2
10. $\ln 50 = 3.9$ **11.** $\ln 2.718 = 1$ **12.** $\ln \frac{1}{2} = -0.7$
13. e **14. a.** $(1.02)^4$ **b.** $(1.01)^8$ **c.** $e^{0.08}$

Written Exercises 5-8, pages 181-183 **1. a.** $e^3 \approx 20$
b. $e^{9.2} \approx 10,000$ **c.** $e^{-1.2} \approx 0.3$ **d.** $e^1 = e$ **3. a.** 2
b. 3 **c.** -1 **d.** $\frac{1}{2}$ **5.** $\ln 2$ **7.** $\ln 10$ **9.** $\ln 1000$

11. $\ln 18$ 13. a. x b. x c. x^2 d. $\dfrac{1}{x}$

15. $x = e^3 + 2$ 17. $x = e^4$ 19. $x = e^3$ 21. $x = 2$
23. $x = e$ 25. $x = 2$ 27. a. $x = 1.0986$
b. $x = 0.5493$ c. $x = 2.0986$ 29. a. 10.25%
b. 10.38% c. 10.5%

31.
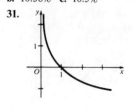

33. $x = \ln 3$ or
$\quad x = \ln 2$
35. $x = \ln(1 + \sqrt{11})$
$\quad \approx 1.4624$
37. $x = \ln 2$
43. a. $e - 1$ or
$\quad \approx 1.718$
b. ≈ 1.052
c. ≈ 1.005

Calculator Exercises, page 183 **1.** 30 years
3. 3.04×10^{64} **5.** e^π

Chapter Test, page 185 **1.** a. 36 b. 27 c. $\frac{9}{10}$ d. 256
3. a. $x = \pm\frac{1}{8}$ b. $y = \frac{1}{16}$ c. $y = -\frac{1}{2}$ d. $x = 27$ or
$x = 1$ **5.** a. 4 b. $\frac{7}{2}$ c. -3 d. $-\frac{1}{3}$
7. a. $2\log M + 2\log N$ b. $\frac{1}{3}\log M - \log N$
9. a. $x = \pm\frac{1}{25}$ b. $x = 5$ **11.** \$4900 **13.** a. 1 b. 4
c. 4

Cumulative Review (Chapters 1–5), pages 186–187
1. a. $(-4, -1)$; $4\sqrt{5}$ b. $y = -\frac{1}{2}x - 3$
c. $2x - y = -7$ **3.** a. $2x - 5y = -22$
b. $5x + 2y = 3$ **5.** $b^2 - 4ac = 30^2$ so the roots are
rational; $(3x - 8)(3x + 2) = 0$; $x = \frac{8}{3}$ or $x = -\frac{2}{3}$
7. $t = -1 \pm 3i$ **9.** $C(3, -6)$; $r = 7$ **11.** $\dfrac{3 + \sqrt{5}}{2}$,
$\dfrac{3 - \sqrt{5}}{2}$; $f(x + 1) = x^2 - x - 1$; $f(-i) = 3i$

13.
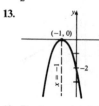

17. a. The only possible
rational roots, ± 5 and ± 1,
do not satisfy the equation;
$P(-2) = 1$ and
$P(-3) = -13$, so there is
an irrational root between
-2 and -3. b. -2.1
19. Roots are $\pm i$, $2 \pm \sqrt{5}$; sum is 4; product is -1;
$a = -4$; $b = -4$; $c = -1$ 21. a. $k = \pm 2\sqrt{2}$
b. $-2\sqrt{2} < k < 2\sqrt{2}$ 23. 23; $\frac{121}{2}$ 25. $h^{-1}(x) = \dfrac{x + 1}{x}$;
$h(x)$ has domain $\{x : x \neq 1\}$ and range $\{y : y \neq 0\}$;
$h^{-1}(x)$ has domain $\{x : x \neq 0\}$ and range $\{y : y \neq 1\}$
27. a. $-\frac{2}{3}$ b. $-\frac{15}{16}$ 29. a. $\frac{5}{3}$ b. $-\frac{1}{3}$ c. 6 31. 1
33. a. 6.12 b. -0.49 35. $\dfrac{\log 5}{\log e} - 2$

Oral Exercises 6–1, page 192 **1.** a. π b. $\dfrac{\pi}{2}$ c. $\dfrac{\pi}{4}$
d. $\dfrac{\pi}{3}$ e. $\dfrac{2\pi}{3}$ f. $\dfrac{4\pi}{3}$ g. $\dfrac{\pi}{6}$ h. $\dfrac{\pi}{180}$ **2.** a. $360°$

b. $180°$ c. $90°$ d. $45°$ e. $135°$ f. $60°$ g. $30°$
h. $150°$ **3.** Examples: a. $370°, -350°$ b. $460°, -260°$
c. $40°, -320°$ d. $355°, -365°$ e. $3\pi, -\pi$ f. $\dfrac{5\pi}{2}$,
$-\dfrac{3\pi}{2}$ g. $2\pi, -2\pi$ h. $\dfrac{5\pi}{3}, -\dfrac{7\pi}{3}$ **4.** a. $60°$
b. $\theta = \dfrac{\pi}{3} + 2\pi n$ **5.** a. 1 b. 2 c. 0.75

Written Exercises 6–1, pages 193–194 **3.** a. $\dfrac{7\pi}{4}$ b. $\dfrac{5\pi}{4}$
c. $\dfrac{\pi}{12}$ d. $-\dfrac{\pi}{4}$ **5.** a. $-\dfrac{2\pi}{3}$ b. $-\dfrac{4\pi}{3}$ c. $\dfrac{5\pi}{3}$ d. 2π
7. a. $-90°$ b. $240°$ c. $-135°$ d. $-30°$ **9.** a. $180°$
b. $-270°$ c. $120°$ d. $210°$ **11.** a. 1.2 b. 0.75
13. a. 1.66 b. 1.92 c. 1.66 d. 2.08 **15.** a. $91.7°$ or
$91°40'$ b. $97.4°$ or $97°24'$ c. $69.3°$ or $69°20'$ d. $75.6°$
or $75°38'$ **17.** $\dfrac{\pi}{6}$ **19.** Examples: a. $140°, -220°$
b. $300°, -420°$ c. $\dfrac{9\pi}{4}, -\dfrac{7\pi}{4}$ d. $\dfrac{4\pi}{3}, -\dfrac{8\pi}{3}$
21. Examples: a. $30', -359°30'$ b. $269°20', -450°40'$
c. $363°21', -356°39'$ d. $475°15', -244°45'$
23. Examples: a. $388.5°, -331.5°$ b. $476.3°, -243.7°$
c. $299.6°, -420.4°$ d. $44.7°, -675.3°$
25. $\theta = (29.7 + 360n)°$ **27.** a. $12{,}600°$ b. 219.91
29. a. 900 b. 15.71 **31.** a. 5256 b. 91.73

Calculator Exercises, pages 194–195 **1.** $12°48'0''$
3. $-18°26'24''$ **5.** $80.43°$ **7.** $280°$ **9.** a. $171.9°$
b. $70.5°$

Oral Exercises 6–2, pages 196–197 **1.** a. π; 2π b. 4π;
12π c. 8; 8 **2.** a. 0.1 km b. The circle is so large
and the arc length so small, it is nearly a straight line.

Review Exercises, page 197 **1.** a. $\dfrac{\pi}{3}$ b. $\dfrac{2\pi}{3}$ c. $\dfrac{5\pi}{6}$
d. $-\dfrac{\pi}{2}$ e. $\dfrac{3\pi}{2}$ f. $\dfrac{\pi}{4}$ g. $\dfrac{3\pi}{4}$ h. $-\dfrac{5\pi}{4}$ i. $\dfrac{7\pi}{4}$ j. $\dfrac{7\pi}{6}$
k. $\dfrac{11\pi}{6}$ l. $-\dfrac{5\pi}{3}$

Written Exercises 6–2, pages 197–200 **1.** b. 3 cm
3. 5 cm; 27.5 cm² **5.** 10 cm; 5 cm **7.** 12 cm² **9.** 2 cm
or 1.5 cm **11.** a. $16{,}200°$ b. 90π c. 810π cm
13. 402,000 km **15.** 1.3×10^6 km **17.** 1885 km²
19. approx. 0.35 km **21.** b. 5 c. 2

Calculator Exercise, page 200 **1.** 2.9×10^{13} km

Oral Exercises 6–3, page 203 **1.** $\dfrac{\sqrt{10}}{10}, -\dfrac{3\sqrt{10}}{10}$
2. $-\dfrac{2\sqrt{5}}{5}, \dfrac{\sqrt{5}}{5}$ **3.** $-\dfrac{4\sqrt{41}}{41}, -\dfrac{5\sqrt{41}}{41}$ **4.** 1; -1
5. $\pm 90°, \pm 270°, \pm 450°$ **6.** a. θ is coterminal with an
angle of $45°$. b. $\theta = \dfrac{\pi}{4} + 2\pi n$

7. $\theta = 270° + n \cdot 360°$; $\theta = n \cdot 180°$ **8.** $\theta = n \cdot 360°$; $\theta = 180° + n \cdot 360°$ **9. a.** decrease **b.** decrease **c.** increase **d.** increase **10. a.** increase **b.** decrease **c.** decrease **d.** increase **11. a.** positive **b.** negative **c.** negative **d.** positive **e.** positive **f.** negative **g.** negative **h.** positive **i.** positive **j.** negative **k.** negative **l.** negative

Written Exercises 6-3, pages 204-205 1. a. 0 **b.** -1 **c.** -1 **d.** 0 **3. a.** 0 **b.** -1 **c.** -1 **d.** 0 **5. a.** II **b.** III **7. a.** $x = \frac{\pi}{2} + 2\pi n$ **b.** $x = \pi + 2\pi n$ **c.** $x = \pi n$ **d.** no such x **9. a.** negative **b.** negative **c.** positive **d.** negative **11. a.** negative **b.** zero **c.** negative **d.** positive **13.** $\sin \theta = \frac{4}{5}$, $\cos \theta = \frac{3}{5}$ **15.** $\sin \theta = \frac{12}{13}$, $\cos \theta = -\frac{5}{13}$ **17.** $\cos \theta = \frac{3}{5}$ **19.** $\sin \theta = -\frac{7}{25}$ **21.** $\cos \theta = -\frac{2\sqrt{6}}{5}$ **23.** $\sin \theta = -\frac{\sqrt{7}}{4}$ **25.** $>$ **27.** $=$ **29.** $=$ **31.** $<$ **33.** $>$ **35.** $\cos 3$, $\cos 4$, $\cos 2$, $\cos 1$ **37. a.** $\left(-\frac{1}{2}, \pm\frac{\sqrt{3}}{2}\right)$

Oral Exercises 6-4, page 209 1. 0.1736 **2.** 0.4787 **3.** 0.8090 **4.** 0.8480 **5.** 0.2924 **6.** 0.2955 **7.** 0.1700 **8.** 0.9249 **9. a.** 10° **b.** 25.1° **c.** 50° **d.** $\pi - 3$ **10.** Examples: **a.** 110° **b.** $-20°$ **11. a.** 320° **b.** 195° **12. a.** $\sin 10°$ **b.** $-\sin 30°$ **c.** $-\sin 40°$ **d.** $\sin 40°$ **e.** $-\sin 15°$ **13. a.** $-\cos 20°$ **b.** $-\cos 2°$ **c.** $\cos 60°$ **d.** $\cos 5°$ **e.** $-\cos 80°$ **14. a.** -0.1392 **b.** 0.9397 **c.** 0.9962 **d.** 0.9962 **15. a.** 0.4169 **b.** 0.4625 **c.** 0.1411 **d.** -0.6536 **16. a.** $\frac{\sqrt{2}}{2}$ **b.** $\frac{\sqrt{2}}{2}$ **c.** $-\frac{\sqrt{2}}{2}$ **d.** $-\frac{\sqrt{2}}{2}$ **17. a.** $\frac{1}{2}$ **b.** $-\frac{1}{2}$ **c.** $-\frac{1}{2}$ **d.** $-\frac{\sqrt{3}}{2}$ **18. a.** $\frac{1}{2}$ **b.** $-\frac{1}{2}$ **c.** $\frac{\sqrt{3}}{2}$ **d.** $\frac{\sqrt{3}}{2}$ **19. a.** $\frac{\sqrt{3}}{2}$ **b.** $-\frac{1}{2}$ **c.** $-\frac{1}{2}$ **d.** $-\frac{\sqrt{3}}{2}$ **20. a.** $\frac{\sqrt{2}}{2}$ **b.** $-\frac{\sqrt{3}}{2}$ **c.** $-\frac{\sqrt{3}}{2}$ **d.** $\frac{\sqrt{3}}{2}$

Written Exercises 6-4, pages 210-211 1. a. 0.47 **b.** 0.72 **c.** 0.42 **d.** 0.47 **3. a.** 0.34 **b.** -0.90 **c.** 0.53 **d.** -0.46 **5. a.** 0.93 **b.** 0.81 **c.** -0.94 **d.** 0.28 **7. a.** $\frac{\sqrt{2}}{2}$ **b.** $-\frac{\sqrt{2}}{2}$ **c.** $-\frac{1}{2}$ **d.** $\frac{\sqrt{3}}{2}$ **9. a.** $\frac{1}{2}$ **b.** $-\frac{1}{2}$ **c.** $-\frac{\sqrt{2}}{2}$ **d.** $\frac{\sqrt{3}}{2}$ **11. a.** $\frac{1}{2}$ **b.** $\frac{1}{2}$ **c.** $-\frac{1}{2}$ **d.** $\frac{\sqrt{2}}{2}$ **13. a.** 1 **b.** $-\frac{1}{2}$ **c.** $-\frac{\sqrt{3}}{2}$ **d.** $-\frac{\sqrt{2}}{2}$ **17. a.** -1 **b.** 0 **c.** $-\frac{1}{2}$ **19.** 60° **21.** 8029 km **23. a.** approx. 1185 km/h **b.** approx. 1451 km/h **25. b.** 0 **27.** $\cos \theta = \cos (-\theta)$

29. $P(\cos t, \sin t)$; $Q(\cos t + \sqrt{16 - \sin^2 t}, 0)$

31.

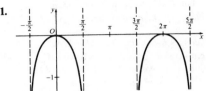

Calculator Exercise, page 212 59.3°

Oral Exercises 6-5, page 216 1. $\theta = 90° + n \cdot 180°$ **2.** $\theta = 90° + n \cdot 180°$ **3.** none; $\theta = n \cdot 360°$; $\theta = 180° + n \cdot 360°$ **4.** $\theta = n \cdot 180°$; $\theta = 45° + n \cdot 180°$; $\theta = 135° + n \cdot 180°$ **5.** π **6.** 2π **7. a.** 33° **b.** -0.6494; 0.6494; -0.6494 **8. a.** -1.0353 **b.** -1.0353 **c.** 1.0353 **9. a.** 11.43 **b.** -1.305 **10.** l_1: 1; l_2: $\sqrt{3} \approx 1.732$ **11.** III **12.** IV **13. a.** $-\frac{3}{5}$ **b.** $-\frac{4}{3}$ **c.** $-\frac{3}{4}$ **d.** $-\frac{5}{3}$ **e.** $\frac{5}{4}$

Written Exercises 6-5, pages 217-218 1. a. $-\sin 15°$ **b.** $\sec 80°$ **c.** $\tan 40°$ **d.** $-\sec (\pi - 2)$ **3. a.** $-\tan 80°$ **b.** $\sec 70°$ **c.** $\cot 5$ **d.** $-\csc (4 - \pi)$ **5. a.** $x = n\pi$ **b.** no such x **c.** $x = \frac{\pi}{2} + 2n\pi$ **d.** $x = \frac{3\pi}{2} + 2n\pi$ **7. a.** $-\frac{1}{2}$ **b.** -2 **c.** $\frac{\sqrt{3}}{2}$ **d.** $-\sqrt{3}$ **9. a.** 1 **b.** -1 **c.** $\sqrt{3}$ **d.** undefined **11. a.** undefined **b.** $-\sqrt{3}$ **c.** 0 **d.** $-\frac{2\sqrt{3}}{3}$ **13. a.** 1 **b.** 0 **c.** $\frac{1}{2}$ **15. a.** -5.671 **b.** -0.1051 **c.** -1.043 **d.** -1.855 **17.** Slide the graph of $\sec x$ 90° to the right. **19.** $\cos x = -\frac{12}{13}$, $\tan x = -\frac{5}{12}$, $\csc x = \frac{13}{5}$, $\sec x = -\frac{13}{12}$, $\cot x = -\frac{12}{5}$ **21.** $\sin x = -\frac{3}{5}$, $\cos x = -\frac{4}{5}$, $\cot x = \frac{4}{3}$, $\csc x = -\frac{5}{3}$, $\sec x = -\frac{5}{4}$ **23.** $\sin x = \frac{2\sqrt{2}}{3}$, $\cos x = -\frac{1}{3}$, $\tan x = -2\sqrt{2}$, $\csc x = \frac{3\sqrt{2}}{4}$, $\cot x = -\frac{\sqrt{2}}{4}$ **25. b.** $\sin^2 x + \cos^2 x = 1$ is true for all values of x. **27. a.** 45° **b.** 168°40′ **29. a.** 149° **b.** 36°50′ **31. a.** domain: $\{x | x \neq \pi(n + \frac{1}{2})\}$; range: all real x; period: π **b.** domain: $\{x | x \neq \pi(n + \frac{1}{2})\}$; range: $\{x | x \leq -1 \text{ or } x \geq 1\}$; period: 2π **33.** $\{x | n\pi < x < \frac{\pi}{2} + n\pi\}$ **35. a.** $\tan x$ **b.** $f(x + p) = f(x)$ **37. b.** 48.2°, 67.1°, 64.7°

Calculator Exercise, page 218 a. $x = 0$ **b.** $x \approx 4.49$

Oral Exercises 6-6, page 222 1. a. $\cos^2 \theta$ **b.** $\sin^2 \theta$ **2. a.** $\tan^2 \theta$ **b.** 1 **3. a.** 1 **b.** $\cot^2 \theta$ **4. a.** $\sin \theta$ **b.** $\cot A$ **c.** $\sin x$ **5. a.** $\sin x$ **b.** 1 **c.** 1 **6. a.** $\cos^2 x$ **b.** $-\cos^2 x$ **c.** $\tan^2 x$ **7. a.** $\cot A$ **b.** 1 **c.** $\cos y$ **8. a.** $\cot B$ **b.** $\sec A$ **c.** -1 **9. a.** 1 **b.** 1

10. a-f. true **Ex. 11-14:** Mult. num. and den. by:
11. a. t b. $\tan A$ 12. a. x b. $\cos \beta$ 13. a. y
b. $\cot \alpha$ 14. a. xy b. $\cos \theta \sin \theta$

Written Exercises 6-6, pages 223-226 1. a. 1
b. $\sin^2 \theta$ c. $-\cos^2 \theta$ 3. a. $\csc^2 A$ b. $\cot^2 A$ c. 1
5. a. $\sec x$ b. 1 c. $\cos x$ 7. a. 1 b. $\sin^2 A$
c. $\cos^2 \theta$ 9. a. $y^2 + x^2$ b. 1 11. a. $a^2 - b^2$ b. 1
13. a. $\dfrac{t}{t^2 + 1}$ b. $\sin \theta \cos \theta$ 15. $\sec A$ 17. 1
19. $\tan x$ 21. 1 23. $\cot x$ 25. 2 27. $\csc \theta$ 29. $\csc y$
31. 1 35. $\sin \theta$ 63. $\sec \theta = \pm \dfrac{1}{\sqrt{1 - \sin^2 \theta}}$
65. $\csc \theta = \pm \dfrac{1}{\sqrt{1 - \cos^2 \theta}}$ 67. 1 69. 1

Oral Exercises 6-7, page 228 1. 66°, 294° 2. 114°, 246° 3. 46°, 134° 4. 226°, 314° 5. 62°, 242° 6. 118°, 298° 7. 6°, 174° 8. 169°, 349° 9. 0.67, 2.47 10. 3.81, 5.61 11. 1.32, 4.96 12. 1.82, 4.46 13. 60°, 300° 14. 225°, 315° 15. 30°, 150° 16. 135°, 315° 17. $\pi + 2n\pi$ 18. $\dfrac{\pi}{4} + n\pi$

Written Exercises 6-7, pages 228-229 1. a. 44.4°, 135.6° b. 224.4°, 315.6° 3. a. 50.2°, 230.2° b. 129.8°, 309.8° 5. a. 78.5°, 281.5° b. 101.5°, 258.5° 7. 70.5°, 289.5° 9. 146.4°, 213.6° 11. 41.8°, 138.2° 13. 0.73, 2.41 15. 1.23, 1.91, 4.37, 5.05 17. 0.83, 2.31 19. 210°, 330° 21. 150°, 210° 23. 45°, 225° 25. 60°, 300° 27. no such θ 29. $\dfrac{\pi}{2}, \dfrac{3\pi}{2}$
31. $\dfrac{2\pi}{3}, \dfrac{4\pi}{3}$ 33. $\dfrac{\pi}{3}, \dfrac{4\pi}{3}$ 35. $\dfrac{\pi}{6}, \dfrac{5\pi}{6}, \dfrac{7\pi}{6}, \dfrac{11\pi}{6}$
37. 0, π 39. $\dfrac{\pi}{2}$ 41. $\dfrac{\pi}{3}, \dfrac{4\pi}{3}$ 43. $\dfrac{\pi}{6}, \dfrac{7\pi}{6}$ 45. $\dfrac{\pi}{4} + \dfrac{n\pi}{2}$
47. $\dfrac{\pi}{4} + \dfrac{n\pi}{2}$

Written Exercises 6-8, pages 231-232 1. 0°, 60°, 300°
3. 30°, 41.8°, 138.2°, 150° 5. 60°, 70.5°, 289.5°, 300°
7. 270° 9. $\dfrac{\pi}{4}, \dfrac{5\pi}{4}$ 11. 0, $\dfrac{\pi}{2}$, π 13. $\dfrac{\pi}{2}, \dfrac{3\pi}{2}, \dfrac{\pi}{3}, \dfrac{5\pi}{3}$
15. $\dfrac{\pi}{4}, \dfrac{5\pi}{4}$ 17. 0.32, 3.46 19. $\dfrac{\pi}{6}, \dfrac{\pi}{2}, \dfrac{3\pi}{2}, \dfrac{5\pi}{6}$ 21. 0, π,
$\dfrac{\pi}{3}, \dfrac{5\pi}{3}$ 23. $\dfrac{\pi}{4}$, 1.11, $\dfrac{5\pi}{4}$, 4.25 25. $\dfrac{\pi}{6}, \dfrac{3\pi}{2}, \dfrac{5\pi}{6}$ 27. 2,
4.28 29. 0.34, 2.8 31. 3 33. $\dfrac{\pi}{6}, \dfrac{\pi}{2}, \dfrac{5\pi}{6}, \dfrac{3\pi}{2}$ 35. $\dfrac{\pi}{3}$,
$\dfrac{2\pi}{3}, \pi, \dfrac{4\pi}{3}, \dfrac{5\pi}{3}$ 37. 1.89, 5.03 39. $\dfrac{\pi}{6}, \dfrac{\pi}{4}, \dfrac{3\pi}{4}, \dfrac{5\pi}{6}, \dfrac{5\pi}{4}$,
$\dfrac{7\pi}{4}$ 41. 5.36, $\dfrac{\pi}{2}$ 43. $\dfrac{\pi}{6}, \dfrac{5\pi}{6}$

Chapter Test, pages 234-235 1. a. 30° b. 135°
c. −420° d. $\dfrac{360°}{\pi} \approx 114.6°$ 3. 15 cm; 157.5 cm²

5. a. −1 b. −1 c. 0 d. 1 7. b. range:
$\{y \,|\, -1 \le y \le 1\}$; period: 2π 9. domain:
$\left\{x \,|\, x \ne \dfrac{\pi}{2} + n\pi\right\}$; range: all real numbers; period: π
11. a. 1.11, 4.25 b. $\dfrac{\pi}{2}, \dfrac{3\pi}{2}$

Oral Exercises 7-1, page 240 1. a. $\sin \alpha = \dfrac{x}{z}$,
$\cos \alpha = \dfrac{y}{z}$, $\tan \alpha = \dfrac{x}{y}$ b. $\sin \beta = \dfrac{y}{z}$, $\cos \beta = \dfrac{x}{z}$,
$\tan \beta = \dfrac{y}{x}$ c. Each equation is true 2. a. sine or cosecant b. cosine or secant c. tangent or cotangent d. cosine or secant 3. Both are correct.
4. a. $\tan \theta = \frac{3}{2}$ or $\cot \theta = \frac{2}{3}$ b. $\cos \theta = \frac{5}{8}$ or $\sec \theta = \frac{8}{5}$
c. $\sin \theta = \frac{7}{4}$ or $\csc \theta = \frac{4}{7}$ d. $\cos \theta = \frac{7}{20}$ or $\sec \theta = \frac{20}{7}$
5. $\sin = \dfrac{\text{opposite}}{\text{hypotenuse}}$, $\cos = \dfrac{\text{adjacent}}{\text{hypotenuse}}$,
$\tan = \dfrac{\text{opposite}}{\text{adjacent}}$, $\cot = \dfrac{\text{adjacent}}{\text{opposite}}$, $\sec = \dfrac{\text{hypotenuse}}{\text{adjacent}}$,
$\csc = \dfrac{\text{hypotenuse}}{\text{opposite}}$

Written Exercises 7-1, pages 240-243 1. $b \approx 7.61$,
$c \approx 16.3$ 3. $d \approx 43.3, f \approx 42.3$ 5. a. $\frac{5}{13}$ b. $\frac{5}{13}$ c. $\frac{5}{12}$
d. $\frac{5}{12}$ e. $\frac{13}{12}$ f. $\frac{13}{12}$ g. $\angle A \approx 22.6°$ h. $\angle B \approx 67.4°$
7. 48.2°, 48.2°, 83.6°; 17.9 sq. units 9. 18.4 cm
11. 126.9° 13. 32.6° 15. $\sin A = \dfrac{\sqrt{t^2 - 4}}{t}$,
$\tan A = \dfrac{\sqrt{t^2 - 4}}{2}$, $\cot A = \dfrac{2}{\sqrt{t^2 - 4}}$, $\sec A = \dfrac{t}{2}$,
$\csc A = \dfrac{t}{\sqrt{t^2 - 4}}$ 17. $\sin A = \dfrac{3}{\sqrt{4t^2 + 9}}$,
$\cos A = \dfrac{2t}{\sqrt{4t^2 + 9}}$, $\tan A = \dfrac{3}{2t}$, $\sec A = \dfrac{\sqrt{4t^2 + 9}}{2t}$,
$\csc A = \dfrac{\sqrt{4t^2 + 9}}{3}$ 19. a. 21.8 m b. 21.9 m
21. 43.3 m 23. $c = a \csc A$ 25. $b = a \tan B$
27. 21.5 m Answers for Exs. 31 and 33 may vary. 31. $x = a \tan \alpha, y = a \tan \alpha \sin \beta$
33. $x = a \sec^3 \alpha, y = a \sec^3 \alpha \sin \alpha$ 39. 3.10 cm²
41. 35.3° 43. $A = nr^2 \sin \dfrac{\pi}{n} \cos \dfrac{\pi}{n}$

Calculator Exercises, page 244 1. a. $\angle RPS \approx 1.29$,
$\angle RQS \approx 1.86$ b. 12 sq. units c. 1.08 sq. units
3. $\tan 15° \approx 0.2679$, $\dfrac{1}{2 + \sqrt{3}} \approx 0.2679$

Oral Exercises 7-2, pages 245-246 1. 10 sq. units
2. 10 sq. units 3. 10 sq. units 4. $\dfrac{5\sqrt{2}}{2}$ sq. units
5. a. $\sin \theta = 0.5$ b. 30° or 150°

Written Exercises 7-2, pages 246–247 **1. a.** 5 sq. units
b. 5 sq. units **3. a.** $3\sqrt{2}$ sq. units **b.** $3\sqrt{2}$ sq. units
7. 158 sq. units **9.** 9 **11.** 30° or 150°
13. $3200\sqrt{2}$ cm^2 **15.** 21 cm^2 **17.** 3.46 cm
19. $K = 6\sin\theta$; domain:
$0° < \theta < 180°$; range:
$0 < K \le 6$ (See diagram.)
21. 13.6 sq. units
23. 55.2
25. area $\approx$ 22.1 sq. units
27. area $\approx$ 6.64 sq. units

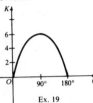

Ex. 19

29. a. area $(\triangle BCD) = \dfrac{\sqrt{3}}{4}x$, area $(\triangle ACD) = \dfrac{\sqrt{3}}{2}x$

b. $\dfrac{\sqrt{3}}{2}$ **c.** $x = \dfrac{2}{3}$

Calculator Exercise, page 247 $\theta \approx 2.5$ radians

Oral Exercises 7-3, page 250 **1.** $\angle B < 90°$ **2.** $b > c$
3. $\dfrac{\sin 60°}{8} = \dfrac{\sin 20°}{x}$ **4.** $\dfrac{\sin 48°}{9} = \dfrac{\sin x°}{5}$
5. $\dfrac{\sin 120°}{7} = \dfrac{\sin 40°}{x}$

Written Exercises 7-3, pages 250–252 **1.** $\angle C = 75°$,
$b = 7\sqrt{6} \approx 17.1$, $c \approx 19.1$ **3.** $\angle C = 15°$,
$a = 4\sqrt{2} \approx 5.66$, $c \approx 2.07$ **5.** $\angle B \approx 39.3°$ or 140.7°,
$\angle C \approx 115.7°$ or 14.3°, $c \approx 4.26$ or 1.17
7. $\angle B \approx 18.9°$, $\angle C \approx 85.1°$, $c \approx 12.3$ **9.** no solution
11. $\angle S \approx 28.8°$, $\angle T \approx 11.2°$ **13.** 15.5 km
15. $13\sqrt{2}$ **17.** $4\sqrt{5}$
21. $K = \dfrac{1}{2}\left(\dfrac{\sin A \sin B}{\sin C}\right)c^2 = \dfrac{1}{2}\left(\dfrac{\sin A \sin C}{\sin B}\right)b^2$

Calculator Exercise, page 252 $\beta \approx 32.0°$

Oral Exercises 7-4, page 255
1. $x^2 = 6^2 + 5^2 - 2 \cdot 5 \cdot 6 \cdot \cos 35°$
2. $\cos x° = \dfrac{5^2 + 6^2 - 7^2}{2 \cdot 5 \cdot 6}$
3. $x^2 = 5^2 + 10^2 - 2 \cdot 5 \cdot 10 \cdot \cos 115°$ **4.** 6.23
5. $\angle Y = 79.6°$ and $\angle X = 50.4°$, or $\angle Y = 100.4°$ and
$\angle X = 29.6°$ (Using the reasoning of Oral Exercise 6,
we reject the first solution.) **6. a.** $\angle X < \angle Z < \angle Y$
b. You may have assumed incorrectly that $\angle Y$ was
acute.

Written Exercises 7-4, pages 255–257 **1.** $c = 7$,
$\angle A \approx 81.8°$, $\angle B \approx 38.2°$ **3.** $\angle Z = 90°$,
$\angle Y \approx 77.3°$, $\angle X \approx 12.7°$ **5.** $r \approx 6.49$, $\angle Q \approx 109.3°$,
$\angle P \approx 20.7°$ **7.** $\angle C = 120°$, $\angle B \approx 27.8°$,
$\angle A \approx 32.2°$ **9.** 90° **11. a.** $\cos C = \frac{3}{5} = 0.6$
b. $\sin C = \frac{4}{5} = 0.8$ **c.** 14 sq. units **13.** 7.07 cm
15. $\sqrt{10}$ cm **17.** 9.74 cm, 13.3 cm **19. a.** $AD = 8$,
$DB = 12$ **b.** 7.35 **21.** $\sin P = \frac{12}{13}$; $\cos P = \frac{5}{13}$ or $-\frac{5}{13}$;
$p = 15$ or 22.5 **25.** $KM = 7$, $JL \approx 7.14$ **27. c.** 7.35

Calculator Exercises, page 257 **1.** 5942 sq. rods
3. 1.81 km

Assorted Exercises 7-2, 7-3, and 7-4, pages 258–260
1. a. area $\approx$ 16.1 sq. units, $p = 5.57$
b. area $\approx$ 26.8 sq. units, $r \approx 15.5$ **3.** 81.8°
7. $\sin A = 0.8$, $\tan A = -\frac{4}{3}$ **9.** 32.7° or 147.3°
11. 19.5° or 160.5° **13. a.** $a = 14$, $e = \frac{120}{7}$, $f = \frac{200}{7}$
b. 49:400 **15.** 7.21 or 13.4 **17.** 5.48 **19.** $\sin x = -\dfrac{\sqrt{26}}{26}$,
$\cos x = \dfrac{5\sqrt{26}}{26}$ **21.** $\sin P = \dfrac{\sqrt{3}}{2}$, $\cos P = -\dfrac{1}{2}$
23. a. $\angle A \approx 38.2°$, $\angle B \approx 81.8°$, $\angle C = 60°$
b. 17.3 sq. units **c.** 4.33 **d.** 4.58 **e.** 4.41 **f.** 4.04
g. 1.73 **25. a.** $c = 17$, $\angle C \approx 126.9°$, $\angle A \approx 25.0°$,
$\angle B \approx 28.1°$ **b.** 36 sq. units **c.** 7.2 **d.** 12.6 **e.** 11.4
f. 10.625 **g.** 2 **27. a.** 3.92, 2.80 **37.** $\frac{6}{7}$

Oral Exercises 7-5, page 264 **1.** 0 **2.** $\frac{\pi}{3}$ **3.** $\frac{\pi}{3}$ **4.** π
5. 0 **6.** $\frac{\pi}{4}$ **7.** 0 **8.** $\frac{1}{2}$ **9.** 1 **10.** $\frac{\pi}{2}$ **11.** $-\frac{\pi}{6}$ **12.** $\frac{\sqrt{3}}{2}$
13. 0.80 **14.** -0.92 **15.** -0.25 **16. a.** $\frac{3}{5}$ **b.** $\frac{5}{3}$
c. $-\frac{4}{5}$ **17. a.** $\frac{b}{a}$ **b.** $-\frac{b}{a}$ **c.** $\frac{a}{c}$

Written Exercises 7-5, pages 264–266 **1.** $-\frac{\pi}{2}$ **3.** $\frac{\pi}{6}$
5. $\frac{3\pi}{4}$ **7.** $-\frac{\pi}{4}$ **9.** $-\frac{\pi}{3}$ **11.** $\frac{\pi}{3}$ **13.** $\frac{2\pi}{3}$ **15.** $-\frac{\pi}{4}$
17. a. -0.73 **b.** 2.42 **c.** 1.41 **19. a.** -1.11
b. -0.46 **c.** -1.57 **21. a.** $\dfrac{2\sqrt{6}}{5}$ **b.** $\dfrac{2\sqrt{6}}{5}$
23. a. $\dfrac{5\sqrt{26}}{26}$ **b.** $-\dfrac{\sqrt{26}}{26}$ **25. a.** $\frac{\pi}{4}$ **b.** $\frac{3\pi}{4}$
27. a. $-\frac{\pi}{3}$ **b.** $\frac{2\pi}{3}$ **29.** It results in an error message.
31. See the graphs of $G^{-1}(x)$ and $H^{-1}(x)$ on page 262.
33. $\dfrac{y}{\sqrt{1 - y^2}}$ **35. a.** true **b.** false. Counterexamples
may vary. For example, $x = \dfrac{3\pi}{4}$ **37.** false.
Counterexamples may vary. For example, $x = 0$.
39. $x = \dfrac{\pi}{2}$
43. $y = \text{Cot}^{-1}\,x$: domain is $-\infty < x < \infty$; range is
$0 < y < \pi$. **45.** $y = \text{Arccsc}\,x$: domain is $x \ge 1$,
$x \le -1$; range is $-\dfrac{\pi}{2} \le y \le \dfrac{\pi}{2}$, $y \ne 0$.

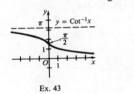

Ex. 43

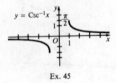

Ex. 45

Answers to selected exercises **13**

Calculator Exercises, page 266 1. Results in error message, because $-1 \le \sin \theta \le 1$ for all θ, but $\pi > 1$. **3.** 0.2164 **5.** 9.129

Written Exercises 7-6, page 269 1. 0.739 **3.** 1.962 **5.** 0.450 **7.** 2.132 **9.** 1.17 **11.** $y = 0.265, y = -0.265$

Chapter Test, page 271 1. 67.4° **3.** 34.1° or 145.9° **5. a.** none **b.** one **7.** 44.4° **9.** domain: $-1 \le x \le 1$; range: $0 \le y \le \pi$

Oral Exercises 8-1, pages 276-277 1. a. 4 **b.** $\frac{1}{2}$ **c.** $f(25) = 1, f(-25) = 0$ **2.** $y = 2f(x)$: 4, 1; $y = \frac{1}{2}f(x)$: 4, $\frac{1}{4}$ **3.** $y = f(2x)$: 2, $\frac{1}{2}$; $y = f(\frac{1}{2}x)$: 8, $\frac{1}{2}$ **4. a.** π, 4 **b.** 4π, 3 **5. a.** 2 **b.** 4 **c.** 6 **6. a.** 1 **b.** 2 **c.** 3 **7. a.** 2 **b.** 4 **c.** 1

Written Exercises 8-1, pages 277-279

1. No
3. Yes; 3, $\frac{1}{2}$
5. a, c.

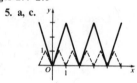

7. a. 2π, 2 **b.** 2π, $\frac{1}{2}$
c. π, 1 **d.** 4π, 1

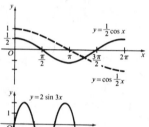

9. a. $\frac{2\pi}{3}$, 1

b. $\frac{2\pi}{3}$, 2

c. 6π, 1

11. a. $\frac{\pi}{2}$, not defined; $\tan 2x$ is undefined at

$x = \frac{\pi}{4} + \frac{n\pi}{2}$. **b.** 2π, not defined; $\tan \frac{1}{2}x$ is undefined at $x = \pi + 2n\pi$. **13. a.** 1, 3 **b.** 365, 5

15. $y = 4 \sin 2x$ **17.** $y = 4 \sin \frac{\pi}{6}t$ **19. a.** 0 **b.** 0, π

c. 0, $\frac{2\pi}{3}, \frac{4\pi}{3}$ **21. a.** $\frac{\pi}{6}, \frac{5\pi}{6}$ **b.** $\frac{\pi}{12}, \frac{5\pi}{12}, \frac{13\pi}{12}, \frac{17\pi}{12}$

c. $\frac{\pi}{3}, \frac{5\pi}{3}$ **23.** $\theta = 10°, 110°, 130°, 230°, 250°, 350°$

25. $\theta = 22.5°, 112.5°, 202.5°, 292.5°$
27. a, b.

29. b. (2, 32) **c.** (4, 96)
33. 3
37. b. No **c.** $x = -p$ or x is any rational number **d.** any irrational number such that $x \ne -p$

Oral Exercises 8-2, page 283 3. a. $x = 3$ **b.** $y = \pm x$; (0, 0) **c.** $x = 0, y = 0$; (0, 0) **4. a.** (i) – (iv) Yes

b. (i) Yes (ii) – (iv) No **c.** (i) No (ii) Yes (iii) – (iv) No **5. a.** $x = 4$ **b.** $x = 1$ **c.** $x = 0$ **6. a.** (2, 1) **b.** $(-1, -5)$ **c.** (0, 7)

Written Exercises 8-2, pages 284-286
1.

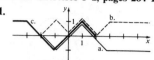

7. c.

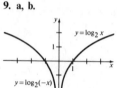

d.

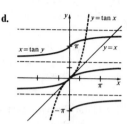

9. a, b.

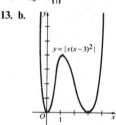

c, d.

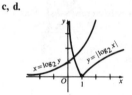

13. b.

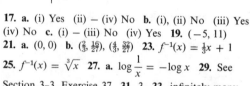

17. a. (i) Yes (ii) – (iv) No **b.** (i), (ii) No (iii) Yes (iv) No **c.** (i) – (iii) No (iv) Yes **19.** $(-5, 11)$
21. a. (0, 0) **b.** $(\frac{2}{3}, \frac{16}{27}), (\frac{4}{3}, \frac{32}{27})$ **23.** $f^{-1}(x) = \frac{1}{3}x + 1$
25. $f^{-1}(x) = \sqrt[3]{x}$ **27. a.** $\log \frac{1}{x} = -\log x$ **29.** See Section 3-3, Exercise 37 **31.** 3 **33.** infinitely many **35.** sin, csc, tan, cot are odd; cos is even. **37.** symmetry in the y-axis **41.** Region enclosed between the graphs has vertices at $(\frac{3\pi}{2}, -\frac{\pi}{2}), (2\pi, 0), (\pi, \pi)$, and $(\frac{\pi}{2}, \frac{\pi}{2})$. Area $= \pi^2$

Oral Exercises 8-3, page 289 2. a. Translate $\frac{\pi}{4}$ units to the left. **b.** Stretch vertically by a factor of 2, translate $\frac{\pi}{4}$ units to the right. **c.** Shrink horizontally by a factor of 2, reflect in the x-axis. **d.** Translate 2 units to the left, 3 units up. **e.** Stretch horizontally by a factor of 2, translate 4 units up. **f.** Stretch horizontally

14　**Answers to selected exercises**

by a factor of 2, stretch vertically by a factor of 3, translate $\frac{\pi}{2}$ units to the right and 4 units up. **3. a.** 2, 4, 6 **b.** 0, 2, 4 **c.** 10, 30, 50 **d.** 0.1, 0.3, 0.5

Written Exercises 8-3, pages 289-293

5.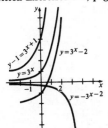

9. d.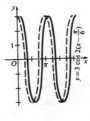

13. d. cot x

15.

19.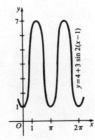

21. $y = 2 \sin\left(\frac{\pi}{2}x\right)$ **23.** $y - 2 = 2 \sin\left(\frac{\pi}{4}x\right)$

25. a. $2(x - 1)^3 - 23(x - 1)^2 + 58(x - 1) + 35 = 0$
b. $2(x + 1)^3 - 23(x + 1)^2 + 58(x + 1) + 35 = 0$
c. $2\left(\frac{x}{2}\right)^3 - 23\left(\frac{x}{2}\right)^2 + 58\left(\frac{x}{2}\right) + 35 = 0$ **29. b.** (0, 7)
c. (1, 7)

31.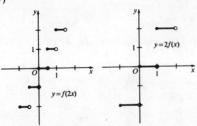

33. They are the same.

35. $f(x) = 0.7 + 0.3\left[\dfrac{x}{0.2}\right]$

37. b. See diagram at right.

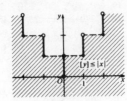

Written Exercises 8-4, pages 295-296 1. $x = -3$, $y = 0$ (See diagram at top of next column.) **3.** $x = 0$, $y = 0$ **5.** $x = \pm 3$, $y = 0$ (See diagram at top of next column.)

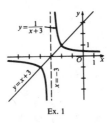
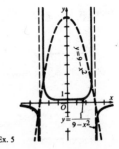

Ex. 1

Ex. 5

7. $x = 0$, $y = 0$ **9.** $x = n\pi$ **11.** $y = 0$ **13.** $x = 4n$, $x = 4n - 1$ **15.** $f(x) = \dfrac{1}{x^2 + x - 12}$

Ex. 7

Ex. 13

17. a. $x = -1$
b. 0
c. 1
d. $y = 1$

e.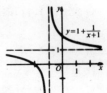

Oral Exercises 8-5, pages 298-299 1. 8 m

2. $y = 1.6 + 1.6 \cos \dfrac{2\pi}{12.4}t$

3. $V(t) = 220\sqrt{2}\sin 60 \cdot 2\pi t$ **4. a.** 400, 500, $\frac{1}{500}$
b. 50, 100,000, $\frac{1}{100,000}$

Written Exercises 8-5, pages 299-302

1. $y = 1.35 + 1.35 \cos \dfrac{2\pi}{12.4}t$

3. $y = 1.6 + 1.6 \cos \dfrac{2\pi}{12.4}t$ **5.** 1.9 m **7. a.** 220, 50, $\frac{1}{50}$
b. 100, 25, $\frac{1}{25}$ **9.** VHF: 30 to 300 MHz; UHF: 300 to 3000 MHz **11. a.** AM **b.** FM **13. a.** 3252.7 km N
b. 0 km **c.** 3252.7 km S **d.** 3252.7 km **15. a.** 12.6
b. Around 3/21 and 9/22 **c.** $y = 12 + 4 \sin \dfrac{2\pi}{365}x$

17. 8.6

21. $T = 15.5 + 12.5 \sin \dfrac{2\pi}{365}(x - 105)$

23. $T = 25 + 3 \sin \dfrac{2\pi}{365}(105 - x)$

Written Exercises 8-6, pages 304-305 For programs in this section, increments of 0.001 were used.
1. 0.346675 **3.** 0.333833 **5. a.** 0.5 **d.** When $n = 4$, the area is about $\frac{4}{5}$. **7.** 2.717

Answers to selected exercises   15

Chapter Test, pages 306–307

1. a. 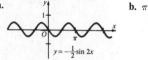 **b.** π

$y = -\frac{1}{2}\sin 2x$

3. **5.**

7. **9.**

11. $y = 2\cos\left(\dfrac{x}{2}\right)$ **13.** $y = 12 + 4\sin\dfrac{2\pi}{365}x$

Oral Exercises 9-1, page 313 1. Example: $\sin 90° = 1$; $2\sin 45° = \sqrt{2}$ **2.** Example: $\cos(90° - 0°) = 0$; $\cos 90° - \cos 0° = -1$ **3.** Example: $\sin(90° - 0°) = 1 = \sin 90° - \sin 0°$ **4.** Example: $\sin(90° + 0°) = 1 = \sin 90° + \sin 0°$ **5. a.** 1 **b.** $\frac{1}{2}$ **c.** -1 **d.** $-\dfrac{\sqrt{3}}{2}$ **6.** $\sin 15° = \sin(60° - 45°)$ or $\sin(45° - 30°)$

Written Exercises 9-1, pages 313–315 1. 1 **3.** 0 **5.** $\sin x$ **11.** $\dfrac{\sqrt{6} - \sqrt{2}}{4}$ **13.** $\dfrac{\sqrt{2} - \sqrt{6}}{4}$ **15.** $\dfrac{\sqrt{2} - \sqrt{6}}{4}$ **17.** $\cos x$ **19.** $\cos\theta$ **21.** 0 **23. a.** $\dfrac{1 - 2\sqrt{30}}{12}$ **b.** $\dfrac{2\sqrt{30} - 1}{12}$ **25.** $\dfrac{3}{5}$ **27.** $\dfrac{56}{65}$ **29.** $\sin x$ **31.** $2\tan\alpha$ **33.** $\sin(x + y)$ **35.** $\tan x + \tan y$ **41.** 0 **47. b.** $BC = \sin\beta$, $CD = \cos\beta$

Oral Exercises 9-2, page 317 1. a. -1 **b.** $-\frac{1}{7}$ **2. a.** 1 **b.** -1 **3.** l_1 and l_2 are perpendicular; $\tan\theta$ is undefined.

Written Exercises 9-2, pages 318–319 1. a. $\frac{7}{4}$ **b.** $\frac{1}{8}$ **3.** 1 **5.** -1 **7.** $\sqrt{3}$ **9.** 3 **11.** $\tan 75° = 2 + \sqrt{3}$; $\cot 75° = 2 - \sqrt{3}$ **13.** $\tan(\alpha + \beta) = 1$ **17.** If $\tan\alpha = 0$ or $\tan\beta = 0$. **19.** $26.6°$, $153.4°$ **21.** $135°$ **23.** $\frac{1}{8}$ **25. a.** $\frac{63}{65}$ **b.** $-\frac{16}{65}$ **c.** $-\frac{63}{16}$ **27.** $\tan 2x = \dfrac{2\tan x}{1 - \tan^2 x}$

Calculator Exercise, page 319 $\angle BAC \approx 70.1°$

Oral Exercises 9-3, pages 322–323 1. $\sin 20°$ **2.** $\cos 30°$ **3.** $\cos 70°$ **4.** $\cos 50°$ **5.** $\sin 6x$ **6.** $\cos 10x$ **7.** $\tan 100°$ **8.** $\tan 80°$ **9.** $\cos^2 x$ **10.** $\cos 2x$

11. $\cos 35° = \cos\left(\dfrac{70°}{2}\right) = \sqrt{\dfrac{1 + \cos 70°}{2}} = \sqrt{0.671}$

12. a. $\dfrac{3}{5}$ **b.** $\dfrac{24}{25}$ **c.** $\dfrac{\sqrt{5}}{5}$

Written Exercises 9-3, pages 323–326 1. $\sin 2A = \frac{120}{169}$, $\cos 2A = \frac{119}{169}$, $\tan 2A = \frac{120}{119}$ **3.** $\sin 2A = \frac{336}{625}$, $\cos 2A = \frac{527}{625}$, $\tan 2A = \frac{336}{527}$ **5.** $\sin 2A = \frac{24}{25}$, $\sin 4A = \frac{336}{625}$ **7. a.** $-\dfrac{23}{25}$ **b.** $\dfrac{\sqrt{15}}{5}$ **9.** $\cos 20°$ **11.** $\cos 40°$ **13.** $\sin 70°$ **15.** $\tan 50°$ **17.** $\cos 80°$ **19.** $\sin 40°$ **21.** $\dfrac{\sqrt{2}}{2}$ **23.** $\dfrac{1}{4}$ **25.** $\dfrac{\sqrt{3}}{2}$ **27. a.** $\dfrac{\sqrt{2} - \sqrt{6}}{4}$ **b.** $-\dfrac{\sqrt{2} - \sqrt{3}}{2}$ **29.** $\frac{24}{25}$ **47.** 2 **49.** $\sin 2x$ **51.** $\sin 2x$ **53.** $\tan^2 x$ **55.** 2 **57.** -2 **59.** $\frac{1}{3}$ **67. c.** 4

Oral Exercises 9-4, page 328 1. Solve for $\sin 2x$ **2.** $\cos 2x = 1 - 2\sin^2 x$ **3.** $2\cos^2 x - 1 = \cos 2x$ **4.** $\sin^2 x - \sin x = 0$; factor. **5.** Divide each side by $\cos x$, $\cos x \neq 0$, and solve for $\tan x$. **6.** Divide each side by $\cos 2x$, $\cos 2x \neq 0$, and solve for $\tan 2x$. **7.** Divide each side by $\cos 3x$, $\cos 3x \neq 0$, and solve for $\tan 3x$. **8.** Solve for $x - 10°$ and then for x. **9.** $\sin 2x$ **10.** $\cos 2x$ **11.** $\sin 4x = 2\sin 2x\cos 2x$ **12.** $\cos 4x = 2\cos^2 2x - 1$

Written Exercises 9-4, page 329 1. $15°$, $75°$, $195°$, $255°$ **3.** $0°$, $120°$, $240°$ **5.** $45°$, $225°$ **7.** $15°$, $75°$, $135°$, $195°$, $255°$, $315°$ **9.** $30°$, $150°$, $210°$, $330°$, $0°$, $90°$, $180°$, $270°$ **11. b.** 0, $\dfrac{\pi}{4}$, $\dfrac{3\pi}{4}$, π, $\dfrac{5\pi}{4}$, $\dfrac{7\pi}{4}$ **13.** $\dfrac{\pi}{6}$, $\dfrac{5\pi}{6}$, $\dfrac{3\pi}{2}$ **15.** $\dfrac{\pi}{6}$, $\dfrac{\pi}{2}$, $\dfrac{5\pi}{6}$, $\dfrac{7\pi}{6}$, $\dfrac{3\pi}{2}$, $\dfrac{11\pi}{6}$ **17.** $15°$, $255°$ **19.** $30°$, $90°$, $150°$, $270°$ **21.** $19.1°$, $199.1°$ **23.** $90°$, $225°$, $270°$, $315°$ **25.** $18.4°$, $108.4°$, $198.4°$, $288.4°$ **27.** $\dfrac{\pi}{4}$, $\dfrac{5\pi}{4}$ **29.** 0, π, $\dfrac{\pi}{6}$, $\dfrac{5\pi}{6}$, $\dfrac{7\pi}{6}$, $\dfrac{11\pi}{6}$ **31.** $\dfrac{\pi}{6}$, $\dfrac{5\pi}{6}$, $\dfrac{7\pi}{6}$, $\dfrac{11\pi}{6}$ **33.** 0 **35.** 3 **37.** 0, $\dfrac{\pi}{3}$, $\dfrac{2\pi}{3}$, π, $\dfrac{4\pi}{3}$, $\dfrac{5\pi}{3}$, $\dfrac{\pi}{6}$, $\dfrac{5\pi}{6}$, $\dfrac{7\pi}{6}$, $\dfrac{11\pi}{6}$ **39.** $\dfrac{\pi}{4}$, $\dfrac{3\pi}{4}$, $\dfrac{5\pi}{4}$, $\dfrac{7\pi}{4}$, $\dfrac{\pi}{2}$ **41.** 0, $\pm\dfrac{\sqrt{3}}{2}$ **43.** 0 **45.** $\pm\dfrac{\sqrt{2}}{2}$

Chapter Test, page 331 1. a. $\sqrt{3}\cos x$ **b.** $\frac{1}{2}$ **3.** 2 **5. a.** $\frac{3}{5}$ **b.** $\frac{7}{25}$ **c.** $\frac{24}{25}$ **d.** $\frac{336}{625}$ **7. a.** $\dfrac{\sqrt{3}}{2}$ **b.** $\sqrt{3}$ **9. b.** $30°$, $150°$, $270°$

16   **Answers to selected exercises**

Oral Exercises 10-1, page 338 **1. a.** $(0, 6)$ **b.** $(0, -3)$
c. $(-2, 0)$ **d.** $(-4, 0)$ **e.** $(0, -1)$ **f.** $(-5, 0)$
g. $(0, 3)$ **h.** $(\frac{1}{2}, 0)$ **2. a.** $(3; 0°)$ **b.** $(8; 270°)$
c. $(4; 180°)$ **d.** $(1; 90°)$ **3. a.** $(4\sqrt{2}, 4\sqrt{2})$ **b.** $(0, 3)$
c. $(2\sqrt{2}, -2\sqrt{2})$ **d.** $(\sqrt{2}, -\sqrt{2})$ **e.** $(-1, -\sqrt{3})$
f. $\left(\frac{\sqrt{3}}{2}, \frac{1}{2}\right)$ **g.** $(-10, 0)$ **h.** $(-\sqrt{3}, 1)$
4. a. $(3\sqrt{2}; 45°)$ **b.** $(4; 60°)$ **c.** $(\sqrt{2}; -45°)$
d. $(1; -30°)$

Written Exercises 10-1, pages 338–340

1.

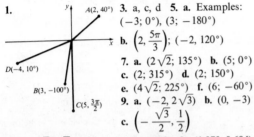

3. a, c, d **5. a.** Examples:
$(-3; 0°)$, $(3; -180°)$
b. $\left(2, \frac{5\pi}{3}\right)$; $(-2, 120°)$
7. a. $(2\sqrt{2}; 135°)$ **b.** $(5; 0°)$
c. $(2; 315°)$ **d.** $(2; 150°)$
e. $(4\sqrt{2}; 225°)$ **f.** $(6; -60°)$
9. a. $(-2, 2\sqrt{3})$ **b.** $(0, -3)$
c. $\left(-\frac{\sqrt{3}}{2}, \frac{1}{2}\right)$

d. $(-\sqrt{2}, \sqrt{2})$ **11. a.** $(0.940, 0.342)$ **b.** $(1.879, 0.684)$
c. $(-0.940, -0.342)$ **d.** $(-0.416, 0.909)$
13. a. $(5; 53.1°)$ **b.** $(\sqrt{5}; 63.4°)$ **c.** $(\sqrt{13}; 123.7°)$

15.
$r = \sin\theta$

17.
$r = 1 - \sin\theta$

$x^2 + (y - \frac{1}{2})^2 = \frac{1}{4}$

$(x^2 + y^2 + y)^2 = x^2 + y^2$

19. $(x^2 + y^2 - 2y)^2 = x^2 + y^2$ (See diagram below.)
21. The line $x = 1$ **23.** Yes. $\theta = 2$: the line passing
through the origin with slope $\tan \approx -2.176$. $r = 4$:
circle with center at $(0, 0)$ and radius 4.
25. Four-leaved rose, has leaves along the axes,
intersect axes at $(0, 0)$, $(0, 1)$, $(1, 0)$, $(0, -1)$, and $(-1, 0)$.

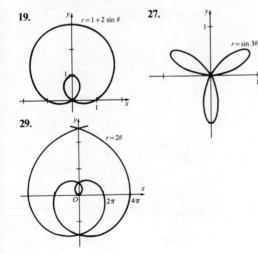

31.
$r = e^\theta$

33. Two-leaved rose, has leaves
along the line $y = x$, passes
through $(0, 0)$, $(\sqrt{2}, \sqrt{2})$,
$(-\sqrt{2}, -\sqrt{2})$.

35. a. b **b.** c **c.** a

Calculator Exercises, page 340 **1.** $(3.606; 56.3°)$
3. $(5; 143.1°)$ **5.** $(2.298, 1.928)$ **7.** $(1.449, 3.728)$

Oral Exercises 10-2, page 343
1. $\sqrt{2}(\cos 45° + i \sin 45°)$ **2.** $\cos 90° + i \sin 90°$
3. $3(\cos 180° + i \sin 180°)$ **4.** $2(\cos 330° + i \sin 330°)$
5. $5i$ **6.** -3 **7.** $2\sqrt{2} + 2i\sqrt{2}$ **8.** $3\sqrt{3} + 3i$
9. $6(\cos 50° + i \sin 50°)$ **10.** $10(\cos \pi + i \sin \pi)$

Written Exercises 10-2, pages 343–344
1. $\sqrt{2}(\cos 135° + i \sin 135°)$ **3.** $2(\cos 60° + i \sin 60°)$
5. $7(\cos 180° + i \sin 180°)$
7. $5(\cos 306.9° + i \sin 306.9°)$ **9.** $-3 - 3i\sqrt{3}$ **11.** 9
13. $10(\cos 90° + i \sin 90°)$; $10i$
15. $4\left(\cos \frac{5\pi}{3} + i \sin \frac{5\pi}{3}\right)$; $2 - 2i\sqrt{3}$
17. a. $4(\cos 60° + i \sin 60°)$;
$2(\cos (-30°) + i \sin (-30°))$; $8(\cos 30° + i \sin 30°)$
b, c. $4\sqrt{3} + 4i$ **19.** $2\sqrt{2}(\cos 45° + i \sin 45°)$,
$2\sqrt{2}(\cos (-45°) + i \sin (-45°))$; $8(\cos 0° + i \sin 0°)$
b, c. 8 **21.** $45°$ **23.** If $z = r(\cos \theta + i \sin \theta)$,
$z^2 = r^2(\cos 2\theta + i \sin 2\theta)$.
25. $(\cos \theta + i \sin \theta)^3 = \cos 3\theta + i \sin 3\theta$

Calculator Exercises, page 344
1. a. $33(\cos 305.1° + i \sin 305.1°)$;
$33.6(\cos 22.8° + i \sin 22.8°)$;
$1109.8(\cos 327.9° + i \sin 327.9°)$ **b.** $940 - 590i$ (by
conversion, $939.3 - 589.2i$)

Oral Exercises 10-3, page 347
1. $4(\cos 90° + i \sin 90°)$ **2.** $8(\cos 135° + i \sin 135°)$
3. $4(\cos (-72°) + i \sin (-72°))$
4. $\cos 360° + i \sin 360°$ **5.** $64\left(\cos \frac{\pi}{2} + i \sin \frac{\pi}{2}\right)$
6. $27(\cos 5\pi + i \sin 5\pi)$
7. a. $\sqrt{2}(\cos 135° + i \sin 135°)$
b. $8(\cos 90° + i \sin 90°)$ **c.** $8i$ **8. a, b.** $-32i$ **c.** b;
$\cos (-450°) = \cos 90°$ and $\sin (-450°) = -\sin 90°$;
this is easier than finding x and n such that
$3150° = (x + 360n)°$.

Written Exercises 10-3, pages 347–349
1. $\cos 30° + i \sin 30°$; $\cos 60° + i \sin 60°$;
$\cos 90° + i \sin 90°$

3. $z = \sqrt{2}(\cos(-45°) + i\sin(-45°))$; $z^{-1} = (\frac{1}{2}, \frac{1}{2})$,
$z^0 = (1, 0)$, $z = (1, -1)$, $z^2 = (0, -2)$, $z^3 = (-2, -2)$,
$z^4 = (-4, 0)$, $z^5 = (-4, 4)$, $z^6 = (0, 8)$, $z^7 = (8, 8)$,
$z^8 = (16, 0)$ **5. a, b.** -8 **7.** z^2, z^4, z^5, z
11. $z = \frac{\sqrt{3}}{2} + \frac{i}{2}$ **15.** real

Written Exercises 10-4, pages 351-352 1. $\pm\frac{\sqrt{3}}{2} + \frac{i}{2}$,
$-i$ **3.** $2, -1 \pm i\sqrt{3}$ **5.** $\pm 2, \pm 2i$ **7. b.** sum = 0,
product = $8i$ **9.** $2, -1 \pm i\sqrt{3}$ **11. b.** $\pm 1, \frac{1}{2} \pm \frac{i\sqrt{3}}{2}$,
$-\frac{1}{2} \pm \frac{i\sqrt{3}}{2}$ **13.** $2, 0.6180 \pm 1.902i$,
$-1.6180 \pm 1.1756i$ **15. a.** 3, 5, 17, 257 **b.** yes, yes,
no, no, yes, yes

Calculator Exercise, page 352 $2.15 + 0.67i$,
$-1.66 + 1.52i$, $-0.49 - 2.2i$

Chapter Test, pages 353-354
1. $(0, -6)$, **3.**
$(-8, 0)$,
$(3\sqrt{2}; 45°)$,
$(2; 270°)$,
$(2; 120°)$

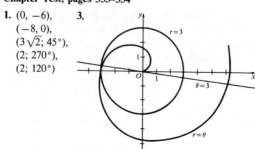

5. a. $z_1 = -2 + 2i\sqrt{3}$, $z_2 = \sqrt{3} + i$,
$z_1 z_2 = -4\sqrt{3} + 4i$ **b.** $-4\sqrt{3} + 4i$
7. a. $r^2(\cos 2\theta + i\sin 2\theta)$

Cumulative Review (Chapters 6-10), pages 355-357
1. a. 1.25 **b.** 10 cm² **3.** Domain: all real nos,; range:
the set of real numbers between 0 and 1, inclusive
5. $\sin\theta = -\frac{2\sqrt{5}}{5}$, $\cos\theta = \frac{\sqrt{5}}{5}$, $\tan\theta = -2$,
$\csc\theta = -\frac{\sqrt{5}}{2}$, $\cot\theta = -\frac{1}{2}$ **7.** 153.4° **9. a.** 2.30,
3.98 **b.** $\frac{\pi}{2}, \frac{3\pi}{2}$ **11.** 18.8 cm **13.** 136 m **15. a.** 53.1°
and 126.9° **b.** 19.7 **17. a.** $\frac{5\pi}{6}$ **b.** $\frac{1}{2}$ **c.** $-\frac{1}{2}$ **d.** $\frac{\pi}{4}$
e. $-\sqrt{5}$ **19.** $\frac{2\pi}{9}, \frac{4\pi}{9}, \frac{8\pi}{9}, \frac{10\pi}{9}, \frac{14\pi}{9}, \frac{16\pi}{9}$
23. amplitude, 150; frequency, 36; period, $\frac{1}{36}$
27. $-\frac{\sqrt{6}+\sqrt{2}}{4}$ **29. a.** $\frac{5 - 12\sqrt{3}}{26}$ **b.** $\frac{119}{169}$ **c.** $\frac{5\sqrt{26}}{26}$
31. $6.6\overline{6}° + 30n$ for $n = 0, 1, \ldots, 11$ **33. a.** two
37. $z_1 = \sqrt{2}(\cos 315° + i\sin 315°)$;
$z_2 = 2(\cos 105° + i\sin 105°)$
39. a. $1024(\cos 60° + i\sin 60°)$ **b.** $512 + 512i\sqrt{3}$

Oral Exercises 11-1, pages 364-365 1. a. horizontal
b. vertices: $(\pm 10, 0)$; foci: $(\pm 6, 0)$ **c.** $\frac{3}{5}$ **d.** 20
e. $\frac{x^2}{100} + \frac{y^2}{64} = 1$ **2. a.** vertical **b.** vertices: $(0, \pm 13)$;
foci: $(0, \pm 12)$ **c.** $\frac{(y-7)^2}{169} + \frac{(x-6)^2}{25} = 1$
3. interior of ellipse with vertices $(\pm 3, 0)$ and foci
$(\pm\sqrt{8}, 0)$; exterior of same ellipse **4. a.** the surface of
an elliptical solid **b.** the interior of an elliptical solid
5. a. $(-5, 8)$ **b.** $a = 2; b = \sqrt{3}; c = 1; e = \frac{1}{2}$

Written Exercises 11-1, pages 365-367 1. vertices:
$(\pm 6, 0)$; foci: $(\pm 2\sqrt{5}, 0)$; $e = \frac{\sqrt{5}}{3}$ **3.** vertices:
$(0, \pm 5)$; foci: $(0, \pm 3)$; $e = \frac{3}{5}$ **5.** vertices: $(\pm 5, 0)$; foci:
$(\pm 4, 0)$; $e = \frac{4}{5}$ **7.** $\frac{x^2}{49} + y^2 = 1$ **9.** $\frac{x^2}{144} + \frac{y^2}{169} = 1$
11. $(0, -3)$, $(-\frac{4}{3}, 1)$ **13.** tangent pt. $(-2, 1)$
15. b. $x^2 + y^2 = 18$ **17. a.** portion of ellipse
$\frac{y^2}{16} + \frac{x^2}{4} = 1$ above x-axis **b.** portion of ellipse below
x-axis **19.** center: $(5, -3)$; vertices: $(0, -3)$, $(10, -3)$;
foci: $(1, -3)$, $(9, -3)$; $e = \frac{4}{5}$ **21.** center: $(3, -5)$;
vertices: $(3 -2)$, $(3, -8)$; foci: $(3, -5 \pm\sqrt{5})$; $e = \frac{\sqrt{5}}{3}$
23. $\frac{(x-3)^2}{25} + (y-2)^2 = 1$; center: $(3, 2)$; vertices:
$(-2, 2)$, $(8, 2)$; foci: $(3 \pm 2\sqrt{6}, 2)$; $e = \frac{2\sqrt{6}}{5}$
25. $\frac{(x+4)^2}{16} + \frac{y^2}{100} = 1$; center $(-4, 0)$; vertices:
$(-4, -10)$, $(-4, 10)$; foci: $(-4, \pm 2\sqrt{21})$; $e = \frac{\sqrt{21}}{5}$
27. $\frac{(x-3)^2}{25} + \frac{(y-7)^2}{16} = 1$
29. $\frac{(x-5)^2}{12} + \frac{(y-5)^2}{16} = 1$
31. $\frac{(x-5)^2}{25} + \frac{(y-6)^2}{36} = 1$
33. $PF_1 + PF_2 = \sqrt{(x-3)^2 + y^2} + $
$\sqrt{(x+3)^2 + y^2} = 10$ **b.** $\frac{x^2}{25} + \frac{y^2}{16} = 1$
35.
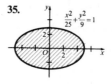
37.
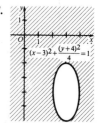
39. b. $x = r\cos\theta, y = r\sin\theta$ **c.** $\frac{\sqrt{3}}{2}$ **41.** The path of
P is an ellipse with equation $\frac{x^2}{4} + y^2 = 1$.

Calculator Exercises, page 368 1. $e \approx 0.016$ **2.** 0.97

Oral Exercises 11-2, pages 373-374 **1. a.** $(\pm 2, 0)$
b. $(\pm\sqrt{13}, 0)$ **c.** $\dfrac{\sqrt{13}}{2}$ **d.** horizontal

e. $\dfrac{x^2}{4} - \dfrac{y^2}{9} = 1$ **f.** $y = \pm\frac{3}{2}x$ **g.** Ex. 1; the hyperbola
is more open. **2. a.** $xy = -8$ **b.** $x = 0, y = 0$
c. $(2\sqrt{2}, -2\sqrt{2}), (-2\sqrt{2}, 2\sqrt{2})$
d. $(x + 2)(y - 1) = -8$ **3. a.** vertical **b.** $(0, \pm 5)$
c. $y = \pm 5x$ **d.** $(0, \pm\sqrt{26})$
e. $\dfrac{(y + 5)^2}{25} - \dfrac{(x - 6)^2}{1} = 1$ **4.** the portion of the
plane not enclosed by the two branches of the
hyperbola $\dfrac{x^2}{4} - \dfrac{y^2}{4} = 1$; the portion of the plane
enclosed by the hyperbola **5.** an ellipse; the sum of
the distances from A to the pencil and from B to the
pencil is fixed. **6.** a hyperbola; the difference of the
distances from A to the pencil and from B to the pencil
is fixed.

Written Exercises 11-2, pages 374-376 **1.** vertices:
$(\pm 1, 0)$; foci: $(\pm\sqrt{2}, 0)$; $e = \sqrt{2}$; asymptotes:
$y = \pm x$ **3.** vertices: $(0, \pm 2)$; foci: $(0, \pm\sqrt{5})$; $e = \dfrac{\sqrt{5}}{2}$;
asymptotes: $y = \pm 2x$ **5.** vertices: $(0, \pm 3)$; foci:
$(0, \pm\sqrt{13})$; $e = \dfrac{\sqrt{13}}{3}$; asymptotes: $y = \pm\dfrac{3}{2}x$

7. $\dfrac{x^2}{36} - \dfrac{y^2}{64} = 1$ **9.** $\dfrac{y^2}{4} - \dfrac{x^2}{4} = 1$

11. a. 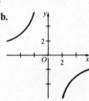 **b.**

11. c. 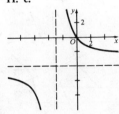 **13.**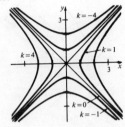

15. Pt. of intersection: $(\frac{41}{10}, \frac{9}{10})$ **17.** Pts. of intersection:
$(-4, 5), (10, -2)$

19. a. 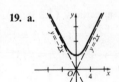 **b.**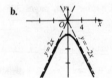

21. center: $(6, 8)$; vertices: $(0, 8), (12, 8)$; foci: $(-4, 8)$,
$(16, 8)$; asymptotes: $\dfrac{y - 8}{x - 6} = \pm\dfrac{4}{3}$; $e = \dfrac{5}{3}$

23. $(y - 1)^2 - (x - 2)^2 = 1$; center: $(2, 1)$; vertices:
$(2, 0), (2, 2)$; foci: $(2, 1 \pm \sqrt{2})$; asymptotes:
$\dfrac{y - 1}{x - 2} = \pm 1$; $e = \sqrt{2}$ **25.** $(x + 4)^2 - \dfrac{(y + 5)^2}{4} = 1$;
center $(-4, -5)$; vertices: $(-3, -5), (-5, -5)$; foci:
$(-4 \pm \sqrt{5}, -5)$; asymptotes: $\dfrac{y + 5}{x + 4} = \pm 2$; $e = \sqrt{5}$

27. $\dfrac{(x - 5)^2}{16} - \dfrac{y^2}{9} = 1$ **29.** $\dfrac{(y - 2)^2}{9} - \dfrac{(x + 4)^2}{16} = 1$

31. a. $|PF_1 - PF_2| = |\sqrt{(x - 10)^2 + y^2} - \sqrt{(x + 10)^2 + y^2}| = 16$ **b.** $\dfrac{x^2}{64} - \dfrac{y^2}{36} = 1$

33. a. 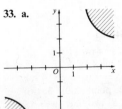 **b.**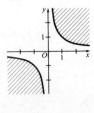

35. **37.**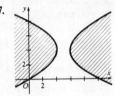

39. b. $\frac{13}{5}$

Calculator Exercise, page 376 $\cosh 0 = 1$; $\sinh 0 = 0$;
$\cosh 1 \approx 1.5$; $\sinh 1 \approx 1.2$; $\cosh (-1) \approx 1.5$;
$\sinh (-1) \approx -1.2$; $\cosh 2 \approx 3.8$; $\sinh 2 \approx 3.6$;
$\cosh (-2) \approx 3.8$; $\sinh (-2) \approx -3.6$; $\cosh \pi \approx 11.6$;
$\sinh \pi \approx 11.5$; $\cosh (-\pi) \approx 11.6$; $\sinh (-\pi) \approx -11.5$

Oral Exercises 11-3, page 380 **1. a.** 1; 1 **b.** 2; 2
c. 5; 5 **d.** 10; 10 **2. a.** (1) $\sqrt{(x + 4)^2 + y^2}$
(2) $\sqrt{x^2} = |x|$ **b.** Set (1) and (2) from Ex. 2a equal to
each other. **3.** $(0, \frac{1}{4})$; $y = -\frac{1}{4}$ **4.** $(0, -\frac{1}{8})$; $y = \frac{1}{8}$
5. $(-6, 4)$; $y = 2$ **6.** $(\frac{1}{4}, 0)$; $x = -\frac{1}{4}$ **7.** $(\frac{1}{4}, 2)$;
$x = -\frac{1}{4}$ **8.** $(-5\frac{1}{12}, 1)$; $x = -4\frac{11}{12}$

Written Exercises 11-3, pages 380-382 **1.** $x = -\frac{1}{4}y^2$
3. $y - 2 = \frac{1}{2}(x - 2)^2$ **5.** $y = -x^2$
7. $x + 2 = \frac{1}{8}(y + 1)^2$ **9. a.** $(0, 0)$; $(0, 2)$; $y = -2$
b. $(0, 0)$; $(2, 0)$; $x = -2$ **11. a.** $(0, 0)$; $(0, -\frac{1}{8})$; $y = \frac{1}{8}$
b. $(0, 0)$; $(-\frac{1}{8}, 0)$; $x = \frac{1}{8}$ **13.** $(2, 1)$; $(2, 2)$; $y = 0$
15. $(4, 7)$; $(\frac{17}{4}, 7)$; $x = \frac{15}{4}$ **17.** $(2, -5)$; $(2, -\frac{39}{8})$;
$y = -\frac{41}{8}$ **19.** $(3, 1)$; $(3, -1)$; $y = 3$ **21.** $(-6, 1)$;
$(-\frac{23}{4}, 1)$; $x = -\frac{25}{4}$

Answers to selected exercises **19**

23. a. 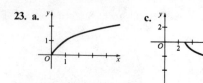 **c.**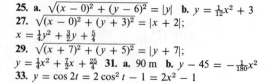

25. a. $\sqrt{(x-0)^2+(y-6)^2}=|y|$ **b.** $y=\frac{1}{12}x^2+3$
27. $\sqrt{(x-0)^2+(y+3)^2}=|x+2|;$
$x=\frac{1}{4}y^2+\frac{3}{2}y+\frac{5}{4}$
29. $\sqrt{(x+7)^2+(y+5)^2}=|y+7|;$
$y=\frac{1}{4}x^2+\frac{7}{2}x+\frac{25}{4}$ **31. a.** 90 m **b.** $y-45=-\frac{1}{180}x^2$
33. $y=\cos 2t=2\cos^2 t-1=2x^2-1$

Calculator Exercise, page 382 $t\approx 9.2$ s; $x\approx 796$

Oral Exercises 11-4, pages 384–385
1. no sol.: $(x+4)^2+y^2=9$; $y^2=x$. 1 sol.:
$(x+3)^2+y^2=9$; $y^2=x$. 2 sols.: $(x+2)^2+y^2=9$;
$y^2=x$. 3 sols.: $(x-3)^2+y^2=9$; $y^2=x$. 4 sols.:
$(x-4)^2+y^2=9$; $y^2=x$. **2.** point $(0,0)$ **3.** two
lines, $y=\pm\frac{1}{3}x$ **4.** no graph **5.** two lines, $y=\pm 2x$
6. no graph **7.** point $(0,0)$ **8.** two lines, $x=3$ and
$y=2$ **9.** two lines, $x=0$ and $y=0$ **10.** two lines,
$y=0$ and $y=3x$ **11.** two lines, $x=0$ and $y=-x$
12. two lines, $y=x$ and $y=\frac{1}{3}x$ **13.** two lines,
$y=-x$ and $y=\frac{1}{5}x$ **14.** no graph **15.** point $(1,-2)$

Written Exercises 11-4, page 385 **1.** ellipse; circle;
$(0,\pm 2)$ **3.** ellipse; parabola; $\left(\dfrac{2\sqrt{17}}{3},\dfrac{16}{9}\right)$,
$\left(-\dfrac{2\sqrt{17}}{3},\dfrac{16}{9}\right)$, $(0,-2)$ **5.** circle; hyperbola; $(4,-3)$,
$(-4,3)$, $(-3,4)$, $(3,-4)$ **7.** hyperbola; parabola; no
points of intersection **9.** hyperbola; parabola;
$(-1,-4)$ **11.** circle; parabola; no points of
intersection **13.** circle; circle; $(-2,\pm\sqrt{5})$ **15.** lines
$y=2x$ and $y=-\frac{1}{2}x$; ellipse; $\left(\dfrac{\sqrt{6}}{3},\dfrac{2\sqrt{6}}{3}\right)$,
$\left(-\dfrac{\sqrt{6}}{3},-\dfrac{2\sqrt{6}}{3}\right)$, $(2,-1)$, $(-2,1)$

17. lines $y=\frac{1}{2}x$
and $y=-\frac{3}{2}x$; parabola;
$\left(\dfrac{1-\sqrt{73}}{9},\dfrac{-1+\sqrt{73}}{6}\right)$,
$\left(\dfrac{1+\sqrt{73}}{9},\dfrac{-1-\sqrt{73}}{6}\right)$,
$(-2,-1)$, $(4,2)$

19.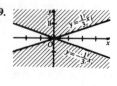

21.

23.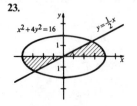

25. $(\pm\sqrt{3},0)$, $(-2,1)$, $(2,-1)$ **27.** 10.6 cm

Calculator Exercise, page 385 $(2.00,3.00)$,
$(3.13,-2.81)$, $(-1.85,3.58)$, $(-3.28,-3.78)$

Oral Exercises 11-5, pages 390–391 **1. a.** A: 4, B: 6,
C: 8, P: $|x-3|$ **b.** A: 8, B: 10, C: 5, P: $|y+6|$
2. a. $\sqrt{x^2+y^2}=2\sqrt{(x-7)^2+(y-2)^2}$
b. $\sqrt{x^2+y^2}=\frac{1}{2}\sqrt{(x+3)^2+(y-4)^2}$
c. $\sqrt{(x+5)^2+(y+1)^2}=|y+6|$
d. $\sqrt{(x+3)^2+(y-4)^2}=2|y+6|$ **3.** a branch of a
hyperbola **4.** The chamber is an ellipse and for any
point P on the ellipse PF_1+PF_2 is constant.
5. a. hyperbola **b.** ellipse **c.** parabola **d.** circle

Written Exercises 11-5, pages 392–395
1. a. $\dfrac{x^2}{4}+\dfrac{y^2}{3}=1$ **b.** $(1,0)$
c. $\sqrt{(x-1)^2+y^2}=\frac{1}{2}|x-4|$; $\dfrac{x^2}{4}+\dfrac{y^2}{3}=1$
3. $\sqrt{(x+6)^2+y^2}=2|x+\frac{3}{2}|$; $\dfrac{3x^2}{25}-\dfrac{y^2}{25}=1$
5. ellipse; $\alpha=\dfrac{\pi}{4}$ **7.** hyperbola; $\alpha=\dfrac{3\pi}{8}$

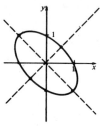

Ex. 5

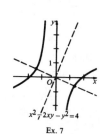

$x^2+2xy-y^2=4$

Ex. 7

9. parabola; $\alpha=\dfrac{\pi}{4}$ **11.** hyperbola; $\alpha=\dfrac{3\pi}{8}$

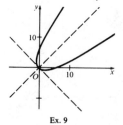

Ex. 9

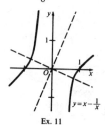

$y=x-\dfrac{1}{x}$

Ex. 11

13. b. (1) $\angle 3\approx 36.9°$; $\angle 4\approx 63.4°$ (2) $\angle 2\approx 26.6°$;
$\angle 1\approx 26.5°$ **23. a.** QO

Chapter Test, pages 396–397 **1. a.** vertices: $(\pm 3,0)$;
foci: $(\pm\sqrt{5},0)$; $e=\dfrac{\sqrt{5}}{3}$ **b.** $\dfrac{(x+3)^2}{9}+(y+1)^2=1$;
vertices: $(-6,-1)$, $(0,-1)$; foci: $(-3\pm 2\sqrt{2},-1)$;
$e=\dfrac{2\sqrt{2}}{3}$

3. a.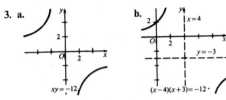

b. $xy = -12$; $(x-4)(x+3) = -12$·

5. a. $(\sqrt{2}, 2\sqrt{2})$, $(-\sqrt{2}, -2\sqrt{2})$ **7. a.** (1) ellipse
(2) hyperbola **b.** $y - 3 = \frac{1}{4}x^2$

Oral Exercises 12-1, page 403 1. a. (1) true (2) false
(3) false (4) true (5) true (6) true **b.** No; in (1) we
add vectors; in (2) we add real numbers. **2. a.** $\overrightarrow{BC}$
b. $|\overrightarrow{BC}|$ or $|\overrightarrow{CB}|$ **c.** $\overrightarrow{BO}$ or $\overrightarrow{OD}$ **d.** $\overrightarrow{AC}$ **e.** $\overrightarrow{AC}$ **f.** $\overrightarrow{AB}$
g. $\overrightarrow{OD}$; $\overrightarrow{AD}$ **h.** $\overrightarrow{DB}$; $\overrightarrow{DC}$ or $\overrightarrow{AB}$ **3.** On the left we add
real numbers; on the right we add vectors. **4.** On the
left we multiply the vector $m\mathbf{v}$ by the scalar k; on the
right we multiply the vector $\mathbf{v}$ by the product of the
real numbers k and m. **5.** because we have not
defined the sum of a scalar and a vector

Written Exercises 12-1, pages 403–407 1. a. true
b. false **c.** true **d.** true **3. a.** $\overrightarrow{RP}$ **b.** $\overrightarrow{SU}$ **c.** $\overrightarrow{TP}$
d. $\overrightarrow{TS}$

5. a. **c.**

7. a. **c.**

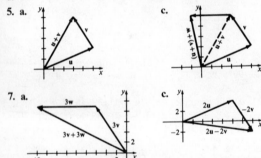

9. a. $\overrightarrow{MN} = a + b$ **b.** $\overrightarrow{QR} = 2a + 2b$ **c.** The line
segment joining the midpoints of two sides of a triangle
is parallel to the third side and is half as long as the
third side.
11. about 440 knots; 5° north of east
13. $|\mathbf{u}| = 10$ N; $\sin \alpha = 0.6$; 37° west of north
15. $|\mathbf{v} + \mathbf{w}| = |\mathbf{v}| + |\mathbf{w}|$; $|\mathbf{v} - \mathbf{w}| = |\mathbf{v}| - |\mathbf{w}|$ when
$|\mathbf{v}| > |\mathbf{w}|$, and $|\mathbf{v} - \mathbf{w}| = |\mathbf{w}| - |\mathbf{v}|$ when $|\mathbf{w}| > |\mathbf{v}|$.
17. a. $\frac{1}{2}\mathbf{v}$ **b.** $-\frac{1}{2}\mathbf{v}$ **c.** $-\mathbf{v}$ **d.** $-\frac{3}{2}\mathbf{v}$ **19. a.** $-\frac{3}{2}\mathbf{v}$ **b.** $\frac{5}{2}\mathbf{v}$
c. $-\mathbf{v}$ **21. a.** $\mathbf{x} + \mathbf{y}$ **b.** $\frac{1}{2}\mathbf{x} + \frac{1}{2}\mathbf{y}$ **c.** $\frac{1}{2}\mathbf{y} - \frac{1}{2}\mathbf{x}$
d. $\frac{7}{2}\mathbf{y} + \frac{1}{2}\mathbf{x}$ **23. a.** $\mathbf{y}$ **b.** $-\mathbf{x}$ **c.** $\mathbf{x} + \mathbf{y}$ **d.** $\frac{1}{2}\mathbf{x} + \frac{1}{2}\mathbf{y}$
e. $\frac{1}{2}\mathbf{y} - \frac{1}{2}\mathbf{x}$ **25. a** $- \mathbf{v} + \mathbf{w}$ **b.** $-\frac{1}{3}\mathbf{v} + \frac{1}{3}\mathbf{w}$ **c.** $\frac{2}{3}\mathbf{v} + \frac{1}{3}\mathbf{w}$
d. $-\frac{5}{3}\mathbf{v} + \frac{2}{3}\mathbf{w}$ **27. a.** $\overrightarrow{AC}$ **b.** $\overrightarrow{AD}$ **c.** $\overrightarrow{BD}$ **d.** $\overrightarrow{AD}$
29. $\angle B = 120°$; $|\overrightarrow{AX}| = 7$ N **31.** 176.9 km; 42.7°
33. b. $L = (4, 3)$; $M = (4, -1)$; $N = (-2, 2)$;
$A = (1, 0)$; $B = (0, 1)$

Oral Exercises 12-2, page 411 1. a. (1, 4); (4, −4)
b. $|\overrightarrow{AB}| = \sqrt{17}$; $|\overrightarrow{CD}| = 4\sqrt{2}$ **c.** $(\frac{17}{4}, 1)$ **d.** (0, −2)
2. a. (2; 150°) **b.** $(-\sqrt{3}, 1)$ **c.** 2 **3. a.** $\overrightarrow{AB} = (2, 1)$;
$\overrightarrow{BC} = (1, 2)$; $\overrightarrow{AC} = (3, 3)$ **b.** $|\overrightarrow{AB}| = \sqrt{5}$; $|\overrightarrow{BC}| = \sqrt{5}$;
$|\overrightarrow{AC}| = 3\sqrt{2}$ **c.** (1) true (2) false **4. a.** (3, 6)
b. (4, 2) **c.** (−2, 2) **d.** (11, 4) **5.** $2\mathbf{v} = (6; 40°)$;
$-\mathbf{v} = (3; 220°)$ **6. a.** (7, 2) **b.** (4, 9)
7. $\overrightarrow{BA} = A - B = -(B - A) = -\overrightarrow{AB} = -(a, b) =$
$(-a, -b)$

Written Exercises 12-2, pages 412–413

1. **3.**

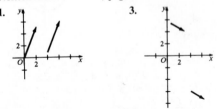

5. a. (5, −7) **b.** (1, 3) **c.** (−1, 3) **7. a.** (6, 3)
b. $3\sqrt{5}$ **c.** (2, 1) **9. a.** (−5, 10) **b.** $5\sqrt{5}$ **c.** (3, 6)
11. a. (6, 3) **b.** $3\sqrt{5}$ **c.** (3, 1) **13. a.** $\sqrt{13}$ **b.** $2\sqrt{13}$
c. $\sqrt{34}$ **d.** $\sqrt{130}$ **15.** $x = 7$; $y = 4$ **17. b.** (5; 53.1°)
19. a. $r \sin \theta$ **b.** $x \approx -8.8$; $y \approx 4.7$ **21.** (11.6; 39.9°)
23. a. $r = 3$; $s = 2$ **b.** The vectors shown are 2(3, 0),
3(1, 2), and the sum of the two vectors, (9, 6). **25.** $\pm\frac{1}{5}$
27. the points of a circle with center (4, 0) and
radius 2 **31. b.** (2, 3)

Calculator Exercise, page 414 37.52

Oral Exercises 12-3, page 418 1. a. (6, 8) **b.** (3, 4)
c. 5 **2. a.** For example, (2, −5), (3, −2), and (4, 1)
b. For example, (1, 3) **c.** 3 **d.** $x = 2 + t$,
$y = -5 + 3t$ **3. a.** $\frac{2}{3}$ **b.** (3, 2) **c.** For example,
$(x, y) = (0, 2) + t(3, 2)$

Written Exercises 12-3, pages 419–421
1. a. $\mathbf{v} = (3, 4)$; $|\mathbf{v}| = 5$ **b.** $\mathbf{v} = (2, 4)$; $|\mathbf{v}| = 2\sqrt{5}$
3. a. (8, 4) **b.** (0, 1) **5.** $x = 2 + 3t$, $y = 3 - t$
7. $x = 2 + t$, $y = 4t$ **9.** For example, $x = 3 + 2t$,
$y = 2 + 4t$; $y = 2x - 4$ **11. a.** a vertical line through
(2, 0) **b.** For example, (0, 1) **c.** The slope is
undefined. **13. a.** $\mathbf{v} = (-3, 2)$; $|\mathbf{v}| = \sqrt{13}$ **b.** When
$t = -1$ at (5, −3)
15. When $t = -2$ at (3, −1);
when $t = 1$ at (0, 2) (See
diagram at right.)
17. a. $(x, y) = (1, 0) + t(2, -4)$;
$x = 1 + 2t$, $y = -4t$
b. $(x, y) = (-2, 3) + t(7, -2)$;
$x = -2 + 7t$, $y = 3 - 2t$
19. a. $(x, y) = (0, 2) + t(1, 3)$;
$x = t$, $y = 2 + 3t$
b. $(x, y) = (3, 0) + t(-3, 1)$; $x = 3 - 3t$, $y = t$
21. $(x, y) = (7, 5) + t(1, 1)$; $x = 7 + t$, $y = 5 + t$

Ex. 15

23. $(x, y) = (\pi, e) + t(1, 0)$; $x = \pi + t$, $y = e$
25. a. $\frac{3}{2}$ **b.** $\frac{3}{2}$ **c.** Each has a direction vector with a slope of $\frac{3}{2}$ and the lines are not coincident. **27. a.** No; yes **b.** No **29.** $(x, y) = (0, 3) + t(8, -6)$
31. $(-4, -5)$ **33.** $y = \frac{1}{4}x^2 + 1$ (See diag. below.)
35. $x^2 + y^2 = 16$ (See diag. below.) **37.** $y = 2x^2 - 1$

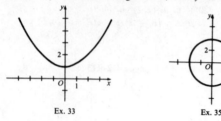

Ex. 33   Ex. 35

Oral Exercises 12-4, page 424 1. $\mathbf{u} \parallel \mathbf{v}$; $\mathbf{u} \perp \mathbf{w}$; $\mathbf{v} \perp \mathbf{w}$
2. a. 2 **b.** 4 **c.** 4 **d.** 4 **3. a.** 5 **b.** 25 **c.** 25
4. a. $5\sqrt{2}$ **b.** $5\sqrt{2}$ **c.** 40 **d.** 0.8 **5.** $\cos \theta = \frac{4}{5}$

Written Exercises 12-4, pages 425-426 1. a. -7
b. 1 **3. a.** 2 **b.** $-\frac{9}{2}$ **5. a.** 13 **b.** 13 **13.** $106.3°$
15. $53.1°$ **19. a.** $\cos C = 0.6$; $\sin C = 0.8$ **b.** 20 square units

Calculator Exercise, page 426 $34.3°$

Oral Exercises 12-5, pages 430-431 1. a. $\sqrt{14}$;
$(\frac{1}{2}, -1, \frac{3}{2})$ **b.** $2\sqrt{14}$; $(4, 2, -1)$ **2. a.** $A = (4, 0, 0)$;
$B = (4, 5, 0)$; $C = (0, 5, 0)$; $D = (4, 0, 3)$; $E = (0, 0, 3)$;
$F = (0, 5, 3)$ **b.** $5\sqrt{2}$ **3.** 8; octants
4. a. $x^2 + y^2 + z^2 = 25$
b. $(x - 1)^2 + (y - 2)^2 + (z - 3)^2 = 25$
5. a. $(5, 9, 4)$ **b.** 0 **6.** 1 **7. a.** For example, $(2, 5, 1)$
and $(8, 12, 9)$ **b.** $x = 2 + 6t$, $y = 5 + 7t$, $z = 1 + 8t$
c. For example, $(6, 7, 8)$ **d.** Both have the same direction vector. **e.** The dot product of the direction vectors is 0.

Written Exercises 12-5, pages 431-433 1. $2\sqrt{6}$;
$(1, 4, -1)$ **3.** $2\sqrt{14}$; $(1, -2, 1)$ **5.** $A = (5, 0, 0)$;
$B = (5, 6, 0)$; $C = (0, 6, 0)$; $D = (5, 0, 4)$; $E = (0, 0, 4)$;
$F = (0, 6, 4)$; $\sqrt{77}$ **7. a.** $(11, 6, 2)$ **b.** -5 **c.** $\sqrt{35}$
9. $\mathbf{u} + \mathbf{v} = (3, 1, 2)$; $\mathbf{u} - \mathbf{v} = (-1, 3, 0)$ **11.** Yes
13. a. $x^2 + y^2 + z^2 = 4$
b. $(x - 3)^2 + (y + 1)^2 + (z - 2)^2 = 4$ **15.** $(1, 2, 3)$; 5
17. $70.5°$ **19. a.** $\vec{AB} = (2, -4, -4)$, $\vec{AC} = (2, -1, 2)$
b. 9 square units **21. a.** $\cos A = \dfrac{\sqrt{3}}{2}$; $\sin A = \dfrac{1}{2}$

b. $\dfrac{\sqrt{3}}{2}$ square units **23. a.** For example,

$x = -2 + 4t$, $y = -t$, $z = 1 + t$ **b.** For example,
$t = 0$: $(-2, 0, 1)$; $t = 1$: $(2, -1, 2)$ **c.** $(-10, 2, -1)$
and $(18, -5, 6)$ are on the line; $(14, -5, 4)$ is not on the line. **d.** $(x, y, z) = (1, 2, 3) + t(4, -1, 1)$

25. $(x, y, z) = (4, 2, -1) + t(2, 1, 3)$; $x = 4 + 2t$,
$y = 2 + t$, $z = -1 + 3t$
29. a. $(x, y, z) = (1, 2, 3) + t(1, 4, -6)$
b. $(x, y, z) = (1, 2, 3) + t(6, 0, 1)$
31. a. $4x + 6y + 12z = 49$ **b.** A plane perpendicular to and bisecting $\vec{AB}$ **33. a.** A plane through A perpendicular to $\vec{AB}$ **b.** $3x + y + 2z = 18$
35. $(7, 2, 2)$, $(-1, 6, -6)$ **37.** $y = 0$; $(-\frac{3}{2}, 0, -6)$
39. b. No

Oral Exercises 12-6, pages 435-436 1. x-intercept, 2;
y-intercept, 6; z-intercept, 3 **2. a.** $z = 4$ **b.** $z = 0$
c. $x = 3$ **d.** $y = 6$ **3. a.** $(2, 3, 4)$ **b.** $(3, 0, -4)$
c. $(0, 0, 1)$ **4.** $x + y + z = 12$ **5.** $6x + 7y + 8z = 0$

Written Exercises 12-6, pages 436-438
1. 　　　　　　　　　 **3.**

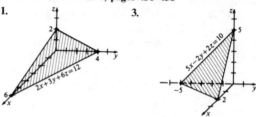

5.

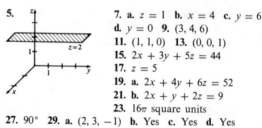

7. a. $z = 1$ **b.** $x = 4$ **c.** $y = 6$
d. $y = 0$ **9.** $(3, 4, 6)$
11. $(1, 1, 0)$ **13.** $(0, 0, 1)$
15. $2x + 3y + 5z = 44$
17. $z = 5$
19. a. $2x + 4y + 6z = 52$
21. b. $2x + y + 2z = 9$
23. 16π square units
27. $90°$ **29. a.** $(2, 3, -1)$ **b.** Yes **c.** Yes **d.** Yes

31. a. For example, $4x + y - 3z = 0$ **b.** For example,
$x - y + z = 0$ **33. a.** $(4, -5, 2)$ **35. b.** $(2, 1, 0)$
37. a. $(x, y, z) = (3, 1, 5) + t(2, 2, 1)$ **b.** $(1, -1, 4)$

c. 3 **39.** $d = \dfrac{|ax_0 + by_0 + k|}{\sqrt{a^2 + b^2}}$

Oral Exercises 12-7, page 441 1. -2 **2.** 12 **3.** 65
4. 31 **5.** 3; 2; 4 **6.** $\begin{vmatrix} 5 & 7 \\ 8 & 6 \end{vmatrix}$; $\begin{vmatrix} 3 & 2 \\ 8 & 6 \end{vmatrix}$; $\begin{vmatrix} 3 & 2 \\ 5 & 7 \end{vmatrix}$ **7.** -3

Written Exercises 12-7, pages 441-442 1. -48
3. -3125 **5.** 435 **7.** 90 **9.** -98 **11.** 105 **13.** No

15. 32 **17. a.** $x = \dfrac{b_2 c_1 - b_1 c_2}{a_1 b_2 - a_2 b_1}$, $y = \dfrac{a_1 c_2 - a_2 c_1}{a_1 b_2 - a_2 b_1}$

Oral Exercises 12-8, page 445

1. $x = \dfrac{\begin{vmatrix} 7 & 4 \\ 8 & 6 \end{vmatrix}}{\begin{vmatrix} 3 & 4 \\ 5 & 6 \end{vmatrix}}$ and $y = \dfrac{\begin{vmatrix} 3 & 7 \\ 5 & 8 \end{vmatrix}}{\begin{vmatrix} 3 & 4 \\ 5 & 6 \end{vmatrix}}$

2. $x = \dfrac{\begin{vmatrix} 4 & 1 & -2 \\ 1 & -1 & 4 \\ 3 & 2 & 7 \end{vmatrix}}{\begin{vmatrix} 3 & 1 & -2 \\ 2 & -1 & 4 \\ 1 & 2 & 7 \end{vmatrix}}, \quad y = \dfrac{\begin{vmatrix} 3 & 4 & -2 \\ 2 & 1 & 4 \\ 1 & 3 & 7 \end{vmatrix}}{\begin{vmatrix} 3 & 1 & -2 \\ 2 & -1 & 4 \\ 1 & 2 & 7 \end{vmatrix}},$

and $z = \dfrac{\begin{vmatrix} 3 & 1 & 4 \\ 2 & -1 & 1 \\ 1 & 2 & 3 \end{vmatrix}}{\begin{vmatrix} 3 & 1 & -2 \\ 2 & -1 & 4 \\ 1 & 2 & 7 \end{vmatrix}}$ **3.** We cannot divide because $D = 0$. **4.** 2 square units

Written Exercises 12–8, pages 445–446

1. $(1, 1)$ **3.** $(1, -2)$ **5.** $\left(\dfrac{1}{a + b}, \dfrac{1}{a + b}\right)$

7. a. no solution **b.** $\begin{vmatrix} a & b \\ d & e \end{vmatrix}$ **9.** 4 square units

11. 8 square units **13.** 6 square units **15.** The points are collinear. **17.** $(3, 2, 1)$ **19.** $(2, 0, 1)$ **21.** 5 cubic units

Written Exercises 12–9, pages 447–448
1. $\mathbf{v} \times \mathbf{u} = (-1, -5, 4)$; $\mathbf{u} \times \mathbf{v} = (1, 5, -4)$; yes
5. $\sqrt{42}$ square units **7.** $(2, 3, 6)$ **9.** $(0, 0, 1)$
11. $5x - 2y + 6z = 10$ **13. a.** $(-1, 4, 1), (1, -4, -1)$
b. $-x + 4y + z = 3$ **c.** $\frac{3}{2}\sqrt{2}$ square units
15. a. $(-4, 4, 2), (4, -4, -2)$ **b.** $2x - 2y - z = 2$
c. 3 square units **17. a.** $\frac{1}{3}\sqrt{5}$ **b.** $\frac{2}{3}$

Calculator Exercise, pages 448–449 320 square units

Chapter Test, pages 450–451 **1.** $3\sqrt{5}$ units; 26.6° north of east (See diagram below.) **3. a.** See diagram below. **b.** $2\sqrt{17}$

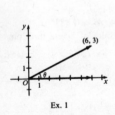

Ex. 1

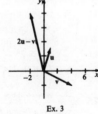

Ex. 3

5. a. velocity $= (2, -1)$; speed $= \sqrt{5}$ **b.** $x = 1 + 2t$, $y = 5 - t$ **c.** when $t = 4$ at $(9, 1)$; when $t = 2$ at $(5, 3)$ **7.** $\frac{3}{5}$ **9.** $(1, 0, -3)$; 4 **11.** -25
13. a. $(-3, 2, 1)$ **b.** $\sqrt{14}$

Oral Exercises 13–1, pages 456–457 **1.** arithmetic; $d = 5$ **2.** geometric; $r = 2$ **3.** neither **4.** arithmetic; $d = -6$ **5.** geometric; $r = -\frac{2}{3}$ **6.** neither **7.** 7, 12, 17, 22; arithmetic **8.** $\frac{2}{3}, \frac{3}{4}, \frac{4}{5}, \frac{5}{6}$; neither **9.** 3, 9, 27, 81; geometric **10.** 7, 19, 31, 43; arithmetic **11.** 8, -16, 32, -64; geometric **12.** 1, 1, 2, 3; neither

13. a. $t_n = 6 + 2n$ **b.** $t_n = 34 - 4n$
14. a. $t_n = 4 \cdot 3^{n-1}$ **b.** $t_n = 24(-\frac{1}{2})^{n-1}$ **15. a.** Yes
b. Yes

Written Exercises 13–1, pages 457–459 **1.** 5, 7, 9, 11; arithmetic **3.** 6, 12, 24, 48; geometric **5.** 2, $\frac{5}{2}, \frac{10}{3}, \frac{17}{4}$; neither **7.** 6, 10, 14, 18; arithmetic **9.** 4, 22, 112, 562; neither **11.** 1, $\dfrac{1}{9}, -\dfrac{1}{243}, -\dfrac{1}{19,683}$; neither
13. arithmetic; $t_n = 4n + 13$ **15.** geometric; $t_n = 8 \cdot (\frac{3}{2})^{n-1}$ **17.** neither; $t_n = \dfrac{n}{n + 1}$ **19.** arithmetic; $t_n = 20n - 28$ **21.** neither; $t_n = 10^n + 1$
23. arithmetic; $t_n = n(a + b) + (a - 3b)$
25. geometric; $t_n = 8(\frac{1}{2})^{n-1}$ **27.** neither; $t_n = n^2$
29. geometric; $t_n = 2^{n - \frac{1}{3}}$ **31.** 129 **33.** 149 **35.** 128
37. $\frac{2}{81}$ **39.** 54 **41.** 38 **43.** 75 **45.** false **47.** 0.4771, 1.0792, 1.6812, 2.2833; $d \approx 0.602$ **49. a.** 3 **b.** 9
51. $x = 3, y = 2$ **53.** $t_n = t_{n-1} + 4$
55. $t_n = t_{n-1} + 2^{n-1}$ **57.** $t_n = t_{n-1} + n$ **59.** $\frac{1}{2}; \frac{1}{4}; \frac{1}{32}$
61. 1, 43, 85, 127, 169, ... **63.** 58, 135, 212, 289, 366
65. any triangle with sides $3d, 4d,$ and $5d$ where d is a positive integer **67. a.** $\frac{13}{2}$; 6 **b.** $\frac{15}{2}$; $5\sqrt{2}$

Calculator Exercises, page 460 **1. a.** 89, 144, 233, 377, 610 **b.** $1.6\overline{18}$, ≈ 1.6179775, $1.6180\overline{5}$, ≈ 1.6180258, ≈ 1.6180371 **c.** The ratios are approaching the golden ratio R. **2.** 1, 1, 2, 3, 5; the sequence seems to be the Fibonacci sequence.

Written Exercises 13–2, pages 462–466 **1.** 210 **3.** 6375
5. 294 **7.** 500,500 **9. a.** $\frac{255}{16}$ **b.** $\frac{85}{16}$ **13. a.** $-28 \cdot 8$
b. 9.5625 **17.** 166,833 **19.** 49,320 **23.** 501 **25.** 320 pages **27.** 305,175,780 **29. a.** ≈ 61 cents per pound
b. \$20,518 **31. a.** 2046 **b.** 19 **33. a.** 1; 4; 9; 16
b. $S_n = n^2$ **35.** $S_n = \left(\dfrac{n(n + 1)}{2}\right)^2$ **37.** $S = \dfrac{n(n^2 + 1)}{2}$
39. a. neither **b.** $t_n = 3n^2 - 3n + 1$ **41.** about \$19,654.58 **43. a.** $(1 + r)A - P$; $(1 + r)^2 A - [(1 + r) + 1]P$; $(1 + r)^3 A - [(1 + r)^2 + (1 + r) + 1]P$

Oral Exercises 13–3, pages 469–470 **1.** 1 **2.** 1 **3.** $\frac{2}{3}$
4. $\frac{8}{5}$ **5.** 1 **6.** 0 **7.** 0 **8.** ∞ **9.** ∞ **10.** 0 **11.** No

Written Exercises 13–3, pages 470–472 **1.** 1 **3.** 1
5. $\frac{3}{8}$ **7.** 0 **9.** 0 **11.** 0 **13.** 0 **15.** does not exist
17. does not exist **19.** ∞ **21.** $-\infty$ **23.** 0 **25.** 1
27. 0 **29.** 1 **31. b.** $\frac{3}{2}$ **33.** e^2 **35. b.** $\frac{1}{4}$ **c.** $\frac{1}{4}$

Calculator Exercises, page 472 **1. a.** 0
b. $\dfrac{1}{\sqrt{n + 1} + \sqrt{n}}$ **c.** 0 **2. a.** 1; 1.5; $1.41\overline{6}$; 1.4142157; 1.4142136 **b.** $\sqrt{2}$ **3. b.** Values approach the golden ratio R **4.** Values approach the golden ratio R

Oral Exercises 13-4, page 475 **1.** 1, $\frac{4}{3}$, $\frac{13}{9}$, $\frac{40}{27}$; $\frac{3}{2}$ **2.** $\frac{1}{2}$, $\frac{1}{4}$, $\frac{3}{8}$, $\frac{5}{16}$; $\frac{1}{3}$ **3.** 1, 4, 9, 16; diverges **4.** 1, 1.1, 1.11, 1.111; $\frac{10}{9}$ **5. a.** $-1 < x < 1$ **b.** $-\frac{1}{2} < x < \frac{1}{2}$ **6. a.** 0.3 **b.** 0.1 **c.** $\frac{1}{3}$ **8.** divergent

Written Exercises 13-4, pages 475-478 **1.** 2 **3.** 16 **5.** $\frac{125}{24}$ **7.** 2 **9.** $\frac{1}{2}$ **11. a.** $-1 < x < 1$ **b.** $\dfrac{1}{1 - x^2}$

13. a. $2 < x < 4$ **b.** $\dfrac{1}{4 - x}$ **15. a.** $x < -2$ or $x > 2$

b. $\dfrac{x}{x + 2}$ **19.** $-\frac{1}{3}$ **23.** $\frac{7}{9}$ **25.** $\frac{400}{9}$ **27.** $\frac{1}{7}$ **29.** $\frac{1}{2}, \frac{2}{3}, \frac{3}{4}, \frac{4}{5}$,

$S_n = \dfrac{n}{n + 1}$; 1 **31.** $\frac{1}{4}, \frac{2}{7}, \frac{3}{10}, \frac{4}{13}$; $S_n = \dfrac{n}{3n + 1}$; $\frac{1}{3}$

33. 56 m **35. a.** 288 square units **b.** $96 + 48\sqrt{2}$ units **37.** 15 **39.** $\frac{100}{9}$ s **41.** $t_3 = 5 + 5i$; $t_4 = 7 + 6i$; $t_5 = 9 + 7i$; $t_{25} = 49 + 27i$; $S_{25} = 625 + 375i$

43. a. $\frac{4}{5} + \frac{2}{3}i$ **b.** $\frac{243}{10} - \frac{81}{10}i$ **45.** $\dfrac{1}{1 + r}$; $\dfrac{r}{1 + r}$ **47.** $\dfrac{2\sqrt{3}}{5}$ square units

Calculator Exercise, page 478 **a.** 5 **b.** 8 **c.** 10

Oral Exercises 13-5, page 480 **1.** $5 \cdot 1 + 5 \cdot 2 + 5 \cdot 3 + 5 \cdot 4$ **2.** $3^2 + 4^2 + 5^2 + 6^2$ **3.** $(-1)^2 + (-1)^3 + (-1)^4 + (-1)^5 + (-1)^6 + (-1)^7 + (-1)^8$ **4.** $1 + \frac{1}{2} + \frac{1}{3} + \frac{1}{4} + \cdots$ **5.** $\sum_{n=2}^{6} n^2$

6. $\sum_{n=1}^{4} \dfrac{n}{n + 1}$ **7.** $\sum_{n=1}^{100} 3n$ **8.** $\sum_{n=1}^{\infty} \dfrac{1}{3^n}$

Written Exercises 13-5, pages 481-482 **1.** $2 + 3 + 4 + 5 + 6 = 20$ **3.** $\frac{1}{1} + \frac{1}{2} + \frac{1}{3} + \frac{1}{4} + \frac{1}{5} = \frac{137}{60}$ **5.** $3^1 + 3^0 + 3^{-1} + \cdots = \frac{9}{2}$

7. $4^{-2} + 4^{-1} + \cdots + 4^2 = \frac{341}{16}$ **9.** $\sum_{k=1}^{5} 4k$

11. $\sum_{k=1}^{25} (4k + 1)$ **13.** $\sum_{k=1}^{\infty} \dfrac{1}{k^2}$ **15.** $\sum_{k=1}^{\infty} \sin kx$

17. $\sum_{t=1}^{4} \log t = \log 1 + \log 2 + \log 3 + \log 4 = \log(1 \cdot 2 \cdot 3 \cdot 4) = \log 24$ **19.** 0 **21.** 625

25. a. $\sum_{k=0}^{5} (-\frac{1}{2})^k$ **b.** $\sum_{k=0}^{5} (-1)(-\frac{1}{2})^k$

27. a. $\sum_{n=1}^{50} (-1)^{n+1}(2n - 1)$ **b.** $\sum_{n=1}^{50} (-1)^n \cdot 2n$

33. 343,400 **35.** 53,130 **37. a.** $\sum_{k=1}^{8} k^2 = 204$ squares in

all **b.** $\sum_{k=1}^{n} k^2 = \dfrac{n(n + 1)(2n + 1)}{6}$ **39.** $\dfrac{n(n + 1)(n + 2)}{6}$
oranges

Calculator Exercises, page 483 **1.** $2^{63} \approx 9.2234 \times 10^{18}$

2. $\sum_{k=1}^{64} 2^{k-1} \approx 1.8447 \times 10^{19}$ **3.** 1.8447×10^{13} bushels

requested; the request was more than 3600 times greater.

Written Exercises 13-6, pages 485-486

1. $\cos \dfrac{\pi}{4} = 1 - \dfrac{1}{2!}\left(\dfrac{\pi}{4}\right)^2 + \dfrac{1}{4!}\left(\dfrac{\pi}{4}\right)^4 - \dfrac{1}{6!}\left(\dfrac{\pi}{4}\right)^6 + \cdots$;

$\sin\left(-\dfrac{\pi}{4}\right) = -\dfrac{\pi}{4} + \dfrac{1}{3!}\left(\dfrac{\pi}{4}\right)^3 - \dfrac{1}{5!}\left(\dfrac{\pi}{4}\right)^5 + \dfrac{1}{7!}\left(\dfrac{\pi}{4}\right)^7 - \cdots$

3. a. $\ln 2 = \ln(1 + 1) = 1 - \frac{1}{2} + \frac{1}{3} - \frac{1}{4} + \cdots$ **b.** The series is defined only when $-1 < x \le 1$.

5. $e^{-x} = 1 - \dfrac{x}{1!} + \dfrac{x^2}{2!} - \dfrac{x^3}{3!} + \cdots$; all real x

7. $\text{Tan}^{-1}2x = 2x - \dfrac{(2x)^3}{3} + \dfrac{(2x)^5}{5} - \dfrac{(2x)^7}{7} + \cdots$;

$-\dfrac{1}{2} < x < \dfrac{1}{2}$

9. $\sin x^2 = x^2 - \dfrac{(x^2)^3}{3!} + \dfrac{(x^2)^5}{5!} - \dfrac{(x^2)^7}{7!} + \cdots$; all real

x **13.** 1 **19.** diverges **21.** converges

Calculator Exercises, page 487 **1.** 2.7183 **2.** 0.5403 **3.** 0.9093

Written Exercises 13-7, pages 489-490

1. For $n = 1$, $\dfrac{1(1 + 1)}{2} = 1$; if

$1 + 2 + \cdots + k = \dfrac{k(k + 1)}{2}$, then

$(1 + 2 + \cdots + k) + (k + 1) =$
$\dfrac{k(k + 1)}{2} + \dfrac{2(k + 1)}{2} = \dfrac{k^2 + 3k + 2}{2} =$
$\dfrac{(k + 1)(k + 2)}{2} = \dfrac{(k + 1)[(k + 1) + 1]}{2}$. **11.** (1) For

$n = 1$, $11^1 - 4^1 = 7$, and 7 is a multiple of 7. (2) If $11^k - 4^k$ is a multiple of 7, then $11^{k+1} - 4^{k+1} = 11 \cdot 11^k - 4 \cdot 4^k = (7 + 4)11^k - 4 \cdot 4^k = 7 \cdot 11^k + 4(11^k - 4^k) =$ the sum of 2 multiples of $7 =$ a multiple of 7. **21.** $a_n = 2^n - 1$

23. $S_n = \left(\dfrac{n(n + 1)}{2}\right)^2$

25. $S_n = \dfrac{n(n + 1)(n + 2)(n + 3)}{4}$

Chapter Test, pages 492-493 **1. a.** geometric; $t_n = 27(\frac{2}{3})^{n-1}$ **b.** arithmetic; $t_n = 7n - 3$ **c.** neither; $t_n = n^2 + 1$ **3. a.** 6 **b.** $\frac{10}{3}$ **5.** $t_n = 2^{n-3}$; $S_n = \frac{1}{2} - (\frac{1}{2})^{n+1}$ **7.** 2550 **9.** $\frac{64}{7}$ **11. a.** $(-1)^0(3 \cdot 0 - 1) + (-1)^1(3 \cdot 1 - 1) + (-1)^2(3 \cdot 2 - 1) + (-1)^3(3 \cdot 3 - 1) = -1 - 2 + 5 - 8 = -6$ **b.** $\dfrac{1}{5} + \dfrac{1}{25} + \dfrac{1}{125} + \cdots = \dfrac{\frac{1}{5}}{1 - \frac{1}{5}} = \dfrac{1}{4}$

13. a. diverges **b.** converges

Oral Exercises 14-1, page 499 **1. a.** 30–35, 36–41, 42–47 **b.** 30–34, 35–39, 40–44 **c.** 30–33, 34–37, 38–41 **2. a.** $\bar{x} = 3.2$, median = 3, mode = 5 **b.** $\bar{x} = 6.875$, median = 7.5, modes = 4 and 8 **c.** $\bar{x} = 2$, median = 2, no mode **d.** $\bar{x} = 2.2$, median = 2.5, mode = 3 **3.** Answers may vary. **4. a.** Yes; if the number of measurements is even, the median is not always an integer. **b.** No

Written Exercises 14-1, pages 499–501 **1.** $\bar{x} = 5.4$, median = 5.5, mode = 6 **3.** $\bar{x} = 14.08\overline{3}$ items, median = 6 items, mode = 5 items

5. a.

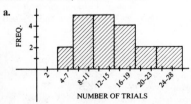

b. median = 13.5 trials

7. a.

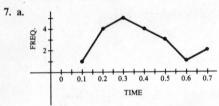

b. $\bar{x} = 0.375$ s, median = 0.35 s, mode = 0.3 s

9. a.

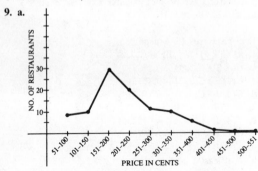

b. $\bar{x} = 222.5$¢, median = 225¢, mode = 175¢
11. a. mean **b.** mode **15.** 82 **17.** $(0, -2)$

Calculator Exercise, page 502 $\bar{x} = 531$

Oral Exercises 14-2, page 505 **1. a.** The heights of the players in team B deviate more from the mean than the heights of the players in team A. **b.** Team B **2.** 3.3 **3. a.** 1 **b.** 2 **c.** 2.5 **d.** -0.5 **e.** 0 **4.** Yes

Written Exercises 14-2, pages 505–506 **1.** $\bar{x} = 7$, $s \approx 1.9$ **3. a.** Row A: $\bar{x} = 3$, $s \approx 1.4$; Row B: $\bar{x} = 13$, $s \approx 1.4$ **b.** mean, 23; standard deviation, 1.4 **5. a.** 2 **b.** -0.5 **c.** 0 **d.** $-\frac{4}{3}$ **7.** Physics students should have

the greater mean score. Scores from students selected at random should have a greater standard deviation.
9. a. 0 **b.** 4 **11.** $\bar{x} = 6$, $s \approx 1.3$ **13.** $\bar{x} = 3.36$, $s \approx 1.78$

Calculator Exercise, page 507 mean, 6; standard deviation, 1.2909944

Oral Exercises 14-3, page 513 **1.** 19.1% **2.** 46.4% **3.** 38.5% **4.** 77% **5.** 2.3% **6.** 84.1% **7.** 91.9% **8.** 91.9% **9.** 75.8% **10. a.** (1) 68.2% (2) 95.4% (3) 2.3% (4) 6.7% **b.** 630 **c.** 570 **11. a.** (1) 47.7% (2) 68.2% (3) 15.9% (4) 0.1% **b.** 1.6 s

Written Exercises 14-3, pages 513–515 **1. a.** 68.2% **b.** 95.4% **c.** 99.8% **d.** 6.7% **e.** 21.2% **f.** 30.3% **3. a.** 2.3% **b.** 1.8% **c.** 4.05% **5.** 12% **7. a.** 2.3% **b.** 15.9% **9. a.** 580 **b.** 430 **11.** 84 is minimum for an A; 75 is minimum for a B. **13. a.** 100% **b.** 2.0074 L **c.** 2.012 L **15.** 11.5%

Oral Exercises 14-4, page 518 **1. a.** $0.2 < p < 0.6$ **b.** $0.1 < p < 0.7$ **2. a.** $0.3 < p < 0.7$ **b.** $0.2 < p < 0.8$ **3. a.** $\bar{p} = 0.5$, $s = 0.05$ **b.** $0.4 < p < 0.6$ **4.** The sample size **5.** No

Written Exercises 14-4, pages 519–520 **1.** $\bar{p} = 0.7$, $s \approx 0.14$, $0.42 < p < 0.98$ **3.** $\bar{p} = 0.36$, $s = 0.048$, $0.264 < p < 0.456$ **5.** $\bar{p} = 0.0625$, $s \approx 0.0121$, $0.0383 < p < 0.0867$ **7.** $\bar{p} = 0.03$, $s \approx 0.0054$, $0.014 < p < 0.046$ **9.** $\bar{p} = 0.09$, $s \approx 0.0286$, $0.004 < p < 0.176$ **11.** $\bar{p} = 0.4$, $s \approx 0.0219$, $0.334 < p < 0.466$ **13.** 4 **15.** 95% **17. a.** $0.537 < p < 0.663$ **b.** $0.505 < p < 0.695$
19. a. **b.** $f(0.5) = 0.25$

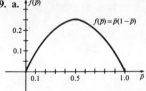

21. $n \geq 10,000$ **23.** $n > 400$

Oral Exercises 14-5, page 524 **1.** c **2.** b **3.** a **4.** d **5.** c; biased language **6.** c **7.** 44%

Written Exercises 14-5, pages 524–526 **1.** Those who wrote may not be representative of the general population. **3.** Those who call the station are often the ones most affected by the issue. **5.** 75, 41, 76, 55, 65 **7.** Answers will vary. **9.** 44.1% **11.** 2400 **13.** 1040 **15.** 47.2%

Oral Exercises 14-6, pages 529–530 **1. a.** 0.6 **b.** 0.8 **c.** -0.6 **d.** -0.8 **e.** 0 **f.** 0.2 **2. a.** positive **b.** approximately zero **c.** positive **d.** positive

Answers to selected exercises **25**

e. negative **f.** negative **3. a.** $\frac{2}{3}$ **b.** $y = 0.444x + 1.222$

Written Exercises 14-6, pages 530-532 1. positive
3. negative **5.** positive **7.** positive

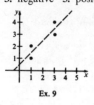

Ex. 9

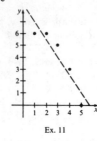

Ex. 11

9. $r \approx 0.89$, $y = 0.995x + 0.51$ **11.** $r \approx -0.93$,
$y = -1.5x + 8.5$ **13. a.** B **b.** C **c.** F
15. $y = 0.04x + 4.82$; about 8.8 metric tons per hectare

Chapter Test, page 533

1.

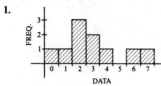

DATA

1. mean $= 3$, median $= 2.5$, mode $= 2$ **3.** -1.5,
-1.0, -0.5, -0.5, -0.5, 0, 0, 0.5, 1.5, 2.0
5. $0.5 < p < 0.7$ **7.** 1800

Oral Exercises 15-1, page 537 1. a. 2 **b.** 6 **c.** 24
2. 362,880 **3. a.** 60 **b.** 180 **4. a.** 8 **b.** 6 **5.** 720

Written Exercises 15-1, pages 538-539 1. a. 120
b. 720 **c.** 5040 **d.** 1 **3.** 120 **5. a.** 17,576 **b.** 13,800
7. 11,880 **11.** 156 **13.** 2 digits: 1,757,600; 3 digits:
17,576,000; 19,333,600 total **15.** 250,000
17. $2^{10} = 1024$ **19. a.** 720 **b.** 720

Calculator Exercise 15-1, page 539 $\log_{10} 9! = 5.5598$;
$\log_{10} 10! = 6.5598$

Oral Exercises 15-2, page 541 1. a. 20 **b.** 10
2. a. 120 **b.** 20 **3. a.** 720 **b.** 120 **4. a.** 24 **b.** 1
5. $P(10, 3) = 720$ **6.** $C(10, 3) = 120$ **7.** The order
matters.

Written Exercises 15-2, pages 542-543
1. a. $P(20, 2) = 380$ **b.** $C(20, 2) = 190$
3. a. $C(10, 5) = 252$ **b.** $P(10, 5) = 30,240$
5. a. $C(200, 3) = 1,313,400$ **b.** $P(200, 3) = 7,880,400$
7. a. $P(8, 3) = 336$ **b.** $C(8, 3) = 56$
9. a. $C(7, 4) = 35$ **b.** $C(7, 3) = 35$ **13. a.** 1 **b.** 1
15. a. $P(12, 6) = 665,280$ **b.** $C(12, 6) = 924$
17. $n = 10$ **21. a.** A must win the last game and three
of the other five games. There are $C(5, 3)$ ways to do
this. **b.** $C(7, 4) + C(7, 4) = 70$

Calculator Exercise, page 543 5.3644738×10^{28}

Assorted Exercises 15-1 and 15-2, pages 543-545
1. $C(100, 3) = 161,700$ **3. a.** 6561 **b.** 5832 **c.** 2000
5. $P(10, 4) = 5040$ **7. a.** 720 **b.** 360 **9.** 62
11. a. $C(8, 4) = 70$ **b.** 35 **c.** 35 **13.** 15
15. a. $P(8, 4) = 1680$ **b.** $P(5, 4) = 120$ **c.** 1560
17. a. 28,800 **b.** 28,800

Oral Exercises 15-3, page 550 1. a. $\frac{1}{52}$ **b.** $\frac{1}{4}$ **c.** $\frac{1}{13}$
d. independent **2. a.** $\frac{1}{4}$ **b.** $\frac{3}{4}$ **3. a.** $\frac{1}{18}$ **b.** $\frac{1}{12}$ **c.** $\frac{5}{36}$
4. $\frac{1}{6}$ **5.** 60% **6.** 0.18 **7. a.** $\frac{1}{2}$; $\frac{1}{2}$ **b.** $\frac{1}{13}$; $\frac{1}{13}$ **c.** Yes
8. a. $\frac{1}{13}$; $\frac{1}{17}$; $\frac{4}{51}$ **b.** No **9.** Yes **11.** $P(A$ and $B) = 0$ so
$P(A$ or $B) = P(A) + P(B)$

Written Exercises 15-3, pages 551-553 1. a. $\frac{1}{4}$ **b.** $\frac{1}{2}$
c. $\frac{3}{13}$ **3. a.** $\frac{5}{36}$ **b.** $\frac{1}{6}$ **c.** $\frac{5}{36}$ **5.** $\frac{1}{4}$ **7.** $\frac{25}{51}$ **9.** 0 **11.** $\frac{4}{13}$
13. $\frac{10}{13}$ **15.** $\frac{25}{204}$ **17.** $\frac{1}{6}$ **19.** $\frac{1}{144}$; $\frac{1}{72}$; $\frac{1}{48}$; 13 **21.** $\frac{3}{4}$
23. a. 0.12 **b.** 0.42 **c.** 0.28 **d.** 0.58 **25. a.** $\frac{147}{400}$ **b.** $\frac{1}{20}$
27. a. $\frac{29}{90}$ **b.** $\frac{29}{147}$ **29.** independent
31. a. $\dfrac{C(13, 3)}{C(52, 3)} = \dfrac{11}{850}$ **b.** $\dfrac{C(26, 3)}{C(52, 3)} = \dfrac{2}{17}$
c. $\dfrac{C(48, 3)}{C(52, 3)} = \dfrac{4324}{5525}$ **d.** $\dfrac{1}{64}$ **e.** $\dfrac{1}{8}$ **f.** $\dfrac{1728}{2197}$
33. a. $\dfrac{4}{12} \cdot \dfrac{3}{11} \cdot \dfrac{2}{10} = \dfrac{1}{55}$ **b.** $\dfrac{3}{12} \cdot \dfrac{2}{11} \cdot \dfrac{1}{10} = \dfrac{1}{220}$
c. $\dfrac{1}{11}$ **35. a.** $\dfrac{2}{7}$ **b.** $\dfrac{5}{7}$ **37. a.** $\dfrac{729}{1000}$ **b.** $\dfrac{5}{18}$

Calculator Exercise or Computer Exercise, page 553
a. ≈ 0.12 **b.** ≈ 0.41 **c.** 23

Oral Exercises 15-4, page 555
1. a. $C(4, 3) \cdot C(5, 0) \div C(9, 3)$
b. $C(4, 2) \cdot C(5, 1) \div C(9, 3)$
c. $C(4, 1) \cdot C(5, 2) \div C(9, 3)$
d. $C(4, 0) \cdot C(5, 3) \div C(9, 3)$
2. a. $C(4, 4) \cdot C(48, 1) \div C(52, 5)$
b. $C(4, 0) \cdot C(48, 5) \div C(52, 5)$
c. $C(13, 4) \cdot C(39, 1) \div C(52, 5)$
d. $C(4, 4) \cdot C(4, 1) \div C(52, 5)$ **3. a.** no aces
b. $P($at least 1 ace$) \approx 0.34$

Written Exercises 15-4, pages 556-557 1. a. $\dfrac{48!5!}{52!}$
b. $\dfrac{48!47!}{43!52!}$ **c.** $1 - \dfrac{48!47!}{43!52!}$ **3.** $P(0$ red$) = \frac{3}{28}$;
$P(1$ red$) = \frac{15}{28}$; $P(2$ red$) = \frac{5}{14}$ **5. a.** $\frac{33}{323}$ **b.** $\frac{352}{969}$ **c.** $\frac{616}{1615}$
d. $\frac{224}{1615}$ **e.** $\frac{14}{969}$ **7. a.** $\dfrac{26!39!}{13!52!}$ **b.** $\dfrac{13!13!13!39!}{6!6!7!7!52!}$
c. $1 - \dfrac{40!39!}{27!52!}$ **9.** $1 - \dfrac{90!96!}{86!100!}$ **11.** $\dfrac{5!8!}{21!}$

Oral Exercises 15-5, page 558 1. 1, 7, 21, 35, 35, 21, 7,
1 **2. a.** 1, 8, 28, 56 **b.** $x^8 + 8x^7y + 28x^6y^2 + 56x^5y^3$
3. $a^3 - 3a^2b + 3ab^2 - b^3$
4. $a^4 + 4a^3b + 6a^2b^2 + 4ab^3 + b^4$

26   **Answers to selected exercises**

5. $a^4 - 4a^3b + 6a^2b^2 - 4ab^3 + b^4$
6. a. $C(6, 2)x^4y^2 = 15x^4y^2$ b. $C(9, 3)x^6y^3 = 84x^6y^3$

Written Exercises 15-5, pages 559-560
1. a. $a^5 + 5a^4b + 10a^3b^2 + 10a^2b^3 + 5ab^4 + b^5$
b. $a^5 - 5a^4b + 10a^3b^2 - 10a^2b^3 + 5ab^4 - b^5$
3. a. $x^7 + 7x^6y + 21x^5y^2 + 35x^4y^3 + 35x^3y^4 + 21x^2y^5 + 7xy^6 + y^7$ b. $x^7 - 7x^6y + 21x^5y^2 - 35x^4y^3 + 35x^3y^4 - 21x^2y^5 + 7xy^6 - y^7$ c. $128x^7 - 448x^6y + 672x^5y^2 - 560x^4y^3 + 280x^3y^4 - 84x^2y^5 + 14xy^6 - y^7$ 5. $1 + \dfrac{3x}{2} + \dfrac{3x^2}{4} + \dfrac{x^3}{8}$
7. $x^{10} + 5x^8 + 10x^6 + 10x^4 + 5x^2 + 1$ 9. 923,521
11. $(a^2)^{100} + 100(a^2)^{99}(-b) + \dfrac{100 \cdot 99}{1 \cdot 2}(a^2)^{98}(-b)^2 +$

$\dfrac{100 \cdot 99 \cdot 98}{1 \cdot 2 \cdot 3}(a^2)^{97}(-b) + \cdots$

13. $(\sin x)^{10} + 10(\sin x)^9(\sin y) + \dfrac{10 \cdot 9}{1 \cdot 2}(\sin x)^8(\sin y)^2 +$

$\dfrac{10 \cdot 9 \cdot 8}{1 \cdot 2 \cdot 3}(\sin x)^7(\sin y)^3 + \cdots$

15. 1.05 17. $C(12, 4) = 495$; $C(12, 8) = 495$ 19. 9th
term 21. a. $C(n + 1, k) = \dfrac{(n + 1)!}{k!(n - k + 1)!}$

b. $C(n, k - 1) = \dfrac{n!}{(k - 1)!(n - k + 1)!}$ if the mayor is

on the committee; $C(n, k) = \dfrac{n!}{k!(n - k)!}$ if not

23. b. $(\cos \theta + i \sin \theta)^3 = \cos 3\theta + i \sin 3\theta$
25. $S_n = 2^n$ 29. $\sqrt{1.04} \approx 1.02$; $\sqrt{0.98} = 0.99$

Oral Exercises 15-6, page 561 1. 3 heads in 4 tosses, 2
heads in 4 tosses, 1 head in 4 tosses, 0 heads in 4
tosses 2. Find $C(4, 3) (\frac{1}{6})^3(\frac{5}{6})^1$ 3. $\frac{1}{6}$ 4. $\frac{1}{2}$

Written Exercises 15-6, pages 562-563 1. a. $\dfrac{1}{8}$ b. $\dfrac{3}{8}$
c. $\dfrac{3}{8}$ d. $\dfrac{1}{8}$ 3. a. $\dfrac{1}{1296}$ b. $\dfrac{5}{324}$ c. $\dfrac{25}{216}$ d. $\dfrac{125}{324}$
e. $\dfrac{625}{1296}$ 5. $\dfrac{25}{216}$ 7. a. $\dfrac{1}{4}$ b. $\dfrac{9}{16}; \dfrac{3}{8}; \dfrac{1}{16}$ 9. $\dfrac{21}{16,384}$
11. P(no odd integers): $\dfrac{8}{125}$, 0; P(exactly 1 odd

integer): $\dfrac{36}{125}, \dfrac{3}{10}$; P(exactly 2 odd); $\dfrac{54}{125}, \dfrac{3}{5}$; P(exactly

3 odd integers): $\dfrac{27}{125}, \dfrac{1}{10}$ 13. $\dfrac{64}{125}; \dfrac{1}{125}; \dfrac{124}{125}$

15. $\dfrac{27}{345}; \dfrac{108}{343}; \dfrac{144}{343}; \dfrac{64}{343}$ 17. $\dfrac{9^9}{10^9} \approx 0.387$

Chapter Test, pages 564-565 1. a. 360 b. 1296
3. $P(12, 3) = 1320$ 5. $\frac{2}{9}$ 7. a. $\frac{2}{9}$ b. $\frac{1}{15}$ c. $\frac{8}{15}$
9. 2.99×10^{-4}
11. $a^{25} + 25a^{24}b + 300a^{23}b^2 + 2300a^{22}b^3$ 13. $\dfrac{9}{256}$

Oral Exercises 16-1, page 573 1. a. 2 b. $\frac{3}{2}$ c. 3
d. ∞ 2. a. 0 b. -1 c. 3 d. 1 3. a. 0 b. -3
c. The right- and left-hand limits are different.
4. a. $\frac{5}{7}$ b. $\frac{3}{4}$ c. $\frac{1}{2}$ d. ∞ e. $-\infty$ f. 8

Written Exercises 16-1, pages 573-575 1. $\frac{3}{4}$ 3. 2
5. $\frac{1}{2}$ 7. 2 9. $\frac{5}{2}$ 11. ∞ 13. $-\infty$ 15. a. 1 b. -1
c. does not exist 17. a. 0 b. undefined c. Limit
does not exist. 19. 2 21. $-\frac{1}{2}$ 23. 0 25. a. 2 b. 4
c. $2x$ 27. a. 1 b. 0 c. ∞ d. $-\infty$ 29. 1 31. 0
33. Limit does not exist 37. 2 39. 1 41. a. 0 b. 0
c. 0

Oral Exercises 16-2, page 577 1. x-intercept 2;
y-intercept $-\frac{2}{3}$; vertical asymptotes $x = 1$, $x = 3$;
horizontal asymptote $y = 0$ 2. x-intercept 0;
y-intercept 0; vertical asymptotes $x = 2$, $x = -2$;
horizontal asymptote $y = 0$ 3. x-intercept -3;
y-intercept 3; no vertical asymptotes; no horizontal
asymptotes 4. x-intercepts 3, -3; y-intercept -1; no
vertical asymptotes; horizontal asymptote $y = 1$
5. no x-intercept; no y-intercept; vertical asymptote
$x = 6$; horizontal asymptote $y = 2$ 6. x-intercept 0;
y-intercept 0; vertical asymptotes $x = 2$, $x = -2$; no
horizontal asymptote

Written Exercises 16-2, page 578

1.
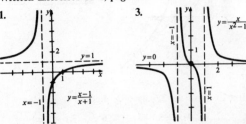

3.
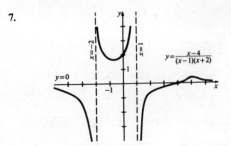

7.
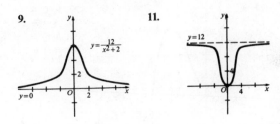

9.

11.

13.

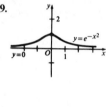

$$y = \frac{x^2 - 2x}{x^2 - 4}$$

$y = 1$

$x = -2$

15.

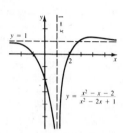

$y = 1$

$$y = \frac{x^2 - x - 2}{x^2 - 2x + 1}$$

17.

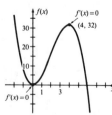

$y = -x$

$$y = \frac{1 - x^2}{x}$$

$x = 0$

19.

$y = e^{-x^2}$

$y = 0$

Oral Exercises 16-3, page 582 **1.** $8x^7$ **2.** $12x^{11}$
3. $-\dfrac{4}{x^5}$ **4.** $21x^6$ **5.** $\dfrac{1}{2\sqrt{x}}$ **6.** $6\sqrt{x}$ **7.** $\dfrac{1}{3\sqrt[3]{x^2}}$

8. $-\dfrac{1}{3x\sqrt[3]{x}}$ **9.** $24x^2 - 14x + 3$

10. $-\dfrac{20}{x^6} + \dfrac{6}{x^4} - \dfrac{9}{x^2}$ **11. a.** -8 **b.** -8 **12.** 32

Written Exercises 16-3, pages 582-583 **1. a.** 7
b. 4.75 **c.** 3.31 **d.** 3 **3.** $10x^4$ **5.** $-6x^2$ **7.** $-\dfrac{6}{x^3}$

9. $\dfrac{\sqrt{5}}{2\sqrt{x}}$ **11.** $7x - 5 + \dfrac{1}{x^2}$ **13.** 3 **15.** -2 **17.** -13

19. $\frac{1}{2}$ **21.** $f'(1) = m, f'(2) = m, f'(3) = m$
23. $y = 12x + 16$ **25.** $y = 2x + 3$

27. a.

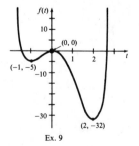

$f(x)$ $f'(x) = 0$ (4, 32)

$f'(x) = 0$

27. b. 0, 4
29. $f(x) = x^4$
31. $h(x) = \frac{1}{2}x^6$
33. $f(x) = 2x^{\frac{3}{2}}$

35. a. $x^n + nx^{n-1}h + \dfrac{n(n-1)}{2}x^{n-2}h^2 + \cdots + h^n$

Oral Exercises 16-4, page 587 **1.** Yes **2.** Maximum
3. Yes **4.** -1 is a local maximum; 2 is a local
minimum. **5.** -2 is a local maximum; 2 is a local
minimum. **6. a.** 0 **b.** 1 **c.** -1

Written Exercises 16-4, pages 587-588 **1.** local
maximum: $(-2, 16)$; local minimum: $(2, -16)$
3. local maximum: $(\sqrt{3}, 6\sqrt{3})$; local minimum:
$(-\sqrt{3}, -6\sqrt{3})$

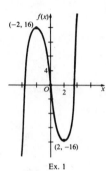

$f(x)$ $(-2, 16)$

$(2, -16)$

Ex. 1

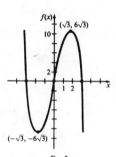

$f(x)$ $(\sqrt{3}, 6\sqrt{3})$

$(-\sqrt{3}, -6\sqrt{3})$

Ex. 3

5. local maximums: $(1, 1)$ and $(-1, 1)$; local minimum:
$(0, 0)$ **7.** local maximum: $(0, 8)$; local minimums:
$(\sqrt{3}, -1)$ and $(-\sqrt{3}, -1)$

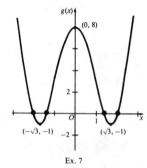

$g(x)$ $(0, 8)$

$(-\sqrt{3}, -1)$ $(\sqrt{3}, -1)$

Ex. 7

9. local maximum: $(0, 0)$; local minimums: $(2, -32)$
and $(-1, -5)$ **11.** local maximum: $(1, 1)$; local
minimum: $(0, 0)$

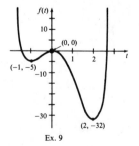

Ex. 9

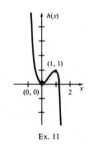

Ex. 11

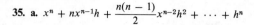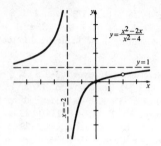

13. $(3, -5)$ **15.** $(a, 0)$ **19. a.** $(1, -5)$

b. $\left(\dfrac{2 - \sqrt{10}}{2}, 5\sqrt{10} - 5\right)$ is local maximum,

$\left(\dfrac{2 + \sqrt{10}}{2}, -5\sqrt{10} - 5\right)$ is local minimum. **21.** local

minimums: $(1, 2)$ and $(-1, 2)$; asymptotes: $y = x^2$ and
y-axis **23.** local minimum: $(\sqrt[3]{\frac{1}{2}}, \frac{3}{2}\sqrt[3]{2})$; asymptotes:
$y = x^2$ and y-axis

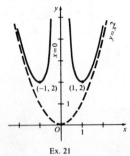

Ex. 21

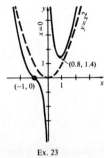

Ex. 23

Written Exercises 16-5, pages 590-592 **3.** 8 sq. units

5. a. $4x^3 - 46x^2 + 120x$ **b.** $\frac{5}{3}$ **7. a.** $(0, 0)$, $(1, 1)$,
$(-1, -1)$ **b.** $\dfrac{\sqrt{3}}{3}$ **9.** 400 **11.** $2\,\text{m} \times 2\,\text{m} \times 2\,\text{m}$

13. b. $\dfrac{256\pi}{3}$ cubic units **15. a.** \$6.60 per mile at 30
mph, \$6.52 per mile at 40 mph **b.** 60 mph

Oral Exercises 16-6, page 595 **1. a.** $48 - 32t$ ft/s;
16 ft/s **b.** -32 ft/s² **2. a.** 2 m **b.** 1 m/s
c. $0 < t < 2$; $t > 2$ **d.** 4 s **e.** 1 m/s; 0 m/s; -1 m/s
f. $0 < t < 1$; $3 < t < 5$ **g.** 2 m **3. a.** 4 m to the left
b. to the left **c.** 6 m **d.** -3 m/s **e.** 5 s

Written Exercises 16-6, pages 595-597
1. a. 32 ft/s   **h.**
b. 32 ft/s
c. 2 s
d. 144 ft
e. $2 < t < 5$
f. 5 s
g. -32 ft/s²

3. a. 2 m **b.** 1 m/s **c.** $0 < t < 3.5$; $t > 3.5$ **d.** 6 s
e. 1 m/s; 1 m/s; 0.5 m/s **f.** $0 < t < 2.5$; 6 s **g.** 3 m
5. a. 7 m to the right **b.** to the right **c.** 8 m
d. 4 m/s **e.** approximately 2.8 s

7.

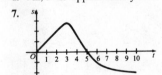

9. a. 2.5 s
b. 40.625 m
c. approx. 2.9 s
11. a. 7.5 mph,
heading west
b. 1:00 pm;
15 m west of
Durham

1. a. 1 **b.** $\frac{1}{2}$ **c.** 1 **d.** $\frac{1}{6}$
3. a. vertical asymptote: $x = 4$; horizontal asymptote:
$y = 1$ **b.** vertical asymptotes: $x = 1$ and $x = -1$;
horizontal asymptote: $y = 1$

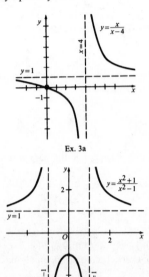

Ex. 3a

Ex. 3b

5. a. $\displaystyle\lim_{h \to 0} \dfrac{f(x + h) - f(x)}{h}$ **7.** Local maximum at
$x = 2$ **9.** 11 m/s; 4 m/s²

Cumulative Review (Chapters 11-16), pages 599-602
1. a. ellipse **b.** parabola **c.** none of these **d.** circle
e. hyperbola **3.** ellipse; center $(1, -3)$; vertices
$(-1, -3)$ and $(3, -3)$ **5.** $x = 4 - y^2$; focus is $(\frac{15}{4}, 0)$;
directrix: $x = \frac{17}{4}$ **7. a.** $5\sqrt{2}$ **b.** $(\sqrt{5}; 63.4°)$ **9.** 53.1°
11. $4x + 7y + z = 17$ **13.** $x = 6$, $y = 7$ **15. a.** 4, 6,
4, 6; neither **b.** 4, 7, 10, 13; arithmetic **c.** 3, 2, 1, 0;
arithmetic **17. a.** 0 **b.** Does not exist **c.** 1 **d.** 0
21. a. $\bar{x} = 23$; median $= 22$; mode $= 31$ **b.** 11.3
23. $\bar{p} = 0.72$; $0.54 < p < 0.9$ **25.** 336 **27.** 53, 130
29. 0.002 **31.** ≈ 0.256 **33.** asymptotes $x = -1$,
$x = 2$, $y = 1$; intercepts: $(1, 0)$ $(3, 0)$ $(0, -\frac{3}{2})$ **35.** local
maximum: $(-\frac{1}{2}, 6)$; local minimum $(\frac{1}{2}, 4)$
37. velocity $= 24s^2 + 16s$ m/s;
acceleration $= 24s + 8$ m/s²

Cumulative Test (Chapters 1-7), page 602 **1.** $-\dfrac{\sqrt{3}}{2}$

3. 2 **5.** -60 **7.** 1 **9.** $\frac{2}{5}$ **11.** $\dfrac{10i\sqrt{3}}{3}$
19. b. $y = \frac{1}{4}x + \frac{7}{4}$ **21.** $x = \pm\sqrt{-1 \pm \sqrt{17}}$;
$f(i) = -34$ **23.** $x = \frac{1}{2}$ **25.** $\dfrac{\sqrt{3}}{4}$; $\frac{1}{2}$ **27.** $\sin x$
29. $-\frac{29}{20}$ **31.** $\angle R = 45°$, $\angle T = 105°$; $\angle R = 135°$,
$\angle T = 15°$

Cumulative Test on Trigonometry (Chapters 6–10), page 603 1. $\dfrac{2\sqrt{3}}{3}$ 3. $\dfrac{\sqrt{3}}{3}$ 5. $\dfrac{13}{12}$ 7. $\sqrt{3} - 2$ 9. $\dfrac{\sqrt{3}}{2}$

11. $\cot\theta$ 13. 13.5 15. a. $x = \dfrac{7\pi}{6}$ or $\dfrac{11\pi}{6}$ b. $x = \dfrac{2\pi}{3}$, $\dfrac{4\pi}{3}$, or 0 19. 166.0° 21. ≈ 22.9 23. $\angle B = 108.2°$

25. $x^2 + (y - 2)^2 = 4$ 27. $z^{10} = -\dfrac{1}{2} + \dfrac{\sqrt{3}}{2}i$; $x_1 \approx -0.1736 + 0.9848i$; $x_2 \approx -0.7660 - 0.6428i$; $x_3 \approx 0.9397 - 0.3420i$

Cumulative Test (Chapters 8–16), page 604 3. $-\dfrac{41}{841}$; $\dfrac{7}{3}$; $\dfrac{-20\sqrt{3} - 21}{58}$ 5. $x = \pi$ 7. $\dfrac{3\sqrt{3}}{2} + \dfrac{i}{2}$; $-\dfrac{3\sqrt{3}}{2} + \dfrac{i}{2}$; $-3i$ 9. $y - 1 = -\dfrac{1}{8}x^2$ 11. $\dfrac{4}{5}$

13. a. 72 b. 728 17. $0.544 < p < 0.736$
19. $C(10, 10)(\tfrac{1}{6})^{10}(\tfrac{5}{6})^0 + C(10, 9)(\tfrac{1}{6})^9(\tfrac{5}{6})^1 + C(10, 8)(\tfrac{1}{6})^8(\tfrac{5}{6})^2$
21. a. -12 b. local minima: (0, 0), (1, 0); local maximum: $(\tfrac{1}{2}, \tfrac{1}{16})$

SAT Sample Questions, pages 611–614 1. C 2. E
3. E 4. D 5. A 6. B 7. E 8. E 9. E 10. B
11. E 12. B 13. D 14. B 15. E 16. B 17. C
18. C 19. D 20. B 21. D 22. D

Mathematics Level I Test Sample Questions, pages 617–619 1. C 2. A 3. D 4. C 5. D 6. D 7. D
8. A 9. E 10. E 11. E 12. B 13. A 14. C
15. E 16. C 17. C 18. E 19. D 20. B

Mathematics Level II Test Sample Questions, pages 619–622 1. A 2. B 3. B 4. A 5. D 6. E 7. D
8. E 9. D 10. C 11. E 12. C 13. C 14. E
15. E 16. D 17. E 18. E 19. E 20. B 21. C
22. B 23. C 24. A 25. C 26. C 27. A 28. E
29. A 30. B

Appendix 1 Exercises, page 647 1. a; Comm. Axiom for Add. 3. 3; Inv. Axiom for Mult. 5. 6; Dist. Axiom 7. 1. Comm. Axiom for Mult.; 2. Iden. Axiom for Mult.; 3. Dist. Axiom; 4. Inv. Axiom for Add.; 5. Thm. 2; 6. Trans. Prop. of Eq.; 7. Inv. Axiom for Add.; 8. Subst. Prop.; 9. Thm. 1

Appendix 2 Exercises, page 650

1.

pos.	neg.	imag.
1	1	0

3.

pos.	neg.	imag.
2	1	0
0	1	2

5.

pos.	neg.	imag.
2	2	0
2	0	2
0	2	2
0	0	4

7.

pos.	neg.	imag.
3	1	0
1	1	2

9. $L = -1$, $M = 2$; $-1 < x < 2$ 11. $L = -2$, $M = 3$; $-2 < x < 3$ 13. $L = -2$, $M = 1$, $-2 < x < 1$ 15. $L = -2$, $M = 3$; $-2 < x < 3$

Appendix 3 Exercises, pages 652–653
1. $s(x) = x^2 + 2x + 2$; $p(x) = 2x^3 + 3x^2 - 2x - 3$; $d(x) = x^2 - 2x - 4$; $q(x) = \dfrac{x^2 - 1}{2x + 3}$, $x \neq -\dfrac{3}{2}$
3. $s(x) = \dfrac{x^2 - x + 3}{x(x - 3)}$, $x \neq 0$, $x \neq 3$; $p(x) = \dfrac{3x - 3}{x(x - 3)}$, $x \neq 0$, $x \neq 3$; $d(x) = \dfrac{x^2 - 7x + 3}{x(x - 3)}$, $x \neq 0$, $x \neq 3$; $q(x) = \dfrac{x^2 - 4x + 3}{3x}$, $x \neq 0$, $x \neq 3$
5. $s(x) = 2x^3 + x^2 + 5$; $p(x) = 2x^5 + 2x^4 - x^3 + 4x^2 + 5x$; $d(x) = 2x^3 - x^2 - 2x + 5$; $q(x) = \dfrac{2x^3 - x + 5}{x^2 + x}$, $x \neq 0$, $x \neq -1$
7.

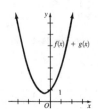

9. increasing 11. neither 13. neither 15. decreasing 17. neither 19. even 21. neither

Appendix 4 Exercises, page 655 1. a. $\Re$
b. $\{(x, y): x^2 + y^2 = 1\}$ c. 2π 3. (0, 1) 5. (0, 1)
7. $(-1, 0)$ 9. $(-1, 0)$ 11. $\cos t = -\dfrac{3}{5}$, $\sin t = \dfrac{4}{5}$
13. $\cos t = -\dfrac{\sqrt{3}}{2}$, $\sin t = -\dfrac{1}{2}$ 15. (0, 0)
17. $\text{cas } 0 = 1$, $\text{san } 0 = 0$; $\text{cas } \dfrac{1}{2} = 1$, $\text{san } \dfrac{1}{2} = \dfrac{1}{2}$; $\text{cas } 1 = 1$, $\text{san } 1 = 1$; $\text{cas } \dfrac{3}{2} = \dfrac{1}{2}$, $\text{san } \dfrac{3}{2} = 1$

Appendix 5 Exercises, page 656 1. a. 3×2 **b.** 3

3. $\begin{bmatrix} 8 & -4 \\ 1 & 4 \end{bmatrix}$ **5.** $\begin{bmatrix} -1 & -12 \\ 8 & 22 \end{bmatrix}$ **7.** $[-2 \quad -3]$

9. $\begin{bmatrix} 15 & -20 \\ -7 & -8 \end{bmatrix}$ **11.** not defined **13.** [5]

15. $\begin{bmatrix} 1 & -36 \\ 18 & 28 \end{bmatrix}$ **17.** not defined **19.** $\begin{bmatrix} -1 & \frac{3}{2} \\ 0 & \frac{1}{2} \\ 2 & 1 \end{bmatrix}$

Appendix 5 Exercises, page 661 1. $\begin{bmatrix} 4 & -7 \\ 5 & -9 \end{bmatrix}$

3. does not exist **5.** $\begin{bmatrix} 3 & -7 \\ -1 & 3 \end{bmatrix}$ **7.** $\begin{bmatrix} \frac{11}{5} & \frac{7}{5} \\ \frac{4}{5} & \frac{3}{5} \end{bmatrix}$

9. $B^{-1}C$ **11.** $X = \frac{1}{2}B$ **13. a.** first row 1, 0; second row
0, 1 **b.** $ax + bz = 0; cx + dz = 1$ **d.** $x = -\dfrac{b}{D}$ and $z =$

$\dfrac{a}{D}$ **15. a.** $\begin{bmatrix} \cos \beta & -\sin \beta \\ \sin \beta & \cos \beta \end{bmatrix}$

b. $\begin{bmatrix} \cos(\alpha + \beta) & -\sin(\alpha + \beta) \\ \sin(\alpha + \beta) & \cos(\alpha + \beta) \end{bmatrix}$ **c.** the rotation about

the origin through an angle $\alpha + \beta$ **17.** $\begin{bmatrix} 5 & -2 \\ 4 & 3 \\ -9 & 6 \end{bmatrix}$